International Symposium on Advanced Materials

(ISAM 2013)

IOP Conference Series: Materials Science and Engineering Volume 60

Islamabad, Pakistan
23 – 27 September 2013

Editors:

S. Qaisar
A Nusair Khan
E. Ahmed Mukhtar

ISBN: 978-1-63439-871-8
ISSN: 1757-8981

Printed from e-media with permission by:

Curran Associates, Inc.
57 Morehouse Lane
Red Hook, NY 12571

Some format issues inherent in the e-media version may also appear in this print version.

Printed by Curran Associates, Inc. (2015)

Institute of Physics
Dirac House, Temple Back
Bristol BS1 6BE UK

Phone: 44 1 17 929 7481
Fax: 44 1 17 920 0979

techtracking@iop.org

Additional copies of this publication are available from:

Curran Associates, Inc.
57 Morehouse Lane
Red Hook, NY 12571 USA
Phone: 845-758-0400
Fax: 845-758-2634
Email: curran@proceedings.com
Web: www.proceedings.com

TABLE OF CONTENTS

IOP Conf. Series: Materials Science and Engineering **60** (2014) 011001 doi:10.1088/1757-899X/60/1/011001

Selected, peer reviewed papers from the

13th International Symposium on Advanced Materials

23-27 September 2013, Islamabad, Pakistan

Editors

Sara Qaisar
A Nusair Khan
Ejaz Ahmed Mukhtar

Organized by: International Symposium on Advanced Materials, Islamabad, Pakistan

Co-organizer: Institute of Space Technology, Islamabad, Pakistan

IOP Conf. Series: Materials Science and Engineering **60** (2014) 011001 doi:10.1088/1757-899X/60/1/011001

Preface

This proceeding is a compilation of peer reviewed papers presented at the 13th International Symposium on Advanced Materials (ISAM 2013) held from September 23-27, 2013, at Islamabad, Pakistan. In my capacity as ISAM-2013 Secretary, I feel honoured that the symposium has ended on a positive note. The ever increasing changes and intricacies that characterize modern industry necessitate a growing demand for technical information on advanced materials. ISAM and other similar forums serve to fulfill this need.

The five day deliberations of ISAM 2013, consisted of 19 technical sessions and 2 poster sessions. In all, 277 papers were presented, inclusive of 80 contributory, invited and oral presentations. The symposium also hosted panel discussions led by renowned scientists and eminent researchers from foreign as well as local institutes.

The ultimate aim of this proceeding is to record in writing the new findings in the field of advanced materials. I hope that the technical data available in this publication proves valuable to young scientists and researchers working in this area of science. At the same time, I wish to acknowledge Institute of Physics (IOP) Publishing UK, for accepting the research papers from ISAM-2013 for publication in the IOP Conference Series: Materials Science and Engineering. The proceeding will be available on the IOP website as an online open access document.

I am profoundly thankful to the Symposium Chairman for his steadfast support and valuable guidance without which ISAM 2013 could not have been the mega event that it turned out to be. My gratitude to all our distinguished participants, session chairs/co-chairs, and reviewers for their active role in the symposium. I appreciate the entire organizing committee for the zest and ardor with which each committee fulfilled its obligations to ISAM. Last yet not the least, my thankfulness goes to all our sponsors for wilfully financing the event.

Dr. Sara Qaisar
Symposium Secretary

ORGANIZING COMMITTEE

Ejaz Ahmed Mukhtar	(Chairman)
Sara Qaisar	(Secretary)
M. Muneeb Asim	(Treasurer)
S. Wilayat Husain	
Anjum Tauqir	
M. Farooque	
A.A. Mazhar	
Ishaat Saboor Khan	
Shaheed Khan	
Iftikhar us Salam	
Zubair Ahmed	
Saleem Akhtar	
Shabbar A. Rizvi	
A. Nusair Khan	
Sajjad Ahmad	

ADVISORY BOARD

Amir Azam Khan	(Malaysia)
F. Ahmad Khalid	(Pakistan)
Mohamad M. Ayad	(Egypt)
Mi Yan	(China)
Muhammad Mujahid	(Pakistan)
S. Wilayat Husain	(Pakistan)
Tahir I. Khan	(Canada)
Yosry Mashal	(Saudi Arabia)
Zhongwu Liu	(China)

SECRETARIAT

Zubair Ahmed	(Incharge)
Shahab Bhatti	
M. Kashif Nasim	

TECHNICAL COMMITTEE

A. Nusair Khan	(Incharge)
Tayyab Subhani	
Mustasim Billah Bhatty	
M. Mansoor Malik	
Haroon Zakir	
Adnan Alvi	

IOP Conf. Series: Materials Science and Engineering **60** (2014) 011001 doi:10.1088/1757-899X/60/1/011001

SPONSORS

Amber Chemical Industries Ltd., Pakistan
Apical Ways, Pakistan
Askari Bank Ltd., Pakistan
Engineering Horizons, Pakistan
Forward Enterprises, China
Hamdard Foundation, Pakistan
Higher Education Commission, Pakistan
M/S OVETEK-P, China
National Bank of Pakistan, Pakistan
National Talent Pool, Pakistan
Pakistan Science Foundation, Pakistan
Peoples Steel Mills Ltd., Pakistan
Professional Systems Ltd., Pakistan

IOP Conf. Series: Materials Science and Engineering **60** (2014) 011002 doi:10.1088/1757-899X/60/1/011002

Peer review statement

All papers published in this volume of *IOP Conference Series: Materials Science and Engineering* have been peer reviewed through processes administered by the proceedings Editors. Reviews were conducted by expert referees to the professional and scientific standards expected of a proceedings journal published by IOP Publishing.

Ageing dependence and martensite stabilization in copper based shape memory alloys

O Adiguzel
Firat University, Department of Physics, 23169 Elazig, Turkey

E-mail: oadiguzel@firat.edu.tr

Abstract. Shape memory alloys exhibit a peculiar property called shape memory effect based on a first order solid state phase transformation, martensitic transformation which occurs in thermal manner on cooling the materials. Martensitic transformation is evaluated by the structural changes in microscopic scale. Copper-based ternary alloys exhibit shape memory effect in metastable beta phase region. These alloys have bcc-based ordered structures at high temperature, and transform martensiticaly to the long-period layered structures on cooling. The material atoms move cooperatively on {110}-type close packed planes of parent phase by means of a shear-like mechanism, and structural and fundamental properties of these alloys are altered by aging in the martensitic state. Therefore, the ageing gives rise to the structural changes in both long and short-range order in material. X-ray powder diffraction studies carried out in a long time interval on copper based shape memory alloys reveal that peak locations and intensities chance with ageing duration in martensitic condition, and these changes lead to the martensite stabilization in the redistribution or disordering manner, and stabilization proceeds by a diffusion-controlled process.

1. Introduction

Martensitic transformations are first order lattice-distorting phase transformations and occur with the cooperative movement of atoms by means of lattice invariant shears in the materials on cooling from high temperature parent phase region. Shape memory effect is based on martensitic transformation, and microstructural mechanisms are responsible for shape memory behaviour. In particular, the detwinning is essential as well as martensitic transformation in reversible shape memory effect [1, 2]. Thermal induced martensite occurs by means of a shear-like mechanism as multivariant martensite in self-accommodating manner and consists of lattice twins. Also, this martensite is called twinned martensite or multivariant martensite.

Shape memory alloys can be deformed plastically in low temperature martensitic condition, and recover the original shape on heating over the austenite finish temperature. The material cycles between the deformed and original shapes on cooling and heating. By applying external stress, the martensitic variants are forced to reorient into a single variant leading inelastic strains, and deformation of shape memory alloys in martensitic state proceeds through a martensite variant reorientation or detwinning of twins. The twinning occurs with internal stresses, while detwinning occurs with the external stresses. The basic mechanism of shape memory effect is schematically illustrated in Figure 1 [1, 2]. The deformed material recovers the original shape on heating over the

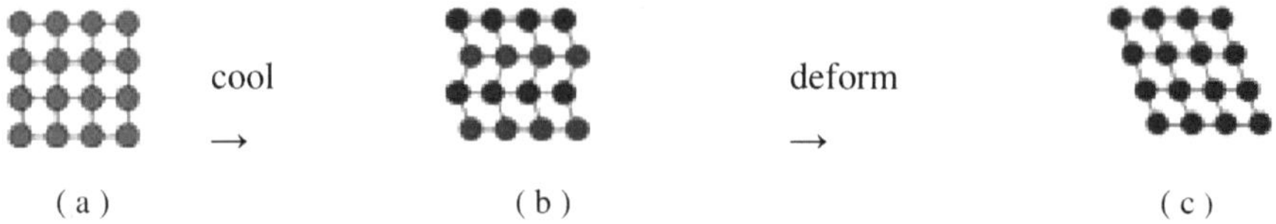

Figure 1. Schematic illustration of the mechanism of the shape-memory effect: (a) atomic configuration on {110} - type planes of parent austenite phase, (b) twinned martensite phase occurring thermally on cooling, (c) detwinned martensite occurring with deformation [1].

austenite finish temperature, and returns to the multivariant martensite structures in irreversible shape memory effect on cooling below the martensite finish temperature; in contrast, the material returns to the detwinned martensite structure in reversible shape memory case.

Shortly one can say that the microstructural mechanisms responsible for the shape memory effect are the twinning and detwinning processes as well as martensitic transformation. Therefore, the detwinning process has great importance for the processing of reversible shape memory effect. Copper based alloys exhibit this property in metastable □□□□ phase region, which has B2 or DO_3-type ordered lattice at high temperature, and these structures martensiticaly turn into layered complex structures with lattice twinning process, on cooling from high temperature austenitic phase region.

Martensitic transformations occur in a few steps with the cooperative movement of atoms less than interatomic distances by means of lattice invariant shears on a {110} - type plane of austenite matrix which is basal plane of martensite. First one is Bain distortion, and second one is lattice invariant shear. Bain distortion consists of an expansion of 26% parallel to the <001>-type axes, called Bain axes, and a compression of 11% parallel to the <110> and <1 $\bar{1}$ 0>-type directions, perpendicular two Bain axes. The compressed directions are [110] and [1 $\bar{1}$ 0] directions for [001] - Bain axes. For the illustration of Bain distortion, an fct cell is delineated in matrix, and this cell undergoes to the corresponding fcc cell with Bain distortion [3]. Lattice invariant shears occur with cooperative movement of atoms less than interatomic distances on {110}-type close packet plans of austenite matrix. The lattice invariant shears occurs, in two opposite directions, <110 > -type directions on the {110}-type basal planes and this kind of shear can be called as {110} <110 > - type mode and has 24 variants in self-accommodating manner [4-6]. These lattice invariant shears give rise to the formation of unusual complex layered structures called long period layered structures such as 3R, 9R or 18R depending on the stacking sequences on the close-packed planes of the ordered lattice.

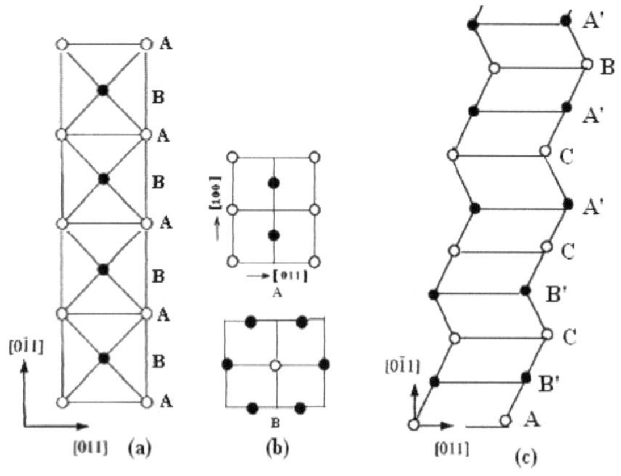

Figure 2. (a) Stacking of (1 $\bar{1}$ 0)$_\beta$ planes viewed from [001]$_\beta$ direction,(b) Atomic configuration on first and second layers of (1 $\bar{1}$ 0)$_\beta$ plane in DO_3 - type structures, (c) inhomogeneous shear and formation of layered structures, stacking sequences of half 18R or M18R unit cell in direction z.

The periodicity and therefore the unit cell are completed through 18 layers in direction z in 18R case. Formation of these layered structures is schematically illustrated in Figure 2.

2. Experimental

In the present contribution, two copper based ternary shape memory alloys were selected for investigation; Cu-26.1%Zn 4%Al and Cu-11%Al-6%Mn (in weight). The martensitic transformation temperatures of these alloys are over the room temperature and both alloys are entirely martensitic at room temperature. Specimens obtained from these alloys were solution treated for homogenization in the β-phase field (15 minutes at 830°C for CuZnAl alloy and 20 minutes at 700°C for CuAlMn alloy), then quenched in iced-brine to retain the β-phase and aged at room temperature after quenching (both alloys). Powder specimens for X-ray examination were prepared by filling the alloys, these specimens were heated in evacuated quartz tubes in the β-phase field (15 minutes at 830°C for CuZnAl and 20 minutes at 700°C for CuAlMn) for homogenization and quenched in iced-brine. These specimens were also given different post-quench heat treatments and aged at room temperature.
X-ray diffraction studies carried out on these specimens, and X-ray diffraction profiles were taken from the quenched specimens using Cu-K_α radiation with wavelength 1.5418 Å.

3. Results and Discussion

X-ray powder diffractograms were taken from CuZnAl and CuAlMn samples. An x-ray powder diffractogram taken from the long term aged CuAlMn alloy sample is shown in Figure 3. This diffractogram has been indexed on the monoclinic M18R basis. On the other hand, electron diffraction pattern were also taken from the samples these alloys, the details were given elsewhere [3, 6]. X-ray diffractograms and electron diffraction patterns reveal that both alloy have the ordered structure in martensitic condition, and exhibit superlattice reflections. A series of X-ray powder diffractograms and electron diffraction patterns were taken from both CuZnAl and CuAlMn alloy samples in a large time interval and compared with each other. It has been observed that electron diffraction patterns exhibit similar characteristics, but some changes occur at the peak locations and intensities on the x-ray diffractograms with aging duration. These changes occur as rearrangement or redistribution of atoms in the material, and attribute to new transitions in diffusive manner [3, 6, 8]. The ordered structure or super lattice structure is essential for the shape memory quality of the material. In the shape memory alloys, homogenization and releasing the external effect is obtained by ageing at β-phase field for adequate duration.
On the other hand, post-quench ageing and service processes in devices affect the shape memory quality, and give rise shape memory losses. These kinds of results lead to the martensite stabilization in the reordering or disordering manner. In order to make the material satisfactorily ordered and to delay the martensite stabilization, copper-based shape memory alloys are usually treated by step-quenching after homogenization. Although martensitic transformation has displacive character, martensite stabilization is a diffusion controlled phenomena, and this result leads to redistribution of atoms on the lattices sites. Stabilization is important factor and causes to memory losses, and changes in main characteristics of the material; such as, transformation temperatures, and x-ray diffraction peak location and intensities.
The martensitic phase in copper-based β-phase alloys is based on one of the $\{110\}_\beta$ planes of parent phase called basal plane for martensite. The (110) basal plane which has a rectangular shape in parent phase is subjected to hexagonal distortion and undergoes a hexagon. This process is schematically illustrated in Figure 4.
The powder specimens were aged at room temperature after quenching process, and many x-ray diffractograms have been taken from both of the alloy samples in a large time interval. Although all of the diffractograms exhibit similar characteristics, some changes have been observed in peak locations on the diffractograms with aging duration. These changes are attributed to new transitions which have diffusive character. It means that some neighbour atoms change locations.

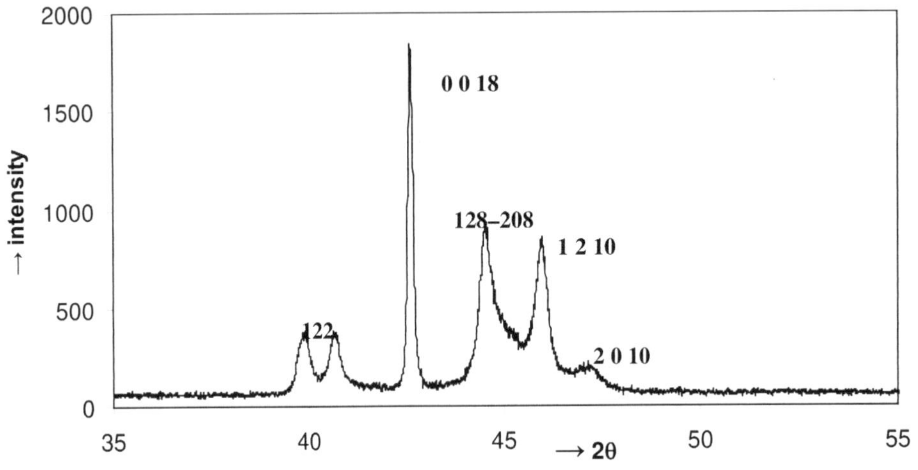

Figure 3. An x-ray diffractogram taken from the long term aged CuAlMn alloy sample.

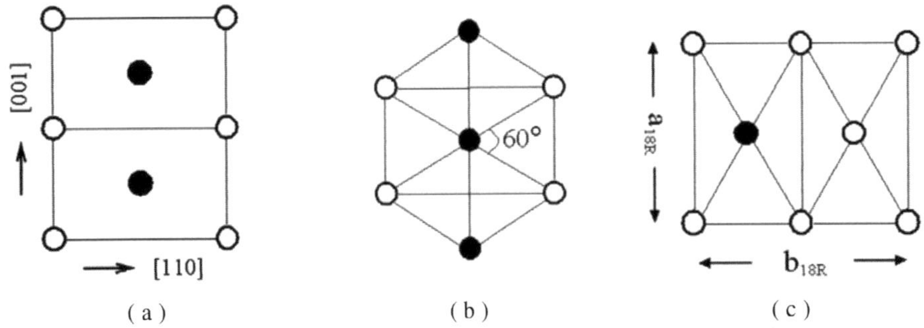

Figure 4. Atomic configuration of the (110) plane of DO_3 - type ordered β-matrix (a) before and (b) after hexagonal distortion, (c) principal basal plane axes of the 18R structure.

In particular, some of the neighbour peak pairs have moved toward each other. It is interesting that miller indices of these plane pairs provide a special relation: $(h_1^2 - h_2^2)/3 = (k_2^2 - k_1^2)/n$ where n=4 for 18R martensite. These plane pairs can be listed as follow; (122)-(202), (128)-(208), (1 2 10) – (2 0 10), (040)- (320). This result can be attributed to a relation between interplane distances of these plane pairs and rearrangement of atoms on the basal plane. In these changes, atom sizes play important role. The different sizes of atomic sites lead to a distortion of the close-packed plane from an exact hexagon and thus a more close-packed layered structure may be expected. In the disordered case, atom sizes can be taken nearly equal, and martensite basal plane becomes an ideal hexagon. Although martensitic transformations are displacive, the post-martensitic transitions have the diffusive character because this transition requires a structural change and this also gives rise to a change in the configurational order. The interplane distances of the diffracted plane pairs are calculated using the following relation in the monoclinic case.

$$\frac{1}{d^2} = \frac{1}{a^2}\frac{h^2}{\sin^2\beta} + \frac{k^2}{b^2} + \frac{1}{c^2}\frac{l^2}{\sin^2\beta} - \frac{2hl\cos\beta}{ac\sin^2\beta}$$

It can be seen with this calculation that dhkl values of the diffracted plane pairs become approximately equal each other in the disordered case, and diffraction peaks overlap each other, considering atom sizes approximately equal each other. In the ordered case, neighbour atom sizes are different, interplane distances become different and each peak appears separately.

Metastable phases of copper-based shape memory alloys are very sensitive to the ageing effects, and any heat treatment can change the relative stability of both martensite and parent phases [8, 9]. Martensite stabilization is closely related to the disordering in martensitic state. Structural ordering is one of the important factors for the formation of martensite, while atom sizes have important effect on the formation of ordered structures [9. 10, 11].

4. Conclusions

It can be concluded from the above results that the copper-based shape memory alloys are very sensitive to the ageing treatments. The diffracted angles and intensities of x-ray diffraction peaks chance with ageing duration in martensitic condition. In particular, some successive peak pairs come close each other. These changes lead to the martensite stabilization in the redistribution or disordering manner, and stabilization proceeds by a diffusion-controlled process. The martensite stabilization is a diffusion controlled phenomena and leads to redistribution of atoms on the lattices sites, although martensitic transformation has displacive character.

The basal plane of martensite turns into a hexagon by means of Bain distortion with martensite formation on which atom sizes have important effect. In case the atoms occupying the lattice sites have the same size, the basal plane of martensite becomes regular hexagon; otherwise the deviations occur from the hexagon arrangement of the atoms. The above mentioned diffracted plane peaks come close each other in the disordered case, and occur separately in the ordered case. The changes in the diffraction angles of the selected plane pairs can be a measure of the ordering degree in martensite.

5. References

[1] MA J, Karaman I and Noebe RD 2010 *International Materials Reviews* 55 257

[2] Liu Y 2001 *Proceedings of SPIE, Smart Materials* 4234 82

[3] Adiguzel O 2007 *Journal of Materials Processing Technology* 185 120

[4] Zhu JJ, Liew KM 2003 *Acta Materialia* 51 2443

[5] Sutou Y et al 2005 *Acta Materialia* 53 4121

[6] Adiguzel O 2012 *Key Engineering Materials* 510-511 105

[7] Pelegrina JL and Romero R 2000 *Materials Science and Engineering A* 282 16

[8] Li Z, Gong S and Wang MP 2008 *Journal of Alloys and Compounds* 452 307

[9] Paidar V 2008 *Materials Science and Engineering A* 481-482 243

[10] Guo YF et al 2007 *Acta Materialia* 55 663

[11] Aydogdu A, Aydogdu Y and Adiguzel O 2004 *Journal of Materials Processing Technology* 153-154 164

Materials with high magnetostriction

R Grössinger[1], R Sato Turtelli[1] and N Mehmood[2]
[1]Institute of Solid State Physics, Technical University of Vienna, Vienna, Austria
[2]National Engineering and Scientific Commission, Islamabad, Pakistan

E-mail: rgroess@ifp.tuwien.ac.at

Abstract. The magnetostrictive properties of cubic Laves phases such as $TbFe_2$, Terfenol D and $SmFe_2$ are summarised. It is shown that $SmFe_2$ can be an interesting candidate as a high magnetostrictive material even in the as cast "dirty" state. Fe-X alloys based on Fe-Ga but also Fe-Al are presented as materials with a medium magnetostriction, however with the big advantage of a magnetically soft material. Generally all nonmagnetic substutional elements which cause an increase of the interatomic distance in the Fe-lattice (such as Ga, Al, Si) cause an increase of the magnetostriction too. In Ni-Ga a reduction of the magnetostriction was found. Another interesting material with high magnetostriction is $CoFe_2O_4$ which is cheap, easy to produce and an oxidic material. Here magnetostriction values up to 400 ppm can be achieved by hydrostatic compaction followed by a field annealing.

1. Introduction

Magnetostriction is an intrinsic property of all kind of magnetically ordered materials. Due to the magnetic exchange the materials change its length (parallel and perpendicular to the external field) as well as its volume. In a single crystal magnetostriction is a tensor property which depends on the symmetry of the lattice. This change of length can be between 10^{-8} and 10^{-3} ($\Delta l/l_0$) depending on the involved elements as well as on the local symmetry. Magnetostriction is based on the coupling of the magnetic moment to the lattice, which means that here the orbital moment **L** of the atom has to be considered. Materials with high magnetostriction are technologically important for sensor or actuator applications.

In this paper different materials with high magnetostriction will be discussed and surveyed. Also a theoretical background of the origin of the magnetostriction will be given. Generally a single crystal is best suited for characterizing the magnetostriction, however often only data of polycrystalline material are available.

2. Rare earth based magnetostrictive systems

Pure rare earth elements (such as Tb, Dy,...) exhibit due to its orbital magnetism generally a huge magnetostriction, however only at low temperatures. Therefore alloys between rare earths and 3d-elements, which have ordering temperatures above room temperature have to be considered. The highest magnetostriction at room temperature was found in $TbFe_2$ as well as in $SmFe_2$ [1,2]. These are cubic Laves phase materials where the easy axis is parallel to the [111] direction. Clark [3] as well as Hathaway and Cullen [4] suggested a microscopic model for this magnetostriction as demonstrated graphically in figure 1. Figure 2 shows as an example the in $TbFe_2$ measured magnetostriction into the [111] direction at different temperatures [1]. At room temperature a value of nearly +2000 ppm is achieved however at external fields of 800 kA/m (~ 10 kOe). This field is too high for technical applications. In order to reduce the magnetocrystalline anisotropy of Tb about 70% Dy is added which leads to a material which is under the trade name Terfenol D ($(Tb_{0.3}Dy_{0.7})Fe_{1.92}$ industrially available. Terfenol D is generally used in a pre-stressed state as well as aligned material. Figure 3 shows room temperature magnetostriction measurements on Terfenol D. With this material a linear magnetostriction of about 1000 ppm is available, however at fields of 5 kOe. H.S.Lim et al [5] report about a plastic bonded Terfenol D which has the advantage that the material can also be used in dynamic applications due to the reduced electrical resistivity. Table 1 compares single crystal magnetostriction values of Fe, Ni and Terfenol D.

Table 1. Single crystal magnetostriction values of Fe, Ni and Terfenol D.

Material	□(100) ppm	□(111) ppm
Fe	20	-21
Ni	-46	-24
Terfenol D	90	1640

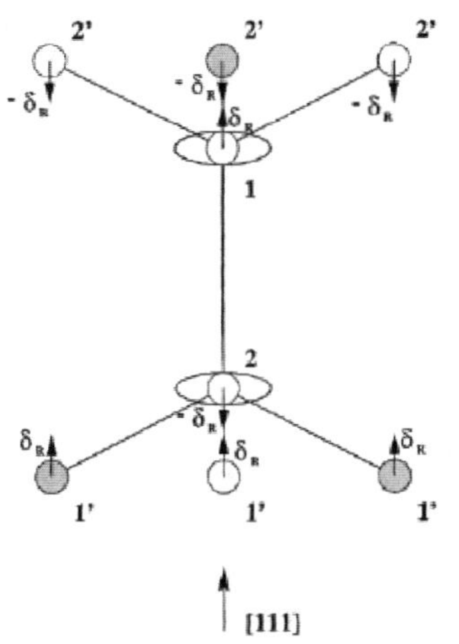

Figure 1. Model of the changed distance between the rare earth atoms (Tb) due to the coupling between the 4f clouds located at “1” and “2” [3,4].

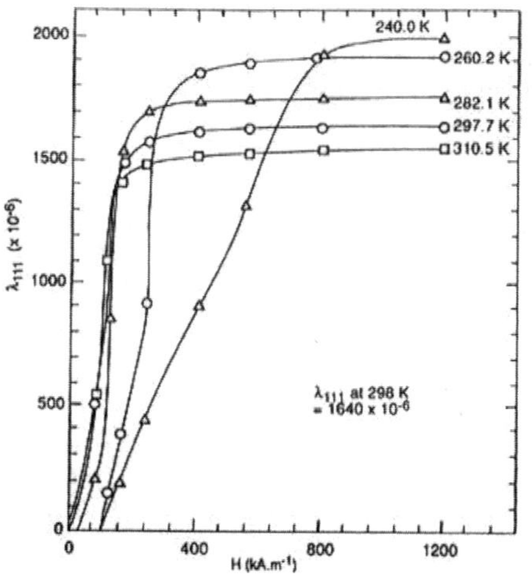

Figure 2. Magnetostriction of a $TbFe_2$ single crystal measured in the [111] direction at different temperatures [1].

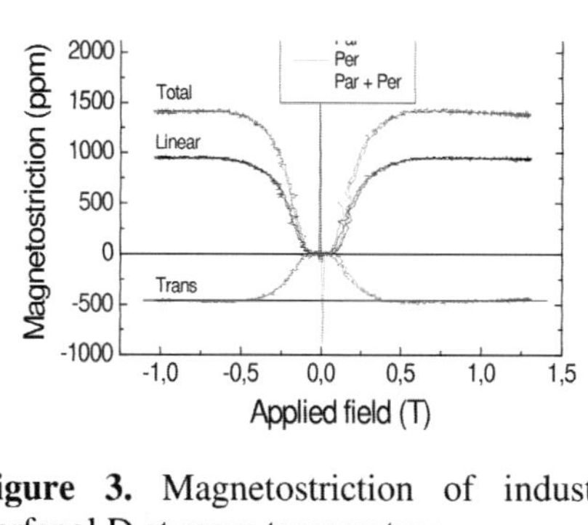

Figure 3. Magnetostriction of industrial Terfenol D at room temperature.

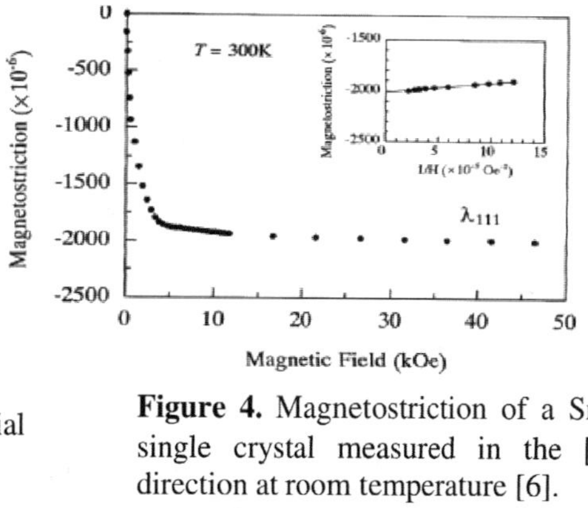

Figure 4. Magnetostriction of a $SmFe_2$ single crystal measured in the [111] direction at room temperature [6].

Even Terfenol D is an excellent high magnetostrictive material (see figure 3), the technical application is limited due to the brittleness of the material as well as due to the high costs of the heavy rare earth elements. Another option is $SmFe_2$ which is also a [111] cubic Laves phase which exhibit a high but negative magnetostriction at room temperature, RT, see figure 4. The easy axis of magnetization is parallel to the [1 1 0] direction at low temperatures, however it changes to the [111] direction at temperatures above 195 K.

$SmFe_2$ shows a typical cubic anisotropy with anisotropy constants $K_1 = 5.3\ 10^6$ and $K_2 = 1.9\ 10^6$ erg/cm^3 at 300 K, however a magnetoelastic part has also to be considered [6]. Magnetostriction constants were determined to be λ_{111} = - 2010 ppm, λ_{100} = -130 ppm and λ_S = -1258 ppm at 300 K. The problem of producing this material is that due to the high vapor pressure of Sm combined with the peritectic point of the formation of $SmFe_2$ in the binary phase diagram it is rather difficult to achieve single phase material. However even just after a melting process (which leads to a "dirty" sample) a rather high magnetostriction of 500 ppm (see figure 5) or more in a polycrystalline state can be achieved. However in the as-cast state, the sample was not homogeneous. The chemical composition of the alloy determined by electron probe microanalysis (EPMA) EDX technique showed that the as cast sample exhibits different compositions at different spots such as: $SmFe_{2.3}$, $SmFe_{1.25}$ and $SmFe_9$. When this material is submitted to a heat treatment at 700°C for 132 h under vacuum with a Ta-foil as getter material, XRD analysis showed that the sample became single phase $SmFe_2$. This single phase $SmFe_2$ presented a crystallite size of 21nm and a lattice constant of 7.426648 Å corresponding to the space group Fd3m. The magnetostriction could be improved significantly – see figure.6.

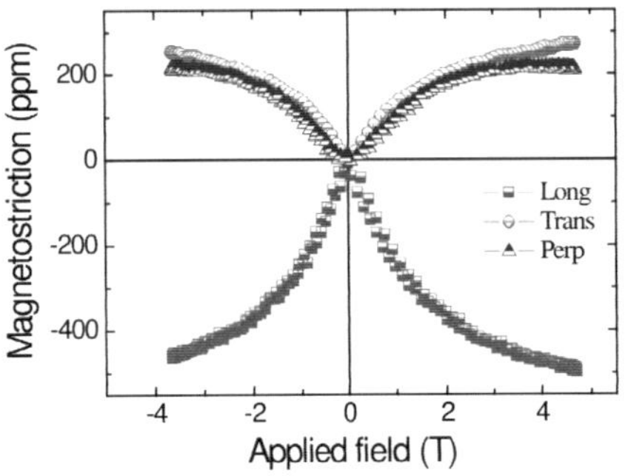

Figure 5. Magnetostriction of a polycrystalline, as-cast "dirty" $SmFe_2$ measured at room temperature.

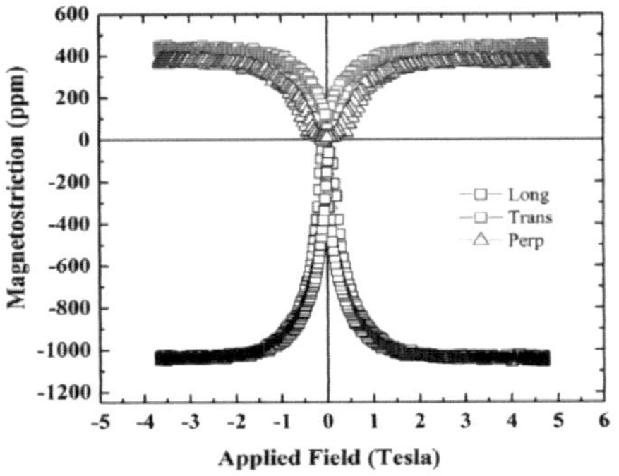

Figure 6. Magnetostriction of a annealed, polycrystalline $SmFe_2$ measured at room temperature.

Also in $SmFe_2$ the magnetocrystalline anisotropy is rather high, which makes it difficult to saturate this material. However producing this material (as well as $TbFe_2$) in a nanocrystalline state might be a way to reduce the anisotropy. This route was already proved on thin film TbFeCo multilayers with excellent soft magnetic properties [7].

2.1.Fe-Ga (Galfenol)

Generally 3d elements (such as Fe,Co or Ni) exhibit a rather low magnetostriction (see table 1), which comes from the fact that in these elements the orbital moment is considered to be quenched. However due to the much lower price and the excellent mechanical properties, especially Fe-based alloys are always of technical interest. Compared to highly anisotropic and corrosive rare earths Fe-based alloys Fe-Ga alloys have the advantage that they are generally magnetically soft and chemically rather stable. In the last years the group of Clark *et al* investigated Fe-Ga and Fe-Ga-Al based alloys [8,9]. It was found that by substituting between 15 and 28 %Ga (or Al) into the Fe lattice causes a significant increase of the magnetostriction [8,9,10]. A similar but smaller effect was found in $Fe_{1-x}Al_x$ [9,10]. In the mean time the effect of substituting many different nonmagnetic elements was investigated [11]. Figure 7 shows 3/2 λ_{100} of the $Fe_{1-x}Ga_x$ system which shows values up to 450 ppm [12].

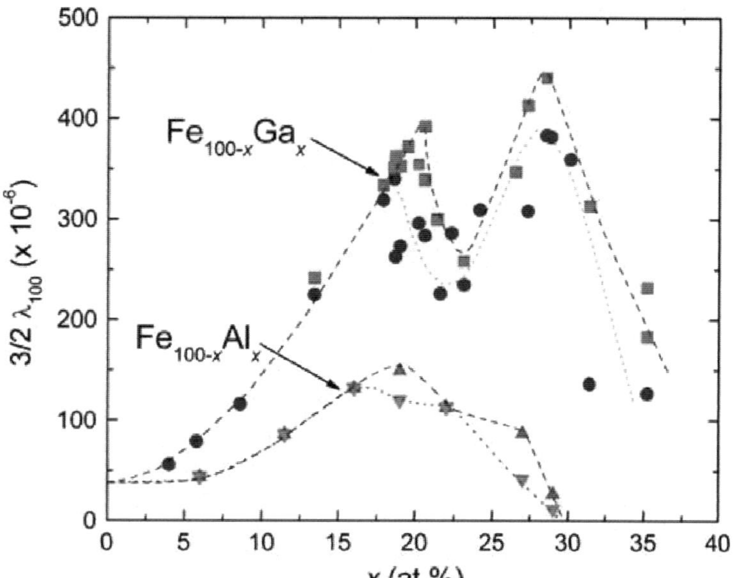

Figure 7. 3/2 λ_{100} of the $Fe_{1-x}Ga_x$ and $Fe_{1-x}Al_x$ systems measured at RT [12].

The fundamental question here was, which mechanism drives the development of a high magnetostriction in these materials. The most important points influencing magnetostriction, beside the involved elements, is the structure as well as the local symmetry at different lattice sides. Most of the Fe-X (X = Al, Si, Ge, Ga) phase diagrams look very similar at the Fe-rich side, exhibiting first a disordered A2 structure which changes at higher X-concentrations to ordered B2 and DO_3 structures. Clark et all have explained the main origin of high magnetostriction in 3d-based Fe-Ga alloys is due to the change of magnetoelastic energy because of the Ga substitution causing a lattice softening. The magnetostriction increase is explained by an increase of the magneto elastic coupling constant due to short range ordering between the Ga–Ga atoms in bcc α-iron (A2 structure) with randomly substituted gallium atoms [13]. Measuring the elastic constants in Fe-Ga and calculating the magnetoelastic energy constants, delivers indeed a peak at 19 at. % Ga [12].

Recently, we investigated the local Fe-Ga atomic structure in highly magnetostrictive α-$Fe_{80}Ga_{20}$ melt-spun ribbons [14] and splat cooled material [15] using Extended X-ray Absorption Fine Structure (EXAFS) analysis at the Fe and Ga K-edges. A careful XRD showed that this ribbon crystallize in the A2 phase. In this work the presence of small Ga clusters could be excluded by EXAFS and XANES, since no first shell Ga-Ga bonds were detected. However, EXAFS analysis of the second coordination shell around Ga clearly provides evidence for the presence of one highly strained (+ 4%) Ga-Ga pair and five Ga-Fe pairs among six crystallographically equivalent <100> atomic pairs. This conclusion supports recent total energy calculations which assign the large magnetostriction in these alloys to the strain caused by the rotation of the magnetization in the vicinity of such defects.

Even polycrystalline material exhibits much higher magnetostriction then pure Fe. Figure 8 shows as an example magnetostriction measurements performed on polycrystalline Fe-Ga samples [16]. It is clear that for a polycrystalline isotropic material the achievable magnetostriction is smaller than λ_{100} (assuming $\lambda_{100} > |\lambda_{111}|$) according to the well-known formula

$$\lambda_s = \frac{1}{5}\left(2\lambda_{100} + 3\lambda_{111}\right) \qquad (1)$$

Fe-Ga based alloys have additionally the advantage that they are magnetically soft, which was demonstrated by hysteresis measurements on ring shaped Fe-Ga samples [17] and is also shown in figure 9. Note that in this case an external field of about 16 kA/m (= 200 Oe) is sufficient to saturate the material.

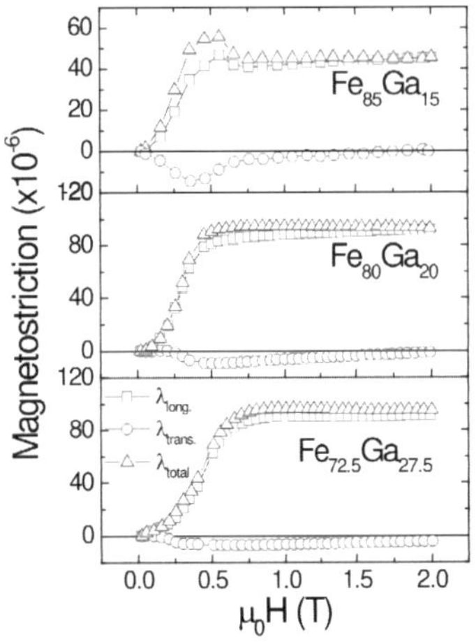

Figure 8. RT magnetostriction measurements on polycrystalline Fe-Ga samples.

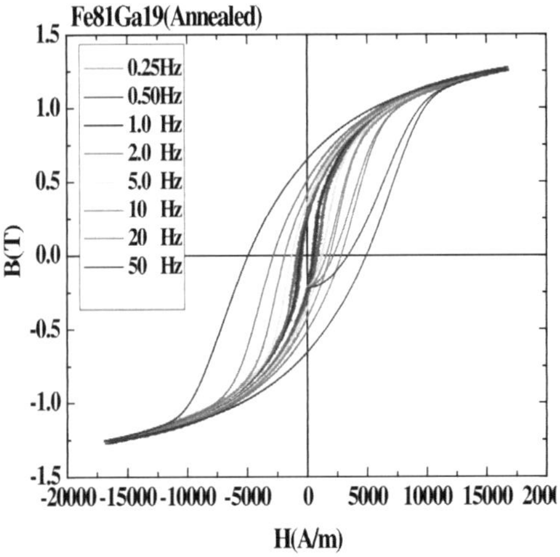

Figure 9. Hysteresys loops measured at different frequencies on annealed polycrystalline $Fe_{81}Ga_{19}$ at RT [17].

2.2.Ni-Ga Alloys

Keeping in view that addition of non-magnetic V, Cr, Al and Ga in Fe increases the magnetostriction of iron as reported by Hall [18] and Clark et al [8,10,19] one has to ask if the same idea will not work in Ni. Therefore two samples of polycrystalline Ni-Ga with the composition $Ni_{98}Ga_2$ and $Ni_{93}Ga_7$ were investigated. Figure 10 shows the magnetostriction (λ_{par}) of $Ni_{98}Ga_2$ and $Ni_{93}Ga_7$ and pure Nickel in the as cast state. It can be seen that after addition of 2 % Ga the magnetostriction of Ni slightly increases from -35 ppm to -39 ppm. However with a higher amount of – 7 at% Ga the magnetostriction value decreases to -10 ppm. The addition of Ga in Ni does not significantly enhance the magnetostriction and for higher Ga values the magnetostriction decreases. This shows that the magnetostriction of the fcc Ni behaves completely different by substituting with the nonmagnetic element Ga.

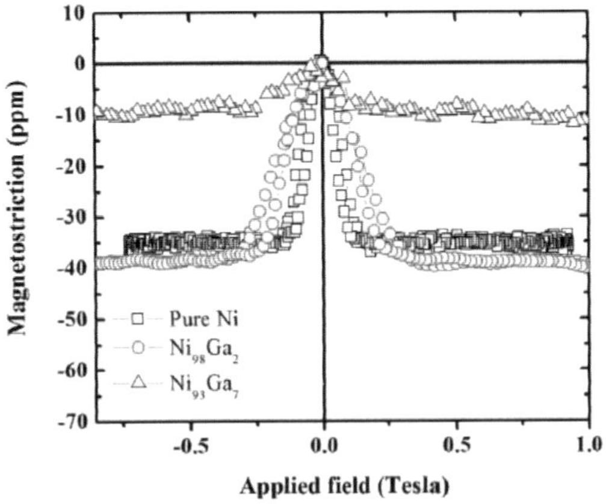

Figure 10. Longitudinal magnetostriction of pure, polycrystalline Ni, $Ni_{98}Ga_2$ and $Ni_{93}Ga_7$ at room temperature

2.3. Fe-Al alloys

The phase diagram of Fe-Al in the Fe-rich region is similar to that of the Fe-Ga system. As in Fe-Ga alloys, the lattice constant increases due to the non-magnetic Al substitution. As can be seen in figure 7, the magnetostriction constant λ_{100} increases with Al substitution up to around 20 at. % Al and reaches a maximum around 150 ppm. The magnetostriction behavior of this system is similar to that of Fe-Ga alloys (see figure 7). In polycrystalline $Fe_{1-x}Al_x$ alloys, we reached a longitudinal magnetostriction of 60 ppm for the sample containing 19 at % Al [20].

The effect of different substituting elements is summarized in table 2. It is evident that the substitution of Ga causes the largest magnetostriction enhancement in the [100] direction. The reason for this is very probable both a structural and electronic one.

Table 2. Magnetostriction at RT of Fe-X single crystals.from R.C.Hall [18]; ** from Landolt Börnstein [21]

Compound-at % in Fe	**□00 (10^{-6})**	**□11 (10^{-6})**
0**	22	-21
13% Ga	153	-16
17% Ga	207	-
16% Al*	86	-2
5% Si**	27	-7
15.6% Cr*	51	-6
15.6% V*	43	-10

2.4. $CoFe_2O_4$

Co-ferrite is a ferromagnetic oxide which is non-metallic. It exhibits from all 3d-element based spinel oxides the highest magnetostriction. The value of the achievable magnetostriction depends strongly on the exact stochiometry as well as on the production methods. Because Co-ferrite is an oxide it is produced by chemical and mechanical methods (such as sol-gel, ball milling etc) which leads finally to a powder material. In order to get bulk samples a pressing and annealing process has to be used. An improvement of the magnetostriction of polycrystalline cobalt ferrite can be obtained by magnetic annealing due to the effect of induced uniaxial anisotropy [22]. On the other hand, it is well known that the magnetic and magnetoelastic properties depend on the processing parameters [23, 24]. Nlebedim *et al.* [24] investigated the influence of powder compaction pressure (87 MPa $\leq$ P $\leq$ 127 MPa) and annealing conditions on the magnetostriction of $CoFe_2O_4$ and found the highest magnetostriction value of -200 ppm for the sample pressed at 127 MPa and annealed at 1350 °C for 24 h.

Later A.Muhamad *et al* investigated the influence of compaction pressure for P > 127 MPa , up to 276 MPA and found a large improvement if the pressed samples are additionally submitted to a field annealing [25]. For the sample pressed at 127 MPa and field annealed, the magnetostriction value was -400 ppm. Figure 12 shows complete measurements of all possible directions between the strain gauge and the hard (easy) axis after field annealing, for example, performed on a sample after a powder compaction pressure of 239 MPa. The open symbols represent a strain gauge placed parallel to the direction of the magnetic field annealing, and filled symbols are results by placing a strain gauge perpendicular to the magnetic field annealing direction. The magnitude of the maximum magnetostriction found by measuring longitudinal magnetostriction, applying the external magnetic field perpendicular to the direction of magnetic field annealing (hard axis), is similar to the transverse and perpendicular magnetostrictions measured by applying the field along the easy axis. Comparing with single crystal data of field annealed cobalt ferrite [26] one gets the impression that λ_{trans} and λ_{perp} are mainly determined by λ_{110}, whereas λ_{long} is given by λ_{100}. So the different single crystal parameter contributions to the "polycrystalline" behaviour are significantly changed by the field-annealing process.

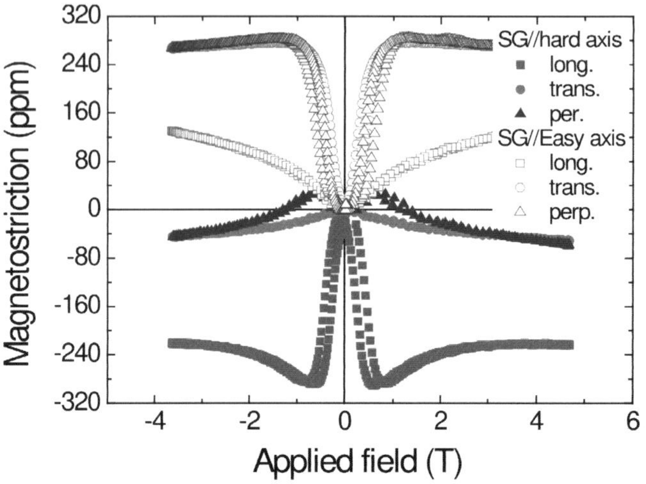

Figure 12. (Colour online) Magnetostriction curves measured after magnetic annealing on a sample pressed at 239 MPa

3. Summary

Materials with high magnetostriction are summarised. The highest magnetostriction was found in $TbFe_2$ and $SmFe_2$, however Terfenol D has the advantage of a reduced magnetocrystalline anisotropy. Rare earth free alloys such as Fe-Ga and Fe-Al exhibit a lower magnetostriction however they are magnetically soft, which means they offer a high magnetostriction at low external fields. Co-ferrite is an oxide material with magnetostriction values between 100 and 400 ppm depending on the production method.

4. References

[1] Clark A E *1980 Ferromagnetic Materials* vol.1, ed E P Wohlfarth (Amsterdam: North-Holland) p 531

[2] Jerems F, Mac Mahon C, Jenner A G and Greenough R D 1999 *Ferroelectrics* 22 333

[3] Cullen J R and Clark A E 1977 *Phys. Rev. B* 15 4510

[4] Hathaway K B and Cullen J R 1978 *J. Appl. Phys.* 49 1975

[5] Lim S H , Kim S R, Kang S Y, Park J K, Nam J T and Son D 1999 *J. Magn.Magn. Mater.* 191 (1999) 113

[6] Samata H, Fujiwara N, Nagata Y, Uchida T and Lan M 1999 *J. Magn.Magn. Mater* 195 376

[7] Duc N H, Richomme F,Tuan N A, HuongGiang D T, Verdier T and J.Teillet 2002 *J. Magn.Magn. Mater* 242 1425

[8] Clark AE, Restorf J B, Wun-Fogle M, Lograsso TA, Schlagel D L 2000 *IEEE Trans. Magn.* 36 3228

[9] Clark A E, Hathaway K B, Wun-Fogle M, Restorf J B, Lograsso T A, Keepens V M, Petculescu G and Taylor R A 2003 *J. Appl. Phys.* 93 8621.

[10] Clark A E, Restorff J B, Wun–Fogle M, Wu D and Lograsso T A 2008 *J. Appl. Phys.*103, 07B310

[11] Grössinger R, Sato Turtelli R and Mehmood N 2008 *IEEE Trans. Magn.* 44 3001

[12] Summers E M, Lograsso T A and Wun-Fogle M 2007 *J. Mater. Sci* 42 9582

[13] Srisukhumbowornchai N and Guruswamy S 2001 *J.Appl. Phys.*90 5680

[14] Pascarelli S, Ruffoni M, Sato Turtelli R, Kubel F and Grössinger R 2008 *Phys. Rev. B*77 184406

[15] Ruffoni M, Pascarelli S, Grössinger R, Sato Turtelli R, Bormio-Nunes C and Pettifer R.F 2008 *Phys. Review Letters* 101 147202

[16] Bormio.Nunes C, Tirelli MA, Sato Turtelli R, Grössinger R, Müller H, Wiesinger G, Sassik H and Reissner M 2005 *J. Appl. Physics* 97 033901

[17] Grössinger R, Mehboob N, Suess D, Sato Turtelli R, and Kriegisch M 2012 *IEEE Trans. Magn.* 48 3076

[18] Hall R C 1960 *J. Appl. Phys.*311037

[19] Cullen J R, ClarkA E, Wun-Fogle M, Restorff J B and Lograsso T A. 2001 *J.Magn.Magn. Mater.*226-230 948

[20] Mehmood N, Sato Turtelli R,Grössinger R and Kriegisch M 2010 *J. Magn. Magn. Mater* 322 1609

[21] Landolt Börnstein "Magnetic Properties of Metals" Group III, Vol. 19, Subvol; Springer Verlag (1994)

[22] Lo C C H, Ring A P, Snyder J E and Jiles D C 2005 *IEEE Trans. Magn.*41 3676

[23] Grössinger R, Mehmood N, Senbaslar G, Muhammad A, Sato Turtelli R, Linert W and Kubel F *Proc. of 20th International Workshop on Rare Earth Permanent Magnets and Their Applications, Crete, Greece, 8–10 September* 2008, p. 214.

[24] Nlebedim I C, Ranvah N, Williams P I, Melikhov Y, Anayi F, Snyder J E, Moses A J and Jiles D C 2009 *J. Magn. Magn. Mater.*321, 2528

[25] Muhamad A, SatoTurtelli R, Kriegisch M, Grössinger R, Kubel F 2012 *J.Appl. Phys.* 111

[26] Bozorth R M, Tilden E F and Williams A J 1956 *Phys. Rev.* 99, 1788

Production and properties of high strength Ni free Zr-based BMGs

M Iqbal[1] and W H Wang[2]
[1]MSG, Physics Division, ATCOP, Pakistan
[2]BMG Group, Institute of Physics, Chinese Academy of Sciences, Beijing, China

E-mail: miqbalchishti@gmail.com, miqbalchishti@yahoo.com

Abstract. Bulk metallic glasses (BMGs) are well known for very attractive physical, mechanical and thermal properties. Zr-based BMGs are used as structural materials in sports goods, electronics, jewelry, medical and aerospace applications. Ni free $Zr_{48}Cu_{36}Al_8M_8$ (M = Nb, Ti and Ta) BMGs are successfully synthesized by Cu mold casting technique. Differential scanning calorimetery (DSC) results show that the $Zr_{48}Cu_{36}Al_8Nb_8$ BMG have good thermal stability, wide supercooled liquid region of 80 K and contain the double stage crystallization. The alloy has fracture strength of 1.953 GPa. Shear angle was measured to be in the range of 43.5±5° for the alloy studied. Vicker's hardness of the BMGs was found to be over 500 H_V for the as cast alloy which enhanced about 11 % more by annealing up to 600 °C/20 min. Intersected shear bands were observed. The observed promising mechanical and thermal properties showed that BMG studied can be used for industrial applications.

1. Introduction

Bulk amorphous materials are new class of materials having better physical, mechanical and thermal properties over their crystalline counter parts [1-3]. Bulk metallic glasses (BMGs) are super cooled liquids having disorder random non-equilibrium structure [4-5]. Zr-based BMGs especially Zr-Cu-Al-Ni quaternary and Zr-Cu-Al-Ni-M (M is a metal) pentanary alloys are widely studied due to their technical importance and are being used for structural and medical applications [6-8]. Ni and Cu are used in Zr based alloys for certain purposes such as to enhance mechanical strength and thermal stability of the super cooled liquid as well as competitive resistance to crystallization during casting process. As enthalpy of mixing (ΔH_{mix}) of Ni-Zr is very high (-44 kJ/mol) as compared to ΔH_{mix} of Cu-Zr phase (-23 kJ/mol) so Ni is avoided to prevent and stop the fast nucleation and rapid growth of Ni-Zr phases. Although a few Ni free Zr-based BMGs have been reported [9-11] but there is no report available on mechanical and thermal properties of $Zr_{48}Cu_{36}Al_8M_8$ (M = Nb, Ti and Ta) BMGs. In this manuscript, results on mechanical and thermal properties of $Zr_{48}Cu_{36}Al_8Nb_8$ BMG synthesized by Cu mold casting technique are reported.

2. Experimental

Alloy buttons of $Zr_{48}Cu_{36}Al_8M_8$ (M = Nb, Ti and Ta) alloys were prepared by using arc melting furnace by taking 50 grams mixtures of 3-4N pure metals of the alloy constituents. Alloy buttons were repeatedly melted at least four times to get the extended chemical homogeneity. Pieces of homogenized alloy buttons were further used for casting of alloy ingots of 1-5 mm thick rods of 70 mm length and sheets of size ~70x10x2 mm^3. Cu mold casting technique was used for materials preparation under pressure of $2x10^{-3}$ Pa. The samples taken from the center of the alloy ingots were characterized by XRD (RIGAKU Diffractometer using Cu $K_{\alpha 1}$ radiation with wavelength $\lambda = 1.54056$ Å) and DSC (Perkin Elmer 7/Pyris DSC and NETZSCH DSC 404 systems). Compression tests were applied on the samples with aspect ratio of ~2 under strain rate of $4x10^{-4}$/sec. Fractures studies were conducted by using scanning electron microscope (SEM) Leo 440i. Analysis of samples was done by energy dispersive x-ray analyzer (EDAX) attached with SEM. For high resolution images, filed emission scanning electron microscope (FESEM) was employed. Density was measured using the "Archimedes principle" at room temperature of ~20 °C. Acoustic studies were conducted and elastic moduli were calculated from the data collected. Low and high temperature differential scanning calorimetry (DSC) was conducted at various heating rates "r" of 5, 10, 20 and 40 K/min. Activation energy was calculated by using Kissinger equation. In order to study crystallization and phase formation, annealing of the samples was done at 350, 450, 550 and 600 °C for 20 minutes in PC controlled tubular furnace with maximum temperature variation of ±1° C by sealing the samples in quartz capsules under high purity Argon. Vicker's hardness of the as cast and annealed samples was measured by using "Everone hardness tester" under appropriate load. Here, results on $Zr_{48}Cu_{36}Al_8Nb_8$ BMG are presented.

3. Results and discussion

XRD results of as cast $Zr_{48}Cu_{36}Al_8Nb_8$ alloy ingot of 5 mm diameter and 70 mm length and sheet of size 70x10x2 mm^3 are shown in figure 1(a). The broad band without any diffraction peaks indicates the amorphous nature of the alloy. Similar patterns were observed for rod samples of 1-4 mm diameter. Average density of as cast samples was found to be 6.9794 g/cm^3 which is comparable with other Zr-based alloys [1]. XRD of the samples annealed at 350 to 550 °C results in the nucleation of $CuZr_2$ phase. XRD pattern of the sample annealed at 600 °C is shown in figure 1(b) which confirmed the formation of $Cu_{10}Zr_7$ stable phase. Atomic radii R_a (in nm) [12] and ΔH_{mix} [13] of the alloy constituents are presented in table 1. It is interesting that no other phase other than $CuZr_2$ was observed. The $CuZr_2$ phase nucleated due to strong interaction of Cu and Zr atoms as enthalpy of mixing of Zr-Cu pair is -23 kJ/mol while ΔH_{mix} of Cu-Al pair is only -1 kJ/mol. The Nb-Zr pair has positive heat of mixing (4 kJ/mol) so Zr and Nb don't react chemically. Actually Nb is added to enhance the mechanical properties. Diffraction peaks in XRD patterns of the samples annealed at 350 to 600 °C indicating the presence of crystalline phases. XRD results confirmed the formation of $CuZr_2$ metastable phase at temperature ≤ 550 °C and stable phase $Cu_{10}Zr_7$ at 600 °C. It is interesting to note that solute atoms (Al and Nb) are not more reactive to solvent atoms (Zr) than Cu atoms because atomic radius of Cu is smaller than all other atoms. So, Cu atoms are more reactive to Zr than Al and Nb atoms. As a result, $CuZr_2$ metastable and $Cu_{10}Zr_7$ stable phases are observed. Al and Cu atoms have positive role in increasing the thermal stability of the Zr-based alloys.

Figure 2(a, b) shows low and high temperature DSC traces of the $Zr_{48}Cu_{36}Al_8Nb_8$ BMG. Low temperature DSC at 10 and 40 K/min shows single stage crystallization while high temperature DSC at 20 K/min shows double stage crystallization at temperature ≥ 600 °C which is in agreement with the XRD results. Exothermic and endothermic reactions occurred are clear in DSC traces.

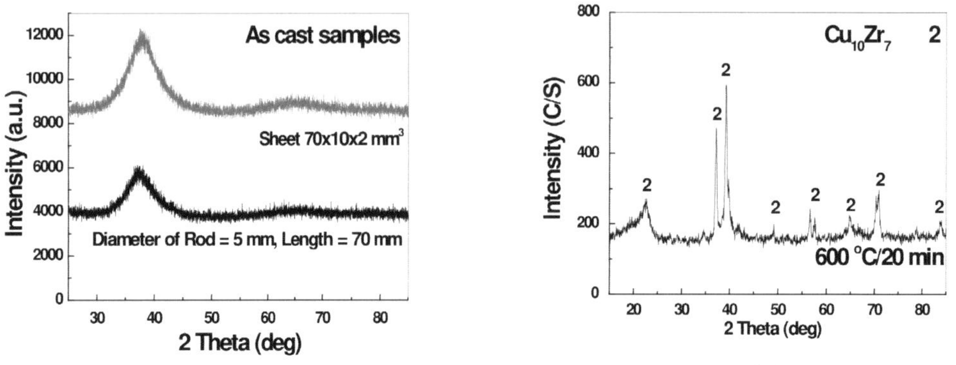

(a) (b)

Figure 1(a-b) XRD patterns of the as cast rod of 5 mm thick and 70 mm long as well as sheet of 70x10x2 mm^3 size (a) XRD of a sample annealed at 600 °C containing $Cu_{10}Zr_7$ stable phase (b)

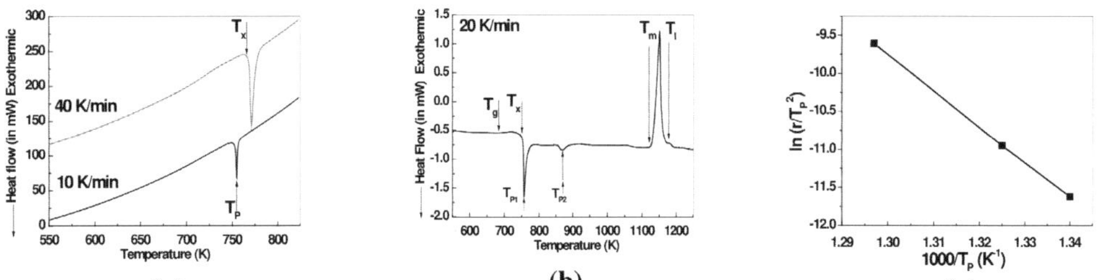

(a) (b) (c)

Figure 2(a-c) Low (a) and high temperature DSC (b) plus Kissinger plot (c) of the alloy studied

Table 1. Mixing enthalpies and atomic radii of alloy constituents

Alloy constituents	Atomic radii R_a (nm)	Mixing enthalpies ($kJmol^{-1}$) Zr	Al	Cu	Nb
		0.16025	0.14317	0.12780	0.14290
Zr	0.16025	-	-44	-23	4
Al	0.14317	-44	-	-1	-18
Cu	0.12780	-23	-1	-	3
Nb	0.14290	4	-18	3	-

Thermal parameters like glass transition temperature T_g, crystallization temperature T_x, peak temperature T_p, melting temperature T_m and liquidus temperature T_l , a key parameter known as "reduced glass transition temperature" $T_{rg1} = T_g/T_m$ and $T_{rg2} = T_g/T_l$, $\gamma = T_x/(T_l+T_g)$ [14], $\delta = T_x/(T_l-T_g)$ [15], $\beta = T_g{*}T_x/(T_l-T_x)^2$ [16] and $\omega = T_g/T_x - 2T_g/(T_g+T_l)$ [17] are also deduced and the results are presented in table 2(a) and table 2(b). In addition some more thermal parameters such as Hruby parameter $K_H = \Delta T_x/(T_m-T_x)$ [18] , Thermal parameter proposed by Saad and Poulain $K_{SP} = (T_p-T_x)(\Delta T_x)/T_g)$ [19], Weinberg parameter $K_W = \Delta T_x/T_m$, $K_{LL} = T_x/(T_g+T_m)$ [20], $K_1 = T_m-T_g$, $K_2 = \Delta T_x$, $K_3 = T_x/T_m$ and $K_4 = (T_p-T_x)(\Delta T_x)/T_m)$ [21] were also calculated and the results are presented in table 2(c). These thermal parameters showed good thermal stability and complete thermal behavior of the alloy studied. The values of thermal parameters of the presently reported $Zr_{48}Cu_{36}Al_8Nb_8$ BMG is better than the results reported by Zhang et al. [9, 10] for Ni free Zr-Fe-Al-Cu, Zr-Cu-Al-Ag and Jiang et al. [11] for $Zr_{100-x-y}(Cu_{5/6}Ag_{1/6})_xAl_y$ BMGs. The reason is substitution of Nb with Fe and Ag. The activation energy "E_{ac}" required for crystallization of an alloy is an important kinetic parameter to determine the thermal stability of amorphous phase. The dependency of crystallization temperature on heating rate was used to determine the associated activation energy (E_{ac}) by means of Kissinger's equation $\ln (r/T_p^2) = -E_{ac}/RT_p + \text{constant}$. Here "r" stands for the heating rate, T_p is the peak temperature in DSC scans, R is the gas constant ~8.3145 J/mol K, with slope $-E_{ac}/R = B$, where B is a constant [22]. Kissinger plot is shown in figure 2(c). The activation energy was calculated to be 392 kJ/mol.

Table 2(a). Thermal parameters (in K) evaluated from low temperature DSC done at various heating rates

r (K/sec)	T_g	T_x	ΔT_x	T_p	E_{ac} (kJ/mol)
5	660	740	80	746.5	
10	670	750	80	755.0	392
40	680	765	80	771.0	

Table 2(b). Various thermal parameters (in K) evaluated from high temperature DSC conducted at heating rate of 20 K/min

r	T_g	T_x	ΔT_x	T_p	T_m	T_l	T_{rg1}	T_{rg2}	γ	β	δ	ω
20	670	750	80	756.5	1119	1175	0.599	0.57	0.407	2.72	1.49	0.17

Table 2(c). Some more thermal parameters (in K) deduced for the alloy studied

r	K_H	K_W	K_{LL}	K_{SP}	K_1	$K_2 = \Delta T_x$	K_3	K_4
20	0.2168	0.714	0.4192	0.0776	449	80	0.67	0.465

Average Vicker's hardness of the as cast and samples annealed at 350, 450, 550 and 600 °C was found to be 524, 542, 570 and 580 H_V respectively. The maximum variation is ±10 H_V. Maximum increase in hardness was found to be ~11 %. The enhancement in hardness is due to nucleation of crystalline phases $CuZr_2$ (metastable) and $Cu_{10}Zr_7$. The hardness at temperature ≥ 550 °C is almost same due to nucleation of stable phase $Cu_{10}Zr_7$. Elastic moduli E, shear modulus G and bulk modulus K were calculated by measuring the acoustic properties of the alloy using the relations, $E = \rho V_s^2[(3V_l^2-4V_s^2)/(V_l^2-V_s^2)]$, $K = \rho(V_l^2 - 4/3\ V_s^2)$ and $G = \rho V_s^2$. Propagation of longitude and transverse acoustic waves are shown in figure 3(a). Time calculated Cal_{t1} and Cal_{t2} for longitudinal and sound waves are fond to be 6.997 and 3.128 μ sec with longitudinal acoustic velocity $V_l = 4.747$ Km/sec and transverse acoustic velocity $V_s = 2.122$ Km/sec respectively. Length and diameter of the samples taken was 5 mm and 7.4 mm respectively having density 6.979 g/cm^3 at 20 °C. The values of elastic modulus E, shear modulus G (modulus of rigidity), bulk modulus K (modulus of compression) and Poisson's ration were found to be 86.434 GPa, 31.43 GPa, 115.37 GPa and 0.375 respectively. The data shows excellent properties of the alloy studied. Compression tests were applied and stress strain curves (SS curves) of $Zr_{48}Cu_{36}Al_8M_8$ (M = Nb, Ti and Ta) are plotted in figure 3(b). Maximum fracture stress "σ_f" was found to be 1.953 GPa for $Zr_{48}Cu_{36}Al_8Nb_8$ BMG which is higher than fracture stress of $Zr_{48}Cu_{36}Al_8Ta_8$ alloy (1.540 GPa) and $Zr_{48}Cu_{36}Al_8Ti_8$ (1.086 GPa) alloys as well as previously reported BMGs **[1]**. It means Nb addition has beneficial role for the improvement of fracture strength. The lower fracture strengths of the alloys containing Ta and Ti are due to their partially crystalline nature. The observed properties are comparable or better than many well-known Zr-based BMGs [2-3].

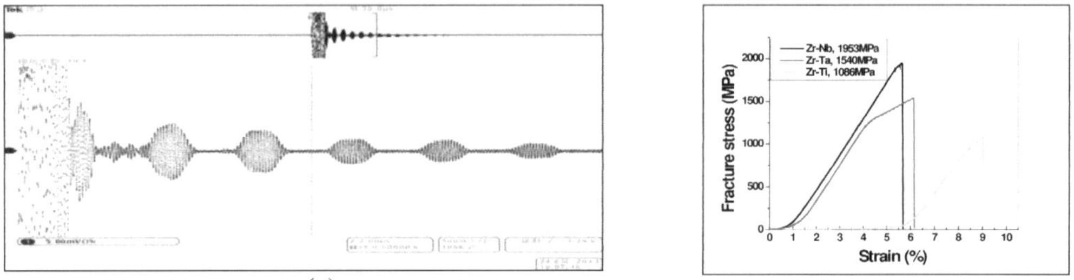

(a) (b)

Figure 3(a-b) Propagation of longitudinal/transverse waves for measurement of acoustic properties of $Zr_{48}Cu_{36}Al_8Nb_8$ BMG **(a)** SS curves of $Zr_{48}Cu_{36}Al_8M_8$ (M = Nb, Ti, Ta) BMGs (b)

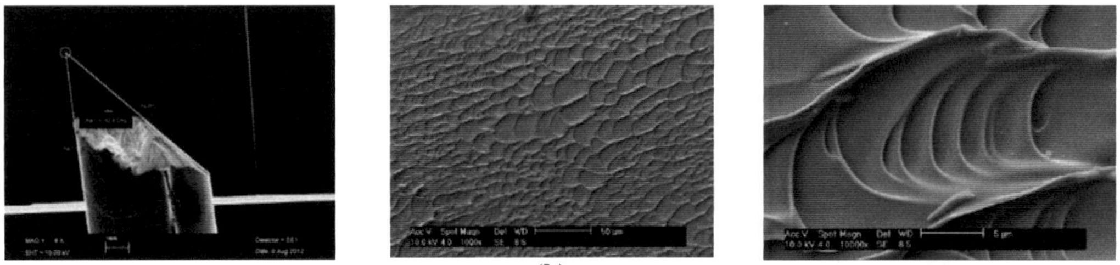

(a) (b) (c)

Figure 4(a-c) SEM images of compression tested samples at low **(a)** and high magnifications **(b)**

Secondary electron images (SEI) taken by SEM/FESEM of $Zr_{48}Cu_{36}Al_8Nb_8$ BMG of the compression tested fractured samples are shown in figure 4(a-c) at low and high magnifications. Figure 4(a) shows shear angel of 43.5° was observed. But the common observation is 45° for compression tested BMGs. Observation of shear angle less than 45° is in agreement with the previous studies [1, 5, 23]. Low shear angles show deviations from Tresca or Von Mises criterion [24]. Intersected shear bands over the cylindrical plain surface were examined. A shear band is usually initiated at a local region where the viscosity or the resistance of the materials to deformation is greatly reduced [25, 26]. Multiple shear bands were usually observed before fracture. There are two hypotheses about the local change in viscosity within the shear bands. The first suggest is that the reduction of viscosity within the shear bands is resulted from the generation and coalescence of free volume and localized heating [25-27]. It may be due to diffusion annihilation of free volume. The second hypothesis is that the local adiabatic heating beyond the glass transition temperature T_g, or even the melting temperature, thus leading to the deformation softening, is the main cause for formation of shear bands in BMGs. Many studies including the present results support this hypothesis. However the precise physical nature of the formation, initiation and propagation of shear bands as well as the possible mechanism for enhancing the plasticity in BMGs is still unclear [25, 26]. Plastic deformation of BMGs occurs in the form of shear bands due to lack of dislocation systems and grain structures. Shear band evolution process is critical to understand the failure mechanism, determining the ductility and improving the reliability of BMGs [27, 28]. The formation of shear band is rate-dependent process and shear banding instability in BMGs has not previously been identified [29]. The density of shear bands in Zr-based alloys increased with strain rate. Strain rate not only affects the strength but also exerts a significant influence on the formation of shear bands in BMGs [25-27]. Figure 4(b, c) shows FESEM images of veins patterns, plain facets of veins, their propagation and formation of liquid droplets within the veins. Veins branching and localized melting was observed in fractured areas. The compressive force is uniaxial and shear is opposite to veins propagation direction. Failure of materials catastrophically is not evident. Mostly the failure is ductile. The promising properties suggest that $Zr_{48}Cu_{36}Al_8Nb_8$ BMG can be used for structural and medical applications. However, further investigations are still needed to improve the ductility of the presently studied BMGs.

4. Conclusions

The $Zr_{48}Cu_{36}Al_8Nb_8$ BMG has good thermal stability and better GFA than the $Zr_{48}Cu_{36}Al_8Ta_8$ and $Zr_{48}Cu_{36}Al_8Ti_8$ metallic glasses. Supercooled liquid region of 80 K and facture stress of 1953 MPa indicate that the alloy studied has better thermal and mechanical properties than other similar quaternary Zr-based alloys. Veins patterns, liquid droplets, low shear angle (43°) and serrated flow observed in fractured samples indicate ductile behavior in Zr-based BMGs containing Nb. The addition of Nb in Zr-based BMGs is beneficial for achieving high fracture strength. Hardness increases due to nucleation of Cu-Zr phases. Activation energy of 392 kJ/mol shows high resistance to crystallization of metastable and stable phases in Zr-base alloys. Wide super cooled region shows good thermal stability of the alloy. XRD, DSC and SEM results are well agreed. The excellent properties indicate that $Zr_{48}Cu_{36}Al_8Nb_8$ BMG is suitable for structural, medical and jewelry applications.

5. Acknowledgements

The first author gratefully acknowledges travel grant and the financial support provided by Third World Academy of Sciences (TWAS). The local hospitality provided by Institute of Physics (IOP), Chinese Academy of Sciences (CAS), Beijing, China is highly appreciated. I am grateful to by Prof. D.Q. Zhao, P. Wen, D.W. Ding, D.P. Wang, Z.G. Zhu, all IOP friends such as Ma Jiang, Bo Huang, Gao, Cui, S.T. Liu and Jiao for providing technical support and best cooperation during my stay at IOP. Thanks to PAEC for permission and support to do this work at IOP, Beijing. Help provided by MSG officials is highly acknowledged. Special thanks to Prof. W.H. Wang, H.Y. Bai and D.Q. Zhao for excellent cooperation. The best cooperation of TWAS Authorities and all concerned official is highly appreciated. The authors are also grateful to Prof. H.F. Zhang and his co-workers for technical support.

6. References

[1] Iqbal M, Sun W S, Zhang H F, Akhter J I and Hu ZQ 2007 *Mater. Sci. Eng. A* 447 167

[2] Wang W H 2012 *Prog. Mater. Sci.* 57 487

[3] Wang W H *Prog. Mater. Sci.* 52 (2007) 540

[4] Iqbal M and Akhter J I 2011 *Adv. Mater. Res.* 326 11

[5] Iqbal M, Akhter J I, Zhang H F and Hu Z Q 2008 *J. Non-Cryst. Solids* 354 3291

[6] Iqbal M and Akhter J I 2010 Key Eng. Mater. 442 41

[7] Telford M 2004 *Mater. Today* 7 36

[8] Inoue A and Nishiyama N 2007 *Mater. Res. Bull.* 32 651

[9] Zhang Q S, Zhang W and Inoue A 2009 *Scripta Mater.* 61 241

[10] Zhang Q S, Zhang W and Inoue A 2006 *Scripta Mater.* 55 711

[11] Jiang Q K, Nie X P, Li Y G, Jin Y, Chang Z Y, Huang X M and Jiang J Z 2007 *J. Alloy. Compd.* 443 191

[12] Senkov O N and Miracle D B 2001 *Mater. Res. Bull.* 36 2183

[13] Takeuchi A and Inoue A 2005 *Mater. Trans.* 46 2817

[14] Lu Z P, Bei H and Liu CT 2007 *Intermetallics* 15 618

[15] Chen Q J, Shen J, Zhang D L, Fan HB, Sun J F and McCartney D G 2006 *Mater. Sci. Eng. A* 433 155

[16] Yuan Z Z, Bao S L, Lu Y, Zhang D P and Yao L 2008 *J. Alloy. Compd.* 459 251

[17] Long Z L, Xie G, Wei H, Peng J, Zhang P and Inoue A 2009 *Mater. Sci. Eng. A* 509 23

[18] Hruby A 1972 *Czech. J. Phys. B* 22 1187

[19] Saad M and Poulain M 1987 *Mater. Sci. Forum* 19 11

[20] Li Y 1999 *J. Mater. Sci. Technol.* 15 97

[21] Nascimento M L F, Souza LA, Ferreira E B and Zanotto E D, 2005 *J. Non-Cryst. Solids* 351 3296

[22] Kissinger H E 1957 *Ana. Chem.* 29 1702

[23] Iqbal M, Akhter J I, Hu Z Q, Zhang, H F, Qayyum A and Sun WS 2007 *J. Non-Cryst. Solids* 353 2452

[24] Donovan P E, 1989 *Acta Metall.* 37 445

[25] Liu L F, Dai L H, Bai Y L and Wei B C 2005 *J. Non-Cryst. Solids* 351 3259

[26] Liu L F, Dai L H, Bai Y L and Wei B C and Eckert J 2005 *Mater. Chem. Phys.* 93 174

[27] Wright W J, Hufnagel T C and Nix W D 2003 *J. Appl. Phys.* 93 1432

[28] Yang B, Morrison M L, Liaw P K, Buchanan R A, Wang G, Liu C T and Denda M 2005 *Appl. Phys. Lett.* 86 141904-1

[29] Dai L H, Yan M, Liu L F and Bai Y L 2005 *Appl. Phys. Lett.* 87 141916-1

Microstructure and fracture mode of a martensitic stainless steel steam turbine blade characterized via scanning auger microscopy and potentiodynamic polarization

D Saidi[1], B Zaid[1], N Souami[2], M Negache[1] and A Si Ahmed[3]
[1]Département de Métallurgie, Division de Technologie du Combustible, Centre de Recherche Nucléaire de Draria, BP 43 Draria, 16000 Alger, Algeria
[2]Département de Spectroscopie, Division des Techniques Nucléaires, Centre de Recherche Nucléaire d'Alger, 16000 Alger, Algeria
[3]Aix-Marseille Université, Im2np, UMR-CNRS 7334, Faculté des Sciences de Saint - Jérôme, F-13397 Marseille Cedex 20, France

E-mail: a.siahmed@univ-amu.fr

Abstract. The microstructure, the fracture and the resistance to pitting corrosion, in a10^{-2}M (NaCl + Na_2SO_4) solution, of steam turbine blades, made of martensitic stainless steel (12% Cr), were analysed prior and after their utilization (during about 100 000 hours). The unused blades display an inter-granular fracture mode whereas a trans-granular one emerges in used blades. The SAM analysis of the fractured surfaces reveals that this change is concomitant with carbon and chromium redistribution from the grain boundaries to the grain interior. The potentiodynamic polarization curves and the SEM-EDS analysis of the pitting sites show that the used blades are less susceptible to pitting corrosion than the unused ones. These results are interpreted as a further qualitative evidence of an evolution of the microstructure leading to the formation of new precipitates.

1. Introduction

Martensitic stainless steels are widely used as steam turbine blades [1-6]. The corrosion resistance is provided by chromium and the high mechanical performances stem from the martensitic structure. The most commonly used alloy is the Cr-12 (12 % at Cr with 1% at C). Additional alloying elements such as Ni, Mo, Si, V and Nb may improve furthermore the performances [5, 7, 8]. In commercial alloys, inclusions and impurities are always present and they can be deliberately introduced for the purpose of achieving better properties [1, 8].

Failure of steam turbine components is associated with crack initiations, from inclusions located at the free surface that could be, in addition, stimulated by pitting corrosion [4, 6]. However, crack initiations within the bulk, from nonmetallic inclusions, have also been identified after 40000 h service, suggesting that the microstructure has evolved [9]. Different carbide transformations have been identified in gas turbine components exposed for 100 000 h at temperatures in the range 673-723 K [10]. Hence, it appears that long term exposures, even at temperatures as low as 623 K, can result in metallurgical transformations having an effect on the mechanical performances. Here, in contrast with steam turbine components, carbon contamination may also interfere.

In this work, we investigate some microstructural changes induced by long-term service (about 100 000 h) of steam turbine blades made of a 12% Cr martensitic stainless steel via the analyses of the fractured surfaces. We will consider the changes occurring in the bulk of the blades (i.e., far from the surface).To this end, samples machined from unused and used blades will be fractured under ultra-high vacuum, within a Perkin Elmer equipment, at nitrogen temperature. The ensuing observations and the punctual analyses of the fractured surfaces will be performed in-situ, via the coupled Scanning Auger Microscopy (SAM). In addition potentiodynamic cyclic polarization measurements will be carried out. Then, the revealed pitting sites will be characterized by means of scanning electron microscope and energy dispersion spectroscopy (SEM-EDS). The motivation is to provide some qualitative evidence of the metallurgical evolutions, occurring within the bulk of the blades, which could spur further investigations aimed at improving the predictions of life span assessment.

2. Materials and Experimental

2.1 SAM analyses

The steel, from which the blades were manufactured, had undergone an austenization heat treatment at 1323 K for 1 h. After quenching in oil, another heat treatment had been performed at 898 K for 1 h. The composition, given in table 1, was determined by means of X-Ray fluorescence and comes out almost identical to that reported in reference [4]. The samples were cut from the central part of the blades and machined to identical dimensions. Then, they were notched for in-situ fracturing and characterization, performed under high vacuum (10^{-10}torr) at nitrogen temperature, within Perkin Elmer equipment (type PHI 600). Sample 1 and 2 will refer, respectively, to the unused blade and the used one. The service conditions, provided by the plant operator, were the time of utilization (about 100 000 h), the steam temperature (near 673 K), the maximum pressure (250 bars) and the spinning speed (3800 rpm).

Table 1: Chemical composition (in wt %) of the blades obtained by means of X-Ray fluorescence

C	Cr	Al	Si	Mn	P	S	N	Mo	Ni	Cu	Ti	V	Fe
0.21	12.4	0.02	0.45	0.74	0.05	0.003	0.01	1.1	0.77	0.12	0.35	0.03	Bal

As mentioned above, the characterizations of the fractured faces (i.e., the micrographs and the Auger spectra) were performed in-situ by means of the SAM. As a result, any surface contamination, during these characterizations, was de facto precluded. The Auger spectra were acquired with Cylindrical Mirror Analyser in derivative mode. We systematically observed the variations of the peak height intensities related to the dN(E)/dE spectra for carbon, chromium, iron and oxygen. The peak intensities were translated into concentrations of elements taking into account their respective sensitivity coefficient. The quantification of the composition is facilitated using peak-to-peak heights of the dN(E)/dE signal corrected by sensitivity factors. The Auger peak-to-peak heights of C (272 eV), O (510 eV), Fe (703 eV), Cr (529 eV) and P (120 eV) were measured.

2. 2. Electrochemical measurements

The solution, 10^{-2}M (NaCl + Na_2SO_4), was prepared from distilled water and analytical-grade reagents. The choice of this solution was furthermore motivated as it appears particularly adapted for the study of pitting corrosion of 12% Cr stainless steel [5]. Incidentally, its composition is close to the actual operating environment of the blades (i.e., because the vapor was obtained from sea water and sodium sulfate was present in the ion exchange resin used in the water purification system).

IOP Conf. Series: Materials Science and Engineering **60** (2014) 012004 doi:10.1088/1757-899X/60/1/012004

The rectangular working electrodes were machined with an exposed surface of 0.7 cm² and then mounted in epoxy resin. Before each anodic polarization, the electrodes surfaces were mechanically polished with silicon carbide paper to 1200 grit, degreased, rinsed and dried.

The electrochemical tests were carried out, at room temperature, in a conventional three-electrode electrochemical glass cell with a saturated calomel electrode, as reference (SCE), and platinum sheet, as counter electrode. The polarization curves were recorded in cyclic potentiodynamic mode, with a potential scan rate of 0.166 mV/s, using a potentiostat/Galvanostat A273 (EG&G Princeton Applied Research) that was connected to a fully compatible IBM computer and controlled by EG&G Softcor M352 corrosion measurement and analysis software. Polarization starts at -250mV under open circuit potential. The samples used in cyclic polarization were observed by means of a Scanning Electron Microscopy (SEM) (Philips environmental: FEI/PhilipsXL30 FEG ESEM). The qualitative punctual analysis of the pitting sites was performed by means of Energy Dispersion Spectroscopy (EDS).

3. Results and discussion

3.1. Observations of the fractured faces

The micrographs of the fractured faces are shown in figure 1 and 2. One can see that the crack path of the fractured sample 1 (unused blade) is dominantly inter-granular and the fractured surface appears smooth. In contrast, the fractured sample 2 (used blade) reveals a trans-granular fracture mode with the presence of facets and cavities. Three samples of each blade were fractured and the dominant fracture mode persisted. Therefore, the reproducibility of the experiments can be anticipated.

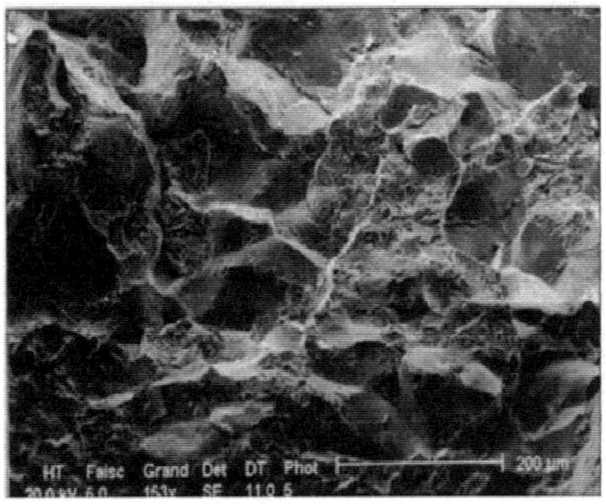

Figure 1. Typical SAM micrograph showing the fractured surface of sample 1(unused blade) displaying the dominance of the inter-granular fracture mode.

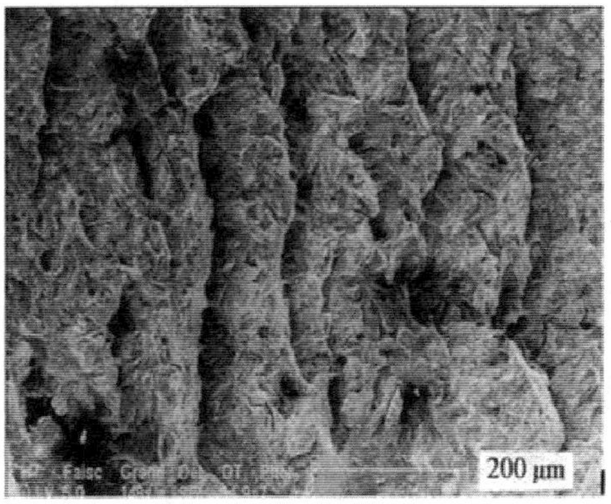

Figure 2. Typical SAM micrograph showing the fractured surface of sample 2 (used blade) displaying the dominance of the trans-granular fracture mode.

3.2. AES line profiles and punctual analyses

The typical Auger spectra acquired from the above fractured surfaces are shown in figure 3 and 4. It seems straight forward to assign the periodicity of the Auger transition to the effects induced by the grain boundaries. Hence, a noticeable carbon and chromium enrichment in the grain boundary region arises in sample 1 (figure 3). On the contrary, sample 2 exhibits a depletion of these elements as seen in figure 4. The analyses were repeated over a few zones and a periodicity was always observed. However, we hasten to point out that the reported period (i.e., the distance between two consecutive depletion or enrichment lines) cannot be interpreted as an average value of the grain size as this would require a statistical treatment. The results of the punctual analyses are shown in figure 5 and 6. The comparison of the spectra of figure 5 (sample 1) with those of figure 6 (sample 2), shows a salient evolution of the intensities of the Auger signals pertaining to C and Cr (located, respectively, at 270 and 490 eV). Clearly, the service conditions have induced a significant redistribution of C from the grain boundaries to the grain interior. The extent of Cr redistribution is however less pronounced.

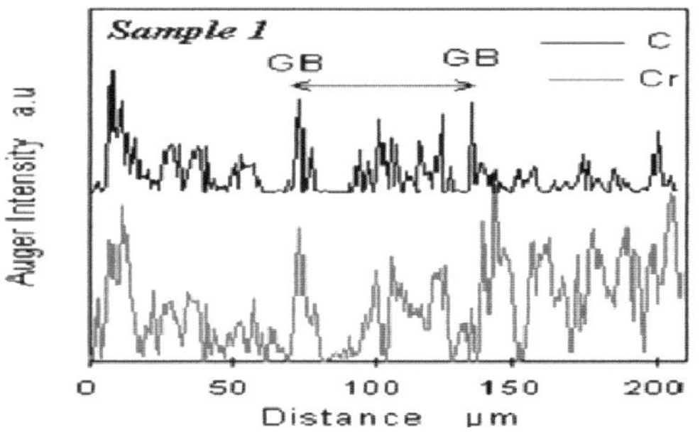

Figure 3. Typical AES line profile of carbon and chromium as a function of distance over the inter-granular fractured surface of sample 1 (unsed blade). The enrichment of carbon and chromium at grain boundaries emerge.

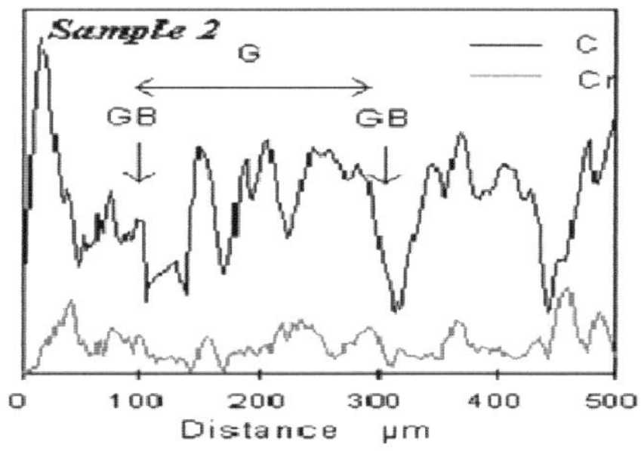

Figure 4. Typical AES line profile of carbon and chromium as a function of distance over the trans-granular fractured surface of sample 2 (used blade). The depletion of carbon and chromium at grain boundaries emerge.

The relative AES peak height intensities of the dN(E)/dE spectra, for sample 1 and 2, allow to estimate the ratio of the concentration of C (and Cr) at grain boundaries (GB) to their respective counterpart within the grain interior (G). The estimates, derived from the Seah formula [11], are given in table 2.

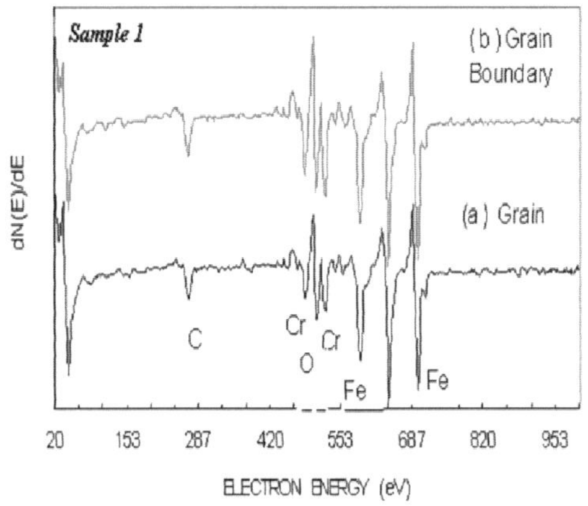

Figure 5. AES spectra of sample 1 (unused blade) corresponding to: (a) the grain interior and (b) the grain boundary

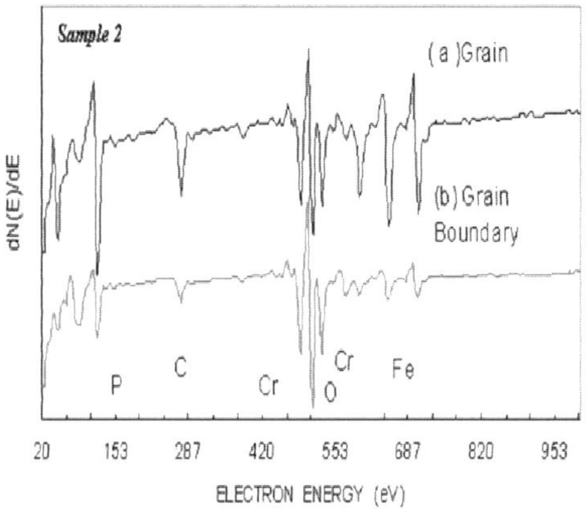

Figure 6. AES spectra of sample 2 (used blade) corresponding to: (a) the grain interior and (b) the grain boundary

Table 2. Ratio of the concentration of carbon (C) and chromium (Cr) at grain boundaries (GB) to their respective counterpart within the grain interior (G) for sample 1 (unused blade) and 2 (used blade) calculated by means of Seah formula [11] and the sensitivity factors.

Steel Type	Sample 1	Sample 2
$X_c(GB)/X_c(G)$	1.19	0.56
$X_{cr}(GB)/X_{cr}(G)$	1.63	1.47

It appears that the grain boundary region in sample 1 is richer in carbon and chromium than the grain interior. The enrichment ratio, resulting from the fabrication process, is found equal to 1.2 (for C) and 1.6 (for Cr). In the case of materials after utilization (sample 2), the grain boundary region is no longer richer in C. Instead, the ratio, $X_C(GB)/X_C(G)$, is drastically decreased (by nearly a factor of 2) to 0.6. However, the reduction seems less noticeable for Cr.

As mentioned above, the analyses were performed over a few different zones that were taken randomly. The degrees of carbon and chromium redistributions kept the same trend. Since a carbon contamination through the external surface of the blades has to be ruled out, due the composition of the blade environment (water vapour), the rather conspicuous carbon redistribution can be assigned, with reasonable confidence, to intrinsic metallurgical transformations that were induced by the service conditions.

3.3. Electrochemical studies (open circuit potential and potentiodynamic cyclic polarization)

The results of figure 7 show the open circuit potential (OCP) versus time obtained with samples 1 and 2 in a 10^{-2}M (NaCl + Na_2SO_4) solution. It can be seen that the OCP curves of both samples decrease with time until reaching their respective quasi-stabilization limit. These curves evolve towards the negative less noble direction. However, the curve of sample 1 (unused blade) lies well below that of sample 2 (used blade). Accordingly, these results could already suggest that the used material is more resistant to corrosion in the 10^{-2}M (NaCl + Na_2SO_4) solution.

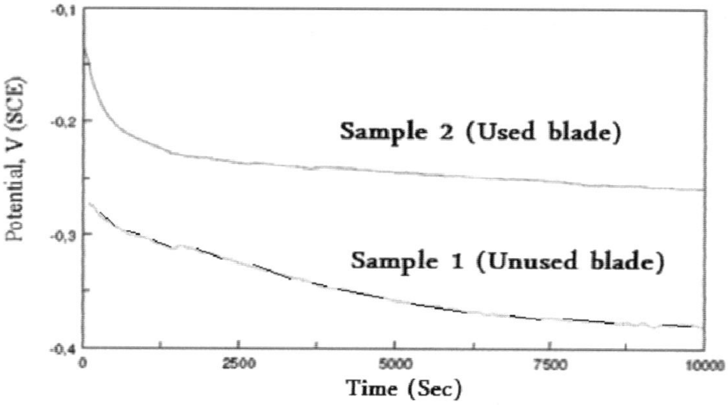

Figure 7. Open circuit potential versus time in a 10^{-2}M (NaCl + Na_2SO_4) solution for sample 1 (unused blade) and 2 (used blade). The potential V (SCE) is measured by taking the saturated calomel electrode as reference (SCE).

The cyclic polarization curves of sample 1 and 2, after exposure in the foregoing solution, are shown infigure 8 and 9. Both curves exhibit spontaneous passivity after the onset of anodic current and a wide passivity domain between the corrosion potential, Ecor, and the pitting potential, Epit, above which stable pits initiate. Figure 8 and 9 also display the reverse anodic scans. The cycle is furthermore characterized by the area of the hysteresis loop and the protection potential Eprot below which no pitting occurs. The electrochemical parameters are given in Table 3.

In comparison with the unused blade, the used material is characterized by:

- nobler pitting and protection potentials;
- a hysteresis loop, of slightly smaller area, that closes at a more positive potential;
- a smaller difference between Eprot and Ecor;
- a greater difference between Epit and Ecor.

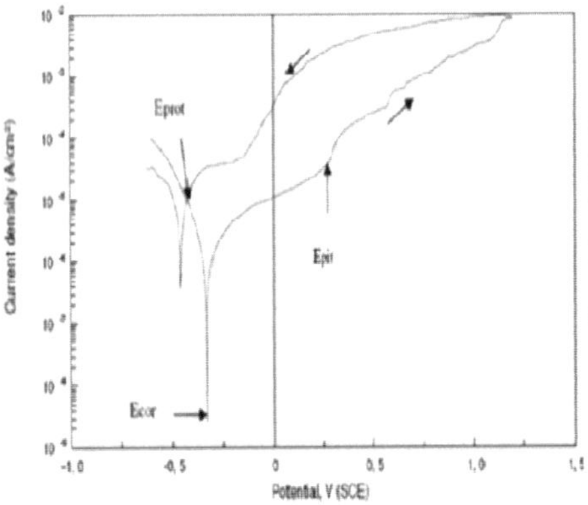

Figure 8. Cyclic potentiodynamic curves of sample 1 (unused blade) measured in 10^{-2}M (NaCl + Na_2SO_4) solution after stabilization of OCP. Epit, Ecor, and Eprot stand for pitting potential, corrosion potential and protection potential.

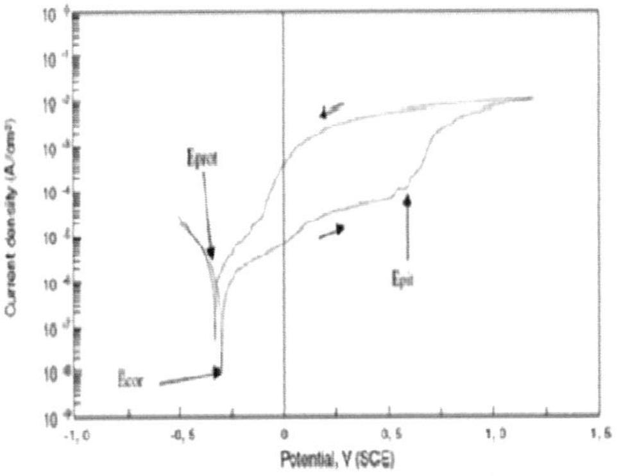

Figure 9. Cyclic potentiodynamic curves of sample 2 (used blade) measured in 10^{-2}M (NaCl + Na_2SO_4) solution after stabilization of OCP. Epit, Ecor, and Eprot stand for pitting potential, corrosion potential and protection potential.

Table 3. Electrochemical parameters obtained from figure 8 and 9. Ecor, Epit and Eprot stand for corrosion potential pitting potential and protection potential

Type Steel	Ecor(mV)	Epit (mV)	Eprot (mV)	Eprot-Ecor (mV)	Epit-Ecor (mV)
Sample 1	-328	296	-424	-96	620
Sample 2	-298	587	-323	-25	885

All the above mentioned characteristics are in accord to express a better resistance to pitting corrosion of the utilized blades [12].

The SEM observations, of the surfaces of the samples after the cyclic potentiodynamic experiments are shown in figures 10 to 12. The qualitative punctual analyses (i.e., the typical EDS spectra) of the pitting sites are also given. Noticeably, the responses of the samples are quite different. First, the pits in the unused blades are much larger. Next, the EDS spectra depend on both the sample and the pit size. For instance, in the case of the unused blades, the Si peak is more noticeable in smaller pits than in the larger ones. Finally, in used blades, the presence of Nb is systematically associated with smaller pits.

4. Discussion

It is recognized that the alloy microstructure can have strong effects on the tendency for pitting corrosion and that the pits more likely initiate at some heterogeneities such as inclusions and

IOP Conf. Series: Materials Science and Engineering **60** (2014) 012004 doi:10.1088/1757-899X/60/1/012004

precipitates [12]. The detailed description of the physical and chemical processes involved during a potentiodynamic cyclic polarization test, in the considered materials, is beyond the scope of this paper. In spite of that, the curves of figure 8 and 9 along with the SEM-EDS analyses can be interpreted as the finger print of the microstructural state as it is also suggested for other types of stainless steels [13, 14].

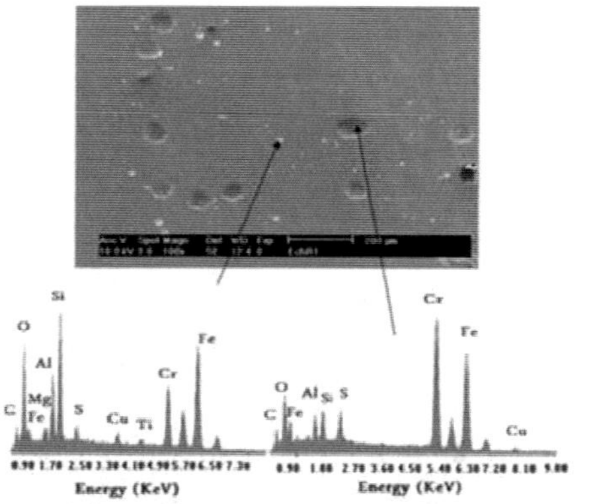

Figure 10. Typical SEM observation of an area Of sample 1 (Unused blade) after a cyclic polarization experiment and EDS spectra from from the pitting sites.

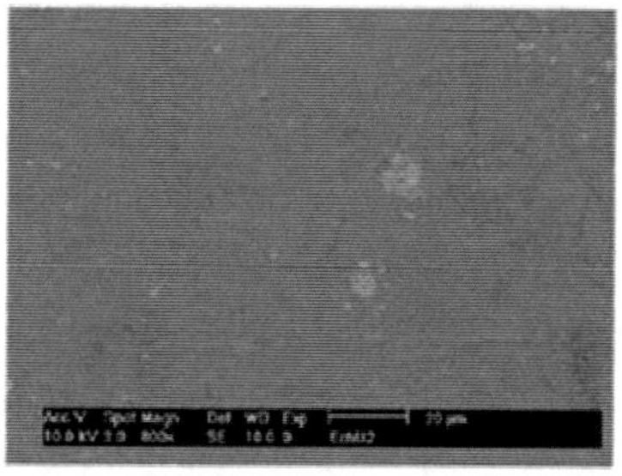

Figure 11. Typical SEM observation of an area of sample 2 (used blade) after a cyclic polarization experiment

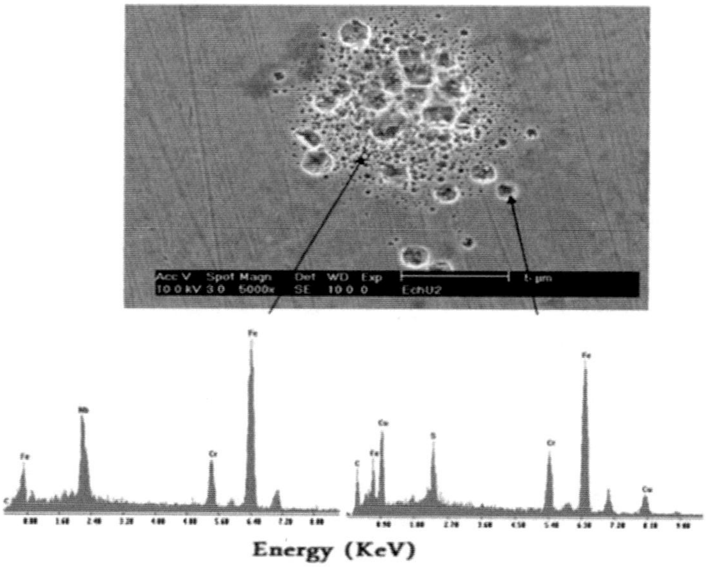

Figure 12. Typical SEM observation of an area of sample 2 (used blade) after a cyclic polarization experiment representing a zone on the left side of figure 11 with a higher magnification and EDS spectra from the pitting sites

Accordingly, the better resistance to pitting of used blades may stem from the microstructural evolution resulting from long time exposure to the service conditions. The differences in the EDS spectra extracted from the pitting sites somewhat reflect the formation of new precipitates or the modification of the vicinities of the existing heterogeneities due, for instance, to fatigue damage accumulation. However, it must be pointed out that these spectra do not display the actual composition of the precipitates or inclusions because the dissolution of elements and the passivation/re-passivation also interfere during the polarization tests [15]. Nevertheless, in the light of the noticeable differences that are mentioned just above, one can connect the spectra to the nature of the heterogeneities that have spawned the pits.

The marked carbon redistribution from the grain boundaries to the grain interior (*cf.* section 3.2), derived from the Auger analyses, highlights that the long-time exposure to the service conditions can lead to changes of the grain boundaries structure. It could also induce small carbide precipitates within the grain such as those linked to NbC whose formation is expected due to the strong affinity of Nb for C. Indeed, the presence of Nb, which is systematically associated with smaller pits (of sub-micron size) distributed within large pit clusters (diameter of about 7 μm), emerges in the EDS spectra (figure 12). This, incidentally, indicates that Nb, although not identified by X-Ray fluorescence (table 2), is present as a trace element.

The change of the fracture mode (from inter-granular in the unused blade to trans-granular in used blades), which is well illustrated by figure 1 and 2, more likely originates from the microstructural evolutions suggested by the preceding interpretations. However, the contribution of fatigue damage is expected to play a triggering role in the occurrence of the trans-granular facture. Indeed, it has been pointed out that fatigue damage can lead to the initiation of micro-cracks in the grain interior at microstructural heterogeneities [16]. Hence, these micro-cracks may favour the dominance of the trans-granular fracture mode that is provoked by the fracture tests. Since the samples were taken from the bulk of the blades (section 2), the foregoing interpretation implies that the failure of steam turbine blades, which is commonly associated with crack initiations at their surface [4, 6], could also originate from cracks located within their bulk [9].

5. Conclusion

This work shows that steam turbine blades made of martensitic stainless steel (12% Cr) undergo metallurgical changes under the service conditions. The change of the fracture mode from inter-granular (in unused blades) to trans-granular (in used blades) is associated with the formation of new precipitates in the grain interior along with the initiation of micro-cracks induced by fatigue damage accumulation.

We acknowledge that further experiments are required for the characterization of the exact nature of the metallurgical transformations. Nevertheless, the micrographs of the fractured faces and the potentiodynamic cyclic polarization curves concur to reflect the evidence of a microstructural evolution. This inference, although qualitative, could be of interest for those who seek refinements of life span assessment of turbine blades.

6. References

[1] Degallaix G, Vogt J B and Foct J 1990 *Mem&Sci Revue de Métallurgie* 87 47

[2] Spachn H, Gangloff R P and Ives M B 1988, in Environment-Induced Cracking of Metals, *Corrosion NACE-10, TX, Houston* p 449

[3] Turnbull A 2008 *Corrosion* 64420

[4] Perkins K M., Bach M R 2005 *Int. J. Fatig.* 27 1508

[5] Pineau A and Castagné J L 1979. *Mém&Sci, Revue de Métallurgie* 6 369

[6] Prabhugaunkar G V, Rawat M S and Prasad C R 1999, *In: 7th International Conference on Shot Peening ICSP-7, Varsovie* p 177

[7] Lewandowski J J, Hippsley C A and Knott J F 1987, *Acta Metall.* 35 1453

[8] Mayes I C and Baker T J 1986 *Materials Science and Technology* 2 133

[9] Karastojkovic Z, Janjuševic Z and Kovacevic Z 2007 in: Low Pressure Turbine Blades, Proceedings of the 3rd International Conference on Deformation Processing and Structure of Materials, September 20 –22, Belgrade, Serbia p 6

[10] Manilova E and Vander Voort G 2004 Materials Characterization 52 27

[11] Seah M P and Dench W A 1979 *Surf Interf. Anal.* 1 2

[12] Frankel G S 1998 Journal of the Electrochemical Society 145 2186

[13] Jeon S, Kim S, Lee I, Park J, Kim K and Kim J Y 2011 *Park Corrosion Science* 53 1408

[14] Cvijović Z and Radenković G 2006 *Corrosion Science* 48 3887

[15] Hara N, Hirabayashi K, Sugawara Y and Muto I 2012 *International Journal of Corrosion* 2012 2206

[16] Meyer S, Brückner-Foit A and Möslang A 2003 *Computational Materials Science* 26 102

Effect of radiation cross-linking on the abrasive wear behaviour of polyethylenes

Rizwan M Gul[1] and Tahir I Khan[2]
[1]Department of Mechanical Engineering, University of Engineering and Technology, Peshawar, Pakistan
[2]Department of Mechanical and Manufacturing Engineering, University of Calgary, Alberta, Canada

E-mail: rgul@nwfpuet.edu.pk

Abstract. This study explores the differences in the dry abrasive wear behavior of different polyethylenes, and compares the effect of radiation cross-linking on the wear behavior. Four different types of polyethylenes: LDPE, LLDPE, HDPE and UHMWPE were studied. Cross-linking was carried out by high energy electron beam with radiation dose of 200 kGy. The results show that in unirradiated state UHMWPE has excellent wear resistance, with HDPE showing comparable wear properties; both LDPE and LLDPE exhibit high wear rate. Cross-linking improves wear rate of LDPE and UHMWPE, however, the wear rate of HDPE and LLDPE increases with cross-linking.

1. Introduction

In general, polyethylenes possess good electrical insulation and chemical properties, and their mechanical properties are determined by the skeletal structure of the polymer chain. The different classes of polyethylenes such as low density (LDPE), linear low density (LLDPE), high density (HDPE), and ultra-high molecular weight (UHMWPE) are being used in a number of diverse applications. These can include house-wares and packaging through to cable insulation and high pressure piping [1-4]. In most cases these polymers are required to show good mechanical properties, particularly at elevated temperatures, and good abrasive wear resistance [5]. As a result, in recent years there has been a growing interest in enhancing the properties of these polymers by inducing cross-linking of the polymer chain through radiation cross-linking; and the effects of irradiation of polyethylenes have been studied extensively [6-8]. In brief, this process uses high energy radiation in the form of high energy electrons or γ-rays which generate free radicals along the polymer chain with the abstraction of hydrogen atoms [9]. These free radicals then react to form a cross-linked network structure with a corresponding change in mechanical properties. Although the cross-linking process can be performed by other methods, radiation cross-linking is a convenient process which does not result in by-products, and produces varying degrees of cross-linking without reformulation. The commercial applications of cross-linked polyethylenes include high temperature applications like insulation on electrical wires and cables, hot water pipes and tubings, heat shrinkable products, and steam resistant food packaging [10-13]. Radiation cross-linked UHMWPE has also been widely used for making orthopaedic implants [14-17].

In this study, the wear resistance under dry abrasive conditions of 4 types of polyethylenes (LDPE, LLDPE, HDPE and UHMWPE) and the effect of radiation cross-linking on wear behaviour has been investigated. The comparative data will assist in materials selection and in exploring the different ways to improve performance. The research also compares and relates the effect of the degree of crystallinity and cross-link density (gel content) induced within the polymer to subsequent changes in the wear resistance properties of these polyethylenes.

2. Experimental

2.1. Materials

The polyethylenes grades selected were: LDPE-EASTMAN 800A (Density ρ 0.918 g/cm^3; Melt Flow Index MFI 1.7 g/10 min), LLDPE-BASF LL 18K FA (ρ 0.919 g/cm^3; MFI 1 g/10 min), HDPE-EASTMAN H6001-A (ρ 0.962 g/cm^3; MFI 8 g/10 min), and UHMWPE-Hoechst Celanese GUR 1050 (ρ 0.930 g/cm^3; MFI <0.01 g/10 min).

For LDPE, LLDPE and HDPE, a Morgan injection moulding press was used for producing rectangular bars 9 mm thick, 25.4 mm wide and 100 mm long. The injection temperature was 218°C, injection pressure 0.4 MPa and clamp pressure 98 kN (for HDPE 133 kN). The mold temperature varied from 65-120°C. In the case of UHMWPE, commercially available bar-stock (ram extruded) material processed by WestLake Plastics, Lenni, PA, USA was used.

Irradiation was carried out using a 2.5 MeV Van de Graff generator at a dose rate of 25 kGy per pass (1 Mrad=10 kGy), at room temperature in air. Eight passes were used to give samples a total dose of 200kGy.

2.2. Characterization

Swelling experiment was carried out according to ASTM test method D 2765 method C to measure the cross-link density (gel content). Two samples (approximately 2-2.5 mg) of each material were dipped in xylene beaker, which was placed in a copper container, and heated on a hot plate. In the case of LDPE, LLDPE and HDPE specimens, xylene was heated to 110°C for 5 hours, while UHMWPE specimens were dipped in xylene at 130°C for 7 hours. The different temperatures were maintained due to the fact that control UHMWPE samples took more time to dissolve and cross-linked UHMWPE samples took more time to equilibrate. The swollen gel was quickly transferred to a weighing bottle, covered and weighed. Samples were then dried in an oven at 85°C for 24 hours and weighed. The degree of swelling was calculated from the ratio of the weights of swollen and dry samples. Flory's swelling equation was used to determine the molecular weight between cross-links and cross-link density. The specific gravity of the diluent was taken as 0.81 Kg dm^{-3} and density of the amorphous polyethylene was taken as 0.862 g/cm^3.

Differential scanning calorimetry (DSC) of the control and radiated samples was carried out on a Perkin–Elmer DSC 7a model to determine the degree of crystallinity and melting points of the different materials. The heating rate was 10°C/min. Results were also used to ascertain any degradation taking place in the samples. The heat of fusion of the materials was compared with the heat of fusion of 100% crystalline polyethylene, taken as 289 J/g.

2.3. Wear test

The wear rate of the control and 200 kGy radiated polymer samples was compared using a pin-on-disc wear testing machine under dry abrasive conditions. The polymer was machined to a cylindrical shape to form the "pin" which had a diameter of 5 mm and length 10 mm. The pin was held against a SiC abrasive paper (600 grit size) attached to the disc using a load of 1 Kg. The test was performed with a sliding speed of 1 m/s for a total sliding distance of 730 m. A new SiC paper was used for each polymer sample tested. No lubricant was used and no attempt was made to clean the abrasive paper from transferred debris. In some tests the pin suffered from extreme wear and a shorter sliding distance had to be used. The wear rate was determined by recording the change in pin weight as a function of sliding distance.

3. Results

Table 1 shows the molecular weight between cross-links and the cross-link density of the four different types of radiated polyethylenes measured by the swelling experiment. Highest cross-link density is achieved in UHMWPE, followed by LLDPE, LDPE and HDPE, respectively. The degree of crystallinity and onset melting point measured by DSC analysis are shown in Table 2. The comparison of the degree of crystallinity of unirradiated samples shows that percent crystallinities of LDPE and LLDPE are in the same range of is 33-34%, while HDPE has the highest crystallinity 67%. UHMWPE lies in between at about 50% crystallinity. Cross-linking does not produce appreciable change in the degree of crystallinity and melting points of LDPE, LLDPE and HDPE. The small variation can be attributed to measurement error. While in the case of UHMWPE appreciable increase in crystallinity is observed.

Table 1. Molecular weight between cross-links (g/mole) and cross-link density (moles/m^3) of the control and cross-linked polyethylenes measured by the swelling experiment

Sample ID	Mol. Wt. between cross-links, g/mole	Cross-link Density, moles/m^3
LDPE 200 kGy	33430 (±10550)	27.1 (±8.57)
LLDPE 200kGy	18700 (±328)	46.1 (±0.81)
HDPE 200 kGy	42420 (±1620)	20.3 (±0.78)
UHMWPE 200 kGy	16220 (±4440)	55.2 (±15.12)

Table 2. Degree of crystallinity (%) and the onset melting point (°C) of the control and cross-linked polyethylenes measured by the DSC analysis

Sample ID	Degree of Crystallinity, %	Onset Melting Temperature, °C
LDPE Control	33.7	104.4
LDPE 200 kGy	32.1	100.6
LLDPE Control	33.4	110.2
LLDPE 200kGy	33.7	110.0
HDPE Control	67.3	128.2
HDPE 200 kGy	68.8	132.3
UHMWPE Control	49.7	130.0
UHMWPE 200 kGy	60.7	132.9

The comparison of the wear behavior of the control unirradiated samples of 4 polyethylenes is presented in Figure 1. A distinct behavior is shown by LDPE and LLDPE in contrast to HDPE and UHMWPE; both the former suffer high wear. The transfer of material on the abrasive paper was also different in the two categories. In LDPE and LLDPE fine flake like wear debris were transferred onto the abrasive paper, but not as a continuous transfer film; the pin suffered from extreme wear and lip formation was observed. Whereas in the case of HDPE and UHMWPE little loose debris were produced leading to smooth and continuous transfer film on the abrasive surface. This play a significant role in reducing further wear of the polymer surface, and provides a counter-surface with low coefficient of friction. Lip formation was not observed in these materials. The comparison of 200 kGy irradiated specimen with the control also exhibit two distinct behaviours (Figure 2). In LDPE and UHMWPE wear resistance improves with irradiation, while in LLDPE and HDPE the wear resistance degrades with irradiation. The most wear resistant material in this study is radiated UHMWPE, and on the other extreme the least wear resistant material is control LDPE.

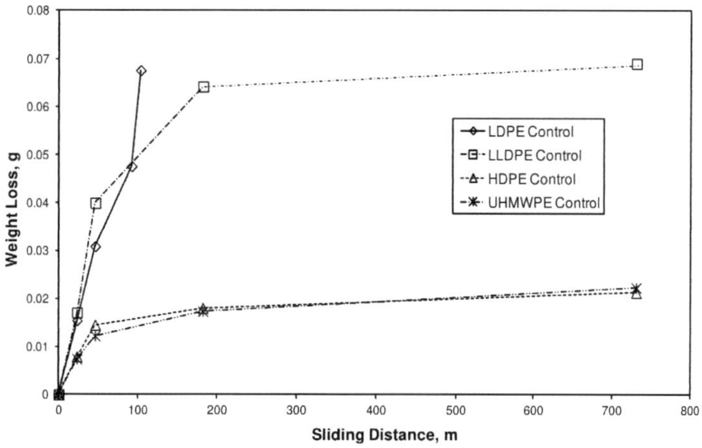

Figure 1. Weight loss (g) versus sliding distance (m) curves for unirradiated (Control) polyethylenes measured from a pin-on-disc wear test under dry abrasive conditions

4. Discussion

To explain the differences in behavior of the different polyethylenes, it is important to understand the skeletal structure of these materials. LDPE has a highly branched structure with crystallinity ranging from 30-40%, as opposed to HDPE which has a linear structure and crystallinity as high as 75% [4, 18]. LLDPE has controlled amount of short branches of the desired length, with little or no long-chain branching making it different from LDPE, however, degree of crystallinity remains about the same. As its name suggest UHMWPE has very high molecular weight linear molecules, which leads to lower crystallinity (typically 45 % [19]) and higher entanglement densities. The differences in behavior are due to the relative amounts of crystallinity and entanglement density, with chain branches and higher molecular weight promoting higher entanglement density and lower crystallinity.

The lower cross-link density achieved in HDPE is due to the fact that cross-linking mainly occurs in amorphous regions which are very small in HDPE. The cross-link densities achieved in UHMWPE are very high as compared to other materials due to its higher molecular weight leading to larger entanglement densities and larger potential cross-linking sites. LDPE show similar behavior but lower cross-link densities; this is due to smaller average molecular weight of LDPE as evident by its large melt flow index (MFI); a smaller MFI indicates a higher molecular weight. As compared to HDPE, LDPE has higher cross-link density at similar radiation dose as also reported in earlier studies [12]. This is mainly due to the lower crystallinity and higher branch content of LDPE as compared to

HDPE. The higher cross-link densities achieved in LLDPE as compared to LDPE can also be due to its higher molecular weight, as shown by its lower MFI.

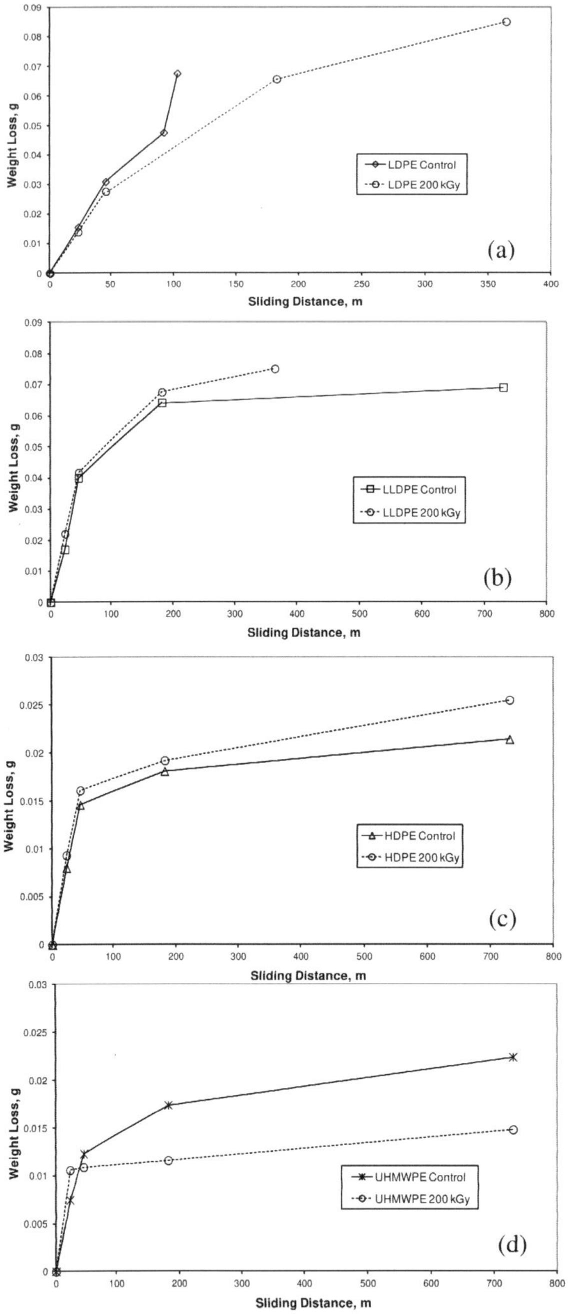

Figure 2. Comparison of the wear behavior of unirradiated (Control) and 200 kGy radiated samples of (a) LDPE, (b) LLDPE, (c) HDPE, and (d) UHMWPE

The DSC analysis confirms that the degree of crystallinity is in range of that expected from literature. The crystallinity and melting point did not changed much with cross-linking in LDPE, LLDPE, and HDPE, which verifies the absence of degradation in molecular weight by oxidation and radiation damage, since any such change would increase the degree of crystallinity. The exception is UHMWPE where appreciably high increase in crystallinity is observed due to degradation in molecular weight. The reason is that free radicals produced during radiation exposure in UHMWPE can survive for a long time due to diminished molecular motion because of its high molecular weight; as oxygen diffuses in the material the radicals are consumed leading to further degradation in molecular weight. The lower molecular weight fragments produced are able to reorganise and crystallize more readily [8]. Since the samples were analysed after two to three years of storage enough time was available for the diffusion of oxygen and subsequent degradation. The technique employed to avoid this problem is to melt anneal UHMWPE after radiation exposure to extinguish any free radicals present and add antioxidents such as vitamin E [20, 21].

The comparison of the wear behavior of control polyethylenes show that LDPE and LLDPE experience very high wear rate as compared to HDPE and UHMWPE, which show comparable and very small wear rate. This observation can be explained on the basis of the smooth transfer film produced in the two later cases leading to reduced friction coefficient and reduction of further wear. The production of this continuous transfer film can be further explained on the basis of high strain to failure and toughness generally observed in HDPE and UHMWPE.

Cross-linking improves the wear resistance of LDPE and UHMWPE, and degrades it in LLDPE and HDPE. Wear mechanisms of polyethylenes can be divided in the following categories; adhesive, abrasive, and surface fatigue wear. Adhesive and surface fatigue wear involves the removal of polymer by the harder counterface asperities after many interactions, while abrasive wear involves the ploughing of polymer by the surface asperities of a harder material and is associated with rough surfaces [22, 23]. Wear test used in this study is likely to produce abrasive wear due to absence of any lubricant and presence of an abrasive counter surface. The decrease in wear rate of radiated LDPE and UHMWPE is due to the formation of extensive cross-link network structure in these cases. This network structure is produced by the cross-linking of large number of entanglements produced by the very high molecular weight of UHMWPE and by the branched structure of LDPE. The cross-links in the network structure will obstruct the formation of wear debris by resisting the uncoiling of the physical entanglement; leading to an improved wear behavior. Due to the linear molecular structure of HDPE the network formation due to cross-linking will not be extensive leading to decreased wear resistance with radiation. The unexpected decrease in the wear resistance of radiated LLDPE, in spite of high cross-link density, can be due to degradation in mechanical properties such as toughness.

5. Conclusions

The results provide a comparison of the properties of different types of unirradiated and radiated polyethylenes. It shows that UHMWPE has the highest cross-linking efficiency, followed by LLDPE, LDPE and HDPE, respectively. The degree of crystallinity does not change appreciably with irradiation since it is carried out at room temperature, except in case of UHMWPE where some degradation in molecular weight due to diffusion of oxygen causes an increase in crystallinity.

Wear test results show that control LDPE exhibits the highest wear rate closely followed by LLDPE. In contrast control HDPE and UHMWPE exhibit very low wear rate due the formation of a continuous transfer film. Radiation cross-linking decreases the wear rate in LDPE and UHMWPE, while increases the wear rate in HDPE and LLDPE. Therefore, cross-linking is only a promising method to improve the wear properties of the former two materials. This effect is due to the formation of extensive network structure aided by the high entanglement density present in LDPE and UHMWPE. Under dry abrasive conditions the most wear resistant material investigated in this study is radiated UHMWPE.

6. Acknowledgment
Authors like to acknowledge help of the following colleagues and students in GIK Institute, Topi, Pakistan: Mr. Fahd N. Khan and Mr. Adeel Raza for conducting the swelling experiment, and Mr. Khawaja Zakaullah for conducting the DSC analysis.

7. References
[1] Dave B 1988 In: *Handbook of elastomers-new developments and technology*, ed A K Bhowmick and H L Stephens (New York: Marcel Dekker) pp 709-728

[2] Amberg L O 1964 In: V*ulcanization of elastomers: Principles and practice of vulcanization of commercial rubbers*, ed G Alliger and I J Sjothun (New York: Reinhold Publishing) pp 286-333

[3] Stern M 2000 *Plastics Engineering* 56 39-43

[4] Anderson J C 1982 *Tribol. Int.* 15(1) 43-7

[5] Thorp J M 1982 *Tribol. Int.* 15(2) 59-68

[6] Feng Y and Z-T Ma 1992 In: *Radiation processing of polymers*, ed A Singh and J Silverman (Munich: Hanser Publishers) p 71

[7] Osagawara M 1993 *Adv. Polym. Sci.* 105 37-79

[8] Choudhury M and Hutchings I M 1997 *Wear* 203-204 335-40.

[9] Venkatraman S and Kleiner L 1989 *Adv. Polym. Technol.* 9(3) 265-70

[10] Cheng S and Kerluke D R 2004 Radiation processing for modification of polymers *Annual Technical Conference of the Society of Plastic Engineering (ANTEC)* pp 2694-99

[11] Torrisi L, Visco A M and Campo N 2007 *J. Mater. Eng. Perform.* 16(1) 97-101

[12] Khonakdar H A, Jafari S H, Wagenknecht U and Jehnichen D 2006 *Radiat. Phys. Chem.*75 78-86

[13] Khonakdar H A, Morshedian J, Wagenknecht U and Jafari S H 2003 *Polymer* 44 4301-09

[14] Suwanprateeb J and Trongtong P 2003 *J. Mater. Sci: Mater. Med.* 14(10) 851-55

[15] Lewis G 2001 *Biomaterials* 22 371-401

[16] Kurtz S M, Muratoglu O K, Evans M and Edidin A A 1999 *Biomaterials* 20 1659-88

[17] Kurtz S M 2004 *The UHMWPE handbook: Ultra-high molecular weight polyethylene in total joint replacement* (San Diago: Elsevier Academic Press)

[18] Mills, C D 1991 *Modern Plastics Encyclopedia' 91* pp 62-3

[19] Coughlan, J J 1986 *Encyclopedia of polymer science and engineering* vol. 6 (New York: J. Wiley & Sons) pp 490-4

[20] Oral E and Muratoglu O K 2011 *International Orthopaedics* 35 215-23

[21] Bracco P and Oral E 2011 *Clinical Orthopaedics and Related Research* 469 2286-93

[22] Marrs H, Barton D C, Jones R A, Ward I M and Fisher J 1999 *J. Mater. Sci.-Mater. Med.* 10 333-42

[23] Unal H, Sen U and Mimaroglu A 2005 *Mater. Design.* 26 705-10

Low magnetostriction in $Fe_{100-x}Mn_x$ (x = 45, 48, 50, 52, 55) alloys

R Sato Turtelli[1], C Grijalva[1], F Kubel[2], D Geist[3], R Grössinger[1], M Kriegisch[1,4], S Sorta[1], M Zehetbauer[3], N Mehmood[5]
[1]Institute of Solid State Physics, Technical University of Vienna, Vienna, Austria
[2]Institute of Chemical Technologies and Analytics, Technical University of Vienna, Vienna, Austria
[3]Research Group Physics of Nanostructured Materials, University of Vienna, Vienna, Austria
[4]Austrian Institute of Technology, Giefinggasse 2, A-1210, Vienna, Austria
[5]National Engineering and Scientific Commission, Islamabad, Pakistan

E-mail: reiko.sato@ifp.tuwien.ac.at

Abstract. Investigation of the magnetostriction in $Fe_{100-x}Mn_x$ alloys with x = 45, 48, 50, 52 and 55, in as-cast, annealed and cold-rolled states was performed. The magnetostriction was measured by two methods: capacitance dilatometer and strain-gauge bridge. A careful structure characterization was made by XRD. For all samples the magnetostriction measured at room temperature was quasi-zero. Additionally, we found an orthorhombic structure for as-cast and annealed samples. These results are in strong disagreement with those reported in literature. Possible reasons for these disagreements are briefly discussed.

1. Introduction

The magnetostrictive materials are defined as materials which undergo a change in shape (length, volume) due to change in the magnetization state of the material. All magnetically ordered materials exhibit a magnetostriction, however the magnitude depends strongly on the involved atoms as well as the symmetry of the lattice. In 3d-elements such as nickel, iron or cobalt, the magnetostriction is between 10 and 50 ppm depending also on the crystallographic direction. In the mid-20 century a new area began with the discovery of giant (around 1000's of ppm) magnetostriction in rare earth alloys, e.g. Terfenol-D, that exhibits magnetostriction values up to 2000 ppm at room temperature in acceptable external fields. Pure rare earth elements or different rare earth based alloys exhibited large magnetostriction, but either at very high magnetic fields, or at cryogenic temperatures, or both. Already a well-known new class of magnetostrictive materials called Galfenol exhibits a rather high magnetostriction of the order of 400 ppm (in single crystals) in iron – 19 % gallium and 450 ppm in iron – 28 % Ga-alloys at room temperature [2], however, in very low external magnetic fields.

Recently, Peng *et al.* [1] obtained an unexpectedly large magnetostriction in antiferromagnetic polycrystalline $Fe_{58}Mn_{42}$ under a moderate applied field of 1 T. They achieved a linear strain of 169 ppm parallel to the field and -581 ppm when the sample is subjected to a compressive stress of 152 MPa. Later, Jingjing Zhang *et al.* [3] obtained an enhanced magnetostriction of 690 ppm at 1350 kA/m in as-cast $Fe_{50}Mn_{50}$ alloy. However a reduced magnetostriction was obtained after annealing the sample at 1100 °C for 24 h, which for this decrease was attributed to the decomposition of the single γ-phase in a mixture of fcc γ-phase, hcp ε-phase and bcc α-phase during isothermal heat treatment. The same authors found a giant magnetostriction of 2100 ppm when the ingot of $Fe_{50}Mn_{50}$ alloy was subjected to 70% cold-rolling and annealed at 600 °C for 1 h [4]. They concluded that the preferred <1 1 0> orientation formed along the rolling direction during the deformation is responsible for the improvement in magnetostriction. Very recently, we found very low magnetostriction, that is around zero, on annealed and cold-rolled $Fe_{100-x}Mn_x$ alloys with x = 38, 42, 46, 50 and 55 [5]. In that work we also showed the susceptibility and the magnetization as function of the temperature and the texture analysis processed through software LABOTEX from the XRD measurements (AXS Bruker D8 diffractometer with GADDS area detector NANOSTAR). The linear increase of the magnetization with applied field and slope increase with temperature indicated that the alloys were antiferromagnetically ordered. Additionally, a very low magnetization value around 1 emu/g was measured at a field of 9 T, at room temperature. In cold-rolled samples with Mn content lower or equal to x = 42, no significant textures have been observed, while for the samples with x > 42 texture components in {110}<1-12> or {011}<01-1> were observed. However, all average longitudinal and transverse magnetostriction values measured on those Fe-Mn alloys were always around zero irrespective of heat treatment, concentration, texture and rolling reduction.

Because of the unclear situation from previous work, we have been motivated to prepare new samples and repeat the magnetostriction investigations in $Fe_{100-x}Mn_x$ alloys with x > 42 where we previously found marked textures [5], that is, x = 45, 48, 50, 52 and 55, in the as-cast, annealed and cold-rolled state. Again, a careful structural characterization was achieved by XRD. Moreover, the longitudinal and transverse magnetostriction was measured at room temperature by two methods: micro-capacitance dilatometer and strain-gauge bridge.

2. Experimental techniques

Ingot of polycrystalline $Fe_{100-x}Mn_x$ alloys with nominal concentrations of x = 45, 48, 50, 52 and 55 were prepared from high purity (> 99.9 %) elements by using high-frequency induction melting (Hüttinger RF Generator, 30kW, 600 kHz) under argon atmosphere in a copper crucible. Each ingot was re-melted three to four times, to assure the chemical homogeneity. Due to the high vapour pressure of Mn, it cannot be excluded that the real Mn-concentrations of samples were smaller than the nominal ones. Parts of each ingot were subjected to 30 %, 50% and 70% cold-rolling at room temperature (the percentages give the relative change in sample thickness due to cold-rolling).

The X-.ray diffraction, XRD, patterns were recorded by means of an Xpert Philips powder diffractometer (Goniometer Philips PW 3050/60) using $CuKa_{1,2}$ radiation in a Bragg Brentano geometry, and a X´Celerator detector. The X-ray generator Philips PW 3040/60 worked at a power of 40 kV and 40 mA and the goniometer was equipped with a graphite monochromator. Diffraction patterns were recorded in the angular range 5° to 135° with a scan step size of 0.02°. Collected data were refined using the Rietveld package TOPAS (Bruker AXS Topas V 2.1) based on the fundamental parameter approach, with diffractometer parameters and wavelength settings adjusted using a LaB_6 standard.

Longitudinal and transverse magnetostriction measurements of as-cast, annealed at 1000 °C for 24 h and cold rolled samples were performed at room temperature by a strain gauge method using an AC-bridge (50-kHz bridge; HBM type KWS 85A1), in a pulsed-field magnetometer, which exhibits a maximum magnetic field of 5 T at pulse duration of 50 ms. It is important to mention that we used a half bridge circuit, where one strain gauge is on the sample and a second strain gauge on a non-magnetic dummy. With this set-up, the effect of the magneto-resistance of the strain gauge is balanced. On annealed $Fe_{48}Mn_{52}$ alloy, the magnetostriction was also measured by micro-capacitance dilatometer method up to 9 T [6].

3. Experimental Results

Figure 1 shows XRD patterns for as-cast $Fe_{100-x}Mn_x$ (x = 45, 48, 50, 52, 55) samples. Our results indicate that the phase present in our as-cast and annealed $Fe_{45}Mn_{55}$, $Fe_{48}Mn_{52}$, $Fe_{50}Mn_{50}$ and

$Fe_{45}Mn_{55}$samples is a single orthorhombic hkl γ-phase instead of cubic fcc γ-phase, as reported in [2, 3]. Additionally, the average Lorentzian crystallite size of all as-cast and annealed samples is very small, i.e. about 50 nm. The sample with highest content of iron $Fe_{55}Mn_{45}$ presents two phases (77 % of α-Fe and 23 % of orthorhombic γ-phase) already in the as-cast state. In all as cast and annealed samples investigated, no significant textures have been observed. In figure 1(a), XRD pattern corresponding to the orthorhombic phase of Fe-Mn was indexed. The lines corresponding to the cubic structure of Fe is indexed in figure 1(e).

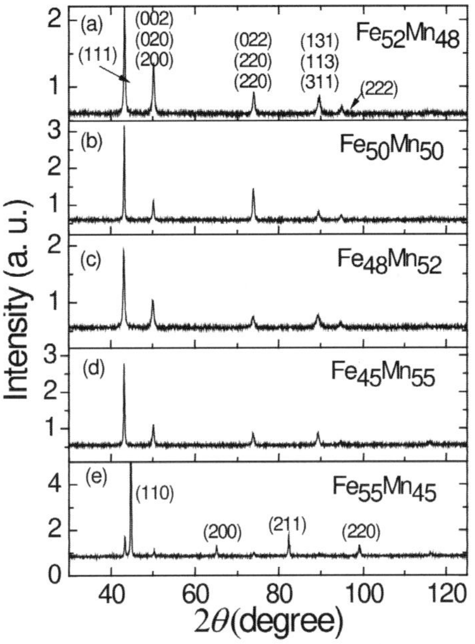

Figure 1. XRD patterns of $Fe_{100-x}Mn_x$ with x = 45, 48, 50, 52 and 55

When the samples were subjected to cold-rolling at room temperature, for the samples rolled at 30% and 70%, phase separation in three phases, namely: α-Fe, Fe-Mn and MnO, occurred. The detailed analysis of the composition, for example for the $Fe_{48}Mn_{52}$ alloy, is shown in figure 2. Due to cold-rolling, the grain size became smaller (an average size about 20 nm) for all samples. For the highest deformation state even an amorphous background was formed (see figure 2(b)). It is also worth to mention the absence of the line corresponding to (222) in the XRD obtained for the sample $Fe_{48}Mn_{52}$ annealed and cold-rolled at 50 %. In figure 2((b), only MnO was indexed (Fe-Mn and Fe were already indexed in the figure 1(a) and figure 1(e), respectively).

In figure 3, all longitudinal and transverse magnetostriction measurements performed on as-cast and annealed single hkl γ-phase samples using a strain-gauge method are given. As can be seen, the average longitudinal and transverse magnetostriction values of all samples are around zero up to an applied field of 5 T. On the other hand, the magnetostriction of the as-cast $Fe_{55}Mn_{45}$ alloy, which contains 77 % of α-Fe and 23 % of the hkl γ-phase, corresponds to that of pure α-Fe, with a saturation magnetostriction around -6 ppm (see figure 4). Therefore for the further studies this sample was not considered.

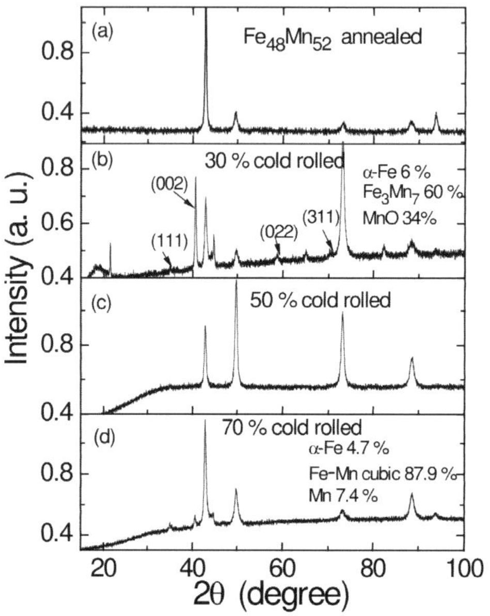

Figure 2. XRD patterns of the heat treated and cold rolled $Fe_{48}Mn_{52}$ alloy with respective compositions

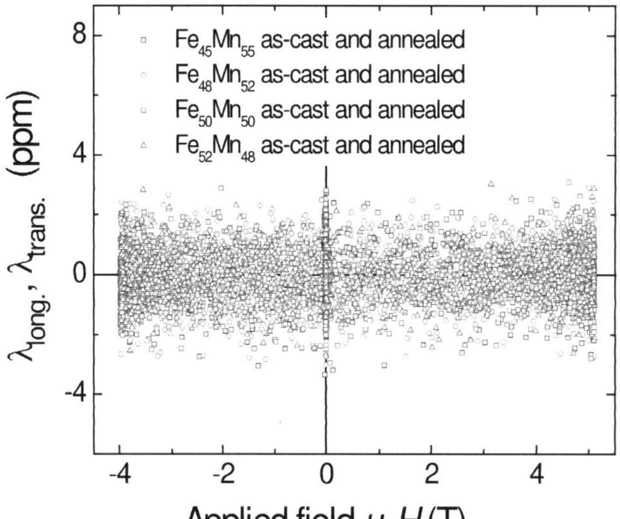

Figure 3. Longitudinal and transverse magnetostriction measured on the as-cast and annealed $Fe_{100-x}Mn_x$ with x =48, 50, 52 and 55

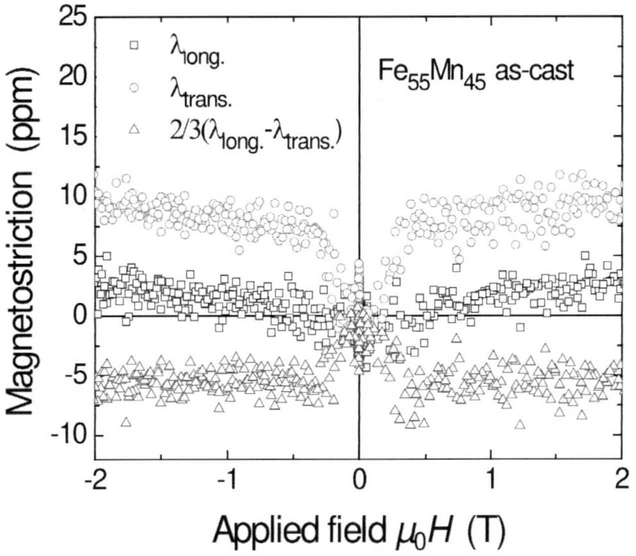

Figure 4. Longitudinal, transverse and saturation (2/3(total)) magnetostriction measured on as-cast $Fe_{55}Mn_{45}$

To investigate the magnetostriction at higher values of the applied field, the longitudinal magnetostriction was measured up to 9 T at 200 K and 300 K on annealed $Fe_{48}Mn_{52}$ alloy using a micro-capacitance dilatometer. The magnetostriction at each temperature was measured three times and the results are shown in figure 5. For both temperatures 200 K and 300 K, the average magnetostriction value lies always around zero even at 9 T. This result is in agreement with the magnetostriction measured with strain-gauge.

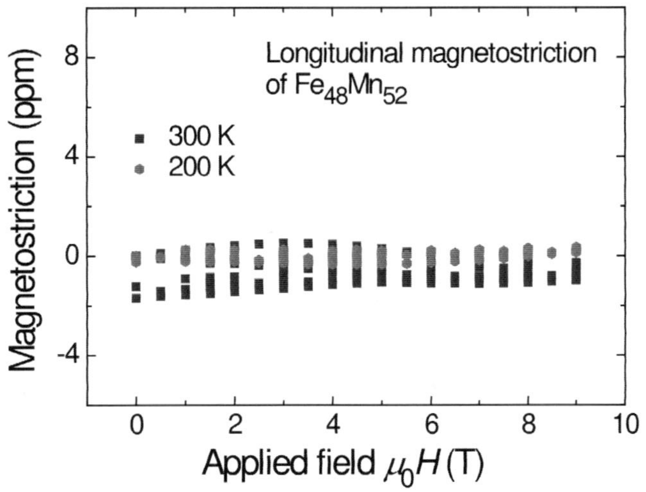

Figure 5. Longitudinal magnetostriction measured with capacitance method on the heat treated $Fe_{48}Mn_{52}$alloy

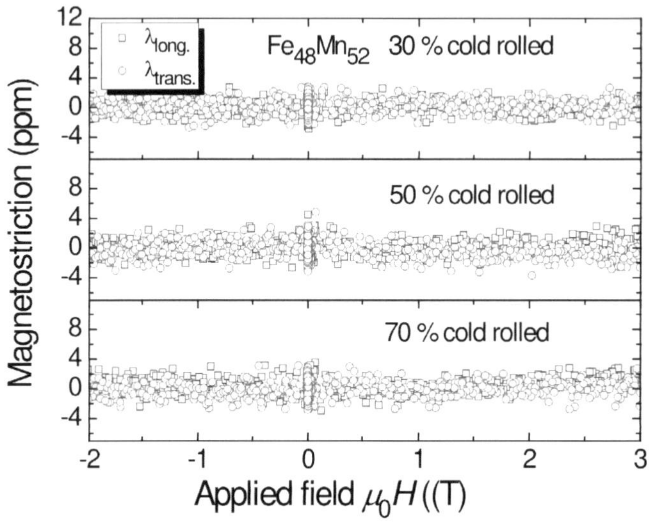

Figure 6. Longitudinal and transverse magnetostriction of annealed cold rolled $Fe_{48}Mn_{52}$ alloy measured with the strain-gauge method

The magnetostriction of all as-cast and annealed cold rolled samples is also around zero up to 3 T, independent of the measuring direction with respect to the rolling direction. Also, annealing at 600 °C for 1 h after the deformation process has no effect to the results. As examples, the longitudinal and transverse magnetostrictions measured on $Fe_{48}Mn_{2}$ which were subjected to cold rolling at 30%, 50% and 70 % are shown in figure 6. These measurements are in agreement with results presented in our previous work [5], however in strong contradiction to published data from the material $Fe_{50}Mn_{50}$ [4].

4. Discussion

All results, structural and magnetostrictive properties, are in disagreement with the results reported in [3, 4]. In our previous work [5] and also in this one, we found very low magnetostriction in$Fe_{100-x}Mn_x$ alloys. One of the reasons for the disagreement could be the difference in structures, that is, our as-cast and annealed samples present an orthorhombic hkl γ-phase while in [2, 3] a cubic fcc γ-phase was reported. For our cold-rolled annealed samples appeared impurities. However, the low magnetostriction value in our Fe-Mn alloys seems reasonable keeping in mind that these antiferromagnetic materials exhibited a very low magnetization (around 1 emu/g at 9 T [5]) and generally, the magnetostriction value in magnetic materials is directly related to the magnitude of the magnetization. It is interesting to note that Fe-Mn alloys are antiferromagnetic materials for a broad concentration range indicating that this ordering occurs due to a band structure effect.

It is worth to mention that measuring magnetostriction with strain gauge, a careful caution must be paid on the magnetoresitience of the strain-gauge employed. The active gauge and dummy gauge must be set parallel and connected to form adjacent sides of a Wheatstone bridge in order to cancel a trivial magnetoresistence change of strain-gauge. The wrong cementation and the wrong set and kind of the gauges lead sometimes to an anomalous increase in resistence with increasing of magnetic field.

Figure 7 shows the longitudinal and transverse magnetostriction of Fe-Mn alloys measured with strain gauge where the magnetoresistance effect was not completely compensated. Generally, the intensity of the measured magnetoresistance varies quadratic with the applied field, and is independent of the alloy' composition of the samples and the direction of the applied field. Therefore, to eliminate the influence of the magnetoresitence on the magnetostriction of the material, the total magnetostriction, λ_{long}-$\lambda_{trans.}$, must be calculated.

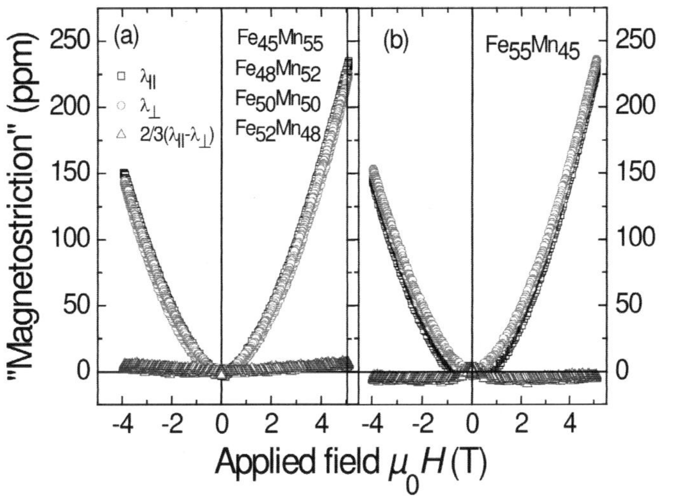

Figure 7. Longitudinal and transverse "magnetostriction" measured in all as-cast $Fe_{100-x}Mn_x$ (x = 45, 48, 50, 52 and 55)

In fact, identical values of longitudinal and transverse "magnetostriction" were obtained for the single hkl γ-phase alloys irrespective of the composition of the samples (see Figure 7(a)). For these samples the total magnetostriction λ_{long}-$\lambda_{trans.}$ is around zero, which is the actual magnetostriction of these samples. For the sample $Fe_{55}Mn_{45}$ alloy, which contains 77% pure Fe, the total magnetostriction λ_{long}-λ_{trans} is typical of that of α-Fe (see Figure 7(b)). These results show that the effect of the magnetoresistence of the strain-gauge on magnetostriction measurements can be canceled measuring the resistence applying a magnetic field longitudinally and transversely. However the influence of the magnetoresistence for fields smaller than 1 T as used in [3, 4] is small.

5. Conclusion

A careful investigation of the structure of $Fe_{100-x}Mn_x$ alloys with compositions x = 45, 48, 50, 52 and 55, in the as-cast, annealed and cold-rolled states was performed refining the XRD data by means of the Rietveld package TOPAS. We obtained a single orthorhombic hkl γ-phase for the as-cast and annealed samples, however three phases (Fe-Mn, α-Fe and MnO) were formed in coexistence when the samples were subjected to cold rolling. The magnetostriction measured by capacitance dilatometer (up to 9 T) and strain-gauge bridge (up to 5 T) was around zero for all samples investigated, irrespective of the composition, annealing & deformation conditions. These results are in strong disagreement with those reported in [3, 4].

6. References

[1] Summers E M, Lagrasso T A, Wun-Fogle M 2007 *J. Mater. Sci.*42 9582.

[2] Peng W Y, Zhang J H 2006 *Appl. Phys. Letters* 89 262501.

[3] Zhang J, Ma T, He A, Yan Mi 2009 *J. Alloys Comp.* 485 510.

[4] Ma T, Zhang J, He A, Yan Mi 2009 *Scripta Mater.* 61 427.

[5] Sato Turtelli R, Bormio-Nunes C, Tiguman K T, Geist D, Panigrahi A, Grijalva C, Sorta S, Grössinger R, Zehetbauer M 2013 Soft Magnetic Materials Conference 21, *IEEE Trans.Mag.*, to be published

[6] Rotter M, Müller H, Gratz E, Doerr M, Loewenhaupt M 1998 *Rev. Sci. Instrum.*69 2742.

IOP Conf. Series: Materials Science and Engineering **60** (2014) 012007 doi:10.1088/1757-899X/60/1/012007

Mechanical property changes in HVOF sprayed nano-structured WC-17wt.%Ni(80/20)Cr coating with varying substrate roughness

Tarek A Ben Mahmud[1], Gobinda C Saha[1,2], and Tahir I Khan[1,3]
[1]Department of Mechanical and Manufacturing Engineering, University of Calgary, Alberta, Canada
[2]Westpower Equipment Ltd., 4451-54th Avenue S.E., Calgary, Alberta, Canada
[3]Qatar Petroleum Chair, Department of Mechanical & Industrial Engineering, University of Qatar, Doha, Qatar

E-mail: tarekbinmahod@yahoo.com

Abstract. Thermally sprayed coatings developed by use of high velocity oxy-fuel (HVOF) process are known for their superior wear characteristics. In many industrial applications, new parts as well as repaired and refurbished parts coated with WC-Co microstructured coatings have shown enhanced erosion-corrosion and abrasive resistant properties when compared with other surface modification technologies such as chrome replacement, fusion welding, and cladding. This research has been further directed towards the development of HVOF technique to deposit dense nanostructured ceramic-metallic composites. The mechanism of plastic deformation, which determines the strength and ductility of materials, in nanostructured materials are different, thereby leading to novel mechanical properties. Various parameters can influence these properties, but the substrate surface preparation by grit blasting before thermal spraying is one critical parameter. The grit blasting process generates a surface roughness, which ensures mechanical anchoring between the coating and the substrate surface. In this work, the sliding wear behavior and microhardness of WC-17wt.%Ni(80/20)Cr cermet coatings deposited onto carbon steel substrates are examined as a function of three different surface roughness values under different loads. The results show that as-prepared surface with different blasting profiles have a direct influence on the surface roughness and wear performance of the coatings. The sliding wear resistance of the coatings increased as the substrate surface roughness increased. The wear depth decreased with increasing surface roughness.

1. Introduction

The oil sands industry in northern Alberta, Canada faces an ongoing challenge from abrasive wear occurring on surfaces of machinery and components that are involved in the heavy oil recovery process. Due to its highly aggressive nature the produced water in this environment causes unpredictable equipment failure leading to complete plant shutdown and reduced production efficiency. Machine lifetime depends on a combination of several factors, particularly on the degree of wear resistance of the components of the machinery.

Traditionally, several techniques have been used to increase the wear resistance by hardening the substrate surface, including heat treatment, hard facing by fusion welding of wear resistant alloys, powder cladding or surface electrodeposition. Up until recently, tungsten carbide (WC) wear protection systems used to resist wear in oil sands applications included bulk sintered and surface engineered products fabricated using processes such as fusion welding, brazing and thermal spraying. Since 1980's WC based thermal sprayed coatings have played an increasing role in protecting metal surfaces from high wear and corrosion environments These included coatings of military and civil aircrafts parts and components (landing gears, propellers and hydraulic actuators etc.) as part of their regular manufacturing and maintenance operations[1]. Atmospheric plasma spraying (APS), detonation spray coating (DSC), and high velocity oxy-fuel (HVOF) are the most commonly used techniques for deposition of WC-Co coatings. The HVOF process provides advantages over other techniques such as higher velocity and lower temperature of in-flight particles, producing coatings with minimal porosity, excellent adherence to the substrate, larger fraction of retained WC and correspondingly improved wear properties [2,3]. In HVOF, feedstock particles are heated and propelled at very high velocities (>1 Mach) resulting from the internal combustion of oxygen and fuel (propylene, kerosene, methane, hydrogen etc.) and from the spray gun special design. The HVOF gas velocity can exceed 2,100 m/sec, pushing the particles with velocities between 400 and 800 m/sec, thereby producing coatings with very high density, low oxidation content, high bond strength, high microhardness and durability, and low residual stress. In the process it does not overheat the powder particle [4]. A successful coating application to its engineering usage is strongly linked to the quality of the adhesion between the coating and the substrate or the previously deposited layers. In HVOF, the adhesion / cohesion is of the mechanical type; surface pits and grooves of a rough surface are filled with the spreading semi-molten material due to the impact pressure. Subsequent solidification leads to mechanical interlocking [5]. The coating must remain bonded to the substrate and this is affected by the prepared substrate surface roughness, which ultimately influences the characteristics like wear resistance, residual stress formation and adhesion strength in the coating [6]. Besides, feedstock powder composition and particle structure exhibit different coating efficiencies during spraying resulting in different coating microstructures. In recent studies it has been shown that the wear resistance of WC-17Co cermet coating increases by reducing the size of the carbide grain size [7]. A thin Co layer on the surface of WC-Co feedstock particle by chemical vapor deposition (CVD) had prevented significant decarburization of WC during HVOF process, thereby improving the sliding wear resistance of the coatings [8]. In this research, a novel 'duplex NiCr alloy coated' WC-17wt.%Ni(80/20)Cr nanostructured cermet powder has been HVOF sprayed to obtain coatings for high wear oil sands applications. The NiCr alloy binder was chosen to increase the corrosion resistance of the coatings. The tribological behavior of these coatings as a function of substrate roughness was investigated by studying changes in wear resistance using a pin-on-plate sliding abrasive tester (as per ASTM G133 standard) and relating this to changes in microhardness of the coatings.

2. Materials and experimental work

2.1. Feedstock powder

In this work, a 'duplex NiCr-coated' nanocrystalline WC grain structured WC-17wt.%Ni(80/20)Cr cermet powder was used. The objective of using this powder particle was to maintain a minimum binder mean free path between the WC grains to achieve a harder coating formation. Furthermore, the powder particles consist of WC grains and 6 wt.% NiCr alloy matrix in the core and a CVD produced external coating consisting of 11 wt.% binder (NiCr). This external coating was designed to reduce the decarburization of WC during the spraying process and reduce the formation of brittle phases in the sprayed coatings. The external NiCr coating was also expected to increase the metallurgical bond between the "splats" which result in the coating on the substrate surface. Figure 1 compares the relationship between the binder mean free path and the hardness of nanostructured and conventional microstructured WC-Co coatings [9]. The chemical composition and size distribution of the powder used in this work are presented in Table 1.

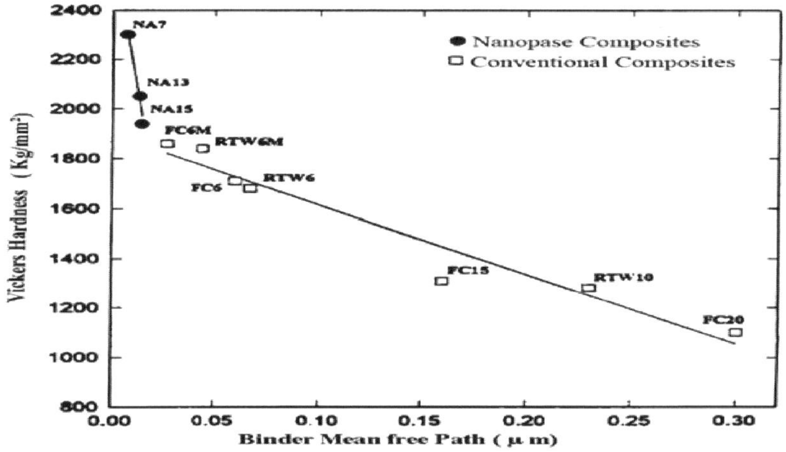

Figure 1. Hardness as a function of binder mean free path of WC-Co cermets [9].

Table 1. Powder composition analysis

Powder	**Element, wt.%**				**Nominal range, μm**
	W	C	Ni	Cr	
WC-17Ni(80/20)Cr	Balance	1.28	20.46	3.34	-22 +10

2.2. Substrate materials

One of the most common steels employed in the oil sands industry is carbon steel AISI 1018, used as the substrate surface in this work. It has been received in the annealed condition with a composition in wt.%: 0.18-C, 0.16-Si and 0.65-Mn. The geometry of this material is cylindrical rod (Ø8 × h12 mm). One cylindrical face was used as the surface for wear tests. Fifteen samples were grouped into three groups, each group having five samples and were named by grit blasting material mesh size numbers 16, 24, and 36, respectively. The samples were degreased with acetone and then submerged in ultrasonic tank using isopropyl alcohol for 15 minutes before surface blasting. Three different blasting grits composed of 97% Al_2O_3 and 3% TiO_2 were used to roughen the substrate surface that was to be coated. The airflow pressure was set at 0.70 MPa, the standoff distance between torch and substrate was set at 100 mm, and the blowing pressure angle was kept 90º. The surface roughness was measured by a Mitutoyo roughness tester (model: SJ-201P). The values of the as-prepared surface roughness with respect to blasting grit sizes are given in Table 2.

Table 2. Relationship between surface blasting grit size and generated surface roughness.

US mesh size #	**Grit size, μm**	**Roughness (R_a), μm**
16	1,190	17
24	707	12
36	500	9

2.3. HVOF spraying

The HVOF spraying process of the engineered nanostructured WC-17wt.%Ni(80/20)Cr powder was carried out at Westpower Equipment Ltd. (Calgary, AB) and the spraying conditions are specified in Table 3. A Sulzer Metco DJ 2700 hybrid diamond jet gun was used. The gun traverse speed was 2,000 mm/minute, and the spray distance was maintained at 254 mm. Approximately 10 μm per pass was deposited to obtain a final coating thickness of 230 μm. By using a coating thickness gauge 3000FX model (Electromatic Equipment Co., NY) the coating thickness was recorded. The substrate surface was preheated before the spray process to maintain the thermal conductivity close between deposited powder particles and the substrate Surface, thereby reducing the generation of residual stresses. The preheating process also guarantees mechanical anchorage at micro-welding sites [10].

2.4. Sliding wear test

Sliding wear tests on the coat surfaces were performed using a custom-designed two-body abrasion wear tester apparatus according to ASTM G133-05 standard. All tests were made at room temperature without lubrication. A multichannel 16-bit 100 kHz data acquisition device USB-1608FS model (microDAQ, OH) was used to gain data and execute a real time analysis of the wear process. The test was conducted under different loads for each grit group number. The sliding speed, stroke length, and run time were 0.029 m/sec, 10 mm and 30 minutes, respectively. Table 4 shows sample organization and test parameters used in the study. Uncoated samples were tested as reference material. Comparison of sliding wear depth in uncoated steel substrates and nanostructured coated materials as a function of applied loads (20, 40, 60 N) is presented. All uncoated and coated samples were cleaned with ethanol and dry compressed air to measure the mass losses before and after sliding wear test. The weight loss measurements were completed by using an electronic analytical balance with a precision of 0.1 mg (model SCIENTECH-ZSA-210) and were recorded before and after the sliding wear tests.

Table 3. The sliding wear test parameters

	Uncoated	Coated		
	Grit No.	16	24	36
Load (N)	20, 30, 40, 50, 60	20, 30, 40, 50, 60	20, 30, 40, 50, 60	20, 30, 40, 50, 60

2.5. Coating characterization

After the sliding wear test, the samples were cut longitudinally by slow speed diamond cutting saw model METTECH-Mark V-seres-600. The surfaces of all samples were prepared by surface preparation techniques using standard metallographic procedure. The final polish was completed at 1 μm diamond paste. The hardness of the coating was performed using Vickers microhardness tester model Micromet-II, Buehler on all coated and uncoated samples on their cross-sections under a load of 300 g. Ten indentations were taken to get an average of hardness value. Coatings' microstructures were characterized by scanning electron microscopy (SEM) model FEI XL30 (Phillips). Coatings' microstructures after the wear tests under 20 N and 60 N loads are presented.

3. Results and discussion

3.1. Characterization of wear depth vs. sliding distance

Figure 2 shows the effect of sliding distance on the worn surface by measuring wear depth of coated and uncoated materials under 40 N loading. All materials have shown similar trends in wear depth plots against the sliding distance. As expected, the maximum wear depth penetration was occurring in the uncoated AISI 1018 material. After the initial wear period (the 'break-in' wear) all samples have reached a more stable and constant wear rate and this can be explained by a transition mechanism from a severe to milder mode of wear. The coatings showed distinct spike in abrasive wear approximately in the first 10 m sliding distance. Furthermore, surfaces blasted with grit number 16 (R_a 17 μm) exhibits the lowest wear depth penetration (about 47 μm). It can be seen that as the prepared surface roughness decreases the wear depth increases. This can be attributed to the strong mechanical bonding between the substrate surface asperities and the coating molten particles generated by the surface blasting prior to deposition of the coating. Figure 3 shows a similar trend in wear depth characteristics obtained from uncoated and nanostructured WC-17wt.%Ni(80/20)Cr coatings under a higher load of 60 N.

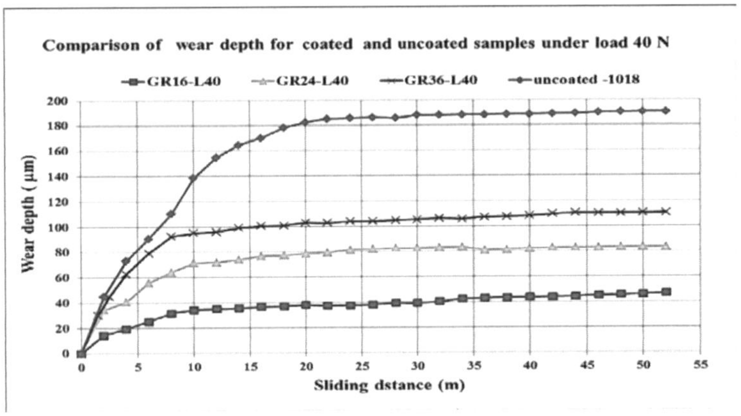

Figure 2. Comparison of sliding wear depth for different surface roughness of coated nanostructured and uncoated carbon steel substrate under 40 N load.

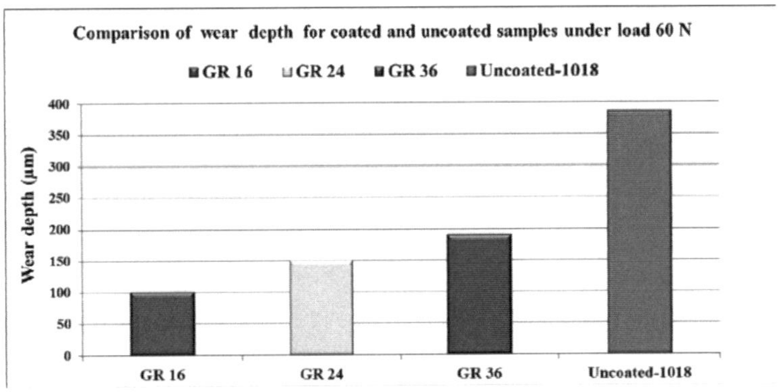

Figure 3. Comparison of sliding wear depth for different surface roughness of coated nanostructured and uncoated carbon steel substrate under 60 N load.

3.2. *Characterization of the mass loss*

Figure 4 shows the abrasive wear mass loss as a function of surface roughness under applied load of 60 N. It is seen that the blasting grit #16 prepared surface experienced less mass loss than compared with the surfaces prepared with blasting grit sizes #24 and #36. This is referring to a good bonding between the coating and substrate surface asperities.

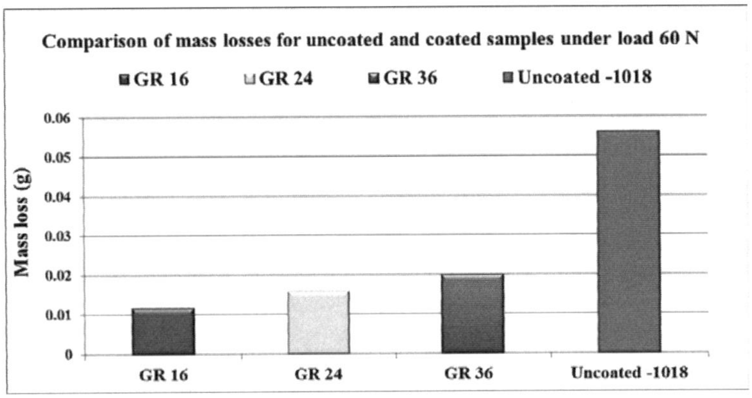

Figure 4. Abrasive wear mass loss as a function of different surface roughness under applied load 60N.

3.3. *Characterization of coating and wear rate*

The results obtained are used to compute the wear rate in terms of distance given by the following equation (1) [11]:

$$Wear\ rate = \frac{\Delta m}{S_d},\quad \left[\frac{g}{m}\right] \quad ....(1)$$

Where Δm is the difference in mass before and after the sliding wear in grams and S_d is the sliding distance expressed in m. Figure 5 shows the wear rate as a function of applied load for coated and uncoated test samples that were prepared with different grit prepared surface roughness. It is seen that the nanostructured coated sample with grit 16 exhibited the best wear resistance than that of the other two nanostructured coated samples prepared with grits 24 and 36, respectively. Further, the low wear rates in all three coated surfaces correspond to the low load (20 N). As the applied load increases the wear rate increases. This phenomenon can be referred to the fact that as the wear run progressed with higher loads at a constant speed there were progressively more debris formation from the worn material (WC-Ni-Cr) leaving them trapped between the coated surface and the abrasive tool. Hence, the wear mechanism shifted from being two-body abrasive wear to three-body abrasive wear process, resulting in an increase in wear rate.

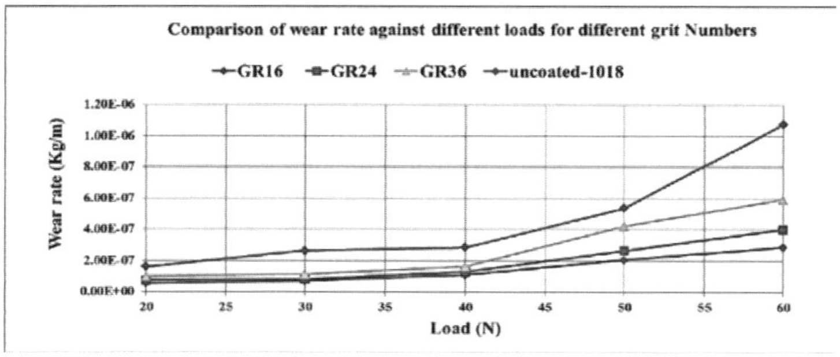

Figure 5. Comparison amongst sliding wear rates on the nanostructured coated and uncoated surfaces prepared by blasting grits (#16, #24, #36) under different loads.

To understand the wear characterization of coated samples, the worn surfaces were observed using the SEM. From the sliding wear test, two groups of coated samples (lowest load 20 N and highest load 60 N) were selected and examined, as shown in figures 6 and 7, respectively. The backscattered SEM micrographs (figure 6) after 30 minutes of tests under the applied load (20 N) all three coated samples (grits 16, 24, 36) retained their bonding with corresponding substrates. In figure 7, however, the protective coatings obtained on prepared surfaces blasted with grits 24 and 36 were visibly pulled out during the tests under the load of 60 N. This can be attributed to the removal of softer binder phase (NiCr alloy) followed by hard WC reinforcement pull out, while this was not the case for worn out sample that was coated on surface prepared by blasting grit 16.

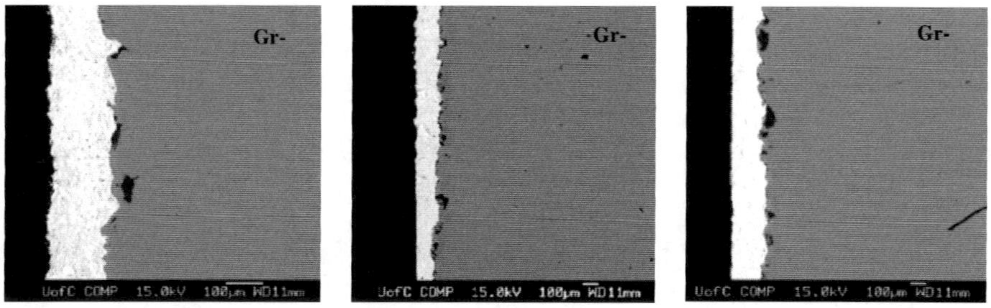

Figure 6. SEM micrographs comparing cross-sectional view of worn surfaces under 20 N load of nanostructured WC-17wt.%Ni(80/20)Cr coatings. The pre-spray surfaces were prepared by grits (16, 24, 36).

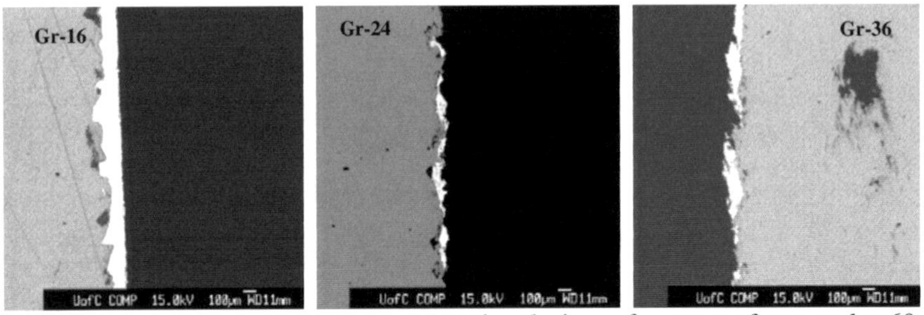

Figure 7. SEM micrographs comparing cross-sectional view of worn surfaces under 60 N load of nanostructured WC-17wt.%Ni(80/20)Cr coatings. The pre-sprayed surfaces were prepared by grits (16, 24, 36).

3.4 Vickers microhardness

Vickers microhardness tests were conducted on the worn surfaces of the coated and uncoated samples to examine the work hardening mechanisms taking place in the coated samples prepared by blasting grits 16, 24, and 36. It is observed that the microhardness remained unchanged and the highest hardness was recorded as 1243 HV0.3.

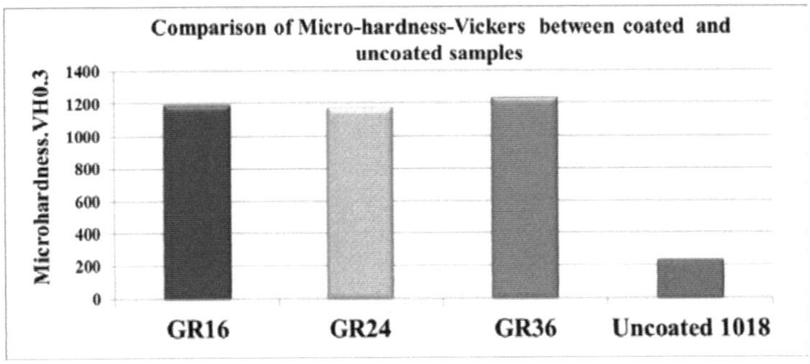

Figure 8. Vickers microhardness in the nanostructured coated and uncoated surfaces prepared by blasting grits (#16, #24, #36) after sliding wear test under 60 N load.

4. Conclusion

In this research it was shown that the novel 'duplex NiCr coated' nanostructured WC-17wt.%Ni(80/20)Cr powder could be HVOF sprayed to obtain coatings with superior wear resistant properties. Furthermore:

The sliding abrasive wear resistance, mass loss and after-wear Vickers microhardness of the nanostructured WC-17wt.%Ni(80/20)Cr coated surfaces were investigated and correlated with the pre-spray substrate surface preparation using three different blasting material profiles (grits 16, 24, and 36). The results showed that the coarse grained surface blasting material (grit #16) contributed to a lower wear rate (high wear resistance), and the highest wear rate took place on coating substrate that was prepared using #36 blasting grit. This could be attributed to the strong mechanical bonding between the substrate surface and the coating semi-molten particles derived during the former prepared surface for sprayed coatings.

- The SEM micrographs showed that the abrasive wear at a load of 60 N for surfaces blasted to produce a fine roughness (grits 24 and 36) was attributed to plastic deformation followed by removal of metallic phase, i.e. NiCr alloy. This resulted in the loss of the WC particles by "pull out".

- The Vickers microhardness measurements from the worn coated samples showed no significant difference in work hardening mechanisms between the three surface grits. However, this aspect of the study requires further investigation.

5. Acknowledgment

Authors acknowledge the financial support provided by the University of Al-Zawia in Libya and Westpower Equipment Ltd. in Alberta, Canada for the continuation of this research study.

6. References

[1] Legg K O Overview of Alternative Technologies for Chromium and Cadmium *Surface Modification Technology 2001: Proceedings of the Fifteenth International Conference on Surface Modification Technologies (ASM International)* pp 235–44

[2] M.R. Dorfman, B.A. Kushner, J. Nerz and A J R 1989 No Title *Proc. 12th Int. Thermal Spray Conf.* ed E I. A. Buclow (The Welding Institute, London,) p 108

[3] B.R. Marple R S L Therm. Spray Technol *Journal of Thermal Spray Technology* 14 67–76

[4] Thorpe, R. Kopech, and H. Gagne N 2000 HVOF thermal spray technology *", Advanced Materials & Processes* 157 27–9

[5] Fauchais P, Fukumoto M, Vardelle a. and Vardelle M 2004 Knowledge Concerning Splat Formation: An Invited Review *Journal of Thermal Spray Technology* 13 337–60

[6] Staia M H, Ramos E, Carrasquero A, Roman A, Lesage J and Chicot D 2000 Effect of substrate roughness induced by grit blasting upon adhesion of WC-17 % Co thermal sprayed coatings

[7] Saha G C and Khan T I 2011 Comparative Abrasive Wear Study of HVOF Coatings Obtained by Spraying WC-17Co Microcrystalline and Duplex Near-Nanocrystalline Cermet Powders *Journal of Engineering Materials and Technology* 133 041002

[8] Mateen A, Saha G C, Khan T I and Khalid F A. 2011 Tribological behaviour of HVOF sprayed near-nanostructured and microstructured WC-17wt.%Co coatings *Surface and Coatings Technology* 206 1077–84

[9] Jia K, Fischer T E and Gallois B 1998 Microstructure, hardness and toughness of nanostructured and conventional WC-Co composites *Nanostructured Materials* 10 875–91

[10] Paredes R S C, Amico S C and d'Oliveira a. S C M 2006 The effect of roughness and pre-heating of the substrate on the morphology of aluminium coatings deposited by thermal spraying *Surface and Coatings Technology* 200 3049–55

[11] Raymond G B 2004 *Mechanical wear fundamental and testing* (Marcel Dekker, Inc)

Thermal properties of TiO_2 films fabricated by atomic layer deposition

Muhammad Rizwan Saleem[1, 2], Seppo Honkanen[1] and Jari Turunen[1]
[1]University of Eastern Finland, Institute of Photonics, P.O. Box 111, FI-80101, Joensuu, Finland
[2]National University of Sciences and Technology (NUST), School of Chemical and Materials Engineering (SCME), Sector H-12, Islamabad, Pakistan

Abstract. Thin, amorphous, high index, dense, low scattering & absorption (low extinction coefficient) and optical grade TiO_2 films of various thicknesses are prepared by Atomic layer deposition (ALD) technique and investigated in terms of thermo-optic coefficient (dn/dT) and temperature dependent density ($d\rho/dT$). The dn/dT and $d\rho/dT$ are calculated by modeling ellipsometric experimental data by developing appropriate optical model such as Cauchy Model. The modeled data was further modeled with Lorentz-Lorenz relation under least-square approach. The dn/dT of TiO_2 films shows negative values for thin and positive values for relatively thicker films and reveals no significant changes in dn/dT and $d\rho/dT$ when film thickness increases more than ~ 200 nm. The coefficient values are calculated for a wavelength range of 380–1800 nm. The average values of $(dn/dT)_{60nm} = -4.2 \pm 0.7 \times 10^{-5}{}^{\circ}C^{-1}$ and $(d\rho/dT)_{60nm} = -6.6 \pm 1.1 \times 10^{-5} gcm^{-3}\ {}^{\circ}C^{-1}$ at wavelength of 640 nm. The reported coefficients values are measured and calculated for TiO_2 film of thickness t = 60 nm and described in detail.

1. Introduction

Thin amorphous TiO_2 films are studied increasingly for sub-wavelength optical structures due to their high index of refraction, and transparency within a broad wavelength region [1–3]. Various techniques such as vacuum evaporation and sputtering have been investigated for the deposition of TiO_2 coatings [4]. The use of Atomic Layer Deposition (ALD) shows promise for low cost and large scale production of thin, high index, dense and amorphous TiO_2 films with high optical quality. These ALD films can be uniformly fabricated on non-flat surfaces having 3D nanoscale structures, with a guided mode resonance filter as an application example [5]. ALD is a unique thin film deposition method based on saturative surface reactions of alternatively supplied precursor vapors [6]. Due to saturative nature of each reaction step, the film growth is self-limiting, provided several practical advantages compared to other deposition techniques: atomic level control of film composition and thickness, uniform coverage of corrugated surface profiles, and large area uniformity.

The refractive index of TiO_2 films deposited by different techniques has been investigated and is found to correlate well with the density of the material [7]. Importantly, the films prepared by the same technique but under different growth conditions exhibit different density and refractive index due to phase changes in their microstructure [8,9]. The density of the deposited TiO_2 films is approximately directly proportional to the corresponding mechanical properties (Young's modulus) [10] and refractive index [11]. For amorphous TiO_2 films the density changes at different temperatures are not caused by the material undergoing a phase change. These density changes are attributed to the varying concentration of hydrogen containing species (H_2O, OH etc.) which adsorb in the voids or residual oxygen vacancies formed during the growth of the films [7,12]. The temperature dependent refractive index of TiO_2 films deposited by different techniques shows a negative thermo-optic coefficient (d*n*/d*T*), i.e., the value of refractive index decreases as temperature increases [13]. The microstructure of TiO_2 thin films depend on the deposition technique and the nature of the substrate. The film properties show variations in film density (porosity), mechanical properties, refractive index, extinction coefficient etc. For example, d*n*/d*T* of TiO_2 films deposited by plasma enhanced chemical vapor deposition [14], deposition by electron beam evaporation [15] exhibit different magnitudes due to the nature of the deposition method being employed.

In this paper we report on thermo-optic and temperature dependent density coefficients of ALD coated TiO_2 films. We fabricated optical grade amorphous TiO_2 films with different thicknesses under the same growth conditions and measured their temperature dependent refractive index and extinction coefficients. The d*n*/d*T* values are evaluated from measured experimental results by modeling with Cauchy model and Lorentz-Lorenz relation for wavelengths 380–1800 nm with a least-square approach. These values are further used to calculate the density of TiO_2 films and corresponding temperature dependent density changes using Lorentz-Lorenz relation. Theoretical calculations of the electronic polarizability at different temperatures are also carried out.

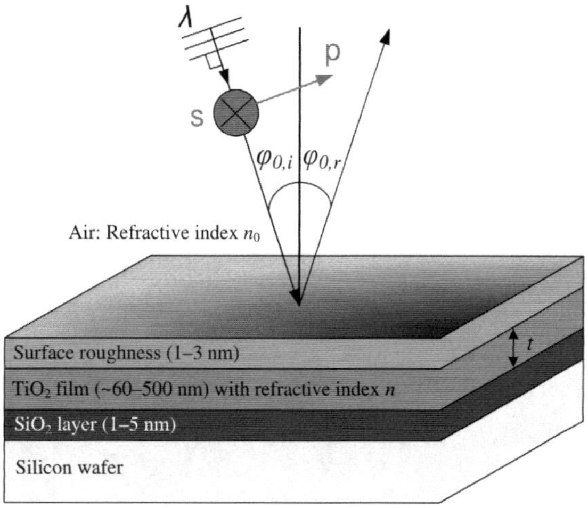

Figure 1. Designed optical-model for Wvase software analysis of TiO_2 films with layered structure

2. Theory

Spectroscopic ellipsometry measurements are represented in terms of ellipsometric parameters Ψ and Δ and related to optical constants of TiO_2 film, shown in Eqs. (1–4). The refractive index data of TiO_2 films is modeled by Wvase32 software by considering layered structure as shown in figure 1. The incident light comes from a residual medium air (layer 0) at an incident angle $\varphi_{0,i}$ and reflected from air-film interface, the part of light transmitted through the film (layer 1) is further reflected from film-substrate medium. The refractive indices of air and film materials are n_0 and n, respectively.

$$\tilde{\rho} = \tan\psi e^{i\Delta} \equiv \frac{\tilde{R}_p}{\tilde{R}_s} \quad (1)$$

where $\tilde{R}_p$ and $\tilde{R}_s$ are p and s polarized pseudo-Fresnel reflection coefficients given as:

$$\tilde{R}_p = \frac{\tilde{r}_{01,p} + \tilde{r}_{12,p} e^{-i2\beta}}{1 + \tilde{r}_{01,p} \tilde{r}_{12,p} e^{-i2\beta}} \tag{2}$$

$$\tilde{R}_s = \frac{\tilde{r}_{01,p} + \tilde{r}_{12,s} e^{-i2\beta}}{1 + \tilde{r}_{01,s} \tilde{r}_{12,s} e^{-i2\beta}} \tag{3}$$

where $\tilde{r}_{01,p}, \tilde{r}_{12,p}$ and $\tilde{r}_{01,s}, \tilde{r}_{12,s}$ are the Fresnel reflection coefficients from air-film and film-substrate interfaces for p- and s-polarized lights, respectively and β is the optical thickness (phase thickness), given as [16]:

$$\beta = 2\pi \frac{t}{\lambda} \sqrt{\tilde{n}^2 - \tilde{n}_0^2 \sin^2 \phi_{0,i}} \tag{4}$$

3. Experimental methods and data modeling

3.1. Experiments

Amorphous TiO_2 films of thicknesses 60, 100, 200, 300, 400, and 500 nm are coated on Si-wafers with <100> orientation by Boneq TFS 500 ALD reactor at a deposition temperature of 120 °C with commonly known $TiCl_4$ and H_2O precursor materials [17]. The film growth rate per ALD cycle was also monitored by measuring the thickness and refractive index with ellipsometer PLASMOS SD 2300, Philips Analytical Technology GmbH. The optical constants of the films were measured by a variable angle spectroscopic ellipsometer VASE manufactured by J. A. Woollam C. The thermal dependence of refractive index of the films is measured by a home made heating assembly directly attached with ellipsometer. The ellipsometry reflectance spectra was measured at incident angles of 65° and 75° with normal to the stage in wavelength range 380–1800 nm with an interval of 20 nm with a beam spot size of 3 mm. The samples were placed firmly on an aluminum hot plate where the temperature was controlled and monitored carefully. The surface temperature of the samples was measured with Convir ST8811 Handheld Infrared Thermometer by Calex Electronics Limited Company with an accuracy of ± 2 °C. The heating rate was 0.5 °C / min with accuracy of ± 0.1 during each measurement interval. The specular and diffused reflectances are measured with PerkinElmer LAMBDA 1050 Spectrophotometer by scanning the wavelength with steps of 1 nm in the visible wavelength range.

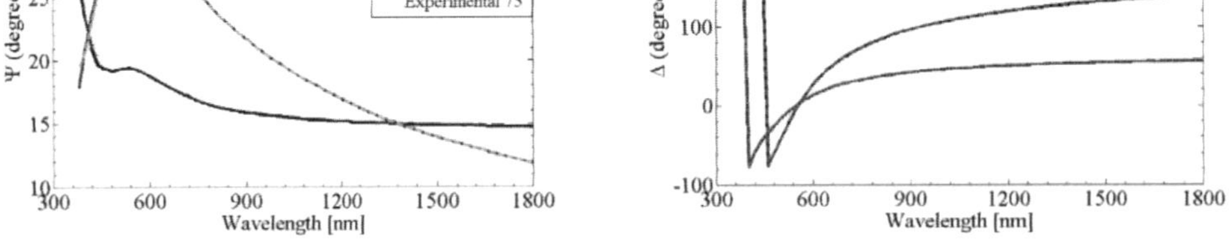

Figure 2. Ellipsometric experimental data of TiO_2 film with thickness t = 60 nm and Cauchy Model fitting. (a) Ψ values and wavelength λ. (b) Δ values and wavelength λ

3.2. Data modeling

The optical constants are obtained by fitting the ellipsometry measured data of Ψ and Δ with Cauchy model using Wvase 32 software at particular film thickness as shown in figure 2. The refractive index data is measured at incidence angles of 65° and 75° under isotropic depolarization conditions. The refractive index data of dielectric amorphous TiO_2 films obtained from Cauchy model is further modeled by Lorentz-Lorenz relation [18], owing to evaluate dn/dT and density via the relation.

$$\frac{n^2-1}{n^2+2}=\frac{4\pi}{3}\alpha_e\frac{\rho N_A}{m} \tag{5}$$

where n is the refractive index of TiO_2 films at room temperature, α_e is the electronic polarizability at optical frequencies in (cm^3), ρ is density in (gcm^{-3}), N_A is Avogadro number 6.023 × 10^{23} (electrons mol^{-1}) and m is the molecular weight 79.9 (g mol^{-1}). The electronic polarizabilities of Ti^{+4}, O^{-2} and TiO_2 molecule are 0.19 × 10^{-24}, 2.4 × 10^{-24} and 5 ×10^{-24} cm^3, respectively [16]. The electronic polarizability connects dipole moment **p** to applied field **E** by the relation [19].

$$\boldsymbol{p}=\alpha_e\boldsymbol{E} \tag{6}$$

Eq. (6) is commonly used for non-polar molecules at high frequencies so that permanent dipole moments do not follow the electric field. The α_e in Eq. (5) is assumed a scalar polarizabilty (isotropic case) due to insignificant difference in the scattered light in various directions.

3.2.1. Index data

To modal the thermal properties of refractive index of TiO_2 films, reciprocal of left hand side of Eq. (5) is calculated from temperatures 25 to 155 °C with interval of 5 °C and wavelengths 380 to 1800 nm for all TiO_2 films thicknesses.All data points follow a parabolic curve after plotting the function $(n^2+2)/(n^2-1)$ against temperature T as shown in figure 3. The calculated data points at 640 nm wavelength are fitted with least square approach with a characteristic average dn/dT= −4.2 ±0.7 × 10^{-5}°C, as calculated from Eq. (8).

$$\frac{n^2+2}{n^2-1}=5.8\times10^{-8}(T)^2+1.8\times10^{-5}T+1.6 \tag{7}$$

Differentiation of Eq. (7) with respect to T gives

$$\frac{\mathrm{d}n}{\mathrm{d}T}=\frac{-1}{6}\frac{(n^2-1)^2}{n}[11.6\times10^{-8}(T)+1.8\times10^{-5}] \tag{8}$$

Here, n is expressed as a function of temperature T and wavelength λ. The dn/dT value is evaluated at each wavelength for TiO_2 film thickness and is repeated for other film thicknesses, similarly.

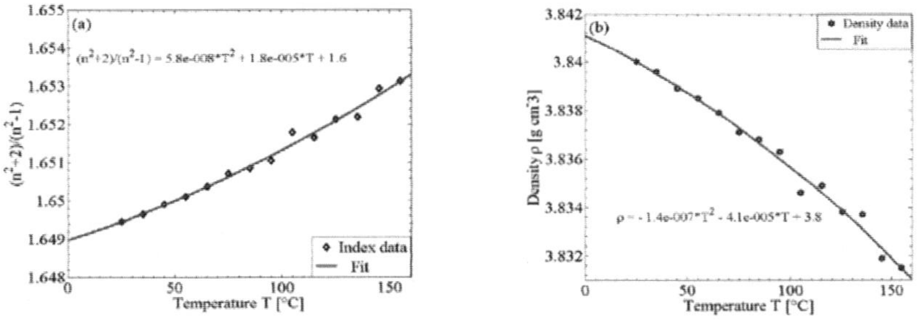

Figure 3. Ellipsometric measured experimentally fitted data of ALD-TiO_2 film with thickness t = 60 nm: (a) index variation with temperature, (b) density variation with temperature.

3.2.2. Density data

It is extensively reported that refractive index is approximately proportional to density of TiO_2 films [20]. From Eq. (5), we calculated the density $\rho(T)$ of TiO_2 films by using index values as a function of temperature. The density values are plotted as a function of temperature in figure 3b, which again shows a parabolic plot after fitting the data points with a fit Eq. (9), an average value of $d\rho/dT = -6.6 \pm 1.1\times10^{-5}$ gcm^{-3}°C^{-1}is computed directly from Eq. (10).

$$\rho = -1.4\times10^{-7}(T)^2 - 4.1\times10^{-5}T + 3.8 \tag{9}$$

Differentiation of Eq. (9) with respect to T evaluates the value of $d\rho/dT$ term.

$$\frac{d\rho}{dT} = -2.8\times10^{-7}(T) - 4.1\times10^{-5} \tag{10}$$

To compute the values of $d\alpha_e/dT$, differentiate Eq. (5) with T

$$\frac{dn}{dT} = \frac{2\pi}{9}\frac{N_A}{m}\frac{(n^2+2)^2}{n}[\alpha_e\frac{d\rho}{dT} + \rho\frac{d\alpha_e}{dT}] \tag{11}$$

use of thermal coefficients dn/dT and $d\rho/dT$ from Eqs. (8) and (10), respectively in Eq. (11) to compute the values of $d\alpha_e/dT$, i.e., about 3.84×10^{-31} cm^3°C^{-1} which in turn give an assessment of the temperature dependent electronic polarizabilities of ALD coated TiO_2 films.

Table 1. Measured temperature dependent refractive index n, extinction coefficient k, and calculated density of ALD-TiO_2 film of thickness 60 nm at wavelength 640nm.

T [°C]	n at 640 nm	K at 640 nm	ρ [g cm^{-3}]
25	2.3705	0.001555	3.8400
35	2.3702	0.001366	3.8396
45	2.3698	0.001262	3.8389
55	2.3695	0.001168	3.8385
65	2.3691	0.001046	3.8379
75	2.3686	0.000982	3.8371
85	2.3684	0.000955	3.8368
95	2.3681	0.000925	3.8363
105	2.3670	0.000781	3.8346
115	2.3672	0.000757	3.8349
125	2.3665	0.000623	3.8338
135	2.3664	0.000459	3.8337
145	2.3653	0.000558	3.8319
155	2.3650	0.000461	3.8315

3. Results and discussion

Figure 3 shows the refractive index and density of ALD-TiO_2 films of thickness t = 60 nm at wavelength of 640 nm. It is evident from the curves that refractive index and density of TiO_2 films are parabolic functions of temperature. The value of thermo-optic coefficient calculated from Eq. (8) depends on the temperature which gives rise small deviation from the linear behavior with temperature. However, for simplicity (linear dependence), and using a single coefficient value at each wavelength, we approximate the thermo-optic coefficient as the average value. Similar approximations are made to the density of ALD-TiO_2 films. Table 1 shows the experimentally measured temperature dependent values of refractive indices, extinction coefficients and calculated density of ALD-TiO_2 films. The decrease in index with temperature correspond to an average value of $(dn/dT)_{60nm} = -4.2 \pm 0.7 \times 10^{-5}$°C^{-1} which is about the same order ofmagnitude reported in (Ref [14]) but an order of magnitude smaller (Ref [15]). This in term explains the dependence behavior of thermo-optic coefficient on nature of deposition technique being used. The packing density of TiO_2 films is an important parameter contributing for the index values. More dense films (less porosity) are demonstrated to have low index gradient than high porosity films. This may be attributed to the presence of hydrogenated species in the pores of ALD-TiO_2 films. On heating water vapors desorb from surface with a result of decrease in optical path of TiO_2 films due to replacement of water by air [21]. Since the ALD films are fabricated through a layer by layer deposition process and inherently exhibit high density than other techniques, the value of index gradient is relatively small. Column 4 of Table 1 shows density of ALD-TiO_2 films which are slightly higher than the highest reported values by ion plating method ([10]).

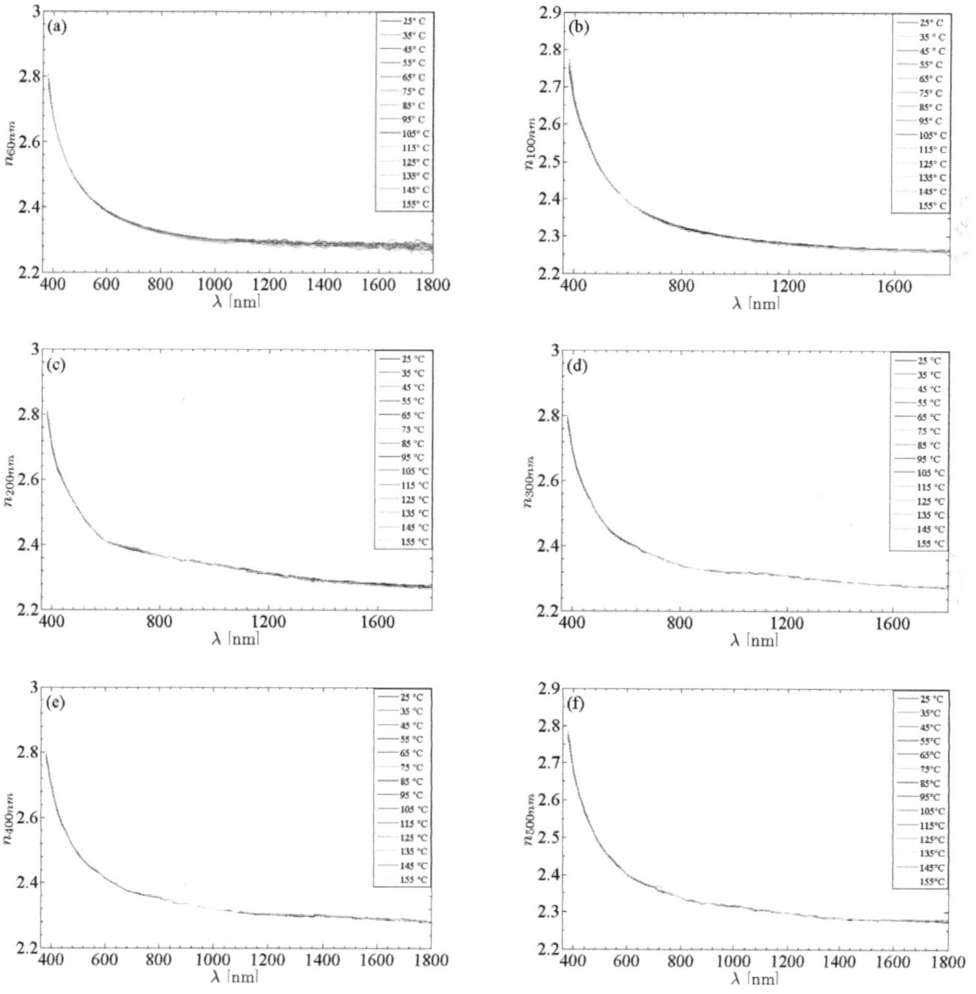

Figure 4. Ellipsometric measured experimental data of temperature dependent refractive index of ALD-TiO_2 films from 25 to 155 °C of thicknesses: (a) 60 nm, (b) 100 nm, (c) 200 nm, (d) 300 nm, (e) 400 nm, (f) 500 nm.

Figure 4 shows the refractive index changes with temperature for TiO_2 films of various thicknesses. It is evident from these dispersion plots that thermal effect on refractive index is very small for these films in the spectral range of 380 to 1800 nm. However, the thermal effect is significant to some extent for films of fewer thicknesses. It is observed that as the film thickness increases, the negative thermo-optic effect becomes less pronounced that makes index gradient with a positive value.

Figure 5 shows the temperature dependent refractive index, density and electronic polarizability of TiO_2 films as a function of wavelength. It depicts that thermal coefficients (dn/dT and dρ/dT) follow dispersion relation and the values become more stable with the increase in film thickness. The values of this coefficient possessed large negative values for thin films (< 150 nm) and tends to attain a decrease in negative dn/dT values as the film thickness increases at intermediate ranges and becomes positive for relatively thicker films (> 200 nm). It is also found that the values of temperature dependent electronic polarizability are so small to contribute significantly to the dn/dT. This may be attributed to the fact that thin films (t = 60 nm) have pores on its surface being initially filled by hydrogen containing species. The pores become empty on rising temperature that results in a decrease of index due to replacement of water molecules with air. Thereby, the effective refractive index of the material decreases up to a few surface nanometer layers that is scanned by ellipsometry in the form of a change of ellipsometric parameters Ψ and Δ. As the film thickness increases, the proportion of the surface porous area decreases in comparison to the overall thickness of the film and results in a decrease in negative index-gradient that makes dn/dT value to approach towards zero or to have a positive value for relatively thicker films. Figure 6 shows the variation of thermal coefficients (dn/dT and dρ/dT) with thickness of the films. It is found that the thermal coefficients approach to positive values when thickness increases from 60 nm and after reaching to a certain critical value of ~ 200 nm, the dn/dT remains constant with further increase in film thickness.

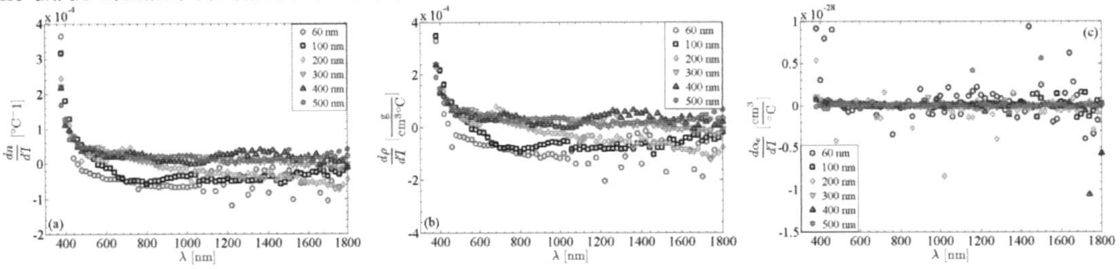

Figure 5. Thermal dependence of material properties of TiO_2 films of various thicknesses: (a) dn/dT, (b) dρ/dT, (c) dα_e/dT.

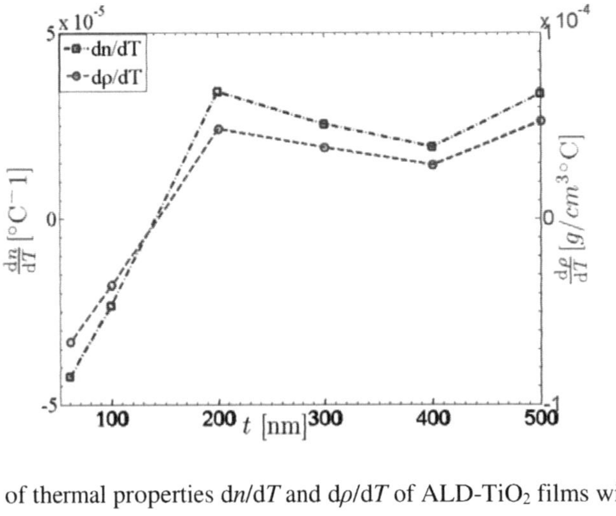

Figure 6. Variation of thermal properties dn/dT and dρ/dT of ALD-TiO_2 films with thickness t

4. Conclusion

Amorphous, high density, optical grade thin TiO_2 films of various thicknesses are coated on silicon substrate by ALD technique. The thermo-optic coefficient and temperature dependent density are calculated for various TiO_2 film thicknesses. These coefficients possess negative values for thinner films and tend to attain positive values for relatively thicker films. The measured dn/dT and $d\rho/dT$ are observed to follow the normal dispersion relation. In addition, the values of these coefficients follow parabolic curves that are obtained after fitting with least square approach. The average values of $(dn/dT)_{60nm} = -4.2 \pm 0.7 \times 10^{-5}$°$C^{-1}$ and $(d\rho/dT)_{60nm} = -6.6 \pm 1.1 \times 10^{-5}$ gcm^{-3} °C^{-1} of ALD-TiO_2 film at wavelength 640 nm are presented in detail while calculated over a wide wavelength range from 380–1800 nm.

5. Acknowledgements

Financial support from Strategic funding of the University of Eastern Finland, Institute of Photonics, Academy of Finland,Tekes and Higher Education Commission HEC (Pakistan), is highly appreciated.

6. References

[1] M. Furuhashi, M. Fujiwara, T. Ohshiro, M. Tsutsui, K. Matsubara, M. Taniguchi, S. Takeuchi, and T. Kawai, "Development of micro fabricated TiO_2 channel waveguides," *AIP advances*1 (2011) 032102.

[2] D. W. Dobbs, I. Gershkovich, and B. T. Cunningham,"Fabrication of a graded-wavelength guided-mode resonance filter photonic crystal," *Appl. Phys. Lett* 89 (2006) 123113.

[3] T. Alassaarela, D. Zheng, L. Huang, A. Priimagi, B. Bai, A. Tervonen, S. Honkanen, M. Kuittinen, and J. Turunen, "Single layer one-dimensional nonpolarizing guided-mode resonance filters under normal incidence," *Opt. Lett.* 36 (2011) 2411.

[4] J. M. Bennett, E. Pelletier, G. Albrand, J. P. Borgogno, B. Lazarides, C. K. Carniglia, R. A. Schmell, T. H. Allen, T. Tuttle- Hart, K. H. Guenther, and A. Saxer, "Comparison of the properties of titanium dioxide films prepared by various techniques," *Appl. Opt.* 28 (1989) 3303.

[5] M. R. Saleem, P. Stenberg, T. Alasaarela, P. Silfsten, M. B. Khan, S. Honkanen, and J. Turunen, "Towards athermal organic-inorganic guided mode resonance filters," *Opt. Express* 19, 24241-24251 (2011).

[6] R. L. Puurunen, "Surface chemistry of atomic layer deposition: A case study for the trimethylaluminum/water process," *Appl. Phys. Rev.* 97 (2005) 121301.

[7] K. Bange, C. R. Ottermann, O. Anderson, and U. Jeschkowski, "Investigations of TiO_2 films deposited by different techniques," *Thin solid films* 197 (1991) 279.

[8] M. Zhang, G. Lin, C. Dong, and L. Wen, "Amorphous TiO_2 films with high refractive index deposited by pulsed bias arc ion plating," *Surf. and Coating Tech.* 201 (2007) 7252.

[9] L. J. Meng, V. Teixeira, H. N. Cui, F. Placido, Z. Xu, and M. P. D. Santos, "A study of the optical properties of titanium oxide films prepared by dc reactive magnetron sputtering,"*Appl. Surf. Sci.* 252 (2006) 7970.

[10] O. Anderson, C. R. Ottermann, R. kuschnereit, P. Hess, and k. Bange, "Density and Young's modulus of thin TiO_2 films," *Fresenius J. Anal.Chem.*358 (1997) 315.

[11] D. Mergel, D. Buschendorf, S. Eggert, R. Grammes, and B. Samset, "Density and refractive index of TiO_2 films prepared by reactive evaporation," *Thin solid films* 371 (2000) 218-224.

[12] G. ketteler, S. Yamamoto, H. Bluhm, K. Andersson, D. E. Starr, D. F. Ogletree, H. ogasawara, A. Nilsson, and M. Salmeron, "The nature of water nucleation sites on TiO_2 (110) surfaces

revealed by ambient pressure X-ray photoelectron spectroscopy," *J. Phys. Chem.*C 111 (2007) 8278.

[13] T. Toyoda and M. Yabe, "The temperature dependence of the refractive indices of $SrTiO_3$ and TiO_2," *J. Phys. D: Appl. Phys.*16 (1983) L251.

[14] S. Wiechmann and J. Müller, "Thermo-optic properties of TiO_2, Ta_2O_5 and Al_2O_5 for integrated optics on silicon," *Thin solid films* 517 (2009) 6847.

[15] G. Gülşen and M. N. Inci, "Thermal optical properties of TiO_2 films," *Opt. Mater.*18 (2002) 373.

[16] H. G. Tompkins and E. A. Irene, *Handbook of ellipsometry*, William Andrew, New-York NY, 2005.

[17] J. Aarik, A. Aidla, H. Mändar, and T. Uustare, "Atomic layer deposition of titanium dioxide from $TiCl_4$ and H_2O: investigation of growth mechanism," *Appl. Surf. Sci.* 172 (2001) 148.

[18] C. Kittel, *Introduction to solid state physics*, John Wiley and Sons, New York NY, 1996.

[19] K. D. Bonin and V. V. Kresin, *Electric dipole polarizabilities of atoms, molecules and clusters*, World Scientific Publishing Co. Pte. Ltd., Singapore, 1997.

[20] C. R. Ottermann and K. Bange, "Correlation between the density of TiO_2 films and their properties,"*Thin solid films* 286 (1996) 32.

[21] S. F. pellicori and H. L. Hettich, "Reversible spectral shift in coatings," *Appl. Opt.*27 (1988) 3061.

Fabrication and properties of CNTs reinforced polymeric matrix nanocomposites for sports applications

A Rasheed and F A Khalid
Faculty of Materials Science and Engineering, GIK Institute of Engineering Sciences and Technology, Topi, KPK, Pakistan

E-mail: atif@giki.edu.pk

Abstract. The polymeric matrix composites have found extensive applications in sports because of high strength to weight ratio, ease of processing, and longer life. This work was carried out to study the properties of different sections of composite field hockey sticks and the influence of carbon nanotubes on their properties. The samples were fabricated by compression molding process. The increase in mechanical properties by the incorporation of carbon nanotubes is correlated with the process parameters to consider enhancement in the overall performance of the stick sections.

1. Introduction

The polymeric matrix composites can be a better choice for production of sports and other items owing to high strength to weight ratio, ease of processing, and longer life [1]. A comparison of the properties of wood, fiber reinforced polymeric matrix composite, and nanocomposite is shown in Table 1. The performance and durability of sports materials is enhanced in recent years by replacing traditional wood with advanced polymeric matrix composites and nanocomposites [1]. In sports items particularly field hockey sticks are now produced by a mix of glass fibers (GF) and carbon fibers (CF) reinforced polymeric matrix composites [2]. The carbon nanotubes (CNTs) are used in advanced polymeric matrix composites to further improve the properties. CNT/epoxy nanocomposites were previously fabricated and studied for creep behavior [3], thermal and mechanical properties [4], morphology [5], cure behavior [6], effect of interfacial chemistry on molecular mobility [7], damping capacity in a broad temperature range [8], doping of nano-particles (titania) for synergistic effects in the multiphase nanocomposites [9], electrical and thermal conduction mechanisms [10], influence of cup-stacked CNTs [11], and effect of silane functionalization on the properties of CNT/epoxy nanocomposites [12].

In glass-fiber/polymer composites, addition of CNTs can increase the cyclic delamination crack propagation resistance significantly. It also shows improvement in interlaminar fracture toughness. The fatigue life can be improved up to three times in case of in-plane cyclic loading. The CNTs can decelerate the delamination crack propagation by CNTs pull-out and fracture, and crack bridging [13]. Glass fiber reinforced epoxy and epoxy based vinyl ester composites were modified with CNTs and studied for interfacial [14] and interlaminar shear strength [15]. CNTs can also improve the Z-axis properties of composite laminates remarkably by direct reinforcing the matrix, fiber bridging mechanism, and toughening effect [15]. The flexural strength of glass fiber reinforced epoxy

Table 1. Comparison of mechanical properties of wood, composite, and nanocomposite used to manufacture field hockey sticks.

Property	**Wood (Mulberry)**	**Composite (GF/Epoxy)**	**Nanocomposite (CNT/GF/Epoxy)**
Density (gcm^{-3})	0.64	1.17	1.20
Flexural strength (MPa)	75	99.8	154.9
Ultimate tensile strength (MPa)	4	5.7	11.3
Young's Modulus (GPa)	4.9	6.2	8.1
Ref.	(Relevant websites)	[12, 14]	[16, 17]

composite modified with SWNTs treated with dispersing agents Volan and BYK-9076 (proprietary) showed enhancement up to 16% mainly caused by better dispersion of SWNTs in the composite [18].

The CNTs have also improved the properties in carbon fiber/polymer composites. Carbon fiber/epoxy hybrid nanocomposites modified with amine functionalized SWNTs were fabricated and studied for crack resistance at cryogenic temperature [19] and quasi-static strength and fatigue properties [20]. Carbon fiber/epoxy hybrid nanocomposites modified with cup-stacked carbon nanotubes (CSCNTs) were fabricated and studied for mechanical properties [21] reporting three times higher fracture toughness than that without CSCNTs [22].

This work was carried out to examine properties of fiber reinforced polymeric matrix composite field hockey sticks and sheet samples produced by compression molding process. The influence of carbon nanotubes on the properties of stick samples is studied. The role of varying volume fractions and number of fiber plies of carbon and glass fiber reinforcement is determined to assess the overall performance of field hockey sticks.

2. Experimental Procedure

Field hockey sticks of specific compositions are presented in Table 2. The fabrication of composite field hockey sticks involves the use of glass fibers and carbon fibers reinforcement in the epoxy matrix. The sticks were fabricated using compression molding process. The fiber orientation in the field hockey stick is shown in Figure 1. The fiber plies were stacked at alternate 0° and ± 45°. The 0° plies impart stiffness while 45° give strength to the stick. The average length of the hockey sticks used in the current study is 95.5 cm (36.5 in). The height of the head is 11.5 cm. The shape of the hockey stick changes along its length. It is circular at the handle while elliptical at the head. The length of major axis changes from 2.5 cm to 5 cm while the minor axis is fixed at 2.5 cm. The wall thickness changes from 1.15 mm (at handle) to 6.10 mm (at head). There are 8 fiber plies in handle while 24 in head. The stick was divided into three portions; initial circular handle (section-I), central elliptical portion (section-II), and the head (section-III). The length of section-I & II is 40 cm while of section-III is 15.5 cm. The epoxy resin used as matrix for composite field hockey sticks modified with CNTs as provided by the suppliers Honlu Technology Co., Ltd. was used to study the influence of

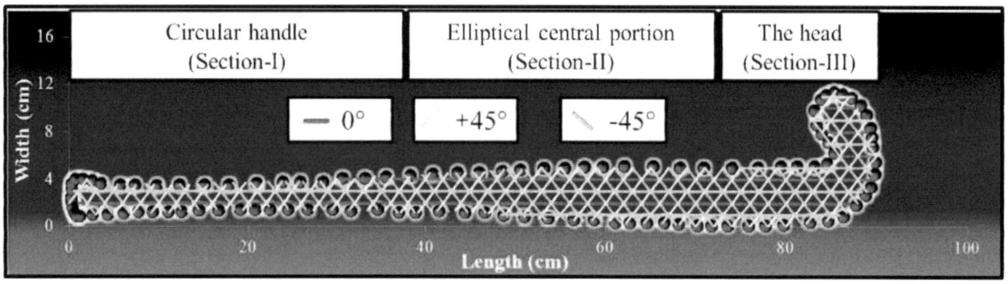

Figure 1. Schematic diagram showing fiber orientation in the composite field hockey stick sample.

Table 2. Compositions of samples fabricated.

Sr.	Compositions		Volume fractions	
	Matrix	**Reinforcement**	**V_f (%)**	**V_m (%)**
Field hockey sticks				
1	Epoxy	CF/GF	47	53
2	Epoxy	CF/GF	49	51
3	Epoxy	CF/GF	46	54
4	Epoxy	CF/GF	52	49
Composite sheets				
5	Epoxy	GWC	60	40
6	Epoxy	GWC/0.4vol% CNTs	61	39

GF = Glass Fibers, CF = Carbon Fibers, GWC = Glass Woven Cloth

CNTs on the properties of composite field hockey sticks. Two composite sheets were fabricated whose composition is provided in Table 2. The processing conditions were maintained to compare results with the field hockey stick samples to assess their properties and performance.

Three point bend test of hybrid nanocomposite samples was conducted to study their flexural stress-strain response. The specimens were prepared according to ASTM Standards D790 and D4476 for rectangular and round specimens, respectively. The specimens were cut using hacksaw and polished along the longitudinal direction. The dimensions were 3.0 mm × 12.7 mm × 70 mm of rectangular specimens and 6.25 mm × 12.5 mm × 120 mm of round specimens. The support span was fixed at 48 mm for rectangular specimens and 100 mm for round specimens. The displacement rate was kept constant at 0.5 mm/min. Five specimens were tested for each composition in longitudinal direction. The test was conducted on Instron 100 kN universal testing machine Model 8501.

3. Results and Discussion

3.1. Field hockey sticks

The flexural stress-strain curves obtained from three point bend test of the samples from section-I, II, and III of the hockey sticks are shown in figures 2-4. The trend shows that the strength increases with increasing volume fraction of carbon fibers.

It can be noted that the trend of increasing strength with carbon fiber volume fraction is not linear which can be explained on the basis of microstructure. The optical micrographs reveal fiber orientation of the transverse sections taken from different locations of the stick samples (Figure 5). Figure 5 (a) shows, mainly, the glass fiber plies while (b) shows the alternate layers of glass fibers and carbon fibers. Porosity was found in the samples. It can be observed that porosity is non-uniform in size and distribution. This non-uniform size and distribution of the porosity result in variation in mechanical properties. The other reason can be non-uniform distribution of glass and carbon fibers. Two optical micrographs taken at same magnification but from different locations of the same stick are shown in Figure 6. Figure 6 (a) shows the optical micrograph of glass fibers while (b) shows carbon fibers. It shows that the fiber distribution is not the same throughout. Another reason could be variation in number of plies and fiber orientation. The optical micrographs of stick head (Sec. III) are shown in Figure 7. The stick is kept underweight throughout the hand lay-up process. It is because that the plies cannot be removed from the stick once applied. However, once the shape of the stick is fixed, the strips of fiber plies can be attached easily to get the desired weight. These strips are unequal in size as difference in weight is different for different sticks. These strips are applied at different regions keeping in consideration the weight balance. These strips can either be of glass fibers or carbon fibers. The fiber orientation in all the strips is also not the same. If these strips are not applied, the required weight would not be achieved which will again produce non-standard samples. Because of all these variables, the trend for strength is non-linear.

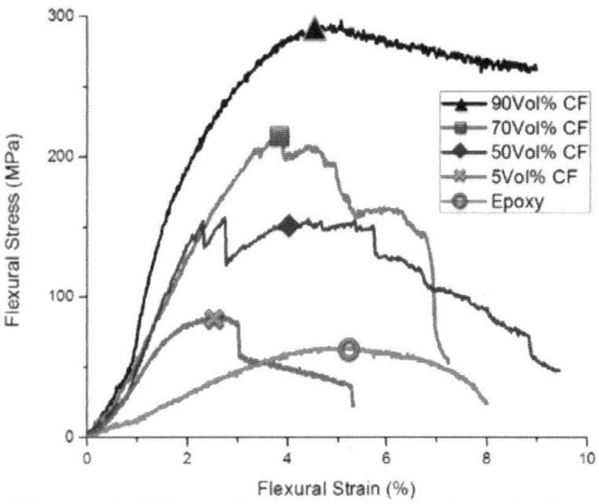

Figure 2. Flexural stress vs flexural strain for stick handle (Sec. I)

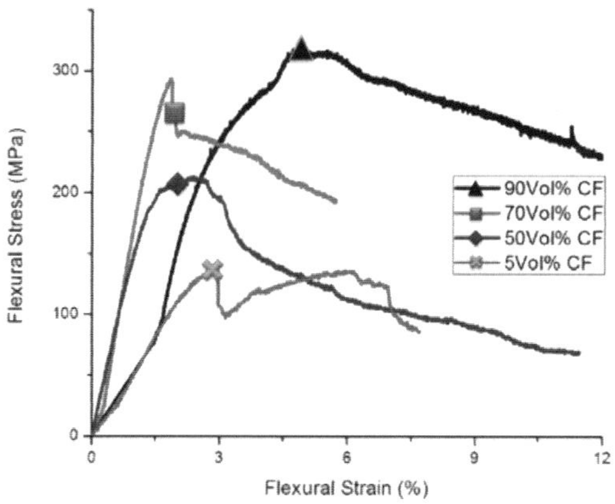

Figure 3. Flexural stress vs flexural strain for stick central portion (Sec. II)

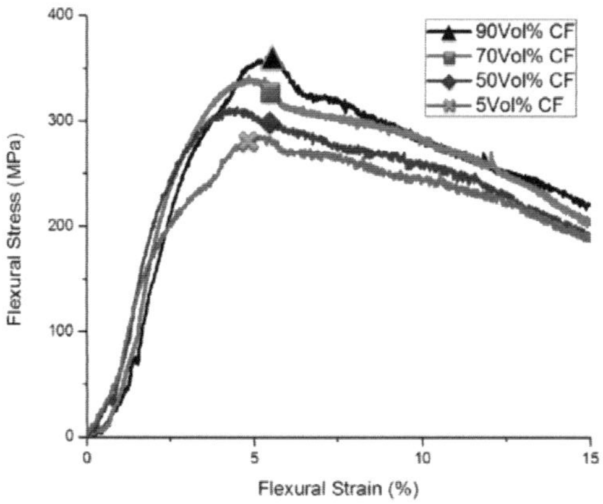

Figure 4. Flexural stress vs flexural strain for stick head (Sec. III)

The influence of carbon fiber volume fraction on flexural strength of different hockey sections is shown in Figure 8. The carbon fibers varied from 5vol% to 90vol% of total fiber reinforcement. There is increase in strength with increasing volume fraction of carbon fibers. The maximum strength is achieved for head (361 MPa) while minimum strength was observed for handle (296 MPa). The number of fiber plies in handle is 8 while 24 in head. The wall thickness of handle (section-I) is 1.15 mm while that of head (section-III) is 6.10 mm. The carbon fibers are concentrated in head to improve the impact strength. It makes head stronger than handle.

3.2. Nanocomposite Sheets

The flexural stress-strain curves obtained from three point bend test of CNTs/GWC/epoxy samples are shown in Figure 9. The flexural strength increased from 54 MPa (neat epoxy) to 342 MPa by making composite with GWC as reinforcement. The strength has further increased to 372 MPa with the incorporation of carbon nanotubes. The CNTs act as additional reinforcement thereby improving the strength. The improvement in properties can further be optimized. The optical micrographs of transverse sections of samples taken from composite sheets without and with CNTs are shown in Figure 10 (a) and (b), respectively. Porosity was found that may have due to air entrapment and relative movement of the fibers and resin during compression molding. Another reason for this porosity can be solvent evaporation. The solvent is added in the resin to reduce its viscosity to help efficient nanotube dispersion. This solvent (along with any other volatiles) gets evaporated while curing at 140 °C leaving behind a huge amount of porosity.

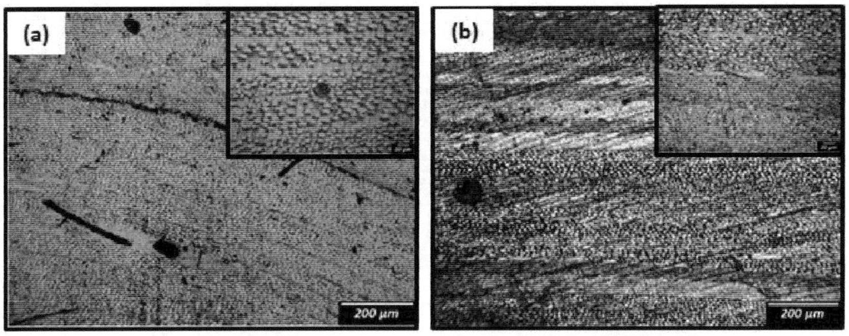

Figure 5. Optical micrographs of stick handle (Sec. I)

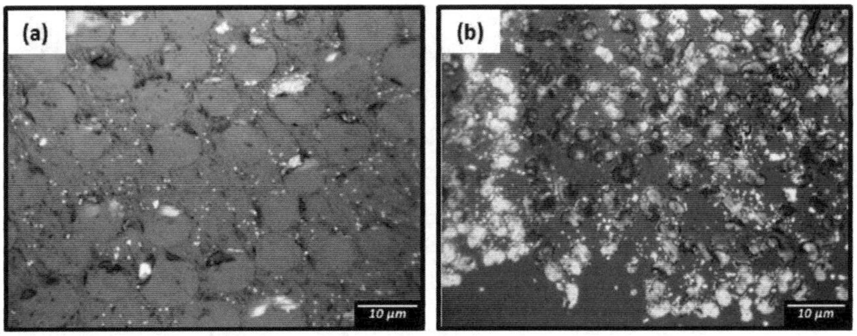

Figure 6. Optical micrographs of stick central portion (Sec. II)

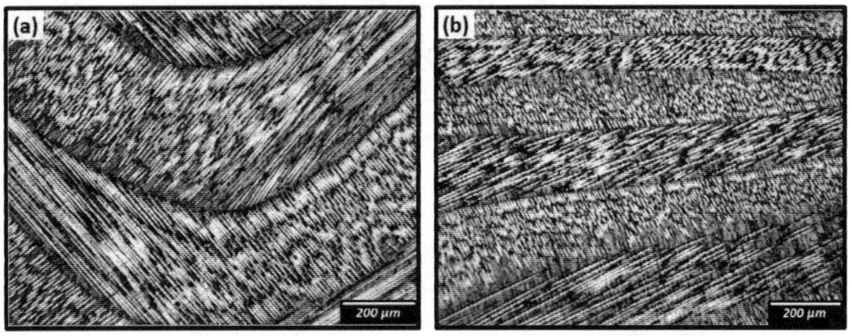

Figure 7. Optical micrographs of stick head (Sec. III)

3.3. Fracture behavior

The SEM image of fractured sample taken from composite field hockey stick is shown in Figure 11 (a, b). Figure 11 (a) shows the region where the fibers are bent to give the required curvature to the stick. The fibers perpendicular to the supports were bearing the load and got fractured. Figure 11 (b) shows the intact fibers above which the matrix has sheared. It shows that the interfacial interactions were not strong enough to transfer the load, to the fibers, completely. The SEM image for composite sheet without CNTs is shown in Figure 11 (c, d). The images show delamination among fiber plies along the longitudinal direction. A similar fracture behavior was observed for composite sheet containing CNTs as shown in Figure 11 (e, f). There is no significant difference observed in the fracture mode with the incorporation of CNTs. However, as the strength was improved by the incorporation of CNTs, the CNTs have delayed the time to fracture by alleviating the stress concentration and by blunting the tip of crack, initiated in the matrix, by CNTs pull-out and fracture i.e. "surface crack bridging" effect.

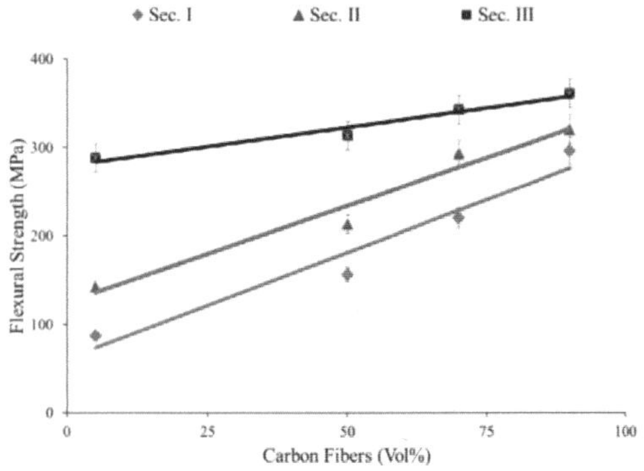

Figure 8. Influence of carbon fiber volume fraction on the flexural strength of field hockey stick samples

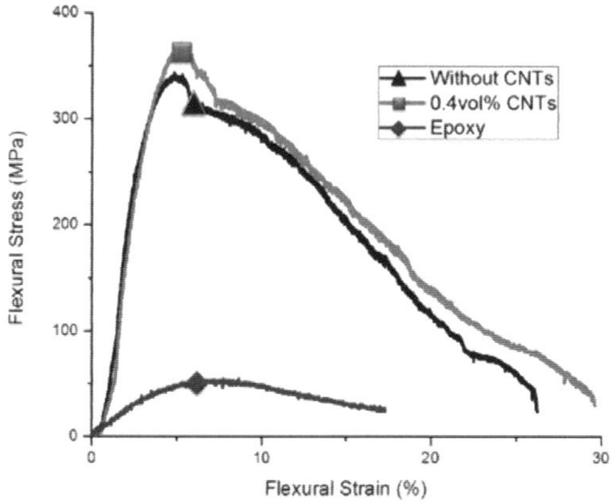

Figure 9. Flexural stress vs flexural strain for CNTs/GWC/epoxy nanocomposites

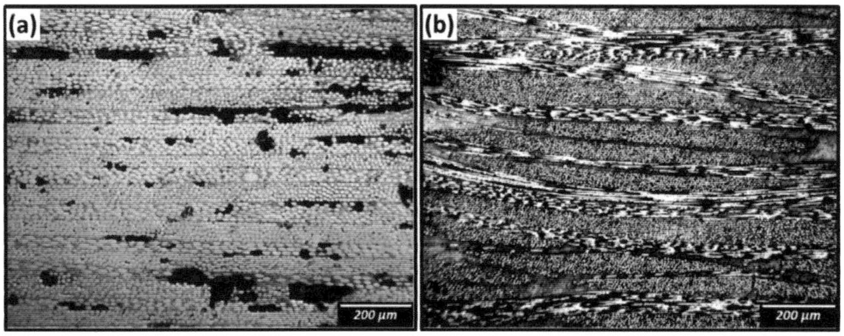

Figure 10. Optical micrographs of transverse sections of nanocomposite sheets; (a) GWC/epoxy, (b) CNT/GWC/epoxy

4. Conclusions

- In this work, glass and carbon fibers reinforced epoxy matrix composite field hockey sticks and sheets of different compositions were fabricated and main results are presented below:
- The fiber distribution is not the same throughout the length of the field hockey stick. The carbon fibers are concentrated in the head while the handle is enriched with glass fibers.
- The strength increases with increasing volume fraction of carbon fibers. The carbon fibers are preferred to increase the strength and to make the stick lightweight.
- The flexural strength increased from 342 MPa to 372 MPa with the incorporation of carbon nanotubes. The CNTs act as additional reinforcement thereby improving the strength.
- Porosity was found that may have due to air entrapment and relative movement of the fibers and resin during compression molding. Another reason for this porosity can be solvent evaporation. . The solvent is added in the resin to reduce its viscosity to help proper nanotube dispersion. This solvent (along with any other volatiles) gets evaporated while curing leaving behind a huge amount of porosity.

- The SEM images of fractured samples showed delamination among fiber plies along the longitudinal direction. There is no significant difference observed in the fracture mode.

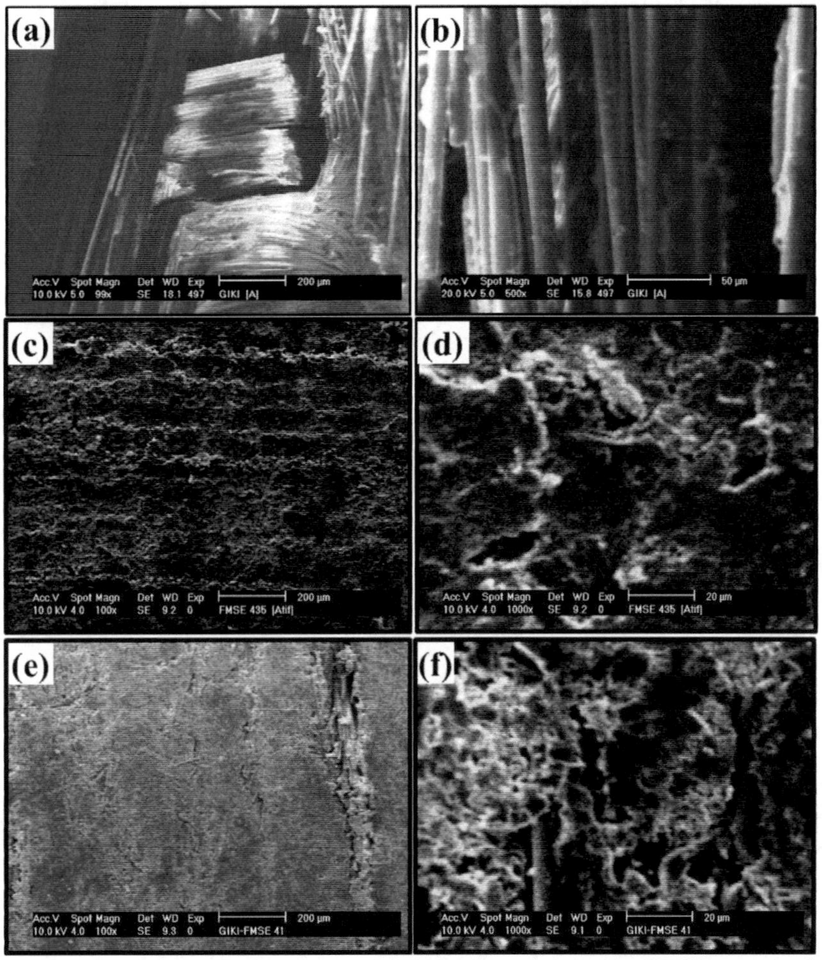

Figure 11. SEM images of fractured surfaces of field hockey stick (a, b), GWC/epoxy (c, d), and CNTs/GWC/epoxy samples (e, f).the incorporation of CNTs

5. Acknowledgements
Mr. Arfan Iqbal (Model sports works, Sialkot) is acknowledged for provision of experimental samples and use of facilities at the plant.

6. References
[1] ASM Handbook, Composites. 2002;21:1071.

[2] Baum CS. Hockey stick formed of composite materials. US Pat 5 217 221 (8 June 1993).

[3] Tehrani M, Safdari M, Al-Haik MS. Nanocharacterization of creep behavior of multiwall carbon nanotubes/epoxy nanocomposite. International Journal of Plasticity. 2011;27(6):887-901.

[4] dos Santos MN, Opelt CV, Lafratta FH, Lepienski CM, Pezzin SH, Coelho LAF. Thermal and mechanical properties of a nanocomposite of a photocurable epoxy-acrylate resin and

multiwalled carbon nanotubes. Materials Science and Engineering: A. 2011;528(13-14):4318-24.

[5] Abdalla M, Dean D, Theodore M, Fielding J, Nyairo E, Price G. Magnetically processed carbon nanotube/epoxy nanocomposites: Morphology, thermal, and mechanical properties. Polymer. 2010;51(7):1614-20.

[6] Abdalla M, Dean D, Robinson P, Nyairo E. Cure behavior of epoxy/MWCNT nanocomposites: The effect of nanotube surface modification. Polymer. 2008;49(15):3310-7.

[7] Abdalla M, Dean D, Adibempe D, Nyairo E, Robinson P, Thompson G. The effect of interfacial chemistry on molecular mobility and morphology of multiwalled carbon nanotubes epoxy nanocomposite. Polymer. 2007;48(19):5662-70.

[8] Auad ML, Mosiewicki MA, Uzunpinar C, Williams RJJ. Single-wall carbon nanotubes/epoxy elastomers exhibiting high damping capacity in an extended temperature range. Composites Science and Technology. 2009;69(7-8):1088-92.

[9] Sumfleth J, de Almeida Prado LAS, Sriyai M, Schulte K. Titania-doped multi-walled carbon nanotubes epoxy composites: Enhanced dispersion and synergistic effects in multiphase nanocomposites. Polymer. 2008;49(23):5105-12.

[10] Gojny FH, Wichmann M. H. G, Fiedler B, Kinloch I. A, Bauhofer W, Windle A. H, Schulte K. Evaluation and identification of electrical and thermal conduction mechanisms in carbon nanotube/epoxy composites. Polymer. 2006;47:2036-45.

[11] Choi Y-K, Gotoh Y, Sugimoto K-i, Song S-M, Yanagisawa T, Endo M. Processing and characterization of epoxy nanocomposites reinforced by cup-stacked carbon nanotubes. Polymer. 2005;46(25):11489-98.

[12] Ma PC, Kim J-K, Tang BZ. Effects of silane functionalization on the properties of carbon nanotube/epoxy nanocomposites. Composites Science and Technology. 2007;67(14):2965-72.

[13] Grimmer CS, Dharan CKH. Enhancement of delamination fatigue resistance in carbon nanotube reinforced glass fiber/polymer composites. Composites Science and Technology. 2010;70(6):901-8.

[14] Godara A, Gorbatikh L, Kalinka G, Warrier A, Rochez O, Mezzo L, et al. Interfacial shear strength of a glass fiber/epoxy bonding in composites modified with carbon nanotubes. Composites Science and Technology. 2010;70(9):1346-52.

[15] Zhu J, Imam A, Crane R, Lozano K, Khabashesku V, Barrera E. Processing a glass fiber reinforced vinyl ester composite with nanotube enhancement of interlaminar shear strength. Composites Science and Technology. 2007;67(7-8):1509-17.

[16] Deng F, Ogasawara T, Takeda N. Tensile properties at different temperature and observation of micro deformation of carbon nanotubes–poly(ether ether ketone) composites. Composites Science and Technology. 2007;67(14):2959-64.

[17] Zhang J, Jiang D. Interconnected multi-walled carbon nanotubes reinforced polymer-matrix composites. Composites Science and Technology. 2011;71(4):466-70.

[18] Yan Z, Yuexin D, Lu Y, Fengxia G. The dispersion of SWCNTs treated by dispersing agents in glass fiber reinforced polymer composites. Composites Science and Technology. 2009;69(13):2115-8.

[19] Kim M-G, Moon J-B, Kim C-G. Effect of CNT functionalization on crack resistance of a carbon/epoxy composite at a cryogenic temperature. Composites Part A: Applied Science and Manufacturing. 2012.

[20] Davis DC, Wilkerson JW, Zhu J, Hadjiev VG. A strategy for improving mechanical properties of a fiber reinforced epoxy composite using functionalized carbon nanotubes. Composites Science and Technology. 2011;71(8):1089-97.

[21] Tomohiro Y, Takahira A, Akiko A, Masaru I, Takashi Y. Fabrication of CNT-Dispersed CFRP using length-controlled CNTs: Measurement of CNT length and characterization of mechanical properties. Tsinghua science and technology. 2009;14:100-4.

[22] Yokozeki T, Iwahori Y, Ishibashi M, Yanagisawa T, Imai K, Arai M, et al. Fracture toughness improvement of CFRP laminates by dispersion of cup-stacked carbon nanotubes. Composites Science and Technology. 2009;69(14):2268-73.

Effect of zirconium addition on welding of aluminum grain refined by titanium plus boron

A I O Zaid
Industrial Engineering Department, University of Jordan, Amman, Jordan

E-mail: a_kilani43@hotmail.com

Abstract. Aluminum oxidizes freely in ordinary atmosphere which makes its welding difficult and weak, particularly it solidifies in columnar structure with large grains. Therefore, it is anticipated that the effect of addition of some grain refiners to its melt before solidification is worth while investigating as it may enhance its weldabilty and improve its mechanical strength. In this paper, the effect of addition of zirconium at a weight of 0.1% (which corresponds to the peretictic limit on the aluminum-zirconium base phase diagram) to commercially pure aluminum, grain refined by Ti+B on its weldability, using gas tungsten arc welding, GTAW, method which was formerly known as TIG. A constant current level of 30 AC Ampere was used because it removes the oxides during the welding process. Metallographic examination of the weldments of the different combinations of Al with Al and Al with its microalloys: in the heat affected zone, HAZ, and away from it was carried out and examined for HAZ width, porosity, cracks and microhardness. It was found that grain refining by Ti+B or Zr resulted in enhancement of the weldment.

1. Introduction

Aluminum and its alloys are versatile and very attractive materials due to their high strength–to-weight ratio, high electrical and thermal conductivities with respect to their weight good corrosion resistance and appearance. Due to these properties Al and its alloys are widely used in engineering and industrial applications particularly in the automobile and air craft industries. Beside these useful properties they have the disadvantage of solidifying in columnar structure with large grain size which tends to deteriorate their mechanical strengths and surface quality [1]. Therefore it became a necessity to grain refine their structure with some grain refiners e.g. Ti, Ti+B, V, Mo etc. to the melt prior to solidification [1,2,3].

Welding is one of the most important means of fabrication and maintenance processes in industry. It is used in joining different alloys in different shapes and applications. The gas tungsten arc welding method, GTAW, which is formerly known as TIG is very often used for welding aluminum, magnesium, titanium and their alloys particularly in welding aluminum and its alloys. It is generally preferred because it produces very high quality welds. Furthermore, the control of distortion which is a major problem in welding of thin sections [4]. To achieve good and strong welds and avoid defects, considerations should be given to the selection of the welding process and the parameters affecting it. In general, the strength and the quality of the weldment are strongly affected by the process parameters which in TIG welding includes: welding current, polarity (DCSP/DCRP), arc voltage (arc length), travel speed, weld joint position, electrode diameter and shielding gas composition and flow rate. The results of the research on the effect of these parameters indicated that they have profound effect on the strength and quality of the weld. Knowledge and control of these variables is essential to consistently produce sound welds of good quality and strength. It is worth mentioning that these variables are not completely independent and changing one requires changing one or more of the others to achieve the desired results, [5,6,7,8,9].

Weld fusion zones typically exhibit coarse columnar structure due to the prevailing high temperature of the weld metal solidification, specially in the case of aluminum and its alloys. This normally results in inferior weld mechanical properties and poor resistance to hot cracking. Hence, it is highly desirable to control solidification structures in welds specially in the HAZ region, which is often very difficult to do so, because of the higher temperatures and the high thermal gradients in the fusion zone. This is unlike the case in casting process of these alloys. Different attempts have been tried for refining the structure in the weld fusion zone, e.g. inoculation with heterogeneous inoculants, microcooler, additions, surface nucleation induced by gas impingement and introduction of physical disturbance techniques such as torch vibrations [10,11,12]. It is worth noting that the use of inoculants for refining the weld fusion zones involves undesirable effects of inoculating elements which tends to adversely affect the mechanical properties of the weld in the HAZ (weld fusion zone), when the inoculants are added to the level which will cause grain refinement. Similarly, the surface nucleation and microcooler additions were also turned down because of the complicated welding setups required and procedures associated with their use. Recently, two techniques namely, magnetic arc oscillation and the current pulsing have gained popularity due to their promising results and the relative ease by which these techniques can be applied to practical and industrial application with only minor modifications to the existing welding setups. The advantages of the pulsed and magnetic arc oscillation welds is claimed to include grain refinement in the fusion zone with reduced width of HAZ which leads to less distortion, control of segregation, reduced cracking sensitivity and reduced residual stresses [13]. Optimization of the pulsed TIG welding process parameters using Taguchi method aiming for increasing the mechanical properties of the weld was investigated and reported [9].To the best of the author's knowledge the effect of addition of grain refiners on the weldability of Al has not been previously investigated. In this paper, the effect of addition of zirconium to aluminum grain refined by Ti+B on its weldability is investigated.

2. Materials, equipment and experimental procedures

2.1. *M*aterials

2.1.1. Base metal

Commercially pure aluminum 99.8% purity of the chemical composition shown in table 1 was used throughout this work.

Table 1. Chemical composition of commercially pure aluminum.

Element	Fe	Si	Cu	Mg	Ti	V	Zn	Mn	Na	Al
Wt. %	0.09	0.05	0.005	0.004	0.004	0.008	0.005	0.001	0.005	Rem.

The commercially available Al-Ti-B ternary master alloy of the chemical composition shown in Table-2 was used for manufacturing the Al- %Ti- %B and its microalloys. High purity zirconium powder was used for manufacturing the Al–3%Zr binary alloy. The crucible and stirring rods used in all the experiments were made of graphite.

2.1.2 Filler and electrode materials

A separate filler made of aluminum with argon as an inert shielding gas, and inconsumable electrode 2.5 mm diameter made of tungsten having conical shape were used in the GTWA welding process.

2.2 *E*xperimental *p*rocedures

2.2.1. Preparation of the Al base metal
The commercially pure bundles of the Al wires were pickled in HNO_3 to remove the oxide layers and any other contamination, and then melted in a graphite crucible inside an electric furnace at about 800°C and then poured to solidify in hollow cylindrical brass rods of 10 mm inside diameter and 55 mm external diameter. Finally, the rods were rolled into sheets of 2 mm thickness , 10 mm width and 240 mm length.

2.2.2 Preparation of the binary master and the microalloys
The Al–Zr binary master alloy were prepared by adding the calculated amount of Zr to the predetermined amount of molten aluminum in the graphite crucible at 850°C for Al–Zr binary master alloy, stirred by the graphite rod for one minute and brought back to the furnace for 20 minutes, brought out and stirred again for one minute and finally spread over a thick cast iron plate to solidify in pieces of thickness less than 5 mm. These master alloys were used for preparing the following three different microalloys, with the chemical composition (wt%): Al-0.05%Ti–0.01%B and Al–0.05% Ti–0.01%B–0.1%Zr. Finally, these prepared microalloys together with the commercially pure aluminum specimen were rolled into sheets of 2 mm thickness, 12 mm width and 240 mm length.

*2.2.3 The welding proc*ess
The tungsten inert gas welding, GTAW, process of the experimental set up shown in figure 1 was used. The single lap joint, figure 2, was used in the welding process.

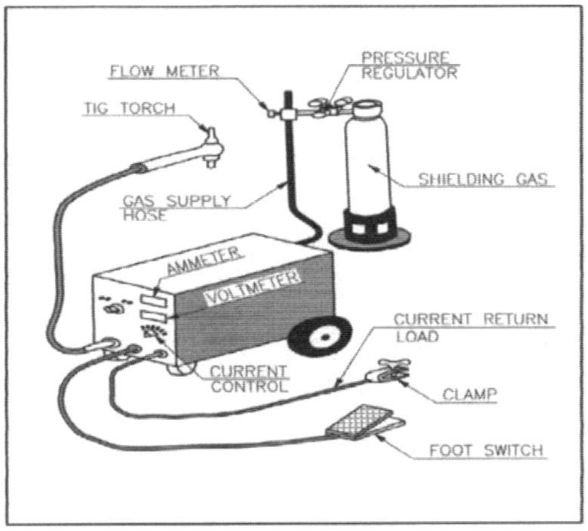

Figure 1. The TIG welding setup.

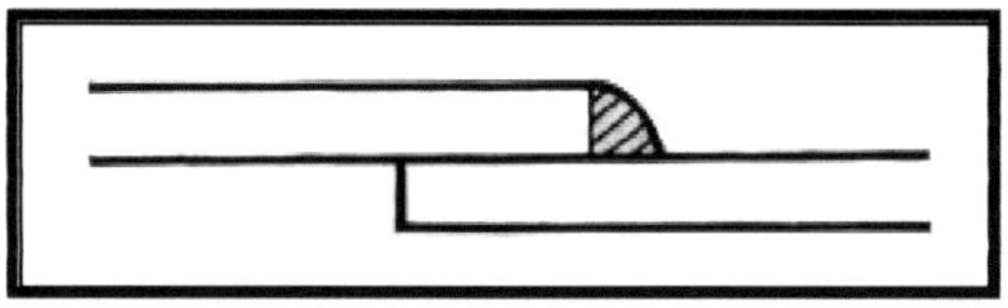

Figure2. The lap Joint.

The welding process started by adjusting the AC current on the welding machine to 30 A which was fixed in welding the different plates. All the specimens were welded using a fixed rate of the shielding gas, constant welding speed, which was calibrated with respect to the motor speed, 0.6 cm/sec, and the current was 30 Ampere AC. The aluminum welding electrode tip is held perpendicular to the welding surface at 5 mm from the workpiece surface. The torch was held perpendicular to the surface of the workpiece and started to heat the starting point of the weld by moving the torch in a circular motion. Once the pool of the molten metal has become bright the torch is ready to be moved along the required weld line.

2.2.4. Hardness testing
The Vicker's microhardness survey were carried out by taking the average of ten values along the HAZ and away from it along the base metal regions, using Digital microhardness tester (model HWDM-3).

2.2.5. Microstructural examination
One specimen of the Al base metal and of each of its microalloys were cut, hot mounted in bakelite ground by different grit size and polished with 1 micron diamond paste. Finally were etched using an etchant of the composition: 1.5% HCl acid, 2.5% HNO3, 0.5% HF acid and 95.5% H_2O ready for grain size and microstructural examination. The photomicrographs were taken on the HAZ region and on the base metal, away from the HAZ regions as shown in figure 3.

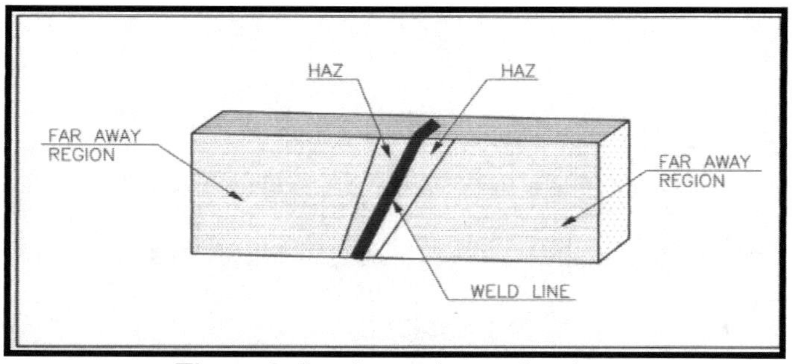

Figure 3. Specimen observed region.

3. Results and discussion
In this section the results are discussed with reference to the effect of addition of either Ti-B or Zr or with Al without addition of any grain refiner. The results and discussion will be dealt with, for welding of Al with each of its microalloys namely Al-Ti-B, Al-Zr and Al-T-Bi-Zr under three headings: First effect on the grain size and second on the hardness and third on the soundness of the weld in the HAZ and in the base metal regions. Also the effect of these grain refiners on the weldability of the three aluminum microalloys together; Al-Ti-B with Al-Ti-B, Al-Zr with Al-Zr and AL-Ti-B-Zr with Al-Ti-B-Zr as compared to the welding of Al with Al under the above three mentioned headings in HAZ and in the base metal regions, is also presented and discussed.

3.1. *Effect of addition of Ti+B and Zr to Al on its grain size*

3.1.1. Effect of welding Al with its microalloys on grain size
It can be seen from the histogram of Fig.4 that the addition of Ti-B to Al on its weldability resulted in 31.25% decrease of its grain size in the HAZ region whereas addition of Zr alone did not affect the grain size in this region. However, addition Ti-B and Zr alone resulted in more increase than Ti-B alone being 37.5% which indicates that their addition is not additive. This agrees with previous findings[1]. Similarly, addition of either Ti-B or Zr to Al in the base region either alone or together resulted in decrease of the grain size in the base. The maximum is in case of addition of both of them together being 38.46%, and followed by Zr addition alone, 26.15%

3.1.2 Effect of Welding Al with its Microalloys on Grain Size
It can be seen from the histogram of figure 4 that the addition of Ti+B or Ti-B + Zr to Al has resulted in decrease of 36.36% in grain size in the HAZ whereas addition of Zr alone did not affect the grain size in this region as compared to the Al with Al. However, addition of both of them together resulted in slight decrease in the HAZ when compared to Ti+B addition alone, only 10%. Furthermore, it can also be seen from figure 4 that in the base metal away from HAZ, addition of Ti+B or Zr either alone or together resulted in increase of the grain size as compared to the HAZ of each combination.

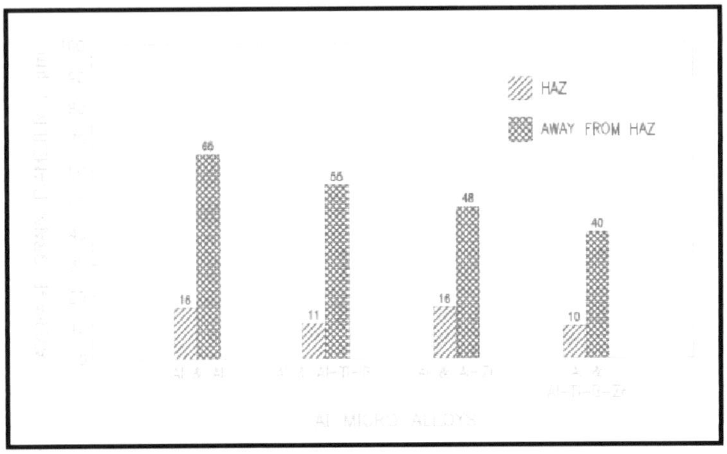

Figure 4. Effect of Ti+B and Zr addition on the grain size in the HAZ and base metals of welding Al with its microalloys.

Comparing the grain size in the base of welding Al with Al or Al with its microalloys Al-Ti-B. Al-Zr and Al-Ti-B-Zr, it can also be seen from Figure 4 that there is high increase in the grain size when compared to the HAZ. However, when comparing the grain size of the base metal (away from HAZ), between welding Al with Al or with its microalloys, it can be seen there is decrease in the welding of Al with the microalloys. The maximum decrease is 40% in welding of Al with Al-Ti-B-Zr microalloy, 18.4 % decrease in welding Al with Al-Ti-B. The fine grain size in the HAZ for welding Al with Al or any of its microalloys Al-Ti-B, Al-Zr and Al-Ti-B-Zr is due to the very high temperature in the HAZ as compared to the base region away from welding Al with the HAZ. However, the decrease of the grain size of the three Al microalloys as compared to Al in the base region is attributed to the refining effect of Ti-B, Zr or Ti-B + Zr when added to Al melt prior to the welding process. This is explicitly illustrated in the photomicrographs of Figure 5.

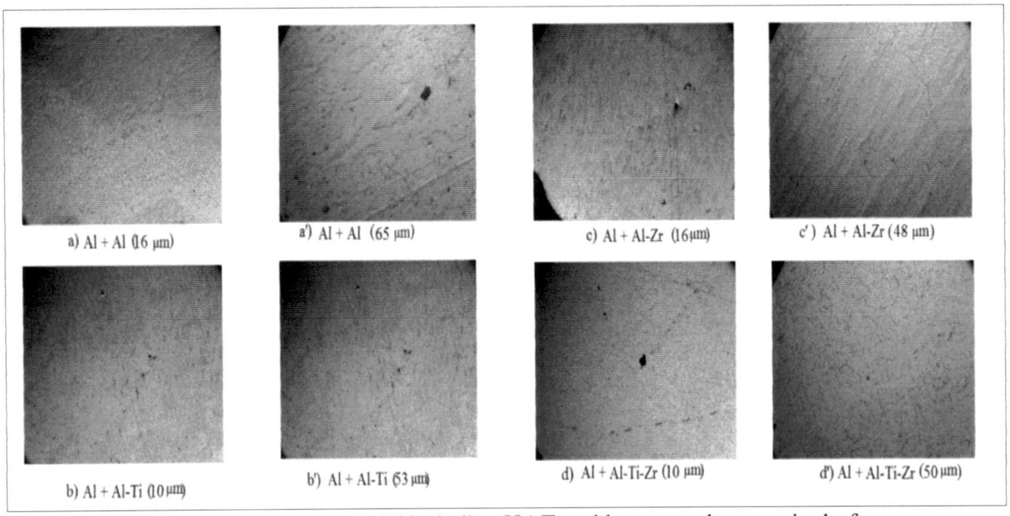

Figure 5. a, b, c, d and a', b', c', d' at HAZ and base metal respectively for welding Al and its microalloys

3.1.3 Effect of welding Al microalloys together on grain size

It can be seen from the histogram of Figure 6 that the grain size in the HAZ and base metal for welding aluminum microalloys together resulted in great reduction in grain size as compared with welding Al with Al being greater reduction in the base metal, for example about 86% in the base metal compared to 37.5 % in the HAZ. This may be due to the grain refinement of the two welded microalloys Al-Ti-B prior to welding, as the HAZ is a combination of both welded strip made of Al-Ti-B. However, welding Al-Zr with Al-Zr did not affect the grain size in the HAZ as compared to the grain size in the HAZ of Al with Al (16 microns). Where a pronounced poisoning (larger grain size) in the base metal is indicated, (from 65 to 80), about 23%. Regarding welding Al-Ti-B-Zr with Al-Ti-B-Zr, i.e. when both Ti and Zr are added together, it produced slightly better refinement than in the case of Al-Ti-B with Al-Ti-B, in the HAZ and no difference in the base metal respectively. This is explicitly illustrated in the photomicrographs of Figure 7 a, b, c, d and a', b', c', d' in the HAZ and base metal respectively.

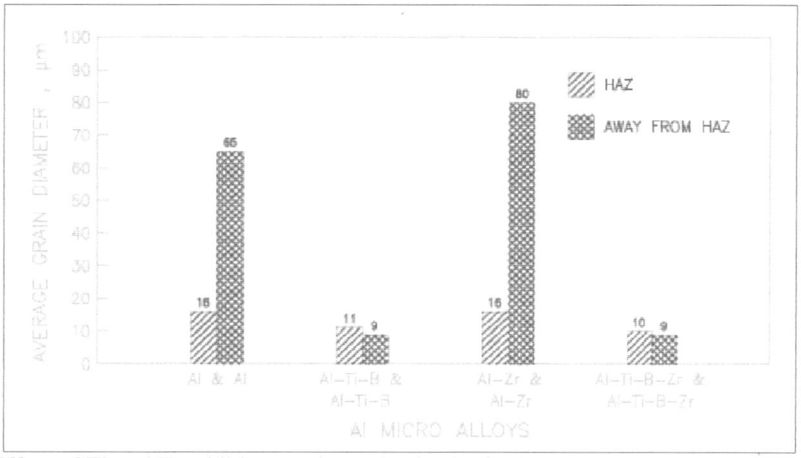

Figure 6. Effect of Ti and Zr addition on the grain size in the HAZ and base metal of welding Al with Al , Al-Ti-B with Al-Ti-B, Al-Zr with Al-Zr and Al-Ti-B-Zr with Al-Ti-B-Zr microalloys.

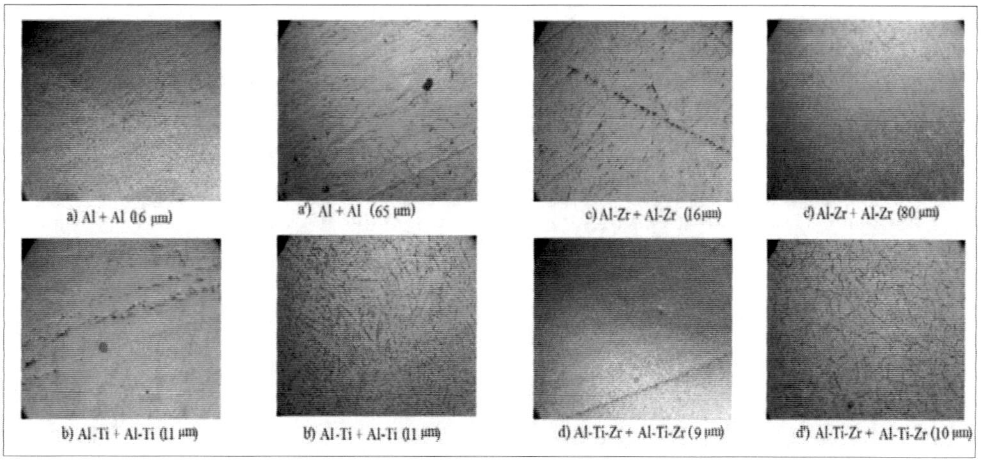

Figure 7. a, b, c, d and a', b', c', d' at HAZ and base metal respectively for welding Al and its microalloys together

3.1.4 . Effect of welding Al with its microalloys on microhardness

The histogram of figure 8 shows that the addition of Ti+B or Zr either alone or together to Al on its hardness welding, resulted in decrease of its Vicker's microhardness in the HAZ region. The maximum decrease is 37.5% in case of welding Al with Al-Ti-B-Zn followed by welding Al with Al-Ti-B-Zr and the minimum decrease in (13%) when. The decrease in hardness is attributed to the grain refinement effect and the existence of the intermetallic phases in the Al matrix and to the softening effect caused by the high temperature in the HAZ region. Regarding the effect on the hardness in the base (away from HAZ) it can be seen that there is relatively very slight increase in case of Ti+B and pronounced increase 26.3% in case of welding Al with Al-Ti-B-Zr and no effect in case of Al wih Al-Zr.

3.1.5 Effect of welding Al microalloys together on microhardness.

The histogram of figure 9 shows that welding the identical microalloys Al-Ti-B, Al-Zr and Al-Ti-B-Zr together resulted in decrease of their hardness both in the HAZ and the base metal regions when compared with the hardness of the corresponding regions of Al with Al and its microalloys with the following decrease percentages in the HAZ 42, 38.6 and 39.7 for Al-Ti-B with Al-Ti-B, Al-Zr with Al-Zr and Al-Ti-B-Zr with Al-Ti-B-Zr respevtively and the following decrease pecentages in the base metal region: 12.3, 8.8 and 14.4 respectively. This is also attributed for the grain refinement of these microalloys prior to welding. The histogram of Fig. 9 indicates that there is slight difference between the hardness of welding the microalloys Al-Ti-B with Al-Ti-B, Al-Zr with Al-Zr and Al-Ti-B-Zr with Al-Ti-B-Zr in HAZ or the base regions. The maximum difference is 6.1 % between Ti and Ti + Zr addition in the HAZ in the base regions respectively. However, a pronounced difference exists in the hardness when comparing the welding of the microalloys with that of Al with Al as shown in figure 9 or with welding of Al with Al-Ti-B, Al-Zr and Al-Ti-B-Zr, as illustrated in figure 8.

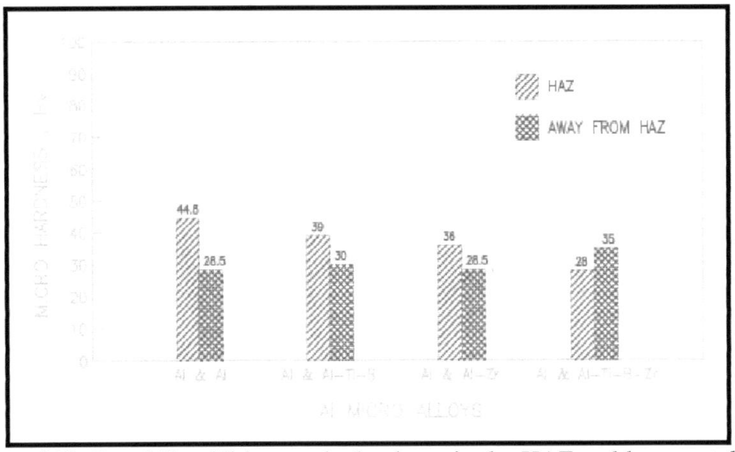

Figure 8. Effect of Ti+B and Zr addition on the hardness in the HAZ and base metal of welding Al with itsmMicroalloys.

IOP Conf. Series: Materials Science and Engineering **60** (2014) 012010 doi:10.1088/1757-899X/60/1/012010

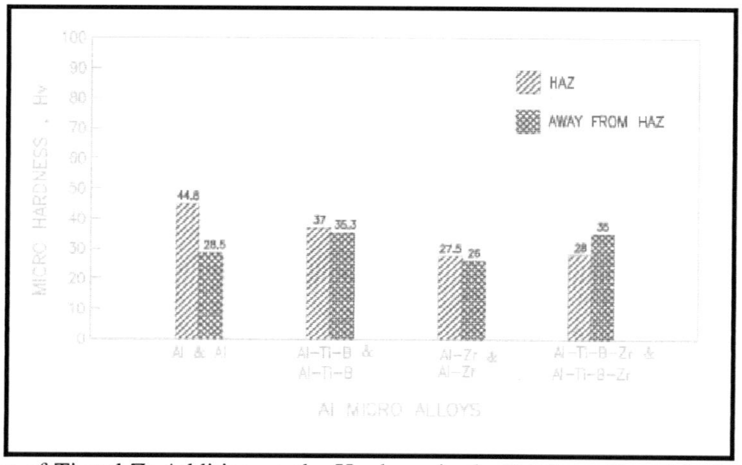

Figure 9. Effect of Ti and Zr Addition on the Hardness in the HAZ and Base Metals of Welding Al with Al , Al-Ti-B with Al-Ti-B, Al-Zr with Al-Zr and Al-Ti-B-Zr with Al-Ti-B-Zr Microalloys.

4. Conclusions

From the investigation carried out in this Paper, the following points are concluded:

- Addition of Ti+B or Zr either alone or together on the weldability of Al resulted in decrease of the grain size both in the HAZ and base regions when welding Al with Al-Ti-B, Al-Zr and Al-Ti-B-Zr as compared to welding of Al with Al except when welding Al-Zr with Al-Zr in the base region.
- Addition of Ti+B or Zr on the weldability of Al resulted in pronounced decrease in the hardness of the HAZ region for welding Al with Al-Ti-B, Al with Al-Zr and Al with Al-Ti-B-Zr and increase in the base metal except when welding Al with Al-Zr as compared to welding of Al with Al. Similarly in hardness in the HAZ but increase the base metal region for welding Al-Ti-B with Al-Ti-B and Al-Ti-B-Zr with Al-Ti-B-Zr and slight decrease in Al-Zr with Al-Zr.
- Effect of addition of Ti, Zr either alone or together on the weldability of aluminum resulted in sound weld: more homogeneous with less voids and segregation, finer grain size and lower hardness both in the HAZ and base metal regions.

5. References

[1] Zaid A.I.O. *Review of the grain refinement of aluminum and its alloys*, (ISAM7 Pakistan, 2001).

[3] Kou S. 2003 *Welding Metallurgy* 2nd ed. John Wiley and Sons Inc.

[4] Din EN 439 1995 *Shielding Gases for Arc Welding and Cutting*.

[5] Simpson R P. 1977 *Weld Journal* 56 67-72.

[6] Gerland G. J.1974 *Metal Construction* 6 121-128.

[7] Prasad R.S. 2001 *Proceedings of National Advances in Materials Processing* Annamalai Nagar, India 176-196.

[8] Kou S. and Le Y. 1986 Weld Journal 65 65-70.

[9] Madusudhan R. G., Gokhale A. A. and R.K. Prasad R.K. Weld microstructure Refinement in a 144 Grade Al-Lithium Alloy 32 4117-4121.

[10] Tseng C. F., Savage W. F. 1971 *Weld J* 50 777-785.

[11] Kumar A., Sundarrajan S. 2009 *Materials and Design* 30 1288-1297.

[12] Kim I. S., Son J. S., Kim I. G. , J.Y. Kim J.Y., Kim O. M. 2003 *Journal of Material Processing Technology* 136.

[13] Cary H. B.1984 *Modern Welding Technology* New Jersey: Prentice Hall.

[14] Davis J. R. 1994 *ASM Specialty Handbook: Aluminum and Aluminum Alloys*, Ohio: ASM International Materials Park.

[15] Tony A 2002 *The American Welder: Weld Journal* 64 77-80.

[16] Kumar A, Sundarrajan S. 2006 *Mater. Manufacturing Process* 21 789-793.

[17] Sundaresan S, Janaki: R. G. H. 1999 *Science Technol Weld Joining* 4 151-160.

[18] Kumar A., Nageshwar R.G.V.S., Sundarrajan S. 2007 *Int. J. Logic Intell. System* 1 85-94.

High temperature oxidation behaviour of nanostructured cermet coatings in a mixed CO_2 – O_2 environment

M A Farrokhzad and T I Khan
University of Calgary 2500 University Dr. NW, Calgary, Canada

E-mail: mafarrok@ucalgary.ca

Abstract. Nanostructured ceramic-metallic (cermet) coatings composed of nanosized ceramic particles (α-Al_2O3 and TiO_2) dispersed in a nickel matrix were co-electrodeposited and then oxidized at 500°C, 600°C and 700°C in a mixed gas using a Thermo-gravimetric Analysis (TGA) apparatus. The mixed gas was composed of 15% CO_2, 10% O_2 and 75% N_2. This research investigates the effects of CO_2 and O_2 partial pressures on time-depended oxidation rates for coatings and compared them to the results from atmospheric oxidation under similar temperatures. The increase in partial pressure of oxygen due to the presence of CO_2 at each tested temperature was calculated and correlated to the oxidation rate of the coatings. The results showed that the presence of CO_2 in the system increased the oxidation rate of cermet coatings when compared to atmospheric oxidation at the same temperature. It was also shown that the increase in the oxidation rate is not the result of CO_2 acting as the primary oxidant but as a secondary oxidant which results in an increase of the total partial pressure of oxygen and consequently higher oxidation rates. The WDS and XRD analyses results showed that the presence of nanosized TiO_2 particles in a nickel matrix can improve oxidation behaviour of the coatings by formation of Ni-Ti compounds on oxidizing surface of the coating which was found beneficiary in reducing the oxidation rates for cermet coatings.

1. Introduction

Nanostructured ceramic-metallic (cermet) coatings are engineered materials produced by incorporating nanosized ceramic particles such as Al_2O_3 or TiO_2 particles into a metal matrix such as nickel. The cermet coatings can be produced using several coating techniques among which co-electrodeposition technique is being commonly used in the industry due to relatively simple equipment and low costs [1-2]. As a result of high demand from the oil and gas production industry for using low cost surface protective coating materials, in recent years there has been an increase in the number of applications where cermet coatings were used to improve mechanical properties, corrosion, and high temperature oxidation resistance of alloyed components [3-6]. The In-Situ Combustion (ISC) process for oil production is a newly proposed unconventional method that utilizes extreme heat (generated by burning the lighter hydrocarbon compounds from the reservoir) to reduce the viscosity of heavier bitumen compounds which then allow an easier extraction from oil reservoirs [7-8]. In the ISC process, hot atmospheric air is injected into the oil reservoir as the feed for an underground combustion process. Due to burning of the light hydrocarbons and combustion reactions in the reservoirs, temperatures can reach as high as 500 °C to 700 °C [7, 9].

The presence of hot oxidizing gases creates a severe environment for metal degradation. In the ISC technique, the presence of high temperature gases, mostly C_xO_y and O_2, at 500°C to 700°C generated in the ignited combustion front gases can lead to the rapid degradation of pipes and tubing used in the oil extraction process [11-12]. Steel alloys are used throughout the oil production industries; however, due to the extreme conditions in the ISC environment, the steel products have limited use. On the other hand, due to the huge volume of tubes and pipes needed for bitumen extraction, the use of high temperature corrosion resistant alloys (CRA's) is uneconomical; therefore, there is a need to develop new types of specialty metal coatings that can withstand the extreme conditions of the ISC. Although electrodeposited nanostructured cermet coatings have shown superior high temperature oxidation resistance in many extreme industrial applications [5], they are not yet fully employed for use as protective coatings in equipment and piping in oil production. As a result, there is a need to develop function specific coatings that can withstand the high temperature oxidation and erosion to which oil production components are subjected to for the ISC process [8]. The analysis of the combustion gases from ISC experiments has shown that CO_2 and/or CO gases can be present in the in-situ combustion produced gas. The CO_2 gas can be found to a maximum level of 15% of volume, whereas CO has only been found in a trace amount in the produced combustion gas [10]. In the high temperature combustion zone, exothermic combustion reactions between oxygen and some light hydrocarbons (that are a part of bitumen chemical composition) create the heat which allows the heavier hydrocarbons to be separated from the sand. The stoichiometric reaction for the combustion of these light hydrocarbon ingredients can be written as [9]:

$$CH_xO_y + (1 - \frac{m}{2} + \frac{x}{4} - \frac{y}{2})O_2 \rightarrow (1-m)CO_2 + mCO + \frac{x}{2}H_2O \qquad \text{(Reaction 1)}$$

Note that reaction 1 is only one of the many stoichiometric reactions that can be written for the in-situ combustion process; but the reactants and products chemicals for all of them are the same chemical compounds. Providing a nickel coating layer onto a steel surface can provide improved oxidation and corrosion resistance at low temperatures, however at elevated temperatures, nickel has a tendency to lose its oxidation resistance and the nickel coating softens (due to the grain growth) resulting in degradation of the coating. To improve the high temperature oxidation resistance of nickel coatings, nano-sized oxide particles can be dispersed into the nickel metal matrix [1-4] using the electrodeposition method. Several studies have shown that the presence of even small amounts of the oxide particles in the nickel matrix can lead to a significant improvement in the oxidation resistance and mechanical properties of the coating [13-19]. In theory, the oxidation growth as a function of time can be linear, logarithmic or parabolic equations [20]. The oxidation process in metals involves the growth of the oxide layer through a diffusion process. The mathematics for this diffusion and oxide layer growth rate is per Fick's first law and can be written as [21]:

$$\frac{dx}{dt} = -D\frac{\partial C}{\partial x} \qquad (1)$$

Where D is the diffusion coefficient and C is the concentration of transferred charged particles. By integration of both sides of the equation, the simple oxidation thickness per function of time is:

$$x^2 = k't + C_1 \qquad (2)$$

Where k' is a parabolic rate of oxidation constant and C_1 is a general constant value. One theory that has been proven to be more accurate and can be applied to a vast majority of the metals is the *Wagner* theory of oxidation based on Fick's diffusion laws. The *Wagner* theory provides a parabolic rate equation in which the oxide scale growth rate on the metal is controlled by diffusion of charged particles (electrons, cations and ions) crossing the oxide-gas scale [22-24]. Similar to the pure metals, the oxidation kinetics of electrodeposited coating can be described by calculating the growth time constant "a" from the following equation [25-27]:

$$\frac{\Delta m}{A} = kt^a + C_2 \qquad (3)$$

Where Δm is the oxidation mass change, A is the surface area, k is the oxidation rate constant, (t) is the time of oxidation, (a) is the growth time constant and C_2 is another general constant value. This research examines the thermal stability and high temperature oxidation behaviour of the double-ceramic cermet coatings at 500°C, 600°C and 700°C in mixed gases composed of O_2, CO_2 and N_2 with a chemical composition similar to that experienced in the in-situ combustion process.

2. Experimental procedure

2.1. Materials

This study used two types of nanosized ceramic powders; alumina (α-Al_2O_3) and titania (TiO_2) purchased from M K Impex Corp. Ltd. The purity for α-Al_2O_3 powder was 99.95% (with traces of Na: 300 ppm, Si: 3.5 ppm, Ca: 1.6 ppm, Fe: 0.2 ppm, and Co: 0.8 ppm). The purity for TiO_2 powder was 98% (with traces of Al: 20 ppm, Ca: 75 ppm, Mg: 65 ppm, Nb: 119 ppm, S: 165 ppm and Si: 102 ppm). The average grain size (measured using TEM) for Al_2O_3 was 20 nm and 50 nm for TiO_2. The substrate material was made from a hot-rolled AISI-1018 carbon steel sheet and the specimens for co-electrodeposition were cut in a rectangular shape (length: 15 mm, width: 6 mm and thickness: 1 mm). The mill scale was removed by mechanical cleaning and the surface was prepared to a 600 grit finish. The surfaces were then cleaned in alkaline solutions (E-Kleen 102-E™ and E-Kleen 129-L™) and finally an acid pickling solution (with 31% HCl) was used to remove any remaining grease or other contaminates.

2.2. Microstructural Characterization

For the grain size analysis of the powders a *Tecani F20-200 kV* (the Netherlands) transmission electron microscope (TEM) in bright-field (BF) mode was used. A *JSM-8200 JEOL* micro-probe (Tokyo, Japan) scanning electron microscopy (SEM) in the back-scattered electron mode (BSE) was used for the images of the cross-sections of coatings and oxidized specimens. The element maps were also produced using the wavelength dispersive spectroscopy (WDS) of the same *JSM-8200 JEOL* micro-probe. A *Rigaku Multiflex* X-ray diffactometer (Japan) was used for the X-ray diffraction spectroscopy of oxidized specimens.

2.3. Co-electrodeposition of coating

The standard *Watt's* bath formula was used for the chemical composition of the electrolyte solution and the concentrations of ceramic particles in the electrolyte solution were obtained from previous research [14, 28-30, 37]. The pH of the electrolyte solutions was 4.0 to 4.2, and the temperature of electrolyte solution was kept between 50 to 55°C. The electrolyte solutions for double-ceramics cermet coatings were also composed of two mixtures of ceramic powders which were called; A1 and B1 coatings. The coatings classification and the compositions of the electrolyte bath solutions are summarized in table 1.

Table 1. Type and concentration of ceramic powders in electrolyte solutions

Coatings	Molar Concentration	Particles in electrolyte [gr/lit]	Stirring Speed [rpm]	Electrolyte Composition
A1	Al_2O_3: 0.25 M TiO_2: 0.25 M	Al_2O_3: 25.5 TiO_2: 20	350	Standard *Watt*'s bath solution: Nickel Sulphate Hexahydrate (1M), Nickel Chloride Hexahydrate (0.2 M), and Boric acid (0.5M), dissolved in distilled water.
B1	Al_2O_3: 0.375 M TiO_2: 0.125 M	Al_2O_3: 38.25 TiO_2: 10	380	

The anode was made of commercial high purity nickel (99.9%). The coatings were produced using a standard DC setup for electroplating. A current density of 2 A/dm^2 was applied for 60 minutes for the electrodeposition of the coatings. Scanning electron microscopy (*JEOL JXA-8200*) was used to measure the average electrodeposited coating thickness. The produced average thickness was measured to be 40±5 μm.

2.4. TGA Oxidation
For thermo-gravimetric analysis, a simultaneous symmetrical thermoanalyser (*Setaram, TGA 16, KEP technologies*, France) was used. The thermo-gravimetric analysis was conducted at 500°C, 600°C and 700°C on specimens coated with A1 and B1 double-ceramics cermet coatings. The sensitivity level for mass change measurement was 1μg and the maximum allowed weight of specimens was 1 gram. The mixed gas consisted of 15% CO_2 and 10% O_2 with the remainder being N_2. The oxidation time was 86,400 seconds (24 hours) for the 500°C and 600°C tests, and 100,000 seconds (27.7 hours) for the 700°C test.

3. Results and discussion

3.1. Characterization of coatings
The appearance and morphology of the coatings deposited for both A1 and B1 ceramic mixture were, in general, similar and a uniform distribution of Al_2O_3 and TiO_2 particles in the nickel matrix was produced for all the coatings. The cross-section of the A1 cermet coatings produced with a current density of 2 A/dm^2 is displayed in figure 1(left). The WDS element mapping of the cross-section of the coatings with respect to nickel, aluminum and titanium elements, shown in figure 1 (middle and right images), also indicated a uniform distribution of dispersed particles in the nickel matrix. The total particle volume fractions in the nickel matrix were 11.8 % for A1 coating and 9.2% for B1 coatings.

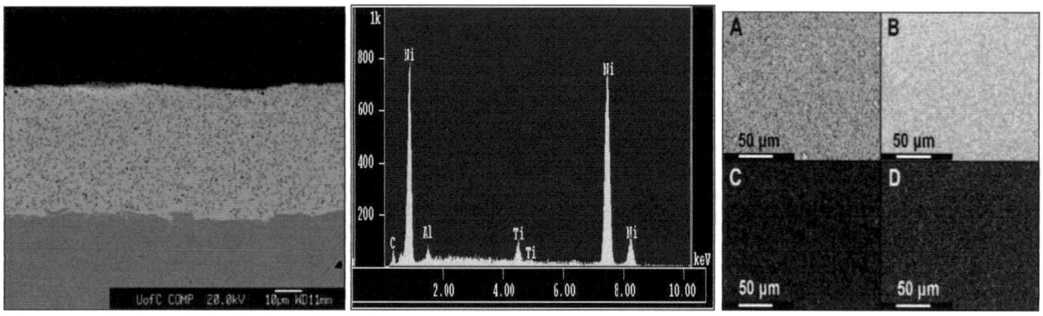

Figure 1. SEM image of cross-section (left) and WDS compositional analysis (middle) of A1 coating. Element maps (right) showing A1coating surface (A), nickel (B), aluminum (C) and titanium (D)

3.2. Thermo-gravimetric analysis of coatings
The oxidation "mass change as a function of time" graphs are illustrated in figure 2. The solid lines are the TGA mass change results for A1 and B1 double-ceramic cermet coating specimens oxidized in the mixed CO_2-O_2 environment. The dotted lines on these plots show the mass change per unit of area of similar coating specimens oxidized in the air. The values for the "oxidation rate constant (*k*)" and "growth-rate time constant (*a*)" for cermet coatings oxidized in atmosphere were calculated in the previous work by the same authors [27].

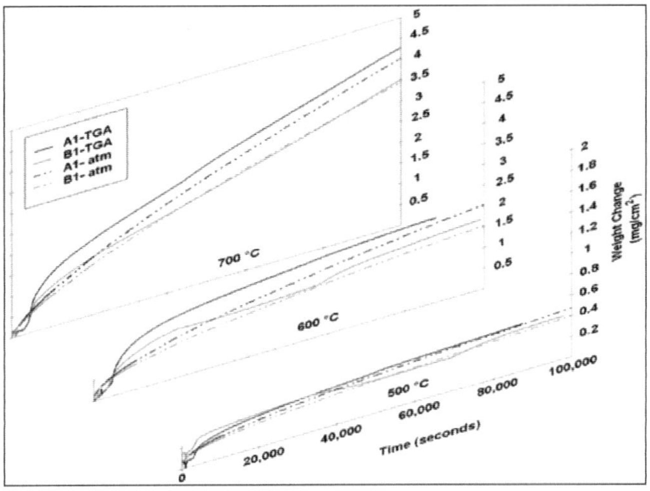

Figure2. Oxidation mass change as a function of time of oxidation for A1 and B1 cermet coatings in a mixed (CO_2-O_2) gaseous environment (solid line) and calculated atmospheric oxidation rate (dotted lines) at 500°C, 600°C and 700°C.

3.3. Discussion on effect of partial pressure of CO_2

Unlike carbon and nitrogen, which do not directly react chemically with nickel, oxygen reacts with nickel as a primary or secondary oxidant and forms nickel oxide. The chemical reaction between oxygen as the primary oxidant and nickel can be written as:

$$2Ni_{(s)} + O_{2\,(g)} = 2NiO_{(s)} \qquad \text{(Reaction 2)}$$

Meanwhile, carbon from CO_2 can act as a secondary oxidant by decomposing into carbon monoxide (CO) and oxygen and then can introduce the additional oxygen into the nickel in which the oxygen first adsorbs and then diffuses into the nickel lattices:

$$Ni_{(s)} + CO_{2(g)} = NiO_{(s)} + CO_{(g)} \qquad \text{(Reaction 3)}$$

The diffusing oxygen ions eventually react with migrating nickel cations to the surface and form the thickness of the NiO layer or they recompose into the pores inside the NiO layer or coating and form oxygen molecules (*Rahmel and Tobolski et al.*) [32]. For the voids that exist near the surface, it can be assumed that the amount of pressure inside the voids and the chemical composition of the gas in the pores is the same as the pressure and the chemical composition of the external atmosphere (i.e. atmospheric pressure and chemical composition for this study). However the chemical pressure and the chemical composition of the gas inside of the pores that is further into the bulk of the coating is not the same as the pressure and chemical composition of the external atmosphere. For calculating the pressure and chemical composition of the diffused ions from oxidizing gases into the pores and voids within the bulk of coating material, it should be noted that the partial pressure of reacting oxygen will be at equilibrium with cations (Ni^+) migration towards the oxide layer and as it is defined by activity of Ni^+ when reaction 2 is written by half cell oxidation equation ($2Ni \rightarrow 2\,Ni^+ + 2\,e^-$):

$$2Ni^+ + 2e^- + O_2 \rightarrow 2NiO \qquad \text{(Reaction 4)}$$

$$\Delta G^\circ = -RT\ln\left(\frac{a_{NiO}^2}{a_{Ni^+}^2 \times a_{O_2} \times a_{e^-}^2}\right) = -RT\ln\left(\frac{1}{a_{Ni^+}^2 \times a_{O_2}}\right) \qquad (4)$$

Where a_{NiO} and a_{e^-} are equal to unity and $a_{O2} = p_{O2}$ [36]. In fact in an equilibrium condition the activity of the nickel cation (a_{Ni+}) can also be considered almost equal to unity at the oxide/gas interface (when it is considered as vapour pressure of the condensed element) but in the bulk of the material, its activity corresponds to the amount of oxygen that is diffused into the bulk of the material. Therefore the equilibrium partial pressure of oxygen as a function of nickel activity becomes:

$$p_{O_2} = \frac{1}{a^2_{Ni^+}} \exp\left(\frac{\Delta G^\circ}{RT}\right) \quad (5)$$

By calculating the Gibbs free energy of nickel oxidation for reaction 2 for the tested temperature, the oxygen partial pressure will be:

$$\Delta G^\circ \approx (2 \times \Delta G^\circ_{NiO}) - (2 \times \Delta G^\circ_{Ni} + \Delta G^\circ_{O2})$$
$$\Delta G^\circ \approx (2 \times \Delta H^\circ_{NiO} - 2 \times T\Delta S^\circ_{NiO}) - (2 \times \Delta H^\circ_{Ni} - 2 \times T\Delta S^\circ_{Ni} + \Delta H^\circ_{O2} - T\Delta S^\circ_{O2})$$
$$\Delta G^\circ \approx 2 \times \Delta H^\circ_{NiO} - 2 \times T\Delta S^\circ_{NiO} - 2 \times \Delta H^\circ_{Ni} + 2 \times T\Delta S^\circ_{Ni} - \Delta H^\circ_{O2} + T\Delta S^\circ_{O2}$$

Where standard values for ΔH°, ΔG°, ΔS° are given in table 2:

Table 2. Standard enthalpies, free energies and entropies at 298K (1 atm) [33]

Substance	ΔH° [kJ/mol])	ΔG°[kJ/mol])	ΔS°[J/mol.°K]
Ni	0	0	30
O_2	0	0	205
NiO	- 244	- 216	39
CO	- 111	- 137	198
CO_2	- 394	- 394	214

Therefore the Gibbs free energy at test temperatures and in the pressure for reaction 2 becomes:

$$\Delta G^\circ \approx (2 \times -244000 - 2 \times 39T) - [(0 - 2 \times 30T) + (0 - 205T)]$$
$$\Delta G^\circ \approx -488000\ [J/mol] + 187\ T\ [J/mol.°K]$$

The amount of partial pressure of oxygen in the bulk of nickel oxide (as a result of diffusion) can be calculated using equation 5:

Table 3. Partial pressure of oxygen in the bulk of nickel oxide

Temperature	ΔG° [J/mol]	p_{O_2}
500°C (773 °K)	- 343499	6.13 x 10^{-24} atm
600°C (873 °K)	- 324749	3.70 x 10^{-20} atm
700°C (973 °K)	- 306049	3.71 x 10^{-17} atm

Nickel will not be oxidized below these pressures shown in table 3 (also called the dissociation partial pressure of Ni) at the given temperature. This also means that the oxygen partial pressure in the bulk of the pores will be increased as a function of temperature and consequently the oxidation rate will increase as was observed in these experiments (see figure 2). Meanwhile the oxygen partial pressure inside the pores in the bulk (for a non-cracked oxide layer) is a small fraction of the oxygen partial pressure for the oxidizing environment (the partial pressure of oxygen in air is approximately 0.21 atm.). This is due to the fact that in an oxidizing surface, the transfer of oxygen from the high oxygen side (the external atmosphere) to a low oxygen side (pores) is a low rate mass transfer process [22, 36]. The diffusion of oxygen in the bulk is assumed to be from adsorption of oxygen molecules at the gas-metal interface followed by a chain of "formation-decomposition-reformation" process where the oxide dissociates and releases oxygen which then migrates inward to low energy sites such as pores or voids around the deposited α-Al_2O_3 or TiO_2 particles [22]. *Sheasby et al.* described that since CO and CO_2 do not react with nickel, the nature of their diffusion is mass transportation through a short-circuit path such as high-angle grain boundaries [34]. From reaction 3 we can calculate the O_2 partial pressure associated with the secondary oxidant (CO_2):

$$Ni_{(s)} + CO_{2(g)} = NiO_{(s)} + CO_{(g)}$$
$$\Delta G° \approx (\Delta G°_{NiO} + \Delta G°_{CO}) - (\Delta G°_{Ni} + \Delta G°_{CO2})$$
$$\Delta G° \approx [(\Delta H°_{NiO} - T\Delta S°_{NiO})+(\Delta H°_{CO} - T\Delta S°_{CO})]-[(\Delta H°_{Ni} - T\Delta S°_{Ni})+(\Delta H°_{CO2} - T\Delta S°_{CO2})]$$
$$\Delta G° \approx \Delta H°_{NiO} - T\Delta S°_{NiO} + \Delta H°_{CO} - T\Delta S°_{CO} - \Delta H°_{Ni} + T\Delta S°_{Ni} - \Delta H°_{CO2} + T\Delta S°_{CO2}$$
$$\Delta G° \approx -244000 - 39T - 111000 - 198T + 30T + 394000 + 214T$$
$$\Delta G° \approx 39000_{[J/mol]} + 7T_{[J/mol.°K]}$$

Although CO and CO_2 do not react with nickel, they can act as the secondary oxidant and release oxygen to react with Ni. The activity of nickel (a_{Ni+}) is about the unity, therefore the calculated ratio between the partial pressures for the CO/CO_2 system are:

$$\frac{p_{CO}}{p_{CO_2}} = a_{Ni^+} \exp\left(\frac{-\Delta G°}{RT}\right) = \exp\left(\frac{-\Delta G°}{RT}\right) \tag{6}$$

Table 4. Partial pressure for CO/CO_2 system

Temperature	**ΔG° [J/mol]**	$\frac{p_{CO}}{p_{CO_2}}$
500°C (773 °K)	44411	9.97 x 10^{-4}
600°C (873 °K)	45111	1.99 x 10^{-3}
700°C (973 °K)	45811	3.47 x 10^{-3}

Table 4 shows that only small amounts of oxygen released from the conversion of CO_2 to CO (reaction 3) become available as the secondary oxidant and the majority of CO_2 remains inactive in the system. The oxygen partial pressure associated with the secondary oxidant at each temperature can be calculated by considering the redox reaction between CO to CO_2:

$$CO_{2\,(g)} = CO_{(g)} + \tfrac{1}{2}O_{2\,(g)} \qquad \text{(Reaction 5)}$$

$$\Delta G° \approx (\Delta G°_{CO} + \tfrac{1}{2}\Delta G°_{CO}) - (\Delta G°_{CO2}) \approx 282000_{[J/mol]} - 87\,T_{[J/mol.°K]}$$

The partial pressure associated with reaction 5 can be written as:

$$p_{O_2} = \left(\frac{p_{CO_2}}{p_{CO}}\right)^2 \exp\left(\frac{-2\Delta G°}{RT}\right) \tag{7}$$

Therefore the partial pressure of oxygen as the secondary oxidant for each test temperature becomes:

Table 5. Partial pressure of oxygen as the secondary oxidant

Temperature	**ΔG° [J/mol]**	p_{O_2}
500°C (773 °K)	214749	9.52 x 10^{-24} atm
600°C (873 °K)	206049	5.55 x 10^{-20} atm
700°C (973 °K)	197349	5.37 x 10^{-17} atm

A comparison between the primary and secondary oxidant O_2 partial pressure is provided in table 6:

Table 6. A comparison for partial pressure for primary and secondary oxidants

Temperature	**Primary O_2 Oxidant Partial Pressure ($2Ni + O_2 = 2NiO$)**	**Secondary O_2 Oxidant Partial Pressure ($Ni + CO_2 = NiO + CO$)**	**Secondary to Primary Partial Pressure Ratio**
500°C (773 °K)	6.13×10^{-24}	9.52×10^{-24}	1.55
600°C (873 °K)	3.70×10^{-20}	5.55×10^{-20}	1.50
700°C (973 °K)	3.71×10^{-17}	5.37×10^{-17}	1.45

From tables 5 and 6 and by comparing the oxygen partial pressure between the primary and the secondary oxidants, it can be concluded that in the presence of a CO/CO_2 oxidizing system, the O_2 partial pressure in the voids and pores in the bulk of the coating material will be greater as compared to a system, which consists of only oxygen. This also means that a slightly higher oxidation rate for a CO_2/CO gas system is expected as compared to an atmospheric system.

3.4. Discussion on effect of partial pressure of O_2

Since reactions 2 and 3 are both thermodynamically possible for the tested temperature in these experiment, the standard state Gibbs free energy ($\Delta G°$) exchanged between the reactants and products should be negative and that means that the equilibrium partial pressure of O_2 is:

$$p_{O_2} = \exp\left(\frac{\Delta G^{\circ}}{RT}\right) \tag{8}$$

Where T is the temperature, p_{O2} is the equilibrium partial pressure of the reaction (or as secondary oxidant, the partial pressure of oxygen from dissolution of CO_2: $CO_2 \rightarrow CO + \frac{1}{2}O_2$). Changes in partial pressure of oxygen can affect the oxidation rate constant (k) of most metals including nickel according to [22, 36]:

$$k' \propto \left(P_{O_2}\right)^{1/n} \tag{9}$$

The value for n is unique to each metal and it can be found by analyzing the oxidation reaction with respect to exchange of ions, cations and electrons in oxidation process when they are written in *Kröger* nomenclature formulation [36]. The growth of the oxide layer in a cation deficient NiO occurs at the gas-oxide layer interface. When the exchange of ions, cations and electrons in oxidation of Ni is written in *Kröger* formulation, it becomes:

$$O_2 + 2Ni = 2NiO + 2V''_{Ni} + 4\,\dot{h} \qquad \text{(Reaction 6)}$$

Where the left side of the reaction is reactants and the right side is the products, V''_{Ni} is the doubly charged cation vacancy (2 for Ni) and $\dot{h}$ is the number of electron holes (4 for Ni). The thermodynamic equilibrium for the exchange of reactants and products at gas-oxide interface when considering the activity of compounds can be written as:

$$K = \frac{[NiO]^2 \times [C_{V''_{Ni}}]^2 \times [C_{\dot{h}}]^4}{[O_2] \times [Ni]^2} = \frac{1 \times [C_{V''_{Ni}}]^2 \times [C_{\dot{h}}]^4}{p_{O_2} \times 1} = \frac{[C_{V''_{Ni}}]^2 \times [C_{\dot{h}}]^4}{p_{O_2}} \tag{10}$$

Where $[NiO]$, $[C_{V''Ni}]$ and $[c_{\dot{h}}]$ are the concentration (activity) of products and $[O_2]$ and [Ni] are the concentration (activity) of the reactants. The activity of oxygen corresponds to the partial pressure of the reactant oxygen. Similarly, the activity of solids (Ni and NiO) is equal to unity. The equation 10 becomes:

$$p_{O_2} \propto [C_{V''_{Ni}}]^2 \times [C_{\dot{h}}]^4 \tag{11}$$

Since for every two cation vacancies in Ni there are four electron holes ($V''_{Ni} = 2\dot{h}$) then:

$$p_{O_2} \propto [C_{V''_{Ni}}]^2 \times [C_{V''_{Ni}}]^4 \propto [C_{V''_{Ni}}]^6 \quad (12)$$

or

$$C_{V''_{Ni}} \propto (P_{O_2})^{1/6} \quad (13)$$

Another possibility for reaction 6 is when an electron hole is associated with one cation vacancy or:

$$\tfrac{1}{2}O_2 = O + V'_{Ni} + \dot{h} \quad \text{(Reaction 7)}$$

And hence equation 13 becomes:

$$C_{V''_{Ni}} \propto (P_{O_2})^{1/4} \quad (14)$$

Recent research has confirmed that both types of defects from reactions 6 and 7 are possible [22]. For an oxidation reaction to progress, the outward flux of Ni cations (J_{Ni+}) should equal inward migration of oxygen cation (or vacancies in Ni), ($J_{O-}=J_{VNi}$) but in the opposition direction, hence Fick's first law (equation 1) becomes:

$$\frac{dx}{dt} = k' J_{Ni^+} = -k' J_{V_{Ni^+}} = -k' \times D_{Ni+} \frac{C''_{V_{Ni^+}} - C'_{V_{Ni^+}}}{x} \quad (15)$$

Where D_{Ni+} is the diffusivity coefficient of cation migration, $C'_{V_{Ni^+}}$ is the vacancy concentration at the metal-oxide (Ni-NiO) interface and $C''_{V_{Ni^+}}$ is vacancy concentration in oxide-gas (NiO-O_2), therefore equations 13 and 14 respectively become:

$$\frac{dx}{dt} = -k' \times D_{Ni+} \frac{\left(p^{1/6}{}_{O_2(NiO-O_2)}\right) - \left(p^{1/6}{}_{O_2(Ni-NiO)}\right)}{x} \quad \text{or} \quad k' \propto \left(p^{1/6}{}_{O_2(NiO-O_2)}\right) - \left(p^{1/6}{}_{O_2(Ni-NiO)}\right)$$

$$\frac{dx}{dt} = -k' \times D_{Ni+} \frac{\left(p^{1/4}{}_{O_2(NiO-O_2)}\right) - \left(p^{1/4}{}_{O_2(Ni-NiO)}\right)}{x} \quad \text{or} \quad k' \propto \left(p^{1/4}{}_{O_2(NiO-O_2)}\right) - \left(p^{1/4}{}_{O_2(Ni-NiO)}\right)$$

Since the oxygen partial pressure at the oxide-gas interface is greater than the partial pressure of diffusing oxygen in oxide-metal interface, therefore we have:

$$k' \propto p^{1/6}{}_{O_2(NiO-O_2)} \quad (16)$$

$$k' \propto p^{1/4}{}_{O_2(NiO-O_2)} \quad (17)$$

The chemical composition of the oxidizing mixed gas used for this experiment was prepared based on the literature review of chemical composition of produced gas in the in-situ combustion which showed that the maximum volume of CO_2 to be 15% and the maximum volume for O_2 to be 10%. Since the TGA equipment applies an atmospheric pressure during the oxidation tests, the equivalent partial pressure of CO_2 to 15% of the total gas volume creates an oxygen partial pressure equivalent to:

15% x 1 atm = 0.15 atm.

Similarly the partial pressure of oxygen to 10% of the total gas volume is equivalent to:

10% x 1 atm = 0.1 atm.

The combined partial pressure from both CO_2 and O_2 is equivalent to 0.25 atm which is greater than the partial pressure of oxygen from dry air in atmospheric conditions (0.21 atm.). Therefore per equations 16 and 17, we should experience an increase in oxidation rate constant (*k'*) close to 3% to 5%. The mass change results in figure 2 showed the correlation between this increases in the oxidation rate constant to the presence of CO_2.

3.5. Microanalysis of oxidized coatings

Figure 3 is SEM images of the cross-sections and surfaces of A1 specimens oxidized at 500°C, 600°C and 700°C. In comparison with oxidation in dry atmosphere for the same temperature and oxidation time, these images showed similar microstructural features of the coatings and oxide layers between oxidized specimens in the mixed gas to the oxidized specimens in dry atmospheric condition [27].

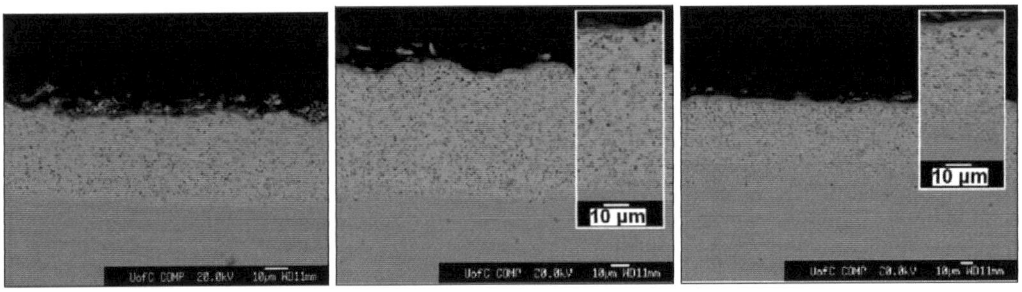

Figure 3. SEM of A1 coatings oxidized in mixed gas (15%CO_2-10%O_2) at : 500°C (left), 600°C (center) and 700°C (right)

3.6. Element maps and XRD analyses of oxidized coatings

Figure 4 represents the element distribution maps for aluminum (Al) and titanium (Ti). These elements maps are based on microprobe WDS area scan analyses from the surfaces of A1 and B1 coatings oxidized in mixed gas (15% CO_2-10% O_2) at 500°C, 600° and 700°C. The element map analyses of non-oxidized coating surfaces were also included for comparison. XRD analyses were conducted on the surfaced of A1 and B1 coatings oxidized in mixed gas at all three temperatures and the graphs and compound analysis results are given in figure 5.

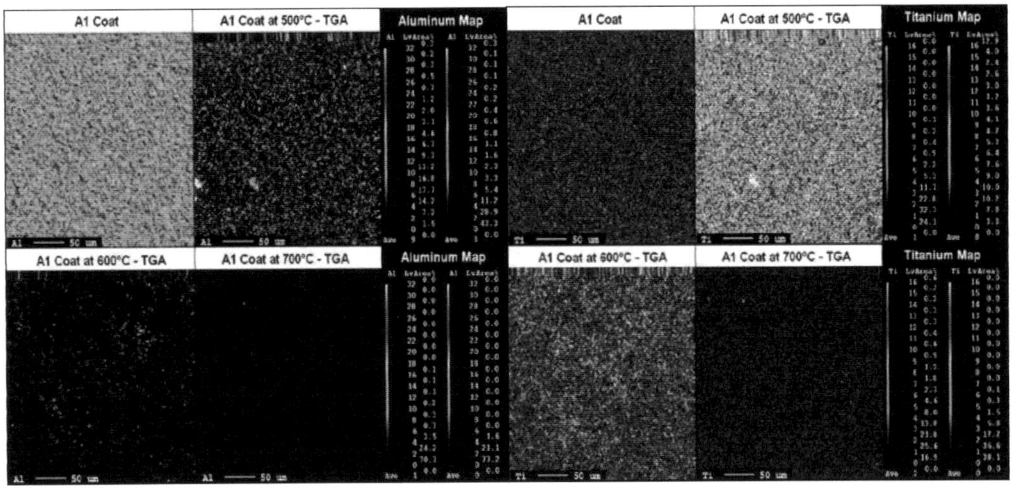

Figure 4. WDS Aluminum and Titanium maps for A1 oxidized in mixed gas (15%CO_2-10%O_2): non-oxidized (top-left), at 500°C (top-right), at 600°C (bottom-left) and at 700°C (bottom-right)

It should be noted that WDS scanning by *JEOL-8200* microprobe was not sensitive enough to detect the presence of carbon; therefore the search for carbides or other carbonic compounds (as a result of oxidation of the coating with CO_2 or CO) was conducted using XRD. Element maps for titanium and aluminum on oxidized B1 coatings in a CO_2 rich mixed gas were similar to the maps from oxidized A1 coatings, therefore only the results for A1 coatings are presented. The aluminum maps showed an outward migration of Ni cations, thus imposing an obscuring effect on aluminum content as a function of temperature, which was seen previously in atmospheric oxidation of A1 and B1 coatings. The element maps for Ti and Al suggest that the TiO_2 particles actively remain on the surface as compared to Al_2O_3 particles (Fig. 4). Oxidized cermet coatings containing dispersed TiO_2 particles showed formation of nickel titanate compounds; $NiTiO_3$ and Ni_3TiO_5 [38-41]. The formation of Ni-Ti compounds could affect nickel oxidation resistance in two ways; firstly, it can have some alloying effects between the dispersed particles and the Ni matrix. *Zeng et al* [42] have argued that formation of the $NiTiO_3$ may contribute to the lower oxidation rates for the alloys, especially at temperatures below 750°C. Secondly XRD results of the oxidized surface showed that these nickel titanate can be found in the oxidized surface, suggesting that unlike the Al_2O_3 particles where upward migration of Ni did not form any Ni-Al compounds and the migrating Ni cations may have bypassed the dispersed Al_2O_3 particles, for TiO_2 particles, the Ni-Ti compounds can affect the number of migrating Ni to the surface. Therefore for coatings with dispersed TiO_2, the migration of Ni cations does not leave the upper part of the coating empty of dispersed particles. Literature review shows that nickel titanate $NiTiO_3$ can be formed in the presence of NiO and TiO_2 at around 600°C [39-40]. However Ni_3TiO_5 is a meta-stable phase that can be found in stable phase at a higher temperature [38, 40-41]. The XRD analyses also confirmed that when the temperature was increased, the frequency of finding these compounds in the oxide layer also increased, which is an indicator of an increase in the quantity of the compounds in the oxide layer. Carbon has an extremely low solubility in the non-molten form of nickel and therefore the carburization of Ni coatings and formation of C-Ni compounds (carbides) does not occur at the temperatures used in this project. The absence of C-Ni carbides or graphite in pure or reacted forms in the oxide layer formed on the coating surfaces was later confirmed with XRD analysis results. Due to the low solubility of C in Ni at low temperatures, the practical data for Ni-C solids is nonexistent, however *Singleton and Nash et al* [35] have constructed a Ni-C phase diagram using computer models and also based on some measured values from other literatures and have concluded that Ni-C at below eutectic can only exist in metastable form (Ni-Ni_3C). The XRD analysis in this research did not detect this metastable form Ni-C system. Additionally, nitrate or nitride compounds of nickel (i.e. nickel (II) nitrate: $Ni(NO_3)_2$) were not found in XRD analysis, therefore it was concluded that the nitrogen (N_2) in the mixed gas does not react with nickel for the temperatures used in this research.

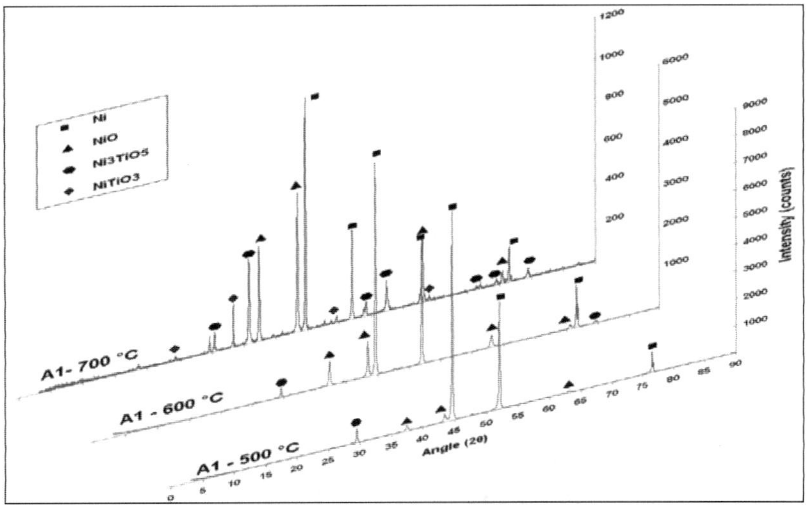

Figure 5. XRD analyses for A1 coatings oxidized in mixed gas at: 500, 600 and 700°C

4. Conclusion

The TGA experiment was designed to study the effect of oxidation of the double-ceramic oxides cermet coatings in a gaseous environment containing CO_2 and O_2. Based on the results, it was showed that the presence of CO_2 in the system can increase the oxidation rate compared to oxidation at atmospheric conditions. It was concluded that the increase in the oxidation rate is not the result of carbon-dioxide acting as the primary oxidant but as a secondary oxidant agent. Carbon does not dissolve in nickel at low temperatures (below the eutectic temperature), but CO_2 can act as the secondary oxidant by decomposing into CO and releasing extra oxygen to the surface of the oxidizing coating. It was found that the additional oxygen from the decomposition will increase the total partial pressure of oxygen in the voids which results in higher oxidation rates. The increase in partial pressure due to the presence of CO_2 was calculated and then correlated to the oxidation rate. The results showed that the oxidation behaviour of the coatings in an atmosphere of CO_2 was similar to that observed under atmospheric conditions. The titanium element maps showed a relative stability and presence of titanium on the surfaced of oxidized coatings which is believed to be caused by formation of Ni-Ti compounds (Ni_3TiO_5 and $NiTiO_3$) which was also confirmed by XRD analyses.

5. Acknowledgements

The authors would like to thank the department of Mechanical and Manufacturing Engineering of the University of Calgary, Alberta, Canada, Natural Science and Engineering Research Council of Canada (NSERC), and Statoil Canada Ltd. for their financial support.

6. References

[1] Sautter F K 1963 *J. Electrochem. Soc.*110, Issue 6, 557–560

[2] Gupta P K, Tiwari A N. and Agrawal B K 1989 *Trans. Japan Instit. Met.* 23 320–327

[3] Saha R K, Haq I U, Khan T I and Glenesk L B 2010 *Key Eng. Mat.* 442 187–194

[4] Saha R K, Khan T I and Glenesk L B and Haq I U 2009 *Ceram. Trans.* 208 37–44

[5] Chang L M, Liu J H and Zhang R J 2011 *Mat. Corrs.* 62, Issue 10, 920–925

[6] Kannan K R, Vasntasree V and Hocking M G 1984, *Coatings and Surface Treatment for Corrosion and Wear Resistance* (New York: Ellis, Horwood) pp 263–275

[7] Moore R G, Laureshen C, Belgrave J M, Ursenbach M, Mehta S A 1995 *Fuel* 74 1169–1175

[8] Roche P 2009 In-Situ Combustion, *New Technology Magazine* (September 2009)

[9] Mamora D D 1993, *Kinetics of In-Situ Combustion*, PhD Thesis, (Stanford University)

[10] Xu H 2001, *In-Situ Upgrading of Heavy Oil*, M. Sc. Thesis, (University of Calgary)

[11] Brooks S, Ferguson J M, Meadowcroft D B and Stevens C G 1978, *Materials and Coatings to Resist High Temperature Corrosion* (London: Applied Science Publishers) pp. 121–138

[12] Brigham W E, Castanier L 1998, *Reservoir Engineering*, Chapter 16, (The National Energy Technology Laboratory -NETL)

[13] Ciubotariu A C, Benea L, Varsanyi M L, Dragan V 2008 *Electrochim. Acta* 53 4557

[14] Saha R K and Khan T I 2010, *Surf. Coat. Tech.* 205 890–895

[15] Gül H, Kilic F, Aslan S, Alp A, Akbulut H 2009 *Wear* 267 976–990

[16] Erb U 1995, *Nanostruture Mater.*6 533–538

[17] Bockris J O M and Razumney G A 1967, *Fundamental Aspects of Electrocrystallization*

(New York: Plenum Press)

[18] Walsh F C.and Herron M E 1991, *Jour. Physic, D: Appl. Phys.* 24 217–225

[19] Qu N S, Chan K C, Zhu D 2004 *Scripta Mater.* 50 1131–1134

[20] Ernest Birchenall C 1982, *High Temperature Corrosion – NACE 6* 3–7

[21] Wahab M A 2005 *Solid State Physics* (Oxford: Alphas Science International) Chapter 6

[22] N. Birks N, Meier G H, Pettit F S 2006, *Introduction to the High-Temperature Oxidization*, 2nd Edition, (Cambridge: Cambridge University Press)

[23] Davis J R 1999 *Heat-Resistant Materials*, (Ohio: ASM International) p.36

[24] Gesmundo F and Viani F 1981 *J. Electrochem. Soc.* 128, Issue 2, 470–479

[25] Susan D F and Marder A R 2002 *Oxid. Met.* 57 131–158

[26] Susan D F and Marder A R 2002 *Oxid. Met.* 57 159–180

[27] Farrokhzad M A , Khan T I 2012 *Key Eng. Mater.*510-511 32–42

[28] Gabe D R 1978 *Principle of Metal Surface Treatment and Protection* 28 (Oxford: Pergamon)

[29] Kim M J, Kim J S, Kim D J, Kim H P 2009 *Met. Mater. Int.* 15, No. 5, 789–795

[30] Kuo S L , Chen Y C, Ger M D and Hwu W H 2004 *Mater.Chem. Phys.* 86 5–10

[32] Rahmel A and Tobolski J, Werkst. U 1965 *Korr* 15 662

[33] West J M 1980 *Basic Corrosion and Oxidation* (Chichester: Ellis Horwood Ltd.)

[34] Sheasby J, Boggs W E and Turkdogan E T 1985, *Mater. Sci.*18 127

[35] Singleton M and Nash P 1989 *Bulletin of Alloy Phase Diagrams*- The C-Ni (Carbon-Nickel) System- 10, No.2

[36] Bradford S A 2001 *ASM Handbooks- Corrosion* 13 (Ohio: ASM International) pp. 62–76

[37] Feng Q, Li T, Teng H, Zhang X , Zhang Y, Liu Ch, Jin J, 2008 *Surf. Coat. Tech.*202 4137–44

[38] Laqua W, Schmalzried H 1982 *NACE 6*, 110–114

[39] Lopes K P *et al* 2009 *J. Alloys Compd.* 468 327–332

[40] QIU Ai-tao *et al* 2011, *Trans. of Nonferrous Met. Soc. China* 21 1808–1816

[41] Taylor D J, Fleig P F, Schwab S T and Page R A 1999 *Surf. Coat. Tech.* 465, pp.120–121

[42] Zeng C L, Li M C, Liu G Q, and Wu W T 2002, *Oxid. Met.* 58, Nos. 1/2, pp.171–184

Characterization of MBE grown diamond structure on Ti/sapphire

M Asghar[1], K Mahmood[1], Noor ul Ain[1], W Stiputra[2] and R Tsu[2]
[1]Department of Physics, the Islamia University of Bahawalpur, 63100, Pakistan
[2]Department of Electrical and Computer Engineering, UNC-Charlotte, 28223, USA

E-mail: mhashmi@iub.edu.pk

Abstract. Diamond crystallites were found on titanium/sapphire surface as a result of evaporation of C_{60} from effusion cell in molecular beam epitaxy (MBE) reactor. Substrate temperature, effusion cell temperature and base pressure of the MBE chamber were set as 1050 °C, 650 °C and 1× 10^{-9} Torr, respectively. Raman scattering of the as-grown layers revealed the existence of diamond and graphite at 1337 cm^{-1} and 1592 cm^{-1}, respectively. Moreover, x-ray-diffraction pattern, micrographs exposed from scanning electron microscopy and optical microscopy also confirmed the evidence of MBE diamond crystallites.

1. Introduction

Diamond and diamond like carbon is an active area of research because it has some exceptional properties such as hardness, low coefficient of friction, chemical inertness, wide electrochemical working potential window, good electron emitting characteristics, high transmittance near infrared region and especially the highest resistivity in it pure form [1-4].

Despite of these promising applications, the growth of pure diamond is however very difficult by the frequently used methods like chemical vapor deposition. The reason behind that is the incorporation of hydrogen in diamond films during CVD growth. In fact, hydrogen has been reported to induce detrimental effects on the tribological and optoelectronic properties of diamond [5]. Single solid source molecular beam epitaxy (MBE) has evolved as a best alternate for the conventional growth methodologies for high purity synthesis of diamond. Molecular beam epitaxy is versatile technique due to ultra-high vacuum conditions, highly controlled conditions and low growth temperature. In this perception we used fullerene (C_{60}) as a single source to grow diamond on titanium surface. Raman scattering revealed a well-defined peak of diamond at 1337 cm^{-1} along with a graphitic line. The apparent shift from the pure diamond line has been discussed with the help of literature.

2. Experimental

MBE chamber (SVT MBE system) with C_{60} effusion cell was prepared for the growth of hydrogen-free diamond on Ti/ c-plane sapphire Al_2O_3 substrate. The substrate was loaded into the MBE chamber. During the 3 hours growth procedure, the chamber pressure, substrate temperature and C_{60} cell temperature were set as 1.4×10^{-9} Torr, 1050 ^{0}C and 650 ^{0}C respectively. It is pertinent to mention here that an amorphous layer of C_{60} was deposited for about half an hour while maintaining the substrate temperature at 400 ^{0}C. Characterization of the as-grown film was performed by following equipment: XRD by PANALYTICAL-Philips, SEM/EDAX by JEOL 6430 and Micro Raman spectroscopy from HORIBA excited by optical source having wavelength 532.14 nm. All the measurements were carried out at room temperature.

3. Results and Discussion

3.1 SEM/EDAX Measurements

Figure 1 displays the SEM/EDAX images of sample taken at different magnifications. The picture clearly demonstrates the crystallite-growth of diamond. The large scale EDAX measurements of the as-grown layer revealed that the concentration (by at %) of carbon in as-grown film is 95% (c.f. inset). Whereas the interior of the same crystallite at higher magnification revealed a 100% carbon at% as depicted by figure 1b. The same crystallite was further investigated using the Raman microscopy.

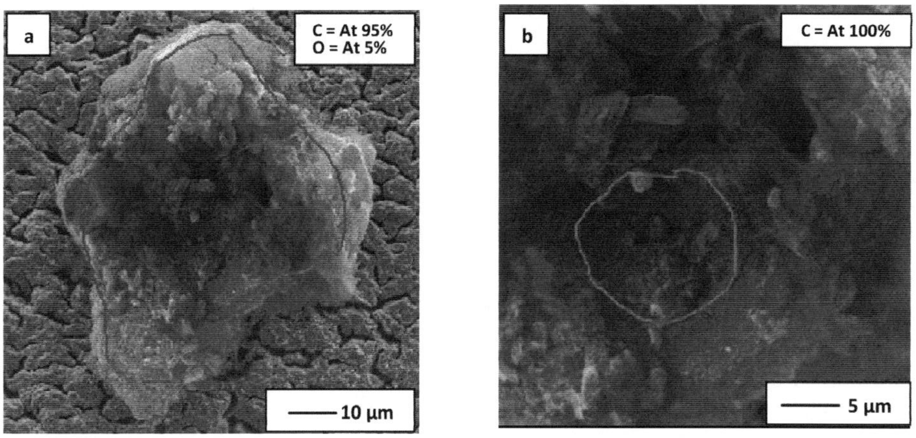

Figure 1. SEM images of MBE diamond crystallite on titanium surface (background) under different magnifications to witness 100% carbon by at%.

3.2 X-ray diffraction

X-Ray diffraction was performed to confirm the phases of diamond. Figure 2 reveals a typical XRD pattern showing diamond peak P5 at 43.21° which is related to the diamond plane (111) [6, 7], the small intensity reflects the amorphousness of diamond cluster. Other XRD peaks P1-P4 at 20.49°, 37.43°, 39.39° and 41.69° are associated with C_{60}, Ti and sapphire, respectively. The most intense signal is from c-plane sapphire indicating the highly oriented and crystalline structure along [0 0 6] plane. Titanium being an orthorhombic system therefore it is hardly observed with the XRD and therefore its small intensity is justified [8]. Ratio of the integrated intensities of diamond peak to carbon peak i.e.$A_{Diamond}/A_{Carbon}$ = ~0.5, indicates that almost half of carbon is available in the chamber has been used for the diamond growth.

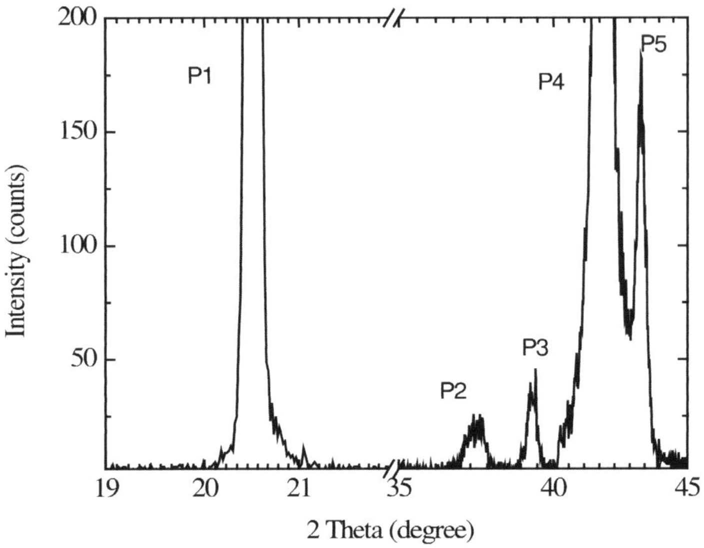

Figure 2. X-ray diffraction pattern of as-grown to demonstrate presence of diamond structure on Ti/Al_2O_3 surface.

3.3 Micro-RAMAN mapping

Diamond is a monoatomic system having first order phonon with the symmetry T_{2g} which is the only Raman active mode. That is why for more than two decades Raman spectroscopy is used for the detection of synthetic and natural diamond in the carbonaceous materials [9].

Micro Raman spectrum of the typical diamond crystallite is shown in figure 3 exposed on the portion of the film optically displayed in the inset. In the figure, we can see two Raman lines at 1337 cm^{-1} and 1592 cm^{-1} respectively. With the information from the literature, we assign 1337 cm^{-1} to diamond and 1592 cm^{-1} line to graphite [10]. A higher frequency shift of the diamond line i.e. from 1332.5 cm^{-1} (theoretical) to 1337 cm^{-1}, is attributed to the compressive stress between the diamond and substrate [11, 12]. Similarly the apparent broadening of the diamond peak is related with the presence of defects sites in the crystal, to be discussed later. We understand stress-induced frequency shift is due to film-substrate mismatch of lattice constant and thermal expansion coefficient [13]. Quantitatively, the stress-induced frequency shift was calculated by following formula:

$$V_s - V_r = \sigma a \tag{1}$$

where ν_s is the measured frequency of Raman line i.e. 1337 cm^{-1}; ν_r is the position of stress-free Raman line i.e. 1332.5 cm^{-1}; a is the stress present in the film and σ is the coefficient of stress induced frequency shift, the reported value of α is 2.88 ±0.17 cm^{-1}/GPa [14]. By incorporating all values in the above equation, the diamond crystal is found to be under a compressive stress of ~1.5-1.7 GPa which is consistent with the reported data [12].

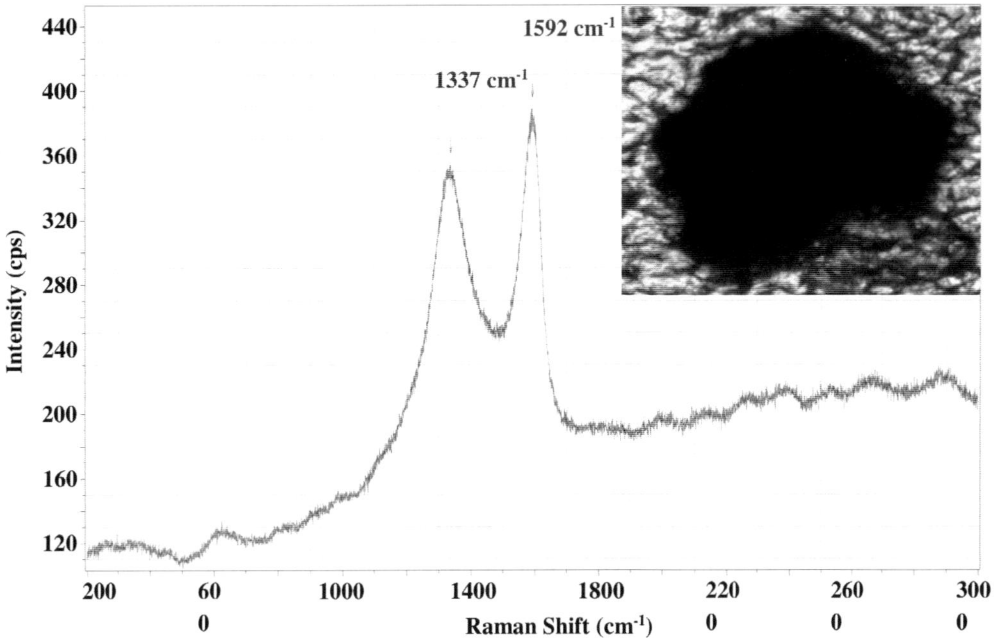

Figure 3. Raman spectroscopy data (Investigated portion is shown in the inset) recorded with argon ion 514.5 nm laser line.

The broadening of Raman peak is also utilized to get information about the crystal quality and stress present in the sample. Generally the sources of Raman line broadenings are crystal disorder and/or crystal size. Since the lifetime of the phonon in the crystal is shortened by scatter at the defect site and hence results in the broadening of the FWHM of the Raman. The observed FWHM of Raman line is 169 cm^{-1}, indicating that the phonon lifetime is very short (50 μSec). Therefore, the broadening of Raman line in diamond (ours) is justified.

Table1. Raman measurements and parameter details

Diamond peak	1337 cm^{-1}
Graphite G peak	1592 cm^{-1}
FWHM of Diamond peak	169 cm^{-1}
FWHM of Graphite G peak	78.94 cm^{-1}
$\frac{I_{Daimond}}{I_{Graphite}}$	0.9071
$Purity\ Index\ (PI) = \frac{I_{Daimond}}{I_{Diamond} + \frac{I_{Carbon}}{50}}$	0.656688
$Figure\ of\ Merit\ (FOM) = \frac{I_{Daimond}}{I_{Diamond} + I_{Carbon}} \times 100$	48.8858 %

Moreover Raman parameters were used to find the quality and purity of the grown crystallites. Using Sails ET. al. [15] method, we calculated the figure of merit (FOM) and quality factor for MBE diamond. Similarly, purity index (PI) was worked out with the help of formula found in Braga et.al and Guo and Chen papers [16, 17]. As-calculated parameters for as-grown diamond crystallite are listed in Table 1.

4. Conclusion

In summery we have successfully grown diamond on titanium/sapphire by Molecular Beam Epitaxy (MBE) using C_{60} as a single solid source. Different diagnostic techniques were employed to verify that the grown film was diamond. Micro Raman spectrum of a carbon rich region of the as-grown layers revealed the existence of diamond and graphite at 1337 cm^{-1} and 1592 cm^{-1}, respectively on Ti/Al_2O_3. The possible reason of the higher-frequency shift of Raman line with respect to the theoretical one (1332.5 cm^{-1}) was compressive stress: ~1.7-1.5 GPa on the diamond crystallites induced by lattice mismatch and/or thermal expansion coefficient of the hetero-substrate. Micrographs exposed from scanning electron microscopy demonstrated well defined crystallites of diamond. The composition of the SEM image was ~ 100% carbon (EDAX data).

5. Acknowledgement

Authors are thankful to Higher Education Commission Pakistan for financial support. The authors are also thankful to UNC-Charlotte, USA for technical support.

6. References

[1] Hupert M, Muck A, Wang J., Stotter J, Cvackova Z, Haymond S, Show Y and Swain G M 2003 *Diamond & Related Materials* 12 1940-1949.

[2] Zhou D, Krauss A R, Qin L C, McCauley T G, Gruen D M, Corrigan T D, Chang R P H and Gnaser H 1997 *J. Appl. Phys.* 82 4546-4550.

[3] Krauss A R, Auciello O, Ding M Q, Gruen D M , Huang Y, Zhirnov V V, Givargizov E I, Breskin Chechen A, Shefer E, Konov V, Pimenov S, Karabutov A, Rakhimov A and Suetin N 2001 *J. Appl. Phys.* 89 2958-2967.

[4] Sharda T, Rahaman, M M, Nukaya Y, Soga T, Jimbo T and Umeno M 2001 *Diamond & Related Materials* 10 561-567.

[5] Shenq Luo Y, Jui-Kang K, Brian Y, James Sung C, Chuang-Wen D and sung T, Tsai J 2001 *Materials Chemistry and Physics* 72 133–135.

[6] Pandey M, Cunha R D and Tyagi A K 2002 *J. Alloys and Compounds* 333 260–265.

[7] Hua P, Xingquan W, Guling Z, Hua C, Guohua L and Size Y 2010 *Appl. Sur. Sci.* 265 6403-6407.

[8] Cullity B D Elements *of x-ray diffraction*, second edition Addison-Wesley Publishing Company Inc. 1978, 337.

[9] Diane Knight S and William White B 1989 *J. Mater. Res.* 4 385-393.

[10] Tuinstra F and Koenig J L 1970 *J. Chem. Phys.* 53 1126-1130.

[11] Denisov V N, Mavrin B N, Serebryanaya N R, Dubitsky G A, Aksenenkov V V, Kirichenko A N, Kuzmin N V, Kulnitskiy B A, Perezhogin I A and Blank V D 2011 *Diamond & Related Materials* 20 951-953.

[12] Easmus R M, Comins J D, Mofokeng V and Martin Z 2011 *Diamond & Related Materials* 20 907-911.

[13] Ahmed A, Bayerlein K, Rosiwal S M, Goken M and Durst K 2011 *Act Materilia* 59 5422-5433.

[14] S A Catledge, Vohra Y K, Ladi R and Rai G 1996 *Diamond & Related Materials* 5 (1996) 1159–1165.

[15] Sails S R, Gardiner D J, Michael B, James S and Don R 1996 *Diamond and Related Materials* 5 589-591.

[16] Braga N A, Cairo C A, Baldan M R, Trava-Airoldi V J and Ferreira N G 2011 *Diamond & Related Materials* 20 31–35.

[17] Liang G and Guohua C 2007 *Diamond & Related Materials* 16 1530–1540.

Properties enhancement and recoil loop characteristics for hot deformed nanocrystalline NdFeB permanent magnets

Z W Liu[1, 2], Y L Huang[1], S L Hu[1], X C Zhong[1], H Y Yu[1] and X X Gao[2]
[1]School of Materials Science and Engineering, South China University of Technology, Guangzhou, 510640, P.R. China
[2]State Key Laboratory for Advanced Metals and Materials, University of Science and Technology Beijing, 100083, P.R. China

E-mail: zwliu@scut.edu.cn

Abstract. Nanocrystalline NdFeB magnets were prepared by spark plasma sintering (SPS) and SPS followed by HD using melt spun ribbons as the starting materials. The microstructure of SPSed and HDed magnets were analyzed. The effects of process including temperature and compression ratio on the microstructure and properties were investigated. High magnetic properties were obtained in anisotropic HDed magnets. The combination of Zn and Dy additions was successfully employed to improve the coercivity and thermal stability of the SPSed magnets. Open recoil loops were found in these magnets with Nd-rich composition and without soft magnetic phase for the first time. The relationship between the recoil loops and microstructure for SPS and HD NdFeB magnets were investigated. The investigations showed that the magnetic properties of SPS+HDed magnets are related to the extent of the aggregation of Nd-rich phase, which was formed during HD due to existence of porosity in SPSed precursor. Large local demagnetization fields induced by the Nd-rich phase aggregation leads to the open loops and significantly reduced the coercivity. By reducing the recoil loop openness, the magnetic properties of HDed NdFeB magnets were successfully improved.

1. Introduction

Anisotropic NdFeB magnets are generally prepared by two processes. One is pressing in a magnetic field followed by sintering; the other is inducing a c-axis orientation by plastic deformation, such as hot deformation (HD) like extrusion, compression or rolling. HDed NdFeB magnets have attracted much attention not only because of their high magnetic properties but also due to their exceptional corrosion resistance, thermal stability, and fracture toughness [1-3]. Up to now, the largest reported value of $(BH)_{max}$ for HDed magnets is 54.4 MGOe, achieved in a $Nd_{13.5}(Fe,Co)_{80}Ga_{0.5}B_6$ alloy[4]. The deformation mechanism, which is not capable of classic plastic flow through dislocation mechanism, is believed to be a combination of stress-assisted grain growth via mass transport and grain boundary sliding [5]. To achieve enhanced magnetic properties, the precursor and the hot deformation processing have to be well optimized for improving the grain structure.

The precursors for the hot deformed magnets are generally prepared by hot pressing or spark plasma sintering (SPS). SPS is known as one of the novel sintering techniques for preparing bulk materials with fine structure [6]. Low sintering temperature and short holding time make it possible to sinter nanocrystalline NdFeB powders into fully dense bulk magnet without significant grain growth [7]. This paper reports our systematic study on the preparation of SPSed and HDed NdFeB magnets. We also observed the open recoil loops in the HDed magnets with RE-rich compositions for the first time. The relationships among the recoil loops, microstructure and coercivity were investigated in detail.

2. Experimental

Spark plasma sintering (SPS) was employed as the pre-process for obtaining nanocrystalline bulk NdFeB precursors from melt spun powders. Commercial melt spun ribbons with nominal composition of $Nd_{13.5}Co_{6.7}Ga_{0.5}Fe_{73.5}B_{5.6}$ or $Nd_{10.15}Pr_{1.86}Fe_{80.41}Al_{1.67}B_{5.91}$ were consolidated to dense isotropic magnets by SPS (SPS-825, Sojitz Machinery Co.) using the optimized processing parameters obtained previously (700 °C/5 min/50 MPa) [8] if not stated otherwise. The SPSed magnets were subjected to hot deformation using a vacuum hot deformation facility (HP-12×12×12, Centorr Vaccum Industries, USA) at various temperatures and compression strain rates. The density for the magnets was measured based on the Archimedes principle. X-ray diffraction patterns were obtained with X-ray diffractometer (XRD, Philip X-pert) using Cu-*Kα* radiation. To characterize the microstructure, the SPSed and HDed magnets were carefully broken off manually from an indentation and the fracture was examined by SEM (Nano430, FEI). Magnetic properties were characterized by physical properties measurement system (PPMS-9 Quantum Design, USA) equipped with a 9T vibrating sample magnetometer.

3. Results and discussion

3.1 Process optimization for hot deformation

The XRD patterns of SPSed and HDed magnets obtained from the surface perpendicular to the pressing direction at the various compression ratios were shown in the Fig. 1. All XRD peaks are attributed to the tetragonal hard magnetic $Nd_2(FeCo)_{14}B$ phase. For the HDed magnets, enhanced (00l) peak densities indicate the c-axis crystallographic alignment. With increasing the compression ratio from 54 % to 80 %, the degree of c-axis orientation increases.

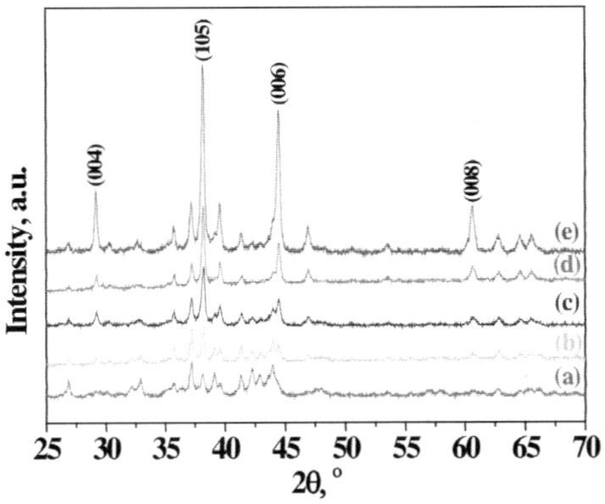

Figure 1. XRD patterns for the SPSed (a) and HDed magets at various temperatures and compression ratios: (b) 750°C-54%; (c) 750°C-68%; (d) 800 °C-73%; (e) 800 °C-80%

Two distinct zones with different grain sizes were noticed in SPSed NdFeB magnets as shown in the Fig. 2 (a) and (b). The coarse grain zones correspond to the particle boundary area of ribbons and the fine grain zones correspond to the interior of the particles. The formation mechanism of the two-zone structure was discussed in our previous paper [8]. The microstructure of the coarse grain zones and fine grain zones are consisted of equiaxed grains for SPSed isotropic magnets. For HDed anisotropic magnets, the two-zone microstructure still exists, as shown in the Fig. 2 (c). It is observed that the microstructure of the fine grain zones consists of small platelet-shaped $Nd_2Fe_{14}B$ grains due to the deformation of the grains, as shown in the Fig. 2(c) and (d). Within each grain the crystallographic c-axis, which is the magnetic easy axis, is perpendicular to the platelet. However, in the coarse grain zones, the microstructure of equiaxed grains was maintained, as shown in the Fig. 2 (c), indicating that large grains are hard to be deformed. The reason can be explained as follows. Compared to the large grains, the fine grains need a shorter diffusion distance for preferential growth of $Nd_2Fe_{14}B$ grains. A larger driving force exists for the growth of small grains [9]. The results imply that the microstructure with large grain size is not beneficial to obtain anisotropic magnet by plastic deformation.

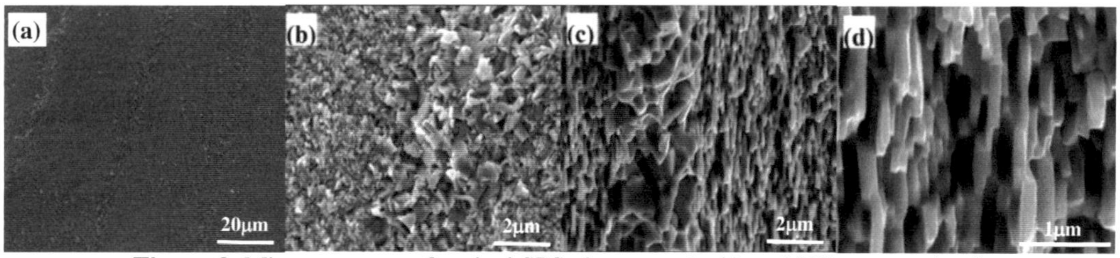

Figure 2. Microstructure of typical SPSed magnet (a, b) and HD magnet (c, d)

Fig.3 shows the microstructure of HD magnets obtained at various deformation conditions. The coarse grain zones in Fig.3 (a), (c), (e), and (g) correspond to the particle boundary area of ribbons and the fine platelet-shaped grain zone correspond to the interior of the particles.[8] With the increase of compression ratio from 54% to 80%, the average width of the coarse grain area increases from ~3.9 to ~5.4 μm, and the average width of the fine grain area decreases from ~11.2 to ~4.1 μm.

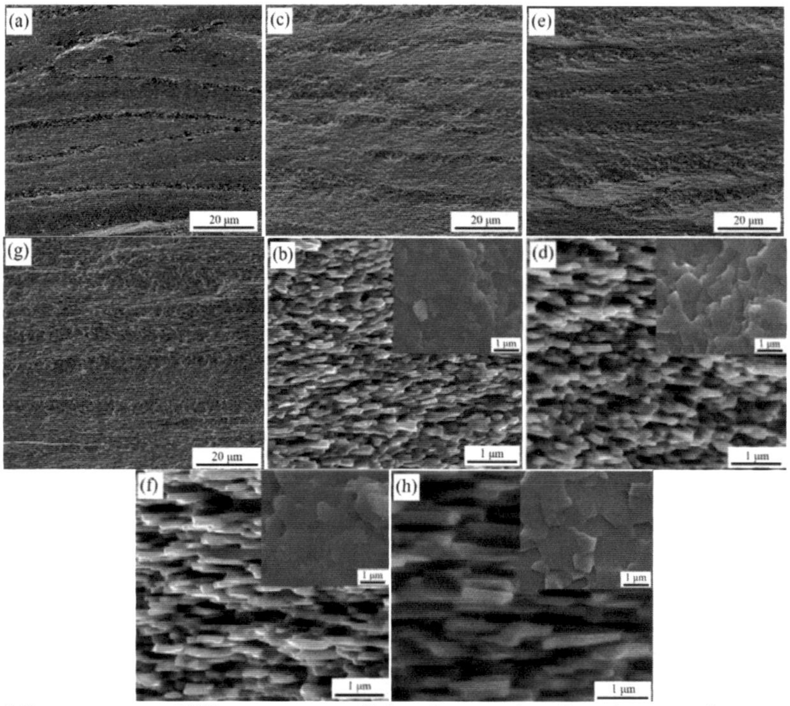

Figure 3. Microstructure of HD magnets parallel to the pressing direction at various compression ratios: (a), (b) 750°C-54%; (c), (d) 750°C-68%; (e), (f) 800 °C-73%; (g), (h) 800 °C-80%; the insets are the microstructure of HDed magnets perpendicular to the pressing direction respectively

Fig. 4 shows the dependences of the average widths of the fine and coarse grain zones on the compression ratios. The width of the coarse grain zone clearly increases with increasing compression ratio. The increase of deformation time and temperature are responsible to the slight increase of coarse grain area width. The preferential growth of $Nd_2Fe_{14}B$ grains under stress perpendicular to the pressing direction is contributed to the decrease of the fine grain area width. The mean dimensions of grain perpendicular and parallel to the pressing direction, defined as *w* and *h* (Fig.4 inset), respectively, in the fine grain zone for HD magnets prepared at various compression ratios are plotted in Fig. 5. Both *w* and *h* increase with the increasing compression ratio, indicating an increasing grain volume due to mass diffusion. The increase value of *w* is attributed to the increasing compression ratio and time. Large compression ratio can enhance grain rotation and deformation. Increase of deformation time leads to more mass transportation caused by diffusion [5]. Both of them can enhance the preferential growth of $Nd_2Fe_{14}B$ grains under stress along perpendicular to the pressing direction, consistent with the XRD results. This microstructure characteristic has an important effect on the magnetic properties.

Fig.5 shows magnetic hysteresis loops at 300K of the SPSed and HDed magnets parallel and perpendicular to the pressing direction at various deformation conditions. As listed in the Table 1, the degree of texture, indicated by the remanence along two directions $(J_r^{//}-J_r^{\perp})/J_r^{//}$, for SPSed magnet is only 0.012, indicating an isotropic behavior. For the magnetic properties of HDed magnets parallel to the pressing direction, with increasing compression ratio from 54% to 80%, the remanence J_r increases from 1.19 T to 1.34 T, indicating the increase of the orientation degree of c-axis texture. At the same time, the coercivity ${}_jH_c$ decreases from 1239 kA/m to 731 kA/m. The different dependences of remanence and coercivity on the compression ratio lead to a maximum of energy product at compression ratio of 73%. The optimal combination of the magnetic properties are J_r = 1.32 T, ${}_jH_c$ = 847 kA/m, and $(BH)_{max}$ = 303 kJ/m^3. With the formation and enhancment of c-axis texture, the remanence along the direction perpendicular to the pressing direction decreases from 0.39 T to 0.23 T.

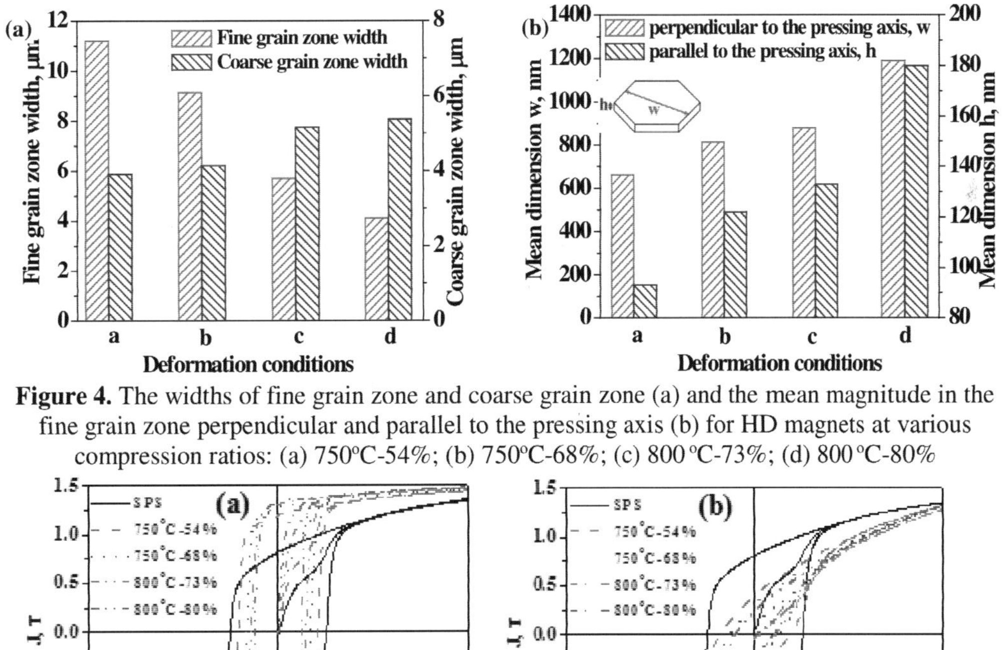

Figure 4. The widths of fine grain zone and coarse grain zone (a) and the mean magnitude in the fine grain zone perpendicular and parallel to the pressing axis (b) for HD magnets at various compression ratios: (a) 750°C-54%; (b) 750°C-68%; (c) 800 °C-73%; (d) 800 °C-80%

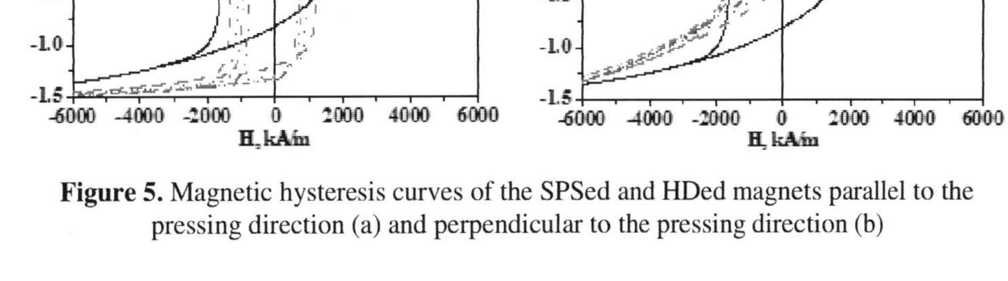

Figure 5. Magnetic hysteresis curves of the SPSed and HDed magnets parallel to the pressing direction (a) and perpendicular to the pressing direction (b)

The magnetic properties are directly related to the microstructure. The dimension *w* of the platelet-shaped grain is related to the remanence and coercivity. Larger values of *w* result in higher remanence, as shown in Table 1. This can be attributed to the improved anisotropy. Table 1 also shows that with the increase of compression ratio, $_jH_c$ decreases. Since the increase of the platelet-shaped grain diameter enhances the irregular degree of grain, which possibly increases the stray field acting on the grains. Therefore, the increased of stray field is one possible reason for the reduction of the coercivity.

Table 1. Magnetic properties and degree of texture of SPSed and HDed magnets parallel and perpendicular to the pressing direction

Process of sample	J_r (T)		$_jH_c$ (kA/m)		$(BH)_{max}$ (kJ/m³)		$\frac{J_r^{//} - J_r^{\perp}}{J_r^{//}}$
	$J_r^{//}$	$J_r^{\perp}$	$_jH_c^{//}$	$_jH_c^{\perp}$	$(BH)_{max}^{//}$	$(BH)_{max}^{\perp}$	
700 °C-SPS	0.82	0.81	1516	1517	116	113	0.012
750°C-54%	1.19	0.39	1239	1095	259	25	0.670
750°C-68%	1.29	0.30	995	773	293	14	0.768
800°C-73%	1.32	0.24	847	638	303	9	0.820
800°C-80%	1.34	0.23	731	585	285	8	0.823

3.2 Compositional modification for SPS magnets

It is still not possible to completely avoid the grain coarsening during SPS, which leads to inevitable decrease in coercivity for SPSed magnets. Hence, maintaining a high coercivity is particularly important for SPSed NdFeB magnets. To improve coercivity, the magnetic powders can be mixed with some metal powders (such as Zn and Cu powders). During SPS, the metal powders can react with the magnetic matrix and presumably diffuse into the neodymium-rich phase surrounding the $Nd_2Fe_{14}B$ grains at high temperature and pressure required for densification, which has been confirmed by electron microprobe analysis [10]. By this method, the enhancement of coercivity has been achieved by Cu and Zn additions. It was found that both elements increased the coercivity of hot-pressed magnets by ~37%. For die-upset magnets, coercivity was increased by ~91% for Zn addition, which is higher than ~75% by Cu addition [11]. Another effective way to enhance coercivity and maintain the low concentration of high cost rare earth element Dy is the development of a core-shell microstructure, which is formed by segregation of Dy along the grain boundary area, by the addition of powdered Dy compounds (oxides, fluorides, or hydride) [12-14]. It was reported that adding 4 wt.% Dy_2O_3 enhanced the coercivity by 37.9% (from 11.6 to 16.0kOe) for sintered $Nd_{15}Fe_{77}B_8$ magnet [12].

Table 2. Room temperature magnetic properties, density and temperature coefficients for SPSed $Nd_{10.15}Pr_{1.86}Fe_{80.41}Al_{1.67}B_{5.91}$ magnets with the optimal powder additions

SPSed Samples	J_r (T)	$_jH_c$ (kA/m)	$(BH)_{max}$ (kJ/m³)	Density (g/cm³)	α (%K⁻¹)	β (%K⁻¹)
Nd-Fe-B	0.77	953	96	7.43	-0.145	-0.518
Nd-Fe-B+0.6 wt.% Zn	0.76	1054	97	7.38	-0.135	-0.454
Nd-Fe-B+2wt.% Dy_2O_3	0.74	1098	92	7.23	-0.132	-0.488
Nd-Fe-B+0.6 wt.% Zn+2wt.% Dy_2O_3	0.73	1142	91	7.22	-0.133	-0.462

We investigated the effects of separated and combined Dy_2O_3 and Zn additions on the magnetic properties, microstructure and thermal stability of SPSed nanocrystalline NdFeB magnets. The magnetic properties of the magnets with different additions are listed in Table 2. The magnets with 0.6 wt.% Zn and 2 wt% Dy_2O_3 additions have $_jH_c$ values about 11% and 15% higher than that of the additive free magnet, respectively. To further improve the properties, the combination of 2 wt.% Dy_2O_3 and 0.6 wt.% Zn additions were employed. The magnetic properties for the magnet with Dy_2O_3+Zn addition are also shown in Table 2. The combined addition considerably increases the magnetic properties, which produced a ~20% increase in coercivity. An optimal combination of J_r=0.73 T, ${}_jH_c$=1142 kA/m, and $(BH)_{max}$=91 kJ/m^3 were obtained. Importantly, the thermal stability of the doped magnets is also improved, demonstrated by the reduced temperature coefficients of remanence (α) and coercivity (β) in Table 2. These SPSed samples can not only work as the isotropic permanent magnets but also can be employed as the ideal precursors for hot deformed anisotropic NdFeB magnets.

3.3 Recoil loop characteristics

As a series of magnetic characteristic curves, recoil loops, measured by removing and reapplying an increased successively reversed field on a previously positively saturated magnetic sample, can provide physical insight into magnetization reversal mechanism. In practice, open recoil loops can cause working point change for permanent magnets. The energy loss associated with the open recoil loops is not negligible, which is disadvantageous to the application of permanent magnets. It is commonly accepted that the open recoil loops results from the exchange decoupled soft phase in the nanocomposite alloys [15]. Generally, single phase alloys have narrow and even closed loops [16]. Recently, McCallum [15] pointed out that the enclosed area in the loops is related to the mean field interaction strength, the distribution of coercivity, and the volume fraction of the soft phase. From the experimental studies, the open recoil loops was also found to be related to the unstable magnetic moments caused by thermal fluctuation [17], strong exchange coupling due to the nanostructure [18], and the inhomogeneity in the magnetic anisotropy [19]. Here we investigated the relationship between the recoil loops and microstructure for hot deformed NdFeB magnets.

For the experiments, SPS was carried out at various temperatures with a pressure of 50 MPa for 5 min. HD was carried out at 750°C for 20 min under a uniaxial stress of 350 MPa. As expected, the density of SPSed magnets increases with increasing SPS temperature. The magnet SPSed at 700 °C has a density of 7.33 g/cm^3, while the density of that SPSed at 650°C is only 7.03 g/cm^3. The low magnification images for these two samples are shown in Fig.6a and Fig.6b. Some pores can be clearly observed for the low density magnet (Fig.6a). Reduced porosity is evident in Fig.6b. The microstructure of SPSed magnet consists of uniform equiaxed grains (insets in Figs.6a and 6b).

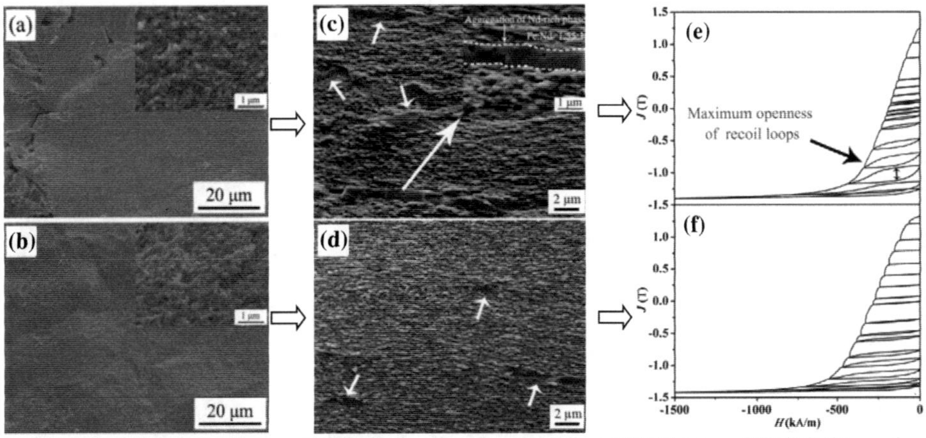

Figure 6. Microstructures for SPSed magnet prepared at 650°C (a) and 700°C (b), and corresponding HDed magnets (c,d) and recoil loops (e, f).

Above mentioned two samples were subjected to HD at 750°C, and the microstructures after HD are shown in Figs.6c and 6d. Platelet-shaped $Nd_2Fe_{14}B$ grains due to deformation are formed. Some strip-shaped grains can be observed, indicated by the arrows. The Fe/Nd atomic ratios detected by EDS indicate that these grains are Nd-rich phase. The existence of porosity in SPSed precursor is considered to be the main reason for Nd-rich phase aggregation in HDed magnets. It can be assumed that liquid Nd-rich phase aggregated in the pores at high temperature during the early stage of HD, and it was squeezed to form some strip-shaped grains under the uniaxial stress during deformation. Comparing Fig.6a with Fig.6b, higher SPS temperature led to an improved density and less porosity, and therefore, it reduced Nd-rich phase aggregation in SPS+HDed magnets, as shown in Fig.6c and Fig.6d. To confirm above assumption, a low porosity SPS magnet with high density close to the theoretical value of 7.50 g/cm^3 was prepared by optimal process and raw powders. After hot deformation, no Nd-rich phase aggregation was observed, as presented in Fig.7a, which is attributed to the elimination of the porosity in the SPSed precursor. Hence, the results demonstrated that the Nd-rich phase aggregation results from the existing porosity in SPSed precursor and the reduced porosity in SPS precursor is beneficial to eliminate Nd-rich phase aggregation.

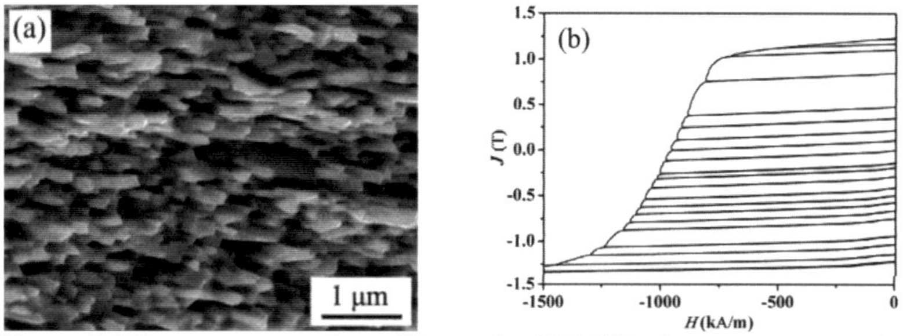

Figure 7. Microstructure and recoil loops for SPS+HDed magnets prepared with 68% deformation ratio using high density (7.50 g/cm^3) SPS precursors

The recoil loops for SPSed and SPS+Hded magnets were obtained. All SPSed magnets had closed loops, similar to the normal single phase NdFeB magnet and RE-rich sintered microcrystalline NdFeB magnets. Interestingly, the recoil loops of hot deformed magnets with same composition are open, as shown in the Fig.6e and 6f, although there is no soft magnetic phase detected by XRD. This type of open recoil loops for Nd-rich NdFeB magnets has never been reported before and is worthy of investigation. On the other hand, the maximum openness of recoil loops does not take place at the coercive field, which is different from that of nanocomposite NdFeB alloys.[20]

Considering the microstructure and magnetic properties, we suggested that strip-shaped Nd-rich phase with dimension up to several micrometers, as shown in Figs.6c and 6d, generate non-negligible local demagnetizing fields. Large demagnetizing factor induced by the Nd-rich phase aggregation along pressing direction (c axis) produces a large demagnetization field. The Nd-rich phase aggregation with various sizes, lager than grains, leads to non-uniform distribution of local demagnetization fields. These demagnetization fields can reduce the local pinning field and lead to the inhomogeneity of magnetic anisotropy. This variation of local magnetic anisotropy in the magnets should be responsible for the formation of open recoil loops. In other words, the open recoil loops arise from the difference in local reversal field due to non-uniform distribution of large local demagnetization fields. Since there exists non-uniformly distributed demagnetization fields, the reverse fields at different parts of the magnets are different, which eventually results in open recoil loops.

Table 3. A summary of the process, density, Nd-rich aggregation, recoil loop characteristics and coercivity for SPS+HDed magnets

SPS process	**HD process**	**SPS precursor density, g/cm^3**	**Nd-rich phase aggregation**	**Recoil loops openness**	**$_jH_c$, kA/m**
650°C-50MPa-5min	750°C-20 min-350 MPa	7.03	Serious	large	226
700°C-50MPa-5min	750°C-20 min-350 MPa	7.33	Medium	Medium	288
Optimal process and raw powders	750°C-20 min-350 MPa	7.50	Not observed	No	995

Above explanation has been verified by experiments. As shown in Figs.6c and 6d, with increasing SPS temperature from 650 to 700°C, Nd-rich phase aggregation was reduced in SPS+HDed magnets. Correspondingly, the openness of recoil loops is also reduced, as presented in Figs.6e and 6f. Furthermore, for the HDed magnets without Nd-rich phase aggregation, the recoil loops is found to be fully closed (Fig.7b). These results confirmed that the openness of recoil loops arises from the aggregation of Nd-rich phase. A more serious distribution of local demagnetization field contributes to the more open recoil loops. The characteristics of recoil loops and magnetic properties for SPS and HDed magnets are summarized in Table 3. The clear relationships between Nd-rich phase aggregation, recoil loops openness and coercivity have been demonstrated. Distribution of large local demagnetization fields not only leads to reduced coercivity, but also results in open recoil loops.

4. Conclusions

Anisotropic nanocrystalline magnets were prepared by SPS followed by hot deformation. Excellent magnetic properties were obtained by optimized process. The microstructure evolution and its influences on the magnetic properties were investigated. Fine and regular grain microstructure is beneficial to deformation and improving coercivity. The increase of stray field is responsible for the decrease of coercivity with increasing compression ratio. Zn and Dy_2O_3 additions have been successfully employed to improve the coercivity and thermal stability of SPSed magnets. The aggregation of Nd-rich phase not only leads to low coercivity, but also results in open recoil loops. The openness of recoil loops is related to the local demagnetization fields induced by Nd-rich phase aggregation. The reduction of the Nd-rich phase aggregation is beneficial to achieve deformation anisotropy and high magnetic properties.

5. Acknowledgements

This work is financially supported by the Natural Science Foundation of China (Grant No. 51174094), Guangdong Provincial Science and Technology Program (Grant No. 2012B091000005), and the State Key Laboratory for Advanced Metals and Materials (Grant No. 2011-ZD05).

6. References

[1] Mishra R K, Panchanathan V and Croat J J 1993 *J. Appl. Phys.* 73 6470

[2] Ma B M, Lee D, Smith B, Gaiffi S, Owens B, Bie H and Warren G W 2001 *IEEE Trans. Magn.* 37 2477

[3] Yue M, Tian M, Zhang J X, Zhang D T, Niu P L and Yang F 2006 *Mater. Sci. Eng. B* 131 18

[4] Saito T, Fujita M, Kuji T, Fukuoka K and Syono Y 1998 *J. Appl. Phys.* 83 6390.

[5] Li L and Graham C D 1990 *J. Appl. Phys.* 67 4756.

[6] Tokita M 1999 *Mater. Sci. Forum* 308-311 83

[7] Anselmi-Tamburini U, JGaray J E and Munir Z A 2005 *Mat. Sci. Eng. A-struct.* 407 24

[8] Liu Z W, Huang H Y, Gao X X, Yu H Y, Zhong X C, Zhu J and Zeng D C 2011 *J. Phys. D: Appl. Phys.* 44 025003

[9] Li L and Graham C D 1992 *IEEE Trans. Magn.* 28 2130.

[10] Fuerst C D and Brewer E G 1990 *Appl. Phys. Lett.* 56 2252

[11] Fuerst C D and Brewer E G 1991 *J. Appl. Phys.* 69 5826

[12] Ghandehari M H 1986 *Appl. Phys. Lett.* 48 548

[13] Yan G L, McGuiness P J, Farr J P G and Harris I R 2010 *J. Alloys Compd.* 491 L20

[14] Kianvash A, Mottram R S and Harris I R 1999 *J. Alloys Compd.* 287 206

[15] McCallum R W 2006 *J. Magn. Magn. Mater.* 299 472

[16] Bollero A, Gutfleisch O, Müller K -H, Schultz L and Drazic G 2002 *J. Appl. Phys.* 91 8159

[17] Rong C B, Liu Y Z and Liu J P 2008 *Appl. Phys. Lett.* 93 042508

[18] Rong C B and Liu J P 2009 *Appl. Phys. Lett.* 94 172510

[19] Choi Y, Jiang J S, Pearson J E and Bader S D 2007 *Appl. Phys. Lett.* 91 022502

[20] Zheng B, Zhang H W, Zhao S F, Chen J L and Wu G H 2008 *Appl. Phys. Lett.* 93 182503

Shear viscosity of two-dimensional strongly coupled complex (dusty) plasmas

Aamir Shahzad[1,2], Arffa Aslam[1], Mariam Sultana[1] and Mao-Gang He[2]
[1]Department of Physics, Government College University Faisalabad (GCUF), Allama Iqbal Road, Faisalabad 38000, Pakistan
[2]Key Laboratory of Thermo-Fluid Science and Engineering of Ministry of Education (MOE), Xi'an Jiaotong University, Xi'an 710049, P. R. China

E-mail: aamirshahzad_8@hotmail.com, aamir.awan@gcuf.edu.pk

Abstract. A molecular dynamics method has been employed of studying shear viscosity for two-dimensional plasma liquids. For the entire range of strongly coupled liquid states, shear autocorrelation functions indicate overall valid viscosity coefficients. A systematic dependence of shear viscosity value on screening strength (κ) is observed for an intermediate and higher Coulomb coupling strengths (Γ). The simulation data indicate that the position of the viscosity value shifts towards higher Γ as κ increases. It is observed that valid viscosity coefficient exists and it is dependent on plasma parameters (Γ, κ). A finite minimum viscosity exists nearly at the same value of Γ where the most extreme super-diffusion was earlier found and is reported for a wide range of coupling and screening parameters.

1. Introduction

In recent years, the transport problems in non-ideal systems of interacting particles are of significant interest in various fields of science and technology. In some cases, the presence of dust particles can be quite beneficial in the efficiency of solar cell where as an undesirable contaminant that can destroy microchips during their plasma enhanced fabrication. The presence and formation of dust in plasma systems has been widely studied both experimentally and numerically [1,2]. The industrial applications and widespread existence of dusty plasmas, combined with a number of their general and standard thermophysical properties, make dusty plasmas extraordinary attractive and interesting subjects for many researchers. Shear processes play a dominant role in determining dynamical properties of many physical, chemical and biological systems. Shear viscosity is one of dynamical property of fluids required to describe shear flows and damping of waves. Shear viscosity is essential for predicting collective mode properties of strongly coupled complex (dusty) plasmas (SCCDPs) plasma, an issue at the forefront of rapidly evolving field of dusty plasma [3,4].

The thermophysical properties of two and three dimensional SCCDPs have been studied in great detail for many decades [1-9]. This work reviews the outcomes of studies carried out in the last two decades, which facilitate us to understand a picture of the variation in the viscosity coefficient over the whole range of plasma parameters of the SCCDPs. The correlation functions of two dimensional (2D) SCCDPs have recently attracted the particular attention to the researchers for studying dusty plasma [3,10,11]. A slow decay $\propto t^{-1}$ of the velocity autocorrelation function (VACF) was observed in the earlier 2D molecular dynamics (MD) measurements of hard disk liquids, showing no diffusive motion [12]. Similar conclusions have been publicized with an additional argument that this asymptotic behaviour should also exist in kinetic contributions of the autocorrelation functions for the shear viscosity and thermal conductivity, regardless of the pair potential [13]. Another MD simulation with observed a t^{-1} decay of the shear autocorrelation function (SACF) in the 2D soft disk liquids [14].

Shear viscosity has been experimentally investigated in 2D dusty plasma liquids (DPLs) monolayer of dust particles [2], and these observations were supported by earlier equilibrium MD (EMD) simulations with 2D DPLs [14]. Soon after, nonequilibrium MD simulations (NEMD) were employed with 2D Yukawa liquids [11] and investigated the viscosity coefficients at lower plasma parameters (Γ). A long standing controversy over these transport observations with conflicting experimental and simulation results in recent 2D Yukawa liquids is our motivation for presented work.

Recently, in a previous paper, the present authors have been studied similar computations with detailed argument on shear stress autocorrelation functions (SACFs) measurements for intermediate and higher Coulomb coupling strengths of 2D SCCDPLs [15]. Our previous results, for SACFs, show that the long time behaviour will determine whether the shear viscosity coefficient is unique or not. In this presented work, we have extend our work for the determination of shear viscosity coefficients (η) for the whole range of plasma parameters (Γ, κ) than those used previously known results for the 2D SCCDPs [11,14].

The purpose of the presented work is to investigate further all possible dependences of the shear viscosity on the screening and Coulomb coupling strengths (Γ, κ), using the same method and the extended set of parameters as in Shahzad and He [15]. Moreover, the aim of this work is to investigate a systematic study of the effects of screening length of the dust particle interaction on shear viscosity and to investigate the existence of the related shear viscosity coefficients of 2D SCCDPLs, by using extensive numerical simulations. This paper is organized as follows: In Section 2, the estimate of the shear viscosity coefficient for Yukawa liquids is analyzed through the equilibrium MD method, and this section also describes the simulation technique. Next, the obtained results are presented and discussed with simulation data reported by other author's in Section 3. The work is summarized in Section 4.

2. Theory and EMD algorithm

A well known Green-Kubo relation (GKR) is commonly used for the shear viscosity coefficients of uncharged particles [16,17]. Equilibrium MD is used for 2D Yukawa liquid simulations [3,15] and it is similar to the 2D monolayer dusty plasma experiments [2] conducted in equilibrium environments. It has been shown that the standard GKRs of liquids apply to the SCCDPs [3-9,11,15]. Currently, the Yukawa (screened Coulomb) potential model is the best known model for simulation of the pairwise interactions between dust particles with each other in laboratory plasma of gas discharge and in many physical systems (biology, medicine, physics of polymers, etc.)

$$\phi(|\mathbf{r}|) = \frac{Q^2}{4\pi\varepsilon_0}\frac{e^{-|\mathbf{r}|/\lambda_D}}{|\mathbf{r}|}, \tag{1}$$

where, r is the interparticle distance, λ_D is the Debye screening length and Q is the charge on a dust particle. In thermodynamic limit, such a Yukawa model can be fully described by two reduced parameters [3,10,11,15]: the Coulomb coupling parameter $\Gamma=(Q^2/4\pi\varepsilon_0)(1/ak_BT)$ and the screening parameter $\kappa\equiv a/\lambda_D$. Here, T is absolute temperature, k_B is Boltzmann's constant, $a=(n\pi)^{-1/2}$ is the Wigner-Seitz radius [7] with n is the equilibrium dust number density. Time scales of interest are regarded as inverse of the plasma frequency $\omega_p=(Q^2/2\pi\varepsilon_0 ma^3)^{1/2}$, where m is the dust particle mass.

To compute the shear viscosity η, we start from the following usual equations

$$\eta = \frac{1}{Ak_BT}\int_0^\infty Z_\eta(t)dt\,, \qquad Z_\eta(t) = \left\langle P^{xy}(t)P^{xy}(t_0)\right\rangle, \tag{2}$$

where $Z_\eta(t)$ is the shear stress autocorrelation function, A is area for a 2D system, T system temperature, and the angular bracket indicates an ensemble average for Yukawa systems. In order to compute the shear viscosity, time series data for the positions (x_i, y_i), velocities (v_{ix}, v_{iy}) and off-diagonal elements of pressure tensor P_{xy} are required and given as

$$P^{xy}(t) = \sum_{i=1}^{N} m v_{ix} v_{iy} - \frac{1}{2}\sum_{i}\sum_{i\neq j}^{N} \frac{x_{ij} y_{ij}}{r_{ij}} \frac{\partial \phi(r_{ij})}{\partial r_{ij}}, \tag{3}$$

where N is the number of particles, m is the dust particle mass and $\mathbf{r}_{ij}=\mathbf{r}_i-\mathbf{r}_j$. The first term of Eq. (3) is kinetic contribution, which depends on particle velocities and the second term is a potential contribution, which depends on the Yukawa pair potential given in Eq. (1). In calculating SACF from MD simulation data, it is used a common method of overlapping time segments. Further details of the correlation functions from MD simulation method, were reported in [3]. The 2D viscosity η is normalized by $\eta_0=nm\omega_p a^2$. All time series data are recorded during MD simulation and used Eqs. (2) and (3) to calculate η.

2.1. Numerical simulation

We start as usual with a system of N=1024-18225 particles enclosed in a simulation box with edge lengths $Lx/Ly=2/\sqrt{3}$, interacting through a pairwise potential of Yukawa type, and the number particle motion is modeled by Newton's equations. The particle number is chosen large enough to be free of the size effects noted for the present work. It is reported in our earlier equilibrium MD simulation estimations that particle number is sufficient to test such system sizes [15]. Therefore, this particle number N is sufficient for simulations that in order to save computer time. Our MD computations are performed in a microcanonical ensemble and a fifth order predictor-corrector algorithm [1,3] is employed to solve the normalized equation of motion of N Yukawa particles. Periodic boundary conditions (PBCs) are maintained in all directions and the minimum image convention of the N Yukawa particles throughout our numerical experiment, for simulation of an infinite system. The imposed periodicity needs to be consistent with the fact that the ground state at low temperatures is the triangular lattice [3,15,17]. When the number of dust particles satisfies N = (integer)^2, or is equal to the number of unit cells, our periodicity satisfies this condition.

We derived the Ewald sums expression of the force, energy and pressure tensor which are implemented in our numerical code. In the Ewald summation method, the interaction potential is divided into two parts, one of which converges rapidly in real space sum and the other converges rapidly in the reciprocal space sum. For higher screening values κ the real space sum part alone gives enough performance and precision. Pair wise Yukawa interparticle forces are summed over a κ-dependent cutoff radius [1,3,7-9]. Additional details of the Ewald sums technique for Yukawa interactions are reported by Shahzad and He [7]. The desired system temperature is attained by periodically synchronize all particles and renormalizes their velocities to the required target value for T and the system reaches thermodynamical equilibrium. Once the system reaches thermodynamical equilibrium, the periodic renormalization of particle velocities is continued and allows the system progress under the constant energy conditions. Production stages are employed between $9.5\times10^6/\omega_p$ and $12.5\times10^6/\omega_p$ time units in the series of data recording of η_0. The simulation time step $0.001/\omega_p \leq dt \leq 0.005/\omega_p$ is maintained the large cell size to allow computing the important data at long times [3,15]. In addition, to reduce statistical noise, two independent equilibrium MD calculations runs are carried out with arbitrarily chosen configurations of the particle positions, for each target temperature Γ, and averaging these measurements to improve the statistics. In this paper, the computations are reported for shear viscosity of 2D SCCDPs over wider range of plasma parameters of ($40 \leq \Gamma \leq 180$) and ($1 \leq \kappa \leq 3$).

3. Results and Discussion

The results obtained for our MD series of simulation of a 2D Yukawa liquid are estimations of shear viscosity showing the indication of superdiffusive motion at intermediate and higher Coulomb coupling parameters, as we earlier reported extreme superdiffusion in Ref [15]. The shear viscosity trends in 2D DPLs depend on both the Coulomb coupling strength and the screening strength $1/\kappa$. These simulation data are for a large system size (N = 18252), but with different Γ, which covers the value of coupling strength, where the viscosity (η_0) has minimum possible value depends on κ, [6,15] up to Γ=180.

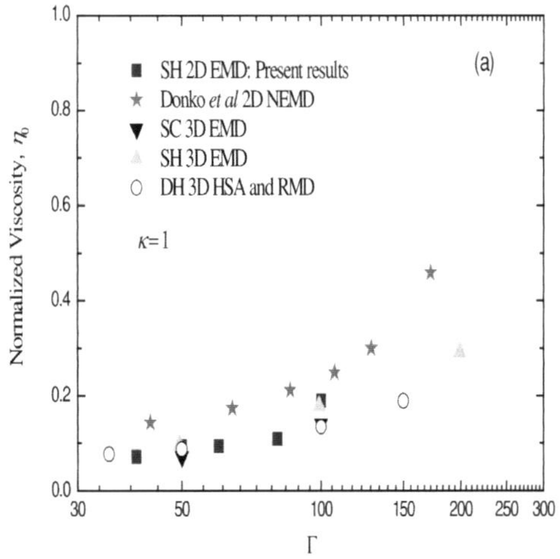

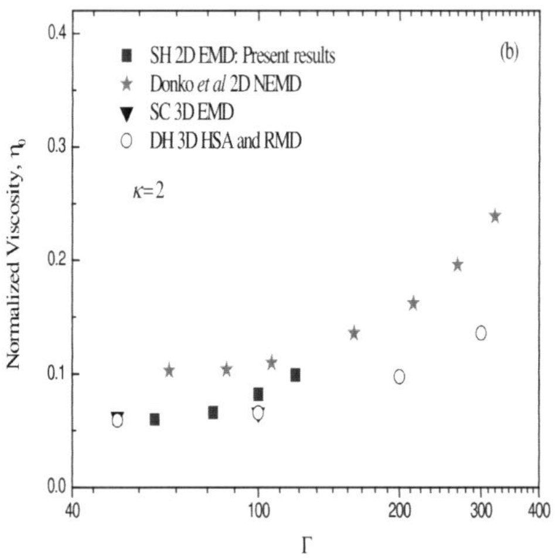

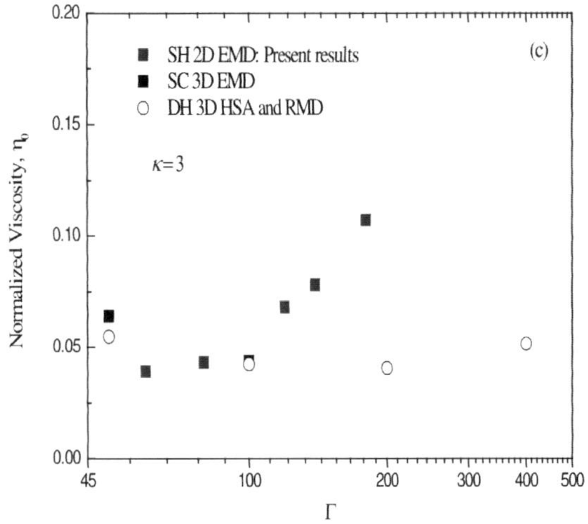

Figure 1. Comparison of results for minimum possible values of shear viscosity as a function of Coulomb coupling Γ for Yukawa liquid, obtained from Green-Kubo relation defined in Eqs. (2) and (3), at (a) κ=1; (b) κ=2 and κ=3 (c). EMD of Shahzad and He: SH 2D EMD: present results, NEMD of Donkó *et al* [11]: Donkó *et al* 2D NEMD, EMD simulation results of Salin and Caillol [4]: SC 3D EMD, EMD calculations of Saigo and Hamaguchi [5]: SH 3D EMD, and Homogenous Shear Algorithm (HSA) and Reverse MD (RMD) of Donko and Hartmann [6]: DH 3D HSA and RMD.

Figure1shows the main results obtained from the EMD method for the different plasma states simulated for the Yukawa liquids at κ=1, 2 and 3, respectively. Parts (a), (b) and (c) of the figure 1 show our calculations for the shear viscosity with N =18252 particles at a long run, as well as the previous results taken from the NEMD simulations of Donkó *et al* [11], Green-Kubo EMD of Salin and Caillol [4], EMD work of Saigo and Hamaguchi by using tensor-product spline function [5] and homogenous shear algorithm (HSA) and reverse MD (RMD) of Donko and Hartmann [6]. A first observation is that η_0 has its minimum possible value, nearly indicating the most extreme superdiffusion, at Γ=40 for κ=1, Γ=60 for κ=2 and Γ=80 for κ=3. The evidence of maximum values of superdiffusion has been discussed in our earlier published work [15]. These results are important because this happens nearly at the same value of Γ where the shear viscosity was earlier to have a minimum value [11,14]. There are different factors to be chosen to obtain the best results by the presented equilibrium MD for shear viscosity. We can vary the strength of screening parameter (κ), system length (run time), the system temperature (Γ) and system size (N) in order to examine how efficiently the EMD algorithm computes the shear viscosities of SCCDPs.

Starting from the values of Γ at which η_0 has lowest values, the present simulations give a definitely lower η_0, compared to results in Ref. 11. The simulations are carried out with N= 18252 for κ = 1, 2 and 3, and it is important that the EMD method gives measurements for η_0 within the limited statistical uncertainties over the wide range of system sizes. It is observed that the presented results of η_0 are slightly lower than and lie more close to the earlier NEMD calculations of Donkó *et al* [11] but our results are slightly higher at region between $180 \leq \Gamma \leq 300$ for the κ=2 case. These investigations are significant because the results happen at the same values of coupling Γ where the $\eta_0 \cong 0.072$ at κ= 1 and $\eta_0 \cong 0.06$ at κ= 2 were earlier shown to have the minimum values of η_0 [11,14]. The presented results for $40 \leq \Gamma \leq 300$ at κ= 1 and $60 \leq \Gamma \leq 180$ at κ= 2 lie between of data sets of 2D NEMD of Donkó *et al* [11] and the results of 3D calculations of η_0 by different authors [4-6]. Our simulation data for viscosity are dependent on the degree of correlations in the 2D system, and the dependence is clearly monotonic. Our second examination from figure 1 is that the η_0 is dependent on κ, like 3D simulation [4-6]. The present simulations show that the minimum possible value of η_0 shifts towards higher Γ as κ increases. The same shifting behavior was also seen in earlier 3D simulations. The minimum value of η_0 decreases with increasing κ. Our simulation trends have a good agreement with the earlier 2D MD of Ref. 14 and NEMD of Ref. 11 simulations. In addition, the presented results are also well matched with the earlier 3D Green-Kubo EMD investigations of Ref. 4, spline function EMD work of Ref. 5 and homogenous shear algorithm (HSA) and reverse MD (RMD) calculations of Ref. 6.

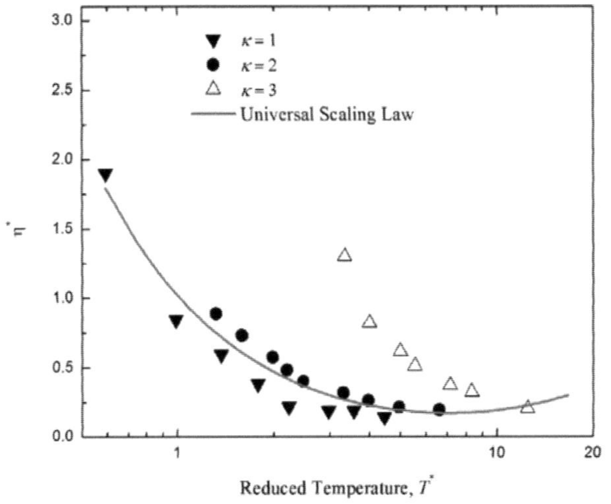

Figure 2. Variation of reduced shear viscosity normalized by Einstein frequency $\eta^*(T^*)$ with reduced temperature T^* for Yukawa system at different κ=1, 2 and 3. The thick line is obtained using the simple Scaling Law: $\eta^*= AT^*+B/T^*+C$, showing the Universal behaviour.

Finally, it is revealed the universal scaling law for the CDPLs. Figure 2 shows the variation of reduced shear viscosity η^*(normalized by the Einstein frequency) at different normalized temperatures

T^* (ratio of the temperature to melting temperature) for κ=1, 2 and 3. It is shown from figure 2 that the η^* is dependent on both the screening and reduced temperature in the SCCDPs. The results obtained for η^* using EMD method provide the suitable trend nearly for all the plasma state points of κ and Γ. The solid line given in figure 2 is obtained using the simple empirical fit of form, $\eta^* = AT^* + B/T^* + C$, with coefficients: A= 0.02287, B= 1.14437, and C= -0.15092, same as in Ref. 5 and 11. This scaling law is found to reproduce appropriately the shape of $\eta^*(T^*)$ curves at three κ values and it gives suitable relationship between η^* and T^* for nearly the whole plasma state points of (Γ, κ), confirming the earlier simulation results [8, 11, 17].

4. Conclusion

Equilibrium MD simulation approach has been employed to determine shear viscosity of a 2D SCCDPs over a wide range of plasma parameters of Γ and κ. New simulations provide more reliable data for Yukawa shear viscosity than earlier known results. The calculations confirm that the minimum possible value of shear viscosity exists at intermediate and higher Coulomb couplings. It is revealed that the prediction of minimum value of shear viscosity helps to understand the fundamental behavior in 2D Yukawa dusty plasma systems. Our data indicate that the shear viscosity dependent on the screening strength and the degree of correlation in 2D Yukawa liquids. Our simulations show that the minimum possible value of viscosity shifts towards higher Coulomb coupling as screening parameter increases. Our shear viscosity calculations illustrate a fair agreement with earlier MD and NEMD simulations. In future work, it will also be of highly interest for analysis to see how quantum effects influence viscosity in 3D and 2D SCCDPs and it is suggested a need for an experiment or simulation.

5. Acknowledgments

This work was sponsored by the China Postdoctoral Science Foundation (CPSF No. 2013M532042). The authors thank Z. Donkó (Hungarian Academy of Sciences) for providing his shear viscosity data of Yukawa Liquids for the comparisons of our simulation results, and useful discussions. We are grateful to the National High Performance Computing Centre of Xian Jiaotong University for allocating of computer time to test and run our MD code.

6. References

[1] Shahzad A and He M-G 2012 *Physics of Plasmas* 19 083707

[2] Nosenko V and J Goree 2004 *Phys. Rev. Lett.* 93 155004

[3] Shahzad A, He M-G and He K He 2013 *Phys. Scr.* 87 035501

[4] Salin G and Caillol J M 2002 *Phys. Rev. Lett.* 88 065002

[5] Saigo T and Hamaguchi S 2002 *Physics of Plasmas* 9 1210

[6] Donko Z and Hartmann P 2008 *Phys. Rev. E* 78 026408

[7] Shahzad A and He M-G 2012 *Plasma Science and Technol.* 14 771

[8] Shahzad A and He M-G 2012 *Contrib. Plasma Phys.* 52 667

[9] Shahzad A and He M-G 2013 *Interaction Contributions in Thermal Conductivity of Three-Dimensional Complex Liquids* AIP Conf. Proc. 1547, 173

[10] Hou L J and Piel A 2009 *J. Phys. A: Math. Theor.* 42 214025

[11] Donko Z, Goree J, Hartmann P and Kutasi K 2006 *Phys. Rev. Lett.* 96 145003

[12] Alderand B J and Wainwright T E 1970 *Phys. Rev.* A 1 18

[13] Ernst M H, Hauge E H and Van Leeuwen J M J 1970 *Phys. Rev. Lett.* 25 1254

[14] Liu B and Goree 2005 J *Phys. Rev. Lett.* 94 185002

[15] Shahzad A and He M-G 2012 *Phys. Scr.* 86 015502

[16] Evans D J and Morriss G P 1990 *Statistical Mechanics of Non-equilibrium Liquids* Academic (London)

[17] Shahzad A and He M-G *2014 Computer Simulation of Complex plasmas: Molecular Modeling and Elementary Processes in complex plasmas* Scholars' Press (Germany)

IOP Conf. Series: Materials Science and Engineering **60** (2014) 012015 doi:10.1088/1757-899X/60/1/012015

Oxidation resistant peroxide cross-linked UHMWPE produced by blending and surface diffusion

Rizwan M Gul[1], Ebru Oral[2] and Orhun K Muratoglu[2]
[1]Department of Mechanical Engineering, University of Engineering and Technology, Peshawar, Pakistan
[2]Harris Orthopaedic Laboratory, Massachusetts General Hospital, Boston, MA 02114, USA

E-mail: rgul@nwfpuet.edu.pk

Abstract. Ultra-high molecular weight polyethylene (UHMWPE) has been widely used as acetabular cup in total hip replacement (THR) and tibial component in total knee replacement (TKR). Crosslinking of UHMWPE has been successful used to improve its wear performance leading to longer life of orthopedic implants. Crosslinking can be performed by radiation or organic peroxides. Peroxide crosslinking is a convenient process as it does not require specialized equipment and the level of crosslinking can be manipulated by changing the amount of peroxide added. However, there is concern about the long-term stability of these materials due to possible presence of by-products. Vitamin E has been successfully used to promote long-term oxidative stability of UHMWPE. In this study, UHMWPE has been crosslinked using organic peroxide in the presence of Vitamin E to produce an oxidation resistant peroxide crosslinked material. Crosslinking was performed both in bulk by mixing peroxide and resin, and only on the surface using diffusion of peroxides.The results show that UHMWPE can be crosslinked using organic peroxides in the presence of vitamin E by both methods. However, the level of crosslinking decreases with the increase in vitamin E content. The wear resistance increases with the increase in crosslink density, and oxidation resistance significantly increases due to the presence of vitamin E.

1. Introduction

Crosslinking by organic peroxides is used commercially to crosslink different types of polyethylenes by blending into the resin before processing [1]. It is a convenient process as compared to radiation crosslinking as it does not require specialized equipment and the level of crosslinking can be manipulated by changing the amount of peroxides added. Peroxide crosslinking by blending was also used to improve the wear behavior of ultrahigh molecular weight polyethylene (UHMWPE) in vitro for orthopedic implants [2]. However, there is concern about the oxidation of these materials due to possible presence of the by-products of peroxide decomposition. Vitamin E has been successfully used to promote oxidative stability of UHMWPE [3]; however, since it is a free radical scavenger, its addition in the presence of peroxides may disrupt crosslinking [4]. We hypothesized that the desired crosslinking level and oxidation resistance can be achieved by peroxide crosslinking of UHMWPE in the presence of vitamin E.

Limiting crosslinking to the surface is desirable, as crosslinking in bulk leads to decrease in mechanical strength and toughness [5]. We also hypothesized that surface diffusion of peroxides followed by heating to decompose the peroxide could surface cross-link UHMWPE.

2. Experimental

2.1. Materials

Blending of Peroxides: Medical grade UHMWPE (GUR 1050) was blended with 0.1 wt% vitamin E. The peroxide was 2,5-di(tert-butylperoxy)-2,5-dimethyl-3-hexyne (P130; Sigma-Aldrich, St. Louis, MO). The peroxide (a liquid at room temperature) was mixed with the vitamin E-blended resin by hand tumbling for 20 minutes. The mixture was compression molded into pucks (10 cm diameter, ~1 cm thickness) under 15-26 MPa for 2 hours. To determine the effect of vitamin E, both virgin and 0.1 wt% vitamin E-blended resins were processed with 1 wt% P130. To determine the effect of consolidation temperature, 0.1 wt% vitamin E-blended UHMWPE was processed with 0.5 wt% P130 at 170, 180, 190, 200 or 210°C.

Diffusion of Peroxides: Medical grade UHMWPE (GUR 1050) was mixed with vitamin E at 0.1wt%, compression molded and cubes (1cm) were machined. Two different types of organic peroxides, dicumyl peroxide (DCP) or P130 were used. Cubes were doped with the peroxide under argon gas and then heated to decompose the peroxides; they were weighed at each step. To determine the effect of doping temperature on cross-linking, cubes (n=3 each) were doped with DCP at 60, 80 or 100°C for 4 hours followed by heating at 130°C for 4 hours. Another set of cubes (n=3 each) were doped with P130 at 80, 100 or 120°C followed by heating at 180°C. To determine the effect of decomposition temperature on cross-linking, cubes (n=3 each) doped with DCP at 80°C for 4 hours were heated at 130 or 140°C. Another set of cubes (n=3 each) doped with P130 at 80°C were heated at 150, 165 or 180°C. Wear testing pins machined from oversized blocks (11×11×15mm) were doped with DCP at 100°C for 4 hours followed by heating at 130°C for 4 hours, and with P130 at 120°C for 4 hours followed by heating at 180°C for 4 hours. Pins had one doped surface without machining maintained as the wear surface.

2.2. Characterization

Crosslink Density: Crosslink density of blended samples was measured by swelling cubes (n=6; 3 mm) in xylene at 130°C. The gravimetric swell ratio was converted to a volumetric swell ratio using the density of polyethylene as 0.94 g/cc and the density of xylene at 130°C as 0.75 g/cc. The cross-link density was further calculated as previously described [6]. In case of diffusion samples surface crosslink density was measured from a surface section (n=6; 1mm deep and 3x3mm) swollen in xylene at 130°C.

Mechanical Testing: Type V tensile specimens (n=6) according to ASTM D638 were stamped out of 3.2 mm thick sections and then tested at 10 mm/min. Only blended samples were tested. The ultimate tensile strength (UTS) was reported, and true elongation at break (EAB) was determined using a laser extensometer.

Wear Testing: Pin-on-disc (POD) wear testing was performed on cylindrical pins (n=3) as previously described at 2 Hz [7] for 1.2 million-cycles (MC). Wear was measured gravimetrically every ~0.16 MC and the wear rate was measured by the weight change from 0.5 to 1.2 MC.

Oxidative Aging: Accelerated aging was conducted on 1 cm cubes (n=3) in a pressure vessel at 5 atm oxygen at 70°C for 14 days. Thin sections (150μm) were microtomed from an inner surface of the cubes, and analyzed using Fourier Transform Infrared Spectroscopy (FTIR) at an interval of 100μm from the surface along the depth of the sample as an average of 32 scans at a resolution of 4 cm^{-1}. An oxidation index was calculated by normalizing the absorbance at 1700 cm^{-1} (1680-1780cm^{-1}) against that at 1370 cm^{-1} (1330-1390cm^{-1}).

3. Results

3.1. Blending of Peroxides

Extensive crosslinking was achieved by adding 1 wt% P130 to both virgin and 0.1 wt% vitamin E-blended UHMWPE (Table 1). But, the cross-link density of the 0.1 wt% blend was less than that of virgin UHMWPE ($p<0.0001$). As the consolidation temperature increased from 170 to 180, 190, 200 and 210oC for 0.1 wt% vitamin E-blended, peroxide cross-linked UHMWPE, the crosslink density decreased slightly from 0.269±0.007 to 0.256±0.003, 0.250±0.004, 0.245±0.003 and 0.242±0.004 moles/m3, respectively. The processing temperature in the range of 170-210oC did not have significant effect on the UTS (Fig 1a) or EAB (Fig 1b).

Table 1. Crosslink density, ultimate tensile strength, elongation at break, and average oxidation index of peroxide-crosslinked virgin and 0.1 wt% vitamin E-blended UHMWPEs with 1wt% P130.

	Virgin UHMWPE	0.1 wt% vitamin E-blended UHMWPE
Cross-link Density, moles/m^3	0.343±0.005	0.301±0.005
Ultimate tensile strength, MPa	32.9±3.8	26.7±3.3
Elongation at Break, %	234±17	251±15
Average Oxidation Index	0.27±0.05	0.04±0.01

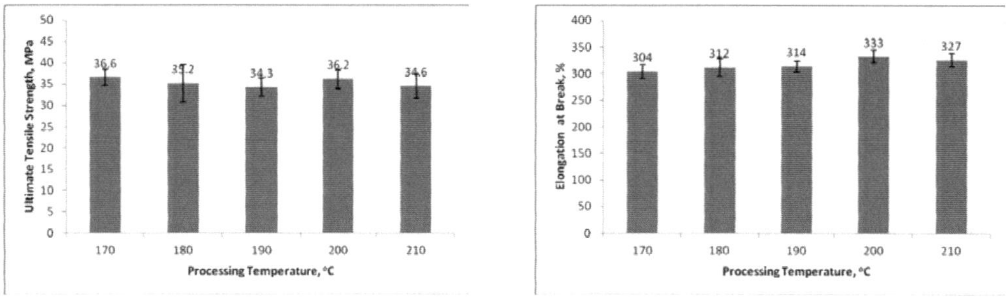

(a) (b)

Figure 1. (a) Ultimate Tensile Strength and (b) elongation at break of 0.5 wt% Peroxide 130, 0.1 wt% vitamin E-blended UHMWPEs at different processing temperatures.

The wear rate of 0.5 wt% P130, 0.1 wt% vitamin E-blended UHMWPE processed at 180 and 210°C was 1.9±0.3 and 1.8±0.2 mg/MC, respectively. The wear rate of control 0.1 wt% vitamin E sample with no peroxide was 13.6±0.1 mg/MC. After accelerated aging, the average oxidation index of 0.1 wt% vitamin E-blended, 1 wt% peroxide crosslinked UHMWPE was 0.04±0.01 compared to 0.27±0.05 for virgin UHMWPE cross-linked with the same amount of peroxide (Fig 2).

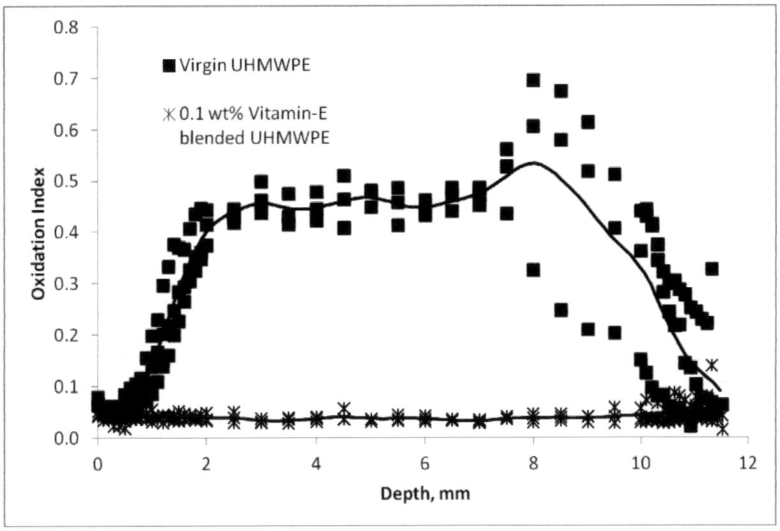

Figure 2. The oxidation index of virgin UHMWPE and 0.1 wt% vitamin E-blended UHMWPE both with 1 wt% Peroxide 130. Three samples of each material were aged in oxygen atmosphere at 70°C under 5 atm pressure for 14 days.

3.2. Diffusion of Peroxides

The crosslink density of DCP-doped samples at 60, 80 and 100°C for 4 hours followed by heating at 130°C for 4 hours was 0.109±0.032, 0.230±0.038, and 0.303±0.020moles/m3, respectively (Fig 3a). Decomposition of the 80°C-doped sample at 140°C increased the cross-link density to 0.280±0.034moles/m3. The crosslink density of P130-doped samples at 80, 100 and 120°C followed by heating at 180°C was 0.123±0.010, 0.243±0.022, and 0.348±0.023moles/m3, respectively (Fig 3b). Increasing the decomposition temperature from 150 to 180°C increased crosslinking (Fig 3b). The weight of DCP-doped cubes (at 60, 80 and 100°C) increased by 0.3, 0.8 and 2.3wt%, respectively (Fig 4a). After heating at 130°C, 0.06, 0.23, 0.59wt% peroxide remained (Fig 4a). The weight of P130-doped cubes (at 80, 100 and 120°C) increased by 0.35, 0.95, and 4.6wt%, respectively (Fig 4b). After heating at 180°C, 0.00, 0.00, and 0.48wt% peroxide remained (Fig 4b).

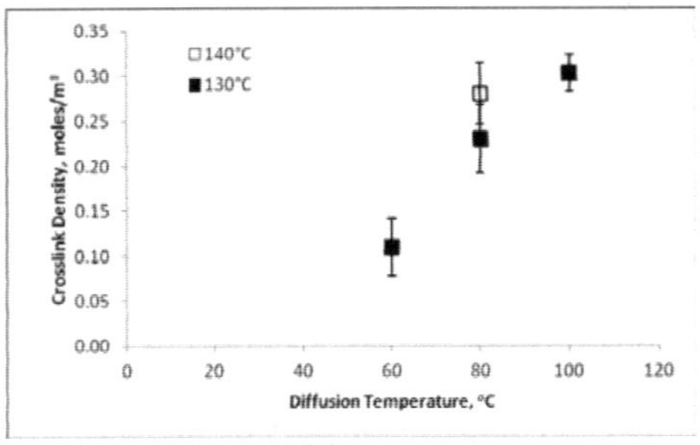

(a)

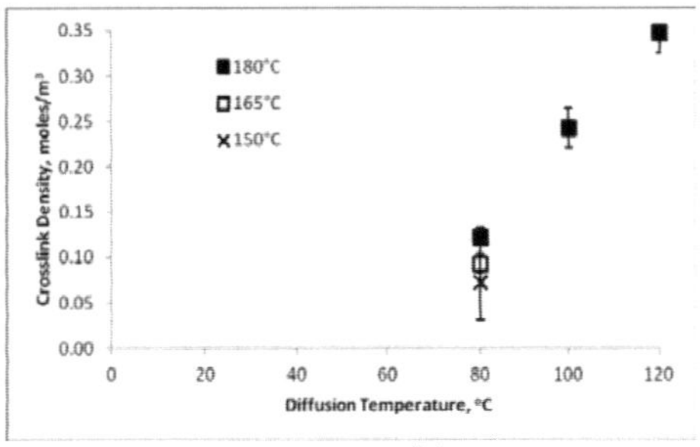

(b)

Figure 3. Crosslink density of peroxide-crosslinked, vitamin E-blended UHMWPEs processed by diffusion of peroxides. (a) Crosslink density as a function of dicumyl peroxide (DCP) doping temperature and subsequent peroxide decomposition temperature; (b) Crosslink density as a function of Peroxide 130 (P130) doping temperature and subsequent peroxide decomposition temperature.

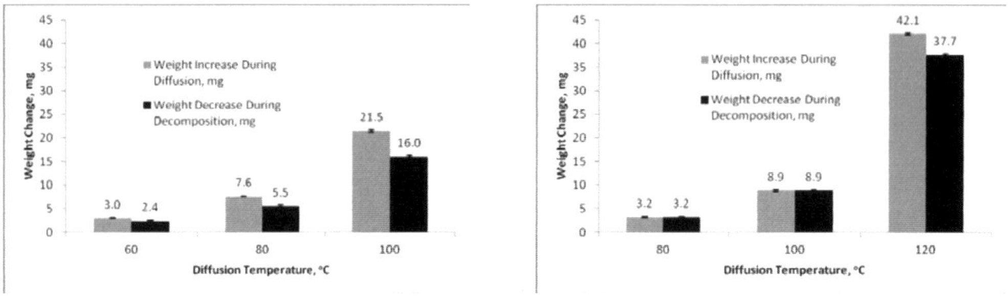

(a) (b)

Figure 4. Weight gain during doping and weight loss during subsequent peroxide decomposition for (a) DCP-doped and (b) P130-doped 0.1 wt% vitamin E-blended UHMWPEs. Doping was performed at the given temperature for 4 hours and decomposition was performed at 130°C for DCP and 180°C for P130.

Wear rate of the 0.1 wt% vitamin E-blended UHMWPE doped at 100°C with DCP and homogenized at 130°C was 1.5±0.9mg/MC. Similar wear rate was observed with the 120°C P130-doped blend that was decomposed at 180°C at 1.6±0.6mg/MC. The wear rate of uncrosslinked 0.1wt% vitamin E-blended UHMWPE was 13.6±0.1mg/MC. After accelerated aging, the average oxidation index (surface 1mm) of DCP cross-linked 0.1 wt% vitamin E/UHMWPE blend (DCP doped at 100°C and decomposed at 130°C) was 0.00±0.062 while that of the virgin sample was 2.81±2.07. The average oxidation index of the P130-doped 0.1 wt% vitamin E/UHMWPE blend (P130-doped at 120°C and decomposed at 180°C) was 2.02±0.05 and that of virgin sample with same doping condition was 3.41±0.11 (Fig 5).

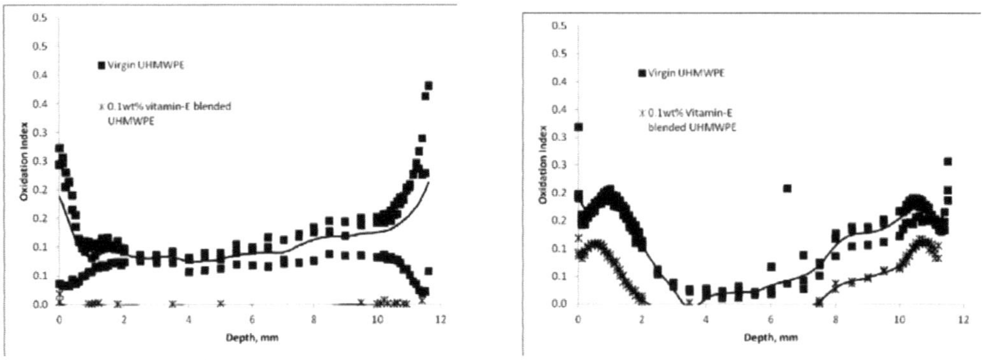

(a) (b)

Figure 5. The oxidation index of virgin UHMWPE and 0.1 wt% vitamin E-blended UHMWPE both crosslinked by diffusion of peroxides. (a) Samples DCP doped at 100°C and decomposed at 130°C and (b) samples P130-doped at 120°C and decomposed at 180°C. Three samples of each material were aged in oxygen atmosphere at 70°C under 5 atm pressure for 14 days.

4. Discussion

Our hypothesis that crosslinking using an organic peroxide can be achieved in the presence of an antioxidant free radical scavenger was confirmed with both blended and diffusion samples. Using blending technique with 1wt% P130, even in the presence of vitamin E, led to high crosslink density. The comparatively lower value of the crosslink density of vitamin E blended UHMWPE as compared to the virgin UHMWPE is presumably due to vitamin E's free radical scavenging ability similar to its effect when using radiation cross-linking [4]. For peroxide cross-linking of polymers, the consolidation has to be optimized to address two main concerns, which are (1) to avoid premature crosslinking that can lead to incomplete consolidation and (2) to ensure stability against oxidation due to peroxide decomposition by-products. The wide range of consolidation temperatures, which resulted in good cross-link density and mechanical properties (Figs 1a and 1b), suggested that P130 was a suitable peroxide for cross-linking UHMWPE with a wide range of processibility. The slight decrease in crosslink density with increasing processing temperature may be due to increased chain scissioning at higher temperatures. Most importantly, the vitamin E-blended, peroxide cross-linked UHMWPE had higher oxidation resistance than the virgin, peroxide cross-linked UHMWPE, which may help enable peroxide cross-linking as a feasible method for fabricating wear and oxidation resistant UHMWPEs for total joint arthroplasty. High levels of cross-linking led to marked improvement in wear rate compared to the uncross-linked UHMWPE, comparable to previously reported wear rates of 100-kGy irradiated and melted virgin UHMWPE (1-2 mg/MC at comparable crosslink density [8]).

Our hypothesis on surface crosslinking that UHMWPE could be cross-linked by diffusing peroxides and elevating the temperature to decompose peroxides tested positive. Surface crosslink density increased with increasing peroxide content, which was increased by the doping temperature (Fig 1a,b). The decomposition temperature (the temperature at which half of the peroxide is decomposed in 1 hour) was 137°C for DCP, which enabled doping and decomposition below the peak melting point of UHMWPE. The decomposition temperature was 152°C for P130, which required decomposition above the melting point of UHMWPE. The crosslinking efficiency of the two peroxides was similar (Fig 1a,b). The achieved crosslink density of peroxide-diffused samples even in the presence of vitamin E was comparable to that of 100-kGy irradiated and melted virgin UHMWPE (~0.26mol/m3), which has excellent wear resistance in vivo [9]. Wear rates of surface peroxide-crosslinked UHMWPEs were lower than their non-crosslinked counterpart, and the wear resistance for the peroxide crosslinked UHMWPEs was dependent on crosslink density. After decomposition, there was a higher amount of DCP remaining compared to P130 (Fig 2a,b), which suggested higher volatility for the by-products of P130. While the presence of these by-products can be a possible source of long-term oxidation of UHMWPE, the negligible oxidation observed after accelerated aging of DCP doped samples supported our hypothesis that vitamin E can improve the oxidation resistance of peroxide cross-linked UHMWPEs. In case of P130, the vitamin E blended sample did show some oxidation on surface but it is appreciably less than the virgin sample, and higher vitamin E content would be required to achieve negligible oxidation. Thus, these results supported our hypothesis that a wear and oxidation resistant UHMWPE could be achieved by cross-linking with both blending and diffusion of peroxides in the presence of vitamin E.

5. Conclusions

Peroxide crosslinking of vitamin E-blended UHMWPE was achieved by both blending and diffusion techniques, resulting in a wear and oxidation resistant crosslinked UHMWPE. The blending method lead to good mechanical properties material and this may present a convenient and economical alternative of crosslinking UHMWPE for total joint implants. Limiting crosslinking to the surface of vitamin E-blended UHMWPE by diffusion and decomposition of organic peroxides is a new technique, providing a feasible alternative of making a wear and oxidation resistant cross-linked UHMWPE for total joint implants.

6. Acknowledgment
The authors acknowledge Harris Orthopaedic Laboratory for providing funds and Fulbright Scholar Program for providing fellowship for Rizwan M. Gul to conduct the research.

7. References
[1] Morshedian J and Hoseinpour P M 2009 *Iranian Polymer Journal* 18(2) 103-28

[2] Gul R M 2008 *J Mater Sci-Mater Med.* 19(6) 2427-35

[3] Bracco P and Oral E 2011 *Clin Orthop Relat Res.* 469(8) 2286-93

[4] Oral E, Godleski Beckos C, Malhi A and Muratoglu O K 2008 *Biomaterials* 29 3557-60

[5] Baker DA, Bellare A and Pruitt L 2003 *J Biomed Mater Res.* 66A 146-54

[6] Oral E, Ghali B, Rowell S, Micheli B and Muratoglu O K 2010 *Biomaterials* 31 7051-60

[7] Bragdon C R, O O'Connor D, Lowenstein J D, Jasty M, Biggs S A and Harris W H 2001 *J Arthroplasty* 16(5) 658-65

[8] Muratoglu O K, Bragdon C R, O'Connor D O, Jasty M, Harris W H, Gul R and McGarry F 1999 *Biomaterials* 20(16) 1463-70

[9] Digas G, Kärrholm J, Thanner J and Herberts P 2007 *Acta Orthop.* 78(6) 746-54

IOP Conf. Series: Materials Science and Engineering **60** (2014) 012016 doi:10.1088/1757-899X/60/1/012016

Multilayer dielectric narrow band mangin mirror

K Ahmed[1], A N Khan, A Rauf and A Gul
Institute of Industrial Control Systems, P. O. Box – 1398, Islamabad

E-mail: drkamalahmed@yahoo.com

Abstract. The design of multilayer stack of dielectric films for narrow band mirror is developed using thin film coating software. The proposed design is materialized by employing thin film coating (PVD) method and reflectance in narrow band spectrum range is achieved. Thickness of high and low refractive index material is taken precisely up to nanometer level. The curved coated substrate is cemented with another K9 matching substrate that forms a Mangin mirror for wavelength 650nm. Narrow band mirrors with reflectivity more than 90% has been produced by properly stacking of 21 layers and advantage of the use of this type of mirror as an interference filter is discussed.

1. Introduction

In mid 60s the use of lasers has been increasing that demand for high quality multilayer dielectric narrow band mirrors [1-3]. Technology evolution for small size diode laser has created further demand of the dielectric narrow band stop filter [4]. New application of narrow band transmission mirrors is now a day that it is a main sophisticated part of the optical sights that is used to reflect the red desired band and transmits the entire visible region. This sight is extensively used in the surveillance, forestry and hunters to target for their desired purposes.

Thin film coating on optical components can be attributed to the following [5]:

- To reduce scattering, absorption and reflection from window/mirrors due to poor polish, strains, wedge and or incorrect mounting angle. Careful cleaning and assembly procedures are necessary.
- Diffraction losses depend upon mirror radius of curvature and geometry. Optical filters or interference filter are devices that selectively transmits or reflects electromagnetic radiation of desired region. These filters are commonly used in optical sights now a day for desired purpose. Dielectric filters are particularly suited for precise scientific work, since their exact colour range can be controlled by the thickness and sequence of coating with different refractive index optical material. They are usually much more expensive and delicate than absorption filter.

In our optical system the minus filter is coated on back surface of the negative Convexo-concave lens that is known as Mangin mirror [6]. This type of mirror was invented in 1876 by French scientist Alphonse Mangin. This mirror reduces spherical aberration affects by refracting the beam twice.

In this report we present the design of narrow band stop filter and PVD method employed to fabricate thin film coating on optical component. The parametric analysis and comparison of design with experimental electromagnetic spectrum is discussed.

2. Multilayer Dielectric Coating

The reflectance R and transmittance T of a stack of non-absorbing, optically homogeneous films deposited upon a transparent substrate of refractive index n_s are a function of the geometrical thickness. A high reflectance can be obtained from a stack of quarter wave dielectric layers of alternate high and low refractive index. The desired wavelength band reflected from all interfaces in the stack of equal phase reach the front surface and combine constructively [7].

The reflection in air or free space is given by:

$$R = \left[\frac{1-\left(\frac{n_H}{n_L}\right)^{2m}\left(\frac{n^2{}_H}{n_S}\right)}{1+\left(\frac{n_H}{n_L}\right)^{2m}\left(\frac{n^2{}_H}{n_S}\right)}\right]^2 \qquad (1)$$

Where n_H and n_L are the indices of the high and low index respectively and m is the number of layers in the stack. The greater number of layers the greater reflectance. Maximum reflectance for a given odd number of layers is always obtained with the high index layers outmost.

If

$$\left(\frac{n_H}{n_L}\right)^{2m}\left(\frac{n^2{}_H}{n_S}\right) > 1 \qquad (2)$$

Then

$$R \cong 1 - 4\left(\frac{n_L}{n_H}\right)^{2m}\frac{n_S}{n^2{}_H} \qquad (3)$$

and

$$T = 1 - R \cong 4\left(\frac{n_L}{n_H}\right)^{2m}\frac{n_S}{n^2{}_H} \qquad (4)$$

The materials used are transparent and the absorption in a multilayer stack is very small.

3. Experimental

Thin film coating technique was employed by using optical design software while n_L: 1.38 and n_H: 1.63 refractive indices were used that correspond to coating materials MgF_2 and Al_2O_3 respectively. The adopted coating scheme was (H 2L)10 0.25H at reference wavelength λ_o: 868nm.

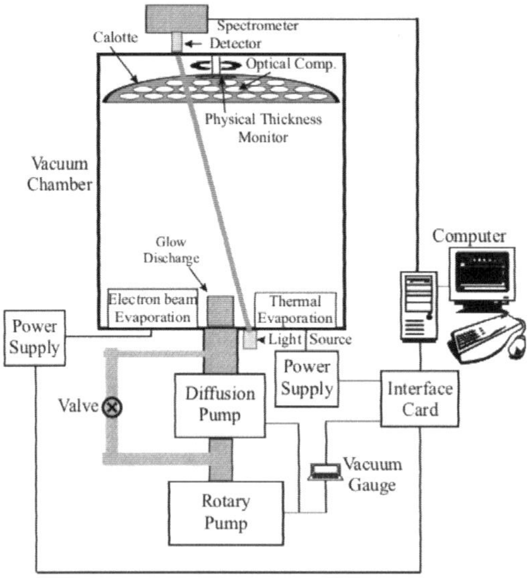

Figure 1. Schematic diagram of thin film coating machine

The high quality polished cleaned optical glass substrate (Diameter: 29mm, thickness: 2.6mm and radius of curvature of coated convex surface: 116.6mm) was used and thin film coating was carried out on coating machine as shown in figure 1. This is a back surface of the optical component that forms a Mangin mirror. Coating was performed at 280°C and pressure of chamber was maintained in the range of 10^{-6}mbar. Before coating, optical components were cleaned in the chamber for half an hour through glow discharge. MgF_2 and Al_2O_3 were evaporated through thermal evaporation and electron beam evaporation techniques respectively. The coated component was cemented in between of other K9 substrates of matching surfaces for protection from environmental effects. These coated components have gone through the environmental test and transmission spectrum of coated sample was recorded on Spectro-Photmeter. The results are shown in figure 2.

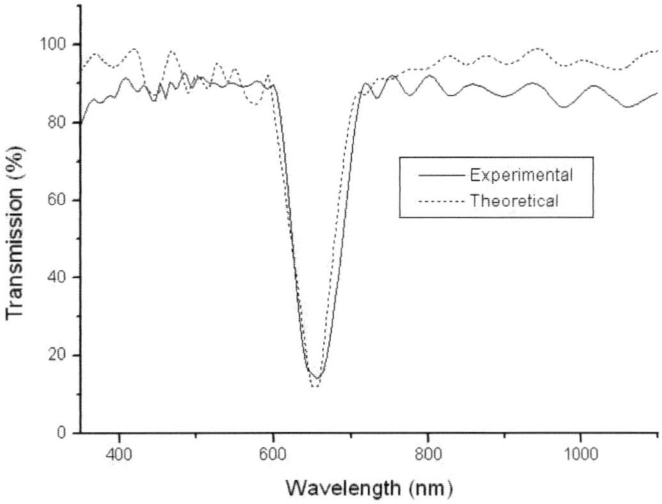

Figure 2. Theoretical and experimental transmission spectrum response

4. Results and Discussions

The colour of coating on polished optical components was reddish yellow and spectral response of theoretical simulation and experimental is shown in figure 2. The theoretical simulation and experimental spectrum response is quite similar. The whole theoretical spectrum shows transmission in both wing of narrow band dip averaged around 95% and experimental spectrum less than 90%. While the narrow band stop dip at wavelength 650 nm of theoretical simulated spectrum is more than the experimental spectrum by factor of 3 to 4 percent. The band width of transmission dip at λ: 650 nm is slightly larger than the simulated value. It is because of chamber environment pressure, temperature and purity of the coating materials. This difference of transmission in theoretical and experimental is because of the absorption factor of dielectric thin film coating material that is quite natural. The band width of narrow band stop filter of calculated and experimental is very similar and shifting of dip is around 5nm. This mirror is a unique and can be used in optical sights as an aiming device in which the surveillance person aiming the target however aiming spot on the target does not appear.

A sight is a common optical instrument non-magnifying reflector that gives the user an aiming point in the form of illuminated red dot [8]. The typical configuration for red dot sight is a tilted spherical Mangin mirror with a red light emitting diode as shown in figure 3. The advantage of Mangin mirror is the red light refracted from the first surface of the mirror that gives narrow reflection angle from reflecting surface and sends the rays parallel towards the eyes of the aiming person. Thus a red light emitting diode (LED) at the focus of collimating optics generates a dot style illuminated reticule.

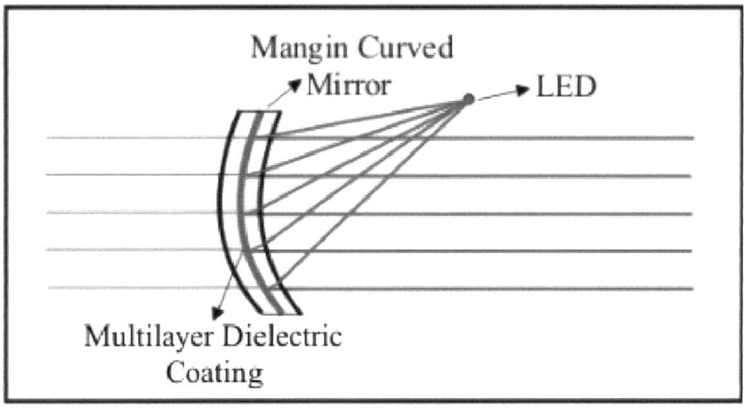

Figure 3. Ray tracing of aiming device.

The sight attached with the weapon remains aligned regardless of eye position. The mirror has partially multilayer dielectric coating to reflect just the desired red wave band and allowing it to pass through most other visible light. The size of the dot generated by LED is controlled by a tiny aperture hole in metal or coated glass placed in front of it.

5. Conclusion

The simulated and experimental spectra are in quite good agreement (figure 2). The multilayer dielectric coating techniques can be implemented in an optical sight to aim the target in the form of red dot in place of old type of reticule. The reticule patterns such as cross hairs or concentric circles can be used but need more complex aberration free Optics however, in aiming dot sights the narrow band Mangin mirror optical system is an aberration free optics. These minus mirrors can produce in large quantity by employing PVD techniques in one go. They are considered to be fast acquisition and easy to use in gun sights for target shooting, hunting and surveillance purposes.

6. Reference

[1] Perry D L 1965 *App. Opt.* 4(8) 987

[2] Baryan H C 1978 US Patent 4106855

[3] Slaughter J M, Medower B S, Watts R N, Tarrio C, Lucatorto T B, Falco C M 1994 Opt. Lett. 19 1786

[4] Young L and Cristal E G 1966 *App. Opt.* 5(1) 77

[5] Lim Y C, Westerwalbesloh T, Aschentrop A, Wehmeyer O, Haindl G, Kleinberg U and Neinzmann U 2001 *App. Phys.* A72 721

[6] Leite Jr. P R, Silva M, Padi E T 2009 *Proc. SPIE* 7338 01

[7] H. A. Macleod 1986 Thin Film optical Filter (2nd edition) Adam Hilger Ltd. Bristol, England

[8] Terebizh V Y 2007 Telesc. Opt. 5 322

Surface engineering glass-metal coatings designed for induction heating of ceramic components

Amir Azam Khan[1] and Jean Claude Labbe[2]
[1]Department of Mechanical and Manufacturing Engineering, Faculty of Engineering, Universiti Malaysia Sarawak, 94300 Kota Samarahan, Sarawak, Malaysia
[2]Institut Universitaire de Technologie (IUT), Université de Limoges, 87060 Limoges, France

E-mail: akamir@feng.unimas.my

Abstract. The term Surface Engineering is of relatively recent origin and use, however, the use of coatings and treatments to render surfaces of materials more suitable for certain application or environment is not new. With the advent of Vacuum Technology, Surface Engineering has gained a whole new impetus, whereby expensive materials with adequate mechanical, chemical and thermal properties are being coated or treated on their surfaces in order to achieve what is called as Surface Engineered materials. The present paper presents an overview of recent achievements in Surface Engineering and gives a detailed view of a specific application where glass-metal composite coatings were deposited on ceramic components in order to render them sensitive to induction heating. Sintered glaze coatings containing silver particles in appropriate concentration can be used for the induction heating of porcelain. Mixtures of glass ceramic powders with silver are used to prepare self-transfer patterns, which are deposited over porcelain. Several configurations of these coatings, which are aesthetic to start with, are employed and heating patterns are recorded. The microstructure of these coatings is discussed in relation to the heating ability by a classical household induction system. The results show that this technique is practical and commercially viable.

1. Introduction

Materials surfaces can be modified at the atomic, micro and macro level in order to obtain, what is called as Surface Engineered materials. This modification can be brought about through Surface Treatment or Surface Deposition process. In each case, surface properties are engineered in a way to obtain specific properties which would otherwise be impossible if only bulk materials are employed. The earliest use of Surface Engineering can be traced back to the Egyptians some 5000 years ago. The use of ink for writing can be considered as the oldest form of surface engineering.

During the last 40 years or so, Surface Engineering has emerged as an avant garde of Materials Engineering and Processing. Novel materials are deposited on substrates through techniques which make use of vacuum or inert environment. This provides solution to engineering problems for severe or demanding applications. The surface modified components are in some cases cost effective and in other cases only available options for specific applications. Examples can be such as Thermal Barrier Coatings (TBCs) on turbine blades [1,2] and zirconia coatings for wear resistance and corrosion protection [3].

An almost unlimited range of processes is available but the choice of a process depends on several parameters such as size of the component, structure of the coating, process cost, deposition rate, environment, pressure, temperature and working conditions etc. For electronic and optical applications, atomic/molecular layer deposition processes such as Physical Vapor Deposition (PVD) and Chemical Vapor Deposition (CVD) for example, are used. Very thin wear resistant coatings can also be obtained using the same techniques. For thick wear resistant coatings for corrosion, wear or thermal protection, a range of thermal spraying techniques can be employed such as High Velocity Oxy Acetylene (HVOF) spraying, Thermal Plasma Spraying (APS, LPPS, TPS), Thermal Arc Spraying and Detonation Gun Spraying. Chemical and/or Electrochemical routes can be employed through Electrodeposition, Electroless Deposition, Paints, Galvanizing, etc.

The present work discusses the application of Surface Engineering in the domain of decorative coatings. These coatings have been used in the form of glaze on porcelain for several hundred years. Glazed glass panels can be found on the windows of ancient buildings in Europe and Asia, some dating back to more than 1000 years. A glaze is a glass coating applied in the form of a glass powder. A heating process converts this fine powder attached to a surface into a molten glass film of a given color. Upon solidification this film converts to a reflecting colored glass which provides several advantages in addition to give an aesthetic value. Glazed surfaces don't absorb water hence can be cleaned easily. Glazed film is composed of a mineral glass which is relatively inert to the environment, hence the life of the substrate is increased. The glazed coatings can also be used for the heating purpose. The Induction heating of food products, for cooking or re-heating purpose, has become a useful and widely employed technique in the recent years. The induction heating is done using a medium frequency (20 to 45 kHz) spiral shaped induction coil, continuously cooled through forced circulation of air. In such type of system, the vessel containing the food is directly heated through induction currents and therefore, should be an electric conductor with appropriate electrical properties and dimensions. It should be placed at some appropriate distance from the induction plate in order to obtain efficient heating through induced currents.

The idea of depositing conducting films on the surface of ceramic substrates (e.g. porcelain utensils), in order to allow their possible use through induction heating, has recently developed a lot of interest and has also resulted in the demand of a number of patents[4,5,6,7,8]. Initially this type of films was developed to help in the use of ceramic utensils in microwave ovens. One such attempt was made to deposit silver coatings for the purpose of induction heating. Many problems such as control of film structure, form and thickness remain to be resolved in order to control heating rate, to obtain required temperatures and to eliminate the risk of fracture in porcelain through rapid heating. Some work has been conducted and published on mixtures of glaze with silver particles, sintered together to produce silver containing glazes [9 , 10]. Further study on combinations of continuous and discontinuous glazed layers of different thickness is presented in the following work. The objective is to obtain as much as possible, a uniform temperature distribution on the surface of the plate, making it safe for use under induction heating.

2. Experimental Results and Discussion

A mixture of glass powder and silver powder is used to produce these films. They are mixed together in an organic solvent to form a self-adhesive, flexible, transfer layer deposited over a backing sheet made up of a release layer and a transfer layer. The glass powder called as flux, is mainly composed of oxides of Pb, Si, Sn and Na, with small amounts of CaO and K_2O. The presence of a flux is necessary for such an application due to aesthetic, hygienic and technical reasons as it transforms to a glaze after sintering. Figure 1 presents the particle size distribution of Ag powder (CERDEC 512 origin Matthey Beyrand, France), measured at the laboratory using X-ray sedigraphy. It can be seen on this figure that majority of the powder particles have sizes ranging between 0.1 to 1 μm. This powder is mixed with a flux (CERDEC 192001 origin Matthey Beyrand, France) to form a paste, in a ratio of 73 weight % Ag – 27 weight % flux. this paste is added in an organic solvent (CERDEC 80612 origin Matthey Beyrand, France) to produce an ink, which is used to produce uniform thickness transfer layers. The overall percentage of Ag, including the flux and solvent comes out to be 57.1 %. The film, after sintering, shows a glassy glazed surface on cooling, with silver particles embedded as a contiguous network inside the glaze. This glaze also helps to accomplish a good contact and adhesion between the silver agglomerates and the porcelain substrate. It protects these particles against wear and therefore helps to increase the life of the deposited film.

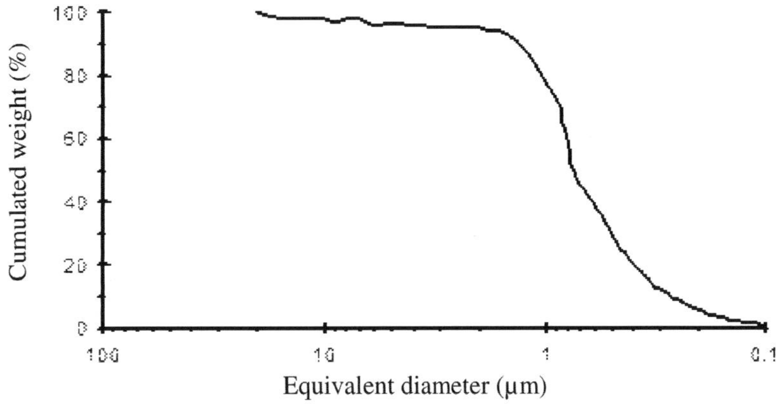

Figure 1. Particle size distribution of the Ag powder (CERDEC 512), measured using X-ray sedigraphy technique (Sedigraph 5000 D, Micromeritics).

The heating plate is constituted of a magnetic induction coil working at medium frequency between 20 and 45 kHz. The transferred film after sintering at around 820 °C on the plate surface gives a thickness of 27 +/- 2 μm but due to segregation of Ag particles the zone containing Ag agglomerates is about 15 +/- 3 μm thick. Figure 2 presents the microstructure of this film. On the same figure it can also be seen that the glassy phase produced by the melting of the flux contains spherical pores. These pores are a typical feature of a glaze on porcelain. Major effort is required to keep these pores as little as possible in number and size to give a good thermal conduction between the deposited layer and the porcelain substrate. This is mainly due to the densification effect and due to the removal of organic components incorporated in the film during processing. The presence of bubble like pores shows that the glaze becomes partially liquid at the sintering temperature. The silver particles form large size agglomerates and migrate to one side of the film, probably due to differences in the density. It should be noted that on the micrograph it appears that the Ag particles/agglomerates are not in contact at some points, but as the film is in three dimensions whereas the micrograph gives a 2 dimensional view, this impression can be misleading.

These initially deposited films were tested on the induction plate for their heating performance. The porcelain plate was of 140 mm diameter and 5 mm thickness. It was placed on the induction plate in such a way that the film was facing upwards. This gives a coupling distance between the film and induction coil of the same order as to a distance between heating coil and film as used under real conditions (about 11 mm). The plate is heated and temperatures at the surface are measured using thermocouples attached at equal distances. The Figure 3 gives the result of temperature variation as a function of time at different distances from the center of the plate.

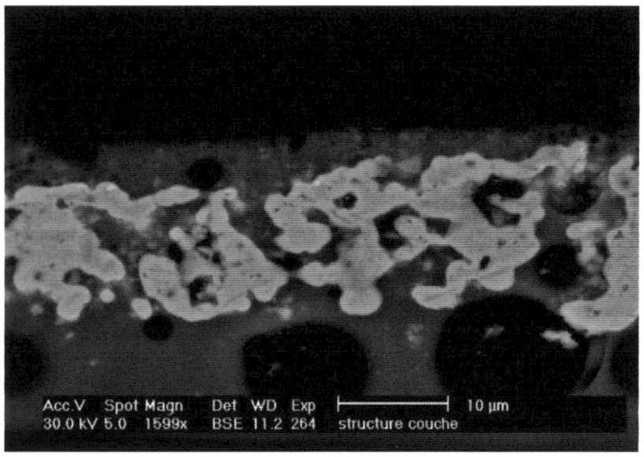

Figure 2 Microstructure of a silver film with an initial thickness of 27 µm, sintered at 820 °C

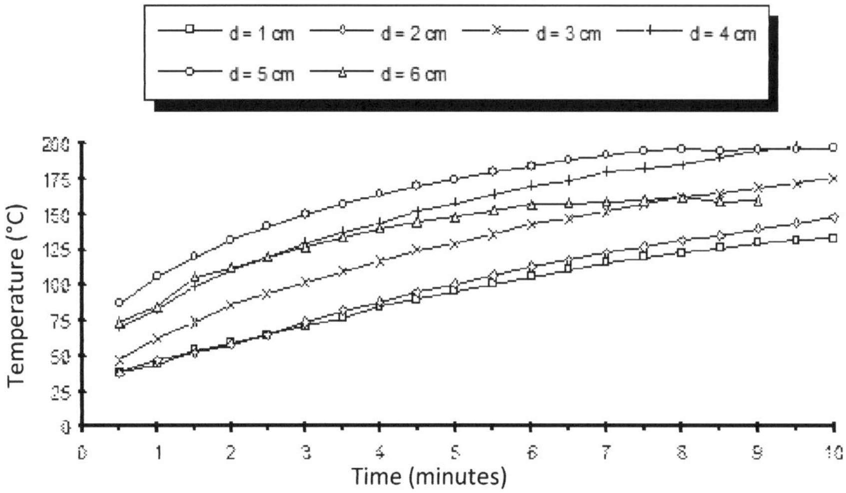

Figure 3. Variation of temperature of a 27 µm thick silver film as a function of time, at different distances from the center of a porcelain disc, heated over an induction plate at half power.

As it can be seen on Figure 3, for a film of 27 +/- 2 µm thickness heated at half power, the temperature stabilizes at about 200 °C at a distance of about 5 cm from the center of the plate. The temperature remains lower at all other points on the surface of the plate. Two important observations are worth noting. The highest temperature is not located at the center of the plate, nor at its edges, but at a distance of 5 cm from the center. Secondly, there is an auto-regulation of temperature, which takes place due to coupling-decoupling of the induction coil with the silver film. It is expected that this would change as the thickness of the film or the induction frequency would change.

2.1 Structure of silver coatings at different thickness and sintering temperature

The simple continuous film with a thickness of 27 +/- 2 µm creates a heating pattern which can give a damaging thermal shock in case the plate is heated without any food in it. It was found that sudden heating of the film can create cracks. Several options were negotiated to reduce this thermal shock. The shape and thickness of the film were manipulated in order to reduce the temperature gradient over the surface of the porcelain. In all cases porcelain substrates with more than 95% relative density were used. The layers were sintered at different temperatures with a heating rate of 5 °C.min-1 and the samples were kept at the sintering temperature for 10 minutes. Films of different thickness and configurations were sintered in order to find a suitable sintering cycle. The Figure 4 shows the microstructure of a 54 +/- 2 μm film. The sintering was carried out at 760 °C. This temperature is barely enough to sinter the glass powder. This gives an advantage that there are no large size pores and the silver particles are not agglomerated as they are in the sample shown in Figure 2. The sample in Figure 2, sintered at 820 °C, contains large size pores due to degassing of solvent at high temperature and silver particles migrate and agglomerate under gravity and under long range surface forces. On first sight this microstructure seems acceptable, but this coating can degrade over time very quickly as glaze is not entirely sintered and attached to the porcelain surface. Moreover, further sintering and shrinkage of the film is possible while heating over induction plate.

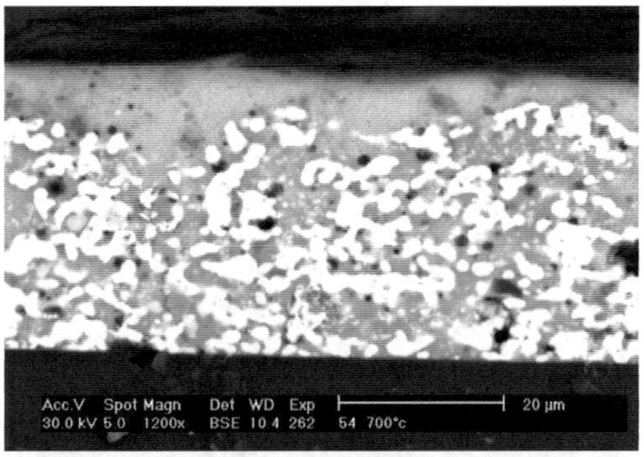

Figure 4. Structure of a 54 ± 2 μm silver film, sintered at 760 °C. The fine structure of silver grains and porosity can be observed inside the film

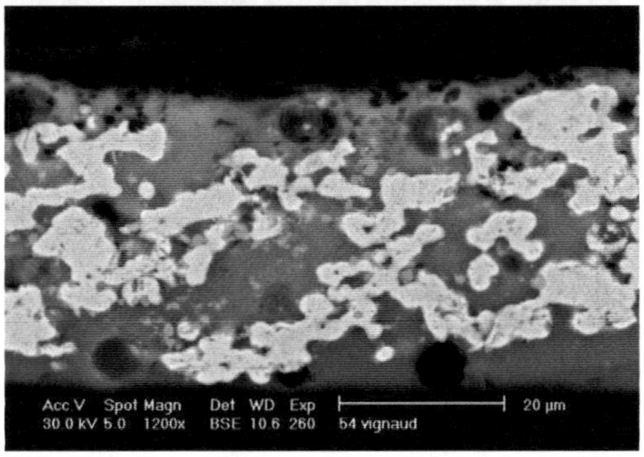

Figure 5. SEM micrograph showing the structure of a 54 ± 2 μm silver film, sintered at 780 °C.

Figure 5 presents the microstructure same film sintered at 780 °C. The difference in microstructure is visible in this micrograph. The silver particles are agglomerated relative to the previous one and there is presence of pores, although these are not of very large size. From point of view of optimization in sintering cycle, this cycle seems to fulfill and optimize the properties of the film.

2.2 Design of silver coatings

Once the sintering and microstructure of the film is optimized, the second stage is to design the film in a manner so as to distribute the temperature as evenly possible. This would create a condition for lowering the thermal shock. Several configurations were investigated, three of them are shown in Figures 6, 7 and 8.

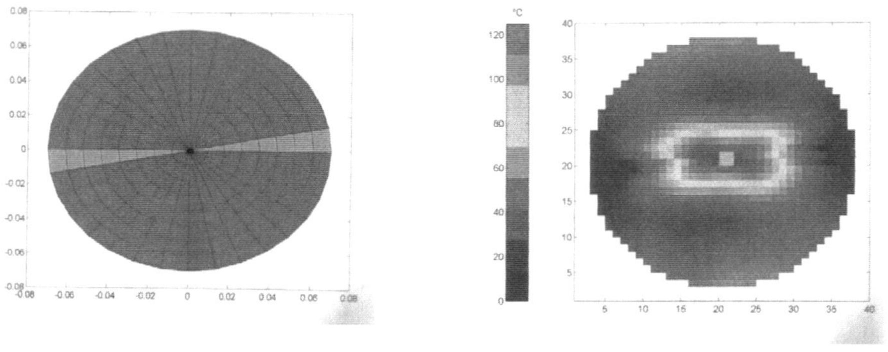

(a) (b)

Figure 6. (a) presents an optional design of glazed layer including silver particles with a cut, in order to change the distribution of induced current and give a different heat distribution for a 27 µm thick layer. (b) shows the temperature distribution on the surface of the porcelain plate for this configuration

The Figure 6 (a) presents a modified design in which a discontinuous layer in two halves is deposited. The layer thickness is 27 ± 2 µm at the time of deposition. The Figure 6 (b) shows the temperature distribution on the surface of porcelain substrate under similar conditions as mentioned under section 1. The areas close to the centre of the substrate now reach higher temperature which is better than the previous continuous layer, but the maximum temperature has dropped from approx. 200 °C to 120 °C. Other configurations are also tried as shown in Figures 7 and 8. By increasing the number of discontinuities within the film, the temperature distribution over the surface is improved but the maximum temperature drops for each discontinuity. The temperature reaches mere 35 °C in the case of a film with 4 discontinuities.

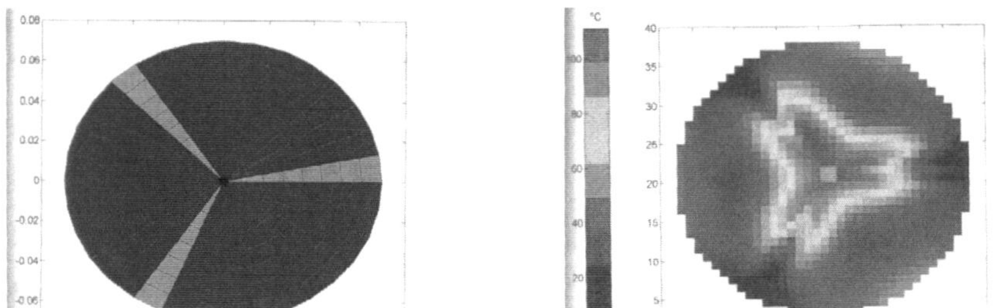

(a) (b)

Figure 7. (a) presents the design of glazed layer with three cuts, within a 27 µm thick layer. (b) shows the temperature distribution on the surface of the porcelain plate

The objective of this exercise is to introduce within a conducting layer, non-conducting zones so that the current flow trajectory can be changed. As it can be seen, this helps to change the heating pattern towards a more regular one, but greatly reduces the heating efficiency. As a general statement we can say that any tentative to bring the current more towards the centre of the substrate would reduce the current intensity, resulting in lower efficiency. The solution is more complex than initially envisaged. The solution lies probably in combining the effect of discontinuities of the film, to improve temperature distribution and increasing the thickness of the film. As shown in previous section (Figures 4 and 5) films of greater thickness can be transferred to the porcelain substrate and sintered successfully. This thing has to be taken with care, as increasing thickness of silver coating increases the cost as well as it is not so good from the aesthetic purpose. Some initial results were obtained for a different design of the film.

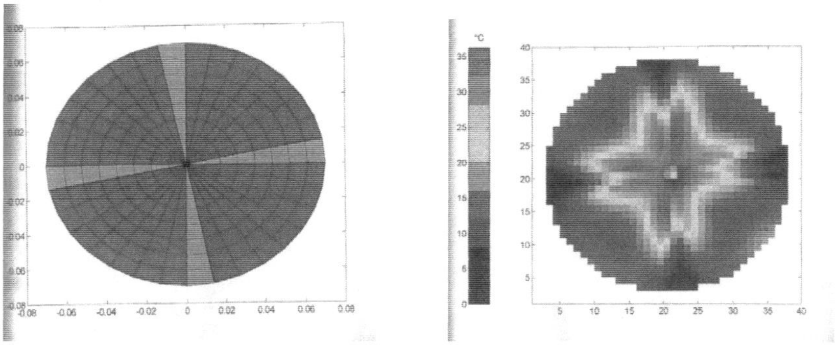

Figure 8. (a) Presents the design of glazed layer with four cuts, within a 27 μm thick layer, (b) shows the temperature distribution on the surface of the porcelain plate.

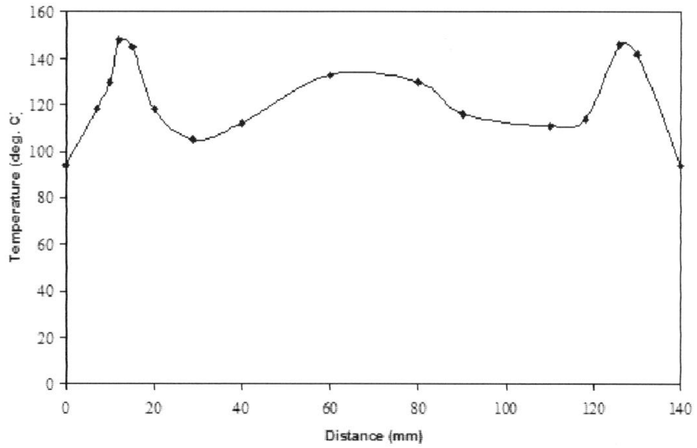

Figure 9. Temperature distribution profile measured along the diameter of a porcelain plate, measured after 6 minutes of heating. Inner coating diameter = 105 mm, thickness 54 ± 2 μm, outer crown between 106 to 140 mm, thickness 36 ± 2 μm.

2.3 Double layer Configuration

A double layer pattern was investigated. It composed of an inner film of 54 ± 2 μm thickness and a diameter of 105 mm and an outer film of inner diameter 106 mm and outer diameter 140 mm. The outer layer thicknesses tried in this case is 36 ± 2 μm. In this design the inner layer ensures the central part of the substrate to remain hot. As the inner film thickness is 54 ± 2 μm, this is also expected to produce sufficient heat for the central part. The outer layer which is less thick (36 ± 2 μm) heats the outside boundary of the substrate. Excellent results are obtained with this configuration, as shown in Figure 9. These results are obtained using an infrared camera (Hughes Inc) after six minutes of heating at half power. This figure shows the temperature distribution over a double layer combination of 54 ± 2 μm thickness inner film and a 36 ± 2 μm thickness outer film. It shows that the temperature is well distributed over the surface with a maximum of 150 °C situated at about 15 mm from the edge. Continued heating checks have ensured that no plate gets cracked, no matter what type of heating is conducted.

3. Conclusion

This work was conducted with an objective to study the induction heating of porcelain substrates and to improve the existing configuration of a single glazed silver layer containing no pattern. The coating microstructure and sintering conditions were also optimized and a suitable sintering cycle was proposed for which the coating exhibits minimum defects and a uniform distribution of silver particles within the glaze. The initial design was based on a continuous silver/glaze film deposited over the substrate. The heating pattern for this continuous layer gave uneven heating of the periphery of the film, with very little heat at the centre of the plate, a characteristic of induction currents. This produced a thermal shock which was damaging for the plate if it was heated empty. Initially an effort was done to use perforated or discontinuous layers in order to achieve better heat distribution but this also resulted in the lowering of the heating efficiency.

The maximum temperature obtained was significantly reduced depending on the pattern used. To further solve the problem two distinct layers of different thickness were employed. The coating was split in two parts, an inner coating of 105 mm diameter, with a slot in between, and an outer crown of 106 mm inner diameter and 140 mm outer diameter. The inner coating thickness was increased to 54 ± 2 μm as no coupling effect was possible for a 27 ± 2 μm film. The outer coating had a thickness of 36 ± 2 μm. By using two films of different thickness it was made possible to regulate the heating rate and to make the temperature distribution more uniform. The results hence obtained have shown that the heat remains relatively uniform. No sample showed any damage despite several heating cycles, empty or full of food.

4. References

[1] D. Stover, C. Funke, 'Directions of the Thermal Barrier Coatings in Energy Applications' *Journal of Materials Processing Technology*, 92-93(1999), pp 195-202

[2] D. Bhattacharyya, M. Targa, J. R. Nicholls, 'Electron Beam Physical Vapor Depositited Zirconia based Ceramics' *Surface Engineering*, Vol 28, No 2 (2012), pp 122-128

[3] A. A. Khan, M Asim, M F. Zafar, 'Development of Calcia Stabilized Zirconia (CSZ) Thermal Barrier Coatings (TBCs) : Characterization and Improvement of CSZ Powder for optimization in coating properties' *Journal of Sc. & Tech. in the Tropics* (JoSTT) Vol 7, (2011), pp 57-64

[4] M. Schneider, L.A. Blanco, 'Ustensile pour cuisson par microondes et couche de transfert pour sa fabrication.' *Commercial Decal. Inc.* French patent demand registered under N° FR 89 14519 to Institut National de la Propriété Industrielle, (6 Nov. 1989).

[5] O. Kunz 'Feeding vessel - Comprising calcined piece made of porcelain or ceramic with glazed layer' *Winterling Porzellan AG.* Deutsche patent demand N° DE 44 144 39 to BundesRepublic Deutschland Deutsches Patentamt C1 (26 April 94).

[6] A. Imano, 'Récipient pour un appareil de cuisson électro-magnétique' *Sony SA, Tokyo,* Japan patent demand published under N° H3-4479 to Institut National Japonais de la Propriété Industrielle (10 Jan. 1991).

[7] Sou-Kuein Chiu 'Film conducteur de la chaleur pour absorber les ondes électromagnétiques et l'énergie des micro-ondes' *Sou-Kuein Chiu, TW.* French patent demand N° FR 2 671 709 to Institut National de la Propriété Industrielle, (17 Jan. 1991).

[8] H. Kawanishi, S. Yano, T. Shibata 'Récipient non métallique destiné à la fabrication d'un appareil de cuisine' *Matsushita S.A. and Tatero Chemical Industries,* Japan patent demand published undr N° H1-157088 to Institut National Japonais de la Propriété Industrielle (20 June 1989).

[9] A.A. Khan, J.C. Labbe, 'Effect of heating cycle on the structure of Ag films deposited over porcelain substrates' *J. Mat. Engg. & Perfor.* (ASM Int.) 7, [6], (1998), 757-60

[10] A.A. Khan, J. C. Labbe, A. Teillet, J. Fouladgar, G. Develey, « Properties and Microstructure of Silver Coatings used for Induction Heating of Porcelain Substrates » Proceedings of 101st Annual meeting of American Ceramic Society, 25-29 April 1999, Indianapolis, USA. Innovative Processing and Synthesis of Ceramics, Glasses and Composites III, Ceramic Transactions 108, pp 493-501, Edited By J P Singh, N P Bansal, K Niihara

IOP Conf. Series: Materials Science and Engineering **60** (2014) 012018 doi:10.1088/1757-899X/60/1/012018

TiAu based shape memory alloys for high temperature applications

Abdul Wadood[1], Hideki Hosoda[2] and Yoko Yamabe-Mitarai[1]
[1]High Temperature Materials Unit, National Institute for Materials Science, 1-2-1 Sengen,Tsukuba, Ibaraki 305-0047, Japan
[2]Precision and Intelligence Laboratory, Tokyo Institute of Technology, 4259-R2-27, Midori-Ku, Yokohama 226-8503, Japan

Email: wadood.abdul@nims.go.jp

Abstract. TiAu (equiatomic) exhibits phase transformaion from B2 (ordered bcc) to thermo-elastic orthorhombic B19 martensite at about 875K and thus TiAu is categorized as high temperature shape memory alloy. In this study, recent research and developments related to TiAu based high temperature shape memory alloys will be discussed in the Introduction part. Then some results of our research group related to strengthening of TiAu based high temperature shape memory alloys will be presented. Potential of TiAu based shape memory alloys for high temperature shape memory materials applications will also be discussed.

1. Introduction

Ti-50Au (atomic %) has B19 orthorhombic martensite at room temperature that transforms to detwinned martensite upon mechanical loading and B2 (ordered bcc) phase at around 875K upon heating. Upon cooling, B2 phase reverts back to B19 orthorhombic martensite phase at around 830K. Transformation from B19 $\leftrightarrow$ B2 is thermo-elastic in nature and Ti-50Au is categorized as high temperature shape memory alloy [1]. Melting point and phase transformation temperatures of Ti-50Au is ~100K higher than Ti-50Pd [2, 3]. Ti-50Ni also known as nitinol is famous shape memory alloy. However as phase transformation temperatures of Ti-50Ni are below 100K, so binary Ti-50Ni has room temperature engineering and biomedical shape memory and super-elastic applications [4].

Phase transforamtion temperatues of Ti-52.6Au, Ti-50Au, Ti-47.4Au, Ti-45Au and Ti-42.5Au (e.g. all compositions in atomic%) are reported by Donkersloot H C et al. in 1970 [5]. Donkersloot H C et al. showed that the phase transformation temperatures were decreased by decreasing Au content from 52.6at.% Au to 42.5% Au. X ray diffraction data at room temperatrue (RT) showed only B19 martensite for Ti-52.6Au, Ti-50Au, Ti-47.4Au and Ti-45Au. However B19+Ti_3Au two phase structure was identified for Ti-42.5Au [5]. Decrease in the austenite finish temperature (A_f) and the martensite finish temperature (M_f) with the decrease in Au content in TiAu from Ti-52.6Au to Ti-42.5Au is given in Figure 1 [5]. Decrease in the austenite transformation temperatures of TiAu per atomic percent decrease in Au content (e.g. A_f and A_s=14K/Au) was found higher than decrease in the martensite transformation temperatures (e.g. M_s, M_f=8K/Au) from Ti-52.6Au to Ti-45Au.

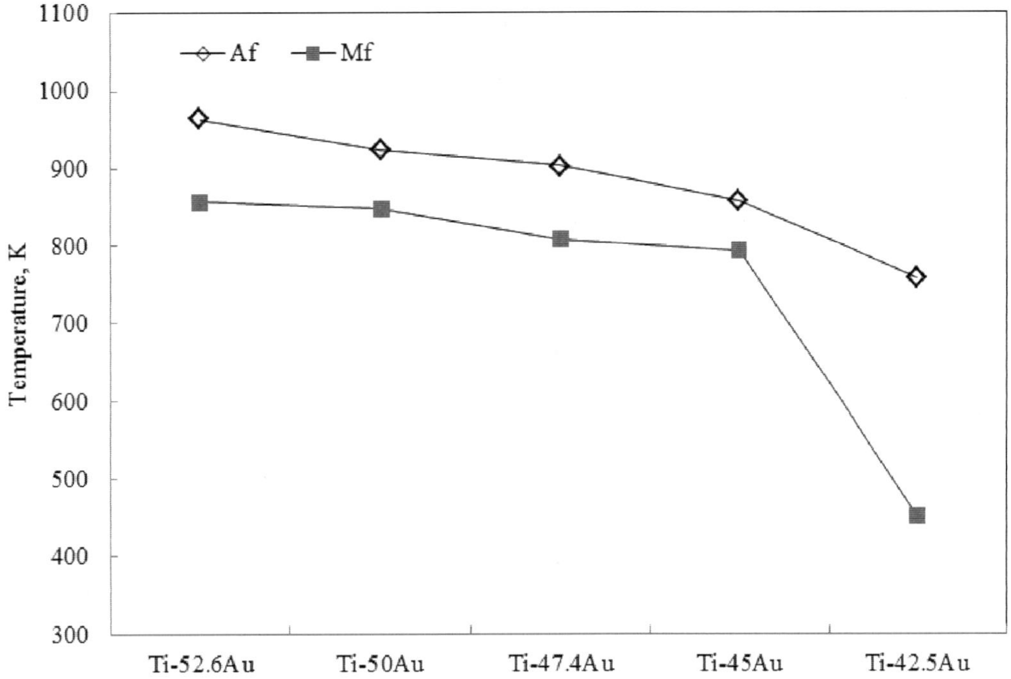

Figure 1. Decrease in A_f and M_f with Au contentof TiAu. Please see Ref. [5] for details.

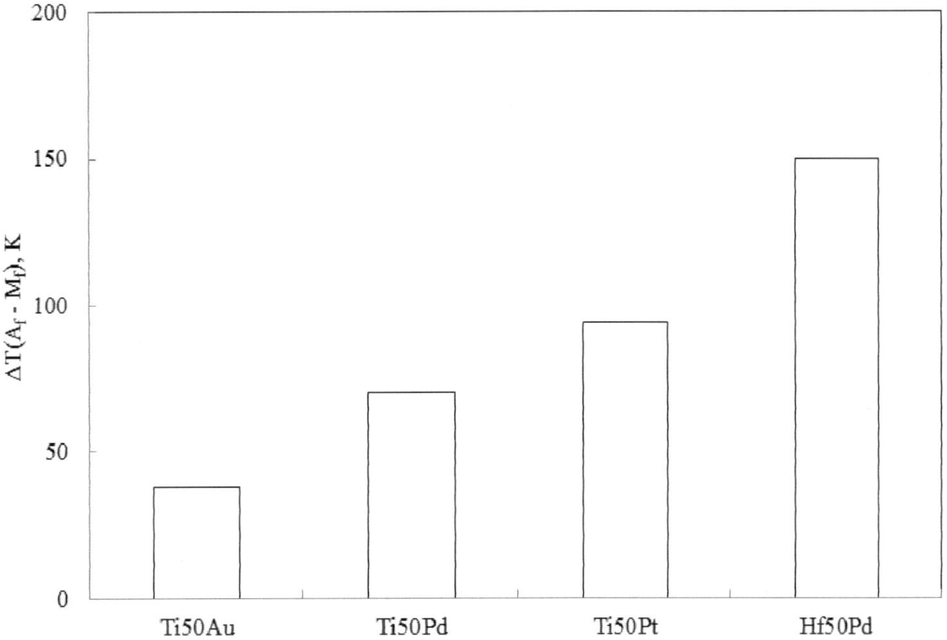

Figure 2. Transformation hysteresis of Ti-50Au [9], Ti-50Pd [5], Ti-50Pt [11] and Hf-50Pd [10].

However the decrease in the transformatoin temperatures from Ti-45Au to Ti-42.5Au was found at much higher rate (A_f=40K/Au, As=58K/Au, M_s=114K/Au, M_f=136K/Au) than decrease in transformation temperatures from Ti-52.6Au to Ti-45Au [5]. As from *X* ray results, B19+Ti_3Au two phase structure was analyzed for Ti-42.5Au compared to Ti-45Au that has only B19 martensite. It was analyzed from Figure 1 that there was abrupt decrease in A_f and M_f for Ti-42.5Au compared to Ti-45Au.This abrupt decrease in A_f and M_f is related to precipitation of Ti_3Au for Ti-42.5Au compared to Ti-45Au that has only B19 martensitic phase at room temperature. This showed that the precipitation of Ti_3Au phase in B19 martensite also affects the phase transformation temperatures [5]. Recently Hosoda-Inamura research group [6] showed that from geometrical point of view, Ti-53Au can exhibit 11% transformation strain comparable with Ti-50Ni shape memory alloy. Such high theoretical transformation strain and high transformation temperature (~875K) of Ti-50Au has attracted our research group interest to further explore this system for high temperature shape memroy materials applications.

Although Donkersloot H C et al. showed the potential of Ti-50Au for high temprature appplications in 1970 [5], however even uptill now, there is not much literature availabe related to Ti-50Au. One reason may be the high cost of Au. However for high temperature shape memory materials applications, functionality of materials is more important than cost. Compare toTi-50Pd and Ti-50Pt high temperature shape memory alloys, cost of Au is comparable with that of Pd and Pt metals. One advantage of Ti-50Au is small temperature hysteresis compared to Ti-50Pd and Ti-50Pt as shown in Figure 2. Small temperature hysteresis is required for shape memory alloys as large temperature hysteresis results plastic deformation [1]. Hosoda-Inamura research group [7] also showed the effects of ternary additions on martensitic transformation temperatures of Ti-50Au and co-related the decrease in M_s with e/a ratio, valence electron number (N_e), Mendeleeve number (N_M) and lattice parameters of martensite phase. Wu S K and Wayman C M [8] showed the martensitic transformation in Ti-50Au and Ti-40Au-10Ni (atomic%) alloys. Both Ti-50Au and Ti-40Au-10Ni has orthorhomic B19 martensite (2H) and both compositions exhibit one way shape memory effect, not two way shape memory effect.

Recently Declairieux C et al. discussed the potential of Ti-50Au as high temperature shape memory alloy [9] as Ti-50Au exhibit stable phase transformation temperatures during thermal cycling, showing thermal stability. Authors of Ref. [9] also showed that Ti-50Au exhibits partial shape memory effect (e.g. 78%) and has potential as high temperature shape memory alloy due to small temperature hysteresis and stability of transformation temperatures during thermal cycles. For comparison, transformation temperature hysteresis of Ti-50Au [9], Hf-50Pd [10], Ti-50Pd [5] and Ti-50Pt [11] is given in Figure 2. It is clear from Figure 2 that Ti-50Au exhibits small temperature hysteresis compared to Ti-50Pd, Ti-50Pt and Hf-50Pd intermetallic alloys. Small temperature hysteresis is beneficial for good shape memory properties [1]. Vermaut P et al. [10] showed that although Ti-50Au exhibits ~950MPa strength at RT and ~600MPa at 50K below *Ms* in B19 martensite phase. However strength of Ti-50Au drastically decreases when it transforms from B19 martensite to B2 phase (~120MPa). So this is an important research challenge to improve the high temperature strength of Ti-50Au in B2 phase region. It is expected that by increasing the strength of Ti-50Au in B19 and B2 phase, its shape memory properties will also improve due to increase in critical stress for slip deformation. So we tried to improve the high temperature strength of Ti-50Au by partial substitution of Ti whose atomic radii is 0.147nm with Zr whose atomic radii is 0.158nm [12] and size misfit parameter [19] between Ti and Zr is 0.075.

In our previours studies, we found that Zr is an effective element to improve the high temperature strength and shape memroy properties of Ti-50Pt [13-14] and Ti-50Pd [15-16] alloys. Both Ti and Zr belongs to same group IV of periodic table, same crystal structure (hcp; α) at room temperature that transforms to β (bcc) phase at 1155K. Melting point of Zr (2130K) is 187K higher than Ti (1943K). Ti-Zr phase diagram shows that both Ti and Zr have complete solubility in solid (α, β) as well as in liquid phase [17]. Au-50Zr intemetallic compound also exists in Au-Zr phase diagram [18]. So due to these properties, we were expecting that partial substitution of Ti with Zr can improve the room temperature to high temperature strength of Ti-50Au in B19 as well as in B2 phase region.

2. Experimental Procedures

Ti-50Au-10Zr (atomic %) alloy ingot was prepared by Ar arc melting method using high purity (99.9%) metals. Then alloy ingot was sealed in a silica tube with small amount of Ar gas and solution-treated at 1273K for 10.8ks followed by ice-water quenching. Sputter etching (dry etching) was performed by glow discharge optical emission spectrometer (GD-OES) using 200Pa pressure, 20W power for 30s sputtering time in Ar and N_2 gas mixture. SEM analysis was conducted of sputter etched Ti-50Au-10Zr samples at an accelerating voltage of 20KV using JEOL JSM-7001F field emission SEM. Electron probe micro analysis (EPMA) was also conducted for composition analysis. For shape memory effect measurement, rectangular samples of 3mm x 3mm x 6mm dimensions were cut from the solution-treated alloy ingot using wheel cutting machine. These samples were then compressed for 5% applied strain at test temperature of 50K below the martensite finish (M_f) temperature using Shimadzu *AG–X* compression test system. Sample's dimensions were measured before compression test, after 5% applied strain and after heat-treatment above the austenite finish temperature. The recovery ratio due to the shape memory effect (R_{SME}) was calculated using following equation:

$$R_{SME}\ \% = 100*(\varepsilon_s - \varepsilon_r)/\varepsilon_s) \qquad (1)$$

Where ε_s is the strain after deformation and ε_r is the residual strain after heat-treatment above the austenite finish temperature followed by air cooling. For high temperature mechanical testing, cyclic loading–unloading compression tests were conducted for 20% applied strain at room temperature, test temperature of 50K below the martensite finish temperature and 50K above the austenite finish temperature using Shimadzu *AG–X* compression test system. Thermo-mechanical analyzer (TMA) was used for phase transformation temperatures between room temperature (RT) to 1073K at a heating and cooling rate of 0.17K/s in air environment.

3. Results and Discussion

Room temperature SEM micrograph of solution-treated and sputter etched Ti-50Au-10Zr alloy is presented in Figure 3. It was analyzed from Figure 3 that Ti-50Au-10Zr intermetallic alloy has two phase structure; twinned oriented martensite (matrix) and second phase at grain boundary as well as inside the matrix

Electron probe micro analysis (EPMA) of Ti-50Au-10Zr is given in Table 1. From this EPMA composition analysis, it was analyzed that the composition of matrix was close to theoretical composition and second phase at grain boundary was analyzed as $(Ti,Zr)_3Au$ type precipitates. Some Ti_3Au second phase was also analyzed inside the matrix. It is not clear either this $(Ti,Zr)_3Au$ second phase was formed during liquid solid transformation or it was formed during solid solid transformation. Transformation temperatures of Ti-50Au (M_s=870K, M_f=859K, A_s=888K and A_f=897K) [9] were decreased by partial substitution of Ti with Zr in Ti-50Au-10Zr alloy (M_s=758K, M_f=741K, A_s=768K, A_f=784K). Decrease in transformation temperatures was ~12K/atomic% partial substitution of Ti with Zr in Ti-50Au alloy. This showed that the role of Zr addition in Ti-50Au is to stabilize the B2 phase structure.

Figure 3. SEM micrograph of Ti-50Au-10Zr. Martensite and second phase at grain boundary was analyzed.

Table 1. EPMA composition analysis of matrix and second phase in Ti-50Au-10Zr.

	Matrix (atomic %)		
Ti	**Zr**	**Au**	**Comment**
39.7	10	50.3	$(Ti, Zr)_{50}Au_{50}$
	Black phase at grain boundary (atomic %)		
Ti	**Zr**	**Au**	**Comment**
69	3.3	27.7	$(Ti, Zr)_3Au$

Ti-50Au-10Zr specimens deformed successfully for 20% applied strain at room temperature, 50K below M_f and 50K above A_f in air environment. This showed that Ti-50Au-10Zr has good ductility. Stress for reorientation of martensite and maximum stress for 20% applied strain at room temperature, 50K below M_f and 50K above A_f of Ti-50Au [10] and Ti-50Au-10Zr are presented in Figure 4 and Figure 5 respectively. It was analyzed from Figure 4 that the stress for reorientation of martensite decreased with increasing temperature for Ti-50Au [10] and Ti-50Au-10Zr. Figure 4 also showed that

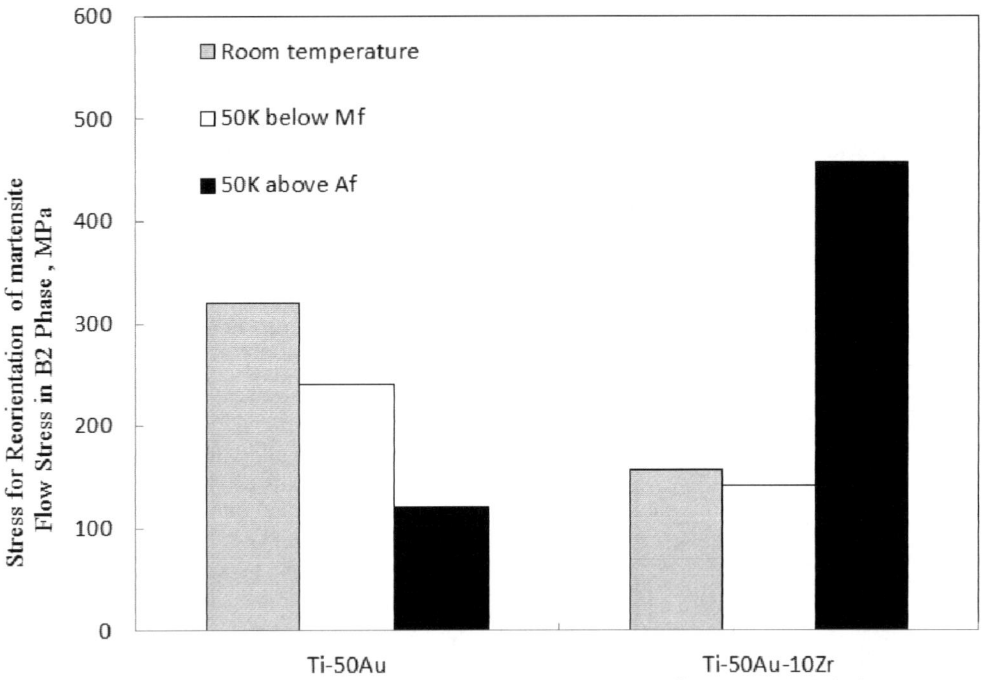

Figure 4. Martensite reoriention stress in B19 and flow stress in B2 phase of Ti-50Au [10] and Ti-50Au-10Zr.

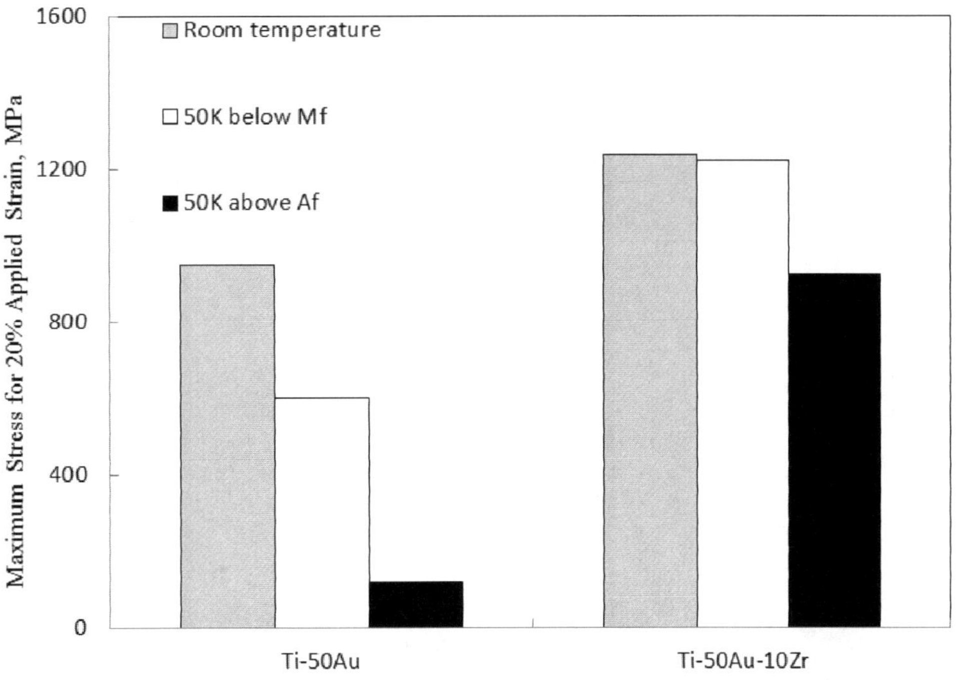

Figure 5. Maximum stress for 20% strain of Ti-50Au [10] and Ti-50Au-10Zr.

Ti-50Au-10Zr exhibited lower stress for reorientation of martensite compared to Ti-50Au [10] at room temperature and at 50K below M_f. This result showed that stress for reorientation of martensite decreased by addition of Zr in Ti-50Au alloy. It was also analyzed from Figure 4 that Ti-50Au exhibits very low flow stress (~100MPa) in B2 phase region, lower than stress required for martensite reorientation in B19 phase. From Figure 5, it was analyzed that Ti-50Au has very low strength in B2 phase region. Shape memory alloys having very low strength in B2 phase compared to their strength in martensitic phase are not suitable for shape memroy materials applications as during mechanical loading, plastic deformation can easily introduce in such alloys in B2 phase region [1]. Figure 4 also showed that Ti-50Au-10Zr has high flow stress value in B2 phase region compared to Ti-50Au [10]. Comparing the stress values in B19 and B2 phase (Figure 4), flow stress value of Ti-50Au-10Zr in B2 phase was found higher than stress required for martensite reorientation in B19 phase region. This is an important research achievement by partial substitution of Ti with Zr in Ti-50Au alloy. Higher flow stress value in B2 phase region compared to stess require for martensite reorientation is beneficial for shape memory effect [1]. It was also analyzed from Figure 5 that the strength of Ti-50Au was improved by partial substitution of Ti with Zr in B19 martensite phase region as well as in B2 phase region.

Ti-50Au-10Zr exhibited 82% recovery ratio due to shape memory effect. Total recovery ratio (elastic recovery +recovery due to shape memory effect) after 5% applied strain for Ti-50Au-10Zr was 90% that was found higher that Ti-50Au that exhibits 78% total shape recovery ratio [9]. This result showed that partial substitution of Ti with Zr in Ti-50Au was found effective in improving the high temperature shape memory properties of Ti-50Au alloy. Still we could not achieve 100% recovery ratio due to shape memory effect and further research is in progress to enhance the shape memory properties of Ti-50Au alloy. Increase in the shape recovery ratio due to shape memory effect for Ti-50Au-10Zr as compared to Ti-50Au showed the increase in critical stress for slip deformation, decrease in residual strain and increase in recovered strain due to shape memory effect.

It should be noted that as martensite finish temperatures of Ti-50Au (859K) [9] and Ti-50Au-10Zr (741K) are well above the room temperature. So B2 phase upon cooling again transforms to twinned oriented martensite and at room temperature, there is single B19 martensite phase structure. As both Ti-50Au and Ti-50Au-10Zr exhibits partial shape memory effect, so some fraction of plastic deformation was introduced during shape memory testing.

4. Conclusions

Recent research and developments related to Ti-50Au high temperature shape memory alloys are discussed. Potential of Ti-50Au intermetallic alloy in terms of thermal stability, phase transformation temperatures, shape memory effect and temperature hystersis is discussed in this paper. Some results of our research group to improve the strength and shape memroy properties of Ti-50Au by partial substitution of Ti with Zr are presented. High temperature strength of Ti-50Au in B19 martensite as well as in B2 phase was improved by partial substitution of Ti with Zr. Partial substitution of Ti with Zr in Ti-50Au was also found effective in improving the high temperature shape memory properties. Transformation hysteresis ($A_f - M_f$) of Ti-50Au (38K) was slightly increased by partial substitution of Ti with Zr in Ti-50Au-10Zr (43K) alloy. Ti-50Au-10Zr has potential for high temperature shape memory materials applications.

5. Acknowledgements

This research is supported by the Japan Society for the Promotion of Science (JSPS) through the Funding Program for Next Generation World-Leading Researchers (NEXT Program) initiated by the Council for Science and Technology Policy (CSTP). Authors are thankful to Nishio Mitsuaki san, Materials Analysis Station NIMS Japan for EPMA of our research samples.

6. References

[1] Ma J, Karaman S, and Noebe R D 2010 *Inter. Mater. Rev.* 55 257

[2] Murray J L 1990 *Binary Alloy Phase Diagrams*, II Ed., Ed. T.B. Massalski, Vol. 1 p 442

[3] Okamoto H 1993 *J. Phase Equilib.* 14 128

[4] Pelton A R, Stockel D and Deering T W 2000 *Mater. Sci. Forum* 327-328 63

[5] Donkersloot H C and Van Vucht J H N 1970 *J. Less Common Metals* 20 83

[6] Hosoda H, Tachi R, Inamura T, Wakashima K and Miyazaki S 2007 *Mater. Sci. Forum* 561-565 1541

[7] Kawamura T, Tachi R, Inamura T, Hosoda H, Wakashima K, Hamada K and Miyazaki S 2006 *Mater. Sci. Eng. A* 438–440 383

[8] Wu S and Wayman C M 1987 *Metallography* 20 359

[9] Declairieux C, Denquin A, Ochin P, Portier R and Vermaut P 2011 *Intermetallics* 19 1461

[10] Vermaut P, Declairieux C, Ochin P, Kolomytsev V, Pasko A, Monastyrsky G, Denquin A and Portier R *J. Alloys Compd.* (2012), doi:10.1016/j.jallcom.2012.02.025.

[11] Yamabe-Mitarai Y, Hara T, Miura S and Hosoda H 2006 *Mater. Trans.* 47 650

[12] Shackelford J F and Alexander W 2001 *Mater. Sci. Eng. Handbook*, Third Edition, chapter 1, CRC Press LLC, Florida, Printed in USA

[13] Wadood A, Takahashi M, Takahashi S, Hosoda H, and Yamabe-Mitarai Y 2013 *Mater. Sci. Eng. A* 564 34

[14] Wadood A, Takahashi M, Takahashi S, Hosoda H and Yamabe-Mitarai 2013 *TMS2013 Annual Meeting Supplemental Proceedings*, San Antonio, Texas USA 949

[15] Kawakita M, Takahashi M b, Takahashi S and Yamabe-Mitarai Y 2012 *Mater. Lett.* 89 336

[16] Arockiakumar R, Takahashi M, Takahashi S and Yamabe-Mitarai Y 2013 *Mater. Sci. Eng. A* 585 86

[17] Hari Kumar K C, Wollants P and Delaey L 1994 *J. Alloys Compd.* 206 121

[18] Okamoto H 2002 *J. Phase Equilib.* 23 111

[19] Yamabe-Mitarai Y, Maruko T, Miyazawa T and Morino T 2005 *Mater. Sci. Forum* 475-479 703

IOP Conf. Series: Materials Science and Engineering **60** (2014) 012019 doi:10.1088/1757-899X/60/1/012019

Frequency behaviour of SPD treated soft magnetic materials

R Grössinger[1], N Mehboob[1] and M Zehetbauer[2]
[1]Institute of Solid State Physics; T.U. Vienna, Wiedner Hauptstrasse 8; A-1040 Vienna, Austria
[2]Research Group Physics of Nanostructured Materials, University of Vienna, A-1090 Vienna, Austria

E-mail: rgroess@ifp.tuwien.ac.at

Abstract. The effect of SPD treatment on the hysteresis properties of selected soft magnetic materials such as Fe3%Si and Fe17%Co steel is presented. The enhancement of the coercivity due to the mechanical deformation can be explained by the magnetoelastic energy The frequency dependence of the coercivity and of the losses can be described using an eddy current based model.

1. Introduction

Soft magnetic materials are characterised by the shape and the most important points (coercivity, remanence, losses) of their hysteresis loop. Figure 1 shows for a comparison the typical hysteresis loops of soft and hard magnetic materials.

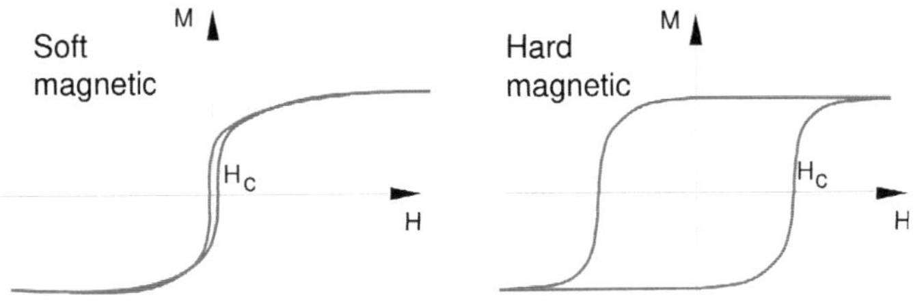

Figure 1. Typical hysteresis loop of soft and hard magnetic materials.

Generally soft magnetic materials are formed by 3d metals such as Fe,Co or Ni respectively 3d based alloys. Here well-known examples are Fe-Si, Fe-Si-Al, Fe-Ni etc (see e.g. [1]).

In general, technically useful soft magnetic materials should exhibit the following intrinsic properties at room temperature: a high saturation magnetization, a low anisotropy and a low magnetostriction. Additionally extrinsic properties such as low coercivity, high permeability and low losses are important. The extrinsic properties are determined by the microstructure such as grain size, different phases, grain boundaries and internal stresses, which can be influenced by the heat treatment but also by plastic deformation.

In this paper, the magnetic properties of some important soft magnetic metallic materials such as Fe-3%Si as well as a standard steel will be given and discussed. In many technical applications these materials are produced or used with stresses added. In order to investigate the effect of stresses on the hysteresis properties, the samples were produced by applying a so-called SPD (Severe Plastic Deformation) treatment [2]. The hysteresis properties of SPD-treated soft magnetic materials were investigated in detail within the PhD thesis of N.Mehboob [3].

2. Experimental

Samples: In the last years, the magnetic properties (frequency dependence of the hysteresis loop, losses) on all kind of soft magnetic materials, such as pure Fe, pure Ni, steels, Fe3%Si and Fe6.5%Si [1,3] but also amorphous and nanocrystalline materials [4,5] were investigated. Within this review we will focus only on Fe-based alloys, i.e. a Fe3%Si (electro-steel) and a Fe-Co steel.

All bulk samples were studied in an "undeformed" state but also after a so-called SPD (Severe Plastic Deformation) treatment which is nowadays a well established method to produce nanocrystalline bulk materials [2]. Specimens have been disc-shaped to a diameter of 10 mm, and from each material a set of four samples was produced. These samples were deformed by High Pressure Torsion ("HPT" being one of the most effective SPD methods [6], see also below) at liquid nitrogen temperature (77 K), ambient temperature (RT = 298 K), and 723 K corresponding to 0.4 T_{hom} (= T_{HPT} /$T_{melting}$) under a force of 500 kN which corresponds to a pressure of about 6 GPa. With the rotation tools, three complete turns were done, thus achieving an average torsional shear strain of about ≡80. One sample from each set was kept undeformed as a reference. Ring-shaped samples were made from the periphery of the thin HPT-deformed discs, which are in the saturation state of deformation and microstructure, and from magnetic point of view the best choice for hysteresis measurements. The HPT processing parameters are given in Table 1.

Figure 2 shows the principle of HPT setup where a high pressure combined with torsion at different temperatures can be applied. The drawing is from the facility used in this work, which is available at the Erich Schmid Institute, Austrian Academy of Sciences, Leoben; Austria with two systems applying at maximum force of 500 kN, and 4MN, respectively. This treatment can be performed between 77K and 800 K. In the periphery of sample (see e.g. Figure 7), a maximum of grain refinement is achieved. Summarizing, this method can cause nanocrystallization [7] as well as a strongly deformed state [8].

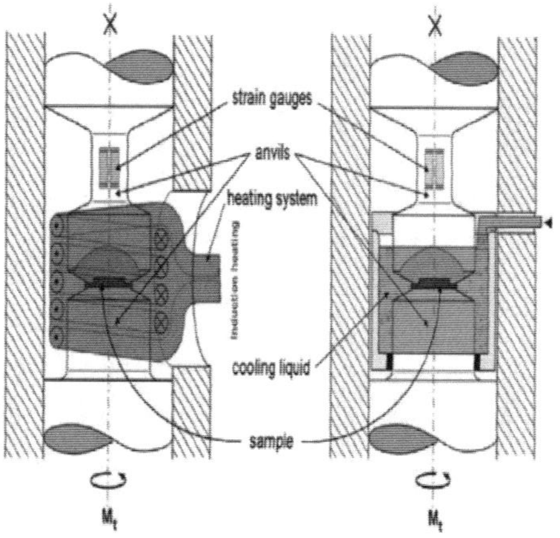

Figure 2. Schematic view of the used HPT equipment for hot (left) and cold (right) processing.

Table 1: HPT parameters as used for pure Fe-3 wt%Si and Fe-17wt %Co samples, T_{HPT} - HPT treatment temperature; n-number of turns; t - thickness of sample before HPT; d_o-outer diameter; d_i-inner diameter of ring-shaped samples. The pressure was about 6 GPa (500kN over a sample area of 80 mm^2), the average torsional shear strain was about 80.

Sample	T_{HPT}	N	t [mm]	d_o [mm]	d_i[mm]
Fe3%Si		0	0,81	7,053	5
Fe3%Si	RT	3	0,638	7,01	5
Fe3%Si	723 K	3	0,637	7,026	5
Fe3%Si	77 K	3	0,653	6,91	5
Fe17%Co			0,781	7,053	5
Fe17%Co	RT	3	0,64	7,01	5
Fe17%Co	723 K	3	0,657	7,026	5
Fe17%Co	77 K	3	0,653	6,91	5

3. Hysteresis measurement

For measuring the frequency dependence of the hysteresis loop (at room temperature) we used a fully automated hysteresigraph which was described in [9, 10] - see figure 3. Its main part is a standard PC with a National Instrument measurement card (NI 6120). The measuring electronics for generating the magnetic field consists of an external power amplifier (KEPCO) for generating the magnetizing current. The whole measurement is handled by a LabView program. The advantages of such a system are: i) the current I(t,f) can be chosen freely as e.g. triangular, sinusoidal one etc, ii) averaging over many loops which improves the signal - noise ratio, iii) automatic demagnetization of the sample, iv) frequency dependent measurements, v) temperature dependent measurements. The software was recently improved in order to measure under constant dB/dt conditions which is necessary to determine the true losses [11]. The hysteresis measurements are usually performed at room temperature; however, other temperatures (as e.g. -30°C or + 100°C) are also possible. All samples were demagnetized before each hysteresis measurement.

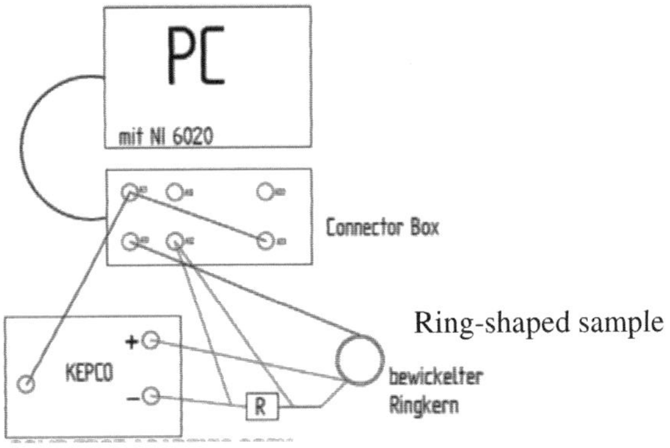

Figure 3. Block diagram of the Labview-automated hysteresis measurement system for ring-shaped samples

The frequency dependence of the hysteresis loop was measured on ring-shaped samples at room temperature applying a triangular field H(t). Using a triangular field allows to determine the relative permeability $\mu_r(H)$ directly from measurements of minor loops using formula (1):

$$\mu_r(H) = \frac{1}{\mu_0} \Delta B(H) / \Delta t \frac{4}{f.H_{max}} \quad (1)$$

4. Results and Discussion

First we want to summarise the intrinsic magnetic properties of pure Fe and compare with Fe3%Si – see table 2. However, the hysteresis properties such as the coercivity, the remanence etc. depend - beside the intrinsic properties - mainly on the actual microstructure such as grain size, defects etc.

Table 2. Room temperature data of the saturation magnetization □M_S, the crystal anisotropy K_i and the magnetostriction □□ of pure Fe compared with the data of Fe3%Si[13].

	□M_S (T)	K_1 (J/m^3)	K_2 (J/m^3)	□□□□□ppm)	□□□□□ ppm)	□□□□□ cm	Density (g/cm^3)
Fe	2,15	4,9.10^4	1,1.10^4	21	-21	9,61	7,86
Fe3%Si	2,0	3,6.10^4	-	23	-4	48,0	7,65

The hysteresis loops of undeformed and HPT-deformed Fe, Fe-3 wt% Si, Fe-6.5 wt% Si and Fe-17 wt%Co (P800) samples were measured. Figure 4 shows - as an example - measurements of Fe3%Si which were performed at room temperature in the frequency range of 0,25 Hz to 1 kHz. The maximum average applied field was about 20000 A/m. It can be seen from the width of the hysteresis loops that the coercivities (H_c) increased significantly with frequency. This behaviour can be simply described using an eddy current based formula (2) of the type:

$$H_C(f) = a + b.\sqrt{f} + c.f \quad (2)$$

The strong frequency dependence of the coercivity (H_c) at high frequencies is understood in iron (which is a well conductive material), due to the increasing effect of eddy currents induced in the material with increasing frequency [12].

It is well known that the addition of non-magnetic silicon lowers the saturation flux but significantly decreases the coercivity. For example the coercivity is reduced up to factor 2,5 with 3 wt%Si as compared to pure undeformed iron whereas the electrical resistivity of pure iron increases about 4,5 times with the addition of 3 wt%Si and 8 times with 6.5 wt%Si [13], thereby reducing the eddy currents. Moreover with increasing Si content also the magnetocrystalline anisotropy constant K_1 [12] increases which is obvious from the width of the hysteresis loops.

The coercivity of all HPT-deformed samples at low frequencies is considerably high as compared to undeformed material: although the grain size is reduced to nano-scale due to the SPD deformation, it is still larger than the exchange coupling length. Moreover, the SPD deformation increases the level of microstresses □□□□as well as of the number of dislocations and other lattice defects which leads to domain wall pinning. Thus the deformation increases the magnetoelastic energy (□□) which results in an increase of the (static) coercivity – see formula (3).

$$H_C = \frac{\sigma.\lambda + \langle K \rangle}{\mu_0 \; M_S} \quad (3)$$

However, at higher frequencies, the effect of eddy current is comparatively smaller in HPT-deformed samples as compared to undeformed ones because the deformation induced dislocations/lattice defects also enhance the electrical resistivity and reduce the permeability of the material.
Figure 4 shows as examples the frequency dependent hysteresis loops as measured on undeformed bulk polycrystalline Fe3%Si and on a RT-SPD processed sample.

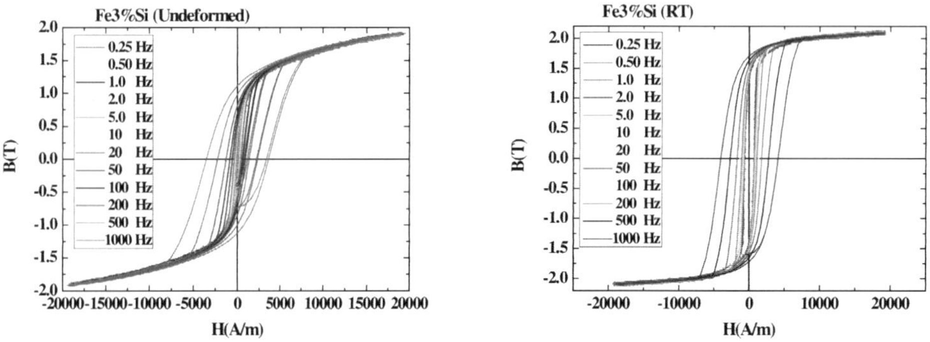

Figure 4. Frequency dependence of the at room temperature measured hysteresis loop of undeformed (left picture) and of RT-SPD deformed (right picture) Fe3%Si.

$H_C(f)$ of FeSi shows a nonlinear relation between the coercivity and the frequency, indicating the importance of eddy currents within the dynamic process (see figure 5). $H_c(f)$-curves for deformed samples are less steep as compared to undeformed samples. The slope of the $H_c(f)$-curve is reciprocally proportional to the square root of the electrical resistivity of the material. This indicates that the resistivity of the material increases with deformation which is in line with the fact that deformation induced dislocations and/or lattice defects enhance the resistivity. Figure 6 shows the frequency dependence of the losses of all SPD treated Fe3%Si samples. The applied field H(t) was triangular. Table 3 shows the initial and the maximum permeability (calculated from formula (1)), the coercivity extrapolated to zero frequency, as well as the ratio B_r/B_{max}. This ratio should be about 0,8 for an isotropic cubic material. The fact that all values are significantly below 0,8 can be explained assuming a strong induced texture. Applying formula (2) gives the fit parameters (a,b,c) shown in table 3 and also achieves the results of a loss analysis using formula (4). This formula is based on a loss separation usual for Fe-Si transformer sheets [14]

$$(BH) = W_{tot} = W_h + W_{class} + W_{exc} = W_h + B_2.f + B_1.f^{0.5} \quad (4)$$

$$W_{tot} = W_h + CB_{max}^{1.5} f^{0.5} + a\frac{A.\pi^2.B_{max}^2}{6\rho} f = W_h + CB_{max}^{1.5} f^{0.5} + aW_C f \quad (5)$$

where W_h represents the frequency independent hysteresis losses, and B_1, B_2 are fitting parameters. $A(m^2)$ is the cross section of the sample, B_{max} gives the maximum induction of the material and ρ is the specific electrical resistivity; a describes a constant which is determined by the shape of the driving field. The excess losses can be describe according to Bertotti [14] with a formula which contains mainly a hypothetic field V_0 which is an important parameter for the damping process:

$$W_{exc} = C.B_{max}^{1.5}\sqrt{f} = 8\sqrt{\frac{G.A.V_0}{\rho}}B_{max}^{1.5}\sqrt{f} \quad (6)$$

The value of Vo can be estimated by using the parameter B_1, the specific electrical resistivity of the material and B_{max}.

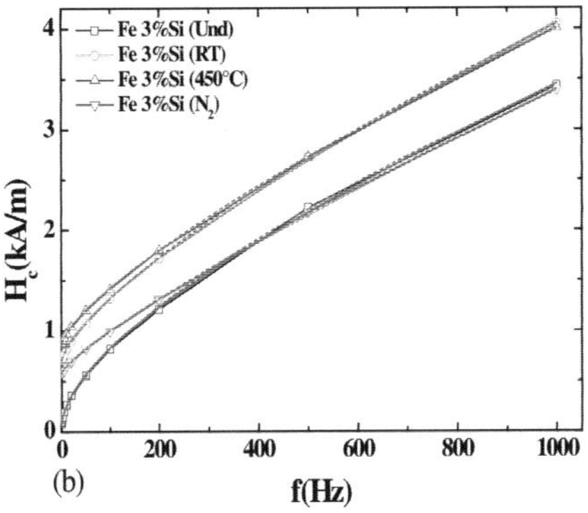

Figure 5. $H_C(f)$ of SPD treated Fe3%Si.

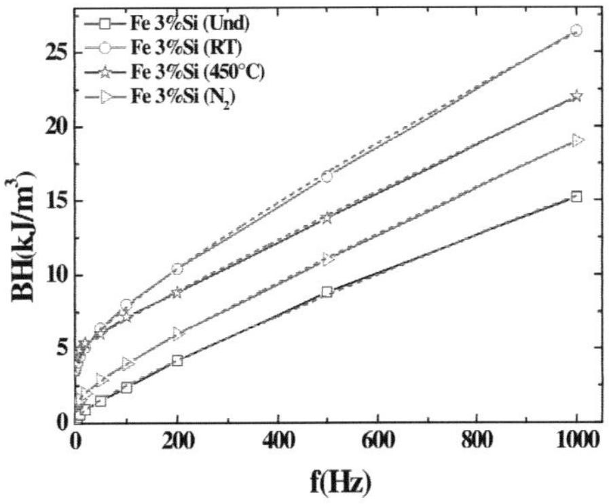

Figure 6. Frequency dependence of the losses of SPD treated Fe3%Si.

Table 3: Initial permeability μ_0, maximum permeability □max determined at the field H(..), zero frequency coercivity H_c (A/m), and the ratio B_r/B_{max}. Polynomial fit parameters (a,b,c) of the frequency dependence of the coercivity for all HPT treated Fe3%Si samples at RT. Fit parameters of the frequency dependence of the losses (W_h, B_1, B_2) using formula (4).

Fe3%S	μ_0	μ_{max} at H(A/m)	H_c (A/m	B_r/B_{ma}	a	b	c	W_h	B_1	B_2	aB_2	V_0
Undef	1000	1000(115)	47	0,15	45	35	1,3	86	146	11	a.35	244
RT	290	1000(710)	290	0,23	659	48	1,8	3392	312	13	a.26	1537
723 K	120	900(920)	842	0,7	830	41	1,9	4393	126	13	a.25	258
77 K	220	650(422)	385	0,4	552	25	2	973	176	12	a.25	503

5. Fe-Co steel

In this material, all deformation states processed at different HPT temperatures correspond to a torsional shear strain of □≈ 75. Crystallite sizes were extremely decreased in the highly deformed states. Deformed microstructures are shown in figure 7 following HPT at 723, 293 and 77 K, respectively [15]. After HPT the mean crystallite size as calculated by the area fraction method decreased from 270 to 83 and 72 nm after HPT deformation at 723, 293 and 77 K, respectively. By comparing the microstructures of Fe-17 wt% Co in figure 7 (b,c), one can clearly see that both the grain size and the grain shape have strongly changed.

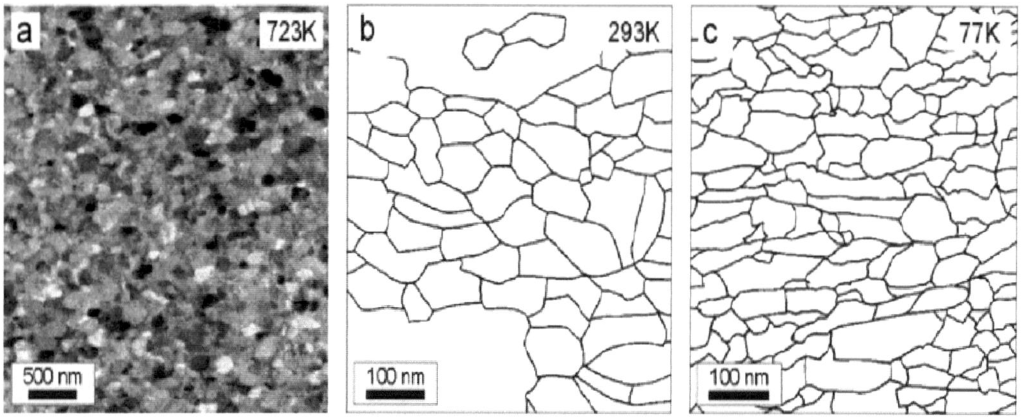

Figure 7. Grain size maps for Fe-17 wt% Co alloy corresponding to HPT temperatures (a) 723 K, (b) 293 K and(c) 77 K [15].

Figure 8 shows for comparison the frequency dependence of the hysteresis loops as measured on undeformed Fe-17wt%Co and on at RT deformed material. Note that the coercivity here is significantly higher than that of Fe3%Si. Figure 9 (left picture) shows the frequency dependence of the coercivity as obtained for all SPD treated steel samples. Figure 9 (right picture) shows the frequency dependence of the losses as obtained for all SPD treated steel samples.

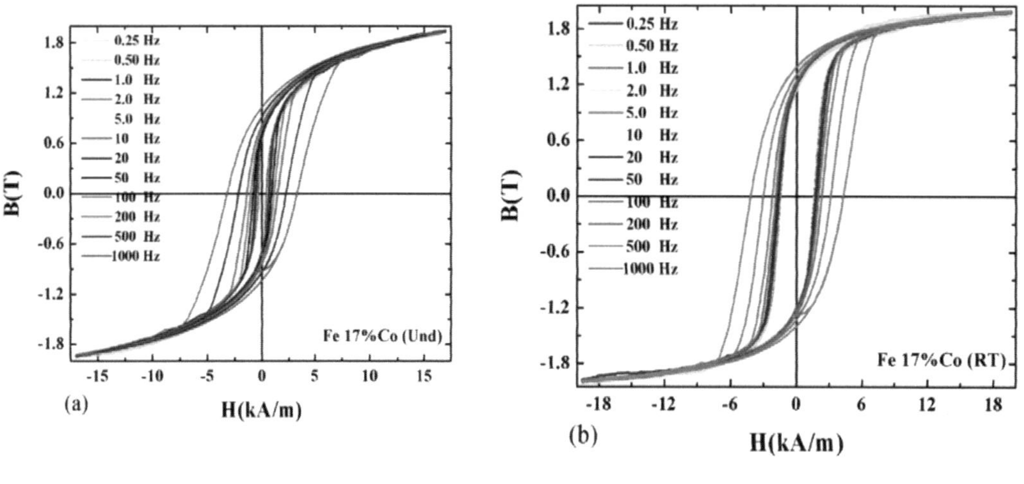

Figure 8. Frequency dependent hysteresis loops of Fe-17wt%Co (a) undeformed, (b) HPT treated at RT

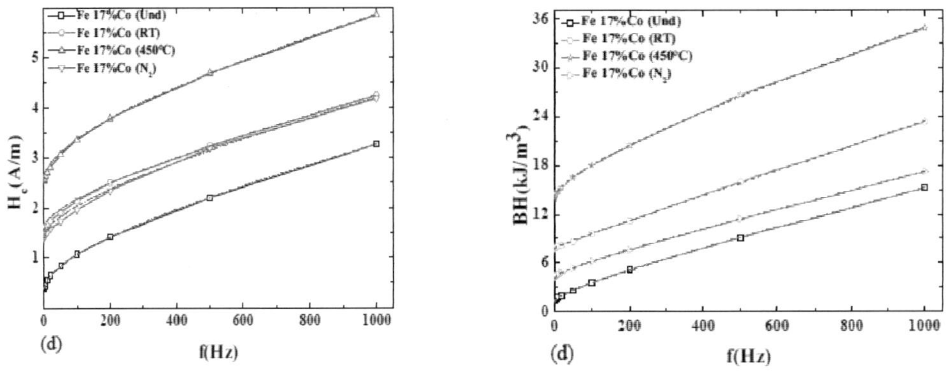

Figure 9. The frequency dependence of the coercivity (left picture) and the losses (right picture) of all HPT-treated Fe-17 wt% Co samples.

Table 4. Results of analysing the frequency dependence of the hysteresis loop (coercivity) applying formula (2) as well as the losses using formula (4), as follows:

Samples Fe-Co	a	b	c	W_h	B_1	B_2	aB_2	V_0
Undef	372	63	1	1283	132	10	a.296	23
RT	1533	22	2	7728	44	14	a.215	4
723 (K)	2510	57	1,6	13760	315	11	a.121	168
77 (K)	1379	31	1,8	4150	108	10	a.245	19

6. Conclusion

Severe plastic deformation (SPD treatment) generally causes an increase in the static coercivity as well as in the hysteresis losses W_h. In particular, the plastic deformation increases the magnetoelastic energy which is responsible for all effects on the hysteresis loop. The frequency dependence of the coercivity as well as of the losses can be described with the same type of formula which is based on the idea that there are two dynamic contributions, one which comes from the classical eddy currents and a second which arises from the movement of the domain walls (excess losses).

7. References

[1] Boll R 1979 *Soft Magnetic Materials-Fundamentals, Alloys, Properties, Products, Applications* (The VACUUMSCHMELZE Handbook; Siemens Aktiengesellschaft/ Heyden & Sons Ltd.

[2] Zehetbauer M, Zhu Y 2009 *Bulk Nanostructured Materials* (Wiley VCH, Weinheim, Germany)

[3] Mehboob N 1988 *Hysteresis Properties of Soft Magnetic Materials* (PhD Thesis, TU Vienna)

[4] Yoshizawa Y, Oguma S, Yamauchi K 1988 *New Fe-based soft magnetic alloys composed of ultrafine grain structure J. Appl.Phys.* 64, 6044-6046

[5] Herzer G 1989 *Grain Structure and Magnetism of Nanocrystalline Ferromagnets* IEEE Trans. Magn. 25, 3327-3329

[6] Pippan R, Scheriau S, Hohenwarter A, Hafok M 2008 *Advantages and Limitations of HPT A Review* Mater.Sci.Forum 584-586, 16–22

[7] Vorhauer A, Rumpf K, Granitzer P, Kleber S, Krenn H, Pippan R 2006 *Magnetic Properties and Microstructure of a FeCo Ferritic Steel after Severe Plastic Deformation* Mater.Sci.Forum 503-504, 299-304

[8] Waitz T, Antretter T, Fischer F.D, Simha N.K, Karnthaler H.P 2007 *Size effects on the martensitic phase transformation of NiTi nanograins* J. Mech. Phys. Sol. 55, 419-444

[9] Oser, P 2008 (*Diploma Thesis* TU Vienna)

[10] Czichos H, Saito T 2006 *Handbook of Materials Measurement Methods* (Ed. L.Smith, Springer,. ch.10)

[11] Antoni M 2013 (*Diploma Thesis* TU Vienna)

[12] Goodenough J.B 2002 *Summary of losses in magnetic materials* IEEE Trans.Magn. 38, 3398-3408

[13] Landolt-Börnstein Group III 1994 Vol 19 "*Magnetic properties of Metals*" (Springer)

[14] Bertotti G 1998 *Hysteresis in Magnetism* (Academic Press in Electromagnetism; Ed. I.M.Mayergoyz)

[15] Valiev R.Z, Ivanisenko Yu.V, Rauch E.F, Baudelet B 1996 *Structure and deformation behaviour of Armco iron subjected to severe plastic deformation* Acta Mater. 44, 4705-4712

Co-ferrite – A material with interesting magnetic properties

R Sato Turtelli[1], M Kriegisch[1,2], M Atif[3] and R Grössinger[1]
[1]Institute of Solid State Physics, Technical University of Vienna, Vienna, Austria
[2]Austrian Institute of Technology, Giefinggasse 2, A-1210, Vienna, Austria
[3]Department of Physics, Air University, PAF Complex E-9, Islamabad, Pakistan

E-mail: reiko.sato@ifp.tuwien.ac.at

Abstract. The development of Co-ferrite nanocrystals, and bulk materials for magnetostrictive applications is reviewed. This includes magnetic and structural properties and also processing conditions which are of special importance to increase the magnetostriction as well as to achieve very fine Co-ferrite nanocrystals. The magnetic properties of a single crystal, where a field induced phase transition was found, are also presented.

1. Introduction

Over the past few years, there has been an increasing interest in Co-ferrites as magnetic nanocrystals (as ferro-fluids, magnetic drug delivery, high density information storage etc [1-3].) and magnetostrictive material (as for actuator and sensor applications [4,5]). This interest is due to bulk Co-ferrite exhibits remarkable properties such as high saturation magnetization, high coercivity, high anisotropy with good mechanical hardness, chemical stability and high magnetostriction [4-9]. Additionally, it is inexpensive and easy to produce. Conventional piezo-ceramics, such as PZT and PMN, and rare-earth based magnetostrictives, such as Terfenol-D, offer large strain values, but have limited mechanical robustness, therefore requires an operation under compressive loading. In contrast, the good mechanical properties of Co-ferrites and the ability to induce anisotropy through stress and field annealing [8,9], enables Co-ferrite to operate in the tensile regime without losing magnetic performance. Although the saturation magnetostriction of cobalt ferrite is lower than that of Terfenol, the slope of the magnetostriction or strain derivative ($d\lambda/dH$) which determines the stress sensitivity of a material is higher. Magnetostriction values of about 400 ppm can be achieved in polycrystalline ferrite at room temperature under the right processing conditions [9].

Cobalt ferrite is a non-metallic ferromagnetic oxide, which exhibits the highest magnetostriction among all 3d-element based spinel oxides. Its magneto-elastic properties rise mainly due to the presence of Co^{2+}-ions in the spinel lattice, which are due to its orbital contribution responsible for the large magnetostriction. The value of the achievable magnetostriction depends strongly on the exact stochiometry as well as on the production methods [10]. An improvement of magnetic properties of polycrystalline Co-ferrite can be obtained by heat treatment, high compaction pressure and magnetic annealing besides adequate production process of the powders [7-10].

This article reviews the development of the nanocrystalline and bulk polycrystalline cobalt ferrites including structural and magnetic properties (such as magnetization, coercive field, magnetostriction). Additionally the importance of different processing conditions on the improvement of magnetic properties of Co-ferrite as well as on nanocrystalisation is shown. In order to investigate the magnetic properties without grain or grain boundary effects single crystals are ideal to come to a fundamental understanding of all kind of magnetic properties. Therefore also magnetostriction and magnetization measurements of a $Co_{0.8}Fe_{2.2}O_4$ single crystal are presented.

2. Co-ferrite

Ferrites based on Fe_3O_4 crystallize in the spinel structure with the general formula $Fe^{2+}Fe_2^{3+}O_4^{2-}$ which allow a large variety of substitution possibilities on the Fe site. When Fe^{2+} ions are substituted by Co^{2+} ions, we obtain a Co-ferrite. For adjusting the magnetic properties for various applications, the cations can be substituted by other elements such as Zn, Mn, etc which go to preferential sites (tetrahedral (site-A) or octahedral (site-B), see figure 1) according to their oxidation state as well as their ionic radius. The unit cell consists of eight tetrahedral places and 16 octahedral sites.

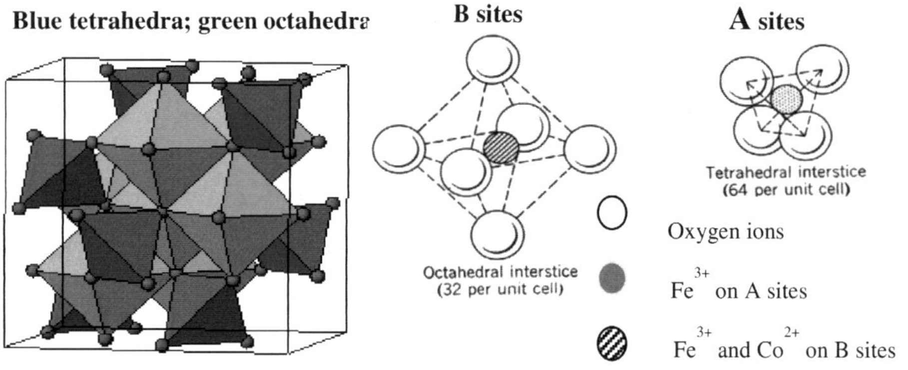

Figure 1. The crystal model of the unit cell for spinel ferrite structure and relative position of the nearest neighbour atoms of the tetrahedral and the octahedral sites

According to the distribution of cations on A and B sites, the spinel Co-ferrites are classified into two categories:

- Inverse spinel ferrites: the eight Co^{2+} filling half of the available octahedral sites, and Fe^{3+} ions are distributed randomly over the available tetrahedral and octahedral sites.
- Mixed spinel ferrites: the A- and B- sites are occupied by the both Co^{2+} and Fe^{3+} ions.

The occupancy of one type of divalent cations in the tetrahedral site is called degree of inversion, *i*. For the $CoFe_2O_4$ ferrite can be written as: $(Co_{1-i}Fe_i)[Co_iFe_{2-i}]O_4$, where the round brackets indicate the tetrahedral and the square brackets indicate the octahedral sites. Thus we have for the inverse spinel $i = 1$ and for the mixed spinel: $0 < i < 1$. As the tetrahedral site is crystallographically smaller than the octahedral, generally the smaller ion usually tends to go on that site.

Magnetically, it is well known that $CoFe_2O_4$ is a ferrimagnetic material. The magnetic moments of cations in the A- and B-sites are aligned antiparallel with respect to one another yielding a total moment of spins with ferrimagnetic ordering. In the inversed ferrites, half of Fe^{3+} with magnetic moment $5\mu_B$ is placed in A-sites and another half in B-sites. Their moments are mutually compensated and the net moment of ferrite is due to the moment of $3\mu_B$ of Co^{2+} in the B-site. The general formula of the net magnetic moment per formula unit of spinel ferrite as a function of inversion degree may be written as: $\mu = \mu_{B\text{-sites}} - \mu_{A\text{-sites}} = \{[3i+5(2-i)]-(5i+3-3i)\}(\mu_B) = (7-4i)(\mu_B)$, where μ is the net moment of the material, $\mu_{B\text{-sites}}$ and $\mu_{A\text{-sites}}$ are moments of cations in B and A-sites, respectively. Thus, the cation distribution affects the magnetic properties of the cobalt ferrite.

3. Nanocrystalline Co-ferrites

Many methods have been developed to prepare nanocrystalline $CoFe_2O_4$, namely: co-precipitation [11], combustion [12], micro-emulsion [13], ion exchange [14], sol–gel [7,15], modified oxalate

IOP Conf. Series: Materials Science and Engineering **60** (2014) 012020 doi:10.1088/1757-899X/60/1/012020

method [16], forced hydrolysis [17,18]. We could obtain very fine particles by means of modified oxalate method [16] and the forced hydrolysis method [18]. The later, it seems to be one of the best methods to obtain very fine nanocrystalline $CoFe_2O_4$, since it allows to prepare nano-particles with an equiaxial shape and narrow size distribution exhibiting high saturation magnetization and coercivity at low temperatures [17,18]. Using this method, nanocrystals of $Co_{1-x}Zn_xFe_2O_4$ (x = 0, 0.2, 0.4) with average particle size of 3 nm could be produced [18] (see figure 2, for example for x = 0.2 sample). Obviously these particles are superparamagnetic at high temperatures as can be seen, for example in susceptibility measurements shown in figure 3. This figure shows also the effect of Zn substitution on the blocking temperature, which changes with Zn content. Additionally, the saturation magnetization, M_s, increases and the coercive field, H_c, decreases with Zn [18]. Detailed investigations of the magnetic and structural properties were also made on annealed samples [19,20].

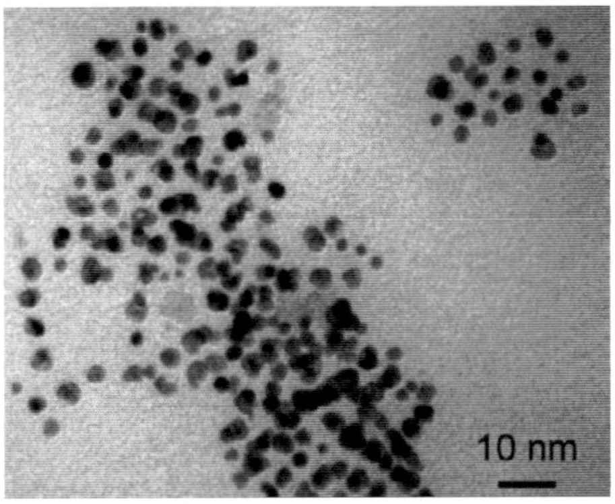

Figure 2.TEM image of $Co_{0.8}Zn_{0.2}Fe_2O_4$ sample [18].

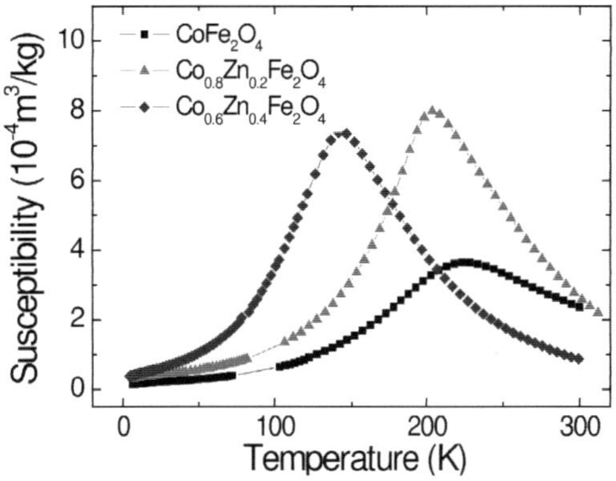

Figure 3. Susceptibility as function of the temperature measured on the $Co_{1-x}Zn_xFe_2O_4$ samples with x = 0, 0.2 and 04 [18].

For the nanocrytalline $CoFe_2O_4$ prepared by modified oxalate method a larger average particles size (8 nm) and also a broad distribution of particles size were obtained [16]. The hysteresis loops measured with applied field up to 9 T at different temperatures are presented in figure 4. The blocking temperature of this material is around room temperature, RT.

4. Magnetic properties of single crystalline $Co_{0.8}Fe_{2.2}O_4$

We investigate the magnetostriction and magnetization of a cobalt ferrite ($Co_{0.8}Fe_{2.2}O_4$) single crystal prepared by a flux method [21], as a function of the temperature applying a magnetic field along the three different crystallographic axis: [100], [110] and [111]. Magnetostriction is an intrinsic property of all kind of magnetically ordered materials. Due to the magnetic exchange, the materials change its shape when a magnetic field is applied. Magnetostriction is based on the coupling of the magnetic moment to the lattice, therefore the orbital moment of the atoms must be considered. In a single crystal, the magnetostriction is a tensor property which depends on the crystal structure and symmetry of the lattice.

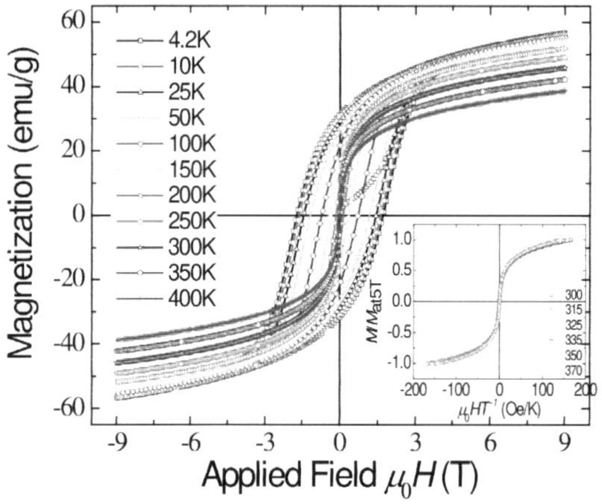

Figure 4. Hysteresis loops of nanocrystalline $CoFe_2O_4$ measured at 4.2 K ≤ T ≤ 400K. The insert shows $M/M_{(at\ 90kOe)}$ vs. H/T which curve is typical of Langevin function that describes a superparamagnetic material.

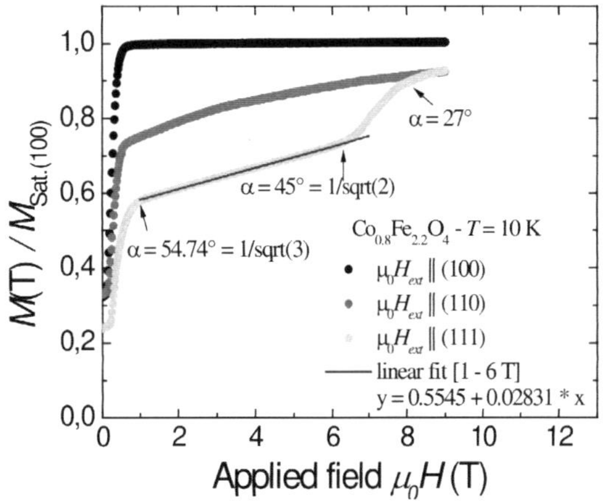

Figure 5. Normalized to $M_s(100)$ magnetization of $Co_{0.8}Fe_{2.2}O_4$ measured at 10 K with field // to the [100], [110] and [111] axis. α is the angle between the applied field and M [21]

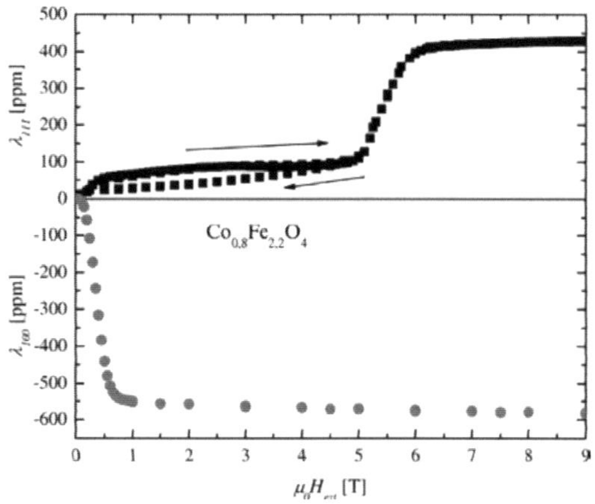

Figure 6. Magnetostriction measurements of $Co_{0.8}Fe_{2.2}O_4$ at 4.2 K and the field applied // to the [100] and [111] axis [21]

Figure 5 shows the applied field, H, dependence of the normalized magnetization, M, measured at 10 K with a H along the three principal axis, and figure 6 shows the magnetostriction, λ, as function of H measured with field along the crystallographic axis [100] and [111], at 4.2 K. The M and the λ measurements revealed that the [100] axis is the easy axis of magnetization of Co-ferrite over the whole temperature range reaching M_s for an applied field $\mu_0H < 1$ T. The other two axis are magnetically hard, which at 9 T are far away from saturation, as shown in figure 5.

We observed unusual behavior in the magnetic hard axis [111] of the single crystal which manifests in a jump of the M and λ curves at a critical field. This jump is visible at $T < 150$ K for both M and λ curves. The jump in λ was observed first time by us [21], while the jump in M curve was already observed by Guillot *et al* [22] in $Co_{1.04}Fe_{1.96}O_4$. For this jump was attributed to a spin-flip phenomenon. However we could prove that the jump in the $M(H)$ curve can be explained as an

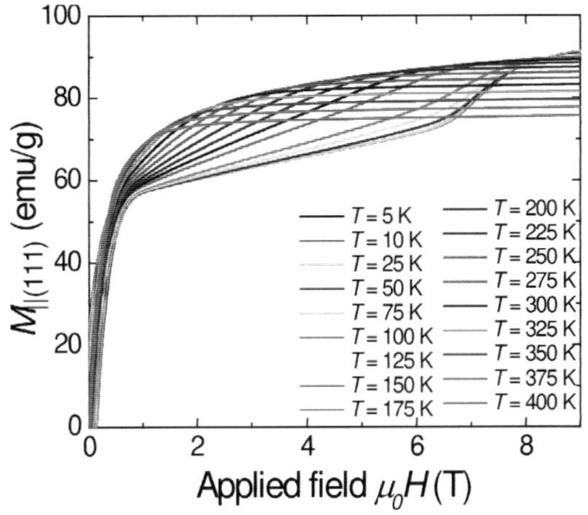

Figure 7. Magnetization vs. applied field along [111] axis measured at different temperatures on $Co_{0.8}Fe_{2.2}O_4$.

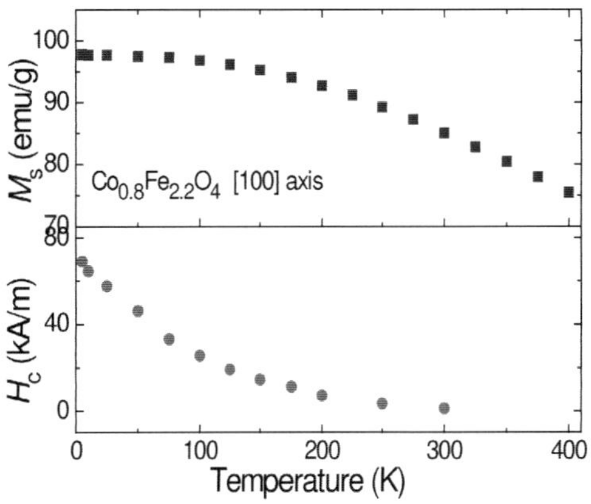

Figure 8. Temperature dependence of M_s and H_c measured on a $Co_{0.8}F_{2.2}O_4$ single crystal, applying the field along the [100] axis.

anisotropy driven transition [21], which is known as first order magnetization process (FOMP). This transition is caused by a rotation of the magnetization vector jumping over an energy barrier. Magnetization as function of the applied field along the [111] axis measured at different T in $Co_{0.8}Fe_{2.2}O_4$ is shown in the figure 7. As can be seen the field where the FOMP occurs, H_t, depends on the temperature. It is worth to mention that H_t, at a certain temperature is very sensitive to direction deviations of the applied field with respect to the [111] axis. Composition fluctuations are also important.

The decrease of the saturation magnetization and coercive field with temperature measured for the magnetic easy axis of $Co_{0.8}Fe_{2.2}O_4$ is shown in figure 8. Due to high anisotropy in the [111] axis, at low temperatures the sample was not magnetically saturated up to 9 T (see figure 7). Although saturation magnetization values are comparable to those of polycrystalline Co-ferrites, the coercive fields of single crystal are much smaller than those measured on nanocrystalline (at low temperatures) and polycrystalline materials, because in a single crystal the influence of grain size and grain boundaries don't exist.

5. Polycrystalline Co-ferrite

As was mentioned in the nanocrytalline section, there are many methods to prepare nanocrystalline Co-ferrite. Generally bulk polycrystalline Co-ferrites are obtained annealing as-cast produced nanocrystalline materials at a convenient temperature and annealing time. Among the preparation methods cited in section 3, for economical reasons the sol-gel and ball-milling method are most common. These methods allow an easy and cheap production process delivering large quantities.

The magnetic properties of polycrystalline Co-ferrite depend strongly on the processing conditions [7-10]. The heat treatment of the samples changes significantly the micro-structural and magnetic properties. Structurally, the grains increase, becoming more homogeneous with increasing annealing temperature, T_a, for a certain time or annealing time, t_a, at certain temperature. With annealing, due to the increasing of grain size, the magnetization and magnetostriction increase and the coercive field decreases. However, these properties become insensitive for larger grains (see figure 9) [7]. From XANES and XES measurements it was evidenced that all investigated samples in [7] exhibit only Co^{2+} and Fe^{3+} cations (no Co^{3+} and Fe^{2+}ions) and adequate Fe-spin states. Additionally, we found that the concentration of Co^{2+} cations in B-sites is larger in BM sample than in SG sample. Therefore the inversion degree in BM sample is higher than that of the SG sample resulting in lower saturation magnetization (see figure 9). These results showed thus the dependence of the cations distributions on the production method.

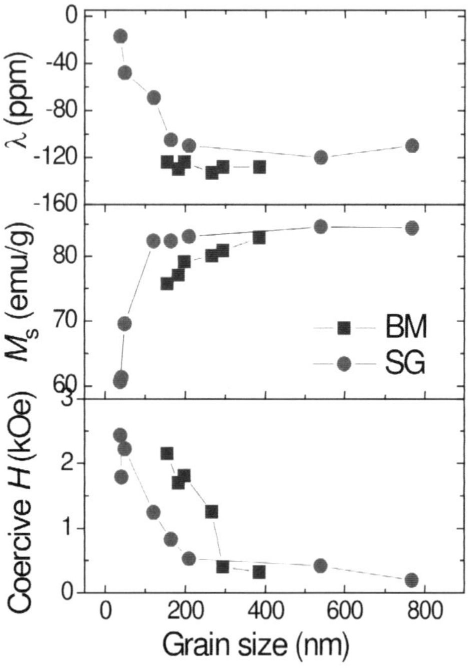

Figure 9. Grain size dependence of the magnetostriction, saturation magnetization and coercive field measured on samples produced by modified sol-gel and ball-milling methods

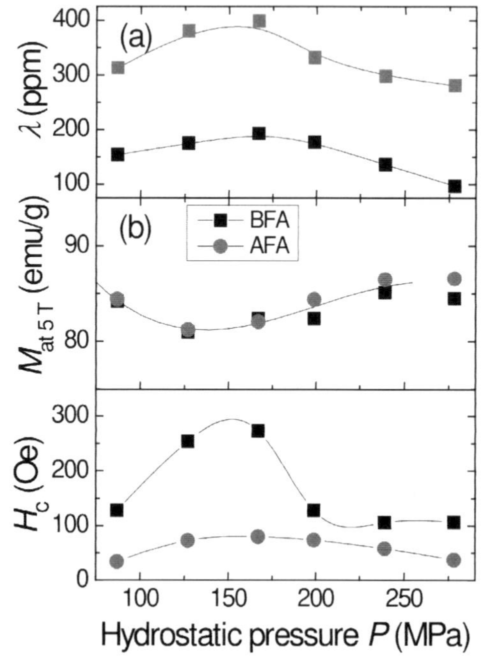

Figure 10. H_c, M at 5 T and λ vs. P: BFA are the data from the samples annealed at 1350 °C for 24 h after the application of P, while AFA are data obtained from the samples BFA field annealed at 300 °C for 3 h.

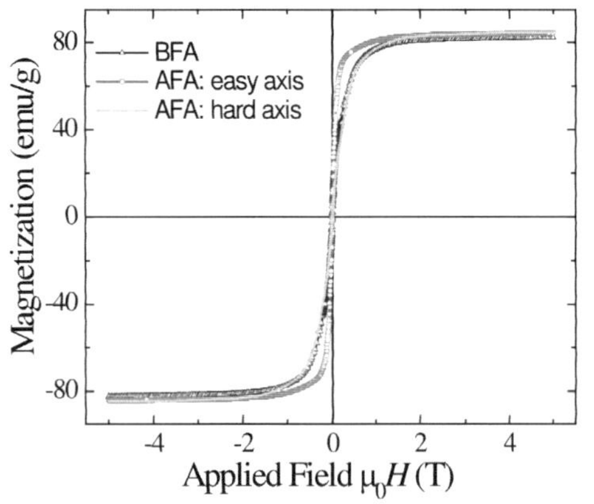

Figure 11. Magnetization curves obtained for the BFA and AFA samples pressed at 199 MPa. In the AFA sample, M was measured along the magnetic easy and hard axis

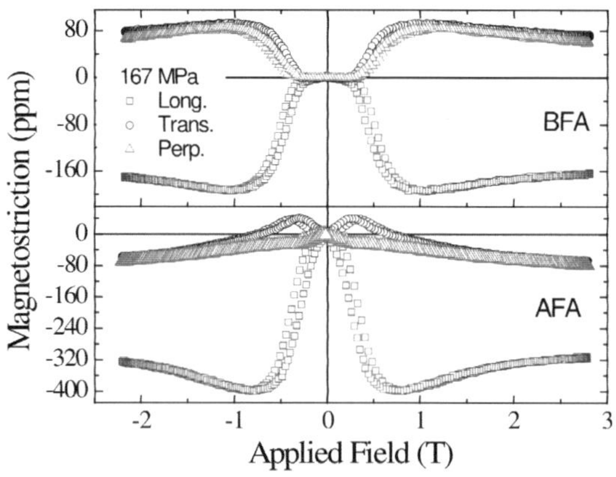

Figure 12. Measurements of the longitudinal, transverse and perpendicular (tri-dimensional) magnetostriction, performed on the BFA and AFA samples pressed at 167 MPA.

An improvement of magnetostriction of polycrystalline cobalt ferrite can be obtained by magnetic annealing due to the effect of induced uniaxial anisotropy [8,9]. Nlebedim *et al.* [8] investigated the influence of powder compaction pressure (87 MPa $\leq P \leq$ 127 MPa) and annealing conditions on the magnetostriction of $CoFe_2O_4$ and found the highest magnetostriction value of 200 ppm for a sample pressed at 127 MPa and annealed at 1350 °C for 24 h. Later we investigated the influence of compaction pressure on magnetostriction also for higher values of P, i. e, for 87 MPa $\leq P \leq$ 278 MPa. We found an optimum compaction hydrostatic pressure at around 160 MPa where a maximum magnetostriction of about 400 ppm in the pressure dependence of the magnetostriction curve was found (see figure 10). This high λ value was achieved when the sample was pressed and annealed at 1350 °C for 24 h and afterward again submitted to a heat treatment at 300 °C for 3 h, but in high magnetic field of 10 T [9]. This field annealing process causes an induced uniaxial anisotropy (see figure 11), which results in a reduction of the coercivity (in the easy axis) as well as a dramatic increase of λ up to 400 ppm along the hard axis (see figure 12). However, the microstructure and the saturation magnetization were not varied with field annealing. In the coercive field as a function of the compaction pressure (see figure 10) and magnetic anisotropy show also maxima values [9]. At this optimum pressure the saturation magnetization takes a minimum value (figure 10). Thus the change in magnetic properties is due to the occurrence of a redistribution of Co and Fe cations between A- and B- sites with pressure once a change in microstructure with P was not observed. These results show that the population of Co^{2+} cations in octahedral environment is a deciding factor for magnetic and magnetostrictive properties The M, H_c and λ vs. P behaviors are similar to those found before magnetic annealing (figure 10) showing once that the controlling factor is the distribution of cations.

The final specification of the magnetic and magnetostrictive properties of the cobalt ferrites is not only determined by the basic composition, but also by structural characteristic, induced uniaxial anisotropy and cations distributions influenced by the route of ferrite production.

6. Polycrystalline $Co_{1-x}Mn_xFe_2O_4$ (x = 0-0.4)

In chapter 3 the effect of Zn substitution on nanocrystalline Co-ferrite was shown. Beside many other possible substitutions of elements, always Mn is interesting to investigate. As was described in [23] the substitution of Mn causes an increase of the lattice constant, however the magnetic properties such as M_s, H_c, the Curie temperature and also the magnetostriction decrease with Mn substitution. Figure 12 shows the magnetostriction measured on polycrystalline $Co_{1-x}Mn_xFe_2O_4$ (x = 0-0.4) at room temperature. It is obvious that with Mn substitution the magnetostriction decreases, because this is accompanied with a decreasing magnetocrystalline anisotropy which causes a reduced coercivity [23].

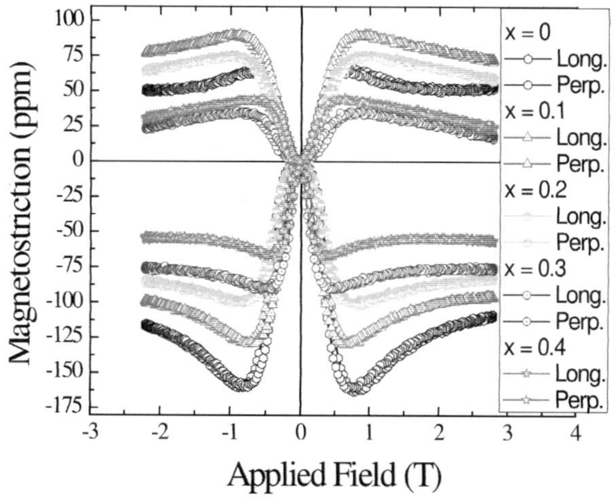

Figure 12. Measurements of the longitudinal and perpendicular magnetostriction, performed on $Co_{1-x}Mn_xFe_2O_4$ (x = 0-0.4)

7. Conclusion

The magnetic properties of polycrystalline Co-ferrite depend strongly on the production process (chemical routes etc., pressure). The annealing process (temperature and time) influence additionally the distribution of the Co^{2+} cations but also the microstructure (grain size). Powder compaction hydrostatic pressure and field annealing cause a significant increase of the magnetostriction, which is due to a reordering of the Co^{2+} ions. Single crystalline Co-ferrite exhibits a FOMP transitions which is also visible in the λ times H curve. Substitutions of Co by other elements such as Zn or Mn cause generally a magnetic softening of the material.

8. References

[1] Raj K, Moskowitz R and Casciari R 1995 *J. Magn. Magn. Mater.* 149 174

[2] Häfelli U, Schütt W, Teller J and Zborowski M 1997 *Scientific and Clinical Applications of Magnetic Carriers Plenum*, New York

[3] Kryder M H 1996 *MRS Bull.* 21 917

[4] McCallum R W, Dennis KW, Jiles D C, Snyder J E and Chen YH 2001 *Low Temperature Physics* 27 266

[5] Chen Y, Snyder J E, Schwichtenberg C R, Dennis K W, McCallum R W and Jiles D C 1999 *IEEE Trans. Magn. 35* 3652

[6] Valenzuela R 1984 *Magnetic ceramics* Cambridge University Press, Cambridge p. 212.

[7] Sato Turtelli R, Atif M, Mehmood N, Kube Fl, Biernacka K, Linert W, Grössinger R, Kapusta Cz and Sikora M 2012 *Mater Chem. Phys. 132* 832

[8] Nlebedim I, Snyder J, Moses A and Jiles D C 2010 *J. Magn. Magn. Mater.* 322 3938

[9] Muhammad A, Sato Turtelli R, Kriegisch M, Grössinger R, Kubel F and Konegger T 2012 *J. Appl. Phys.* 111 013918

[10] Grössinger R, Mehmood N, Senbaslar G, Muhammad A, Sato Turtelli R, Linert W and Kubel F 2008 *Proc. of 20th International Workshop on Rare Earth Permanent Magnets and Their Applications,* Crete, Greece, 8–10 September 2008, p. 214 (ISBN:978-960-86733r-r6-6).

[11] Pannaparayil T and Komarneni S 1989 IEEE Trans. Magn. 25 4233

[12] Yan C H, Xu Z G, Cheng F X, Wang Z M, Sun L D, Liao C S and Jia T J 1999 *Solid State Commun.* 111 287.

[13] Moumen N, Veillet P and Pileni M P 1995 *J. Magn. Magn. Mater.* 149 67

[14] Ziolo R F, Giannelis E P, Weinstein B A, O'Horo M P, Ganguly B N, Mehrotra V, Russell MW, Huffman D R 1992 *Science* 257 5067

[15] Cheng F X, Peng Z Y, Liao C S, Xu Z G, Gao S and Yan C H 1996 *Solid State Commun.* 107 471

[16] Sato Turtelli R, Atif M, Krippel J, Grössinger R, Kubel F and Linert W 2012 *International J. Mater. Research* 09 1163

[17] Hanh N, Quy O K, Thuy N P, Tung L D and Spinu L 2003 *Physica B: Condensed Matter* 327, 382.

[18] Duong G V, Sato Turtelli R, Hanh N, Linh D V, Reissner M, Michor HH, Fidler J, Wiesinger G and Grössinger R 2006 *J. Magn. Magn. Mater.* 307 313

[19] Duong G V, Sato Turtelli R, Nunes W C, Schfler E, Hanh N, Grössinger R and Knobel M 2007

Journal of Non-Crystalline Solids 353 805

[20] Sato Turtelli R, Duong G V, Nunes W, Grössinger R and M. Knobel 2008 *J. Magn. Magn. Mater.* 320 e339

[21] Kriegisch M, Ren W, Sato Turtelli R, Müller H, Grössinger R and Z. Zhang 2012 *J. Appl. Phys.* 111 07E308.

[22] Guillot M, Ostorero J and Marchand A 1988 *Z. f. Physik B Condensed Mater.* 71 193

[23] Atif M, Sato Turtelli R, Grössinger R and Kubel F 2013 *J. Appl. Phys.* 113 153902

IOP Conf. Series: Materials Science and Engineering **60** (2014) 012021 doi:10.1088/1757-899X/60/1/012021

Cavity closure during compression between semi-closed dies using superplastic tin-lead alloy

Adnan I O Zaid[1] and Mohammad M Al-Tamimi[2]
[1]Industrial Engineering Department, University of Jordan, Amman, Jordan
[2]Industrial Engineering Department, University of Jordan, Amman, Jordan

E-mail: a_kilani@hotmail.com, mhn_iie2@hotmail.com

Abstract. Superplasticity is a feature of a material or alloy, which allows the material to deform plastically to an extremely large strain at low values of stress under certain loading conditions of strain rate and temperature. Eutectic tin-lead alloy is a practical material for research investigations as it possesses a superplastic behavior at room temperature and low strain rate which makes it a useful tool in simulating the ordinary engineering materials at high strain rate and temperature, and has been extensively used as a model material. In this paper, superplastic tin-lead alloy was used at room temperature to simulate the closure of cavities in steels at high temperatures in the hot region under dynamic loading (high strain rate) under the effect of compressive loads using semi-closed dies (modified dies) with 45° inclination and compare the results from these dies with those of flat platens (open dies) published previously. Hollow specimens having different values of bore diameter (D_b) to outer diameter (D_{out}), of the same height and volume were investigated under 40% height reduction. The cavity closure for each specimen was determined. Comparison is made between flat platens and semi-closed dies regarding cavity closure based on bore diameter, bore volume, reduction percentage in bore diameter and reduction percentage in bore volume, at the 40% reduction in height. It was found that modifying the platens (45° inclination) resulted in lower values of bore diameters and volume i.e. higher values of reduction in bore diameters and volumes percentages irrespective of the value of bore diameter and the ratio of D_b/D_{out}.

1. Introduction

Cast ingots normally include some defects, for example; porosity and voids, inclusions and entrapped gasses which tend to reduce their mechanical properties and surface quality. Therefore, it is necessary to get rid of these defects. One recommended method is the free upsetting test for the cast ingots, and it is becoming customary to subject the casts, while hot, into compressive stresses. In the early studies on cavity closure during compression, friction between the workpiece and die interface was considered the main parameter in determining the force, energy, flow of metal and the changes in the cavity dimensions (enlargement or shrinkage) [1,2]. The ring test was also used to determine the changes in the coefficient of friction during compression at room and high temperatures [3,4], and at different strain rates [5]. Void growth in 6061 – aluminum alloy under triaxial stress system was investigated and reported using notched tensile test specimens, to maintain the state of triaxial stress system [6]. The effect of local deformation conditions on the cavitation behavior and cavity growth during open-die hot forging of Ti-6Al-4V was reported by Nicolaou and Semiatin [7]. The effect of strain path reversal on microstructure evolution and cavitation during hot torsion testing of Ti-6Al-4V is given by the same authors. They found that the cavity growth persisted as long as there was a flow-stress difference in the regions surrounding a given cavity [8]. The effect of stress and strain path on cavity closure during hot working of an alpha/beta titanium alloy was investigated by Nicolaou *et al.* Their work indicated that the rate of cavity closure decreases with increasing deformation. This suggests that deformation at very large strains may be required to totally heal damage or complete cavity closure [9].

1.1. Superplasticity and Superplastic Materials

Superplasticity is a feature which allows the material to deform to large strains at low stress if the material is deformed under certain loading conditions. The elongation may be of the order of a few hundred percent up to 2000% [10]. Superplasticity can be induced either in materials possessing a stable, ultra-fine grain size at the temperature of deformation ≥0.4 Tm, where Tm is the absolute melting point or in those subjected to special environmental conditions [11]. Superplastic materials are a unique class of metals that have the ability to undergo extra ordinary large tensile ductility where elongations of 200% and more are experienced. The requirements necessary to achieve superplastic behaviour are: i) fine and equiaxed grain size, usually less than 10 microns. ii) Testing temperature that is greater than or approximately half the absolute melting temperature of the subject material. iii) Controlled strain rate [12]. This shows the importance of using superplastic material for investigating cavity closure in ordinary engineering materials during plastic deformation in the forging process. Taylor *et al.* have investigated the effect of void size and spacing on the ductility and flow stress of superplastic Pb-Sn tensile specimens using experimental testing and Finite Element model [13]. Recently, Ridely *et al.* have studied and reported the effect of strain rate path on cavitation in superplastic aluminum alloy 5083. The alloy was tested under two different conditions: one in uniaxial tension and the other under constant strain rate. They observed continuous nucleation of cavities in both test conditions. Furthermore, they found that the cavities were associated with inter- metallic particles as observed under metallographic examination. The analysis of their experimental data led to the conclusion that the growth of cavities was controlled primarily by plastic deformation, although diffusion may play a role in the early stages of growth [14]. The deformation behavior of the fine grained tin–lead eutectic alloy was investigated by Soliman and Al-Seif using compression specimens at initial strain rates in the range 10^{-5} to 10^{-1} s^{-1} [15]. Recently, Mulholland *et al.* have conducted set of parametric experiments on a superplastic material (eutectic tin–lead alloy) with one or more pre-drilled holes in each specimen. They found an increase in ductility with the number of holes up to 10 holes followed with a decrease thereafter [16].

1.1.1. Mechanical Behaviour of Superplastic Materials

Unlike the mechanical behaviour of an ordinary engineering metal or alloy where the flow stress is function of strain, strain rate and temperature. The flow stress of a superplastic material is mainly affected by strain rate at a specified temperature and is strain independent. Steady state flow of structurally superplastic materials invariably shows an increase in flow stress with increasing strain rate. The eutectic tin-lead alloy (61.9% tin and 38.1% lead by weight) has extensively been used as a model material to investigate the deformation characteristics of superplastic materials and ordinary engineering materials at high strains and hot loading conditions [11, 17, 18].

1.2. Effect of Strain Rate on cavity Closure

Recently, Zaid and Al-Tamimi have used superplastic tin-lead alloy at room temperature to simulate the closure of cavities in steels at high temperatures in the hot region under dynamic loading (high strain rate) under the effect of compressive loads using flat platens (open dies). Hollow specimens having different values of bore diameter (D_b) to outer diameter (D_{out}), of the same height and volume were investigated under different values of height reduction percentages ranging from 20% to 80% , and the percentage of cavity closure at each reduction percentage was determined and discussed. It was found that the cavity closure percentage increases or decreases at slow rate for reduction percentage in height less than 40% and increases more rapidly for reduction percentages in height above this value [18].

2. Materials, Equipment and Experimental Procedures

2.1 Materials

High purity, 99.99%, tin and lead supplied in granular form were used for the production of the superplastic tin-lead alloy. Graphite crucibles were used for melting and graphite rods were used for stirring.

2.2 Equipment and experimental procedures

The experimental procedure was carried out in the following sequence: preparation of the superplastic tin-lead alloy, carrying out the extrusion process on the cast eutectic alloy to increase its degree of superplasticity, preparation of specimens from the extruded alloy for the compression test between the semi-closed dies at 40% height reductions and finally sectioning the different specimens along their axis by wire cutting and finally tracing their profiles using the Profilometer, (contracer).

2.2.1 Preparation of the superplastic tin-lead alloy

The preparation process started by placing the pre-calculated amount of pure Tin granules inside a graphite crucible, charged into an electric furnace at 523 ^{o}C. After 10 minutes the crucible was taken out of the furnace and the calculated amount of lead (Pb) granules were introduced to the crucible and mixed with Tin (Sn). The crucible was then returned to the furnace for 10 minutes, taken out of the furnace, and the melt was stirred for one minute and poured to solidify in a thick brass hollow cylinder. It is worth noting that the amount of tin and lead were calculated to satisfy the eutectic point of 61.9% tin and 38.1% lead by weight.

2.2 Extrusion process

The extrusion process was carried out at different billets from the prepared tin-lead alloy to increase their degree of superplasticity. The extrusion process was carried at 25 ^{o}C which corresponds to a homologous temperature of 0.65 i.e. hot extrusion, using five extrusion dies of different openings to produce hollow cylindrical specimens of bore to outer diameter, (D_b/D_{out}), of 0.1, 0.2, 0.3, 0.4 and 0.5 at extrusion ratios of 6.6, 6.4, 6.0, 5.6 and 5. All the extrusion tests were carried out at cross head speed = 5mm/min, and were kept in a freezer at -10 °C to avoid grain growth. Figures 1 and 2 show the designed and manufactured dies in the assembled and exploded conditions.

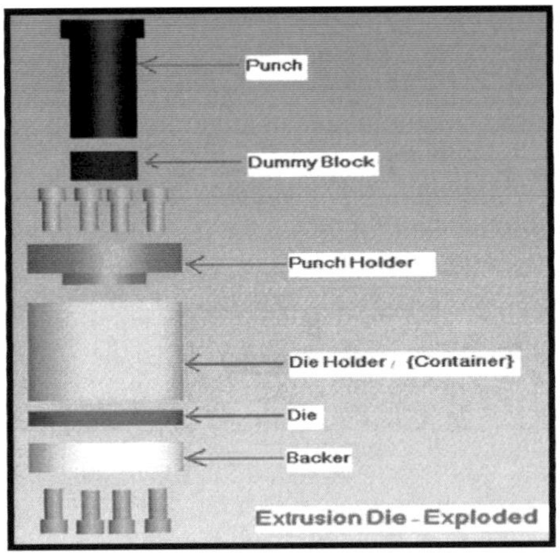

Figure1. Photograph of the designed and **Figure 2.** Extrusion die - exploded manufactured dies.

2.3 Compression between two 45° semi-closed dies test

The compression test between two "45° semi-closed dies" for each specimen was performed using GUNT HAMBURG WP 310 Hydraulic Universal Material Tester of 50 kN capacity at cross head speed of 5 mm/min, which corresponds to strain rate of 4.17 $*10^{-3}$ s^{-1} at the end of compression (40% reduction in height). The die design allows only 40 % height reduction on specimens of different (D_b/D_{out}) namely 0.1, 0.2, 0.3, 0.4 and 0.5 but all having the same solid volume. After the test, the actual bore volume of each specimen was measured by filling water inside the bore using 1-ml Insulin Syringe having an accuracy of ±0.005ml).

2.4 Tracings of the specimens profiles

Five different specimens of 20mm height and with different (D_b/D_{out}) ratios were sectioned using a wire cutting process, after which the deformation of the bore was traced using Shadomaster R122 BM Profilometer at a magnification of 10.

3. Results and Discussion

All the following results were obtained for hollow cylindrical specimens of the same volume but of different bore diameters and outside diameters. Furthermore, all the specimens were compressed between the same semi-closed dies with 45° inclination at cross head speed of 5 mm/min, which corresponds to strain rate of 4.17 $*10^{-3}$ s^{-1}, which falls within the range of the rate sensitivity of the tin-lead alloy i.e. when it behaves superplastically.

3.1 Comparison between Flat Platens and Semi-Closed Dies

In this section, comparison is made between flat platens and semi-closed dies with 45° inclined platens based on bore diameter, bore volume, reduction in bore diameter percentage and reduction in bore volume percentage, only at 40% reduction in height. It can be seen from the histograms of Figures. 3, 4, 5 and 6 that modifying the platens (45° inclination) resulted in lower values of bore diameters and volume and higher values of reduction percentages in bore diameters and volumes respectively irrespective of the value of bore diameter and the ratio of D_b/D_{out} . It can also be seen, that the difference is more pronounced at smaller bore diameters and smaller ratios of D_b/D_{out}. This means the plastic flow in the radial direction in the radial direction towards the center is more pronounced in case in case of small bore diameters and smaller ratios of D_b/D_{out}, i.e. lower friction at the interference as compared to higher ratios of D_b/D_{out} , which agress with the previous work of the ring test [1,2]. Furthermore, it can be seen from Fig. 5 that at D1 (D_b/D_{out} = 0.1) at 40% reduction in height the diameter closure is 65% in the semi-closed , where is the diameter closure in case of the flat platens die is about 19%. This is explicitly shown in Fig. 6 that at D1 (D_b/D_{out} = 0.1) the volume cavity closure is almost completely closed in the semi-closed dies (about 92%) at 40% reduction in height, where is the volume cavity closure in case of the flat platens die is about 58% for the same D_b/D_{out} at the same reduction in height.

3.2 Comparison between Flat Platens and Semi-Closed Dies on the Shape Changes of the Cavity

Comparing the effect of modifying the flat platens to 45° semi-closed die with the flat platens dies when compressing identical specimens at the same reduction in height percentage namely 40%, it can be seen from Figure 7 (a) to (e) inclusive, that the semi-closed dies produced more cavity closure than the flat platens for all the specimens irrespective of their D_b/D_{out} ratio. Furthermore, in case of the flat platens, the bore increased in diameter in certain regions and reduced in others, where as in the case of semi-closed dies reduction of the bore diameter was observed along the whole height of the specimens and being maximum at the middle section of the height except for the specimen D5 (D_b/D_{out} = 0.5), Figure 7 (e), where the bore increased in some regions and reduced in others.

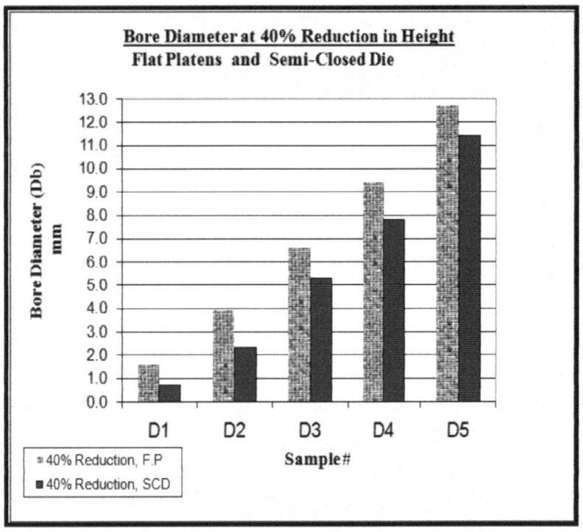

Figure 3. Bore diameter after compression (40% reduction in height) flat platten and 45° semi-closed dies

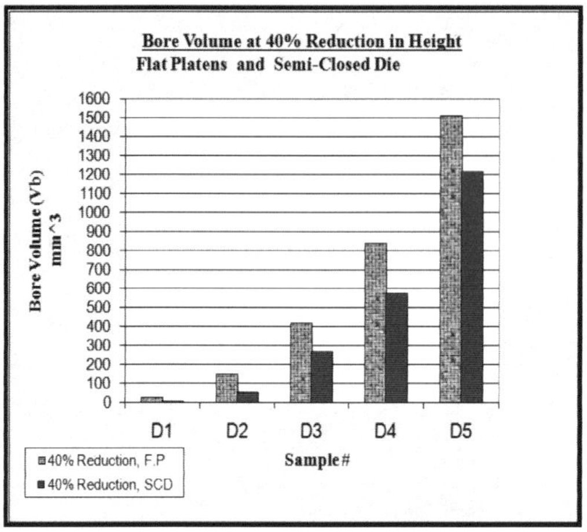

Figure 4. Bore volume after compression (40% reduction in height) flat platten and 45° semi-closed dies

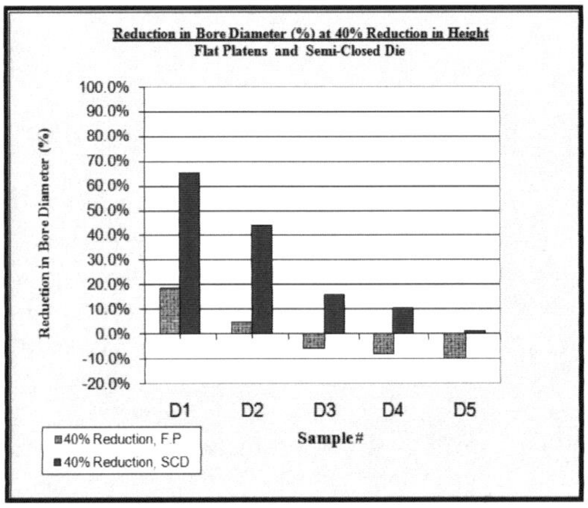

Figure 5. Reduction in bore diameter percentage after compression (40% reduction in height) flat platten and 45° semi-closed dies

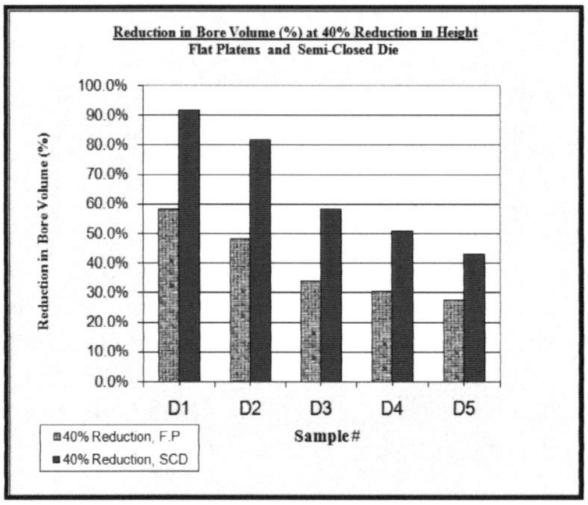

Figure 6. Reduction in bore volume percentage after compression (40% reduction in height) flat platten and 45° semi-closed dies

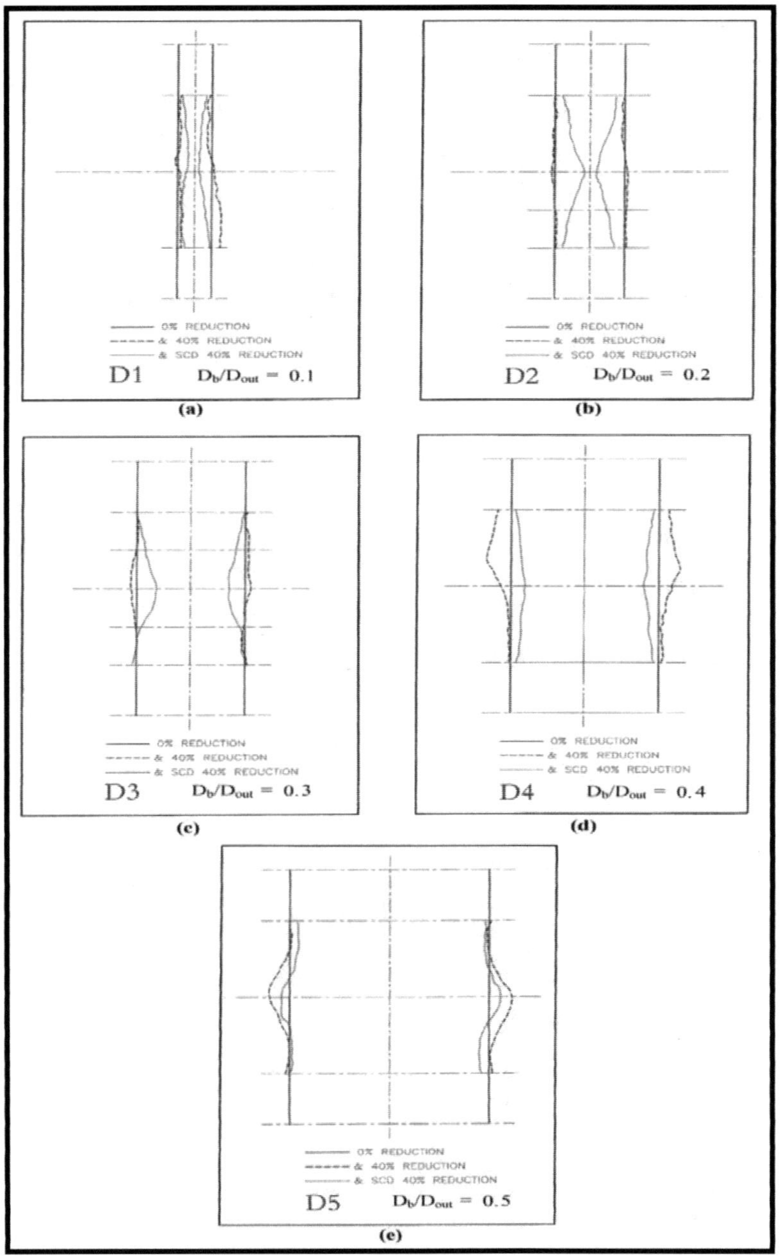

Figure 7. Tracings of the sectioned specimens D1, D2, D3, D4 and D5 along their axes at 0 and 40 % reduction in height using flat platens and 40% reduction in height using semi-closed dies (a), (b), (c), (d), and (e) at [D_b/D_{out}] of 0.1, 0.2, 0.3, 0.4, and 0.5 respectively

4. Conclusions

The following points may be concluded:

- Superplastic tin-lead alloy can be used successfully, under certain conditions, as a model material to simulate the behavior of metals at high strain rate and hot conditions in flat and semi-closed dies forging.
- Cavity closure does not depend only on the end friction at the die-workpiece interface as repeatedly reported in the literature, but also depends on other parameters, e.g. strain rate, the geometrical shape of the platens if flat or at angle, reduction in height and bore diameter to the outside diameter ratios of the specimen, (D_b/D_{out}) .
- Using semi-closed dies during compression by modifying the platens by an angle (45° inclination) resulted in lower values of bore diameters and volume i.e. higher values of reduction in bore diameters and volumes percentages respectively irrespective of the value of bore diameter and the ratio of D_b/D_{out} .

5. References

[1] Male A T and Cockcroft M G 1965 A Method for the Determination of the Coefficient of Friction of Metals Under Conditions of Bulk Plastic Deformation, *Journal of the Institute of Metals*, 93 38-46.

[2] Wiegels H and Herbertz R 1981 Plotting of Stress-Strain Curves at a High Deformation Rates in the Cylinder Compression Test with Low and High Friction *Stahl U. Eisen* 101 1127-1132.

[3] Slater R A C ,Johnson W and AkuS Y 1968 Fast Upsetting of Short Circular Cylinders of Plain Medium Carbon (0.55%) Steel at Room Temperature , *9th International M.T.D.R. Conference* University of Manchester.

[4] Male A T 1966 Variations in Friction Coefficients of Metals during Compressive Deformation *Journal of the Institute of metals* 194 121-125.

[5] Jain S C and Bramely A N 1968 Speed and Frictional Effects in Hot forging *Proc, Inst. Mech. Engrs*182 783-788.

[6] Agarwal H Gokhale A M, Grahamb S and Horstemeyer M F 2003 Void Growth in 6061-Aluminum Alloy under Triaxial Stress State *Materials Science and Engineering,* A341 35-42.

[7] Nicolaou P D and Semiatin S L 2005 An Analysis Of Cavity Growth During Open-Die Hot Forging Of Ti-6Al-4V *Metallurgical And Materials Transactions. A* 36A 1567-1574.

[8] Nicolaou P D and Semiatin S L 2007 Effect of Strain-Path Reversal on Microstructure Evolution and Cavitation during Hot Torsion Testing of Ti-6Al-4V *Metallurgical And Materials Transactions. A* 38A 3023-3031.

[9] Nicolaou P D, Goetz R L and Semiatin S L 2008 Effect of Stress and Strain Path on Cavity Closure during Hot Working of an Alpha/Beta Titanium Alloy, *Metallurgical And Materials Transactions. A* 39A 659-665.

[10] Kalpakjian S and Schmid S R 2002 *Manufacturing Processes for Engineering Materials* (4th ed.), New Jersey: Pearson Education In.

[11] Zaid A I O 2004 Superplasticity and Superplastic Materials: Applications and Recent Developments *Proc of 2nd International Conference on Frontiers of Advanced Engineering Material*, Lahor, Pakistan 37-44.

[12] Pilling J and Ridely N 1989 Superplasticity in Crystalline Solids, *The Institute of Metalls*, London, UK 102-157.

[13] Taylor M B, Zbib H M and Khaleel M A 2002 Damage and Size Effect During Superplastic Deformation *International Journal of Plasticity* 18 415-442.

[14] Ridley N, Bate P S and Zhang B 2007 Effect Of Strain Rate Path On Cavitation In Superplastic Aluminum Alloy *Materials Science and Engineering* A463 224-230.

[15] Soliman M S and Al-Seif K A2005 Compression Testing of Superplastic Lead–Tin Eutectic Alloy at Room Temperature, *The Arabian Journal for Science and Engineering*. 30 177-188.

[16] Mulholland M, Khraishi T Shen Y L and Horstemeyer M 2006 Void growth and interaction experiments: Implications to the optimal straining rate in superplastic forming *International Journal of Plasticity* 22 1728–1744.

[17] Soliman M S 1995 Effect of Strain Rate and Grain Size on the Ductility of Superplastic Pb-62% Sn Eutectic at Room Temperature *Scripta Metallurgica et Materialia* 33 919-924.

[18] Zaid A I O and Al-Tamimi M M 2011 Effect of Strain Rate on Cavity Closure During Compression Between Flat Platens Using Superplastic Tin-Lead Alloy *12th International Symposium on Advanced Material* Islam Abad, Pakistan. 37-44.

Large-scale oxide nanostructures grown by thermal oxidation

Z W Liu, M L Zhong, C M Tang
School of Materials Science and Engineering, South China University of Technology, Guangzhou, 510640, P.R. China

E-mail: zwliu@scut.edu.cn

Abstract. Large scale oxide nanostructures of CuO, Fe_2O_3, Co_3O_4, ZnO, etc. were prepared by catalyst-free thermal oxidation process in atmosphere using pure metal as the starting material. Various single crystalline nanostructure arrays, including nanowires, nanobelts, nononeedles, nanoflakes, and nanowalls were obtained. These nanostructures can be grown from bulk materials, like foils or sheet, or from the microsized metal powders and the pre-deposited metal film. The growth time, temperature and substrate have important effects on the morphology, size and distribution of the nanostructures. Different from V-S or V-L-S mechanisms, the growth of nanostructure is found to be based on the metal ion diffusion process. The gradual oxidation process of the metals was clearly demonstrated. The properties of these nanostructures including gas sensing, magnetism, photoluminescence, and field emission were extensively investigated.

1. Introduction

Nanostructures with large surface area and possible quantum-confinement effect exhibit distinct mechanical, electronic, magnetic, optical and thermal properties. Some of the most important candidates in the family of nanomaterials are metal oxide nanostructures, which have great potential in addressing some fundamental scientific issues on low dimensionality and applications thereof. Oxide nanostructures can be grown by various methods such as chemical routes, template ways and by electro-spinning techniques. Future applications in commercial devices require further development of the growth method with the advantages of mass-production, efficiency, controllability, and catalyst-free. In addition, it should be able to synthesize the materials at relatively low temperatures, and to assemble the nanomaterials onto various substrates for further characterization and applications [1].

As early as the 1950s, the Cu, Fe, Zn, Ni and Mn oxides whiskers or filaments were prepared by annealing the corresponding metals in air at specific temperatures [2,3]. Currently, the thermal annealing method has been well developed and attracted much more attention due to its simplicity, low cost, and use of standard equipment. This synthesis can be carried out without a catalyst and sometimes even without any gas control. Such a chamber free approach is suitable for future large-scale production and for a variety of metals. In addition, the growth temperature for this technique is much lower than that used in normal vapor-based method. However, a clear understanding of the growth process and the properties of the nanostructures is still in need. In our recent work, various oxide nanostructures were synthesized by thermal oxidation and a systematic study on the growth and functional properties were carried out.

2. Synthesis of metal oxide nanostructures by thermal oxidation

Thermal oxidation methods can grow the metal oxide nanomaterials directly on the surface of bulk metals like foils, plates, or wires, as well as on the thin film or nanoparticles prepared by other processes. The purity, grain size and surface roughness of the precursors have effects on the nanostructure growth. The temperature, typically between 300 to 800°C, is a key factor governing the growing process, which can be provided by hot plate, furnace, oven or other alternative heating systems such as flame, resistive heating. Various heating sources can provide different temperature ramp rates from about 5°C/min to100°C/s. The oxidation time can be varied from a few seconds to dozens of hours or more in some case. The oxidizing atmosphere can be dry or wet air, pure oxygen, a mixture of oxygen with argon, nitrogen, water vapor, etc. and it can be static or flow through the oxidation system. In addition, the background pressure can be ambient atmosphere or less than 1 atm, which also affects the growth process. The simplest thermal oxidation is heating the starting metals in air. The typical set-up of our hot-plate technique for thermal oxidation is shown in Fig.1. Various nanostructured oxides have been prepared from bulk, powder and film precursors.

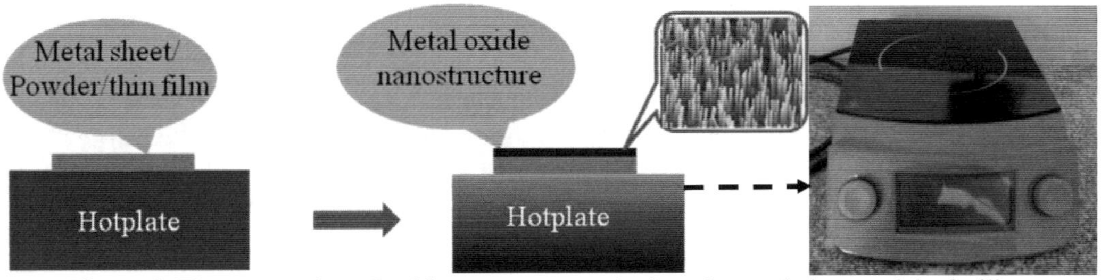

Figure 1. The preparation of oxide nanostructures by hot-plate technique and the facility

2.1 CuO nanowires

CuO is a p-type semiconductor with a narrow band gap of 1.4 eV, and is also an antiferromagnetic material. It has been used as a basic material in high-T_c superconductors and also has potential application in solar cells, field emission emitters, electronic cathode materials, catalysts and gas sensors [4-6]. Using the hotplate technique, we have grown single crystalline CuO nanowires with good quality [7], as shown in Fig.2. Typically, the nanowires were synthesized by heating copper in various forms at 400~500°C in air for 24-96 h. It was found that the growth temperature has a distinct influence on the diameter, while the growth time has an important effect on the length and density of nanowires (Fig.3). Kumar et al.[8] also obtained similar results, which are related to the growth mechanism. The substrate can also affect the nanostructure growth, as shown Figs.2. The different morphologies were obtained for the oxidations of Cu foil, grid, powders and film. Love et al. [9] prepared the CuO nanowires by heating the Cu powders with average particle sizes of 1, 10 and 50μm in a box furnace and found that the nanowire coverage also depends on the particle size.

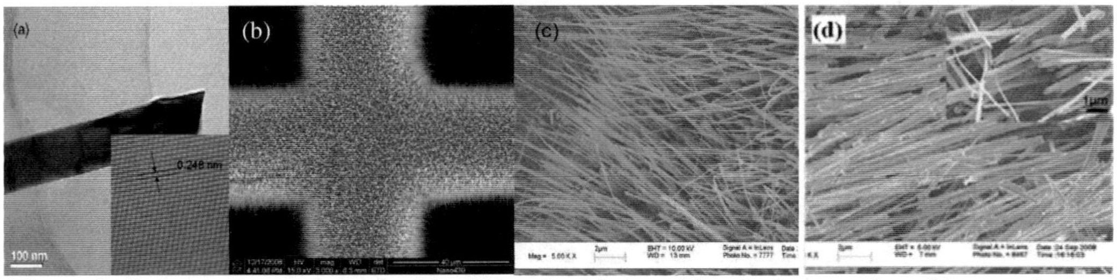

Figure 2. CuO nanowires synthesized by thermal oxidation: (a) a single nanowires; (b) nanowires on Cu grid; (c) nanowires on Cu foil (450°C); and (d) nanowires on Cu plate at 480 °C

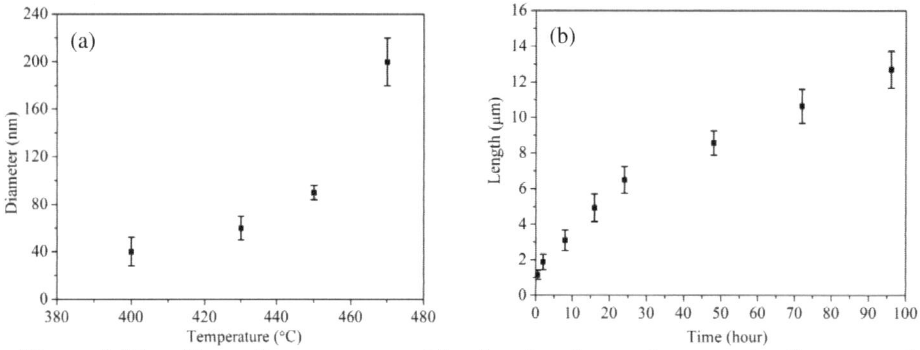

Figure 3. The effects of temperature and heating duration on the nanowire diameter and length (a: duration of 72 h; b: heating temperature of 430 °C).

2.2 Fe_2O_3 nanowires, nanoneedles and nanoflakes

Hematite (α-Fe_2O_3) is another important semiconductor metal oxide with small band gap (E_g=2.1 eV), high corrosion resistance and low cost. Yu et al. [10] synthesized large scale α-Fe_2O_3 nanoflakes by heating the Fe foil in air using a hotplate. All flakes grew perpendicularly to the surface of the foil and have very sharp tips with mean diameter of 17 nm. Most nanoflakes are single crystalline with growth direction of [110]. The temperature has a district influence on the morphology. Zheng et al. [11] also synthesized ultra-sharp α-Fe_2O_3 nanoflakes on the hotplate at a very low temperature of 260°C and found that the nanoflakes become broader with increasing temperature. Three-dimensional urchin-like α-Fe_2O_3 structures were also formed via the thermal oxidation of Fe powders in air at ~280 to 480°C [12]. We prepared α-Fe_2O_3 nanostructures on Fe foils, powders and films (Fig.4) and found that the heating temperature, duration, substrate and the form of the precursor have important effects on the morphology, size and distribution of the nanostructure [13]. For the nanostructured grown on foil, it was found that there was wire-like structures at the low temperature (<430°C), flakes-like structures at 480°C, and no noticeable nanostructure coverage above 530°C. The obtained nanostructures were demonstrated single crystalline.

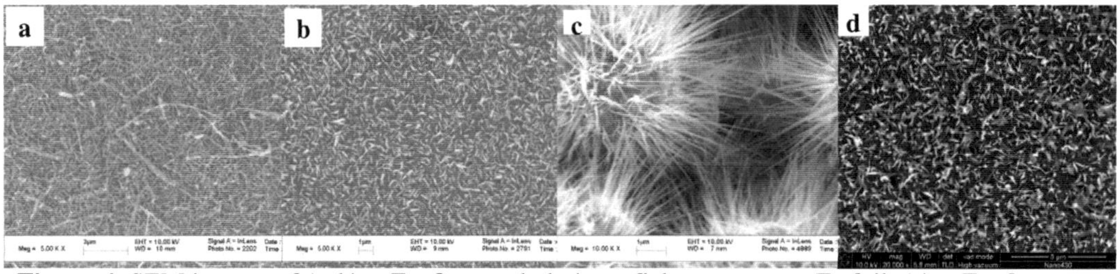

Figure 4. SEM images of (a, b) α-Fe_2O_3 nanobelts/nanoflakes grown on Fe foil, (c) α-Fe_2O_3 nano-needles grown on Fe powders, (d) α-Fe_2O_3 nanobelts grown on Fe thin film

2.3 Co_3O_4 and MnO_2 nanostructures

The Co_3O_4 nanostructures have also been synthesized by direct heating the cobalt foils in air conditions. Yu et al. [14] prepared Co_3O_4 nanostructures by directly heating Co foil on a hotplate in ambient conditions at 200-350°C for 1min to 12h. They obtained single-crystalline Co_3O_4 nanowalls with several to tens nanometers in thickness and hundreds nanometers in length. They suggested that the morphology of the nanowall be controlled by the growth temperature and duration. We also synthesized Co_3O_4 and MnO nanowall or nanoflakes on Co and Mn foils, but no 1-dimensional nanostructures were observed (Fig.5).

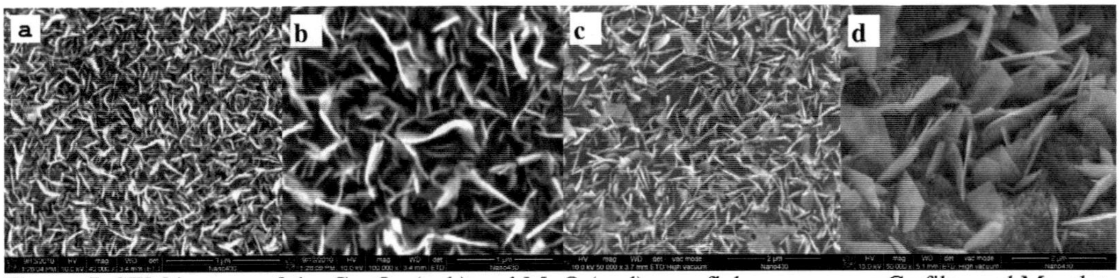

Figure 5. SEM images of the Co_3O_4 (a, b) and MnO (c, d) nanoflakes grown on Co film and Mn plate

2.4 Other oxide nanostructures

ZnO nanowires were also successfully synthesized by oxidation zinc foils, films in air [10,15]. We oxidized granular sputter deposited Zn films at 450-600°C [16] and ZnO nanowires were found to grow from individual Zn nanograins. Results showed that the nanowires preferably grow from relatively porous Zn film and a small amount of oxygen flow has beneficial effect. The single crystal nanowire obtained at 600 °C had a mean diameter of <50 nm with good structural quality.

Excepting the nanostructures of the monary metal oxide nanostructure, the hybrid nanostructures can also be produced when the precursors of binary alloys are employed. Zhu et al. [15] synthesized ZnO/CuO hybrid nanostructures by directly heating Cu-Zn alloy (brass) on a hotplate in ambient condition. With the increase in Zn concentration in brass, the dominant products transition from CuO nanowires to ZnO nanostructures. Both CuO and ZnO nanoflakes are obtained from the brass substrates with the intermediate ZnO content. It is worth noting that although CuO and ZnO are formed simultaneously on the same substrate, doping of Cu in ZnO or Zn in CuO was observed. Hence, this technique has potential to provide preparations for complicated oxide nanomaterials.

3. The growth mechanism for thermal oxidation process

The morphology and crystalline structure of the as-prepared oxide nanostructures are determined by the crystalline structure, purity, grain size, surface roughness of the starting metal. Also, the reaction parameters, such as temperature, atmosphere composition, oxygen partial pressure, heating systems, have significant effects. To control the growth of oxide nanostructures for various applications, their growth process and growth mechanism have to be fully understood. Numerous observations have been performed and have led to different theories regarding to the mechanisms for the spontaneous oxide nanostructure growth via thermal oxidation.

1D metal oxide nanostructures growth is usually described by either VS or VLS mechanism [17]. In VS process, the key steps are the evaporation of the precursor at high temperature followed by condensing of vapour and growth of the nanostructure in low temperature region. However, the temperature for thermal oxidation (typically 300-800°C) is far less than the melting points of the metals and their oxides, and the low and high temperature regions are difficult to distinguish. In VLS process, nanostructures grow with a catalyst (metal droplet) at the tip and the diameter of the structure determined by the size of the metal droplet is uniform. However, no metal catalyst droplet was observed in the experiments and the diameter of the nanostructures decreased from bottom to tip. Therefore, both VS and VLS mechanisms should not be responsible for the thermal oxidation growth.
We investigated carefully the structure and growth of various oxide nanostructures based on our hot plate technique. X-ray diffraction investigations showed various products obtained for oxidizing each metal. During thermal oxidation of Cu, both Cu_2O and CuO were observed. For the oxidation of Fe, both Fe_3O_4 and Fe_2O_3 were also obtained. TEM results showed that the nanowires are single phase of CuO or Fe_2O_3. The cross-section of the oxidized product was investigated by FESEM with energy dispersive X-ray spectroscopy (EDS). A three layer structure of Cu_2O film/CuO film/CuO nanowires for obtained sample is finally established for thermal oxidation of Cu, as shown in Fig.6a. The oxidation process of Cu includes two steps, i.e. $4Cu+O_2 \rightarrow 2Cu_2O$ and $2Cu_2O+O_2 \rightarrow 4CuO$. Based on the Cu-O phase diagram, both CuO-Cu_2O and Cu_2O-Cu phase equilibriums exist at high temperature.

The equilibrium oxygen pressures at CuO-Cu_2O and Cu_2O-Cu interfaces, $P_{CuO\text{-}Cu2O}$ and $P_{Cu2O\text{-}Cu}$, can be calculated thermodynamically according to the equation $P_{\text{equilibrium}} = \exp(-2\Delta G_0/RT)$, where ΔG_0 is the free energy of formation of CuO or Cu_2O, R the gas constant (8.31 J/mol K) and T the absolute temperature. Based on the values of ΔG_0 for Cu_2O and CuO [18], $P_{CuO\text{-}Cu2O}$ and $P_{Cu2O\text{-}Cu}$ can be given by 3.2×10^{-7}–1.6×10^{-3} and 3.3×10^{-17}–1.7×10^{-10} atm, respectively, in the temperatures range of 400-500 °C. The environmental oxygen partial pressure (0.21 atm) is higher than $P_{CuO\text{-}Cu2O}$ and $P_{Cu2O\text{-}Cu}$. Therefore, both CuO and Cu_2O can be formed in the present case. Thermodynamically, the CuO phase will be formed in the late stage of the oxidation due to the difference in above two equilibrium oxygen partial pressures. Indeed, from the experimental observation, the Cu_2O phase is the main product during the first stage of the thermal oxidation of copper, whereas the Cu_2O-CuO phase equilibrium becomes important afterwards. In addition, it has been suggested that several oxides including copper oxide formation follows a parabolic law [19], i.e., $l^2=k\cdot t$, where l is the thickness of CuO film, k is growth rate, and t is the time. If we consider the thickness as the nanowire length, from Fig. 3, the obtained values for the growth rate constant, k, were about 3×10^{-15} m^2/min for T=430°C, which indicates that the growth of nanowire prepared by thermal oxidation follows the same mechanism that control the copper oxide film formation.

Very recently, Gonçalves et al. [20] proposed that the CuO nanowires are formed as a result of rapid, short-circuit diffusion of the Cu ions across grain boundaries and/or defects in a Cu_2O underlayer. Based on their proposal and our experimental results, a proposed nanowire growth model is given in Fig. 6c [7]. Initially a Cu_2O layer is formed, and due to the interface compressive stress such layer is highly porous and defective. Further oxidation requires that Cu atoms reach the film/air interface by two pathways: lattice and grain-boundary diffusion. The first pathway leads to a continuous increase of the Cu_2O and CuO layers, whereas the second results in the formation of the nanowires. The time dependence of the nanowire length, observed in Fig. 2 from 400 to 470 °C, can be directly attributed to grain-boundary diffusion of Cu ions toward the Cu_2O-CuO interface since the variations in the nanowire densities and diameters during growth at various temperatures are small. The morphology of the nanowires is determined by the microstructure of the underlying Cu_2O layer because it is within this intermediate layer where the "short-circuit" diffusion of Cu atoms takes place. Similarly, for oxidation of Fe or Co, Fe→Fe_3O_4→Fe_2O_3 or Co→CoO→Co_3O_4 were demonstrated by both theoretical and experimental results. The cross section image of oxidized Fe is shown in Fig.6b.

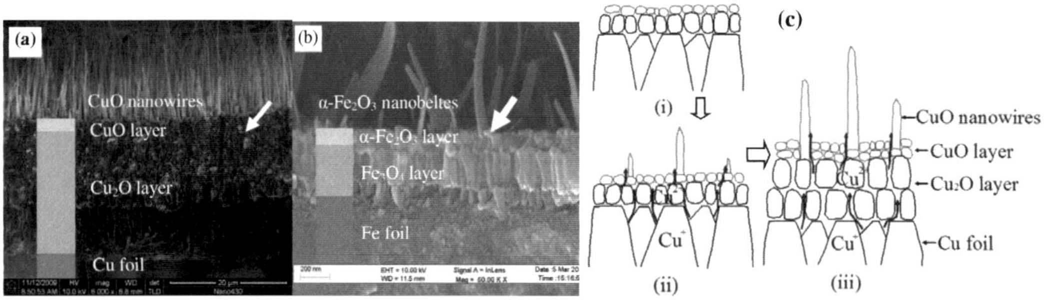

Figure 6. Cross section SEM images for the Cu and Fe oxides (a, b) prepared by thermal oxidation and the schematic for thermal oxidation mechanism (c).

4. Properties and applications of the nanostructures

4.1 Gas sensing properties

A very simple gas sensor prototype was built in this work [7], shown in Fig. 7a. The CuO nanowires together with CuO and Cu_2O underlayers were used as the sensing materials. The current provided by a stable power resource was introduced through two silver electrodes from the top and bottom of the sample. This sensor was used to test ethanol concentrations of 50-1000 ppm at 100 °C. Fig. 7b shows the resistances at various time for several ethanol concentrations. It indicates that the gas sensor has a very short reaction time, typically less than 30s. The resistance increases drastically with the increasing ethanol concentration. For a resistance of ~0.12 KΩ in air, when the ethanol concentration increases from 50 to 1000 ppm, the resistance increased from ~50 to >500 KΩ. By removing the sensor from ethanol atmosphere, it is found that, for a concentration of 1000 ppm, recovery time is less than ~120 s. The testing results are also found repeatable and reliable. As we know, CuO is a *p*-type semiconductor, and the electron injection emanated from the reaction between the reducing gas and the negatively charged surface oxygen increases the sensor resistance by decreasing the major charge carrier (hole) concentration. The high sensitivity of the prototype in a wide range of ethanol concentration makes it very practical for applications in many fields. For example, our prototype are very good for drunk driving detection working at ethanol concentration range of 200~800 ppm. Similarly, we have found that the Fe_2O_3 nanowires are also suitable for gas sensor with high sensitivity.

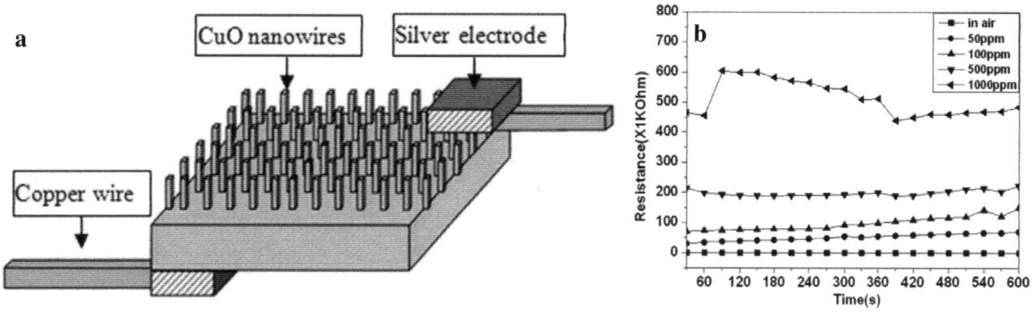

Figure 7. Schematic diagram of a sensor fabricated from CuO nanowires and a layer of CuO and the resistance in air and at different ethanol concentration at 100 °C

4.2. Magnetic properties

The magnetic properties of the prepared Fe_2O_3, Co_3O_4 and MnO nanostructures have also been investigated. As we all know, the Morin transition is a magnetic phase transition in α-Fe_2O_3 hematite where the antiferromagnetic ordering is reorganized from being aligned perpendicular to the c-axis to be aligned parallel to the c-axis below TM. A change in magnetic properties takes place at the Morin transition temperature. Fig.9 shows the *M*(T) curves of α-Fe_2O_3 nanowires measured in the temperature range of 5-300K. The transition temperatures of 131 and 126 K were observed for the nanowires prepared at 430 and 480°C, respectively, which are much lower than T_M = 263 K for the bulk α-Fe_2O_3. Similar results have been found by other researchers [21,22]. It has been suggested that the reduction in T_M in the hematite nanostructures result from lattice strain and defects [21] or uncompensated surface spins and/or shape effects of the nanowires [22].

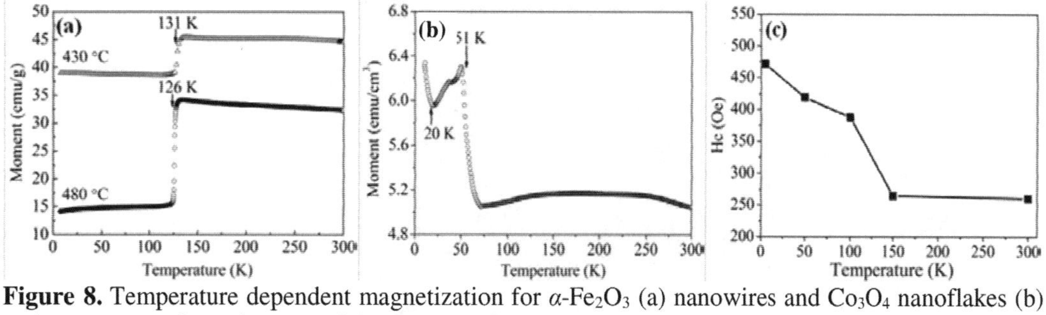

Figure 8. Temperature dependent magnetization for α-Fe_2O_3 (a) nanowires and Co_3O_4 nanoflakes (b) and temperature dependent coercivity for α-Fe_2O_3 nanowires (c)

4.3. Optical properties

The optical properties of ZnO nanowires have been well studied. Here we investigated the photoluminescence of α-Fe_2O_3 nanowires. As we know, in bulk α-Fe_2O_3, there is no PL due to magnetic and thermal relaxations, but in nanostructures, PL may occur for quantum confinement effect, the reduction of hyperfine-field or magnetic interactions, and the self-trapped state formation due to the enhanced electron–phonon interactions [23]. Zou et al. observed strong photoluminescence in Fe_2O_3 nanopareticles with size less than 20 nm [23]. A PL peak at 578 nm (2.14 eV) was also evident forα-Fe_2O_3 nanowires by Han et al. [24] Fig.9 shows the PL spectra for α-Fe_2O_3 nanowires synthesized on Fe foils at different temperatures under the 400 nm line exitations. Different from previous work, a PL peak at 791~795 nm (~1.56 eV) was found. Since the band gap of bulk α-Fe_2O_3 is 2.10 eV, the appearance of the peak in infrared light range indicated an unclear mechanism. Generally, the PL of 1-D α-Fe_2O_3 results from the enlarged Fe-O bond, which leads to magnetic coupling of neighbouring Fe^{3+} ions. Fig.9 also show that the samples prepared at 380°C and 430°C with different morophologies have significant different PL intensities, which is related to the size dependent quantum confinement effect, since the sample prepared at 380 has shown high length/width ratio.

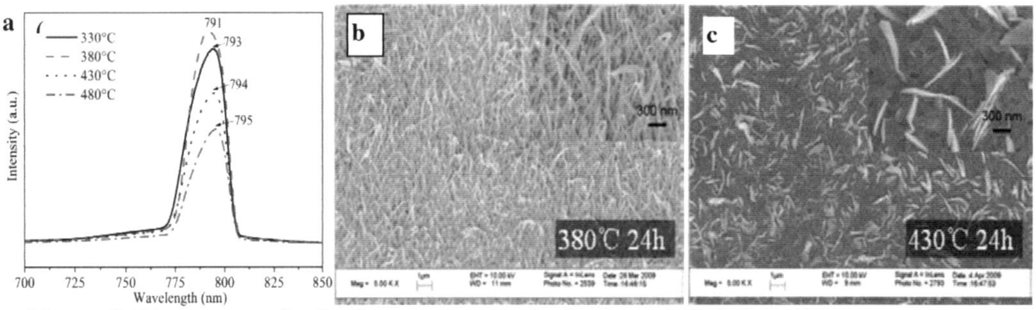

Figure 9. PL spectrum of α-Fe_2O_3 nanowires synthesized on (a) Fe foils at different temperatures

4.4 Field emission

The 1-D nanomaterials are suitable for field emission. We tested the field emission properties of α-Fe_2O_3 nanoneedles obtained by hot plate technique under a vacuum of 5.1×10^{-5} Pa. Fig.10 shows the structures and current density vs electric field curves. The nanoneedls prepared on Fe foil has the threshold field of 14.5 V/μm corresponding to the current density of 10 μA/cm^2. The current density is 17.4 μA/cm^2 when the electric field is 16.20 V/μm. The nanoneedls prepared on Fe/Si substrate has the threshold field value of 13.3 V/μm corresponding to the current density of 10 μA/cm^2. For the nanoneedles prepared on the Fe/ITO substrate, relatively low threshold field of 9.3 V/μm was obtained. Rackauskas et al. [25] investigated various nanowires produced by thermal oxidation and found the CuO nanowires showed the best FE characteristics. The threshold field value corresponding to the current density of 0.01 mA/cm^2 was ~4 V/μm. The maximum current density was ~100 mA/cm^2.

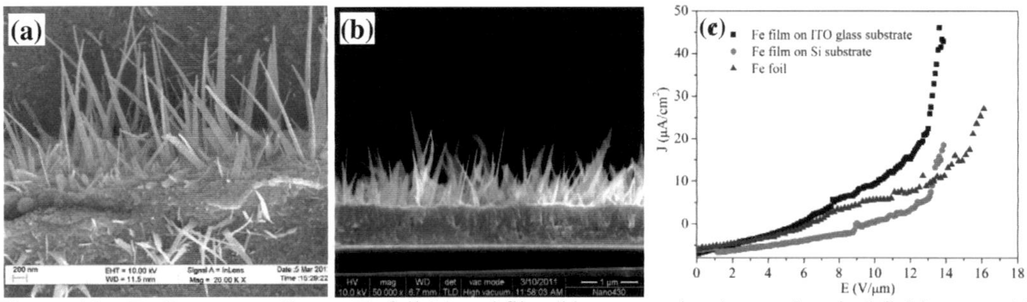

Figure 10.α-Fe_2O_3 nanoneedles on Fe foil (a) and film (b), current density vs electrical field curves (c)

5. Summary

The nanostructures of metal oxides including CuO, α-Fe_2O_3, Co_3O_4, ZnO etc. have been synthesized via thermal oxidation. This process is very simple, cost-effective and suitable for mass production in large-scale applications. The nanostructures can be grown on the surface of bulk metals, as well as the thin film or nanopartices. The heating temperature and composition of the oxidizing atmosphere have distinct effects on the morphology and structure. The length can be determined by the thermal duration, whereas the diameter can be determined by the substrate temperature. The VS and VLS mechanism cannot explain the thermal oxidation growth process and a short-circuit diffusion mechanism was proposed. In addition, the grown nanostructures show some interesting physical properties and have potential applications for magnetic, optical, field emission and gas sensing devices. It is expected that new oxide nanostructures with more complicated structure and composition and novel properties may also be produced by this simple process.

6. Acknowledgements

This work is supported by the Program for New Century Excellent Talents in University (NCET-11-0156) and the Research Fund for Returned Chinese Scholar.

7. References

[1] Xia Y N, Yang P, Kim F and Yan H 2003 *Adv. Mater.* 15 353
[2] Arnold S M and Koonce S E 1956 *J. Appl. Phys.* 27 964
[3] Takagi R 1957 *J. Phys. Soc. Japan* 12 1212
[4] Chowdhuri A, Sharma P, Gupta, V, Sreenivas K and Rao K V 2002 *J. Appl. Phys.* 92 2172
[5] Hsieh C T, Chen J M, Lin H H and Shih H C 2003 *Appl. Phys. Lett.* 83 3383
[6] Anandan S, Wen X G and Yang S H 2005. *Mater. Chem. Phys.* 93 35
[7] Zhong M L, Zeng D C, Liu Z W, Yu H Y, Zhong X C and Qiu W Q 2010 *Acta Mater.* 58 5926
[8] Kumar A, Srivastava A K, Tiwari P and Nandedkar R V 2004 *J. Phys.: Cond. Matt.* 16 8531
[9] Love C J, Smith J D, Cui Y H and Varanasi K K 2011 *Nanoscale* 3 4972
[10] Yu, T. Zhu Y, Xu X, Yeong K S, Shen Z, Chen P, Lim C T, Thong J T L and Sow C 2006 *Small* 2 80
[11] Zheng Z, Chen Y, Shen Z, Ma J, Sow C, Huang W and Yu T 2007 *Appl. Phys. A* 89 115
[12] Hsu L C, Yu H C, Chang T H and Li Y Y 2011 *ACS Appl. Mater. Interfaces* 3 3084
[13] Zhong M L, Liu Z W, Zhong X C, Yu H Y and Zeng D C 2011 *J. Mater. Sci. Techn.* 27 985
[14] Yu T, Zhu Y, Xu X, Shen Z, Chen P, Lim C T, Thong J T L and Sow C H 2005 *Adv. Mater.* 17 1595
[15] Zhu Y, Yu T, Cheong F C, Xu X, Lim C T, Tan V B C, Thong J T L and Sow C H 2006 *Adv. Funct. Mater.* 16 2415
[16] Liu Z W, Yeo S W and Ong C K 2007 *J. Mater. Sci.* 42 6489
[17] Pan Z W, Dai Z R and Wang Z L 2001 *Science* 291 1947
[18] Brady G S, Claucer H R and Vaccari J A 2002 *Materials Handbook* 15th edn. (New York: McGraw-Hill)
[19] Atkinson A 1985 *Rev. Mod. Phys.* 57 437.

[20] Gonçalves A M B, Campos L C, Ferlauto A S and Lacerda R G 2009 *J. Appl. Phys.* 106 034303

[21] Diaz-Guerra C, Perez L, Piqueras J and Chioncel M F 2009 *J. Appl. Phys.* 106

[22] Kim C H, Chun H J, Kim D S, Kim S Y, Park J, Moon J Y, Lee G, Yoon J, Jo Y, Jung M, Jung S I and Lee C J 2006 *Appl. Phys. Lett.* 89 223103

[23] Zou B S, Huang W, Han M Y, Li S F Y, Wu X C, Zhang Y, Zhang J, Zhang J S, Wu P F and Wang R Y 1997 *J. Phys. Chem. Solids* 58 1315

[24] Han Q, Liu Z H, Xu Y Y, Chen Z Y, Wang T M and Zhang H 2007 *J. Phys. Chem. C* 111 5034

[25] Rackauskas S, Nasibulin A G, Jiang H, Tian Y, Kleshch V I, Sainio J, Obraztsova E D, Bokova S N, Obraztsov A N and Kauppinen E I 2009 *Nanotechnology* **20** 165603,

Thermal performance of alumina filler reinforced intumescent fire retardant coating for structural application

Faiz Ahmad[1], Sami Ullah[2], Wan Farhana bt Mohammad[1] and M Farth Shariff[1]
[1]Department of Mechanical Engineering, University Technology PETRONAS, Perak, Malaysia
[2]Research Center for CO_2 Capture, Department of Chemical Engineering, University Technology PETRONAS, Perak, Malaysia

E-mail: faizahmad@petronas.com.my

Abstract. In the modern construction, fire safety has significant consideration for the protection of people and assets. Several intumescent fire protection systems are in practice and have constrain of releasing toxic gases on degradation forms an insulating char layer protecting underlying substrate. An intumescent coating expands many times of its thickness on exposure to fire and protect the underlying substrate from fire. This study presents the results of thermal performance of an intumescent fire retardant coating (IFRC) developed for structural application. IFRC was developed using expandable graphite (EG), ammonium poly phosphate (APP) and melamine (MEL), epoxy resin Bisphenol-A (BPA) and hardener triethylenetetramine (TETA) were used as a binder as a curing agent. Char expansion of IFRC was measured by furnace fire test at 450°C, thermal performance was measured using a Bunsen burner at 950°C and temperature of substrate was recorded for 60 min at an interval of two min. Results showed that IFRC containing 3wt% alumina showed char expansion X19. After one hour exposure of coating to heat, substrate temperature recorded was 154°C. X-ray Diffraction (XRD) results showed the presence of high temperature compounds present in the char of coating, considered responsible to reduce the penetration of heat to the substrate.

1. Introduction

Steel is a commonly used material in the construction of our regular infrastructures almost existing in all construction of buildings, vehicles, airplanes and oil platforms. Steel acquire good strengths and ductility while also being incombustible, however steel loses its physical properties when subjected to heat at high temperature such as in the case of fire. Heat treated carbon steel begins to lose its mechanical properties at temperatures above 400°C and continues to reduce in strength at steady rate up to temperature of 800°C [1].

Intumescent coating is a type of passive fire protection system which commonly used to protect steel against fire. The coating comprises of three active ingredients, usually ammonium polyphosphate-APP, expandable graphite and melamine-MEL [2]. The three main ingredients are linked together with a binder which then reacts together at higher temperatures. The result of the reaction is the expansion of the coating and the formation of a foamed char. The resulting foamed char acts as a thermal barrier and help reduces heat transfer to the coated material.

The binders for intumescent coatings mainly based from organic binders which have good expanding effect and char structure. Although having a good expanding effect and char structure, organic binders release solvent, toxic gas and smoke in a fire [3]. Comparing with inorganic intumescent coatings, the amount of toxic gas emission and smoke upon heating is relatively low. However, the inorganic intumescent coating only protects against fire at low temperature and are vulnerable to water and moisture [4, 5].

The objectives of this research work are to develop an intumescent coating formulation with added mineral based fillers of talc and alumina and to investigate the coating performance on the heat shielding effect, char expansion, structure, and water resistivity.

2. Materials and Methods

Ammonium polyphosphate (APP), expendable Graphite (EG), melamine (MEL), boric acid (BA), alumina, bisphenol A epoxy resin BE-188 (BPA) and Tetraethylene tetramine hardener and steel plate. The intumescent ingredients were mixed with their weight percentage composition homogeneously for 30min using Ultra Turrax mixer. The formulation NF (No filler) acts as a control with APP-EG-Boric acid-epoxy-hardener. The formulation was coated manually on the steel substrate. The coating was allowed to dry for 24 hours. Then, Bunsen burner and furnace oven was used for fire test. FEEM images of char formation were taken.

Table 1. Weight percentage of IFRC

Sample	BPA	TETA	APP	EG	Mel	ZB	Filler
NF	44.44	22.22	11.11	5.56	5.56	11.11	-
T1	43.94	21.72	11.11	5.56	5.56	11.11	1 (Talc)
T2	43.44	21.22	11.11	5.56	5.56	11.11	2 (Talc)
T3	42.94	20.72	11.11	5.56	5.56	11.11	3 (Talc)
T4	42.44	20.22	11.11	5.56	5.56	11.11	4 (Talc)
T5	41.94	19.72	11.11	5.56	5.56	11.11	5 (Talc)
A1	43.94	22.22	11.11	5.56	5.56	11.11	1 (Alumina)
A2	43.44	21.72	11.11	5.56	5.56	11.11	2 (Alumina)
A3	42.94	21.22	11.11	5.56	5.56	11.11	3 (Alumina)
A4	42.44	20.72	11.11	5.56	5.56	11.11	4 (Alumina)
A5	41.94	19.72	11.11	5.56	5.56	11.11	5 (Alumina)

3. Analysis and Characterization

Heat shielding effect: Bunsen burner was used as a fire test source by using the standard UL94. K-type thermocouple connected to AMS-850 data logger was used to measure the temperature in front and the backside of the substrate. The flame temperature of Bunsen burner is 800°C.
Field Emission Scanning Electron Microscope (FESEM): The charring layer and the morphological structures of the char formation were observed and analyzed using SUPRA Instrument FESEM.
X-Ray Diffraction: The composition of residual char of the intumescent coating after the fire test was analysed by XRD and measurements were performed on a Diffractometer Bruker AXS D8 Advance using Cu Kα radiation and a nickel filter (k = 0.150595 nm) in the range ($10 < 2\theta < 90$).
Static immersion test: Static immersion test is considered as a standard method to evaluate water

resistivity of coating using gravimetric method. Coating formulations were immersed in distilled water at 25°C and weighed every 24 h. The samples then were dried at 40°C for 1 h in furnace oven. Weight change percentage was calculated using Eq. (1), and the result was plotted against time.

$$\Delta W = \frac{W_2 - W_1}{W_1} \times 100\% \qquad (1)$$

ΔW weight loss, %;
W_1 sample weight before water immersion, g;
W_2 sample weight after water immersion, g.

4. Results and Discussion

4.1. Heat shielding effect

In this study, 11 samples with different compositions of intumescent ingredients were tested. The result after testing shows that formulation A5 which contains 5% of alumina fillers gives the best intumescent effect.

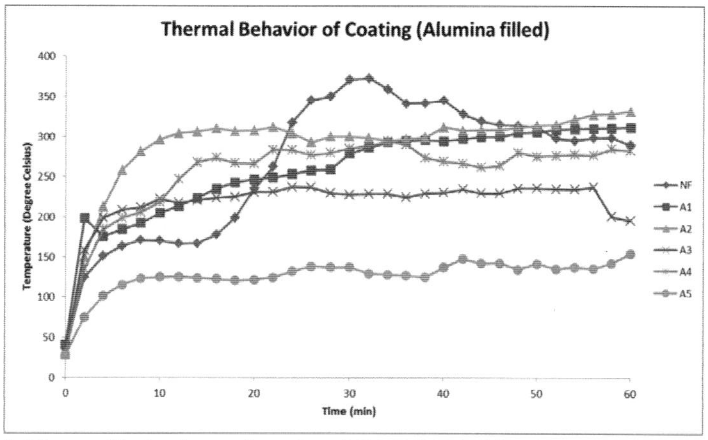

Figure 1: Thermal behaviour of coating APP-EG-BA-Alumina-Epoxy and hardener.

The temperature time curves for heat shielding of fire retardant coating with added alumina fillers are illustrated in Figure 1. The coating with no added filler has the backside substrate temperature of 290°C after 60 min. Fig 2, after 60 min, formulation A1 has the backside substrate temperature of 312.2°C, formulation A2 has the backside substrate temperature 332.0°C, formulation A3 has the backside substrate temperature of 196.0°C, formulation A4 has the backside substrate temperature of 282.8°C and formulation A5 has the backside substrate temperature of 154.6°C. From the alumina filled result, formulation A5 has the lowest backside substrate temperature followed by A3, A4, A1 and A2.

From the result, formulation A5 gives the best intumescent effect as it contains the highest percentage of alumina. Higher percentage of alumina gives better strength to the char formation thus giving the better intumescent effect. Alumina have outstanding physical ability such as high mechanical strength at room temperature and high melting point of 2050°C [6] hence higher percentage of alumina presence would enhance the heat shielding capability of the coating.

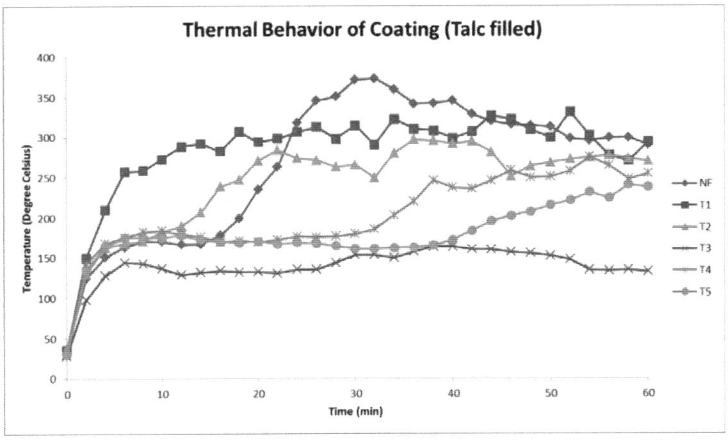

Figure 2: Thermal Behaviour of Coating APP-EG-BA-Talc-Epoxy and hardener.

The temperature time curves for heat shielding of fire retardant coating with added talc fillers are illustrated in Figure 2. The coating with no added filler has the backside substrate temperature of 290°C after 60 min. Fig 3, after 60 min, formulation T1 has the backside substrate temperature of 293.9°C, formulation T2 has the backside substrate temperature 270.1°C, formulation T3 has the backside substrate temperature of 132.4°C, formulation T4 has the backside substrate temperature of 254.2°C and formulation T5 has the backside substrate temperature of 237.7°C. From the heat shielding result of talc filled coating, formulation T3 has the lowest backside substrate temperature followed by T5, T4, T2 and T1. 3% filled talc formulation, T3, produces the best intumescent effect after A5 formulation. This is because presence of talc reinforces the mechanical stability of the intumescent protective shield [4]. A study conducted by Almeras et al. [7] on the effect of talc fillers suggested a chemical reactivity between talc and intumescent ingredients that may affect the performance of the formulation. The study demonstrates that the ratio percentage of talc and intumescent ingredients must be optimized to produce the best intumescent effect. This explains why 3% filled talc performed better than the higher filled talc formulation.

This proved that the addition of talc and alumina as fillers enhanced the fire retardant effect. The presence of alumina and talc provide better strength to the char structure and better mechanical stability thus maintaining the thermal barrier longer before rupturing.

4.2. Field Emission Scanning Electron Microscope (FESEM)

The FESEM micrograph images of char showing surface microstructure for sample A5 and T5 were obtained. Sample A5 and T5 were selected for testing as A5 gives the best intumescent effect and T5 as a direct comparison. The char obtained are from fire testing conducted earlier.

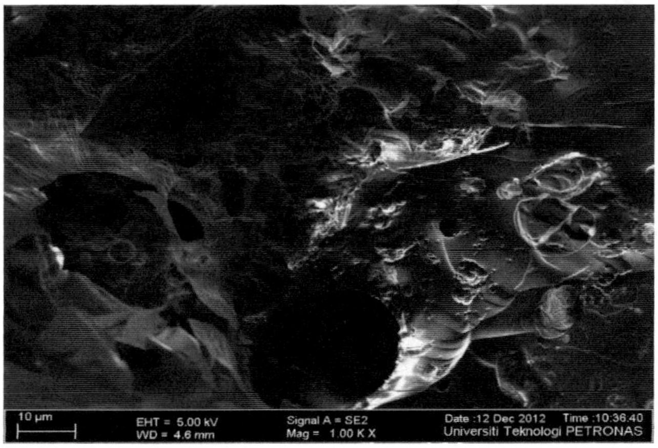

Figure 3: FESEM micrograph A5 450°C char at 1000X.

Figure 3 and figure 4 showed the holes and voids in FESEM image of formulation A5 and T5 respectively. The amount voids observed from formulation T5 is slightly lower compared to formulation A5. From Figure 5, no voids were observed due to the absence of filler which subsequently reduce the strength of the char. The holes were formed from evolution of trapped gas by blowing agent when the coating is subjected to fire. The voids act as a thermal barrier which prevent heat transmission to the substrate [7]. Voids also prevent diffusion of gaseous degradation products to the combustion zone and prevent oxygen diffusion to the surface polymer [8].Fillers were believed to improve the efficiency of intumescent coating. This is down to the plate like microstructure of the fillers which consequently provide stabilizing effect on the cell structure of char foam [9].

This proved that the addition of talc and alumina fillers produced more voids compared to the no filled intumescent coating. Higher amount of voids produced means higher strength of char structure. This subsequently contributes to better intumescent effect.

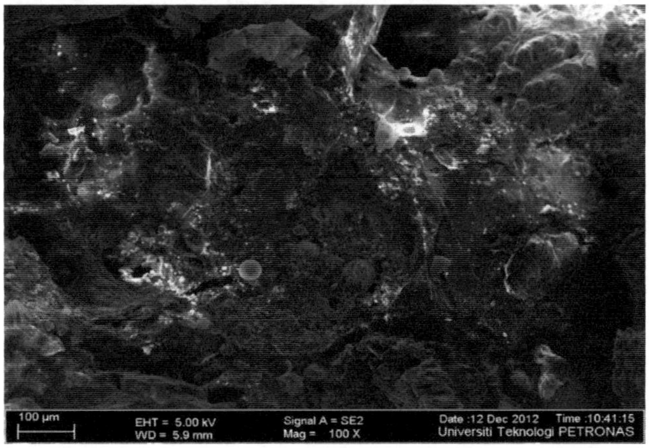

Figure 4: FESEM micrograph T5 450°C char at 100X.

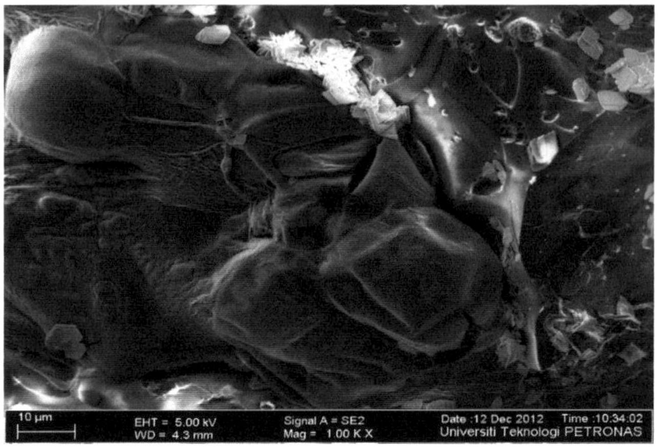

Figure 5: FESEM micrograph NF 450°C char at 1000X.

4.3. X-Ray Diffraction Analysis

After the residue char of the intumescent coating was gradually oxidized at high temperature, only some amorphous carbon and inorganic materials remained. The inorganic materials might be the main protecting layer at later stages of burning via XRD analysis [2]. Figure 6 and 7 shows the X-Ray diffraction peak of T5 formulation (5% talc) and A5 formulation (5% Alumina) with d-spacing value. Figure 6, it shows that after the addition of 5% talc, the d-spacing value obtained at 9.21968 and 2.98847 which is assigned to be Mg_2P4O_{12} and $Si_5O(PO4)_6$. Previous study performed by Levchik and co-workers [10] found that in XRD analysis, the residue of the reaction of talc and APP was ammonium $Si(NH_4)_2P4O_{13}$ at 300°C. At higher temperature, all the nitrogen group loss and the residue of magnesium cyclotetrapolyphosphate $Mg_2P_4O_{12}$, and silicon oxomonophosphate, $Si_5O(PO_4)_6$ was obtained. The overlapping peak at 3.8896, 3.51456 and 3.17264 is assigned to be titanium dioxide (TiO_2) and titanium pyrophosphate (TiP_2O_7). This residue can enhance char structure and protect the underlying substrate at higher temperature [11].

From Figure 7, the XRD analysis showed the existence of graphite, boron oxide (B_2O_3) and borophosphate (BPO_4). The dehydration of boric acid resulted in boron oxide which is a very hard glass. Besides, the formation of borophosphate was due to the reaction between ammonium polyphosphate and boric acid. When combining APP and boric acid, a good mechanical resistance is observed because borophosphate is a hard material. That resulted in providing the good structural properties of the char whereas phosphorous allows a good adhesion of the char to the steel plate [12].

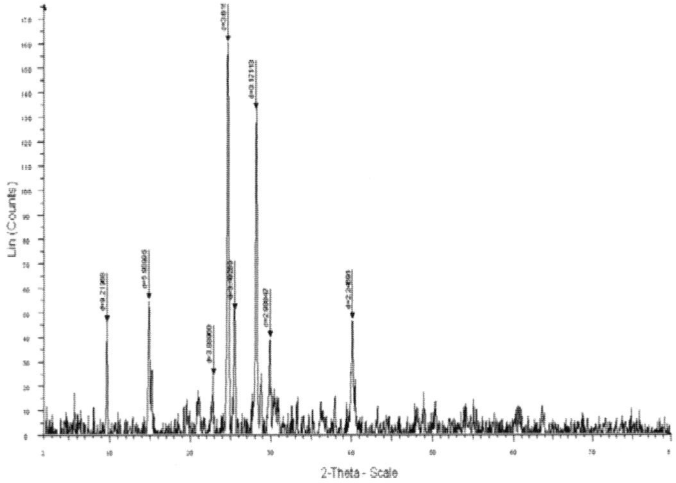

Figure 6. XRD curve of T5 (Talc 5%)

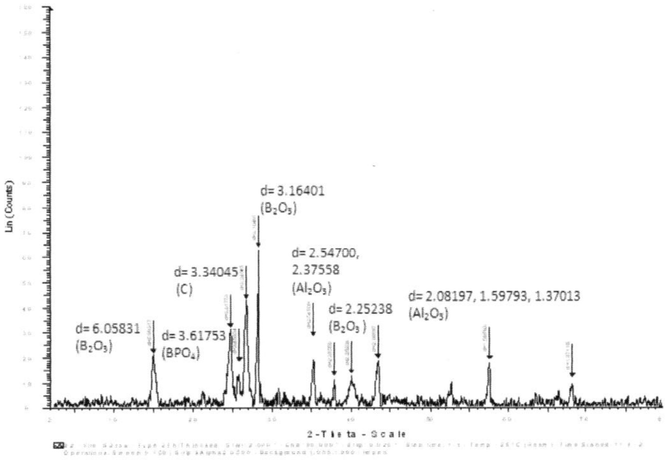

Figure 7. XRD curve of A5 (Alumina 5%)

4.4. Static immersion test

Weight change – time curves of all formulations are shown in Figure 8 and 9. Looking at the graph of alumina filled coating, weight change percentage of all the coating has increased dramatically from the first day until the fourth day of immersion where the percentage declined slightly. Equilibrium is not achieved in the initial stage of immersion therefore coatings continue to gain weight [13]. The increased weight change percentage is due to the absorption of water by the coating and the slight declination is due to the weight loss of the coating from formation of pores and damages to the surface [14]. The formation of pores allowed more water to be absorbed and the trend continue to increase until it reached physical and chemical equilibrium. Formulation A2 shows the highest water absorption rate of 0.167% and formulation NF has the lowest rate of absorption of 0.032%.

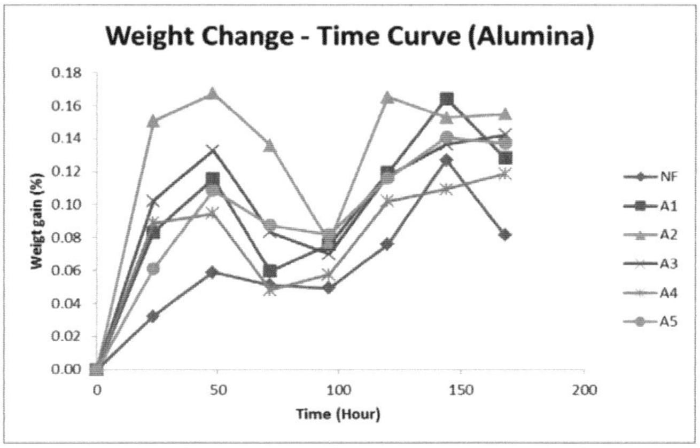

Figure 8. Weight change time curves of alumina filled coating.

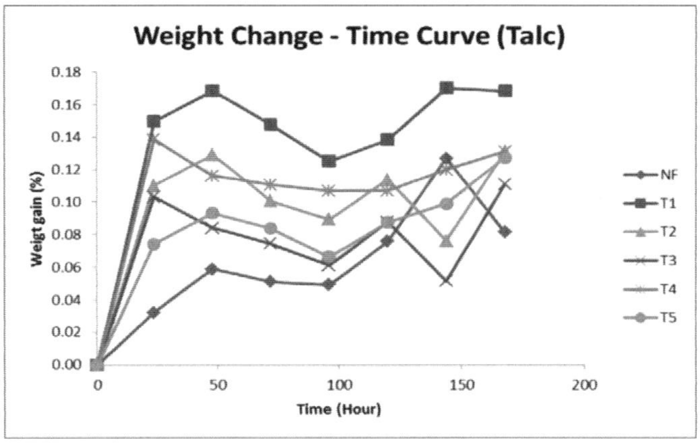

Figure 9. Weight change time curves of talc filled coating.

Based on the weight change time curves of talc filled intumescent coating, the trend of the curves are almost similar with alumina filled coating but with more settled curves. Formulation T1 has the highest water absorption of 0.170%. The talc filled intumescent coating does not produce pores and surface defects as alumina filled coating thus reaching physical and chemical equilibrium after 48 hours of testing.

5. Conclusion

The fire retardant time of EG-APP-Mel-B.A coating with 5% wt alumina showed the best intumescent effect and the temperature of the substrate was 154.6°C after 60 min. FESEM result showed improved formation of voids by addition of talc and alumina. Thus, the rate of heat transfer to the steel substrate can be reduced therefore giving better intumescent effect. Static immersion test show that alumina does not improve the water resistivity of the coating while talc improved the water resistivity by achieving equilibrium faster after just one day of immersion. Talc filled coating also does not produce pores and surface defects compared to alumina. Thus, the water resistivity is improved by the addition of talc.

6. References

[1] M. Jimenez, S. Duquesne and S. Bourbigot, "Intumescent fire protective coating: Toward a better understanding of their mechanism of action," Thermochimica Acta, vol. 449, pp. 16-26, 2006.

[2] S. Ullah, F. Ahmad and P. S. M. M. Yusoff, "Effect of boric acid and melamine on the intumescent fire-retardant coating composition for the fire protection of structural steel substrates," Journal of Applied Polymer Science, vol. 128, pp. 2983-2993, 2013.

[3] Ullah S., F. Ahmad. and P. S. M. Megat-Yusoff, "Effect of Boric Acid with Kaolin Clay on Thermal Degradation of Intumescent Fire Retardant Coating," Journal of Applied Sciences, vol. 11, pp. 3645-3649, 2011.

[4] K. Langille, D. Nguyen and D. E. Veinot, "Inorganic Intumescent Coatings for Improved Fire Protection of GRP," Fire Technology, vol. 35, pp. 99-110, 1999/05/01 1999.

[5] J. Green, "Mechanisms for flame retardancy and smoke suppression - A review," Journal of Fire Sciences, vol. 14, pp. 426-442, 1996.

[6] J. Green, "A Review of Phosphorus-Containing Flame Retardants," Journal of Fire Sciences, vol. 14, pp. 353-366, 1996.

[7] X. Almeras, M. Le Bras, P. Hornsby, S. Bourbigot, G. Marosi, S. Keszei, and F. Poutch, "Effect of fillers on the fire retardancy of intumescent polypropylene compounds," Polymer Degradation and Stability, vol. 82, pp. 325-331, 2003.

[8] M. Wladyka-Przybylak and R. Kozlowski, "The thermal characteristics of different intumescent coatings," Fire and Materials, vol. 23, pp. 33-43, 1999.

[9] Faiz Ahmad, S. Ullah and M. S. Hamizol, "To Study the Effect of Aluminium Trihydrate and Fumed Silica on Intumescent Fire Retardant Coating," Journal of Applied Sciences, vol. 12, pp. 2631-2635, 2012.

[10] A. R. Horrocks, "Developments in flame retardants for heat and fire resistant textiles - The role of char formation and intumescence," Polymer Degradation and Stability, vol. 54, pp. 143-154, 1996.

[11] Z. Wang, E. Han and W. Ke, "Effect of nanoparticles on the improvement in fire-resistant and anti-ageing properties of flame-retardant coating," Surface and Coatings Technology, vol. 200, pp. 5706-5716, 2006.

[12] Z.-y. Wang, E.-h. Han and W. Ke, "Fire-resistant effect of nanoclay on intumescent nanocomposite coatings," Journal of Applied Polymer Science, vol. 103, pp. 1681-1689, 2007.

[13] T. A. Roberts, L. C. Shirvill, K. Waterton and I. Buckland, "Fire resistance of passive fire protection coatings after long-term weathering," Process Safety and Environmental Protection, vol. 88, pp. 1-19, 2010.

[14] Z. Wang, E. Han and W. Ke, "An investigation into fire protection and water resistance of intumescent nano-coatings," Surface and Coatings Technology, vol. 201, pp. 1528-1535, 2006.

Multi-passes warm rolling of AZ31 magnesium alloy, effect on evaluation of texture, microstructure, grain size and hardness

J Kamran, B A Hasan, N H Tariq, S Izhar and M Sarwar
Department of Metallurgy and Materials Engineering, Pakistan Institute of Engineering and Applied Sciences, Islamabad, Pakistan

E-mail: jawedkamran@hotmail.com

Abstract. In this study the effect of multi-passes warm rolling of AZ31 magnesium alloy on texture, microstructure, grain size variation and hardness of as cast sample (A) and two rolled samples (B & C) taken from different locations of the as-cast ingot was investigated. The purpose was to enhance the formability of AZ31 alloy in order to help manufacturability. It was observed that multi-passes warm rolling (250°C to 350°C) of samples B & C with initial thickness 7.76mm and 7.73 mm was successfully achieved up to 85% reduction without any edge or surface cracks in ten steps with a total of 26 passes. The step numbers 1 to 4 consist of 5, 2, 11 and 3 passes respectively, the remaining steps 5 to 10 were single pass rolls. In each discrete step a fixed roll gap is used in a way that true strain per step increases very slowly from 0.0067 in the first step to 0.7118 in the 26th step. Both samples B & C showed very similar behavior after 26th pass and were successfully rolled up to 85% thickness reduction. However, during 10th step (27th pass) with a true strain value of 0.772 the sample B experienced very severe surface as well as edge cracks. Sample C was therefore not rolled for the 10th step and retained after 26 passes. Both samples were studied in terms of their basal texture, microstructure, grain size and hardness. Sample C showed an equiaxed grain structure after 85% total reduction. The equiaxed grain structure of sample C may be due to the effective involvement of dynamic recrystallization (DRX) which led to formation of these grains with relatively low misorientations with respect to the parent as cast grains. The sample B on the other hand showed a microstructure in which all the grains were elongated along the rolling direction (RD) after 90 % total reduction and DRX could not effectively play its role due to heavy strain and lack of plastic deformation systems. The microstructure of as cast sample showed a near-random texture (mrd 4.3), with average grain size of 44 μm & micro-hardness of 52 Hv. The grain size of sample B and C was 14μm and 27μm respectively and mrd intensity of basal texture was 5.34 and 5.46 respectively. The hardness of sample B and C came out to be 91 and 66 Hv respectively due to reduction in grain size and followed the well known Hall–Petch relationship.

1. Introduction

Magnesium is a very important advanced structural material and has received significant research attention for automobile aerospace and electronic applications due to its low density (1.7 g/cm^3, which is 2/3rd of aluminum & 1/5th of steel), high specific strength and adequate specific stiffness [1-3]. The combination of low density and reasonable strength of magnesium helps in weight reduction of components for automobiles and aerospace applications [4]. A lot of research work has been carried out to gain an in depth understanding on the deformation behavior of wrought Mg alloys as compared to cast alloys that normally have low strength and poor ductility. The use of wrought magnesium alloys in the form of sheet is increasing in recent years. In this regard Mg-Al-Zn type alloy AZ31 has got marvelous industrial applications for sheet production by rolling [5].

One of the most critical issues currently limiting applications of Mg based alloys is their poor formability & plastic flow characteristics at room temperature because of HCP structure and limited available active slip systems [6]. Therefore, the forming processes of AZ31 alloy are carried out at elevated temperatures in order to avoid edge as well as surface cracking. Another obstacle to the formability of the magnesium sheet production is the formation of strong basal texture, which leads to a greater anisotropy and affects the enhancement of its press formability [7]. It is well known that the formability of magnesium alloys can be improved significantly by changing the (0002) basal texture [8,9]. Therefore, many researchers are working to weaken the basal texture either by using severe thermomechanical processes (TMP) or by modifying the well known wrought alloys by addition of alloying elements. Literature reports that some TMP have been proven successful to weaken the basal texture like equal channel angular pressing(ECAP) [10,11], cross-roll rolling [12], differential speed rolling(DSR) [13,14], cyclic extrusion compression (CEC) [15], friction stir process(FSP) [16,17] and repeated unidirectional bending(RUB) [18,19]. Though these processes are helpful in controlling the basal texture, yet all of them are not used for the purpose of manufacturing of sheets.

Present study is an effort to increase the formability of AZ31 through conventional rolling carried out in steps with different number of rolling passes in each step. The roll gap in each pass was fixed in a way that the resulting true strain increases very slowly. In this regard warm temperature range (250-350°C) was selected due to the fact that deformation in warm forming temperature range is now well known to generate dynamic re-crystallization (DRX), deformation twins, and dislocation slips. The main objective of this work is to establish the optimum process parameters for multi-pass warm rolling which can help in improving the formability of the AZ31 magnesium alloy by improving (1) basal texture (2) microstructure (3) grain refinement and (4) mechanical properties.

2. Experimental

The Mg–3Al-1Zn (wt. %) magnesium alloy (abbreviated as AZ31) was prepared from high purity elements Mg (99.99%), Al (99.99%) and Zn (99.99%). The charge was melted in an iron crucible using an electrical resistance furnace, in the presence of high purity Argon gas to protect the melt from oxidation. After stabilization at 780 °C for about 5 min, the melt was poured into a copper mold with dimensions (80 × 75 × 8 mm). After cooling in Ar atmosphere the cast ingot was demolded and three samples A, B & C were collected from the ingot from the location shown in figure 1, their dimensions are listed in table 1. The chemical analysis was carried out by inductively coupled plasma atomic emission spectroscopy (ICP), results are shown in Table 2.

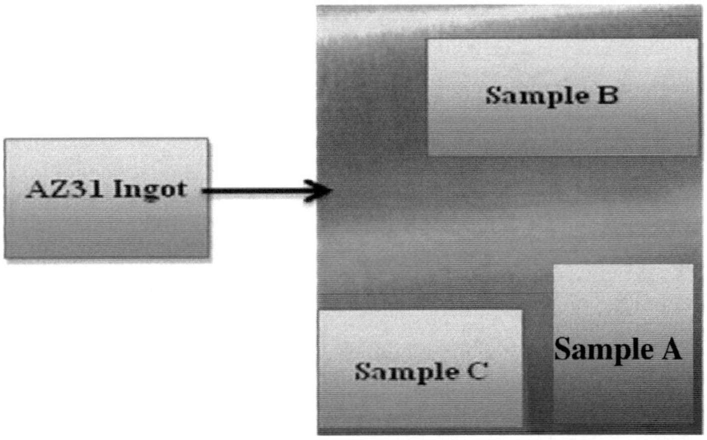

Figure 1. Location of the samples taken from warm rolling.

Table 1. Dimensions of two samples before and after multi pass warm rolling.

Samples	Initial Dimensions (mm)			Final Dimensions (mm)			Reduction (%)
	Length	Width	Thickness	Length	Width	Thickness	
Sample B	33.12	17.79	7.76	315	23.63	0.78	90
Sample C	24.89	18	7.73	88.73	29.11	1.16	85

Table 2. Chemical Composition of AZ31 magnesium alloy (wt %).

Materials	Mg %	Al %	Zn%	Fe ppm	Cu ppm	Mn ppm	Ni ppm
Pure Mg	Balance	ND	ND	ND	ND	13	ND
AZ31	Balance	3.0	1.02	51	ND	20	ND

Sample A was retained in the as-cast condition for determination of texture, microstructure, grain size and hardness. A unidirectional multi-pass warm rolling (250-350 °C) of as-cast samples B and C was carried out in JOLIOT rolling machine without any lubricant with roll diameter of 124 mm and roll length of 150 mm. Rolling speed was fixed at 27 rev/min. Samples B & C were preheated at 400 °C in air for 2 hours in induction heating furnace prior to the start of rolling. After two hours the furnace was switched off and samples were allowed to cool in furnace as soon the temperature was dropped to 350 °C, multi-pass warm rolling was started at 300 °C with an intermediate heating time of 5 minutes according to the scheme given in Table 3. The step numbers 1, 2, 3 and 4 consist of 5, 2, 11 and 3 passes respectively while the next 5 steps consist of only 1 pass each. In each discrete step a fixed roll gap was used in a way that value of true strain per step might increase very slowly from 0.0067 in the first step to 0.7118 in the 26th step.

The plane of examination for microstructure and macro-texture examinations of as cast sample was that of solidification direction (SD) whilst in case of rolled samples same study was performed on the plane containing rolling direction (RD) & transverse direction (TD). The samples for microstructure examination were ground by using SiC papers in the sequence of 1200-grit, 1500 grit and 2000 grit. Polishing was carried out by means of solution of alumina 0.05 μm and ethanol. After final polishing the samples B & C were etched by using acetic picral with the composition: acetic acid 10 ml, acetic picral 4.25g, distilled water 10ml, and ethanol 70 ml. The microstructures were analyzed using OLYMPUS optical microscope (OM). Texture evaluation was carried out by X-ray diffraction (XRD) technique using a Bruker D8 diffractometer with Cukα radiation. The hardness measurements were performed by Struers Duramin micro-Vickers apparatus under the load of 1.96 N for the duration of 10 seconds.

Table 3. The multi pass warm rolling schedule for Sample B and C

No of Steps	No of Passes	The Pass Number	Reduced Thickness (mm)	True Strain (%)	Total thickness reduction in a step [Δx] (mm)	Reduction per pass [Δx/no of passes] (mm)
0	0		7.76	0		
1	5	1-5	7.5	0.034	0.26	0.052
2	2	6-7	6.89	0.085	0.61	0.305
3	11	8-18	5.1	0.301	1.79	0.162727
4	3	19-21	4.23	0.187	0.87	0.29
5	1	22	3.79	0.11	0.44	0.44
6	1	23	3.25	0.154	0.54	0.54
7	1	24	2.67	0.197	0.58	0.58
8	1	25	1.94	0.32	0.73	0.73
9	1	26	0.952	0.55 0.712	0.988	0.988
10	1	27	0.44	0.772	0.512	0.512
Total	**27**		**7.32**	**2.872**	**7.32**	

3. Results and Discussion

Figure 2(a), (b) and (c) shows the external appearance, microstructure and basal planes texture of as cast sample A of AZ31 magnesium alloy respectively. The as-cast ingot microstructure was a typical non-equilibrium cast structure which consists mainly of α Mg matrix (light region) and second phase precipitates (dark region).The microstructure of as cast sample shows an average grain size of 44 μm. The second phase precipitates distribute continuously along the grain boundaries. Some black spots were also observed at grain boundaries as well as within the grains. As-cast microstructure also showed few twins, however their volume fraction was very low. Figure.2(c) shows random distribution of basal poles in as cast condition with a mrd of 4.3.

Figure 2(d), (e) and (f) shows the external appearances, microstructure and associated basal planes texture of sample C respectively after total thickness reduction of 85%. The rolling schedule is provided in table 3. The micrograph of this sample shows an equiaxed grain structure [figure. 2(e)] with an average grain size of 27 μm evaluated by linear intercept method. It is envisaged that these refined equiaxed grains are mainly induced by the dynamic recrystallization (DRX) involved during the repeated pre-heating for about 5 min during multi-pass warm rolling [24]. Another interesting result of sample C is that optical micrograph exhibits twin free grains. Such twin free and equiaxed microstructure is the characteristics of annealed samples followed by deformation either by rolling or extrusion etc. Twin free grains microstructure mostly produce a strong basal texture of as-rolled plate due to the fact that twining is one of the most important deformation mechanism involved in magnesium alloys [20]. Basal pole figure shown in [figure. 2(f)] revealed that the basal planes are trying to reorient their c-axis parallel to the compression axis. Once most of the c-axes are parallel to the compression axis by repetitive rolling, {1012} twinning will be suppressed because {1012} twinning is an extension twin which is active under loading perpendicular to the compression axis. Thus, the possibility of the occurrence of deformation twinning as an additional deformation mode will decrease with an increase in the number of rolling passes during multipass warm rolling [21].

Similar relation of equiaxed microstructure and basal texture was observed while studying the role of tension twins on warm deformation behavior of AZ31 Mg alloy [20]. However, in the present case the intensity of basal texture could only be lowered down to 5.46 mrd.

The splitting of basal poles from normal direction (ND) towards the rolling as well as transverse direction (TD) is shown in figure 2(f). The resulting double peak texture represents a clear tendency for a broad c-axes distribution in the RD and TD, which supports the findings of Li *et. al.* [22] who related it with improved formability. They reported that the deformation was carried out primarily by basal slip assisted by prismatic and <c + a> pyramidal slip. This splitting of basal poles in front of and behind the RD after final rolling pass on sample B reflects indeed a significant improvement in the sheet formability/rollability by decreasing the basal poles intensity [23-24]. Similar double peak basal poles figures were observed by Kohzu *et. al.* during combination of rolling and high-temperature annealing of AZ31 alloy [25].

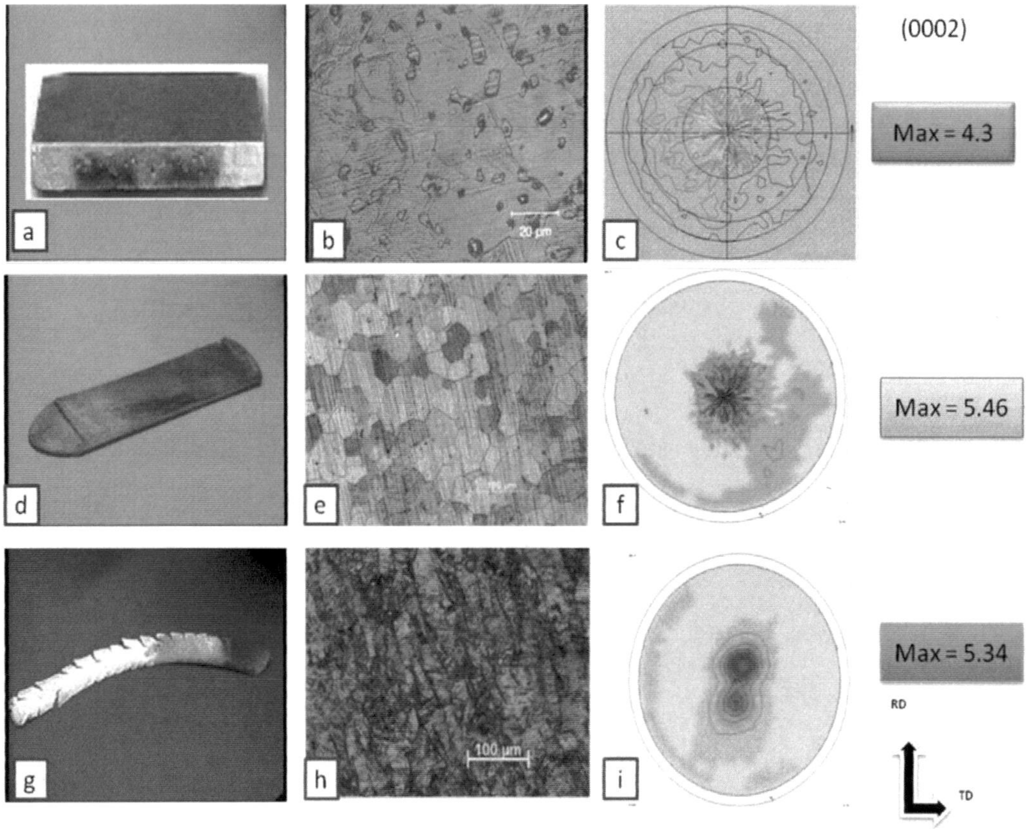

Figure 2. External appearance, microstructures and associated basal pole texture of AZ31 magnesium alloy before multipass rolling and after rolling with a total of 85 % and 90% thickness reduction: (a, b,c) As cast Sample A (d,e,f) 85 % rolled sample C, (g,h,i) 90% Rolled sample sample B.

The microstructural features shown in figure 2 are due to the fact that a high degree of deformation gives rise to a high fraction of DRX structure. Moreover, examination of the microstructural development in AZ31 for different rolling reductions per pass revealed that rolling reductions larger than 53% at warm rolling temperature result in a high fraction of shear banding [25]. It is well known that a shear band is one of the microstructural features that evolve under inhomogeneous deformation [26]. Since conventional rolling is not a complete homogeneous deformation process, therefore, it is very difficult to prevent shear bands formation at high strains (ε =0.772 %). it is however, reported that the formation of shear bands can be delayed by mutli-axial deformation [27,28] which claims a much more homogeneous microstructure during deformation. Therefore, the formation of shear bands during deformation is of great importance to the formability of Mg alloys.

The decrease in the intensity of the {0002} basal plane with the increase of the reduction ratio per pass may be due to delayed formation of shear bands. These shear bands are highly localized deformed regions in which more random grain orientations and thus weaker texture are developed as shown in basal pole figure 2(i). The other aspect of heavy reduction per pass is reflected into fine grain size and high hardness values as shown in figure 3. There is a great difference in grain size and microhardness of both the samples B & C. It appears that repetitive rolling at a smaller reduction ratio per pass up to total of 85% produced an equiaxed and coarse grains structure during DRX in both the samples up to 26th rolling pass according to rolling schedule shown in Table 3. However, as the % age reduction per pass was increased with true strain ε = 0.772 %, a drastic change in microstructure, grain size, grain shape and texture intensity was observed. These findings are in good agreement with previous studies [29].

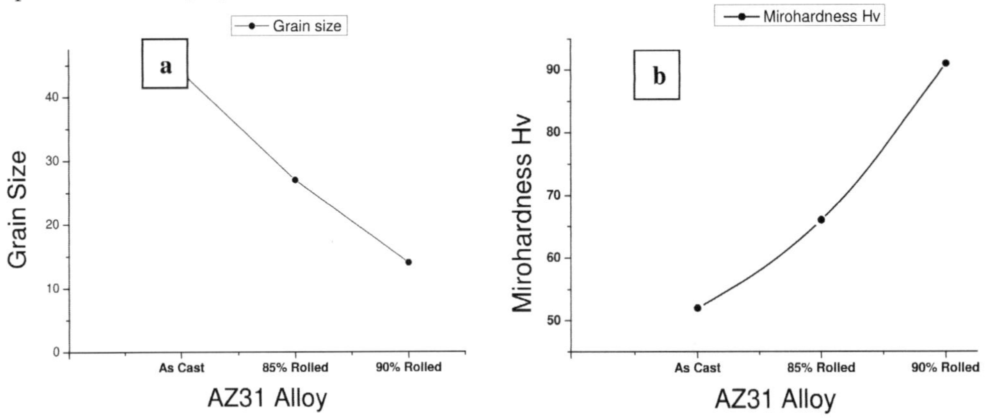

Figure 3. (a) The variation of grain size as a result of multipass warm rolling after total reduction of 85% and 90%. (b) effect of grain size on microhardness after same reduction.

The present results of grain size and experimentally measured micro-hardness are also well in agreement with the well known Hall-Petch relation (for grain size vs hardness plot) as shown in figure 3. The reduction in grain size of sample B is almost half of sample C. Accordingly the hardness showed an increase of up to 91 Hv during last rolling pass at high reduction per pass at ε = 0.772. The grain size distribution becomes more homogeneous and the average grain size becomes finer (14 μm) with an increase in the reduction ratio per pass in 27th pass as shown in figure. 1(h). Researchers are in agreement with the fact that a homogeneous and fine-grained structure could be obtained by using larger reductions. Moreover, it is reported that weaker texture results from greater reductions [30, 31]. Microstructure of this sample consists of elongated grains after total reduction of 90 %. The basal pole texture intensity of sample B is a bit lower than that of samples C as shown in figure 2(i). It seems that in sample B most of the grains are trying to align their c-axes parallel to the compression axis as a result of a very high strain during the 27th pass. The increase of dislocation density resulting from high reduction in final pass seems to be the main reason for strain hardening due to which the dislocation pileup increases and eventually leading to the failure of sample B. It is well established that the evolution of a very high density of uniformly distributed dislocations provide very high strain hardening. Internal stress fields cause the initiation of non-basal slip [32] which might have lowered the basal planes texture of sample B.

4. Summary and Conclusions

It is concluded that both samples B and C exhibited a similar manner up to 85% total reduction in 26 passes. During the 27th pass and reaching a value of true strain of 0.772 %, a drastic change in grain size and grain morphology of sample B was observed and the sample failed. The formation of edge and surface cracks seem not due to strong basal texture but due to a very heavy, 53% thickness reduction, in this pass. In short present study shows that formability can be increased through multi pass warm rolling. Adopting this strategy helped in successful rolling of AZ31 with conventional procedures due to the role of DRX. However, from an industrial point of view the economic feasibility of a long processing time may be taken into account.

5. References

[1] Lee C D 2006 *Met. Mater. Int.* 12 377

[2] Yim C D, Wu G and You B S 2007 *Mater. Trans.* 48 2778

[3] Kang S H, Lee Y S and Lee J H 2008 *J. Mater. Proc. Tech.* 201 436

[4] Easton M, Beer A, Barnett M R, Davies C, Dunlop G, Durandet Y, Blacket S, Hilditch T, and Beggs P 2008 *JOM* 60 57

[5] Spigarelli S, Mehtedi M E, Cabibbo M, Evangelista E, Kaneko J, Jäger A and Gartnerova V 2007 *Mater. Sci. Eng. A* 462(1–2) 197-201

[6] Kim H L and Chang Y W 2011 *Met. Mater. Int.* 17 563 (2011)

[7] Yi S B, Bohlen J, Heinemann F, Letzig D 2010 *Acta Mater.* 58 592–595

[8] Mukai T, Yamanoi M, Watanabe H and Higashi K 2001 *Scripta Mater.* 45 89–94

[9] Chino Y, Lee J, Sassa K, Kamiya A and Mabuchi M 2006 *Mater. Lett.* 60 173–176

[10] Tong L B, Zheng M Y, Hu X S, Wu K, Xu S W, Kamado S and Kojima Y 2010 *Mater. Sci.Eng. A* 527 4250–4256

[11] Kim W J, Hong S, Kim Y S, Min S H, Jeong H T and Lee J D 2003 *Acta Mater.* 51 3293–3297

[12] Chino Y, Sassa K, Kamiya A and Mabuchi M 2006 *Mater. Sci. Eng. A* 441 349–356

[13] Huang X S, Suzuki K, Watazu A, Shigematsu I and Saito N 2009 J. *Alloys Comp.* 470 263–268

[14] Lee J B, Konno T J and Jeong H G 2010 *J. Alloys Comp.* 499 273–277

[15] Chen Y J, Wang Q D, Roven H J, Karlsen M, Yu Y D, Liu M P and Hjelen J 2008 *J. Alloys Comp.* 462 192–200.

[16] Lee H W, Lui T S and Chen L H 2009 *J. Alloys Comp.* 475 139–144

[17] Darras B M, Khraisheh M K, Abu-Farha F. K and Omar M A 2007 *J. Mater. Proc. Technol.*191 77–81

[18] Huang G S, Xu W, Huang G J, Li H C and Song B 2009 *J. Mater. Sci. Technol.* 25 365–369

[19] Zhang L, Huang G S, Zhang H, Song B 2011 *J. Mater. Proc. Technol.* 211 644–649

[20] Park J H, Kim H L, Jung J E and Chang W Y 2013 *Met. Mater. Int.* 19 389-398

[21] Yim C D, Seo Y M and You B S 2009 *Met. Mater. Int.* 15 683-688

[22] Li X, Al-Samman T, Mu S and Gottstein G 2011 *Mater. Sci. Eng. A* 528 7915-7925

[23] Kohzu M, Hironaka T, Nakatsuka S, Saito N, Yoshida F, Naka T, Okahara H and Higashi K 2007 *Mater. Trans.* 48 64-768

[24] Kohzu M, Nakatsuka S and Higashi K 2008 *Mater. Trans.* 49 2096-2099

[25] Kohzu M, Kii K, Nagata Y, Nishio H, Higashi K and Inoue H 2010 *Mater. Trans.* 51 749-755

[26] Yim C D, Seo Y M and You B S 2009 *Met. Mater. Int.* 15 683–688

[26] Lee C S and Duggan B J 1994 Acta Mater. 42 857

[27] Miura H, Yu G and Yang X 2011 *Mater. Sci. Eng. A* 528 (2011) 6981

[28] Liu W C, Li X Y and Meng X C 2009 *Scripta Mater.* 60 768–771

[29] Humphreys F J and Hatherly M 1995 *Recrystallization and Related Annealing Phenomena* (NewYork: Pergamon) p. 363-392

[30] Sun H, Liang S and Wang E 2009 Trans. Nonferrous Met. Soc. China, 19, 349–354

[31] Jeong H T and Ha T K 2007 J. Mater. Proc. Tech. 187–188 229–561

[32] Muránsky O, Carr D G, Barnett M R, Oliver E C and Sittner P (2008) *Mater. Sci. Eng. A* 496 14-24

Electronic and thermoelectric properties of nanograting layers

A Tavkhelidze
Ilia State University, Cholokashvili Ave. 3-5, Tbilisi 0162, Georgia

E-mail: avtotav@gmail.com

Abstract. Recently, new quantum features have been observed and studied in the area of nanostructured layers. Nanograting (NG) on the surface of the thin layer imposes additional boundary conditions on electron wave function and forbids some quantum states. Electrons, rejected from the forbidden quantum states, have to occupy states with higher energy. In the case of semiconductor materials, electrons rejected from the valence band have to occupy empty quantum states in the conduction band. Such increase in conduction band electron concentration n can be termed as geometry-induced doping or G-doping. G-doping is equivalent to donor doping from the point of view of the increase in n. However, there are no ionized impurities. This preserves charge carrier scattering to the intrinsic semiconductor level and increases carrier mobility. G-doping involves electron confinement to NG layer. Here, we investigate the system of multiple NG layers forming a series of homo junctions. Si and GaAs homojunctions were studied and G-doping levels of 10^{18}-10^{19} cm^{-3} were obtained. We also study a system composed of NG layer and an additional top layer forming periodic series of p–n junctions. In such system, charge depletion region develops inside the NG and its effective depth reduces, becoming a rather strong function of temperature T. Consequently, T-dependence of chemical potential magnifies and Seebeck coefficient S increases. We investigate S in the system of semiconductor NG layer having abrupt p–n junctions on the top of the grating. Analysis made on the basis of Boltzmann transport equations shows dramatic increase in S. At the same time, other transport coefficients remain unaffected by the junctions. Calculations show one order of magnitude increase in the thermoelectric figure of merit ZT relative to bulk material.

1. Introduction

Developments in nanotechnology allow fabrication of densely packed, periodic structures [1-4]. Recently, ultra-short period nanopore arrays and nanogratings have been obtained by block copolymer lithography [1, 2]. Another method for nanograting (NG) fabrication is multi-beam interference lithography. Using these techniques, gratings with 10-nm [3] and even sub-10 nm pitch [4] were fabricated. At the same time, it has been shown that NG dramatically improves thermoelectric [5] and electron emission properties [6] when grating pitch becomes comparable with electron de Broglie wavelength. This is owing to special boundary conditions imposed by NG on electron wave function. Supplementary boundary conditions forbid some quantum states, and density of quantum states (DOS) reduces (in all bands). Electrons rejected from NG-forbidden quantum states have to occupy empty states with a higher energy. Fermi energy increases and electronic properties of NG layer change. In the case of semiconductor materials, electrons rejected from the valence band (VB) occupy empty quantum states in the conduction band (CB). Electron concentration n in the CB increases which can be termed as geometry-induced electron doping or G-doping. G-doping is equivalent to donor doping from the point of view of increase in n and Fermi energy. However, there are no ionized impurities. This maintains charge carrier scattering to intrinsic semiconductor level and increases carrier mobility with respect to the donor-doped layer of same electron concentration. G-doping is temperature independent as it originates from layer geometry and no ionized impurities are involved.

Other methods of doping without impurities are well known modulation doping and recently introduced polarization doping [7]. Both are 2D in their nature. However, a 3D approach of modulation doping was introduced in [8] to improve thermoelelectric characteristics of nanocomposites [9]. Influence of periodic structures on electronic properties has been studied in related geometries such as periodic curved surfaces [10, 11], nanotubes [12], cylindrical surfaces with non-constant diameter [13] and strain-driven nanostructures [14].

Electron confinement to the NG layer is needed to obtain G-doping. The layer can be made free standing but in practice it is usually sandwiched between wide band gap layers. NG layer thickness is fundamentally limited by the requirement of having quantum properties. However, thin layers have low optical absorption. Layer thickness also limits lateral charge and heat transport.

Figure 1 shows a cross section of a single NG layer. The grating has depth a and period $2w$. To make a comparison, we choose reference layer as plain layer with thickness H such that it has the same cross section area.

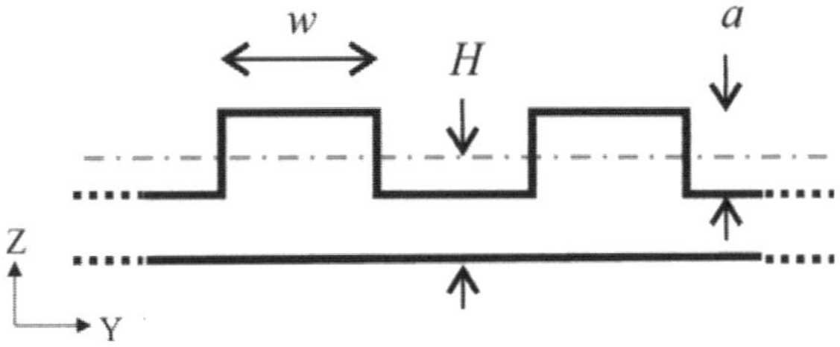

Figure 1. Cross section of nanograting layer.

Nanograting imposes additional boundary conditions on electron wave function and forbids some quantum states. The DOS in energy $\rho(E)$ reduces [15] with respect to the reference well

$$\rho(E) = \rho_0(E)/G \tag{1}$$

Where $\rho_0(E)$ is the DOS (number of quantum states within the unit energy region and in the unit volume) in a reference well and $G = G(H, w, a) > 1$ is the geometry factor. With increasing NG depth a DOS reduces.

Nanograting reduces DOS in all bands. Electrons rejected from the NG-forbidden quantum states have to occupy empty quantum states with higher E. In a semiconductor, electrons rejected from the VB have to occupy empty (and not forbidden by NG) energy levels in CB (figure 2). Electrons reject from low energy levels and occupy high energy ones. During this process, Fermi energy increases from $E_F^{(0)}$ to E_F. To simplify the presentation, in figure 2, we presume that T=0 (T is absolute temperature) and energy levels are equidistant on the energy scale (geometry-induced energy level shift is also ignored).

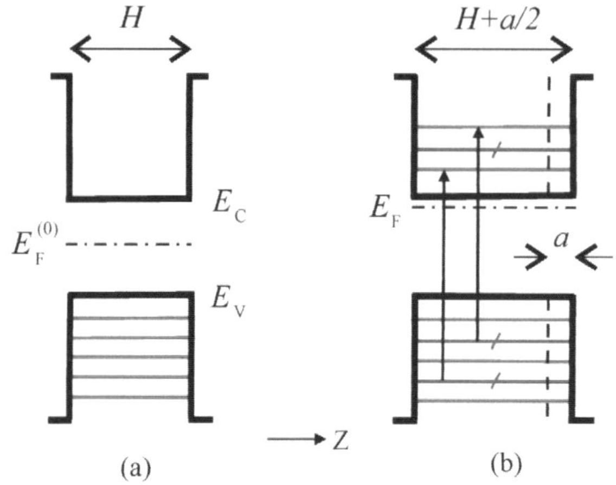

Figure 2. Energy diagrams of a) reference semiconductor quantum well and b) NG layer. Green lines depict occupied energy levels. Red lines depict NG-forbidden energy levels.

We will employ (1) to calculate the DOS and investigate G-dependence of n and E_F. The density of the NG-forbidden quantum states is

$$\rho_F(E) = \rho_0(E) - \rho_0(E)/G = \rho_0(E)(1-1/G). \quad (2)$$

To determine the number of rejected electrons n_r (per unit volume), (2) should be integrated over the electron confinement energy region.

$$n_r = \int_{con} dE\, \rho_F(E) = (1-1/G) \int_{con} dE\, \rho_0(E) = (1-1/G)\, n_{con}. \quad (3)$$

Where, $n_{CON} = \int_{con} dE\, \rho_0(E)$. Here, we assume that electron confinement takes place only in narrow energy intervals inside which G is energy-independent. In most cases, thin layer is grown on a semiconductor substrate and confinement energy regions are band offsets.

To obtain electron confinement in multiple NG layers transitional barrier layers are required. We regard cases of wide band gap material forming heterojunction, and the same material forming homojunction. In heterojunction case, confinement regions are band discontinuities (both type-I and type-II band alignments are considered). In homojunction case, confinement regions originate from barrier layer donor doping.

Here, we investigate G-doping in multiple NG layers. Such layers are quasi-3D and have improved optoelectronic and thermoelectric characteristics with respect to a single layer. Multiple NG layers consist of a replicated structure including main and barrier layers forming a series of hetero or homo junctions. Main layer is thicker than barrier layer and plays a leading role in carrier transport and optical absorption. Barrier layer forms electron confinement energy regions. It also contributes to carrier transport and optical absorption. Both main and barrier layers have NG geometry.

One of the objectives of this work is to calculate n and in multiple NG layers and find out how they depend on layer thicknesses and material properties. First, we introduce G-doping in a single NG layer. Next, we calculate n and in a homojunction multilayer structure. Subsequently, we calculate the same for a heterojunction multilayer structure. Finally, the possibility of realization of such structures and their advantages for optoelectronic and thermo-electric devices are discussed. An analysis was made within the limits of parabolic band, wide quantum well and degenerate electron gas approximations.

Thermoelectric materials are characterized in terms of dimensionless figure of merit *ZT*. Here, *T* is the temperature and *Z* is given by $Z = \sigma S^2 / (\kappa_e + \kappa_l)$, where *S* is the Seebeck coefficient, σ is electrical conductivity, κ_e is electron gas thermal conductivity, and κ_l is lattice thermal conductivity. The difficulty in increasing *ZT* is that materials having high *S* usually have low σ. When σ is increased, it leads to an increase in κ_e, following Wiedemann–Franz law, and *ZT* does not improve much. Another approach is to eliminate the lattice thermal conductivity by introducing vacuum nanogap between the hot and cold electrodes [16–18] and using electron tunnelling. Cooling in such designs was observed in [19] and theoretically studied in [20]. However, vacuum nanogap devices appear extremely difficult to fabricate. In this work, we present a solution that allows large enhancement of S without changes in σ, κ_e, and κ_l. It is based on having series of p–n junctions on the top of the NG.

It is another objective of this work is to calculate *ZT* of NG with series of p–n junctions and compare it with Z_0T of reference nanograting (RNG), in which $G \neq G(T)$, and bulk material. The reduced figure of merit ZT / Z_0T is calculated and μ dependences are presented for such traditional thermoelectric materials as Si and Ge. Analysis was made within the parabolic bands approximation and the abrupt junction's approximation. Since only heavy G-doping was considered, we neglected the hole contribution in transport.

2. Electronic properties of multiple homojunction NG layer

We begin from calculating *n* in a homojunction NG layer, as it is relatively straightforward. Homojunction is formed between intrinsic main layer and donor-doped barrier layer. Let us consider a case when all layers have a plain geometry (NG is not formed yet). Figure 3a shows an energy diagram of such structure (typically obtained by homoepitaxal growth) [21]. Charge depletion regions form inside the both *i*-type main layer and n-type barrier layer. Band edges curve, shaping confinement regions in the main layer VB and in the barrier layer CB. Electrons with energies $E_V - \Delta E_{con}^{(0)} < E < E_V$ are confined to main layers and ones with energies $E_C - \Delta E_{con}^{(0)} < E < E_C$ are confined to barrier layers. Here, we assume that both main and barrier layers are relatively thick so that electron wave functions do not overlap and we can ignore the mini band formation.

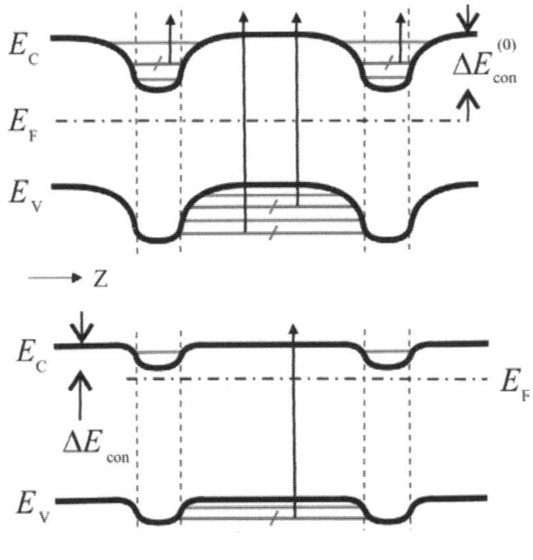

Figure 3. Homojunction structure energy diagrams: a) before NG formation, b) after NG formation. Green lines depict occupied energy levels. Red lines depict NG-forbidden energy levels.

2.1 Influence of Nanograting

Next, we fabricate NG on the surface of all layers simultaneously (fabrication of such structure will be discussed later in this work). NG forbids some energy levels (red lines) and induces G-doping depending on geometry factor value. We introduce two geometry factors G_m and G_b for the main and barrier layers separately (layers have different geometry factors owing to a different thickness).

If $G_m > 1$ some electrons reject from the VB of main layer (figure 3). If $G_b > 1$ some electrons reject from the CB of barrier layer and occupy higher energy levels in the same band. In the beginning, let us regard the case $G_m > 1$ and $G_b = 1$. As $G_m > 1$, the Fermi level increases in main layer, which is equivalent to shifting band edges down on the energy scale. As $G_b = 1$, the bang edges remain intact in barrier layer. With the downward shift in main layer, VB confinement region will shrink and fewer electrons will reject to the CB. This will continue until some equilibrium value of rejected electrons is achieved. In other words, band edge downward shift provides negative feedback to the G-doping process. Next, let us regard the case $G_m > 1$ and $G_b > 1$. Here, barrier layer band edges also shift down. Owing to the downward shift in barrier layer, confinement energy regions widen. This increases the number of rejected electrons in main layer. In other words, barrier layer G-doping provides positive feedback to main layer G-doping. Finally, we have a combination of negative feedback originating from main layer and positive feedback origination from barrier layer. In the case of overall negative feedback G-doping of two layers will saturate and equilibrium value of ΔE_{con} will be attained (figure 3b). Last depends on G_m and G_b and semiconductor material parameters.

2.2 Electron confinement energy region

To find ΔE_{con} we have to calculate electron concentrations in both main and barrier layers first. Applying (2) to the main layer we find the number of rejected electrons

$$n_r = (1 - 1/G_m) \int_{E_V}^{E_V - \Delta E_{con}} dE\, \rho_{0V}(E) \ , \tag{4}$$

where, $\rho_{0V}(E)$ is the initial (to NG) DOS in the main layer VB. Let us regard the case when the rejected electron number is much more than the initial electron number in CB $n_r >> n_0$. Then, the CB electron concentration will be merely n_r. Further, we use the following well known expression for valence band DOS to calculate n_r [22]

$$\rho_{0V}(E) = 2\pi^{-1/2} N_V (k_B T)^{-3/2} (E - E_V)^{1/2} \ . \tag{5}$$

Here, N_V is the effective VB density of states. Inserting (5) in (4) and integrating, choosing zero of energy scale as $E_V = 0$ we find

$$n_r = (4/3\sqrt{\pi}) N_V (1 - 1/G_m)(\Delta E_{con} / k_b T)^{3/2} \ . \tag{6}$$

Further, we find Fermi levels in both layers using corresponding electron concentrations. According to the well known expression for degenerate semiconductor [23] Fermi level in main layer can be found from

$$E_F - E_C^{(m)} = k_B T\left[\ln(n_r / N_C) + 2^{-3/2}(n_r / N_C)\right] \tag{7}$$

where, N_C is the effective CB density of states. CB electron concentration of barrier layer does not change during NG formation, as there are no electrons transferred from the VB to the CB. However, Fermi level in the barrier layer moves up on an energy scale owing to NG-induced reduction in N_C

$$E_F - E_C^{(b)} = k_B T\left[\ln(G_b n_d / N_C) + 2^{-3/2}(G_b n_d / N_C)\right]. \tag{8}$$

Here, n_d is barrier layer initial electron concentration (obtained by donor doping). Energy diagram (figure 3) shows that $\Delta E_{con} = E_C^{(m)} - E_C^{(b)}$. Subtracting (7) from (8) and inserting in the last expression we find

$$(\Delta E_{con}/k_B T) = \ln(G_b n_d / n_r) + 2^{-3/2}\left[(G_b n_d - n_r)/N_C\right]. \tag{9}$$

Inserting (6) in (9) and depicting $\eta \equiv \Delta E_{con}/k_B T$ we get following nonlinear equation for η

$$\eta + \frac{3}{2}\ln\eta + 2^{-3/2}\frac{4}{3\sqrt{\pi}}\frac{N_V}{N_C}\frac{G_m - 1}{G_m}\eta^{3/2} - \ln\frac{3\sqrt{\pi}G_m G_b n_d}{4N_V(G_m - 1)} - 2^{-3/2}\frac{G_b n_d}{N_C} = 0 . \tag{10}$$

Equation (10) was solved numerically to find ΔE_{con} using two geometry factor and n_d, N_C, N_V values for Si and GaAs materials.

2.3 Electron concentration and Fermi level

Numerical solutions of (10) were inserted in (6) to find n_r. Next, we found the Fermi level of the main layer by inserting values of n_r in (7). Formula (8) was used to find a Fermi level position in barrier layer. Table 1 shows the results for geometry factors G_m =1. 02, G_b = 1.1 and *T*=300 K.

Table 1. Electron concentration and Fermi levels in main and barrier layers for Si and GaAs materials. Energy was measured from corresponding layer CB edge.

Material	n_d **[cm^{-3}]**	ΔE_{con} **[meV]**	n_r **[cm^{-3}]**	$E_F^{(m)}$ **[meV]**	$E_F^{(b)}$ **[meV]**
Si	3 x 10^{18}	45	6 x 10^{17}	103	58
	1 x 10^{19}	64	1 x 10^{18}	89	25
	3 x 10^{19}	87	1 x 10^{18}	77	-10
	1 x 10^{20}	126	2.8 x 10^{18}	62	-64
GaAs	1 x 10^{18}	47	3.2 x 10^{17}	3.4	-44
	3 x 10^{18}	86	8 x 10^{17}	-29	-115
	5 x 10^{18}	119	1.3 x 10^{18}	-52	-172
	1 x 10^{19}	197	2.8 x 10^{18}	-110	-297

For barrier donor doping levels 10^{18}-10^{20} cm^{-3} and G values rather close to unity ($G_m = 1.02$, $G_b = 1.1$) main layer G-doping levels of 10^{17}-10^{18} cm^{-3} were obtained. We chose this range of G-dropping levels, as they are frequently used in applications. Higher and lower G-doping levels can be obtained as well. Table 1 shows that one order higher donor doping n_d (10^{20} cm^{-3}) is needed in Si with respect to GaAs (10^{19} cm^{-3}) to obtain a G-doping level 2.8 10^{18} cm^{-3}. It is convenient to give an interpretation of this result using terms of negative and positive feedback. Negative feedback in main layer is weaker for Si material as it has higher n_0 value with respect to GaAs. Owing to high n_0, band edges shift down less rapidly with increasing n_r, following well known logarithmic dependence $\Delta E_F = k_B T \ln(n_r / n_0)$. As ΔE_F is positive in our case, higher n_0 results in weaker negative feedback. Weaker negative feedback in the main layer requires weaker positive feedback in the barrier layer to obtain equilibrium. Positive feedback is weaker in highly donor-doped barrier layers (as shifting band edges require more rejected electrons). This explains why higher n_d is required in Si barrier layer with respect to GaAs.

As figure 3 shows there are potential barriers for electrons in the CB and holes in the VB of multilayer NG. They are affecting carrier transport in Z direction. However, barrier height is ΔE_{con} which is of the order of a few $k_B T$ as table 1 indicates (except for very high level of donor doping. $k_B T$ =26 meV for T=300 K). Charge carriers easily overcome such obstacles (thermionic emission) and their influence can be ignored. In the case of high values of ΔE_{con} barriers may block carrier transport in Z direction. To avoid this, high level of G-doping should be obtained not by increasing donor doping of barrier layers but by increasing of G_m and G_b values.

During numerical calculations, we kept $E_F^{(m)}$ and $E_F^{(b)}$ in proximity of corresponding CB edges to stay within the limits of degenerate electron gas approximation (so that (7) and (8) are valid). For lower doping levels, other expressions should be used instead of (7) and (8).

2.4 G-doping for high electron mobility applications

Realization of G-doping is attractive from the point of view of high mobility applications. Let us use a multi junction solar cell [24] as an example to estimate G-doping improvement of its characteristics. In this device, window, emitter and tunnel junction layer doping levels are roughly 10^{18} cm^{-3}. At this level of donor doping, ionized impurities reduce electron mobility by a factor of 4 in GaAs [25] (at T=300 K) and by the factor of 10 in Si [26]. Most types of solar cells use transparent conductive oxides with doping levels of 10^{20}-10^{21} cm^{-3}. At this donor doping level ionized impurities reduce electron mobility by the factor of 30-50 in GaAs. Thus, using G-doping in these layers can dramatically improve characteristics of solar cells.

Multiple NG layers can be realized by epitaxial growth on the top of lead grating, previously formed on the substrate. Such structure has been modelled [27] and fabricated [28] on GaAs substrate using epitaxial growth technique. Layers, grown using this technique, have diverse geometry compared to single NG layer. They have gratings from both sides and interfaces resemble sine instead of square. With increasing number of layers sine amplitude reduces and finally one gets plain layers. Geometry factor and G-doping level reduce with increasing of layer number.

It is challenging to obtain high G values or considerable DOS reduction in both NG and alike geometries. However, as table 1 show, G-doping required values are quite close to unity (especially for the main layer). The obtaining of such values in multiple NG systems seems to be straightforward.

The above analysis was made using the assumption of G energy independence (3). Usually, electron confinement energy regions are small (tens or hundreds of meV) and this assumption is valid. However, the NG layer can be made free standing or grown on very wide band gap substrate. In these cases, confinement energy intervals are much wider and care should be taken when assuming *G* energy independence.

3. Charge and heat transport in the NG layer with p-n junctions

Here, we propose a system composed of NG layer and an additional layer on the top of the ridges forming periodic series of p^{+}–n^{+} junctions (figure 4). In such system, charge depletion region develops

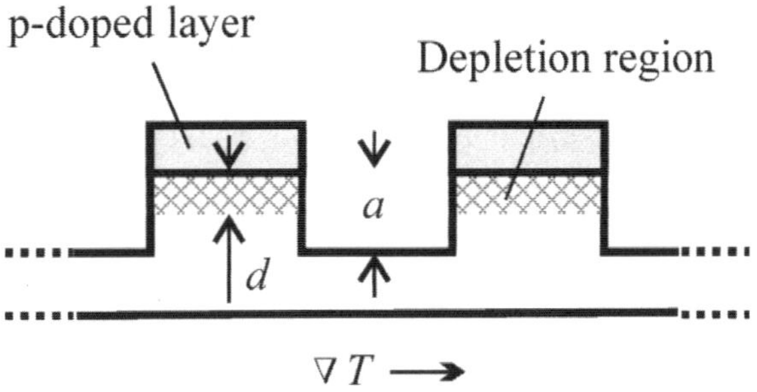

Figure 4. NG with series of periodic junctions grown at the top of the ridges.

inside the ridges and effective ridge height reduces, becoming a rather strong function of temperature *T*. Consequently, *T*-dependence of chemical potential μ magnifies and Seebeck coefficient *S* increases. Depletion region width $\mathrm{d}(T)$ depends rather strongly on temperature. The ridge effective height $a_{\mathrm{eff}}(T) = a - \mathrm{d}(T)$ and consequently the geometry factor *G* becomes temperature-dependent, $\mathrm{G} = \mathrm{G}(T)$. All parameters of NG layer including μ become stronger functions of *T* than it will be in an NG layer layer without the junctions. Seebeck coefficient and thermoelectric figure of merit increase.

Analysis was made using the Boltzmann transport equations. First, we calculate $\nabla\mu$ for the system of NG layer and junctions and express it as

$$\nabla\mu = \nabla\mu_0 + \nabla\mu_{\mathrm{J}}\,. \tag{11}$$

Here, $\nabla\mu_0$ is the chemical potential gradient in an RNG (NG layer without p-n junctions) and $\nabla\mu_{\mathrm{J}}$ is introduced by the junctions. Next, we insert $\nabla\mu$ in Boltzmann transport equations and calculate *S* for the system of NG layer and junctions, expressing it as $S = S_0 + S_{\mathrm{J}}$, where S_0 is Seebeck coefficient of RNG and S_{J} is introduced by the junctions..

Cross-section of the system of NG and periodic junction is shown in figure 4. We assume that there is a temperature gradient ∇T in the Y-dimension. Consequently, depletion depth depends on the Y-coordinate, and geometry factor gradient ∇G appears in the Y-direction. Presence of ∇G and ∇T modifies the electron distribution function and causes electron motion from the hot side to the cold side. This motion is compensated by thermoelectric voltage. Let us write Boltzmann transport equations [29] for the system of RQW and periodic series of junctions

$$J = \mathbf{L}^{11}\left(\mathcal{E} + \nabla\mu/e\right) - \mathbf{L}^{12}\,\nabla T \text{ and } J^{Q} = \mathbf{L}^{21}\left(\mathcal{E} + \nabla\mu/e\right) - \mathbf{L}^{22}\,\nabla T\,. \tag{12}$$

Here, J is the electric current density, J^Q is the heat current density, $\mathbf{L}^{ij}$ are coefficients, $\mathcal{E}$ is the electric field, and e is the electron charge. Within the parabolic bands approximation, $\mathbf{L}^{ij}$ are the functions of integrals of type [30].

$$\Omega^{(\alpha)}(E) = \int_{-\infty}^{+\infty} dE\left(-\partial f_0/\partial E\right)\rho(E)\tau(E)\mathrm{v}_y^{\,2}(E-\mu)^{\alpha} \tag{13}$$

where f_0 is the electron distribution function, $\tau(E)$ is the electron lifetime, v_y is the electron velocity in the Y-direction, and α = 0, 1, 2. Let us find how NG affects $\Omega^{(\alpha)}(E)$. The NG does not change dispersion relation and consequently v_y. It reduces density of states (3D density) G times, i.e. $\rho(E)=\rho_0(E)/G$ and increases the transport lifetime G times [31], i.e., $\tau(E)=G\tau_0(E)$ (the latter follows from Fermi's golden rule). Here, $\rho_0(E)$ and $\tau_0(E)$ are the density of states and carrier lifetime, respectively, in the case $a = 0$. Consequently, product $\rho(E)\tau(E)\mathrm{v}_y^{\,2}$ in the NG layer is the same as in the conventional wide quantum well of the same width (3D case) and in the bulk material. This product does not depend on G and consequently it becomes independent of depletion depth. The NG changes the distribution function f_0, since it increases μ. The NG influences integrals $\Omega^{(\alpha)}(E)$ by changing μ alone. Therefore, integrals are the same as in the bulk material having the corresponding chemical potential

$$\mathbf{L}^{ij} = \mathbf{L}_0^{ij}. \tag{14}$$

where $\mathbf{L}_0^{ij}$ are the coefficients of bulk material having the value of μ obtained by the conventional doping (for instance).

Further, we have to find μ for the system of NG layer and junctions and insert it in (12) together with (14). Chemical potential of degenerated semiconductor ($-2 < \mu^* < 2$ where $\mu^* = \mu/k_BT$) can be written as [32] (note that frequently Fermi energy is used instead of chemical potential in semiconductor literature)

$$\mu = k_BT\left[\ln\left(n/N_{NG}\right)+2^{-3/2}\left(n/N_{NG}\right)\right]. \tag{15}$$

where n is the electron concentration and N_{NG} is the effective conduction band density of states of NG layer. In the case of heavy G-doping ($n >> n_i$, where n_i is intrinsic concentration), we can use (3) for n. Density of states reduces G times in NG layer, i.e. $N_{NG} = N_C/G$, where N_C is the CB effective density of states for bulk semiconductor. Inserting this and (3) in (15), we get

$$\mu = k_BT\left\langle \ln\left[n_{con}(G-1)/N_C\right] + 2^{-3/2}\left[n_{con}(G-1)/N_C\right]\right\rangle. \tag{16}$$

Further, we rewrite (16) as

$$\mu = \mu_0 + k_BT\left\langle \ln\left[(G-1)/(G_0-1)\right] + 2^{-3/2}\left(G-G_0\right)n_{con}/N_C\right\rangle \tag{17}$$

where G_0 is constant. Introduction of $\mu_0 \equiv \mu(G_0)$ defines reference material as n$^+$-type semiconductor with electron concentration of $N_D = n_{CON}(G-1)$, or RRQW having constant geometry factor $(\partial G/\partial T) = 0$ and $G = G_0$. Next, we calculate the gradient of (10) taking into account that $N_C \propto T^{3/2}$. The result is

$$\nabla\mu = \nabla\mu_0 + \theta\nabla T \ . \tag{18}$$

$$\text{where } \theta \equiv k_B\left[\ln\frac{G-1}{G_0-1} - \frac{1}{2}\xi(G-G_0) + T\left(\frac{1}{G-1}+\xi\right)\frac{\partial G}{\partial T}\right] \tag{19}$$

and $\xi \equiv 2^{-3/2} n_{con}/N_C$. Inserting (18) and (14) in Boltzmann equations, we find charge and heat currents in the system of NG layer + Junctions

$$J = \mathbf{L}_0^{11}(\mathcal{E} + \nabla\mu_0/e) - (\mathbf{L}_0^{12} - \mathbf{L}_0^{11}\theta/e)\nabla T \text{ and} \tag{20a}$$

$$J^Q = \mathbf{L}_0^{21}(\mathcal{E} + \nabla\mu_0/e) - (\mathbf{L}_0^{22} - \mathbf{L}_0^{21}\theta/e)\nabla T \ . \tag{20b}$$

Further, Seebeck coefficient, electrical conductivity, and electron gas heat conductivity can be found from (20a) and (20b) in a conventional way.

$$S = \left(\mathbf{L}_0^{12} - \mathbf{L}_0^{11}\theta/e\right)/\mathbf{L}_0^{11} = S_0 - (\theta/e) \ , \tag{21}$$

$$\kappa_e = \mathbf{L}_0^{22} - \mathbf{L}_0^{21}(\theta/e) - \left(\mathbf{L}_0^{12} - \mathbf{L}_0^{11}\theta/e\right)\mathbf{L}_0^{21}/\mathbf{L}_0^{11} = \kappa_{e0} \ , \tag{22}$$

$$\sigma = \mathbf{L}_0^{11} = \sigma_0 \ . \tag{23}$$

Electrical and thermal conductivity in the system of NG layer and junctions remain unaffected (with respect to RNG layer or bulk material having the same μ value) by a series of junctions, and S change according to (21). To calculate θ and then S, we have to find $\partial G/\partial T$ first (19).

Such performance of transport coefficients has following physical interpretation. The geometry factor G is a rather strong function of temperature. Consequently, T-dependence of chemical potential magnifies and Seebeck coefficient, having approximate value $S \approx -(1/e)(\partial\mu/\partial T)$, increases. Electrical conductivity $\sigma = \Omega^{(0)}$ is G independent since, the reduction of state density leads to proportional increase of lifetime, according to Fermi's golden rule (resulting in (14)). In a degenerate electron gas (n^+-type semiconductor) Wiedemann–Franz law works well. Thus, the thermal conductivity κ_e follows σ and is also G independent.

4. Seebeck coefficient of NG layer with p–n junctions

It is reasonable to calculate Seebeck coefficient of NG layer with junctions relative to RNG layer, in which geometry factor is constant $(\partial G/\partial T) = 0$. This will allow comparison of dimensionless figure of merit ZT with Z_0T, using similarity of electric (23) and heat conductivities (22). In (17)–(19) G_0 is arbitrary. To simplify the comparison, we choose G_0 so that NG layer with junctions and reference RNG layer have the same μ value. Equation of chemical potentials leads to $G_0 = G$ (16). Inserting this in (19) and further inserting the obtained result in (21), we obtain for Seebeck coefficient of NG layer with junctions

$$S = S_0 - \frac{k_B T}{e}\left(\frac{1}{G-1} + \xi\right)\left(\frac{\partial G}{\partial T}\right) . \tag{24}$$

3D Seebeck coefficient of reference RNL do not differ from Seebeck coefficient of the bulk material, and within the parabolic bands approximation it is equal to [33]

$$S_0 = \frac{k_B}{e}\left[\frac{r+5/2}{r+3/2}\frac{F_{r+3/2}(\mu^*)}{F_{r+1/2}(\mu^*)} - \mu^*\right] . \tag{25}$$

Here, r refers to scattering parameter and it is assumed that electron lifetime $\tau(E) \propto E^r$, $\mu^* = \mu / k_B T$ is the reduced chemical potential, and $F(\mu^*)$ are Fermi integrals. In the case of no impurities (G-doping) and low energies, acoustic phonons are responsible for electron scattering and $r = 0$. It should be noted here that NG layer has G times more width [31] with respect to the conventional layer. As 3D $\rho(E)$ in wide quantum wells tends to $\rho(E)$ of the bulk material of the same width, using (28) for RNG is a good approximation. However, in the case of thin layers oscillatory behaviour of transport coefficients should be considered [34].

To find the ratio Z / Z_0, we use relations (22) and (23) and obvious relation between lattice thermal conductivities $\kappa_l = \kappa_{l0}$, all together leading to

$$Z / Z_0 = \left(S/S_0\right)^2 . \tag{26}$$

Figure 5 shows the dependence $S_0(\mu)$ according to (25). The $S(\mu)$ in the same figure is determined by first, calculating $S(G)$ by inserting $\partial G / \partial T$ (which was calculated in [5]) in (24), and then (16) was used to convert X-axis so that $S(\mu)$ was obtained from $S(G)$. The ratio $Z(\mu) / Z_0(\mu)$ in the same figure is calculated according to (26). We present μ dependences as they allow the understanding of possible μ ranges, within which real devices can operate without changing sign of $(\partial G / \partial T)$ (see reference 5) and consequently S sign.

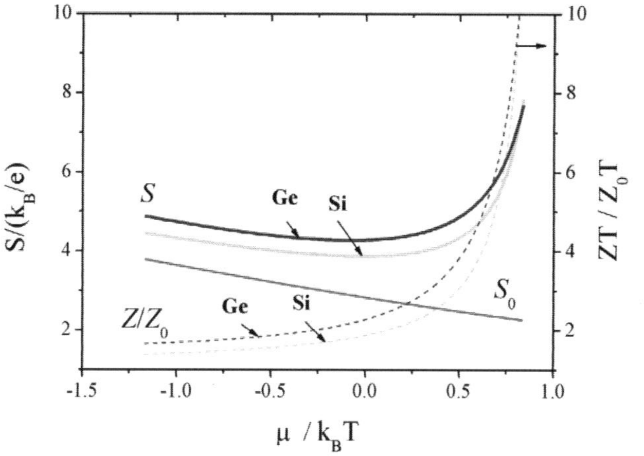

Figure 5. $S(\mu)$ – solid lines, $S_0(\mu)$ – dotted line, and $Z(\mu)/Z_0(\mu)$ – dashed lines (belong to right Y-axis). Dependences $S(\mu)$ and $Z(\mu)/Z_0(\mu)$ are calculated for Si and Ge materials.

Traditional thermoelectric materials Si and Ge are chosen as examples. Dependences are plotted for the following parameters: Material Si, $n_{CON} = 4.5\times10^{18}$ cm^{-3}, a = 20 nm, $\gamma = 3.3$ (γ is reduced chemical potential of p-type layer), $\beta = 0.1$ ($\beta = a_{eff}/d$), T = 300 K; Material Ge, n_{con} = 1.4×10^{18} cm^{-3}, a = 35 nm, $\gamma = -1$, $\beta = 0.25$, T = 300 K. For both Si and Ge, we choose n_{con} so that, for value G=10, μ was close to the optimum value $\mu_{OPT}/k_BT = r + 1/2$ [33]. It should be noted here that NG layer exhibits quantum properties at G times more widths with respect to the conventional quantum well.

Figure 5 shows large enhancement of figure of merit in p+–n+ junction NG layer both for Si and Ge materials. The dependences are quite identical for materials having rather different band gaps (1.12 eV for Si and 0.66 eV for Ge). This shows that despite band gap value entering expressions for depletion depth (Reference 5) and $(\partial G/\partial T)$, almost similar $S(\mu)$ and $Z(\mu)/Z_0(\mu)$ dependences can be obtained by matching such parameters as β and γ.

5. Conclusions

Geometry-induced electron doping (G-doping) was investigated in multiple NG layers. They were composed of main and barrier layers forming a series of isotype homo junctions. Barrier layers were used to form electron confinement energy regions. Both main and barrier layers were n-type. Barrier layers were donor-doped to obtain electron confinement and induce G-doping in main layers. Such parameters as electron concentration, Fermi level and confinement region width were calculated for this system. Main layer G-doping level of 10^{17}-10^{18} cm^{-3} was obtained at barrier layer donor doping of 10^{18} -10^{20} cm^{-3} and geometry factor values G_m=1.02, $G_b = 1.1$. One order higher donor doping was required in Si with respect to GaAs to obtain the same G-doping level. For high electron mobility it was preferable to have high values of geometry factors. Electron mobility was higher in G-doped layers with respect to donor-doped layers of same doping level. G-doping opens prospects for new quasi-3D optoelectronic and thermoelectric systems. At the same time G-doping is temperature independent and can be used to extend working temperature ranges of cryogenic and power electronics.

Thermoelectric transport coefficients were investigated in the system of NG layers and periodic series of p–n junctions at the top of the ridges. Analysis was made on the basis of Boltzmann transport equations. It was shown that the Seebeck coefficient increases considerably. At the same time, electrical and thermal conductivities remain unaffected by the series of junctions. This allows large enhancement of thermoelectric figure of merit. Dependence of Seebeck coefficient on geometry factor G and junction parameters was investigated and the analytical expression was obtained. Seebeck coefficient changes sign for some value of G. Dependences of S and ZT on chemical potential were presented for p–n junction NG layers. Calculations show one order of magnitude increase in thermoelectric figure of merit with respect to the bulk material.

6. Acknowledgements

Author thanks Ismat S. Shah and Juejun Hu for useful discussions. Apparatus received from EU TEMPUS project 530278-TEMPUS-1-2012-1-DE-TEMPUS-JPHES was used for preparing this work for publication.

7. References

[1] Andreozzi A, Lamagna L, Seguini G, Fanciulli M, Schamm-Chardon S, Castro C and M Perego M 2011 Nanotechnology 22 335303

[2] Xia G, Jeong S-J,Kim J E, Kimand B H, Koo C-M and Kim S O 2009 Nanotechnology 20 225301

[3] Paivanranta B, Langner A, Kirk E, David C and Ekinci Y 2011 Nanotechnology 22 375302

[4] Bergmann K, Danylyuk S V and Juschkin L 2009 J. Appl. Phys. 106 073309

[5] Tavkhelidze A 2009 Nanotechnology 20 405401

[6] Tavkhelidze A, 2010 J. Appl. Phys. 108 044313

[7] Simon J, Protasenko V, Lian C, Xing H and Jena D 2010 Science 327 60-64

[8] Yu B, Zebarjadi M, Wang H, Lukas K, Wang H, Wang D, Opeil C, Dresselhaus M,Chen G, and Ren Z 2012 Nano Lett. 12 20772082

[9] H. J. Goldsmid 2010 Introduction to Thermoelectricity (Springer)

[10] Ono S, Shima H 2010 Physica E 42 1224-1227

[11] Kartashov Y V, Szameit A, Keil R, Vysloukh V A, and Torner L 2011 Optics Letters 36 3470

[12] Gupta S and Saxena A 2011 J. Appl. Phys. 109 074316

[13] Fujita N 2004 J. Phys. Soc. Jpn. 73 3115-3120

[14] Ortix C, S. Kiravittaya S, Schmidt O G, and van den Brink J 2011 Phys. Rev. B. 84 045438

[15] Tavkhelidze A and Svanidze V 2008 Int. J. Nanosci. 7 333

[16] Despesse G and Jager T 2004 J. Appl. Phys. 96 5026

[17] Wachutka G and Gerstenmaier Y C 2006 proc. Int. Conference on Mixed Design (MIXDES Poland) p 48

[18] Goldsmid, H J 2003 in Thermoelectrics, 2003 Twenty-Second International Conference on ICT pp 433- 438

[19] Hishinuma Y, Geballe T H, and Moyzhes B Y 2003 J. Appl. Phys. 94 4690

[20] Zeng T 2006 Appl. Phys. Lett. 88 153104

[21] A. Sanders et al 2011 Nanotechnology 22 465703

[22] Sze S M and Ng K K 2007 Physics of Semiconductor Devices (New Jersey: Wiley-Interscience)

[23] Joyce W B and Dixon R W 1977 Appl. Phys. Lett. 31 354

[24] Kirk A P 2010 Solar Energy Materials & Solar Cells, 94 2442

[25] Casey H C Jr, Panish M B 1978 Hetrostructure lasers (Academic, New York)

[26] Bulucea C 1993 Solid-State Electron. 36 489

[27] Ohtsuka M and Suzuki A 1993 J. Appl. Phys. 73 7358

[28] Pickrell G W, Xu C F, Louderback D A, Lin H C, Fish M A, Hindi J J, Simpson M C, Guilfoyle P S, Zhang Z H and Hsieh K C 2004 J. Appl. Phys. 96 4050

[29] Ashkroft N E and Martin N D 1976 Solid State Physics (Saunders College Publishing, NY)

[30] Mahan J D and Sofo J O, 1996 Proc. Natl. Acad. Sci. USA 93 7436

[31] Tavkhelidze A, Svanidze V 2008 Int. J. of Nanoscience 7 333

[32] Joyce W B and Dixon R W 1977 Appl. Phys. Lett. 31 354

[33] Bhandari C M, Rowe D M 1995 CRC Handbook of Thermoelectrics, ed Rowe D M (CRC Press), p 43

[34] Rogacheva E I, Nashchekina O N, Grigorov S N, Us M A, Dresselhaus M D and Cronin S B 2002 Nanotechnology 13 1

Experimental and numerical study of fatigue crack propagation in a thick-walled cylinder under cyclic hoop stress

I Salam[1], M A Malik[2], M Abid[3] and M Farooque[1]
[1]Institute of Industrial Control Systems, Islamabad, Pakistan
[2]Department of Mechanical and Aerospace Engineering, Air University, Islamabad, Pakistan
[3]Faculty of Mechanical Engineering, GIK Institute of Engineering Sciences and Technology, Topi, Pakistan

E-mail: iftikhar.salam@gmail.com, drafzaalmalik@yahoo.com, mabid00@hotmail.com

Abstract. In present work, after finding the anisotropy resulting in dissimilar properties in different orientations of a thick-walled cylinder, experimental and numerical study was performed to reveal the fatigue crack growth behavior of the cylinder under cyclic hoop stress. Fatigue crack growth experiments were conducted on middle tension M(T) samples prepared in an orientation to simulate the hoop stress on the cylinder. The tests were conducted under constant amplitude loading at R ratio 0.1. The fatigue crack growth data was compiled and applied to simulate and predict the crack growth process using two dimensional parametric finite element technique. The fatigue crack propagation was simulated, based on linear elastic fracture mechanics and stress intensity factor determination. Both the experimental and numerical results of crack growth data, at stress levels of 10 to 40 per cent of the yield stress of the material, were found in close agreement. The disparity observed was concluded in the range of statistical scatter in the experimental data. The crack growth rate and the fatigue life of the samples obtained from the experiments and the simulation were also in good agreement at all the stress levels analyzed.

1. Introduction

Reliability of materials and structures in the form of thick-walled cylinders (TWC) is of critical importance. Thick-walled cylinders in form of boilers, gun barrels, pipelines and high-pressure containers are essential structural members for many industries including power, chemical, armament, and food processing industries [1, 2]. In many applications the cylinders are prone to cyclic stresses during their normal operation and large internal pressures produce high tension hoop stresses along the inner surface of the cylinder. The latter may result in the nucleation of the internal surface cracks which propagate due to cyclic action of high-pressure pulses. If the primary crack growth mechanism is slow, the cracks will be detected during routine maintenance by non-destructive testing (NDT) so that corrective measures can be taken before crack growth moves into a high risk regime. After the crack reaches a critical size, the ultimate failure may be catastrophic and result at pressures which are even lower than the design capacity of the cylinder. For this reason, it is quite necessary to analyze the crack propagation behavior in TWC under cyclic hoop stress to ensure the integrity of the cylinder against the fatigue failure. Many researchers have done valuable work to analyze the TWC under fatigue loading [3-5].

The three major fatigue life methods used in design and analysis are the stress-life method, the strain-life method, and the linear-elastic fracture mechanics (LEFM) method [6]. The numerical techniques, based on linear-elastic fracture mechanics [7] are frequently used to establish fatigue failure criteria. Many researchers used numerical approaches for the study of fatigue crack propagation [8-10] and a crack tip node-release scheme was suggested in Ref. [9] for the same.

In present work, experimental and numerical study was performed to reveal the fatigue crack growth behavior of the cylinder under cyclic hoop stress. Since the full-scale fatigue crack growth test of the thick-walled cylinder subjected to internal cyclic pressure involves a significant amount of time and cost, M(T) samples taken from the cylinder in the transverse direction (CR) were alternatively used for the fatigue crack growth simulation [10].

The fatigue crack growth experiments were conducted on M(T) specimens, prepared in CR direction, at various stress levels. The experimental work was supplemented through modeling and simulation with the help of commercially available software. The data collected from the experiments was utilized to predict the fatigue life of the samples with the help of numerical technique based on LEFM. The finite element analysis (FEA) was conducted using ANSYS structural analysis software [11] with two dimensional finite element model and a code in ANSYS Parametric Design Language (APDL) was written to replicate the crack growth process. The results obtained from the two techniques are compared.

2. Experimental procedure

2.1 Fatigue crack growth test - Sample preparation

Fatigue crack growth tests were performed on middle tension M(T) specimens. Samples were cut from the cylinder in CR orientation, figure 1. The inner and outer diameters of cylinder were 100 mm and 150 mm, respectively. This orientation refers to the standard ASTM E 399 which provides the crack plane orientation code for bar and hollow cylinder. In this orientation the samples were perpendicular to the longitudinal axis of the cylinder 'L' while the notch direction was such that the crack propagation was along the radial direction 'R' as shown in figure 1. A TWC under pressure experiences three principal stresses. Among these stresses the tangential or hoop stress is the maximum and has about twice the value of the axial stress [12]. Hence the samples prepared for the experimental study were in the orientation that experiences the maximum stress present in the cylinder. Samples were machined according to the standard ASTM E 647 and the dimensions of the final M(T) sample are shown in figure 2. The machined samples were subjected to mechanical grinding and subsequently fine polishing.

2.2 Testing procedure

Fatigue crack growth experiments were performed on a servo-hydraulic testing machine in accordance with ASTM standard E 647. Tests were conducted in the tension-tension mode under constant amplitude loading with R ratio 0.1. A sinusoidal waveform was applied at a loading frequency of 10 Hz. Tests were conducted at stress levels of 10 to 40 per cent of the yield strength of the material. A total of 12 samples were tested and fatigue crack growth data and SN curves were achieved. Crack length was measured with the help of a traveling microscope, at a magnification of 100x. Tests were conducted in air at a temperature 20°C and approximately 50 per cent relative humidity.

3. Finite element modeling

Fatigue crack growth analysis was performed using ANSYS structural software by repeatedly loading the geometry, recording stress intensity factor K_I at crack tip, advancing the crack and then unloading. Two dimensional finite element analysis of M(T) sample geometry was conducted using four-noded quadrilateral elements under plane-strain conditions. The symmetry in loading and geometry of the M(T) sample was taken advantage of and a solid model for a quarter section of the sample was created in the ANSYS pre-processor. Figure 3 shows the geometry of the quarter model. The quarter model has an initial crack length 'a' of 3 mm and $a/W = 0.4$, where W is the sample half width.

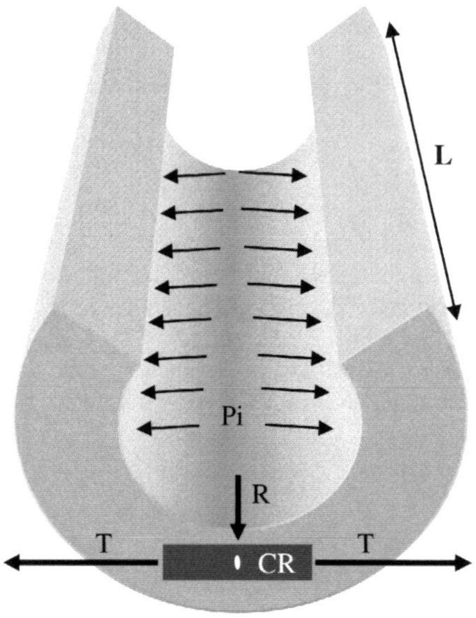

Figure 1. Orientation of the CR sample representing tangential (T) and radial (R) stress under internal pressure (Pi)

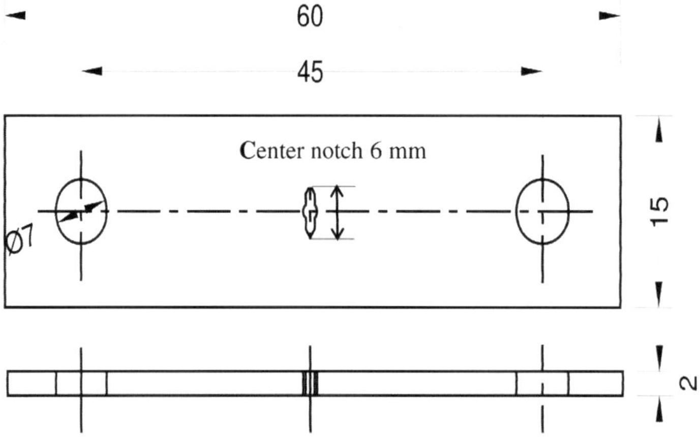

Figure 2. Dimensions (in mm) of the M(T) sample

3.1 Element selection and meshing

The M(T) sample was modeled using two dimensional four-noded, PLANE42 solid elements. The element possesses two degrees of freedom at each node, translation in the nodal x and y directions and does not have rotational degrees of freedom. It also has plasticity, creep, swelling, stress stiffening, large deflection, and large strain capabilities [13]. Other researchers have also conducted fatigue crack growth

analysis by two-dimensional finite element analyses of CT and M(T) geometries. They used the four-noded quadrilateral elements and three-noded triangular elements for the simulation [14-17].
In order to predict the crack growth from the crack tip, the crack advancing region was mapped meshed [17]. The spacing between the consecutive nodes allowed the crack to advance in steps of equal size by node-release scheme [18]. The mesh was optimized by correlation with selected experiments prior to detailed analysis. For the model used in this study the total number of elements and nodes were 12000 and 12261, respectively.

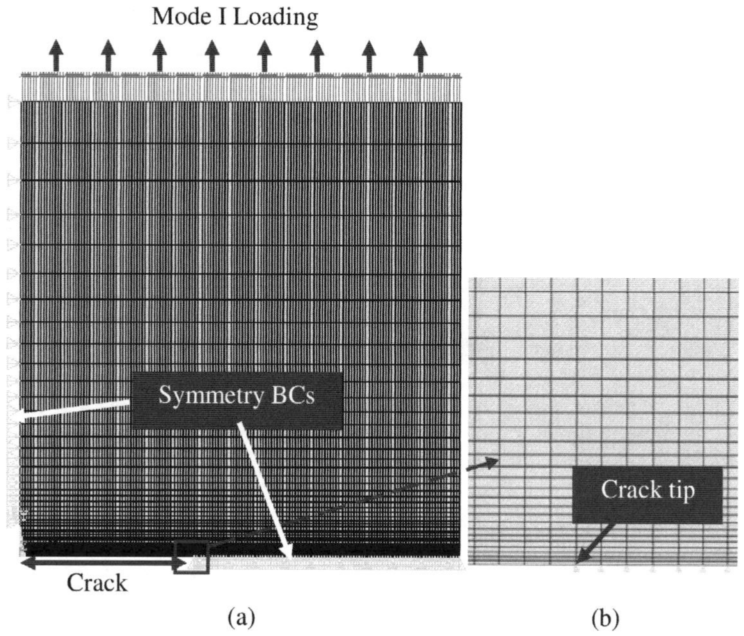

Figure 3. (a) Quarter model of the M(T) sample – element plot with the applied boundary conditions and (b) enlarged crack tip region showing mapped meshing

3.2 Boundary conditions and solution
The boundary conditions applied on the quarter model are shown in the figure 3. The model was constrained applying symmetry boundary conditions along the left and the bottom edges. A 3 mm long crack was modeled by removing the symmetry constraints from 0 to 3 mm along the x direction at the bottom edge, thus providing the crack tip node at 3 mm. The model was loaded by applying tractions at the upper edge in the y direction, simulating mode I loading. After applying the boundary conditions and getting the solution, the value of K_{Imax} at the crack tip was obtained, thereby ΔK calculated. This value of ΔK was used along with the experimental data to obtain the crack growth rate, using the Paris equation. The crack size was increased by releasing the crack tip node, which was equal to the distance between the two consecutive nodes along the line of crack advancement. The number of cycles to move to the next node (one step) was calculated using crack growth rate, and the process was repeated. During crack propagation, ΔK value was monitored and the process stopped when the parameter attained the fracture toughness of the material. In order to validate the FEA results, analysis was conducted by simulating loads similar to those applied during the experimental tests.

4. Results and discussion

4.1 Fatigue crack propagation

After finding the anisotropy in the material, experiments were conducted to reveal the fatigue crack growth behavior of the cylinder in CR direction which shows inferior mechanical properties. In addition, the numerical simulation of fatigue crack growth was performed using Paris law and linear elastic fracture mechanics. The element size in the crack growth region was fixed to 0.05 mm [19, 20]. The detailed FE crack growth analysis was performed at a constant crack growth rate of 0.05 mm/cycle. Figure 4 shows the increasing von-Mises stress at the crack tip node as the crack grows from 3 mm to 4.5 mm under a stress of 40 per cent of the material's yield strength. Von-Mises stress value is normally used in both fatigue and static load design of such cylinders. The stress intensity factor was determined at the crack tip node by defining the path and using KCALC command. Figure 5 shows the plots for the crack length versus the number of cycles and compares the data obtained from the two techniques at different stress levels. The plots provide the number of cycles in the crack growth region and do not include the cycles to initiation the crack. The data covers the entire range from start of the crack at the notch up to the specimen failure.

It was concluded from the results that crack grows faster at higher stress level and vice versa, this observation was validated by literature [6]. Comparison of experimental and FE results shows close agreement at all stress levels analyzed. The small deviation of the FE results from the experimental is within the statistical scatter that was observed in the experimental findings.

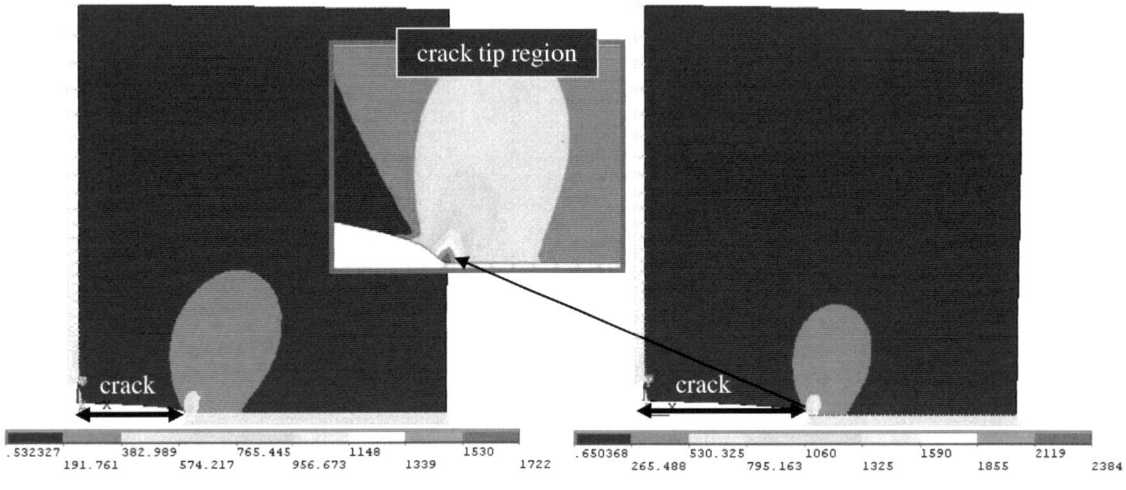

(a) (b)

Figure 4. Increasing von-Mises stress (MPa) as the crack grows from a) 3 mm to b) 4.5 mm under a stress of 40 per cent of the material's yield strength. Inset shows enlarged crack tip region.

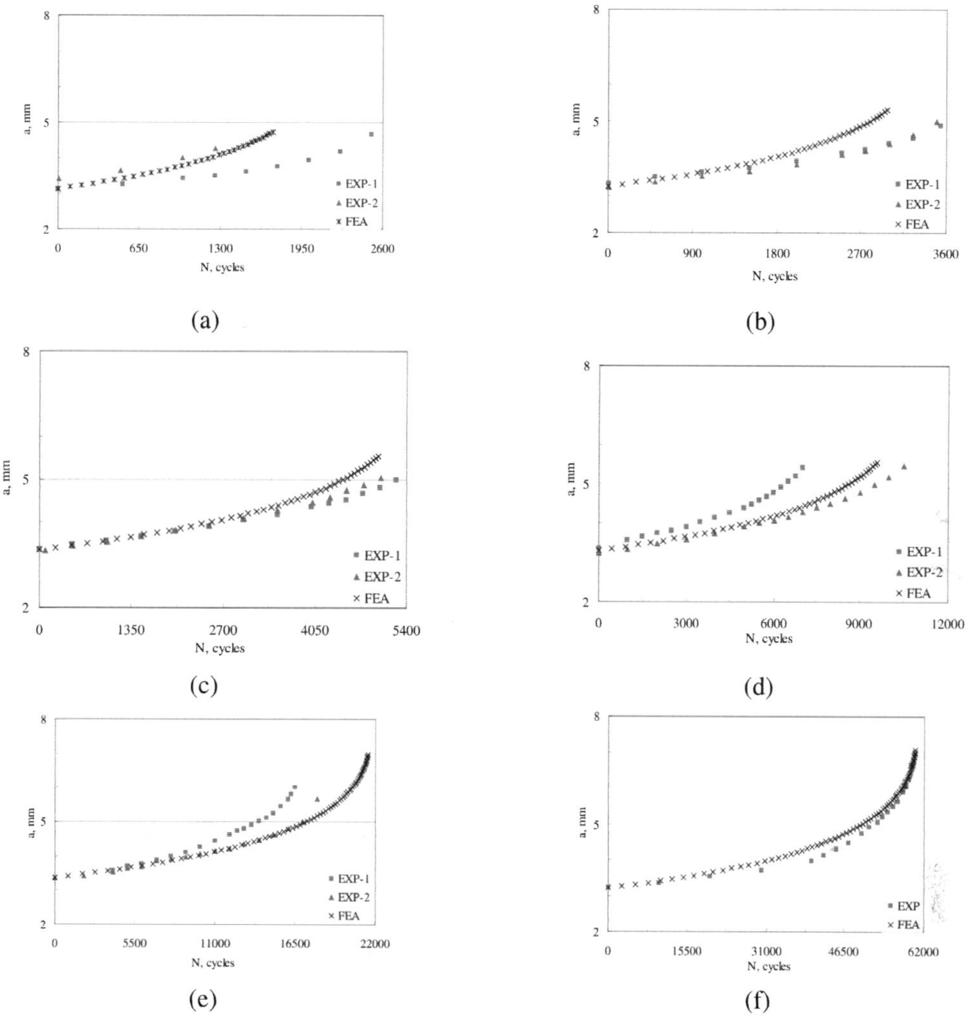

Figure 5. Experimental and FEA results of the crack length versus the number of cycles in the crack growth region at stress levels in per cent of the yield strength: a) 40 b) 35 c) 30 d) 25 e) 20 f) 15 and g) 10. Legend shows the experimental data (EXP-1 and EXP-2) and the FEA results

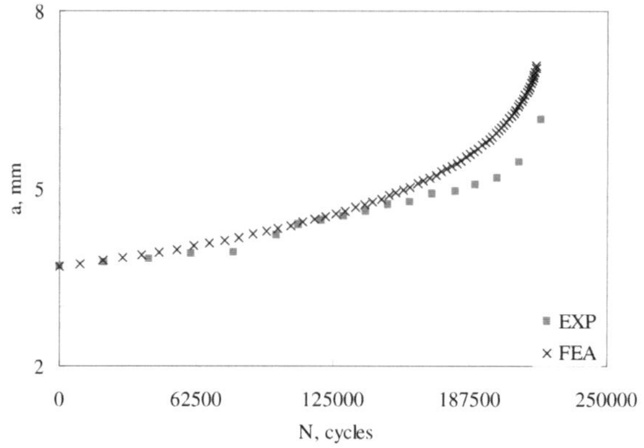

Figure 5. *contd.* (g)

4.2 Fatigue crack growth rate – Experimental vs. FEA

Figure 6 shows the fatigue crack growth rate versus ΔK plot. The experimental and FEA data is plotted to compare the results obtained from the two methods. The experimental data on log-log scale shows almost linear relation. The experimental results show that the crack growth rate increases almost linearly with ΔK. The experimental values of the Paris constants, m and C, were determined by curve fitting of the data through Excel. The values of m obtained from the experimental data and that used for the numerical analysis are 3.40 and 3.35, respectively. A minor adjustment in the constant m was carried out to optimize the modeling results. The value of C was 4E-11 in both the cases. A smooth crack growth rate was achieved in FEA using Paris equation. The fatigue crack growth rate determined by the FEA lied within the upper and lower bounds of the crack growth rate achieved from the experimental data. It can be seen from the plot that the crack growth rates obtained from the two techniques are in close proximity and the predicted results are within the bounds of the experimental data. This concluded that at all the stress levels analyzed the crack growth rate follows the Paris equation and remains in the region II of the crack growth curve.

4.3 Fatigue life analysis

The experimental and the FE results of ΔS versus N_f are shown in the plots in figure 7. ΔS vs. N_f relations for the two data sets was obtained through curve fit and are also given in the figure. The number of cycles to failure includes the number of cycles to initiate the crack and its growth up to the specimen failure. The number of cycles to crack initiation was incorporated in the FE results, from the experimental data. The fatigue life analysis results obtained from both the techniques shows that the fatigue lifetime increases as the stress range decreases. It can be seen that the results obtained from the two techniques are in good agreement at all the stress levels analyzed. Thus the simulation model and the crack growth technique used in the present study provide excellent results and can be utilized satisfactorily in the prescribed stress range.

5. Conclusions

Following conclusions can be drawn on the basis of the present study:

- The finite element technique combined with LEFM and Paris law could predict the fatigue crack growth life of the cylinder.
- Both the experimental and numerical results of crack growth data at the tested stress levels were found in close agreement. However, any difference observed was concluded in the range of statistical scatter in the experimental data.
- SN curves showed that the fatigue life of the samples obtained from the experiments and that predicted from the simulation were in good agreement at all the stress levels analyzed.

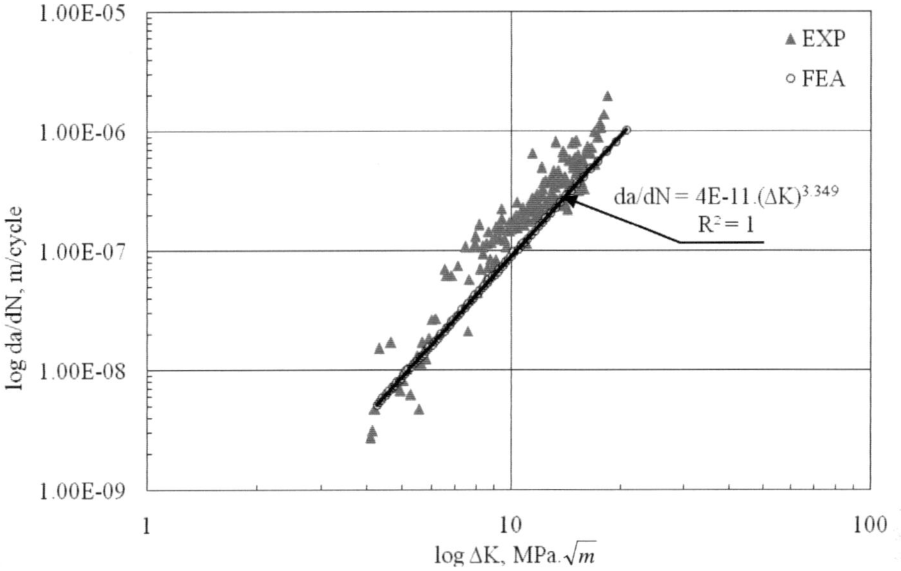

Figure 6. Comparison of the predicted crack growth rate with the experimental observation. Legend shows the total experimental data (EXP) and the FEA results at 10 per cent of the yield strength

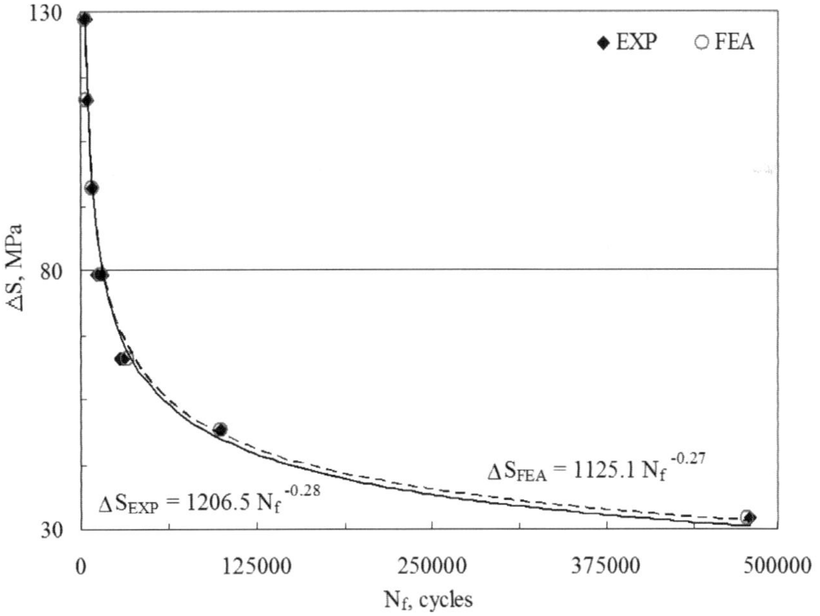

Figure 7. SN curves; Experimental (EXP) versus predicted (FEA) fatigue life

6. References

[1] Alegre J M, Bravo P and Preciado M 2007 Fatigue behavior of an autofrettaged high-pressure vessel for the food industry *Eng. Fail. Anal.* 14 396–407

[2] Shlyannikov V N 2000 Fatigue shape analysis for internal surface flaw in a pressurized hollow cylinder *Int. J. Pressure Vessels Piping* 77 227–34

[3] Rees D W A 1989 Fatigue crack growth in thick -walled cylinders under pulsating internal pressure *Eng. Fract. Mech.* 33(6) 927–40

[4] Tomkins B 1973 Fatigue failure criteria for thick-walled cylindrical pressure vessels *Int. J. Pressure Vessels Piping* 1(1) 37–59

[5] Su B and Bhuyan G S 1998 Elasto-plastic fracture properties of an all-aluminium gas cylinder with different cracks *Int. J. Pressure Vessels Piping* 75(12) 879–86

[6] Shigley J E, Mischke C R and Budynas R G 2004 Mechanical Engineering Design. 7th ed. McGraw-Hill Companies Inc. New York

[7] Newman Jr J C and Ruschau J J 2007 The stress-level effect on fatigue-crack growth under constant-amplitude loading *Int. J. Fatigue* 29 1608–15

[8] Newman Jr J C and Armen Jr H 1975 Elastic–plastic analysis of a propagating crack under cyclic loading *J. AIAA* 13 1017–2023

[9] Newman Jr J C 1977 Finite-element analysis of crack growth under monotonic and cyclic loading, ASTM STP 637 Philadelphia PA 56–80

[10] Koh S K and Na E G 1999 Fatigue crack growth life of thick-walled cylinders with an external radial crack *Int. J. Fatigue* 21 135–46

[11] *ANSYS 8.1 user's Manual, Structural Analysis Guide* 2004, Houston, Swanson Analysis system Inc.

[12] Staat M and Vu D K 2006 Limit loads of circumferentially flawed pipes and cylindrical vessels under internal pressure *Int. J. Pressure Vessels Piping* 83 188–96

[13] *ANSYS Inc. ANSYS elements manual* 2004, 7th edition, Houston, Swanson Analysis system Inc.

[14] Solanki K N, Daniewicz S R and Newman Jr J C 2003 Finite element modeling of plasticity-induced crack closure with emphasis on geometry and mesh refinement effects *Eng. Fract. Mech.* 70 1475–89

[15] Skinner J D 2001 Finite Element Predictions of Plasticity-Induced Fatigue Crack Closure In Three-Dimensional Cracked Geometries *MS thesis* Mississippi State University USA

[16] Solanki K N 2002 Two and three-dimensional finite element analysis of plasticity-induced fatigue crack closure - A comprehensive parametric study *MS Thesis* Mississippi State University USA

[17] Lei Y 2008 Finite element crack closure analysis of a compact tension specimen *Int. J. Fatigue* 30 21–31

[18] Stoychev S and Kujawski D 2008 Crack-tip stresses and their effect on stress intensity factor for crack propagation *Eng. Fract. Mech.* 75(8) 2469–79

[19] Jiang Y and Feng M 2004 In *Fracture Methodologies and Manufacturing Process*, July 25–29, San Diego, California, ASME PVP-Vol. 474, PVP2004-2297 23–31

[20] Ding F, Feng M and Jiang Y 2007 Modeling of fatigue crack growth from a notch *Int. J. Plasticity* 23 1167–88

Effect of Zn/O contents ratio on the structural, optical and electrical properties of MBE grown ZnO

M Asghar[1], K Mahmood[1], F Malik[1], S Rabia[1], Samaa BM[1] and M A Hasan[2]
[1]Department of Physics, The Islamia University of Bahawalpur, Pakistan
[2]Department of Electrical and Computer Engineering, UNC-Charlotte, NC USA

E-mail: mhashmi@iub.edu.pk

Abstract. In this study, the strong dependence of structural, optical and electrical properties of ZnO on the concentration of Zn/O has been observed. Four samples of ZnO with Zn/O-ratio of 1.08, 1.12, 1.17 and 1.22 were grown by molecular beam epitaxy. The x-ray diffraction (XRD) pattern demonstrates that intensity of (002) peak and crystallite size decreases as the Zn/O ratio increases in the samples, suggesting that the crystallinity of the samples degraded. In Raman spectra the E_2^{high} mode shifts towards higher frequency with increasing the concentration of Zn in films. According to electrical measurements, the V_O-Zn_i act as donor defects in ZnO because the carrier concentration increases and resistivity decreases with increasing the Zn/O ratio in the samples.

1. Introduction

ZnO has potentials applications in electronic and optical devices due to its wide band gap (3.37) and rich chemistry of intrinsic defects [1-3]. To enhance the performance of such devices, the basic understanding of defect chemistry of ZnO is essential. ZnO has two type of intrinsic defects; donors and acceptors. O vacancies (VO), Zn antisites and Zn interstitials (Zni) are considered to be donor levels and Zn vacancies (ZnO), O-interstitials act as acceptor levels [4]. Therefore control of density of intrinsic defects in ZnO has fundamental importance. Different methods and growth techniques are used to modulate the density of intrinsic in the grown films. MBE is very versatile and ultra high vacuum technique for the control growth of semiconductor materials. Literature suggested that one can precisely control the concentration of source materials by using MBE. On the other hand the structural, optical and electrical properties of ZnO are strongly depends upon the concentration of intrinsic defects. For example intrinsic n-type of conductivity of ZnO is related to intrinsic donors like V_O, Zn_i and Zn_i-V_O donor complex [5-7]. Similarly the defect luminescence from ZnO is also related with intrinsic defects but its origin is still controversial [8, 9]. The roots of hurdles in the way of p-type conductivity of ZnO are also supposed to be compensation effect of intrinsic donor defects. Therefore a detailed study of intrinsic defects in ZnO would be very interesting and can provide better understanding of film quality and its optical and electrical properties.

In this paper, the effects of Zn/O ratio on the structural, optical and electrical properties of ZnO grown by molecular beam epitaxy, have been investigated by Energy dispersive x-ray spectroscopy, X-ray diffraction, Raman spectroscopy and Hall measurements. The crystal quality of films degrades with increasing Zn/O ratio in the samples. Furthermore Zn_i-V_O complex is the major source of carrier concentration

2. Experimental

Four layers of bulk ZnO each having Zn/O ratio 1.08, 1.12, 1.17 and 1.22 (hereafter referred as D, C, B and A) were grown on 3 inch diameter p-type silicon (111) wafers by means of molecular beam epitaxy (MBE). The detail can be found in reference [6]. Characterization of ZnO films were carried out by the following equipment: XRD by PANALytical X'pert, SEM/EDAX by JEOL, Hall measurements by Ecopia 3000 and Raman/PL spectroscopy by Horiba using He-Cd laser of excitation wavelength 325 nm. All the measurements were performed at room temperature.

3. Results and Discussion

XRD patterns of the as grown samples A, B, C and D confirmed hexagonal structure of ZnO (see Figure 1). Three distinct peaks were observed at angles (2θ) 34.47^0, 36.2^0 and 72.4^0 from all samples associated with the (002), (101) and (004) planes of ZnO, respectively [10]. Peak (002) being the dominant among the others, indicates that the preferable direction of the growth is along this plane i.e. c-plane.

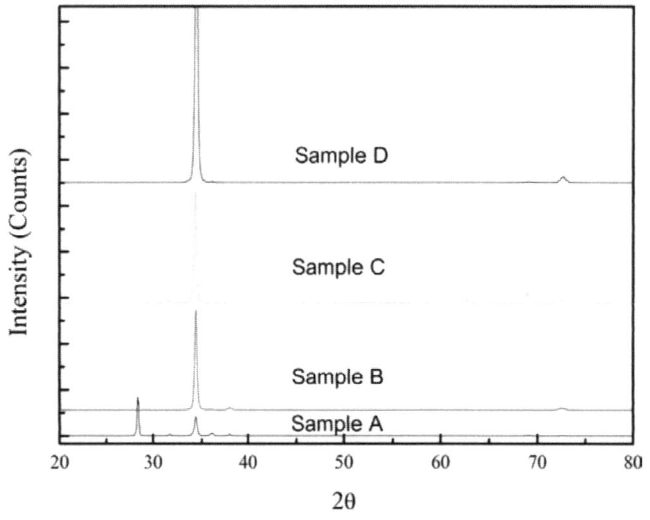

Figure 1. Typical XRD pattern of MBE grown samples with different Zn/O ratio

The position of (002) peak remains constant for all the samples but the intensity of (002) peak decreases with increasing Zn/O ratio in the samples. The lattice constants 'a' and 'c' of the wurtzite structure of ZnO can be calculated using relations given below [11].

$$a = \sqrt{\frac{1}{3}}\frac{\lambda}{sin\theta}$$

$$c = \frac{\lambda}{sin\theta}$$

$$D = \frac{0.9\lambda}{\beta cos\theta}$$

For (002) plane calculated values are a= 3.12 and c= 5.15 A^0 which agree with JCPDS data.
The crystallite size was calculated by using Scherer's formula [1];

Where D is crystallite size, λ is wavelength of x-ray, ß is full width at half maximum of (002) peak.

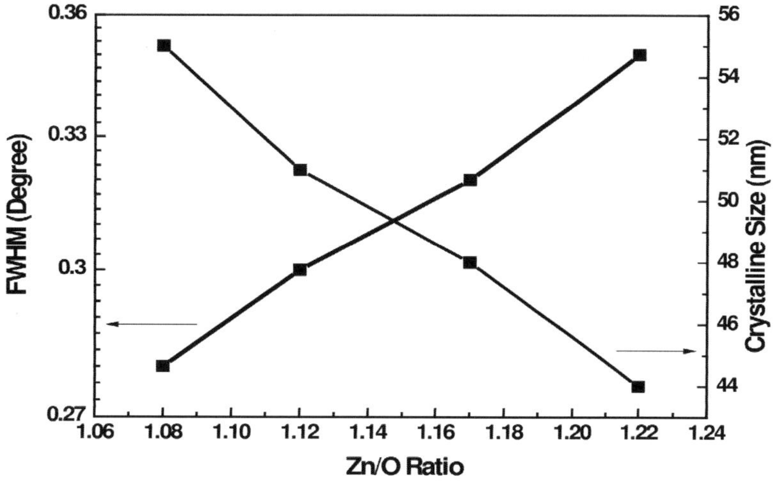

Figure 2. Effect of Zn/O ration on FWHM of (002) peak and crystalline size of MBE grown samples

Figure 2 shows the relation of FWHM and crystallite size with Zn/O ratio in the samples. The FWHM of ZnO (002) peak for samples A, B, C and D is 0.35^0, 0.32^0, 0.30^0 and 0.280^0 respectively, showing the good crystallinity of films. We observed that with increasing the Zn/O ratio in the samples, the FWHM of samples increases and crystallite size decreases.

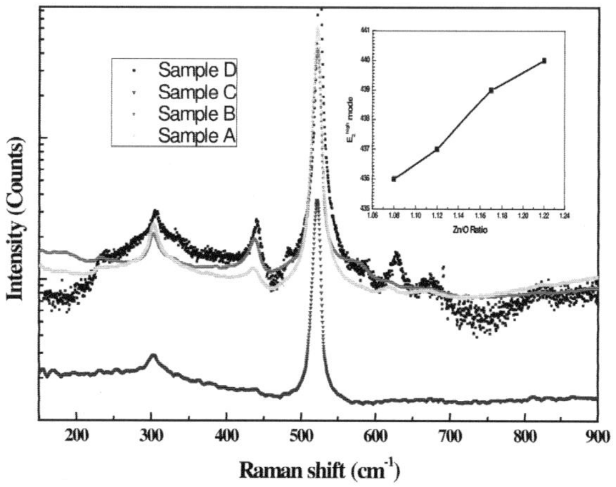

Figure 3. Typical Raman spectra of MBE grown samples with different Zn/O ratio. Inset shows the shifting of E_2^{high} mode towards higher frequencies of ZnO with Zn/O ratio

This means that the crystalline quality degrades with the increase of Zn/O ratio in the samples. The degradation of crystalline quality probably related with stress in the films with increasing the density of Zn_i-V_O complex defects due to increase of Zn/O ratio. The presence of Zn_i-V_O defects in the samples is confirmed from the Raman spectroscopy measurements.

To support the presence of complex donor defects, we performed Raman spectroscopy as well. The ZnO crystal structure belongs to the space group C46v, and the group theory analysis predicts the zone-center optical modes; A1 + 2B1 + E1 +2E2. The A1, E1 and the two E2 modes are Raman active, while the B1 modes are forbidden in Raman scattering. Furthermore, the A1 and E1 modes are polar: their vibrations polarize the unit cell, which results in the creation of a long-range electrostatic field. This field results in the splitting of A1 and E1 modes into longitudinal optical (LO) and transverse optical (TO) components, thus creating the A1 (LO, TO) and E1 (LO, TO) modes.

Figure 3 displays the typical Raman spectra of MBE grown samples with different Zn/O ratio. The inset of figure 3 shows the effect of Zn/O on the E_2^{high} Raman mode of ZnO for samples A, B, C and D. The spectrum shows a non-polar optical phonon mode E_2^{high} for all samples [12]. The theoretical calculations by Tsuboi and Wada [13] predicted the frequency of E2high mode of pure ZnO to be 433 cm-1. We observed this mode at 436 cm-1, 437 cm-1, 439 cm-1 and 440 cm-1 for samples D, C, B and A respectively. The shift of E_2^{high} phonon frequency is directly related to the stress in film. Haung et al. [14] pointed out that under a compressive stress the E_2^{high} shifts up. We observed an up shift of 3 cm-1, 4 cm-1, 6 cm-1 and 7 cm-1 for sample D, C, B and A, respectively. This up shift indicates a compressive stress in the film. The reason of this stress may be the formation of Zn_i-V_O complex defects in the films with increasing II/VI ratio.

Table. 1 Effect of Zn/O ratio on FWHM, Crystalline size, intensity of E_2^{high} mode, carrier concentration and resistivity

Zn/O Ratio	**FWHM (Degree)**	**Crystalline Size (nm)**	**Position of E_2^{high} mode (cm^{-1})**	**Carrier concentration (cm^{-3})**	**Resistivity Ohm-cm**
1.08	0.28	65	436	$5x10^{16}$	1.44
1.12	0.30	60	437	$3x10^{18}$	0.9
1.17	0.32	48	439	$4x10^{17}$	0.08
1.22	0.35	44	440	$2.2x10^{19}$	0.009

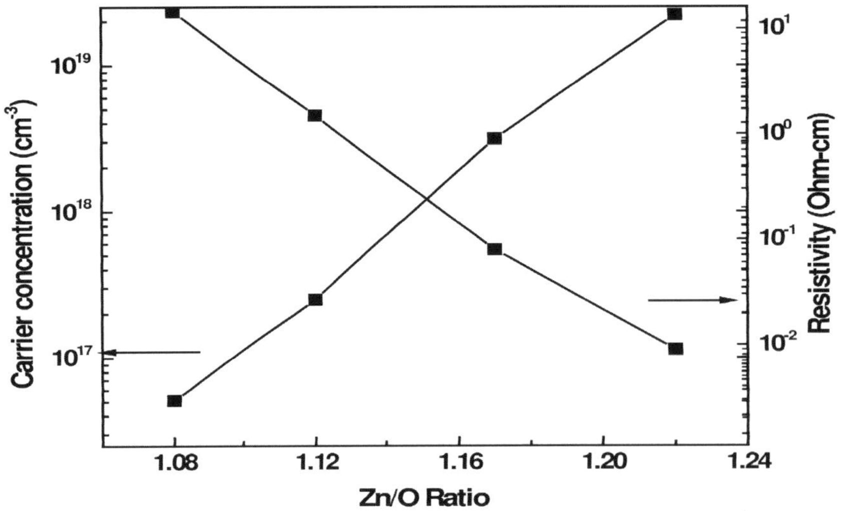

Figure 4. The plot shows the dependence of carrier concentration and resistivity on Zn/O ratio. The graph is evident that with increasing Zn/O ratio, the carrier concentration increases and resistivity decreases.

Figure 4 shows the effect of Zn/O ratio on carrier concentration and resistivity of samples. With increasing Zn/O ratio in the samples the carrier concentration increases and resistivity of samples decreases. The carrier concentration increased from 5.0×10^{16} cm-3 to 2.2×10^{19} cm-3 and resistivity decreased from 1.44 Ω cm to 0.009 Ω cm for samples with increasing Zn/O ratio from 1.08 to 1.22 respectively. The detail of carrier concentration and resistivity can be shown in table 1. The carrier mobility was also calculated and shows that mobility decreased from 7.5 to 1.4 cm^2/V.Sec as the Zn/O ratio increased from 1.08 to 1.22. As ZnO has hexagonal structure with half of the tetrahedral sites are occupied by Zn atom whereas all the octahedral sites are empty, hence there are a plenty of sites for ZnO to accommodate intrinsic (Zn-interstitials, O-vacancy and/or Zn-antisite) [15] and extrinsic defects. Therefore with increasing Zn/O ratio, the densities of both donor defects (Zn_i and V_O) might be increased. Thereafter Zn_i form a complex with V_O vacancy which is donor complex in ZnO also demonstrated by Asghar et al [7]. Therefore the increased of carrier concentration with increasing Zn/O ratio is understandable.

4. Conclusion

We have deposited four samples with Zn-concentration of 52%, 53%, 54% and 55% by molecular beam epitaxy. The structural, optical and electrical properties of grown films are strongly depending on contents ratio of Zn/O. It is concluded that structural and optical properties degrades with increasing Zn/O ratio because the density of V_O-Zn_i donor complex increased

5. References

[1] Asghar M, Mahmood K, Ali A, Hasan M-A, Raja M Y A, Hussain I, Willander M 2011 *ECS transactions* 35 (6) 149-154

[2] Fan J.C, Sreekanth K.M, Xie Z, Chang S.L, Rao K.V 2013 *Progress Mater. Sci.* 58 874

[3] Asghar M, Mahmood K, Faisal F, Hasan M-A 2013 *J. Phys: Conf. Ser.*439 012030

[4] Schmidt-Mende L, Judith L. MacManus-Driscoll 2007 *Materialstoday* 10 40

[5] Janotti A, Van de Walle C.G 2009 *Rep. Prog. Phys.* 72 126501

[6] Asghar M, Mahmood K, Hasan M-A 2012 *Key Eng. Mater.* 510-511 227-232

[7] Asghar M, Mahmood K, Ferguson I.T, Yasin R, Tsu R, Xie Y-H 2013 *Semicond. Sci. Technol.* 28 105019

[8] Bharati P, Aohammad A, Devi S.K, Manoranjan G, Dhirendra B 2010 *Adv. Funct. Mater.* 20 1161

[9] Potzger K, Shengqiang Z.J, Grenzer M.H, Fassbender 2008 *J Appl. Phys. Lett.* 92 182504

[10] Tneh S.S, Hassan Z, Saw K.G, Yam F.K, Hassan H.A 2010 *Physica B* 405 2045

[11] Shakti N, Gupta P.S 2010 *Appl. Phys. Research* 2 1

[12] Zou C.W, Wang H.J, Yi M.L, Li M, Liu C.S, Guo L.P, Fu D.J, Kang T.W 2010 *Apply. Surf. Sci.* 245 2453-2457

[13] Tsuboi M, Wada A 1968 *J. Chem. Phys.* 48 2615

[14] Huang Y, Liu M, Zeng Y, Liu S 2003 *Mater. Sci. Eng B* 97 111

[15] Behera D, Acharya B.S 2008 *J. Lumin.* 128 1577-1586

Effect of copper addition at a rate of 4 % weight on the machininability of ZA-21Al cast alloy by CNC milling

S M A Alqawabah[1] and A I O Zaid[2]
[1]Mechanical Engineering Department, Tafila Technical University, P. O. Box 13720, Amman 11942, Jordan
[2]Industrial Engineering Department, University of Jordan, Jordan

E-mail: safwan_q@yahoo.com, a_kilani43@hotmail.com

Abstract. Little work is published on the effect of copper addition to zinc-aluminium ZA-21Al alloy on its surface quality machined by milling. In this paper, the effect of copper addition at a rate 4 % weight to the ZA-21Al alloy on its hardness and surface quality is investigated. It was found that the addition of 4% Cu resulted in 18.3% enhancement in microhardness whereas the mechanical characteristics were reduced (softening) about 14.5% at 0.2% strain. It was found that the best surface finish for this alloy before copper addition ZA21 was achieved at a feed rate of 100 mm/min and 1.25 mm depth of cut whereas the best surface finish for ZA21-4% Cu was achieved at feed rate 250 mm/min, 1600 rpm cutting velocity and 1.25 mm depth of cut.

1. Introduction

The demand for high quality and fully automated production notes in industry utilizing the fully automated processes focuses attention on dimensional accuracy and surface quality. These normally affect product appearance and function during their lives services. For these reasons it is important to maintain consistent tolerance and surface quality. Among several industrial materials removal processes, milling is a fundamental machining operation. End milling is the most common metal removal operation encountered. It is widely used in a variety of manufacturing industries including the aerospace and automotive sectors, where quality is an important factor in the production of slots and dies. Several factors influence the final surface quality of the end milling operation. Independent factors such as tool geometry, spindle speed, feed rate, and depth of cut that control the cutting operation can be setup in advance, whereas dependant factors such as tool wear, and type of chip formed are uncontrolled. One should develop techniques to predict the surface roughness of a product before milling in order to evaluate the quality of machining parameters such as feed rate or spindle speed for keeping a desired surface. Zn-based alloys have a number of advantages over traditional bearing materials [1, 2, 3]. These advantages can be summarized as high resistance to wear, excellent castability and low cost [4, 5, 6]. Among the monotectoid-based alloys, the best mechanical and wear performance was obtained with the Zn–40Al–2Cu alloy [7, 8]. Selection of cutting tools and cutting conditions represents an essential element in process planning in machining. This task traditionally carried out on the bases of the experience of the process planner with the help of the data from machining handbooks and tool catalogs. Process planners continue to experience great difficulties due to lack of performance data on the numerous new commercial cutting tools with different materials, coating geometry and chip- groove configurations for high wear resistance and effective chip breaking.

The main objective of this present work is to investigate the effect of copper addition at a rate of 4 % to ZA21 on the surface quality of the machined surfaces by the milling process under different cutting conditions of speed, depth of cut, and feed rate.

2. Materials and experimental procedures

2.1 Materials

High purity, zinc, copper powder and commercially pure aluminum were used in preparing the ZA21alloy and the ZA21-4%Cu. The chemical analysis of the prepared alloys ZA21 and ZA21- 4 % Cu are shown in tables 1 and 2 respectively. The density of ZA21 is (6.23) gmcm^{-3} and the melting point ranges from 380-386 °C.

Table 1. Chemical Analysis of ZA21 Alloy, (by weight)

Element	Al	Mg	Fe	Pb	Cd	Sn	Zn
%, Wt	20.3-21.3	0.03-.08	0.1	0.005	0.004	0.003	Bal

Table 2. Chemical Analysis of ZA21-4% Cu Alloy, (by weight)

Element	Al	Cu	Mg	Fe	Pb	Cd	Sn	Zn
%, Wt	20.3-21.3	4	0.03-.08	0.1	0.005	0.004	0.003	Bal

2.2 Experimental procedures

2.2.1 Preparation of the ZA21-4% copper

The ZA21 alloy was prepared by melting the pre-calculated amounts of the 99.99 % Zn powder and the commercially pure aluminum under caryolite flux in a graphite crucible and stirred with graphite rod at 550 °C. The melt was kept at this temperature for 5 minutes and then poured to solidify in a brass mould of 50*50*25 mm manufactured for this purpose as shown in figure1, where the machining set up is shown in figure 2.

Figure 1. Brass mould for producing the square section of milling specimens

Figure 2. Machining process

The ZA21- 4% Cu was prepared by adding the precalculated amount of the high purity copper powder into the ZA21 melt at 550 °C and kept in the crucible inside the furnace for 5 minutes and finally stirred with graphite rod before it was poured to solidify in the thick brass die. Test specimens from each alloy, namely the ZA21 and the ZA21-4 %Cu were prepared for the microhardness, mechanical behavior and metallurgical examination tests.

2.2.2 Machining and surface quality tests

The machining tests on both ZA21 and ZA21-4%Cu were carried out on CNC milling machine type (KM3000) Excel of 0.1 μm resolution. The tests were carried out at different speeds, depth of cut and feed rates as shown in Table 3, under dry condition. A total of 64 machining runs using a 9 mm diameter cutter with four edges, after each run. The average surface roughness of the machined surface, after cleaning, was measured at two locations from which the average surface roughness, Ra, was determined, noting that the measurements were taken close to but not at the edges to avoid grain size at the edges, The average roughness (Ra) measurement was based on cut off distance = 0.8 and ISO 13565 (Rk), after each machining test. All milling tests were performed under dry cutting conditions. The path of the cutter with respect to the surface of the work piece is shown in figure 3.

Table 3. Cutting Conditions for both the ZA-21 and ZA-21-4% Cu

Depth of cut (t) (mm)	0.5	1.25	2	2.75
Cutting speed (v) (rpm)	750	1300	1600	2000
Feed rate (f) (mm/min)	100	150	200	250

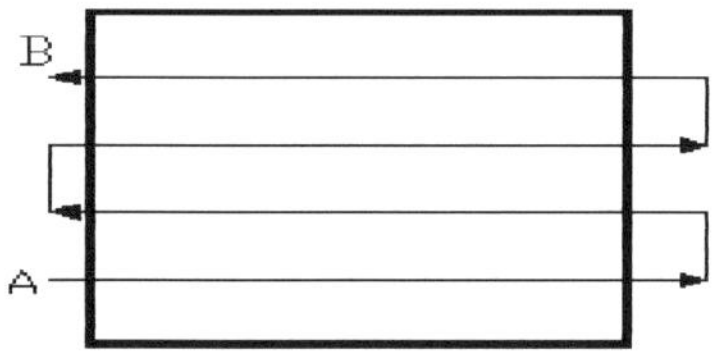

Figure 3. Path of the cutting tool from A to B

3. Results and discussion

In this section, the effect of copper addition on the microstructure, microhardness, mechanical strength and the machining parameters on the surface quality produced from CNC milling of ZA21 cast alloy before and after 4 % Cu will be presented and discussed.

3.1. Effect of copper addition on the microstructure of ZA21 cast alloy

Microphotographs of figure4 shows the optical microphotographs of ZA21 and ZA21- 4%Cu, it can be seen that the microstructure consists of primary aluminium-rich α zinc-rich η and copper-rich ε phases. This is in consistence with the previous work of other researchers,[2,3,4].

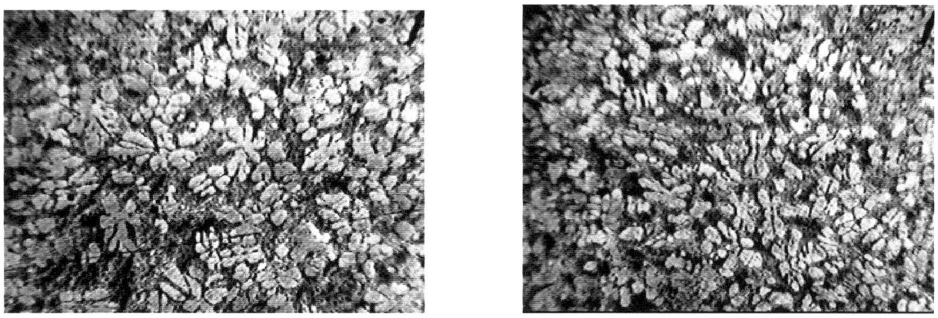

Figure 4. Photomicroscan of ZA21 and ZA21-4%Cu alloys at magnification of 250x

The microstructure of the ternary Zn–40Al–2Cu alloy consisted of aluminium-rich α zinc-rich η and copper-rich ε phases [9]. Other studies point out that the microstructure of the binary monotectoid alloy (Zn-40Al) consisted of aluminum-rich α dendrites and zinc-rich η phase in the interdendritic regions. Addition of copper produced copper-rich ε phase particles, but had no effect on the dendritic structure of the alloys. It was observed that the ε phase was formed in the interdendritic region of the alloys containing more than 2% Cu. This type of microstructure seems to be ideal for bearing materials which usually have two phases, one of which is hard and the other one soft, [4]. In zinc-aluminum-copper alloys, the aluminum-rich α phase having a face-centered cubic (FCC) crystal structure exhibits excellent ductility [10,11,12]. The zinc-rich η phase having a hexagonal-close-packed (HCP) crystal structure with a large (c/a) ratio.

3.2 Effect of copper addition on the microhardness and mechanical characteristics of ZA21 cast alloy
It obvious from figure 5 that the addition of 4 % copper resulted in enhancement in the microhardness however, 18.3 % have been achieved. The results are consistence with the previous work which was observed that the hardness of the alloys increased continuously with increasing copper content up to 5%. Microhardness of the aluminium-rich α phase was also affected by the copper content in a manner similar to that of the tensile strength. It was found that the wear loss of the alloys decreased with increasing copper content and reached a minimum at 2% Cu for a sliding distance of 700 km. However, the coefficient of friction and temperature due to frictional heating were found to be generally less for the copper containing alloys than the one without the element. The effect of copper on the wear behaviour of the alloys was explained in terms of their microstructure, hardness, tensile strength, percentage elongation and microhardness of the a phase [13]. It can also be seen from figure 6, that the mechanical properties after 4 % Cu addition were decreased, however the structure become softer, the maximum reduction is about 14.5 % at 0.2% strain. This result is in consistence with the previous work, which indicates that the tensile strength increased with increasing copper content up to 2%, but above this level the strength decreased as the copper content increased further [13], due to the structural softness of the morphological change of (α+η) phase from fine lamellar structure to coarse equiaxed grains [14]. According to hall-pitch equation, the strength of microstructure with lamellar or equiaxed grains morphology is inversely proportional to the grain size or interlamellar spacing. On the other hand the softness is related to two opposite defects; solid solution hardening of α phase and weakling effect through cracking tendency to ε phase, the later is more effective at copper counted beyond 3 % addition [15].

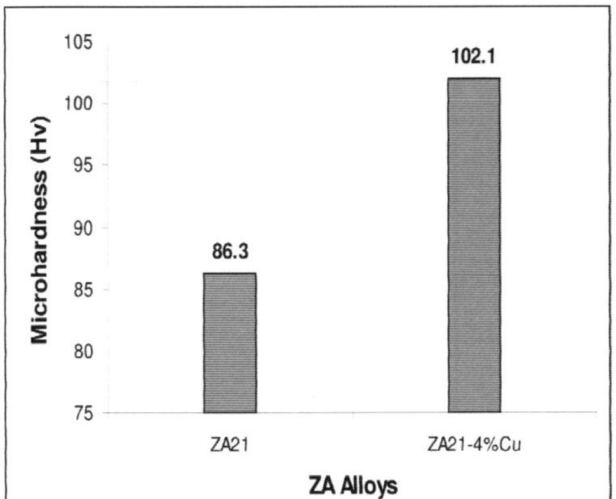

Figure 5. Microhardness of ZA alloys

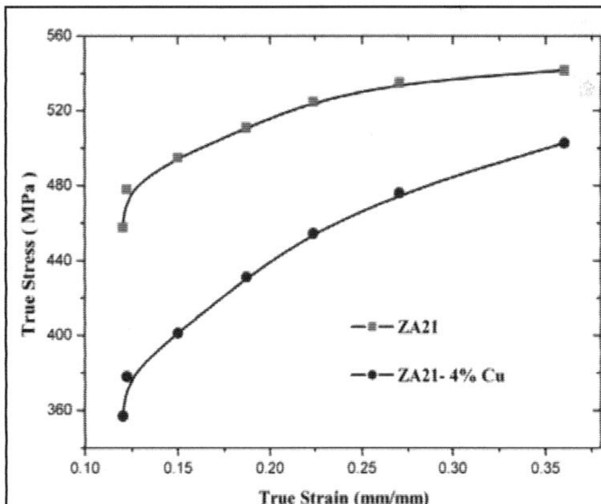

Figure 6. True stress-true strain for ZA21 and ZA21-4% Cu

3.3 Effect of copper addition on surface roughness of ZA-21 Alloys
The effect of cutting parameters on the surface roughness of machine surfaces of the ZA21 alloy without and with addition of 4% copper was investigated by varying the main cutting parameter namely the rotational cutting speed from 750 rpm to 2000 rpm at different depth of cut ranging from 0.5 to 2.75 mm and different values of feed rat ranging from 100 mm/min to 250 mm/min and the

surface roughness was measured at each cutting condition. The results are presented in Fig.7 to Fig.22 inclusive. It can be seen from these figures that increasing the rotational cutting speed from 750 rpm to 2000 rpm at low feed rate, 100 mm/min, and small values of depth of cut, 0.5 mm, the surface roughness of the machined surfaces of both alloys decrease up to 1600 rpm beyond which it starts to increase of the cutting speed. Furthermore, it can be seen that addition of 4% copper to ZA21 alloy resulted in better surface quality i.e. lower values of surface roughness at all cutting speeds. Also increasing the depth of cut from 0.5 to 2.75 mm did not change the trend. These figures also indicate that variation of feed rate is more effective than variation of depth of cut, which agrees with the findings of previous researchers [16]. On the whole the produced surface roughness within the limitation of experimental work of speed and depth of cut resulted in surface roughness ranging from 0.5 to 1 micrometer except at 750 rpm for the ZA2-4%Cu alloy and from 0.9 to 1.4 micrometer for the ZA21 alloy i.e. without copper addition. These values of surface roughness even at highest values fall within the acceptable value in industrial applications. Finally, it can be concluded that, within the limitations of experimental work, addition of 4 % Cu to ZA21 alloy resulted in better surface quality at all cutting speeds, depth of cut and feed rates, which may be attributed to the grain refining caused by the copper addition which also resulted in improvement of its hardness, both of which reduces surface roughness i.e. improves surface quality. This agrees the findings reported in references [14,15].

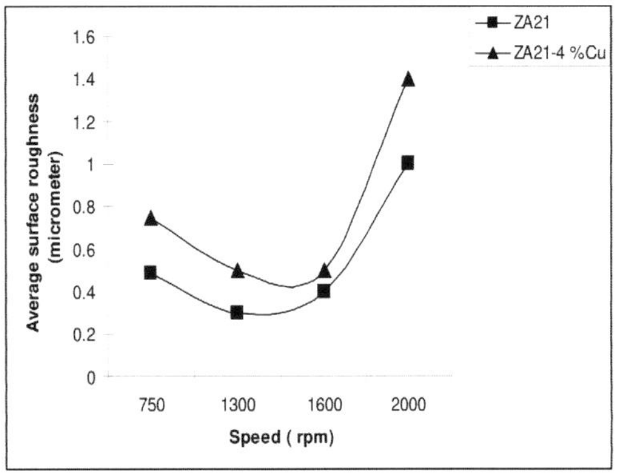

Figure 7. d = 0.5 mm, F=100 mm/min

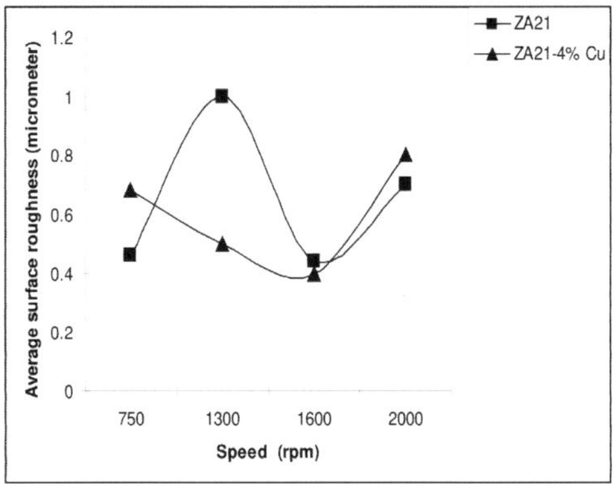

Figure 8. d = 1.25 mm, F=100 mm/min

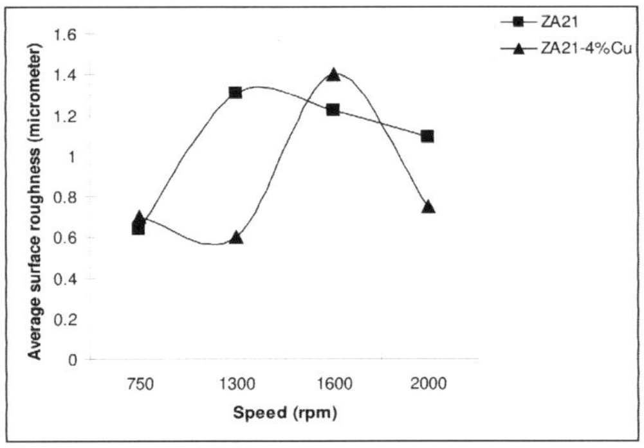

Figure 9. d = 2 mm, F=100 mm/min

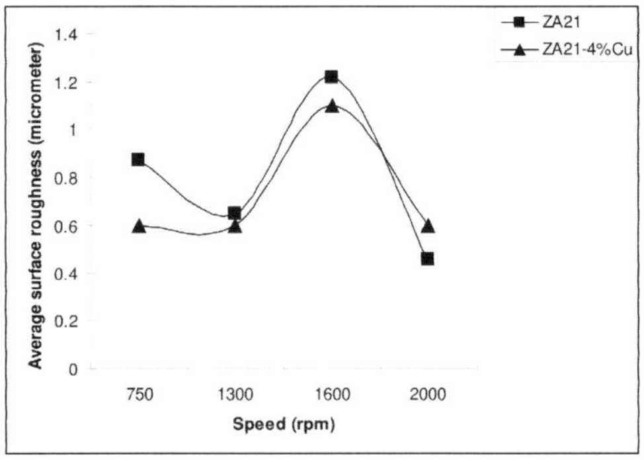

Figure10. d = 2.75 mm, F=100 mm/min

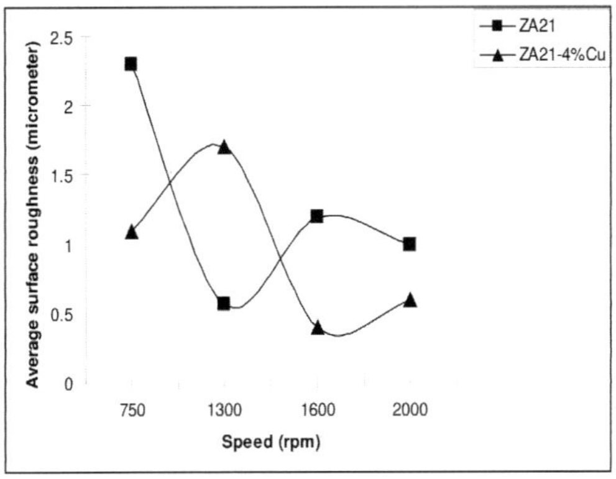

Figure 11. d = 0.5 mm, F=150 mm/min

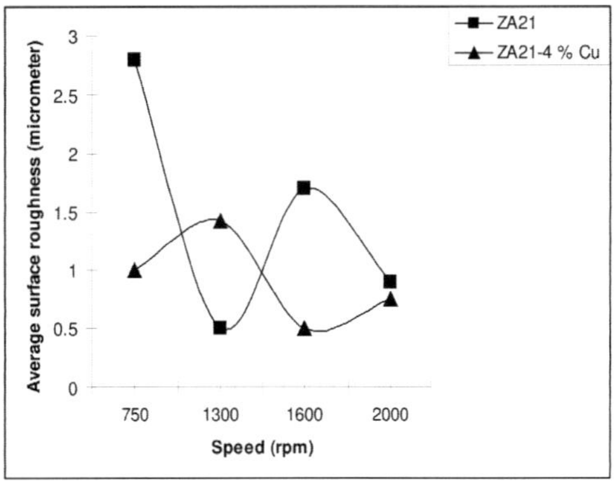

Figure 12. d = 2mm, F=150 mm/min

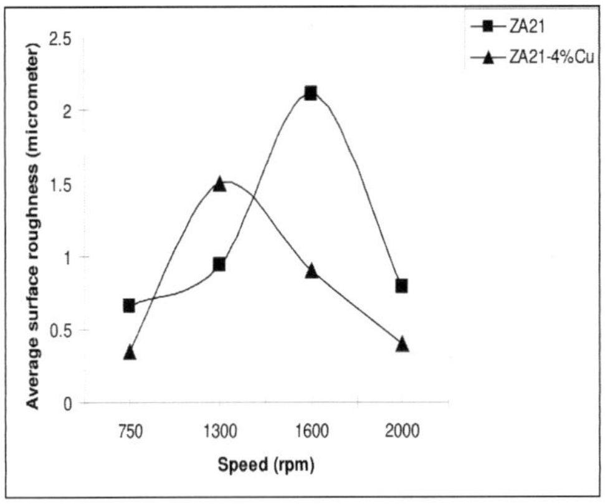

Figure 13. d = 2.75 mm, F=150 mm/min

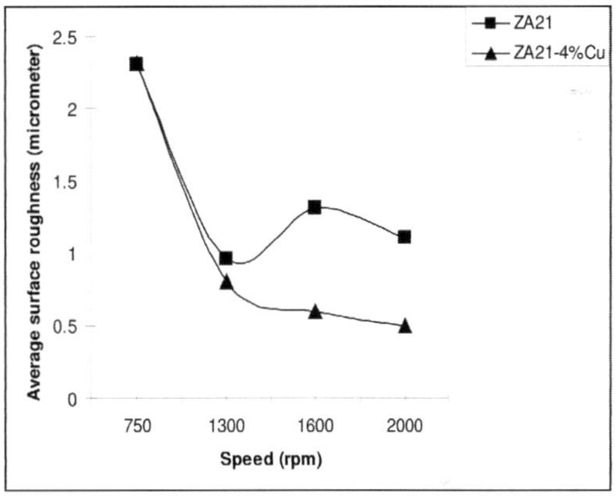

Figure 14. d = 2.75 mm, F=150 mm/min

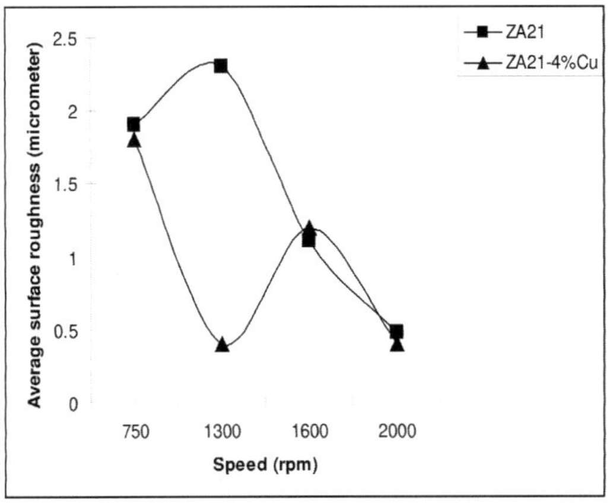

Figure 15. d = 0.5 mm, F=200 mm/min

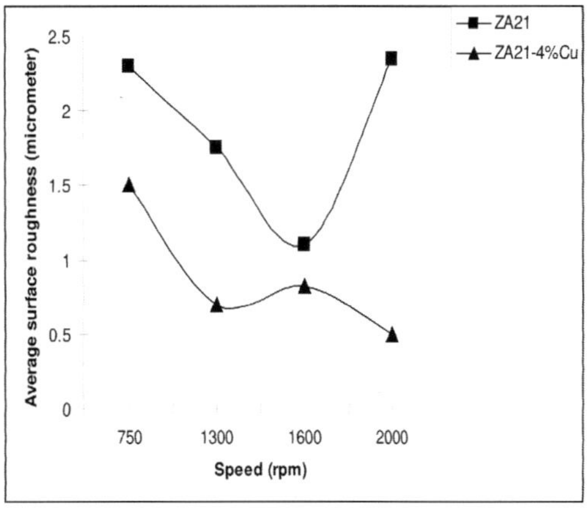

Figure 16. d = 1.25 mm, F=200 mm/min

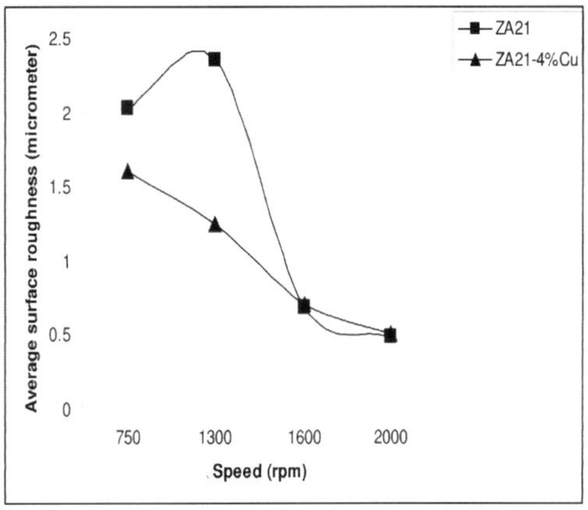

Figure 17. d = 2 mm, F=200 mm/min

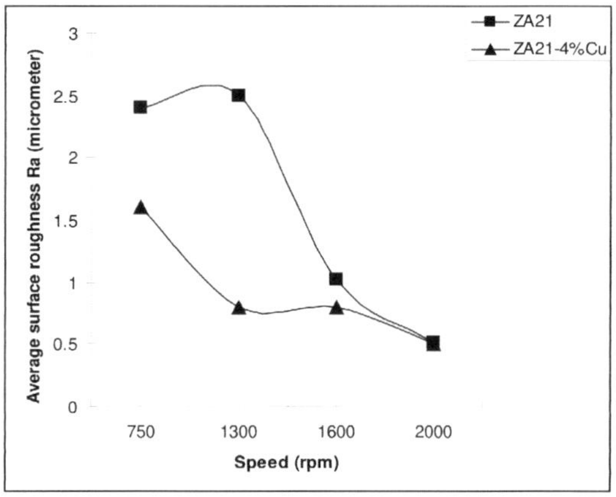

Figure 18. d = 2.75, mm, F=200 mm/min

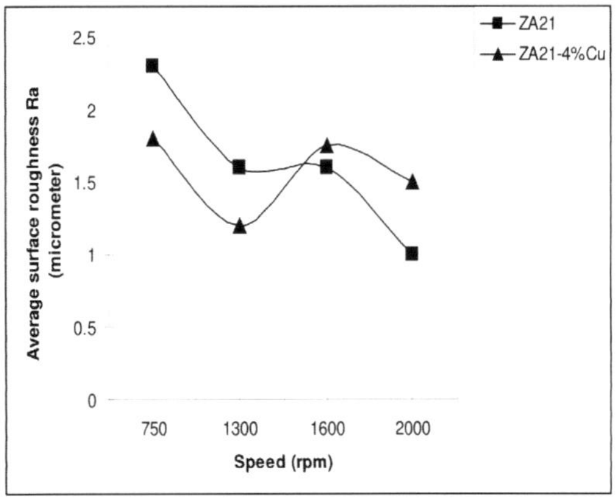

Figure 19. d =0.5mm, F=250 mm/min

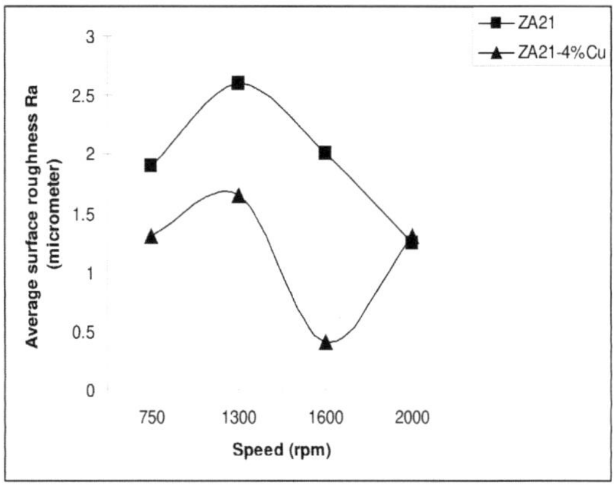

Figure 20. d = 1.25 mm, F=250 mm/min

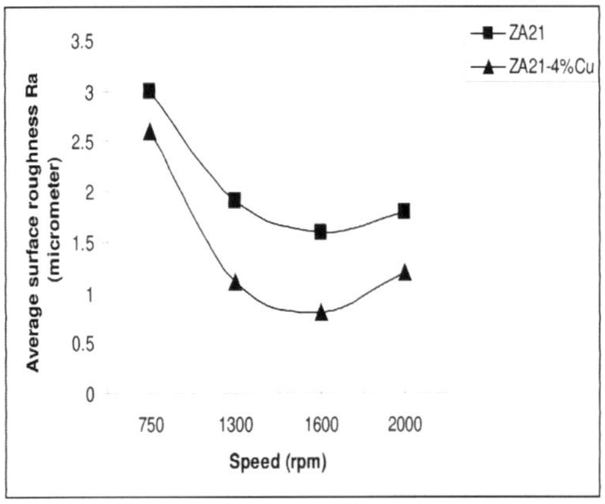

Figure 21. d = 2. mm, F=250 mm/min

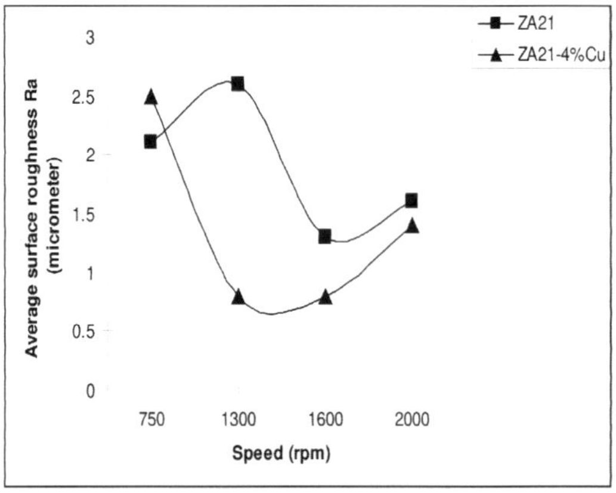

Figure 22. d = 2.75 mm, F=250 mm/min

3.4 Effect of copper addition on the characteristics of chip

It can be seen from figure 23(a) and (b) that the formed chip in the ZA21 alloy before and after addition of the 4% Cu is of the discontinuous type. This is expected, because the type of chip in the milling process is function of the process and not affected by process parameters. Furthermore, the discontinuous chip formed in case of AZ21-4% Cu is finer than that of the alloy before the addition. This may be attributed to the increase in the hardness of the alloy after copper addition.

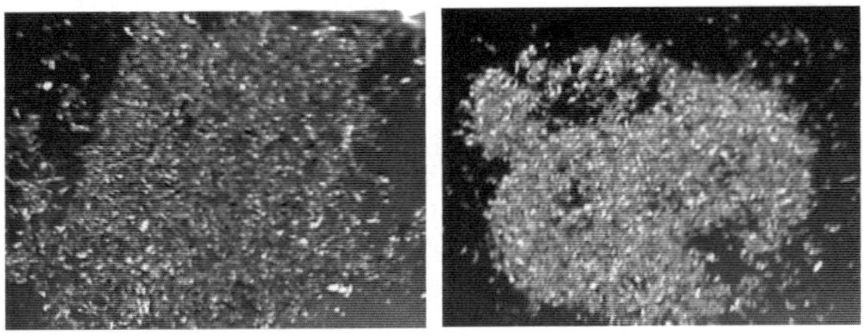

Figure 23. Type of chips, (a) ZA21, (b) ZA21-4 %Cu

4. Conclusions

It can be conclude the following points

- The mechanical strength was decreased at 4 % copper addition, the structure become softer.
- The microhardness increased at 4% copper addition this may attributed to the intermetallic compounds Al_2Cu.
- It can be concluded that the best surface finish for ZA21 was achieved at feed rate 100 mm/min and 1.25 mm depth of cut.
- It can be concluded that the best surface finish for ZA21-4% Cu was achieved at feed rate 250 mm/min, 1600 rpm cutting velocity and 1.25 mm depth of cut.
- The chip in general was discontinuous; the formed chip after copper addition was finer than the ZA21.
- 6-increase in surface roughness had occurred at higher values of feed rates.

5. References

[1] Murphy S and Savaskan T 1984 *Wear* 98 151-161.

[2] Savaskan T and Murphy S 1987 *Wear* 116 211-224.

[3] Lee P P, Savaskan T and Laufer E 1987 Wear 117 79-89.

[4] Savaskan T, Purcek G, and Murphy S 2002 *Wear* 252 693–703.

[5] Prasad B K, Pathwardhan A K and Yegnesvaran A H 2001 *Can. Metall. Q* 40 193.

[6] Prasad B K, Pathwardhan A K and Yegnesvaran A H 1996 *Metall. Mater.Trans.* A 27A 3513.

[7] Savaskan T and Aydıner A 2004 *Wear* 257 377-388.

[8] Savaskan T, Turhal M S and Murphy S: 2003 *Mater. Sci. Technol* 19 67-74.

[9] Savaskan T 1980 The structure and properties of zinc-aluminum based bearing alloys. Ph.D. thesis, University of Aston Birmingham.

[10] Purcek G, Savaskan T , Kucukomeroglu T and Murphy S 2002 *Wear* 252:894-901.

[11] Pandey J P and Prasad B K 1997 *Z Metallkd* 88 739–43.

[12] Prasad B K, Patwardhan A K, A.H.Yegneswaran A H 1997 *Z Metallkd* 88 333-338.

[13] El-tayeb N S M, Low K O, and Brevern P M 2007 *Journal of material processing technology* 186 272-278.

[14] Jur J H, Seong K D, Kim J M, Kim K T and Jung W J 2007 *Alloys and compound* 435 311-314.

[15] Yasin A and Savaskan T 2009 *Tribology international* 42 176-182.

[16] Zaid A I O and Dawud A K 2004 *Effect of cutting parameters on the quality of the machined surface of tool steel D2* ICAMET Kuala Lampur, Malasia, p.165-170.

Effect of cold rolling and annealing on the grain refinement of low alloy steel

Ejaz Ahmad, F Karim, K Saeed, Tanvir Manzoor and G H Zahid
MD, PINSTECH, Post Office Nilore, Islamabad, Pakistan

Abstract: A low alloy steel containing 0.06 wt% C, 1.5 wt% Mn and 0.1 wt% V was given 85% cold rolling reduction. The aim of the rolling reduction was to induce energy due to stresses and distribution of carbides, used for grain refinement in subsequent annealing. The rolled specimens were heat treated at various temperatures from 590°C to 650°C for different lengths of soaking times ranging from five minutes to two hours, to promote the process of re-crystallization. At temperatures with long soaking times the re-crystallization process is expected to be completed with minimum of grain coarsening due to carbide distribution, especially vanadium carbides. A smooth drop in hardness with increase in annealing times was observed which may be due to recovery from stressed conditions during process of re-crystallization. Texture observations supported the re-crystallization process as the preferred orientation of (200) plane in rolled condition was successively reduced with annealing temperatures. Tensile properties observations of two hour annealing times at 590°C to 650°C clearly demonstrated that ductility increased at all annealing temperatures with maximum gain at 625°C and strength is decreased.

1. Introduction

The improvement in mechanical properties, especially the yield strength creates certain attraction in grain refinement. It is well known that yield strength, σ_y, is increased by refining the grain size according to Petch relationship [1]:

$$\sigma_y = \sigma_0 + k_y d^{-1/2} \qquad (1)$$

where σ_0 is yield strength for material containing no grain boundaries, d is the grain diameter and k_y is an experimentally determined coefficient.

Severe deformation with high plastic deformation techniques has gained interest in recent years for recrystallization leading to grain refinement in low alloy steels and aluminum alloys [2-5]. Transformational grain refinement (TGR), in which austenite to ferrite transformation is used to produce refined ferrite structure in steel, was first introduced by Hodgson *et al* [6-8] and got a refined ferrite microstructure of 1 μm in a substantial fraction of its volume. Priestner and Ibraheem [9] used similar process to achieve grain sizes less than 1.5 μm through thickness of a Nb-microalloyed steel rolled to approximately 2.5 mm thickness. They also applied cold rolling after TGR process followed by annealing to promote re-crystallization for the achievement of ultrafine grain refinement. In the present work starting with ferrite and pearlite microstructure a heavy cold rolling reduction of 85% was given to get maximum stored energy used for re-crystallization on subsequent annealing. A heavy deformation like cold rolling provides store energy which acts as driving force for re-crystallization on annealing [10]. In cold rolling the dispersed cementite and vanadium carbides are expected to control the grain coarsening of the re-crystallized grains.

2. Experimental Work

The vanadium micro-alloyed steel was provided in the form of hot rolled and annealed slab with chemical compositions listed in the Table 1.

Table 1. Chemical Compositions of Steel in wt. percent

C	Si	Mn	P	S	Al	N	Cr	Mo	Cu	Ti	V
0.056	0.53	1.46	0.017	0.005	0.032	0.009	0.09	0.03	0.08	0.01	0.1

The initial optical metallographic observations revealed that microstructure was mainly ferrite and pearlite, with minimum banding of prior deformation, shown in Fig.1.

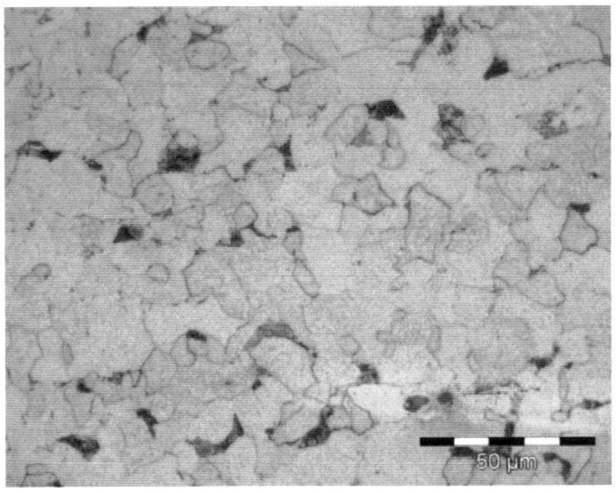

Figure 1. Ferrite and pearlite with equaiaxed microstructure of the steel

After pickling a cold rolling reduction of 85 percent was given to get final thickness of 1 mm steel plate.

Four sets of annealing at temperatures of 590°C, 600°C, 625°C and 650°C with different holding (soaking) times ranging from 5 minutes to two hours were given to promote re-crystallization. For Vickers hardness measurement after annealing average of five reading at different positions of the specimens were taken.

For optical metallographic observations the specimens after grinding polishing were etched in 2% natal solution. The re-crystallized microstructures of the annealed specimens were examined on Olympus 3X-51 microscope equipped with image analyzer and DP-70 digital camera. Line intercept method was employed for grain size measurement. In this method concentric circles of known diameter were superimposed on the image of the microstructure to get the intercepts on the grain boundaries.

For the texture measurement X-ray diffractometer, model "Dmax-III A" of Rigaku Company of Japan, equipped with monochromator and Cu K_{α} radiation was used.

Tensile testing was performed on SANS Universal testing machine with cross head speed of 2mm/min.

3. Results and Discussion

3.1 Cold rolling and annealing for re-crystallization

In cold rolling ferrite and pearlite (or ferrite plus carbide aggregate) microstructure developed previously dissociates and carbides are scattered to play roll in prevention of grain coarsening during subsequent heat treatment. The microstructure revealed after cold rolling is shown in Fig. 2.

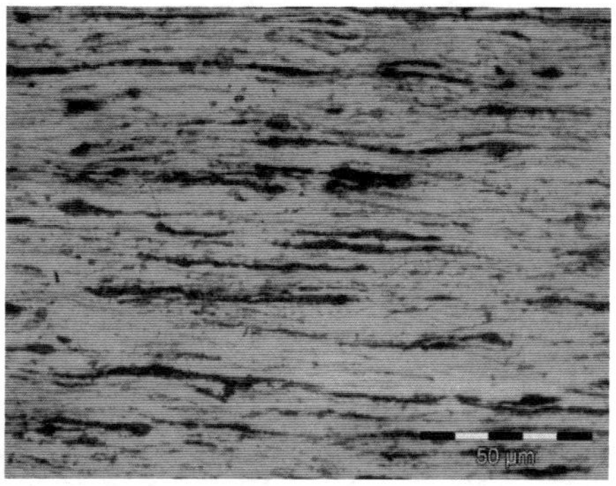

Figure 2. Microstructure after 85% of cold rolling reduction

The aspect ratio of the grains changed along the rolling direction providing a banded morphology with the strain energy used for re-crystallization. A large stored energy resulting from deformation provides large density of sites for nucleating re-crystallization [10]. The energy of deformation is expected to play role in re-crystallization during annealing and scattered carbide especially vanadium carbide (VC) may control the grain coarsening.

The grain refinement by static recrystallisation after cold rolling depends upon recrystallisation temperature, degree of prior deformation and the grain size prior to deformation [11]. Starting with grain size of 8.4 µm of as received material, the cold rolled specimens were annealed at different temperatures to promote static re-crystallization. The effect of annealing on hardness for different holding times after 85% reduction is shown in Fig. 3.

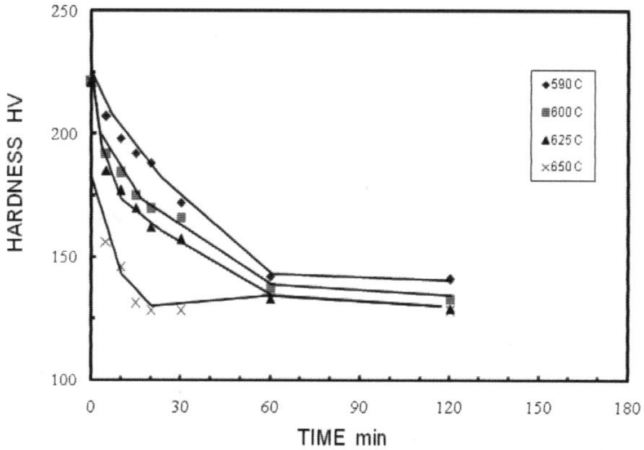

Figure 3. Effect of annealing at different temperatures and holding times on hardness after 85% of rolling reduction

The hardness dropped smoothly with soaking times at all annealing temperatures and this effect is more enhanced at 650°C. The decrease in hardness with annealing times and temperatures are due to the process of recovery and at higher temperature of 650°C it may be completed in 20 minutes. Apart from time and temperature of annealing complete recovery from deformation depends upon the type of deformation applied prior to annealing process. Although equal channel angular extrusion is more effective than cold rolling but accumulated strains introduced by these processes were effectively used in the intercritical annealing for grain refinement of ferrite and martensite [12-13]. The microstructures developed after two hours of annealing at 590°C, 600°C, 625°C and 650°C are shown in Fig. 4.

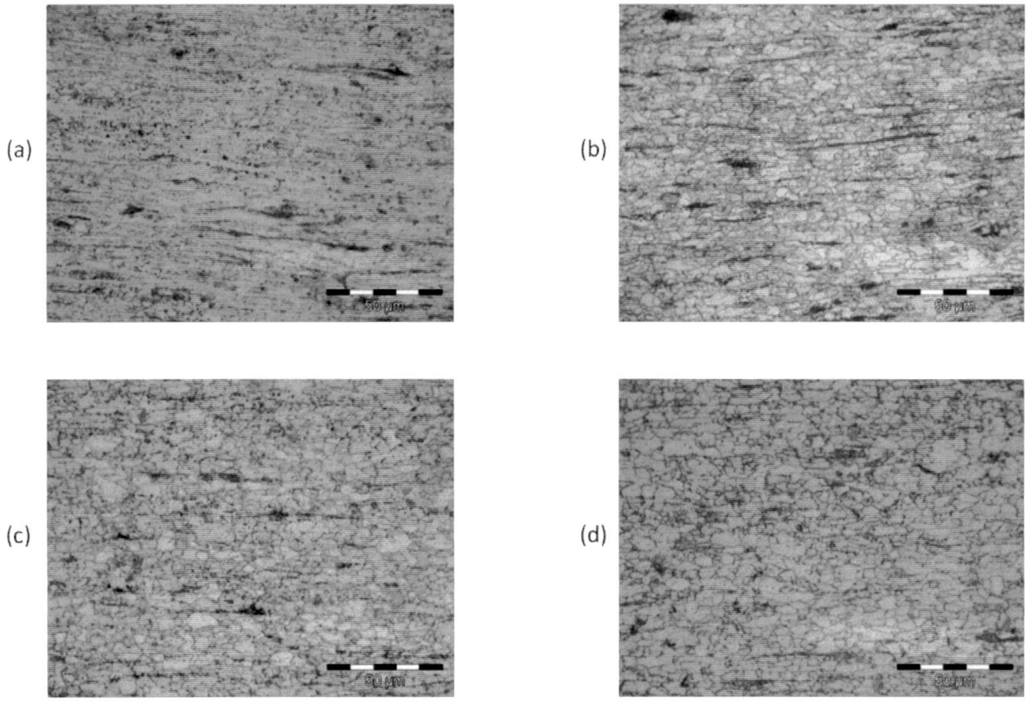

Figure 4. Microstructure developed after cold rolling and annealing for 2 hours at (a) 590°C (b) 600°C (c) 625°C and (d) 650°C

At low annealing temperature of 590°C the process of re-crystallization has started but still some elongated particles are also visible even after two hours of annealing, shown in Fig. 4(a). There is no appreciable drop in hardness between one and two hours of annealing, Fig. 3, predicting the recovery process has completed within or before this period of time. Therefore, the annealing temperature of 590°C may be low enough to mobilize the uncompleted process of re-crystallization by the store energy within the specified period of recovery time. At 600°C the microstructure Fig. 4(b) is mostly equi-axed in nature with some small grains of approximately 1μm are visible. Similarly the grain refinement process looks more efficient with some coarsening of grains in limited localized areas. However, the average grain size obtained after two hours of annealing at all temperatures of rolled specimens was 4.4 μm and minimum of 3.9 μm grain size at 650°C. A noticeable grain size reduction of about 46% was obtained after re-crystallization during process of heat treatment.

IOP Conf. Series: Materials Science and Engineering **60** (2014) 012029 doi:10.1088/1757-899X/60/1/012029

3.2 Deformation Texture

The preferred orientation or textures developed by forming process with severe plastic deformation is called Deformation Texture. In a polycrystalline aggregate where grains have their own random preferred orientation rotate during plastic deformation. As a result grains undergo slip and rotate in a complex fashion to set a new nonrandom orientation which is determined by the imposed forces. The texture development after cold rolling and changes after process of annealing are shown in Fig. 5.

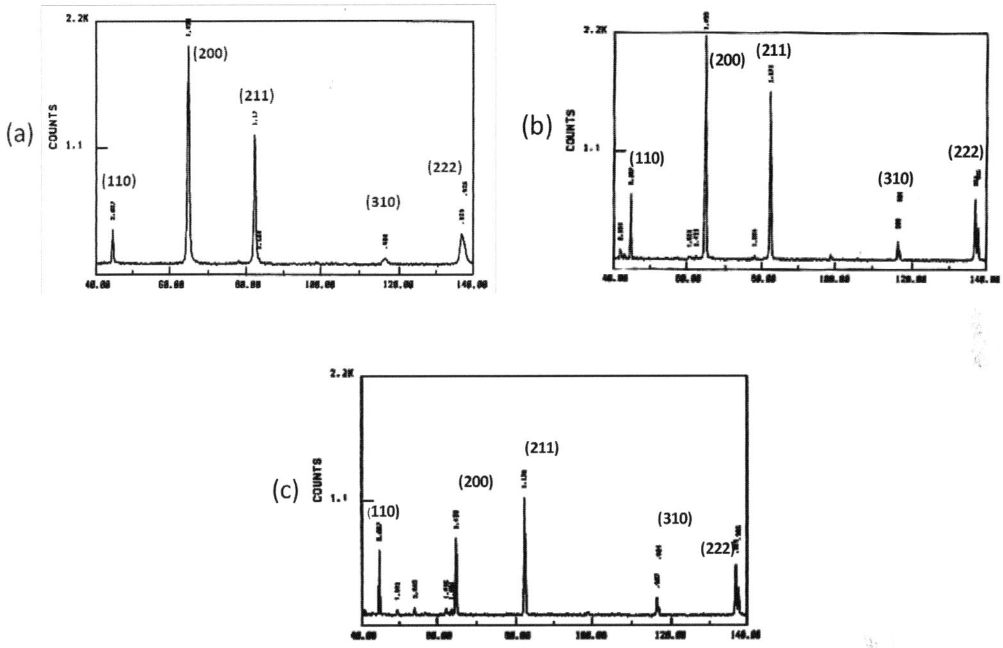

Figure 5. (a) Texture development after cold rolling. Changes in texture after annealing for two hours at (b) 590°C and (c) 650°C

The deformation texture developed by plastic deformation or cold rolling can be assessed by the evaluation from the XRD pattern shown in the present case of 85% reduction sample as Fig. 5(a). The preferred orientation or texture was developed after cold rolling along the (200) plane. After two hours of annealing at 590°C, Fig. 5(b), no appreciable changes of intensities along (200) and (211) planes were observed. There is noticeable decrease in intensity along (200) plane after two hours of annealing at 650°C, Fig. 5(c). The intensity together with random intensity, texture coefficient of (200) plane and texture developed (%) are described in Table 2.

Table 2. Effect of Annealing Temperature for two hours on Texture coefficient along (200) plane

Annealing Temperature ^{0}C	$I_{(200)}$ %	$I_{R\,(200)}$ %	Texture Coefficient	Texture %
Cold rolled	100	20	2.63	163
590	100	20	2.04	104
600	87	20	1.27	27
625	69	20	1.09	9
650	68	20	1.11	11

To calculate the texture coefficient (T.C) the normalized intensities were divided by the average of all the normalized intensities according to the equation:

$$T.C_{(hkl)} = \frac{I_{(hkl)} / Ir_{(hkl)}}{\frac{1}{n}\sum I_{(hkl)} / Ir_{(hkl)}}$$

where

$I_{(hkl)}$ is the measured integral intensity of a given (hkl) reflection in the specimen

$I_{r(hkl)}$ is the calculated theoretical intensity for the same (hkl) reflection in a randomly oriented specimen

n is the total number of rections

When the cold worked metal is annealed and re-crystallized new grains have their own preferred orientation with re-crystallization or annealed texture [14]. A texture of 163% was developed after cold rolling, shown in Table 2 which gradually reduced by increasing the annealing temperature. At 590 °C after two hours of annealing the texture reduced to 104%, reflecting the partial re-crystallization, observed in Fig.4 (a). The reduction in texture percentage along (200) is prominent at higher temperature assessed in the metallographic observations of Fig.4 (b) to (d) and texture reduced to 11% at 650°c for two hours of annealing.

3.4. Tensile Properties

The stress strain curves of the specimens after two hours of annealing at 590°C, 600°C, 625°C and 650°C are shown in Fig. 6.

Tensile testing results of as cold rolled specimen are also added for comparison. In cold rolled condition the tensile strength was increased to 946 MPa with only 4.4% of uniform strain. The expected increase in strength and hardness, as shown in Fig. 3 before the start of annealing process, may be due to introduction of dislocations in the microstructure during heavy plastic deformation in cold rolling. The interaction of these dislocations in tensile deformation raised the tensile strength at the expense of ductility.

After two hours of annealing all the specimens showed sharp decrease in tensile strength and increase in ductility. The decrease in strength may be due to recovery from stressed conditions and release of stored energy used for grain refinement. In annealed conditions when refined microstructure played its role in raising the yield strength, a maximum gain was achieved after annealing at 600°C. The micro-structural observation after two hours of annealing at 600°C, Fig. 4(b), clearly shows re-crystallization and increased population of micron or submicron sizes of grains. The Petch relationship [1]: in equation 1 described an inverse relationship of yielding with grain size of the materials. In tensile testing the plastic deformation of all annealed specimens progressed with minimum of work hardening which is also observed previously by Hodgson et.al [7] in ultrafine ferrite steel.

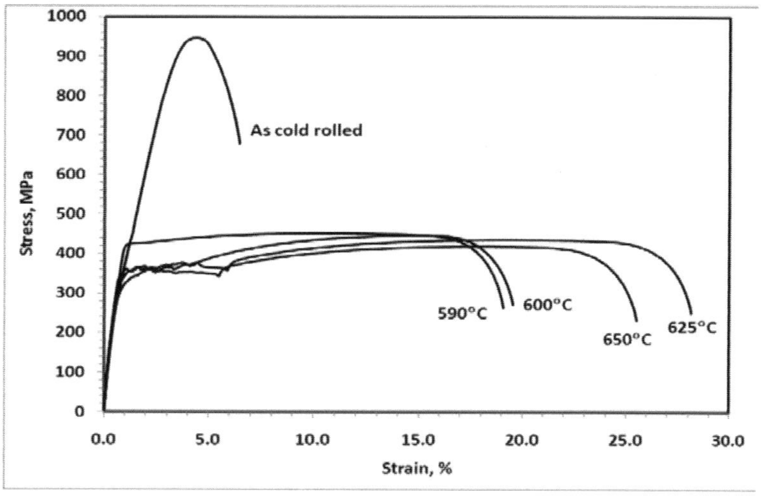

Figure 6. Stress-strain curve after cold rolling and annealing for 2 hours at, 590°C, 600°C, 625°C and 650°C

The tensile strengths of the annealed specimens were not sharply decreased with increasing the annealing temperatures. This may be due to absence of abnormal grain growth during annealing observed previously by Priestner and Ibraheem [9] is steel containing niobium. The present studies the grain coarsening process looks to be effectively controlled by the vanadium precipitates.

4. Conclusions

A low alloy steel containing 6.06 wt% C was given 85% cold rolling reduction to induce effective strain used for grain refinement in subsequent annealing. The cold rolled specimens were annealed at different temperatures with different soaking times. The results are concluded as follow:

- The optical metallography of the rolled specimen showed elongation of the grains and increase of aspect ratio along the rolling direction. The microstructural observations of the annealed specimens clearly revealed the re-crystallizations. An overall grain size reductions of 46% was obtained after annealing. No abnormal grain growth was observed by increasing time and temperature of annealing.
- A texture along (200) plane, as preferred orientation after cold rolling, was developed. The cold rolling texture of 163% was gradually reduced with increase of annealing temperatures. Annealing at 650°C reduced the texture to 11%, reflecting the re-crystallization along the new preferred orientations.
- The Annealed specimens with refined microstructure showed typical tensile properties of ultra-fined steels with lack of excessive work hardening and increased ductility. The tensile strength dropped after annealing and re-crystallization due recovery from stressed conditions. However, increased fracture strains reflect their improved formability essentially required for structural applications of steel.

5. References

[1] N.J. Petch, J.I.S.I., vol.173, (1953) 25-28.

[2] I. Schindler, M. Janošec, E. Mistecký, M. Rùžička L. Čížek, L. A. Dobrzański, S. Rusz and P. Suchánek, Archives Mat. Sci. Eng. Vol. 36(1), 2009, pp. 41-47.

[3] R. Priestner, Y M Al-Horr and A K Ibraheem, Mater. Sci. and Technol., Vol. 18, 2002, pp. 973-980.

[4] P B Prangnell, J R Bowen and P J Apps, Mat. Sci. Eng. A Vol. 375-377, 2004, pp. 178-185.

[5] Y. D. Huang, W. Y. Yang Z. Q. Sun, J. Mat. Proc. Tech., Vol. 134, 2003, pp. 19-25.

[6] P.D. Hodgson, M.R. Hickson and R.K. Gibbs, Mater. Sci. Forum, 1988, vol. 63, pp284-286.

[7] P.D. Hodgson, M.R. Hickson and R.K. Gibbs, Scripta Mater., 1999, vol.40,pp.1179-1184.

[8] M.R. Hickson, P.J. Hurley, G.L. Kelly, P.D. Hodgson and M.R. Barnett, 2nd Symposium on Supermetal, 1-2 Nov. 1999, Tokyo, Japan, Jap. Res. And Dev. Centre for Metals (JRCM) 1999, pp.203-213.

[9] R. Priestner and A.K. Ibraheem, Mater. Sci. and Technol., 2000, vol. 16, pp.1267-1272

[10] F J Humphreys, M. Hatherly, Re-crystallization and Related Annealing Phenomena, ELSEVIER Ltd The Boulevard, Langford Lane Kidlington, Oxford, UK, 2004

[11] A. K. Ibrahim, R. Priestner, J. R. Bowen, P. B. Prangnell and F. J. Humphreys, Ironmaking and Steelmaking, 2001, vol. 28, pp 203-208.

[12] K.T Park, S. Y. Han, B. D. Ahn, D. H. Shin, Y. K. Lee and K. K. Um, Scripta Materialia 51 (2004), pp909-913.

[13] F. Silva, N. I. A. Lopes and D. B. Santos, Materials Characterization 56 (2006) pp. 3-9.

[14] B. D. Cullity, Elements of X-Ray Diffraction, Second Edition, 1978.

Stress corrosion cracking behavior of quenched and tempered 2.25Cr 1Mo steel

T Manzoor, E Ahmad, M Sarwar and M Iqbal
ATCOP, P. O. Box 1509, Islamabad, Pakistan

Abstract. For stress corrosion cracking (SCC) study of 2.25Cr-1Mo steel, specimens were initially austenitized at 925°C for 2 hours and quenched in ice brine solution. With this process of heat treatment martensite was formed and banded morphology in the microstructure developed during previous processing was also removed. Tempering at 575°C for 1h was employed to introduce ductility in the quenched martensite of the specimens. The effects of temperature variations on SCC behavior showed a successive decrease in strength and ductility of the steel. Theses variations in tensile properties were correlated with scanning electron microscopic (SEM) examinations. The SEM results clearly revealed that mostly transgranular cracks initiated at the edges of the tensile specimens grown perpendicular to loading axes, and their intensity is increased with the testing temperatures.

1. Introduction

Chromium-Molybdenum ferritic steels (generally referred as Cr-Mo steels) have excellent high temperature properties, formability, weldability and resistance to stress corrosion cracking and other forms of corrosion [1, 2]. These features make Cr-Mo steels the extensively used family of engineering materials for moderately high temperature applications over several decades, such as conventional power and petroleum industries for fabrication of pressure vessels, boiler tubes and coal liquefaction reactors [1-3]. With increasing demand for suitable materials for supercritical and ultra-supercritical steam generators for power plants having much improved efficiencies, Cr-Mo steels have attracted great interest for their potential application in low-temperature parts of such steam generators that will be expected to operate at higher than usual temperatures.

Among Cr-Mo class of steels with specific applications, 2.25Cr-1Mo has its importance in heat treated conditions. In quenched and tempered states of heat treatment these steels show range of mechanical properties due to carbide precipitation specially chromium carbides [4-5]. The nuclear industry considers 2.25Cr-1Mo steel as strong candidate for structural material of the parts due to their better resistance to swelling than stainless steel [6-7]. In the present studies a simulated environment similar to nuclear application of the material was developed to assess the changes or deterioration in the mechanical properties. The optical and scanning electron metallographic observations were added to support the results of these findings. The chemical composition of the 2.25Cr-1Mo steel is given in Table 1.

Table 1. Chemical Compositions of Steel in wt. percent

C	Si	Mn	P	S	Cr	Mo	Fe
0.14	0.14	0.56	0.009	0.003	2.38	0.89	Bal.

The as received material was provided in the form of bars having 12x12x100mm dimensions. The initial optical metallographic studies revealed ferrite and pearlite with banded morphology of microstructural contents, shown in Fig. 1.

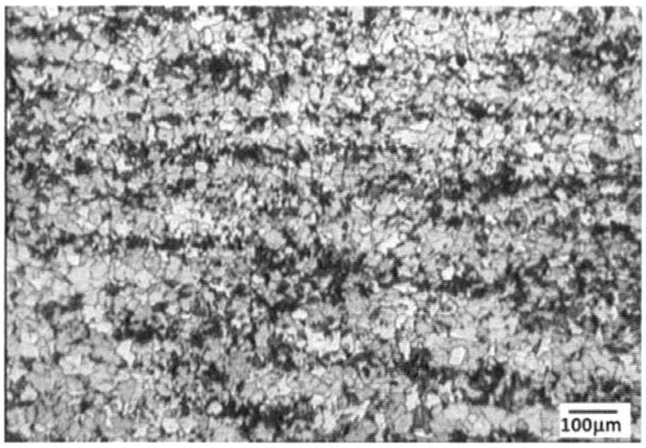

Figure 1. Microstructure of as received material showing banded morphology

The specimens from as received bars were austenitized at 925°C for 2h followed by ice brine quenching. The specimens were further heat treated by tempering for one hour at 575°C.

For stress corrosion cracking behavior studies, slow strain rate tests (SSRT) was carried out using CORTEST testing machine in PWR simulated environment. Normally the pressure vessel in PWR works at about 290–310°C under a pressure of about 15 MPa. The coolant contains Boric acid and LiOH, which are instantaneously added into the coolant system as per requirement to control fission reactivity and pH, respectively. These conditions were simulated in the laboratory using demineralized water (DMW) having a room temperature pH. Boric acid and LiOH were added to the DMW in amounts giving 100 ppm of boron and 7 ppm of Li in the solution. The tensile samples (25 mm gauge length, 5 mm gauge diameter) used in this study, shown in Fig. 2, were machined from the quenched and tempered ferritic steel test bars.

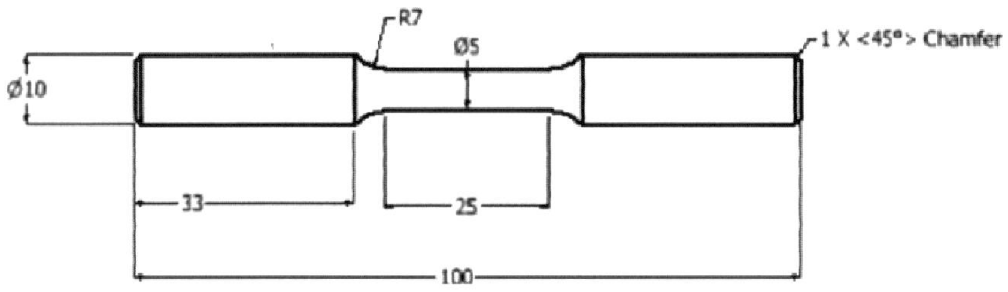

Figure 2. Tensile specimen used for SCC testing

The slow strain rate tests (SSRT) were performed on as received and quenched samples in air at room temperature while SCC testing of quenched and tempered samples was carried out in the autoclave having simulated solutions at temperatures of 75, 150, 200 and 275°C. All tests were performed at an initial strain rate of $1e^{-6}s^{-1}$. The microstructural analysis of the secondary cracks and fracture surface morphology were studied using Lieca optical microscope and Phillips scanning electron microscope (SEM).

2. Results and Discussion:

2.1. Development of Microstructure after Heat Treatment

The microstructures developed after austenitizing at 925°C followed by quenching in brine solution, and tempering after quenching are shown in Fig. 3 (a). and (b) respectively.

(a) (b)

Figure 3. Microstructure of (a) as quenched from 925°C and (b) after one hour tempering at 575°C

The lath or plate-like morphology of martensite is expected in 0.5 wt.% carbon steel after quenching from the austenite region[8]. The morphology of quenched martensitic microstructure appeared in Fig. 3 (a) looks like small laths with lenticular plates. The hardness of the quenched martensite is Hv = 485. The specimens were tempered at 575°C before proceeding towards SCC testing. The aim of tempering was to introduce ductility in the quenched specimens and to compare the SCC behaviors in both of these conditions. The microstructure revealed after tempering is shown in Fig. 3(b). The carbon diffused out in the form of carbides and in the presence of Cr in the composition, chromium carbides would have been formed. A considerable drop in hardness of Hv = 347 was observed after tempering attributed to the diffusion of carbon entrapped at interstitial positions after quenching. The reduction in hardness after tempering at 575°C indicates that softening effect due to diffusion of entrapped carbon is dominant than temper embrittlement. However, change in toughness which is true indicator of temper embrittlement effects can not be assessed from the reduction in hardness after tempering.

In steels containing chromium, two types of carbides are often formed: Cr_7C_6 (trigonal in nature) and $Cr_{23}C_6$ (complex cubic). As chromium is weaker carbide former than vanadium, carbides of $Cr_{23}C_6$ are formed only when their contents increased more than 7% and Cr_7C_6 carbides are formed at1% of chromium level in the steel [8]. However, $Cr_{23}C_6$ were observed in 2.25Cr-1Mo steel after quenched and tempered conditions [9].

2.2. Tensile Properties at Different Temperatures in Simulated Environment:

True stress-true strain curves of specimens to onset of necking in quenched and tempered conditions tested at different temperatures in simulated environment are shown in Fig. 4.

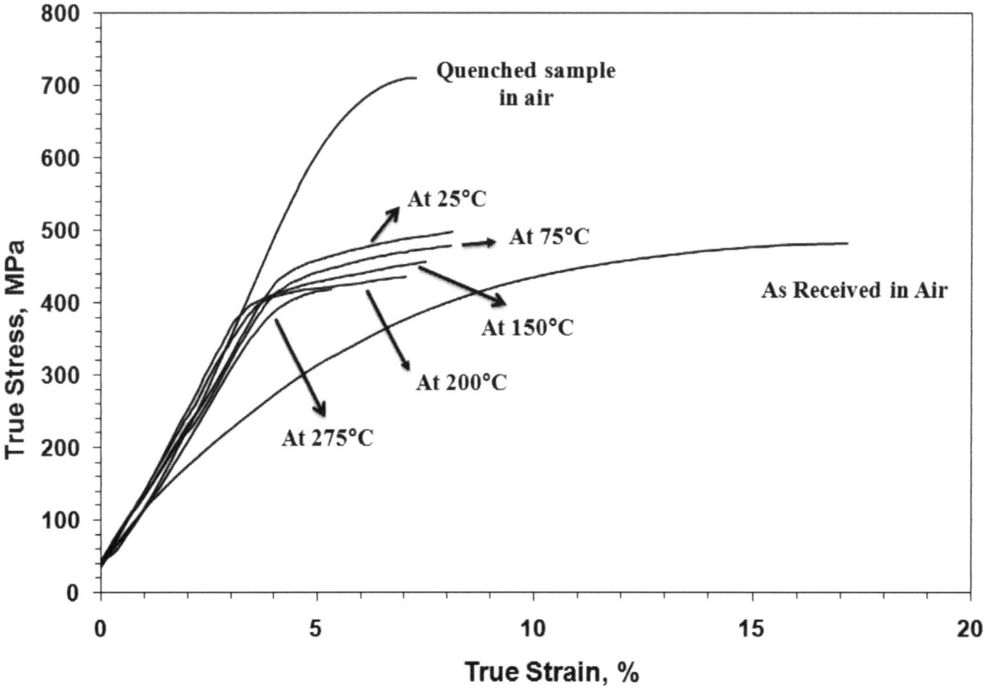

Figure 4. True stress-true strain curves of specimens to onset of necking tested at different temperatures in simulated environment

The tensile curves in as received conditions with ferrite & pearlite microstructure and as quenched condition with un-tempered martensite microstructure, tested at room temperature, are included for comparison. A smooth yielding behavior with absence of sharp yield point was observed in all the specimens. The tensile properties of these specimens are shown in Table 2.

Table 2. Tensile properties

Specimen Code	0.2% Proof Stress (MPa)	Max. True Stress (MPa)	True Uniform Strain (%)
75°	414	480	8.11
25°C	433	498	8.13
150° C C	382	457	7.5
200°C	388	436	7
275°C	378	419	5.35
As Received in air	187	482.4	17.15
Quenched in air	606	711	7.3

In as quenched condition the rise in strength at the expense of ductility is characteristic of hard martensitic microstructure.

The simulated environment testing at 25, 75, 150, 200 and 275°C resulted in successive decrease in maximum true stresses and maximum true strains with rise in testing temperatures, shown in Table 2. This decreasing trend of strength with rise in testing temperatures is shown as block diagram in Fig. 5.

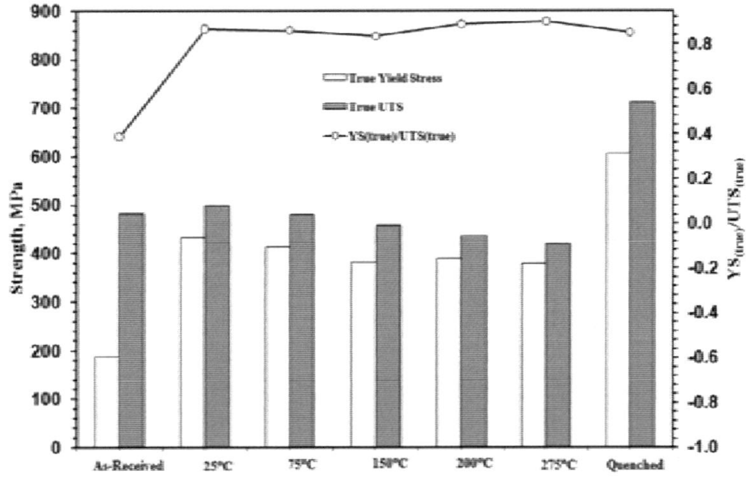

Figure 5. Strength Levels of 2.25Cr-1Mo Steel at Different Temperatures

Also included in Fig. 5 is the yield to tensile strength ratio, predicting the work hardening of the specimens. Although yielding in as received specimen was started very early at 187 MPa of stress but it gained a maximum true stress of 482 MPa due to excessive work hardening in this condition. In quenched and tempered conditions maximum work hardening was achieved for sample tested at 150°C and it decreased with further increase in testing temperature. In SCC the specific environment is very importance, and only very small concentrations of certain highly active chemicals can produce catastrophic cracking [10]. Considering this environmental fact and its application in present SCC testing, the cracks would have been initiated at the surface, and at higher temperatures their intensity increased and they propagated more efficiently. Therefore, inter-granular and trans-granular forces have lost the strength to raise the work hardening component with increasing testing temperatures. This decrease in work hardening resulted in losing ductility due to limited extension of stress-strain curves till start of necking in tensile deformation.

2.3. Stress Corrosion Cracking (SCC) in Simulated Environment

SCC is a combined action of stress and a corrosive environment leading to the formation of a crack which would not have developed by the action of the stress or environment alone. The PWR simulated environmental condition created for the present studies in the laboratory are quite effective for the enhancement of SCC. At room temperature (25°C) near the tensile fracture surface (Fig. 6) little micro-cracks on the edges were observed.

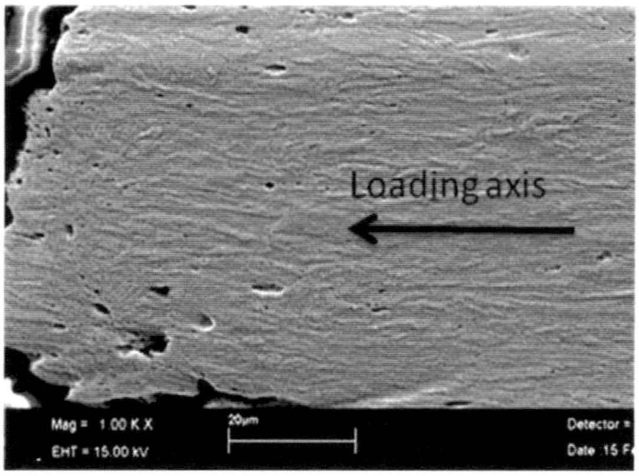

Figure 6. Microvoid formation and deformation along tensile axes near fracture surface on sample tested for SCC at 25°C.

The excessive microvoid concentration and severe plastic deformation of the grains along the tensile axes in the necked portion predict the ductile failure with an increase in the ductility.

Trans-granular cracking was observed at 75°C. At higher testing temperatures micro-cracks initiated from the edges of the specimens trans-granularly and inter-granularly, and propagated inward due to SCC, shown in Fig. 7.

(a)

(b)

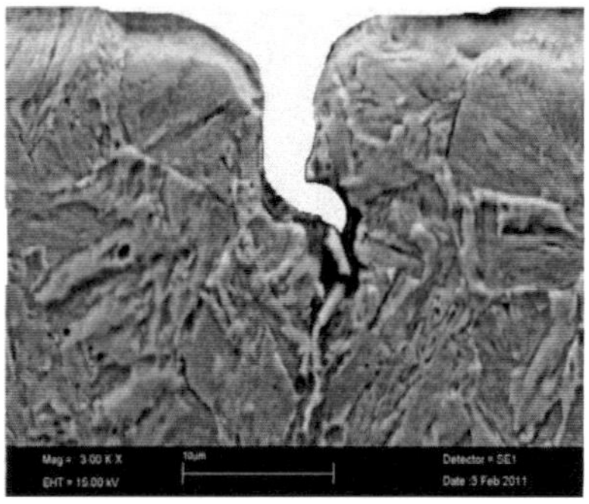

Figure 7. (a) SCC at 75°C showing transgranular cracking perpendicular to tensileaxis. (b) SCC at 150°C showing intergranular cracking perpendicular to tensile axis.

The severity of SCC increased with testing temperatures and branches of cracks from the initial source advanced with inter-granular zigzag movement as shown in Fig. 8.

(a)

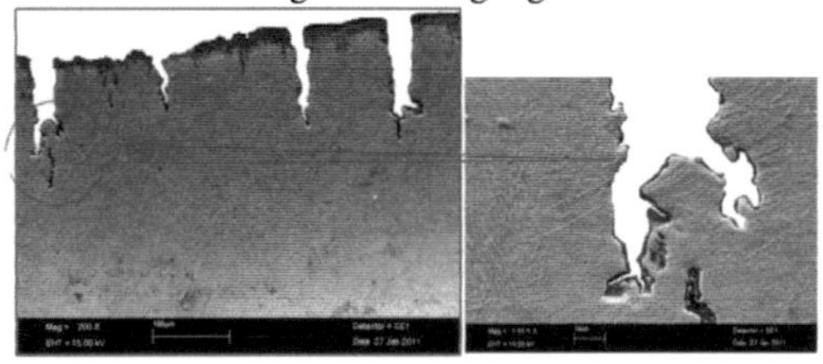

(b)

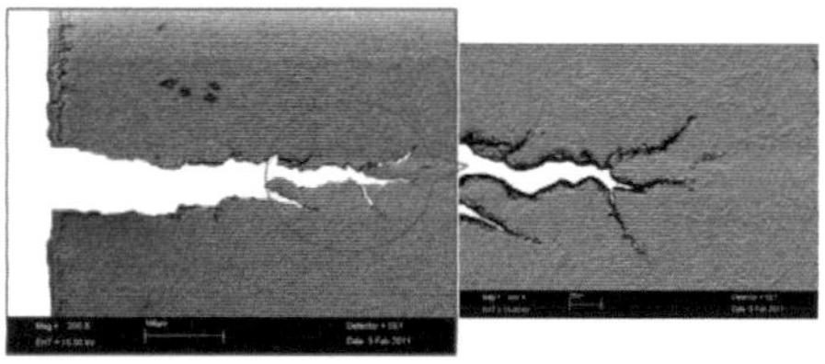

Figure 8. (a) SCC at 200°C showing inter and transgranular crack branches normal to loading axis. (b) SCC at 275°C showing inter and transgranular crack branches normal to loading axis.

The inter-granular cracking looked a prefer mode of fracture in the present case at higher temperatures. This may be due to weakening of the forces or the cohesion strength of the grain boundaries due to their embrittlement. Two micro-mechanisms controlled the changes in mode of fracture due to long term thermal exposure and embrittlement:

- Temper embrittlement caused by growth of carbides, precipitated at the grain boundaries changed the transgranular cleavage from dislocation triggered to carbide induced [11, 12].
- Intergranular embrittlement caused by diffusion of impurities and carbide coarsening at grain boundaries [13, 14].

At 75°C in the present studies, there is no considerable indication of precipitation along the grain boundaries and the crack propagated trans-granularly. However, at higher temperatures the mode gradually changed from trans-granular to inter-granular. There is clear indication of inter-granular growth at 275°C with excessive precipitation along the grain boundaries to follow the crack path.

3. Conclusions

Stress Corrosion Cracking (SCC) behavior was studied at 25, 75, 150, 200 and 275°C, in quenched and tempered states of 2.25Cr-1Mo steel. The results are concluded as follow:

- Initially the steel with banded ferrite and pearlite microstructure was austenitized at 925°C and full martensite was obtained by quenching in iced brine solution. Tempering for one hour at 575°C reasonably reduced the hardness due to diffusion of carbon from entrapped interstitial positions.
- Tensile testing in simulated environment clearly demonstrated that severity of SCC certainly increased with increasing the testing temperatures. A smooth decrease in tensile strength and ductility was observed by increasing the testing temperatures. Similarly the work hardening property was also decreased due to enhancement of simulated environment effect at higher temperatures.
- The microstructural observations revealed that at room temperature the material remained ductile and tensile deformation progressed with void formations in the necked region. The SCC initiated from the edge of the specimen at 75°C trans-granularly, but at higher temperatures mode of SCC changed to inter-granular. Precipitation along the grain boundaries at 275°C promoted the SCC to progress inter-granularly.

4. References

[1] A.K. Khare (ed.), Ferritic Steels for High Temp. Appl.: ASM Int. Conf., Warren, PA, 1981.

[2] S.F. Pugh and EA. Little (eds.), Proc. BNES Int. ConfFerritic Steels for Fast Reactor Steam Generators, Pub: BNES, London, 1978.

[3] T. Wada and G.T. Eldis, Ferrinc Steels for High Temperature Applications: ASM Int. Conf. (A.K Khare (ed.)), Warren, PA, 1981, p.343-62.

[4] M.A. Islam, J.F. Knott, and P. Bowen, J. Mat. Eng. Per, 2004, 13: p. 600-606.

[5] Jai Won Byeon and SookInKwun, Materials Transactions, Vol. 44, No. 6 (2003) p. 1204-1208.

[6] A. L Pitner, S. L. Hecht and R. G. Trenchard "Nonswelling Behaviour of HT9 Alloy Irradiated to High Exposure", WHC-SA-1967-FP (1993).

[7] F. A Smidt, Jr., P. R. Malmbery, J. A. Spague and J. E. Westmoreland, "Swelling Behaviour of Commercial Ferritic Alloys, EM-12 and HT-9, As Assessed by Heavy Ion Bombardment", ASTM STP 611, 227 (1976).

[8] R. W. k. Honeycombe:, Steels Microstructure and Properties, published by Edward Arnold, 1987.

[9] S. H. Kim, W S Ryu and Il H Kuk, J. Korean Nuclear Society, 1999, !3(6), p. 561-571.

[10] ASM International, Metals Handbook (Desk Edition) Chapter 32 (Failure Analysis), American Society for Metals, (1997) pp 32-24 to 32-26

[11] Man, J. Holzmann, M., Vlach, B.: *Zváranie (Welding)*, 32, p. 3, (1983).

[12] Holzmann, M., Dlouhý, I., Vlach, B. & Krumpos , J.: *Int. J. Pres. Ves. and Piping,* 1996, p.68, 99.

[13] Briant, C. L., Banerji, S. K.: *Int. Metals Review*, 23, 1978, p. 164.

[14] Kameda, J.: *ActaMetall*, 34, 1986, p. 867.

TiO_2 and Al_2O_3 promoted Pt/C nanocomposites as low temperature fuel cell catalysts for electro oxidation of methanol in acidic media

Rabia Naeem[1], Riaz Ahmed[1, 2] and Muhammad Shahid Ansari[1]
[1]Department of Chemistry, Quaid-i-Azam University, Islamabad, Pakistan
[2]COMSATS, Shahrah-e-Jamhuriat , G-5/2, Islamabad, Pakistan

E-mail: riaz_ecfc@yahoo.com

Abstract. Carbon corrosion and platinum dissolution are the two major catalyst layer degradation problems in polymer electrolyte membrane fuel cells (PEMFC). Ceramic addition can reduce the corrosion of carbon and increase the stability of catalysts. Pt/TiO_2, Pt/TiO_2-C, Pt/Al_2O_3 and Pt/Al_2O_3-C catalysts were synthesized and characterized. Electrochemical surface area of Pt/TiO_2-C and Pt/Al_2O_3-C nanocomposite catalysts was much higher than the Pt/TiO_2 and Pt/Al_2O_3 catalysts. Peak current, specific activity and mass activity of the catalysts was also determined by cyclic voltammetry and were much higher for the carbon nanocomposites. Exchange current densities were determined from Tafel plots. Heterogeneous rates of reaction of electro oxidation of methanol were determined for all the catalysts and were substantially higher for titania catalysts as compared to alumina added catalysts. Mass activity of Pt/TiO_2-C was much higher than mass activity of Pt/Al_2O_3-C. Stability studies showed that addition of ceramics have increased the catalytic activity and durability of the catalysts considerably.

1. Introduction

In the modern society there is enormous demand for clean energy [1]. Non-renewable energy sources like fossil fuels (e.g. coal, petroleum, natural gas) are not sustainable for long period of time but also cause undesired pollution [2]. In the search of clean and efficient energy sources, fuel cells are emerging as alternate power sources particularly for transport [3]. Mostly efficiency of fuel cells depends on catalysts. For application in fuel cells the catalysts usually comprise of metallic nano particles (NPs) dispersed on suitable nano-support [4]. Most commonly employed nano-material supports are various forms of carbon that include Vulcan carbon (XC-72R), graphite, carbon nanotubes, and carbon nanofibers [4, 5]. Major drawbacks of carbon supported catalysts are corrosion and electrochemical surface area loss. When carbon support degrades and catalyst particle size grows then electrochemical surface area and performance are reduced [6]. Poisoning of the catalyst by CO is another problem. Ceramic oxides are emerging as catalyst supports [7], are resistant to corrosion and help in removal of adsorbed intermediates during methanol oxidation. Among metal oxides, besides others, TiO_2 [8], ZrO_2 [9], Al_2O_3 [10] and WO_3 [11] have been successfully employed as support material.

In this work Pt/TiO_2, Pt/TiO_2-C, Pt/Al_2O_3 and Pt/Al_2O_3-C have been synthesized and characterized. Vulcan carbon was functionalized by oxidative treatment. Electrochemical surface area, specific catalytic activity, exchange current, rates of methanol oxidation and durability studies of the catalysts were done and are compared.

2. Material and Methods

2.1. Pretreatment of Vulcan carbon

The Vulcan carbon XC-72 employed had the specific surface area $254 m^2 g^{-1}$ and average particle size 30nm, respectively. Vulcan carbon was purified by refluxing with conc. HCl and ultra-sonification at 60 ^0C. It was filtered, washed and dried at 110 ^0C. Functionalization of the purified Vulcan carbon was done by ultra-sonication method in 4:1 v/v mixture of 8M HNO_3 and H_2O_2. Titania and Alumina of analytical grade were used as received.

2.2. Synthesis of Pt/Vulcan carbon

Pt/Vulcan carbon (10 wt% Pt) catalyst was synthesized using $NaBH_4$ as reducing agent by conventional reduction method under argon atmosphere at room temperature. In a three-neck round bottom flask, aqueous solution of $PtCl_4$ was added and stirred under argon atmosphere along with Vulcan carbon; the mixture became colorless from golden indicating adsorption of metal ions on the Vulcan carbon. $NaBH_4$ solution in the solvent mixture was drop wise added into the above dispersion. 4 hours stirring resulted formation of black suspension which was filtered and washed several times with de-ionized water till neutral pH of the filtrate. The residue was vacuum-dried for 24 hours at 110°C.

2.3. Synthesis of Pt/TiO₂-C and Pt/Al₂O₃-C Nanocomposite

Pt/TiO_2-C Nano composite catalyst was prepared with two step–methods. In the first step TiO_2 and Vulcan carbon were added into water: 2-propanol (3:1 v/v) mixture (the solvent) and ultra sonicated for 1 hour. The black solid was filtered, washed with deionized water, and dried in vaccum oven for 24 hours at 110°C. TiO_2-C is a required composite support. In second step (10 wt% Pt) was loaded on this TiO_2-C mixed/composite support by conventional co-reduction method in which $NaBH_4$ used as a reducing agent. The resulting black suspension of Pt/ TiO_2-C was filtered and washed several times with de-ionized water till neutral pH of the filtrate. The residue was vacuum-dried for 24 hours at 110°C. Pt/Al_2O_3-C nancomposite catalysts was also prepared by the same procedure. For the preparation of Pt/ TiO_2 and Pt/Al_2O_3 catalysts no carbon was added and other procedure was the same.

2.4 Electrochemical measurements of catalysts

Autolab PGSTAT 12 Eco Chemie (Netherlands) equipped with three-electrode system was employed for electrochemical studies under argon atmosphere at ambient temperature. Graphite electrode having exposed polished tip of surface area 0.785 cm^2 served as working electrode. For uniform mounting on the electrode, 20 mg catalyst was ultrasonically dispersed in 1 mL of the aqueous 2-propanol in a test tube for 30 minutes and 60 µL of the resulting suspension was spread on the tip. After evaporation of the solvent, 10 µL Nafion solution (5%) was applied to the tip and finally dried in air. Nafion was used as an ion exchange binder. Cyclic voltammograms were measured in the potential range from -1.0 to +0.5 V vs. SCE in the aqueous medium containing 0.2 M CH_3OH (Sigma-Aldrich, 99.7%) and 0.5 M KOH (Fluka, 99%).

3. Results and Discussions

3.1. Electrochemical surface area of catalysts

All the catalysts exhibit the typical electrochemical characteristics of polycrystalline Pt, that is, the evolution of currents for hydrogen desorption in the forward sweep and for the reduction of surface oxide and hydrogen adsorption on the reverse sweep [12]. Fig.1.and Fig. 2. Shows the characteristic cyclic voltammograms (CVs) measured in 1M H_2SO_4 at scan rate (υ) 50 $mV.s^{-1}$ from potential range -0.2 to 1.0V on Pt/TiO_2-C, Pt/TiO_2, Pt/Al_2O_3-C, Pt/Al_2O_3 and Pt/C catalysts.

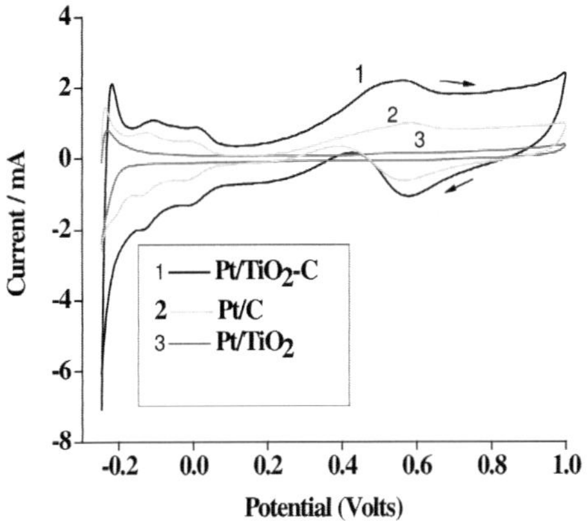

Figure 1. CVs measured in H_2SO_4 on Pt/TiO_2-C Pt/TiO_2 and Pt/C catalysts

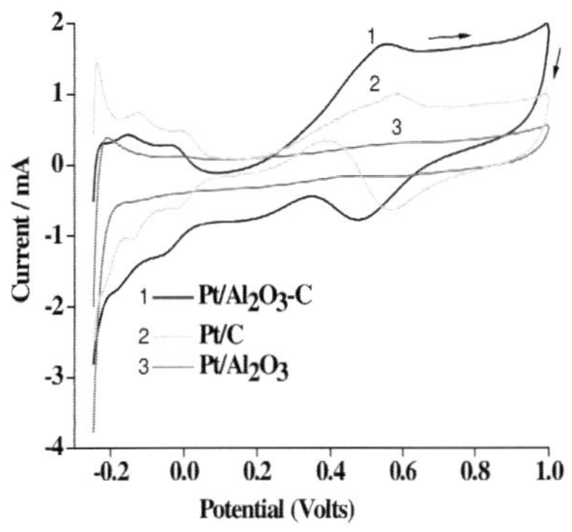

Figure 2. CVs measured in H_2SO_4 on Pt/Al_2O_3-C, Pt/Al_2O_3 and Pt/C catalysts

Hydrogen was adsorbed in the cathodic direction and then oxidized in the anodic direction, and the catalyst modified electrode showed well-defined hydrogen adsorption–desorption region. In forward scans, desorption of hydrogen take place from -0.25 to 0V and in reverse scan, reduction of hydrogen take place from 0 to -0.25V which shows adsorption of hydrogen on the catalyst surface. In reverse scans, oxygen reduction peak was observed at 0.47–0.48V. The specific electrochemical surface area (ESA) of Pt (sometimes referred to as the real Pt catalyst surface area) was calculated by integrating the CV curves in the hydrogen adsorption region by assuming that the charge per real cm^2 of Pt with monolayer adsorption of hydrogen is 210 $\mu C/cm^2$ [13]. Electrochemical surface area (S_{ESA}) may be evaluated from CV by determining the amount of charge transfer (Q_{cat}) in hydrogen adsorption region by applying equation

$$S_{ESA} = \frac{Q_{cat}}{Q_{Pt} \cdot L} \quad (1)$$

Where Q_{cat} is the charge for hydrogen adsorption , Q_{Pt} is the charge for a monolayer adsorption of hydrogen on polycrystalline Pt, which is 210 C (cm^2 Pt$)^{-1}$. L is the mass of platinum loaded on the electrode. Values of Q_{cat} and S_{ESA} are directly proportional to electrocatalytic activity. Knowing the real surface area of the catalysts, one may calculate the roughness factor (RF) by taking the ratio of real surface area to usual geometric area (or apparent area) of an electrode. The comparison of data of real surface area and roughness factor of different catalysts is given in table.1. It can be seen that electrochemical surface area of the composite supports is much higher than the normal carbon support for both the catalysts. Among the two composite catalysts electrochemical surface area of titania is much higher than alumina which indicates that titania supports are better than alumina in acidic media.

Table 1. Electrochemical surface area from CVs in 1M H_2SO_4 on various catalysts

Catalysts	Mass, Pt/mg	Q_{pt-H}/ C	S_{ESA}/cm^{2}·mg^{-1}	Roughness Factor
Pt/C	0.16	4460	132	27.10
Pt/TiO$_2$	0.12	1080	32	6.551
Pt/TiO$_2$-C	0.16	9480	282	57.50
Pt/Al$_2$O$_3$	0.24	1030	30	6.25
Pt/Al$_2$O$_3$-C	0.16	5330	159	32.339

3.2. Electrooxidation of methanol on catalysts.

Catalytic activities of the catalysts for methanol oxidation (MO) were studied in acidic medium using cyclic voltammetry. Fig. 3 and Fig. 4. show the corresponding voltammograms measured in 1 M H_2SO_4 having 1 M CH_3OH at scan rate (υ) of 50 mV·s^{-1} . In anodic sweep methanol oxidation gives a well defined forward anodic peak, while in cathodic sweep MO again gives a reverse anodic peak, which indicates that MO is irreversible process. Table.2 presents various parameters which were calculated directly related to the electro catalytic activity. The performance of Pt/*TiO$_2$*-C is demonstrated by its highest values of peak current (i_p), specific activity (I_s) and mass activity (I_m).

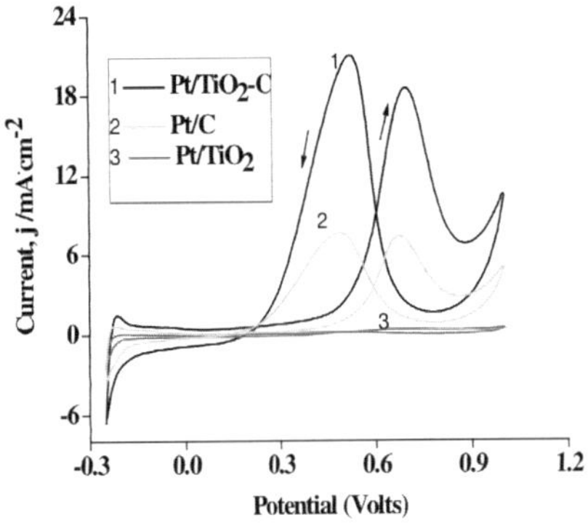

Figure 3. CVs on Pt/TiO$_2$-C, Pt/TiO$_2$and Pt/C, catalysts having 1M H_2SO_4 and 1M CH_3OH

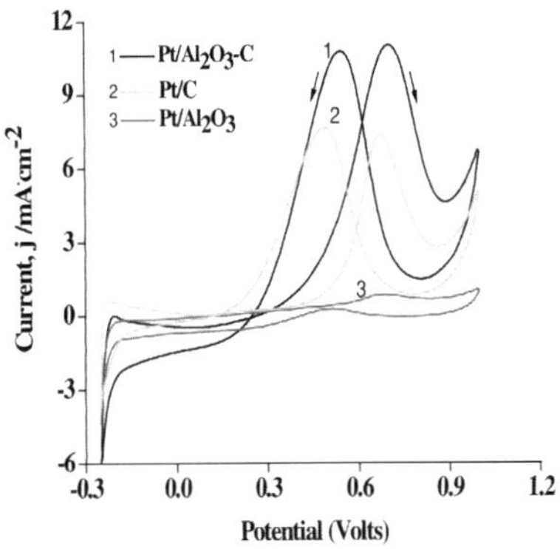

Figure 4. CVs on Pt/Al$_2$O$_3$-C, Pt/Al$_2$O$_3$and Pt/C atalysts having 1M H_2SO_4 and 1M CH_3OH

Table 2. Various parameters calculated from CVs (in 1M CH_3OH + 1M H_2SO_4) at 298 K.

Catalysts	Peak potential Ep/V	Peak Current Ip/(mA)	Specific activity I_s/mA·cm^{-2}	Mass activity I_m/mA·mg^{-1} Pt	Rate Constant k_{het}/cm·s^{-1}* 10^{-5}
Pt/C	0.670	4.87	6.20	30.4	1.62
Pt/TiO_2	0.678	0.11	0.14	1.00	1.14
Pt/TiO_2-C	0.690	11.0	14.0	68.7	7.31
Pt/Al_2O_3	0.685	0.17	0.21	0.71	0.72
Pt/Al_2O_3-C	0.692	6.20	7.90	38.7	2.33

3.3. Comparison of Polarization Curves.

A polarization curve is the most important characteristic of a fuel cell and its performance. Here the polarization curves have been used for Tafel plots. Current –potential data was obtained by analyzing the forward anodic peaks for methanol oxidation. Polarization curves for titania and alumina are given in figs.5 and fig.6.

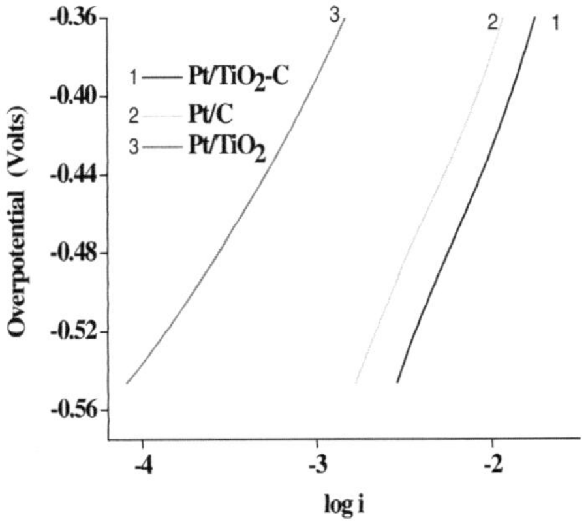

Figure 5. Polarization curves for MO on Pt/TiO_2-C, Pt/$TiO_2$and Pt/C.

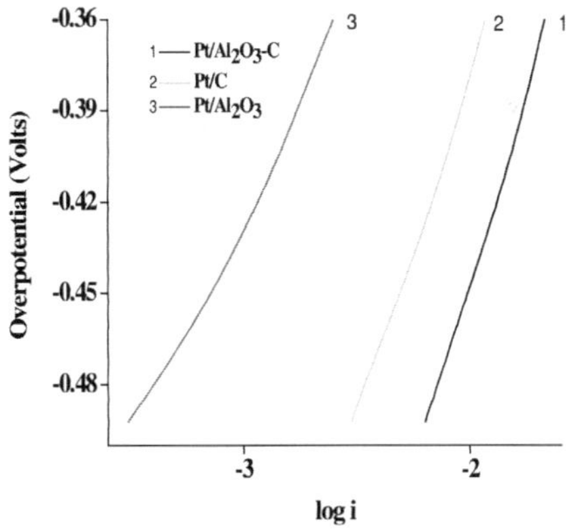

Figure 6. Polarization curves for MO on Pt/Al_2O_3-C, Pt/Al_2O_3 and Pt/C.

Tafel equation can be written as [14].

$$\Delta V_{act} = a + b \log i \qquad (2)$$

Where $a = -2.3\ (RT/\alpha F) \log i_0$ $\quad b = 2.3\ RT/\alpha F$

If voltage-current relationship is plotted in a logarithmic scale, the main parameters, a, b, and i_0 are easily detected. The catalytic activities of catalysts can be evaluated by comparing the data of exchange current density i_0. The value of i_0 can be evaluated from the intercept of the ΔV_{act} vs log i plots [14].The values of exchange current densities for the composite catalysts are compared in table.3. It can be seen that exchange current for the composite catalysts are much higher than the carbon support alone. Among the composites titania has much higher exchange current than others.

Table.3 Polarization data evaluated from Tafel plots in 1M CH_3OH + 1M H_2SO_4

Catalysts	Tafel's slope"b" /(V decade $^{-1}$)	αn_α	Intercept of E vs Log ί	ί° mA·cm^{-2}
Pt/C	0.178	0.331	1.049	5.65
Pt/TiO_2-C	0.334	0.177	1.249	18.7
Pt/Al_2O_3-C	0.281	0.211	1.991	12.2

3.4. Comparison of heterogeneous rate constants.
Heterogeneous rate constants (k_s) for the MO at temperature T can be determined using the Nicholson-Shain equation:[15-17]

$$i_p = 0.227\, nFAC_o k_s \cdot exp[(\alpha n_\alpha \cdot F/RT)(E_p\text{-}E^o)] \quad (3)$$

Where i_p is the peak current, n is the number of electrons transferred, A/cm^2 is the area of the electrode, C_o/mol·cm^{-3} is the bulk concentration of the reactant and k_S /cm·s^{-1} is the standard heterogeneous rate constant. F and R are the Faraday and gas constants. Reinmuth (18) showed that for an irreversible reaction , the current flowing at the root of the wave is independent of the rate of voltage scan. Thus above equation can be written as

$$i_p = 0.227\, nFAC_o k_s \cdot exp[(\alpha n_\alpha \cdot F/RT)(E_p\text{-}E_i)] \quad (4)$$

where E_i is the potential at the root of the wave. A plot of ln (i_P) vs (E_p-E_i) at different scan rates will give rate constant from the intercept and αn_α from the slope. As αn_α decreases the voltammogram will be more drawn out. The rate constants for oxidation of methanol were calculated for all the catalysts and are given in Table-2. It can be seen that rate constant for Pt/TiO_2-C catalyst is much higher than the others. The rate constant k_s/cm·s^{-1} value of $7.31 \cdot 10^{-5}$ at 298 K for Pt/TiO_2-C is respectively 6.5 and 4.5 times higher than those obtained for Pt/TiO_2 and Pt/C in acidic medium. While on the other hand Pt/Al_2O_3-C have rate constant $2.33 \cdot 10^{-5}$ cm·s^{-1} which is also higher than Pt/C catalysts which are normal commercial catalysts. Here, in Table-2 values of peak current (I_p), specific activity (I_s) and mass activity (I_m) are given for the catalysts Pt/TiO_2-C, Pt/Al_2O_3-C, Pt/TiO_2, Pt/Al_2O_3 and Pt/C. The comparison of all synthesized catalysts Table-2 corroborated that Pt/TiO_2-C catalyst showed 4.5 times better electrocatalytic activity than that of Pt/C , 6.4 times from Pt/TiO_2, 3.1 times from Pt/Al_2O_3-C and 10.1 times from Pt/Al_2O_3. From rate constants it is also shown that composite catalysts are much better than the commercial catalysts. It further shows that catalytic activity of Pt/TiO_2 is near to that of commercial type of catalysts. Thus ceramics alone can also be used as supports which will enormously increase the durability of the polymer electrolyte membrane fuel cells and these fuel cells can also be used for the production of hydrogen.

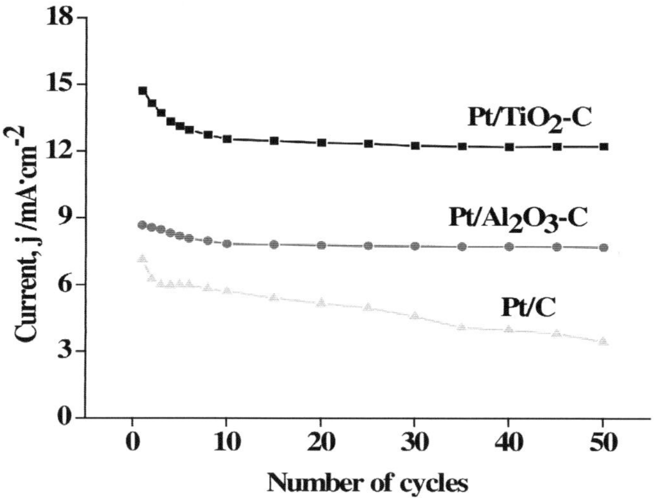

Figure 7. Catalyst durability plots showing specific activity (j) vs. number of cycles in a medium containing 1 MH_2SO_4 and 1 M CH_3OH (scan rate 50 $mV{\cdot}s^{-1}$).

3.5. Durability study of Catalysts.

Figure 7, shows catalyst durability plots when subjected to fifty cycles at scan rate of 50 $mV{\cdot}s^{-1}$. Promoted mechanisms (hydrogen and oxygen spill over effect) has been observed in case of ceramic based nanocomposites, in which CO_{ads} on Pt sites can spill on neighbor ceramic (TiO_2 and Al_2O_3) particles adsorbing OH species. The rapid oxidation will form more free Pt sites covered by CO. So, durability depends on the mechanism of the MO occurring at the catalyst surface. Fig. 7 shows the durability of Pt/TiO_2-C, Pt/Al_2O_3-C and Pt/C catalysts in term of specific activity obtained for 1M CH_3OH + 1M H_2SO_4 at scan rate 50mV.s^{-1} for fifty numbers of cycles in a single run. In all these catalysts peak current decreased for first ten cycles and then up to fifty cycles it stabilized for the composite catalysts and there was decrease for carbon supported catalyst. Although the I_s values increase for all the three catalysts, the plot for Pt/TiO_2-C not only lies much higher but it also becomes horizontal demonstrating a consistent efficiency with time. Decrease in the catalytic efficiency may be attributed to poisoning of the Pt active sites however, presence of TiO_2 and Al_2O_3 in support improved the efficiency and stability of the catalyst. The catalytic activity of carbon supported catalyst showed decrease with increasing the number of cycles, may be due to poisoning of active sites of the catalysts

4. Conclusions

Ceramic-carbon supported composite fuel cell catalysts have been developed with enhanced catalytic activity. Titanium oxide and aluminum oxide inclusion in Carbon support improves the electro catalytic activity and stability of the catalysts because metal oxide in supports exhibit a promising effect on metal and have role in removing poisoning intermediates like CO from the active sites of the catalyst. Durability study has shown that ceramic based composite catalysts are not only efficient but are also durable. Ceramic-based nanocomposites have higher electrochemical surface area and higher catalytic activity when evaluated for exchange current and rate of reaction in acidic medium. Pt/TiO_2-C showed higher performance than Pt/Al_2O_3-C, Pt/TiO_2, Pt/Al_2O_3 and Pt/C catalysts. Pt/TiO_2-C has maximum efficiency in acidic media. Pt/TiO_2 has better efficiency than Pt/C. Pt/TiO_2 has good efficiency and is a potential catalyst to replace the commercial carbon based catalysts which are suffering from corrosion and degradation problems. It can be concluded that the catalyst supports for metal catalyst have an important role in electrodes development for the low temperature fuel cells.

5. References

[1] Apergis, N.; Payne, J. M. Energy Economics. (2011).

[2] Munuswamy, S.; Nakamura, K.; Katta, A. Renew. Energy *36* (2011) 2978–2983.

[3] Haung, S.Y.; Ganesan, P.; Park, S.; Popov, P. N. J. Am. Chem. Soc. 131, 2009, 13898–13899.

[4] Kumar, S. S. M.; Hidyatai, N.; Herrero, J. S.; Irusta, S.; Scott, K. Int. J. Hydrogen Energy, 2011, 36, 5453–5465.

[5] Guo, D. -J.; Qiu, X. -P.; Chen, L. -Q.; Zhu, W. -T.; Carbon, 2009, 47, 1680–1685.

[6] Garzon, F.H., Davey, J.R., Borup, R.L. ECS Meeting Abstracts, Vol.MA ,2005-02,2223

[7] Antonili, E.; Gonzalez, E.R.J. Appl. Catalysis A, 2009, 365, 1–19.

[8] Haung, S. –Y.; Ganesan, P.; Popov, B. N.; J. Appl. Catal. B: Enviromental, 2011, 102 71–77.

[9] Djerad, S., Geiger, B., Schott, F.J.P., Kureti, S., Catal. Commun. 2009, 10, 1103–1106.

[10] Luo, M. F.; Lin, W.R.; Wen, W. H.; Chang, B.W. Surf. Sci. 2008, 602, 3258–3265.

[11] Anotonili, E.; Gonzalez, E.R. J. Appl. Catal. B: Enviromental, 2010, 96, 245–266.

[12] Qian, Y.; Wen, W.; Andock, P.A.; Jiang, Z.; Hakim, N.; Saha, M.S.; Mkherjee S. J. Phys Chem C 2008, 112, 552-557.

[13] Yan, S., Zhang, S., Lin, Y., Liu, G. J. Phys. Chem. C 2011, 115 , 6986–6993.

[14] Barbir, F. 2005, PEM Fuel Cells: Theory and Practice, Elsevier Academic Press, Amsterdam,

[15] Khan, A.S.A., Ahmed, R., Mirza, M.L., Turk. J. Chem. 2008, 32, 743–753.

[16] Nicholson, R.S., and Shain. I. Anal. Chem; 1964, 36, 70,

[17] Bard, A. J., Faulkner, L.R., 2001, Electrochemical methods, Fundamentals and applications, J. Wiley & Sons, New York, pp.231, 251.

[18] Reinmuth, W.H., Anal. Chem. 1960, 32, 1891.

Nano-composite coatings with improved mechanical properties and corrosion resistance by thermal spraying

X C Zhong, Z W Liu, Y S Wu, M T Liu and D C Zeng
School of Materials Science and Engineering, South China University of Technology, Guangzhou 510640, China

E-mail: xczhong@scut.edu.cn

Abstract. This paper reports the synthesis and characterization of nanostructured coatings. To improve the mechanical properties and oxidation resistance of the materials, two new types of nanostructured coatings including CoNiCrAlY-$MoSi_2$ and Ni60-TiB_2 were designed. The nanocrystalline feedstock powders were prepared by high energy ball milling (HEBM). The particle size, morphology and grain size of the feed stocks were investigated. The preparation, microstructure, mechanical properties, and anti-oxidation behavior of the nanostructured CoNiCrAlY-$MoSi_2$ and Ni60-TiB_2 coatings are presented. With a lamellar and compact structure, the optimized nano-composite CoNiCrAlY-$MoSi_2$ coatings is metallurgically bonded with the substrate. It exhibits low porosity, high fracture toughness and excellent thermal shock resistance. The nanostructured Ni60-TiB_2 composite coatings also exhibited better mechanical properties and wear-corrosion resistance than those of its conventional counterpart. This work is expected to play an important role in the preparation and application of high performance nanostructured coatings.

1. Introduction

In the field of surface engineering, thermal spray coating is widely used as overlay to protect components against wear, corrosion and high temperature oxidation. Thermal spraying can provide thicker coatings (>20μm) over a larger area at higher deposition rates as compared to other coating processes such as electroplating, physical and chemical vapor deposition. The microstructure and properties of the coating depend on the thermal and momentum characteristics of the impinging particulate [1], which are determined by both the spraying methodology and the type of feedstock materials employed.

The variations of thermal spraying are distinguished as flame spraying (FS), detonation spraying (DS), wire arc spraying (WAS), plasma spraying (PS), high velocity oxy-fuel coating spraying (HVOF), and et al., among which PS and HVOF are the most popular and effective technique used to prepare metal matrix composite coatings (MMCs), such as WC–Co [2], Cr_3C_2–NiCr [3] or TiB_2–NiCr [4,5], due to its flexibility, low cost and high velocity.

Because of the catastrophic oxidation of WC at high temperature, WC-based coatings are usually used below 450 °C [6,7]. Although Cr_3C_2-based coating can be satisfactorily applied at the temperatures up to 900 °C, its hardness and wear resistance are worse than those of WC-based coating [5]. Thus, development of novel types of coating with excellent properties at high temperature is in need.

TiB_2 has high hardness (34 GPa), low density (4.52 g/cm^3) and excellent oxidation resistance up to 1000 °C [4,5,8,9]. In recent work, TiB_2- containing MMCs such as Fe(Cr)–TiB_2 and Ni(Cr)–TiB_2 coatings have been prepared by HVOF with the excellent abrasive resistance [4,5,8]. However, the high brittleness and low fatigue resistance for TiB_2 have become the main restrictions for the conventional TiB_2-based coatings. In addition, with a high melting point (2030°C), moderate density (6.24g/cm^3) and excellent high temperature oxidation resistance, $MoSi_2$ has also become a promising high temperature protection coating for the superalloys. However, the occurrence of "Pesting" phenomenon in the temperature range from 400 to 600°C and the large difference of thermal expansion coefficient between coating and substrate severely limit its widespread application. To solve above problems, we suggest a new approach to improve the properties of TiB_2- or $MoSi_2$-containing coatings by achieving nanostructure in these films.

In this works, CoNiCrAlY–$MoSi_2$ and Ni60–TiB_2 nano-composite feedstocks were prepared by high energy ball milling. CoNiCrAlY–$MoSi_2$ nano-composite coating was deposited on superalloy GH4169 substrate by plasma spraying technology. Ni60–TiB_2 composite coatings were prepared by HVOF. The microstructure and properties of the two types of coatings were investigated in details.

2. CoNiCrAlY-$MoSi_2$ nano-composite coatings

CoNiCrAlY-$MoSi_2$ with mass ratio of 1:1 powders were produced by mechanically mixing for 12h and the nanocomposite feedstock powders were prepared by ball milling the mechanically mixed powders for 5 to 40 h using a high energy planetary ball miller. The parameters of ball milling include the rotational speed of 300 rpm and ball/powder mass ratio of 15:1. The CoNiCrAlY-$MoSi_2$ nano-composite coatings were sprayed onto the superalloy GH4169 substrates by Praxair 7700 type atmospheric plasma spray syetem. Spraying powder, Spraying distance, main gas (Ar) flow rate, auxiliary ga (He) flow rate and gun moving speed were regulated as follows: 30kW, 70cm, 45 l/min, 5l/min and 500mm/s, respectively[10].

2.1 *Microstructures of the starting powders*

Figure 1a [11] shows the morphology of the mechanical mixing powders, which consists of the CoNiCrAlY powder with smooth surface and spherical and the $MoSi_2$ power with small particle size. When the milling time increases to 30 h, it is difficult to distinguish between CoNiCrAlY particles and $MoSi_2$ particles, because both particles have nearly spherical morphology with similar particles size (Figure 1b). The cross-sectional SEM image of powders milled for 30 h is shown in figure 1c. It can be seen that a large number of dark gray particles ($MoSi_2$) are uniformly and dispersedly distributed in a light-colored matrix (CoNiCrAlY).

Figure 2a shows the XRD patterns of the CoNiCrAlY-$MoSi_2$ powders milled for different times. The results show there is no new phase formed during the entire milling process, suggesting that neither impurity precipitation nor obvious oxidation phenomenon happens after high energy ball milling under argon atmosphere.

Figure 2b shows that the D_{50} (the median diameter D_{50} corresponding to the cumulative distribution value of 50%) value of CoNiCrAlY-$MoSi_2$ mixing powders gradually reduces, suggesting a reduction of powder granularity with the increase of milling time. While the D_{50} value of powders milled for 40 h does not change much, revealing the crushing effect from grinding balls to powders is no longer significant after ball milling for 30 h.

The grain size of $MoSi_2$ phase milled for different times is shown in figure 2c. It is obvious that the grain size of $MoSi_2$ phase approximates to 53 nm only after milling for 5 h. Along with the increase of milling time, the value tends to decrease though the rate of reduction becomes slow, and it seems to be stabilized after milling for 30 h. It is worth noting that although the high energy ball milling can reduce the powder grain sizes to nanoscale, the particle sizes still remain micron scale.

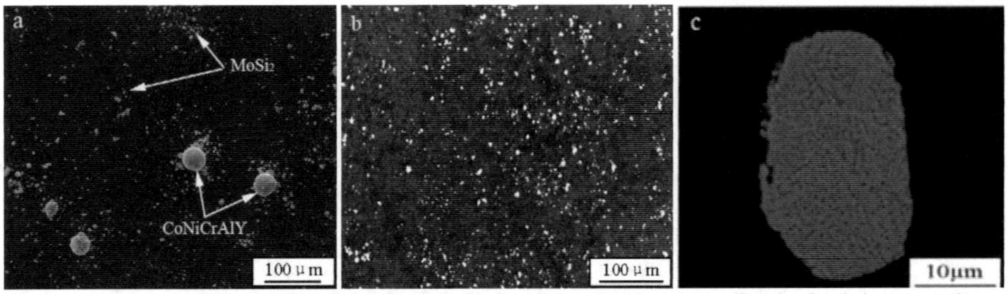

Figure 1. Morphology of CoNiCrAlY-$MoSi_2$ powders [11]: (a) mechanical mixing powders; (b) the powder milled for 30 h; (c) Cross-sectional images of powders milled for 30 h.

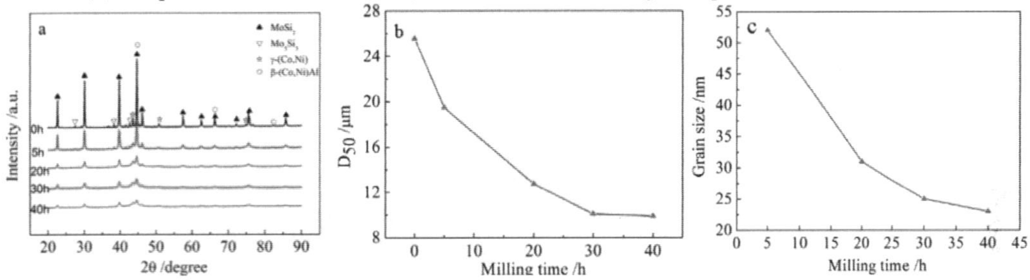

Figure 2. XRD patterns of CoNiCrAlY-$MoSi_2$ powders milled for different time (a); the milling time dependence of the median diameter D_{50} for CoNiCrAlY-$MoSi_2$ powders (b) and the milling time dependence of the grain size for $MoSi_2$ powders (c) [11].

2.2 *Microstructure of as-sprayed coatings*

Figure 3a shows that feedstock powders were overlapped to form a clear "stacking" shape configuration of $MoSi_2$-CoNiCrAlY nanostructured coatings. And nanoscale feedstock powders displayed a good molten effect, but a small amount of semi-molten or unmolten particles is still visible in the local magnification, as arrows pointed in the inset of figure 3a. The presence of these particles allows the coating has a high bonding strength and good wear resistance [12]. The cross-section image of the coating is shown in figure 3b, a typical thermal spray nanostructured coating with unique layered dense structure and less porosity is evident. Figure 3c shows the result of X-ray energy dispersive spectrum (EDS) line scanning of Co element. There is a high Co content in the regions of substrate near the coating, while there is no Co element in the original substrate matrix. This indicates that Co atoms diffuse into the substrate matrix through the interface during the thermal spraying. The diffusion behavior led to a certain metallurgical bonding between coated layer and the substrate, thus it should significantly improve the binding force between them.

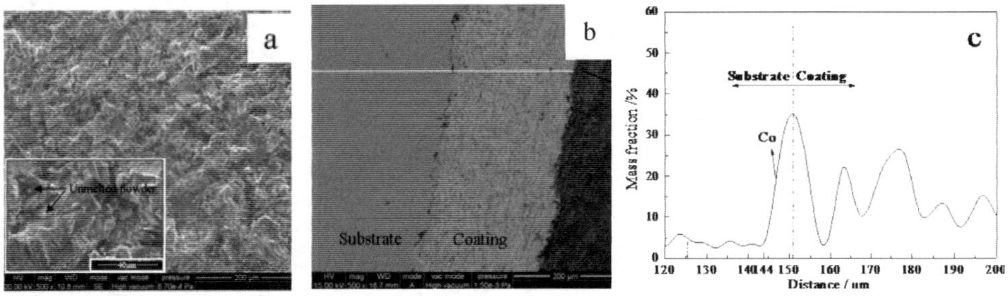

Figure 3. Microstructure of coating and line scanning pattern: (a) surface morphology; (b) cross-section image; (c) distribution spectrum of Cobalt line scanning

2.3 *Mechanical properties of as-sprayed coatings*

2.3.1 *Microhardness and fracture toughness*

With a lamellar and compact structure, the optimized CoNiCrAlY-$MoSi_2$ nanostructured coating is metallurgically bonded with the substrate. It exhibits low porosity (2.31±0.65%) and high fracture toughness (5.27 $MPa{\cdot}m^{1/2}$, 50% higher than that of $MoSi_2$). The average values of microhardness is about 854.4±157$HV_{0.1}$[11].

2.3.2 *Cycle oxidation behavior*

The cycle oxidation kinetics curve of CoNiCrAlY–$MoSi_2$ coating measured under 500°C is displayed in figure 4a. The mass of the coating does not increase continuously in the process of oxidation. In other words, an incubation is needed for the linear increase and the incubation is about 36h. After oxidizing for 48h, the mass gain is substantially constant, the final mass gain is just 0.083mg/cm^2, indicating that the coating has good low-temperature oxidation resistance. Figure 4b shows that there is no "pesting" phenomenon occurring for the composite coating oxidizing at 500°C for 120h, indicating nano-composite coating can improve density and effectively suppress the low-temperature "pesting" phenomenon of $MoSi_2$.

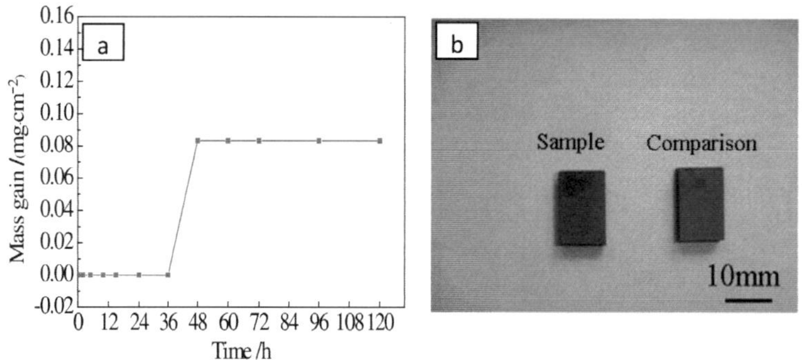

Figure 4. Oxidation kinetics curve of coating (a) and macrograph of coating before and after oxidation (b) [10]. (Note: Comparison sample is the original smaple with coatings before oxidation.)

2.3.3 *Thermal shock behavior*

Figure 5 shows the macrographs of coating surface after different thermal shock cycles. The initial sample surface was gray metallic luster, and there is no apparent unmelted particles. When the thermal shock cycle reach to 13 times, the sample surface begins to darken and a small area is peeling at the corners (figure 5b), which is due to larger stress concentration at the corners of rectangular specimen. Microcracks begin to appear in the center of specimens when the thermal shock testing recycled 18 times (figure 5c), and the crack gradually extended with increasing the number of cycles. When the testing is cycled 22 times, the coating is separated from the substrate and showed an apparent warp (figure 5d). Continued to cycle 27 times, large area peeling appears in the coating along the direction of crack propagation (figure 5e). Therefore, the thermal shock fracture initiation life of the CoNiCrAlY-$MoSi_2$ nanostructured coating is 18 times, thermal shock failure life is 27 times.

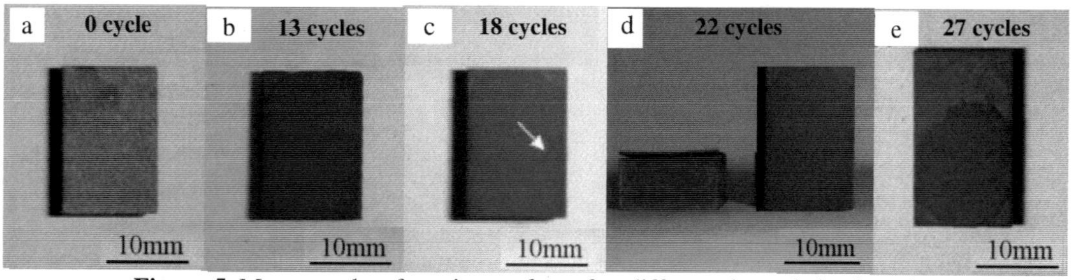

Figure 5. Macrographs of coating surface after different thermal shock cycles

3. Ni60-TiB_2 nano-composite coatings

Ni60–40wt.%TiB_2 powders were prepared by mechanically mixing for 12 h. The nanocomposite feedstock powders were produced by ball milling the mechanically mixed powders for 5 to 40 h using a high energy planetary ball miller under argon atmosphere. The parameters of ball milling include the rotational speed of 400 rpm and ball/powder mass ratio of 10:1. The Ni60–TiB_2 composite coatings were sprayed onto the 45 carbon steel substrates by HVOF thermal spray system (Metallisation MET-JET II). During the HVOF spray, propane and nitrogen were used as the fuel gas and the powder carrier gas, respectively. The spraying parameters were detailed in Ref.[13].

3.1 Microstructures of the starting powders

The SEM micrograph of the mechanically mixed powders (figure 6a) presents the spherical shape of Ni60 spray powders and the small polygonal shape of TiB_2 particles. These TiB_2 particles may reduce the fluidity of mechanically mixed powders. Figure 6b shows the surface morphology of a single milled particle with spherical structure. It is clear that the milled particle is an aggregate composed of many smaller spherical particles with mean size of 50–100 nm. Figure 6c displays the cross-sectional microstructure of the as-polished ball milled powder. Obviously, the ball milled powders are composed of two regions with different contrast which were identified as γ-Ni metallic matrix (white region) and TiB_2 ceramic reinforcing phase (dark gray region). The fine TiB_2 ceramic reinforcing particles evenly dispersed in the γ-Ni metallic matrix.

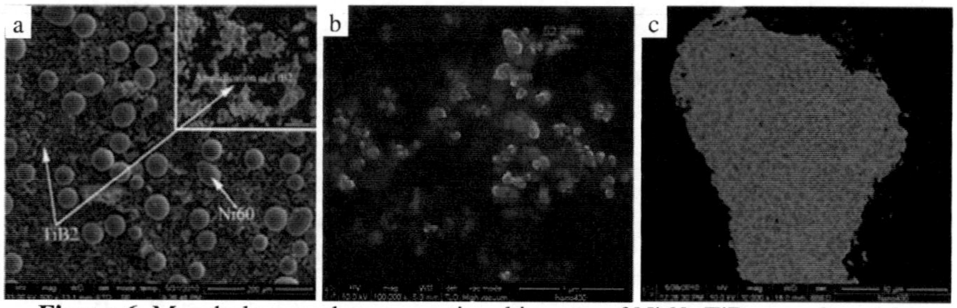

Figure 6. Morphology and cross-sectional images of Ni60–TiB_2 powders: (a) Mechanically mixed powders; (b) and (c) Ball milled powders.

Figure 7 [12] shows the XRD patterns of the Ni60-TiB_2 powders milled for different time and the milling time dependence of the grain size for TiB_2 in Ni60-TiB_2 powders. The results shows that no new phase formed during the whole high-energy ball milling process and the milled powders exist in the form of polycrystalline structure composed of nano-sized grains.

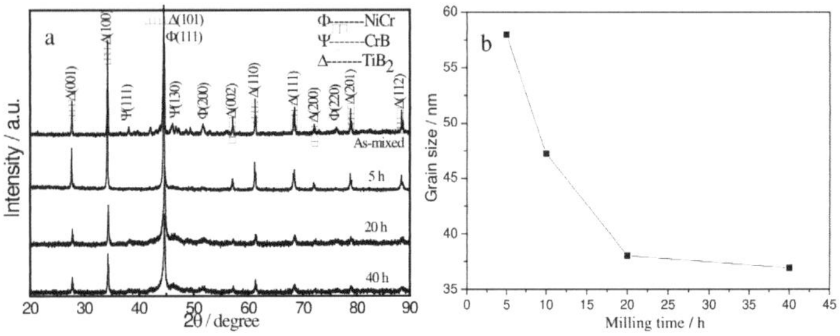

Figure 7. XRD patterns of Ni60-TiB_2 powders milled for different time (a) and the milling time dependence of the grain size for TiB_2 in Ni60-TiB_2 powders (b) [12].

3.2 Microstructure of as-sprayed coatings

Figure 8 shows the microstructure of the Ni60-TiB_2 composite coatings by HVOF with mechanically mixed and ball milled powders, respectively named conventional and nanostructured composite coating. The nanostructured coating exhibits a compact and uniform structure. The fine TiB_2 particles are homogenously distributed in γ-Ni metallic matrix, as shown in figure 8a. However, the microstructure of the conventional coating is not uniform. Micro-sized TiB_2 particles accumulated around the γ-Ni metallic matrix and are separated by the γ-Ni lamellae (figure 8b).

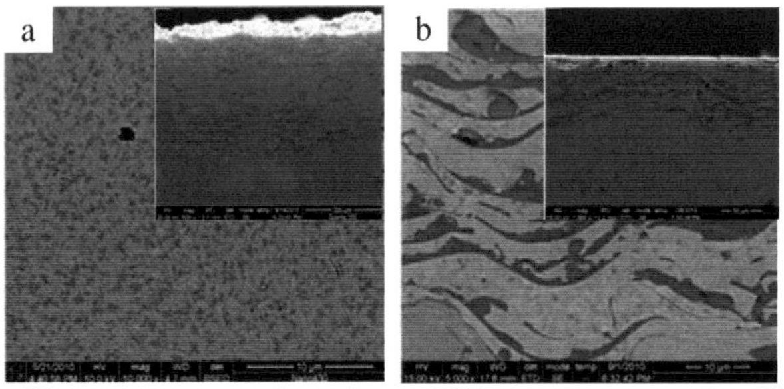

Figure 8. Cross-sectional microstructures of Ni60-TiB_2 composite coatings:
(a) Nanostructured coating;
(b) Conventional coating.

The bright field image in figure 9a presents a uniform, nanoscale microstructure of the composite coating. With an average value of about 45.7 nm, the maximum grain size is around 72.4 nm. As shown in figure 9b, the diffraction patterns are also typical polycrystalline rings which are characterized the crystallographic plane diffractions of γ-Ni and TiB_2. It is evident that the nanostructure characteristics of both metal matrix γ-Ni and ceramic reinforcement particles TiB_2 were successfully preserved during HVOF spraying.

3.3. Mechanical properties of as-sprayed coatings

3.3.1. Microhardness and fracture toughness

Table 1 shows the results of average porosity, roughness, microhardness and fracture toughness of the Ni60-TiB_2 composite coatings. Low porosity was obtained in the nanostructured coating, agreeing well with the microstructure shown in figure 8. The microhardness and fracture toughness of the nanostructured coating are also higher than those of the conventional coating. They are increased by 84.2% and 62%, respectively. The reason for property improvement can be attributed to the microstructural homogenization, grain refine strengthening and dispersion hardening of the nanosized

Figure 9. TEM images of Ni60-TiB_2 nano-composite coating:
(a) bright field image;
(b) diffraction pattern. [12]

TiB_2 particles in the composite coating sprayed by HVOF using ball milled powders.

Table 1. Properties of the Ni60-TiB_2 composite coatings prepared by HVOF

Coating type	Porosity (%)	Microhardness ($HV_{0.3}$)	Fracture toughness ($MPa.m^{1/2}$)	Roughness (μm)
Nanostructured	0.96±0.18	1102±43	3.5±0.2	2.8±0.2
Conventional	1.18±0.25	680±109	1.9±0.4	3.6±0.3

3.3.2. Sliding wear behavior

Average sliding wear coefficients under different applied loads are presented in figure 10 for nanostructured and conventional coatings. Obviously, the coefficient of the nanostructured coatings decrease with increasing applied load, while the sliding wear coefficient of the conventional coating first increases and then decreases with increasing applied load. Figure 11 displays the average volume loss of the two different types of coatings increases with increasing applied loads. It is clear that the nanostructured coating has better dry sliding wear resistance than the conventional coatings.

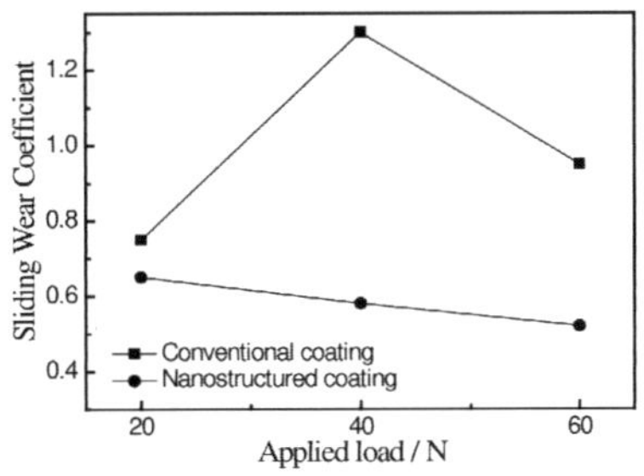

Figure 10. Sliding friction coefficient of Ni60-TiB_2 composite coatings under different applied loads.

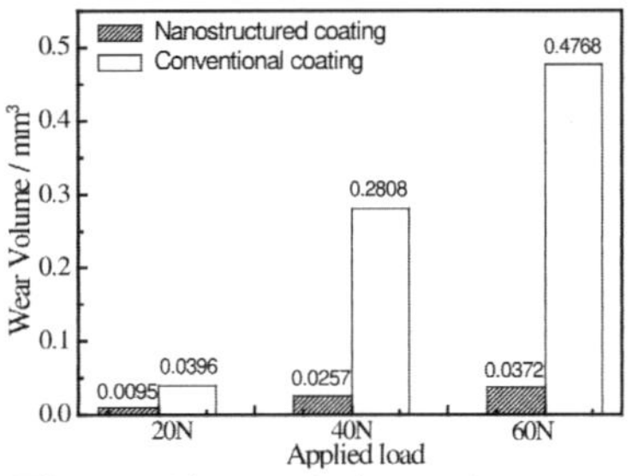

Figure 11. Variations of wear rate of Ni60-TiB_2 composite coatings under different applied loads.

3.3.3. Cycle oxidation behavior

Figure 12 [13,14] presents the cycle oxidation kinetic curves of two types of composite coatings at 600°Cand 800°C in static air. As expected, the cycle oxidation mass gain of the nanostructured composite coating is less than that of conventional composite coating. Both cycle oxidation kinetic behaviors of the composite coatings approximately follow the parabolic law. This indicates that the oxidation process is controlled by the diffusion mechanism. The oxidation rate constant of the nanostructured composite coating is about 1 order of magnitude lower than that of the conventional coating, which indicates that the nanostructured coating has better cycle oxidation resistance than conventional coating.

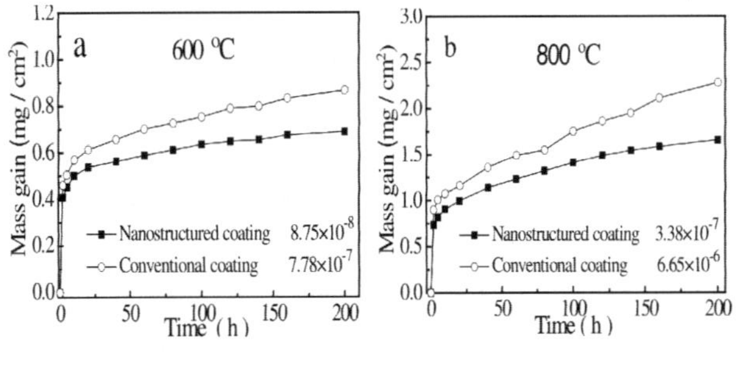

Figure 12. Cycle oxidation kinetics of the Ni60-TiB_2 composite coatings [13,14].

Figure 13 shows the cross-sectional microstructure of the composite coatings oxidized at 800°C. An adherent and thin oxide film with no crack formed on the nanostructured composite coating, which is reasonable for the low oxidation mass gain during cycle oxidation. For the conventional coating, the oxide film is mainly composed of TiO_2 [15] and is thicker than that of the nanostructured coating.

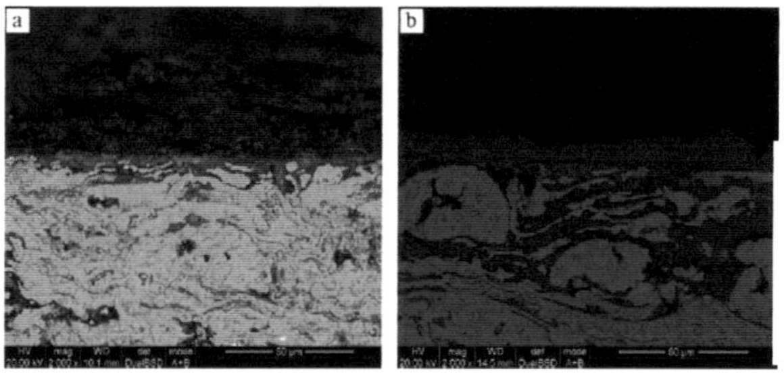

Figure 13. Cross-sectional images of the Ni60-TiB_2 composite coatings [16]: (a) nanostructured composite coating; (b) conventional composite coating.

4. Summary

The synthesis and characterization of the nanostructured feedstock powders prepared by high energy ball milling and the nano-composite coatings prepared by plasma spray and HVOF are introduced. The results show that mechanical milling can be effectively used to synthesize nanostructured powders. The milled powders exist in the form of polycrystalline structure composed of nano-sized grains with the hard phase (nano-$MoSi_2$ or nano-TiB_2) uniformly dispersed in (CoNiCrAlY or NiCr) binder phase matrix. The prepared feedstock exhibits spherical morphology, uniform particle size and good thermal stability, which is expected to be used directly thermal spraying after sieving slightly.

Using nanostructured powders as feedstock powders, CoNiCrAlY-$MoSi_2$ and Ni60-TiB_2 nano-composite coatings have been successfully prepared and have shown improved performance. The as-sprayed nanostructured coatings are metallurgically bonded with the substrate and have compact structure and low porosity. CoNiCrAlY-$MoSi_2$ nanostructured coatings with a lamellar and compact structure exhibit low porosity, high fracture toughness and excellent thermal shock resistance, and the "pesting" phenomenon of $MoSi_2$ does not occur. The Ni60-TiB_2 nanocomposite coating exhibited better mechanical properties and wear-corrosion resistance than those of its conventional counterpart. Available experimental evidence suggests that nanostructured coatings have strong technological potential, and rapid industrial growth is expected over the next decade.

5. Acknowledgements

The authors gratefully acknowledge the financial support provided by the Guangdong Provincial Science and Technology Program (Grant Nos. 2011B010300017, 2010A090200077 and 2009A090100045), and the New Century Excellent Talents in University (Grant No.NCET-11-0156).

6. References

[1] Pawlowski L 1995, *Wiley, England*

[2] Stewart D A, Shipway P H, McCartney D G 1999 *Wear* 225-229 789-798

[3] Mohanty M, Smith R W, De Bonte M, Bonte M D, Celis J P, Lugscheider E 1996 *Wear* 198 251- 266

[4] Lotfi B, Shipway P H, McCartney D G, Edeis H 2003 *Wear* 254 340-349

[5] Horlock A J, McCartney D G, Shipway P H, Wood J V 2002 *Mater. Sci. Eng.* A 336 88-98

[6] Berger L M, Saaro S, Naumann T, Kasparova M, Zahalka F 2008 *J. Therm. Spray Technol.* 17 (3) 395-403

[7] Bjordal M, Bardal E, Rogne T, Eggen T G 1995 *Surf. Coat. Technol.* 70 (2-3) 215-220

[8] Jones M, Horlock A J, Shipway P H, McCartney D G, Wood J V 2001 *Wear* 251 1009-1016

[9] Cutler R A 1991 *Ceramics and Glasses*, ASM International, Ohio

[10] Liu M T, Zhong X C, Liu Z W, Zeng D C, Li Z, Zhang G Q *J. Mater. Eng.* (in pressed)

[11] Liu M T, *Preparation and properties of* $MoSi_2$*-CoNiCrAlY nano-composite coating on GH4169 alloy by plasma spraying* [M] Guangzhou: South China University of Technology. 2013

[12] Wu Y S, *Preparation, wear resistance and corrosion resistance of the nanocrystalline Ni60-*TiB_2 *coating sprayed by high velocity oxygen fuel* [D]. Guangzhou: South China University of Technology. 2011

[13] Wu Y S, *Zeng* D C, Liu Z W, Qiu W Q, Zhong X C, Yu H Y, Li S Z 2011 *Surf. Coat. Technol.* 206 1102-1108

[14] Wu Y S, Qiu W Q, Yu H Y, Zhong X C, Liu Z W, *Zeng* D C, Li S Z 2011 *Chin. J. Mater. Res.* 25(4) 391-398

[15] Wu Y S, Qiu W Q, Yu H Y, Zhong X C, Liu Z W, *Zeng* D C, Li S Z 2011 *Appl. Surf. Science* 257 10224-10232

Synthesis and characterization of Nafion/SiO_2 - MO_x (M= Ti, Zr, W) nanocomposite membranes by sol-gel reaction for fuel cells

Ambreen Shahzadi, Riaz Ahmed and Muhammad Siddiq
Department of Chemistry, Quaid-i-Azam University, Islamabad, Pakistan

E-mail: riaz_ecfc@yahoo.com

Abstract. Development of efficient and durable cation exchange membranes for fuel cells is important particularly a method for inhibiting the volatility of water at higher temperature is a crucial issue. Nafion composite membranes were prepared by impregnation with SiO_2, $SiO_2 - TiO_2$, $SiO_2 - WO_3$, and $SiO_2 - ZrO_2$ by immersing these in-situ sol-gel solution and stirring. The sol-gel solution mixture served to swell the pores of the membrane to maximize the adsorption of the precursor solution. The composite membranes have been characterized by X-ray diffraction, scanning electron microscopy, water uptake, ion exchange capacity and conductivity. XRD studies showed the crystallinity and particle sizes of oxides. FTIR showed the bond formation in oxide mixtures. SEM provides information about the morphology of the particles. Water uptake increased gradually from pure membrane to SiO_2, $SiO_2 - TiO_2$, $SiO_2 - WO_3$, and $SiO_2 - ZrO_2$ particles added membranes and was maximum for $SiO_2 - ZrO_2$ added membrane. Composite membranes exhibited higher water uptake, ion exchange capacity, conductivity and can improve the efficiency and durability of PEM fuel cells considerably.

1. Introduction

Polymer electrolyte membrane fuel cells are very important energy sources particularly in transportation [1-2]. With the increase in temperature efficiency of fuel cells should increase but proton conductivity of mostly used membranes decreases and it is one of the major obstacles to increase the efficiency of fuel cells. Mostly used cation exchange membranes in fuel cells are Nafion membranes [3] Nafion membranes are expensive, unstable at higher temperatures and are proton conducting when soaked in water and their operational temperature is upto 80 ^{0}C – 90 ^{0}C only. It has been reported that gas crossover of nafion also increases with temperature [4]. Considerable efforts are being done to develop membranes for operation at higher temperatures [5]. Higher temperatures have many advantages including faster electrode kinetics and better tolerance to impurities in fuel gas [6-7]. At higher temperature the proton conductivity of Nafion membrane decreases due to dehydration because proton conduction in the membrane depends on the dissociation of protons from SO_3H group in the presence of water [8-9]. Development of nanocomposites by inserting inorganic oxides into Nafion which may retain relative humidity in the membranes even at temperatures at or above 100^{0}C [10-12]. A lot of work has been done to improve the Nafion membranes for use at higher temperatures [13-15]. Incorporation of metal oxides nanoparticles into Nafion to increase its protonic conductivity and operation at higher temperature have been investigated by adding silica, zirconia and titania etc. [5, 15-16]. It has been shown that addition of silica particles by in situ sol-gel reaction enhanced the performance of membrane [17]. It has also been reported that addition of zirconia and titania enhanced the performance of the membrane and could be used at 110^{0}C [18-19].

In this study nanoparticles SiO_2,, SiO_2-ZrO_2, SiO_2-TiO_2 and SiO_2-WO_3 were impregnated onto the nafion membrane via in situ sol-gel technique. The nanocomposite membranes were characterized for water uptake, ion exchange capacity, FTIR, XRD, SEM, and conductivity.

2. Materials and Methods

2.1. Chemicals

Commercially available nafion resins, nafion membrane, Tetraethylorthosilicate ,Tetrabutyl-n-titanate, Zirconium chloride and tungstate were used as received without further treatment.

2.2. Equipment

Fourier transforms infrared (FTIR) spectrometer, Model No, FTSW 300 MX, manufactured by BIO-RAD was used to record Infrared (IR) spectra.

Scanning Electron Microscopy was carried out using JSM 5910, JEOL Japan. X-ray diffraction pattern of all samples were obtained at room temperature by using X-ray Diffractrometer (3040/60 X'pert PRO) using Ni-filtered cu-Kα radiations (40 k V, 30mA). For elemental analysis, energy dispersive X-ray (EDX) spectrometer EDX-720/800 Hs/900 Hs was used. Conductivity was measured using two probe technique.

2.3. Synthetic procedure for SiO_2-MO_x (M=Ti, Zr, W) nanocomposite membranes by in-situ sol gel method

Received Nafion membrane was boiled in 3% H_2O_2 solution for 1 hour. That membrane was rinsed with deionized water at 80^{0}C for 30 minutes. Membrane was immersed in 1 M H_2SO_4 solution for 1 hour at 80^{0}C. Membrane was rinsed with deionized water and was dried in vacuum oven for 6 hours. After the treatment Nafion membrane was dipped into precursor solution of Tetraethyl Orthosilicate: Methanol (3:2 vol%). Membrane was washed with methanol and was dried at 80^{0}C. Using similar protocol SiO_2-MO_x (M=Ti, Zr, W) nanocomposite membranes were synthesized using Tetrabutyl-n-titnate, Zirconium chloride and tungsten solution as precursors followed by same procedure as mentioned for SiO_2-nanocomposite membrane.

2.4. Water uptake measurements

Water uptake values of the Nafion nanocomposite membranes were determined. Different pieces of the nanocomposite membrane were prepared using the following steps: membranes were heated at 105 ^{0}C for 1 h and then at 50 ^{0}C for 24 h in an oven; then, the membranes were cooled down to room temperature weighed immediately and then soaked in deionized water at room temperature for 24 h. The membrane was removed, gently wiped with tissue paper, and weighed again. The water uptake was calculated by following formula: The water uptake (WU%) =W_{wet}-W_{dry} /W_{dry}

2.5. Ion exchange capacity measurements

The IEC of the SPI membrane was measured by the titration method. First dried membrane sample was immersed in 1.0M NaCl solution at 50 ^{0}C for 48 hours to free the H^+ ions (the H^+ ions in the membrane were replaced by Na^+ ions). Then the H+ ions released from the membrane were titrated with 0.01M NaOH using phenolphthalein as indicator. The titrated IEC of the membrane was determined by:

$$IEC = C_{NaOH} . V_{NaOH} / W_P$$

Where V_{NaOH} is the titrated volume of NaOH, C_{NaOH} is the concentration of the NaOH and Wp is the weight of the dried membrane [9].

3. Results and Discussions

3.1. Synthesis of anocomposite membranes

Infiltration of Nafion membranes with SiO_2 and MO_2 has been studied [20]. In this work Tetraethylorthosilicate solution was added to the methanol solution containing Nafion membrane. SO_3^- H^+ group catalyzes the sol-gel reaction and TEOS molecules migrate to the sulfonic acid group clusters. Then the membrane is cured at 100 ^{0}C and condensation of SiOH groups takes place [21].

3.2. FTIR studies of Nafion nanocomposite membrane synthesized by sol gel method

IR spectrum of Nafion membranes showed absorption band at 1195.0cm^{-1} which corresponds to C-F asymmetric stretching in Nafion membrane. Peak at 1142.9 cm^{-1} corresponds to C-F symmetric stretching and at 538.0 cm^{-1} band corresponds to C-F bending [22]. In Nafion /SiO_2 nanocomposite membrane an absorption band at 792.0 cm^{-1} corresponds to Si-O-Si symmetric stretching. Peak at 1026.6cm^{-1} corresponds to S-O symmetric stretching as shown in Fig.1. By observation of nafion/SiO_2-TiO_2 nanocomposite membrane an additional peak was observed at 969.1cm^{-1} which corresponding to Si-O-Ti vibration.1213.6 cm^{-1} corresponds to C-F asymmetric stretching. While peak at 1146.0 cm^{-1} corresponds to C-F symmetric stretching as shown in Fig.2. IR spectrum of Nafion/SiO_2-ZrO_2 nanocomposite membrane showed an additional peak at 919.3cm^{-1} corresponds to Si-O-Zr stretching in Nafion composite membrane. Another absorption peak appeared at 1010.6 cm^{-1} corresponding to C-F stretching as shown in Fig.3. IR spectrum of nafion /SiO_2-WO_3 nanocomposite membrane showed presence of W-O-W vibration group at 726.0cm^{-1}.C-F bending appeared at 558.1 cm^{-1}.Peak at 624.5 cm^{-1} showed C-S stretching.

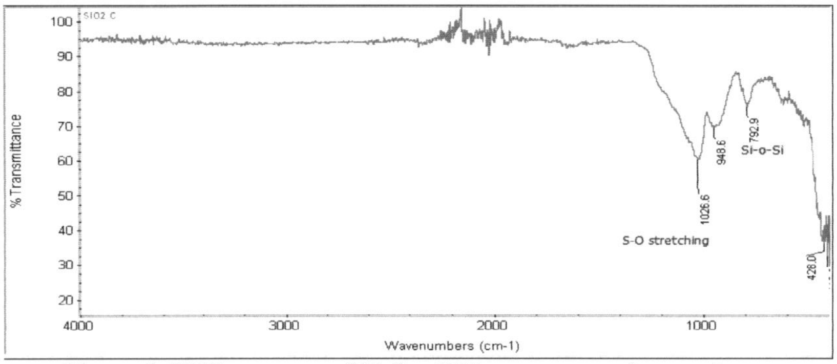

Figure 1. FTIR spectrum of Nafion/SiO_2 nanocomposite membrane

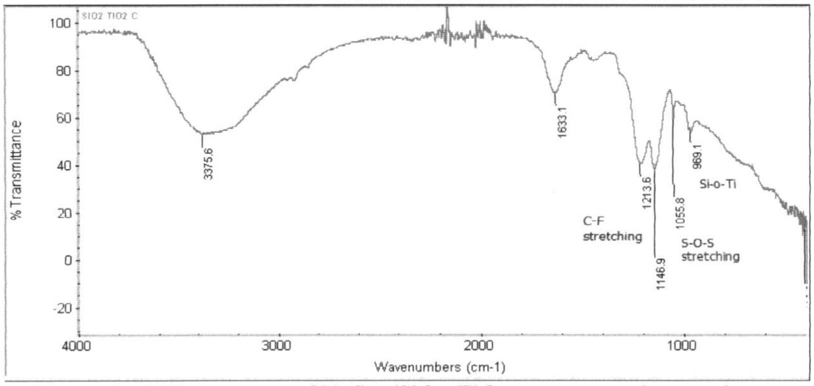

Figure 2. FTIR spectrum of Nafion/SiO_2-TiO_2 nanocomposite membrane

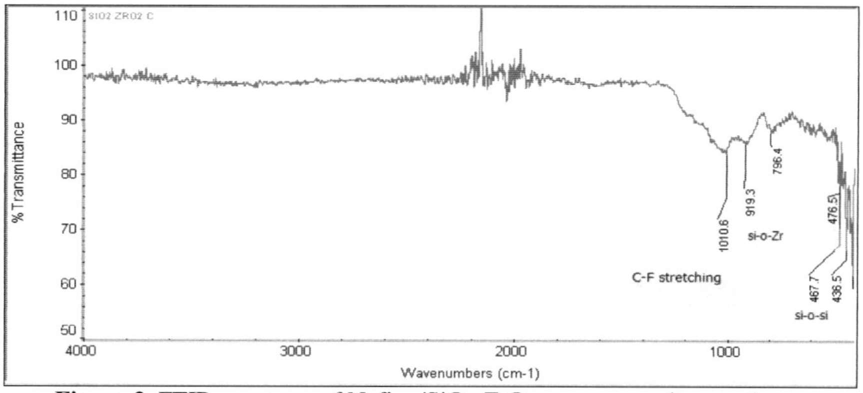

Figure 3. FTIR spectrum of Nafion/SiO_2-ZrO_2 nanocomposite membrane

3.3. XRD analysis

Fig.4 shows the XRD spectra of protonated dry Nafion-115 membrane. It is nearly similar to the XRD reported by [23] for the dry nafion-115 membrane. The broad first peak is indicating the crystalline scattering of the polyfluorocarbon chains in Nafion membranes and their overlapping with the X-ray scattering from the amorphous region of the membrane at lower Bragg angles [24, 25]. XRD spectra for the nanocomposite membrane for Nafion-SiO_2-ZrO_2 is shown in Fig.5. The peaks are nearly the same but first peak is suppressed and the second peak has been enhanced. Intensity of the original peak is lower in the case of nanocomposite membranes. Nearly similar peaks were obtained for all the other nanocomposite membranes. Nanoparticles have not shown their specific peaks which means that the particles are in small size and also may be in amorphous form.

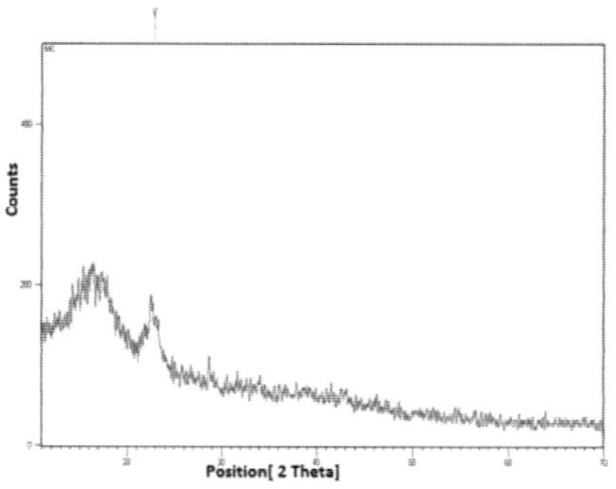

Figure 4. Protonated dry nafion membrane

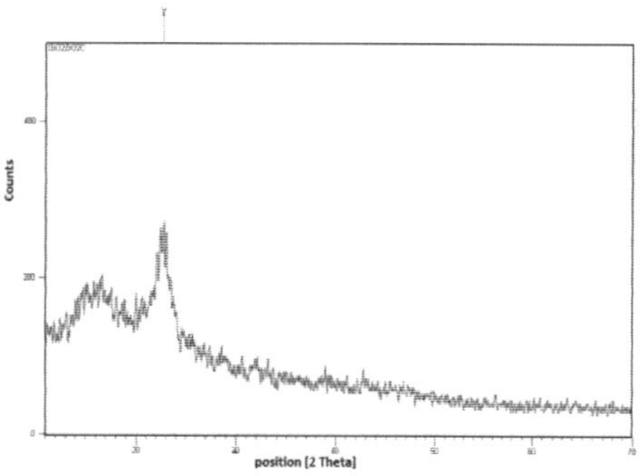

Figure 5. Nafion +SiO_2+ ZrO_2

3.4. SEM and EDX Analysis of nanocomposite membranes

SEM image of pure membrane is shown in Fig.6, and of SiO_2-ZrO_2 nanocomposite membrane (by sol gel method) in fig.7, and of SiO_2-TiO_2 in Fig.8 are shown . It can be seen easily that particles had been adsorbed on the surface of Nafion membrane and are just showing physical interaction with membrane. SEM images of SiO_2-TiO_2 Nafion membrane is indicating the impregnation of binary oxides on surface of Nafion membrane that are also indicating physical interaction and it has been confirmed by FTIR spectrum. The nanoparticles in both the cases can be seen in the micrographs. The particles penetrate into the membrane and have also some interaction with the membrane as confirmed by FTIR but still can be seen in the micrographs. EDX analysis shows that pure Nafion membrane has peaks for carbon fluorine and sulfur only Fig.9, but the EDX for SiO2-ZrO_2 shows clearly the additional peaks of silica and zirconia Fig.10. In Fig.11 is shown the EDX for SiO_2-TiO_2 and it shows the peaks of Titania and silica.

Figure 6. Nafion membrane

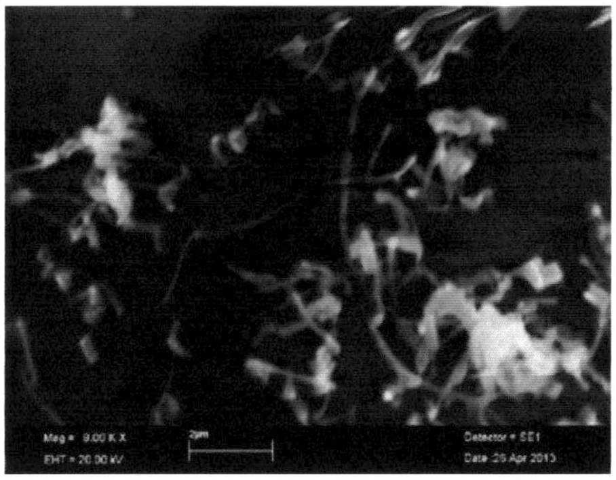

Figure 7. Nafion+SiO_2+ZrO_2

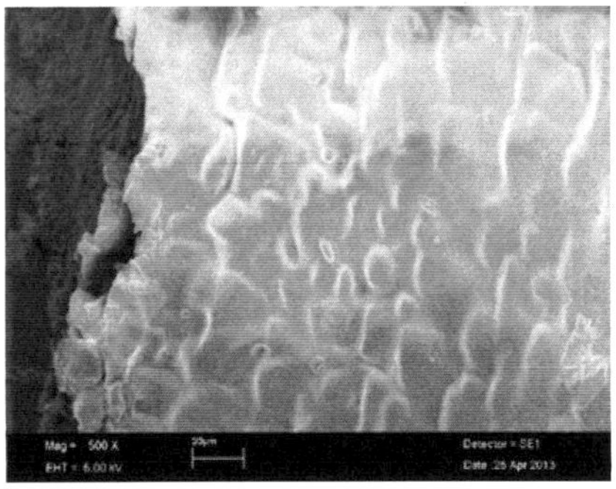

Figure 8. Nafion+SiO_2 +TiO_2

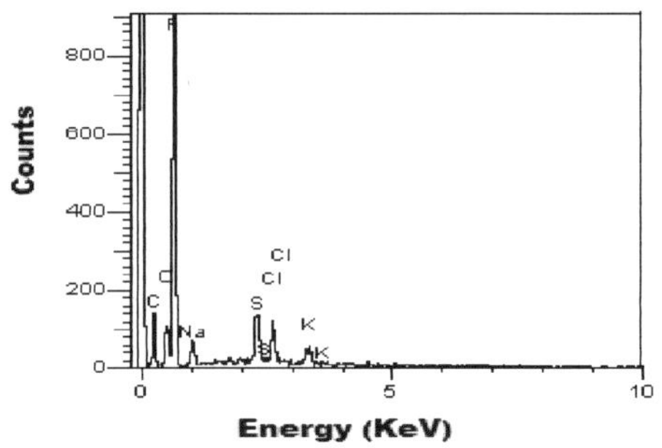

Figure 9. EDX of Nafion membrane

3.5. Water uptake

Water uptake of Nafion/SiO_2 - MO_x (M= Ti, Zr, W) nanocomposite membranes synthesized by sol gel method has been listed in Table 1. By taking under consideration the water uptake values of nanocomposite membranes it was observed that Nafion- SiO_2 membrane showed water uptake 20.7% higher than Nafion membranes. Nafion- SiO_2- WO_3 membrane showed water Uptake 53.7% higher

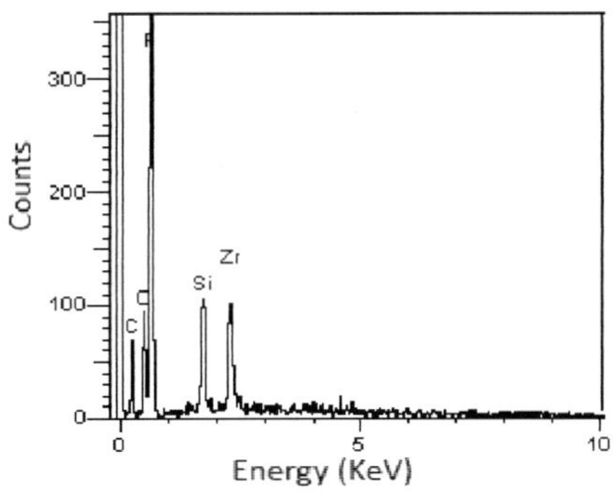

Figure 10. EDX of Nafion $+SiO_2+ ZrO_2$

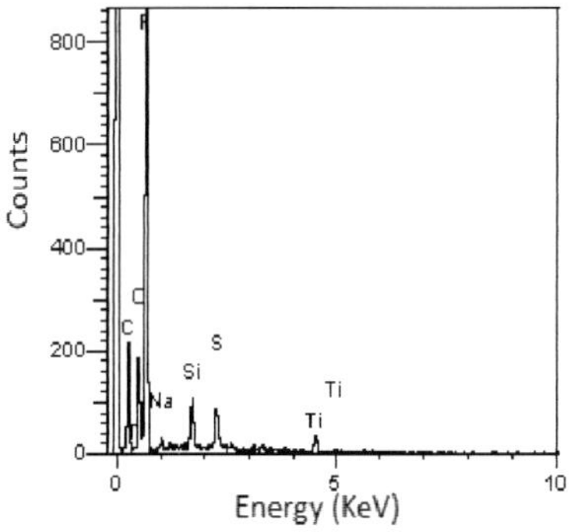

Figure 11. EDX of Nafion/SiO_2-TiO_2

Table 1. Water uptake by nanocomposite membranes

Serial number	Commercially prepared membranes	Water uptake (%)
1	Nafion	43%
2	Nafion-SiO_2	63.7%
3	Nafion- SiO_2-WO_3	96.7%
4	Nafion- SiO_2-TiO_2	109.24%
5	Nafion- SiO_2- ZrO_2	121.52%

than Nafion membrane. Nafion- SiO_2- TiO_2 membrane showed water uptake 66.24% higher than Nafion membrane. Nafion- SiO_2- ZrO_2 membrane showed water uptake 78.52 % higher than Nafion membrane. The increase in water uptake can be due to the hydrophilic nature of the acidic inorganic additives within the pores of Nafion membrane. The increase in acidity and surface areas of nanoparticles may also be responsible for enhancement of water uptake.

With increase in water uptake from silica to tungsten to titania to zirconia nanocomposite there was an increase in acidic strength. On the basis of results obtained, it can be concluded that incorporation of nanosized acidic inorganic with higher surface areas in Nafion membrane, enhances its water uptake properties.That is a key design objective for high temperature membranes. The incorporation of inorganic additives in PEMs is more beneficial at higher temperatures and lower RHs. Higher water uptake and enhanced acidity also increase proton conductivity, which would result into better fuel cell performance under both hot and dry conditions [26].

3.6. Ion exchange capacity

Ion exchange capacity of Nafion/SiO_2 - MO_x (M= Ti, Zr, W) nanocomposite membranes synthesized by sol gel method and by solution casting method has been measured. It was observed that Nafion-SiO_2- ZrO_2 sol–gel was the most acidic membrane. This membrane exhibited highest water uptake followed by Nafion-SiO_2- TiO_2, Nafion-SiO_2- WO_3 and Nafion-SiO_2. Ion exchange measurements are in agreement with the water uptake results [15].

3.7. Conductivity measurements

Resistance of Nafion nanocomposite membranes was measured using resisitivity meter has been shown in Fig.12. It was observed that resistance of nanocomposite membranes was decreased at high temperature. It was also observed that Nafion-SiO_2-ZrO_2 nanocomposite membrane has low resistance as compared to other nanocomposite membranes. Nafion-SiO_2-ZrO_2 membrane has better water uptake, better ion exchange capacity and now better conductivity. Thus addition of SiO_2-ZrO_2 improves the properties of the nafion membrane considerably.

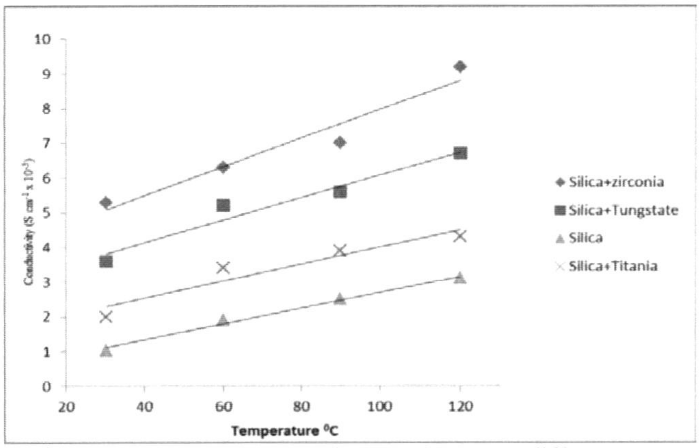

Figure12. Conductivity of composite membranes at different temperatures.

4. Conclusions

Impregnation of nafion membrane with nanoparticles by in situ sol-gel method improved the properties of the membrane considerably. Membranes were impregnated with SiO_2, $SiO_2 - TiO_2$, $SiO_2 - WO_3$, and $SiO_2 - ZrO_2$ particles by sol-gel method. Water uptake by the nanocomposite membranes increased considerably and it was maximum for $SiO_2 - ZrO_2$ impregnated membrane. Thus the membranes will not be dehydrated even at higher temperature and performance of membranes and consequently the efficiency of fuel cells will increase considerably. Ion exchange capacities were also measured and these also showed similar trend. XRD spectra was typical of the dry protonated membrane and absence of peaks for the nanoparticles shows that particles are of small size and of amorphous nature. FTIR spectra of the membranes were taken which showed bond formation between the nanoparticles and membrane which indicates that nanoparticles will not be easily leached out of the membrane. SEM and EDX showed the presence of nanoparticles impregnated onto the membranes. Conductivity of the nanocomposite membranes increased with temperature which shows that the membranes will perform better at higher operational temperatures of the fuel cells. Thus the nanocomposite membranes will improve the efficiency of fuel cells.

5. References

[1] Wang Y, Chen K. S, Mishler J. Cho S. C., Adroher X. C. 2011 Applied Energy 88 981–1007.

[2] Hoogers G(ed.), 2003 Fuel Cell Technology Handbook, CRC Press, London.

[3] Peighambardoust, S.J. Rowshanzamir, S. Amjadi M. 2010 Int. J hyd. energy 35 9349-9384.

[4] Cheng, X. Zhang, J. Tang, Y. Song, C. Shen, J. Song, D. Zhang, J. 2007 J Power Sources, 167, 25-31.

[5] Zhang H. Shen P. K., Chem. Rev. 2012, 112, 2780-2832

[6] Zhang, J. Tang Y. Electrochimica Acta, 2007, 52, 5095-5101.

[7] Pathak A, Panda AB, Tarafdar A, Pramanik P. 2003 J. Indian Chemical Society;80:289-96.

[8] Renaud S, Ameduri B. 2005 Progress in Polymer Science; 30:644-87

[9] Shao P L, Mauritz K. 1996 Journal of Polymer Science, Part-B e Polymer Physics; 34:873-82.

[10] Park K T, Jung U H, Choi D W, Chun K, Lee H M, Kim S H. 2008, Journal of Power Sources 177: 247-53.

[11] So S Y, Kim S C, Lee S Y 2010. Journal of Membrane Science; 360:210-6.

[12] Sacc A, Carbone A, Passalacqua E, D'Epifanio A, Licoccia S, Traversa E, et-al. 2005 Journal of Power Sources;152:16-21.

[13] Jung, G.B. Weng, F.B. Su, A. Wang, J.S. Yu, T.L. Lin, L. Yang, T.F. Chan. S. 2008 Int. J. of Hydrogen Energy, 33, 2413-2417.

[14] Devrim Y , Erkan S , Bac¸ N., Eroglu, I. 2012 Int. J. of hydrogen energy 3 7 1 6 7 4 8-1 6 7 5 8.

[15] Jalani, N. H. Dunn K., Datta R. 2005, Electrochimica Acta 51 , 553–560.

[16] Singaram, V. Hyoung, J.K. Sang, Y.L. 2007, Journal of Power Sources, 167, 325-329.

[17] Jiang, R. Kunz, H.R. Fenton, J.M. 2006, J. Membrane Science, 272, 116-124.

[18] Amjadi, M. Rowshanzamir, S. Peighambardoust, S.J. Sedghi, S. 2012, J. Power Sources, 210, 350-357.

[19] Mohammadi, Gh. Jahanshahi, M. Rahimpour, A. 2013, Int. J. Hydrogen Energy, 38, 9387-9394.

[20] Klein L C, Daiko Y, Aparicio M, Damay F, 2005, Polymer, 46, 4504-4509.

[21] Mauritz K A, 1998, Mater Sci Eng, C6, 121-33.

[22] Cele N. Ray S.S. 2009, Macromolecular Materials and Engineering; 294:719–738.

[23] Yang C. Srinivasan S, Bocarsly A.B., Tulyani, S. Benziger J.B. 2004, J.Membrane Science, 237, 145-161.

[24] Staiti, P. Arico A.S. Baglio, V. Lufrano F, Passalacqua E.E., Antonucci, V.V. 2001, Solid State Ionic; 145:101–107.

[25] Jiang S.H. Xia K.Q. Xu G. Macromolecules 2001; 34:7783–7788.

[26] Jongwon L. Cheol-Woo Y., Keon K. Bull, Korean Chemical Society 2012; 33(5) 1788-1790.

Synthesis of silver nanoparticles by silver salt reduction and its characterization

Muhammad Muzamil, Naveed Khalid, M Danish Aziz and S Aun Abbas
NED University of Engineering & Technology, Karachi, Pakistan

E-mail: muzamil_vir@hotmail.com, naveed_nawab@hotmail.com, mdanishaziz@hotmail.com, aunabbas@live.com

Abstract. The wet chemical method route by metal salt reduction has been used to synthesize nanoparticles, using silver nitrate as an inorganic salt, aldehyde as a reducing agent and amino acid as a catalyst. During the reaction aldehyde oxidizes to carboxylic acid and encapsulates the silver nanoparticles to prevent agglomeration and provide barrier in the growth of particle. The existing work produces particles using lab grade chemical, here the presented work is by using industrial grade chemicals to make the process more cost & time effective. The nano silver powder has been studied for their formation, particle size, shape & compositional analysis using Scanning Electron Microscope (SEM) equipped with EDS. The particles size distributions were analyzed by Laser Particle Analyzer (LPA), structure & morphological analysis using x-ray diffraction (XRD) and Fourier-transform-infrared Spectroscopy (FTIR) confirmed the stabilization of particles by coating of carboxylic group. These studies infer that the particles are mostly spherical in shape and have an average size between 70 to 350 nm.

1. Introduction

In the Synthesis of Silver Nanoparticles, Top down & Bottom up are the two fundamental synthesis approaches for nanoparticles. Top down approach can be regarded as a process starting with a bulk material and then breaking it into small size. Typical processes include attrition, milling, lithography, mechanical & chemical manipulation tools [1].

The main advantages are universality and low cost. Even soft organic matter (e.g. grass) can be ground by first freezing it in liquid nitrogen. The main disadvantages are irregular surface structure, polydispersity of the final particles, and the introduction of many defects. The smaller the particle, the worse is the contamination.

In bottom up approach, organic & inorganic structures are created atom by atom, molecules by molecules, cluster by cluster. This technology synthesizes materials by from atomic or molecular species via by chemical reactions. In this way biological & chemical reactors create conditions for special growth and materials self-assembled together to form nano complex structure [1].

In this synthesization of silver nanoparticles the process is based on bottom up approach, which basically relies on the chemical reduction of metal salts, electrochemical pathways, or the controlled decomposition of metastable organometallic compounds in solution. Control over the growth of the primarily formed nanoclusters depends upon the rate of mixing, two extreme situations yield the desired monodispersity; ultrarapid mixing of very concentrated solutions, and ultraslow mixing of very dilute solutions, and their agglomeration is effected by the use of a variety of stabilizers, in the form of donor ligands, polymers, and surfactants [2].

In the precursors of process, propanal is known to be a weak reducing agent & it has no significant reducing effect on $AgNO_3$, metallic salt, in the absence of a catalyst such as diethyl amine (DEA). Without using DEA, propanal takes quarter of a day to reduce silver salt but in the presence of catalyst reaction will complete in just 10 minutes. The silver precipitates can also be formed without using propanal as in the case of silvering of mirror when ammonia (NH_3) reacts with silver nitrate ($AgNO_3$) and forms ammonium nitrate but the problem is that an excess amount of ammonia (NH_3) cause these silver precipitates to get dissolves in aqueous solution. DEA has a tendency to produce hydroxyl ion which reacts with aldehydes group and forms carboxylic acid which releases an electron in the reaction that will reduce silver ion. The carboxylic acid encapcules the newly formed silver particles stops the further growth of particles & prevent it from agglomeration with other particle.

The formed particles have very high surface energy due to high surface area but energy can also be reduce through surface relaxation, surface restructuring, surface adsorption of chemical & physical species and compositional segregation of impurities [3].

Protective layer plays a vital role in stabilizing a nanoparticle from further agglomeration. There are two types of approaches which are commonly used to protect the particle from agglomeration. Electrostatic stabilization is caused by the Columbic repulsion between particles, caused by the electrical double layer formed by ions adsorbed at the particle surface and the corresponding counter ions [2].

Steric stabilization also called polymeric stabilization which is widely used in stabilization of colloidal dispersions. During the reduction of metal salt formation of carboxylic acid offers two advantages particularly when narrow size distribution is required. It absorbs on the surface of particle provide stabilization & diffusion barrier in the growth of particle.

2. Experimental Work

The starting precursors for synthesis are Silver Nitrate, propanal and diethyl amine. The aqueous solutions with different concentrations of silver nitrate and propanal were mixed together in a calculated volume of deionized water having TDS (total dissolved solid) of 0~5mgs/liter by mixing through magnetic stirrer till the solution became homogenized. A pre calculated milli molar aqueous solution of diethyl amine (DEA) was added to it quickly and stirred vigorously (Conc. of chemical used in synthesis procedure mentioned in Table.1). The color change in solution is observed after 10 to 20 minutes, from black to brown and then finally a dark gray colored higher concentrated Silver Nitrate precipitates were obtained and a yellowish-green colored suspension was obtained for lower concentration of silver nitrate.

The experiment was repeated with argon environment of 0.1~0.2 atm, in a glove box to homogenize the solution and observed that cluster of particles suspended in a solution. After filtering the mixture and washing the residue repeatedly 2 to 3 times with distilled water, the residue was collected and dried in a vacuum furnace at 50°–60°C.

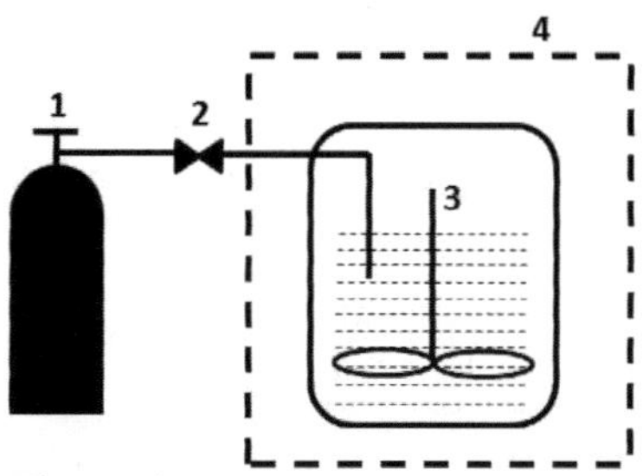

Figure 1. Schematic diagram of the experimental system, (1) Argon cylinder, (2) Control Valve, (3) Magnetic stirrer & (4) Glove Box

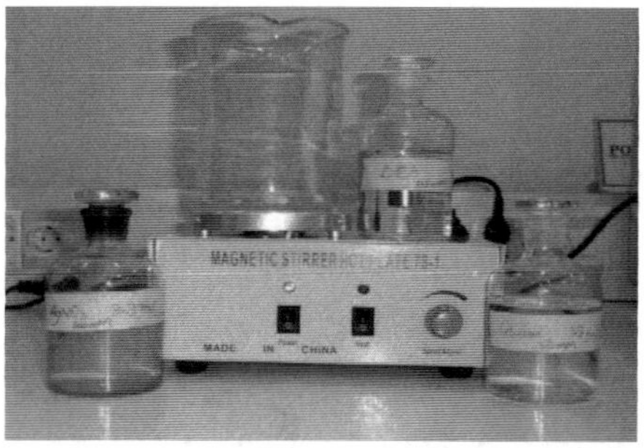

Figure 2. Physical Experimental Setup

This experiment consisted on 2 stages. In the first stage all the mixing of precursors were at ambient conditions of the lab at 30°C by just stirring the solution at 20~30 RPM & in the second stage a controlled environment is provided in the glove box by maintaining the inside working pressure & a jet of argon gas is provided to homogenize & stabilize the particle by removing the trapped oxygen in the solution. Physical arrangement of setup is shown in Figure 2.

Table 1. Concentrations Employed in Experiments for Different Samples

S.No.	Conc. $AgNO_3$ (mM)	Conc. Propanal (mM)	Conc. DEA (mM)	Particle size SEM (±10nm)	Mean (nm)
1	150	60	80	350-300	325
2	150	60	80	325-275	300
3	100	50	85	300-250	275
4	100	50	90	300-225	260
5	75	45	100	200-150	175
6	75	40	100	200-150	175
7	50	40	120	150-100	125
8	50	35	120	150-70	110

The concentration of $AgNO_3$ & DEA has an inverse relation between them. When the concentration of $AgNO_3$ goes on increasing & the concentration of DEA goes of decreasing the outcome is large particles size with wide distribution & vice versa.

3. Results & Discussions

The scanning electron microscope is frequently used for topographical imaging of nanoparticles. These images are typically used for investigating the size, shape, impurities & stabilization of particles. In Sample preparation, nanoparticles attached on aluminum stub using double-sided carbon tape for sample mounting, remove loose sample material by gently knocking the holder to a hard surface. Transfer the stub with sample directly to the SEM for topographical imaging. The characterization was carried in FEI SEM Quanta 200.

SEM images for the *Sample No.7* (Table 1) are shown in *figure 3* at magnification *10,000X* & *figure 4* at magnification at *20,000X*. The secondary electrons results indicate that particles are nearly spherical identical in shape, dispersive & crystalline in structure. The diameters of the polycrystalline particles can be measured by using the scale represented in the each micrograph & the calculated range is written with each below mentioned figure.

Elemental compositional analysis has also been done through EDS (Energy Dispersive Spectroscopy) to find out the purity of silver nanoparticles. Since there is a huge difference in atomic number of oxygen & silver, atomic number contrast of these two element is a primarily tool in identifying the weight percentage. Results are summarized in *Table 2*; shows *98.06%* by weight & *88.11%* by atom silver nanoparticles & just *1.96%* & *11.89%* respectively oxygen were present. Graphically the results are represented in *Graph.1* which is an automated generated graph by built-in software of SEM shows the highest counts of silver. The presence of oxygen shows that some particles got oxidized & agglomerate by forming silver oxide, but it's a very minute quantity, rest of them are still stabilized.

Quantitatively the particle size distribution is being best evaluated through *Laser Particle Analyzer* characterization. To find the distribution, first run the standard solutions of known size distribution, the detectors record the energy scattered & absorbed at particular angle & makes scattering patterns then compares the values of samples with these standard values. BT-9300H Better Size Instrument had been used for particle size analysis.

Eight different samples were being run to find out the mean particle distribution of silver particles (*as shown in Table 1*).The recorded distribution of each individual sample is shown in the histogram

(*Graph 2*), the result are on the basis of volume circulates in the chamber, for different samples assume that circulating pumps took 100% volume to exposed chamber of x-rays.

Figure 3. Sample of Silver Nanoparticle at Magnification 10,000X & Size 100-150nm.

Figure 4. Sample of Silver Nanoparticle at Magnification 20,000X & Size 100-150nm.

Table 2. Compositional Weight Percentage of EDS

Element	Weight %	Atomic %
O K	1.96	11.89
Ag L	98.04	88.11
Totals	100.00	100.00

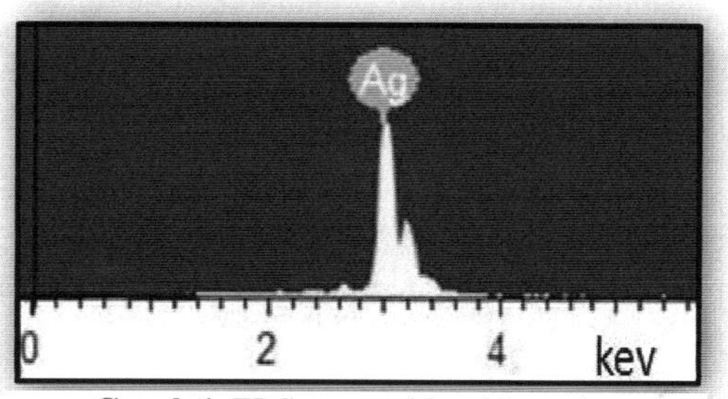

Graph 1. EDS compositional Intensity

For *Sample 1* in Histogram (Graph 2), Result shows out of 100%, 50% in 300-310 nm, 5% in 310-320nm, 30% in 320-330nm, 40% in 330-340nm & 5% in 340-350nm particles of silver were present, same sequence of interpretation will also be followed for the distribution of particles in sample *2,3,4,5,6,7 & 8.*

Fourier Transforms Infra Radiation (FTIR) serves as the molecular fingerprint for the identification organic materials. This technique is extremely useful for the confirmation of the presence of carboxylic acid on the silver nanoparticles as a coating. The obtained FTIR spectrum of the organic material present in the sample compared for best matches with libraries of spectra that have been cataloged for known materials. There are five broad peak can be seen in the *figure 5* of silver particle spectra at *3469, 2370, 1653, 1571, 1506* cm^{-1}.

The first peak at 3469 & 1571 cm^{-1} must be assigned to hydroxyl group [8], because in the oxidation some of the hydroxyl group remains unreactive & also some moisture get absorb on the highly reactive surface of nanoparticles. The peak at 2370cm^{-1} belongs to (CH) stretching band of aldehyde group [9], some solution not succeeded to oxidize to carboxylic acid & the peak at 1550 & 1506cm^{-1} was concluded to be Nitrate from $AgNO_3$, which is starting metal precursor, when we compared the standard FTIR spectra of silver nitrate.

The C=0 absorption is one of the most characteristic in the spectrum. Ketone is considered the root compound while the aldehyde & carboxylic are the derived structures. The characteristic peak of carboxylic acids is present at1653 cm^{-1} is shown in the enlarged spectra of *figure 6*. The carboxylic acid frequency range is 1700-1300 cm^{-1}. They are unique in signature, so far as the hydroxyl group has direct interaction with the carbonyl group, by the formation of a stable hydrogen-bonded structure [7].

In the enlarged spectra *Figure 6*, when this section compared with standard spectra all major & minor peaks should be matched, the carbonyl peaks are present with the hydroxyl group at 1571cm^{-1}, which confirm the presence of carboxylic acid on the silver particle surface.

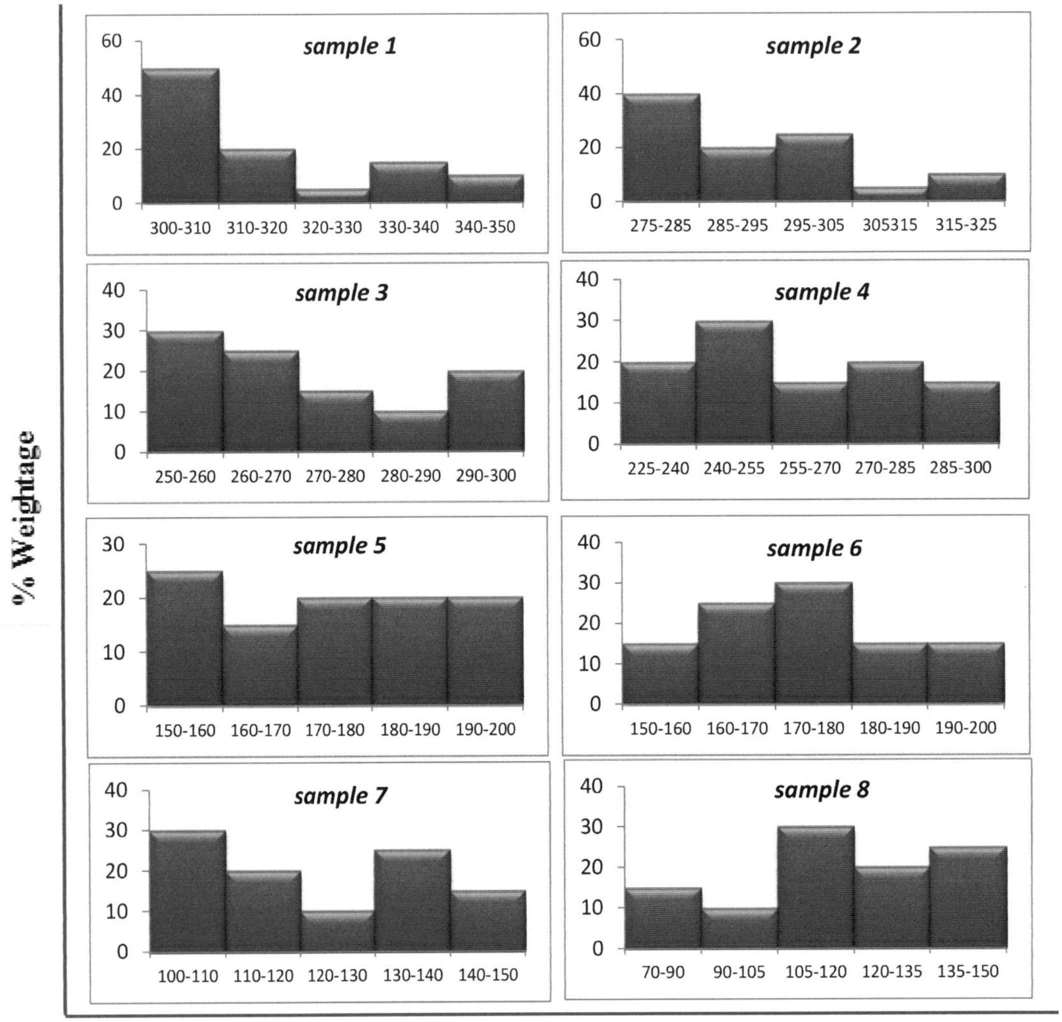

Particles size (nm) distribution

Graph 2. Histogram of Silver Nanoparticles distribution of 8 samples.

The structure of prepared silver nanoparticles has been investigated by X-Ray Diffraction (XRD) analysis. The dry powder of the silver nanoparticles is used for XRD analysis. The diffracted intensities were recorded from 20° to 80° at 2 theta angles. The XRD pattern of the sample, prepared by the presented chemical method is shown in the *Figure 7*.

All diffraction peaks correspond to the characteristic crystalline Face Centered Cubic (FCC) silver lines compared with the standard powder diffraction card of Joint Committee on Powder Diffraction Standards (JCPDS), silver file No. 04–0783. No peaks from other phases are detected, indicating the high purity of silver metal. *Table 3* shows the experimentally obtained X-ray diffraction angle, *d* spacing and the intensity count of Ag specimen.

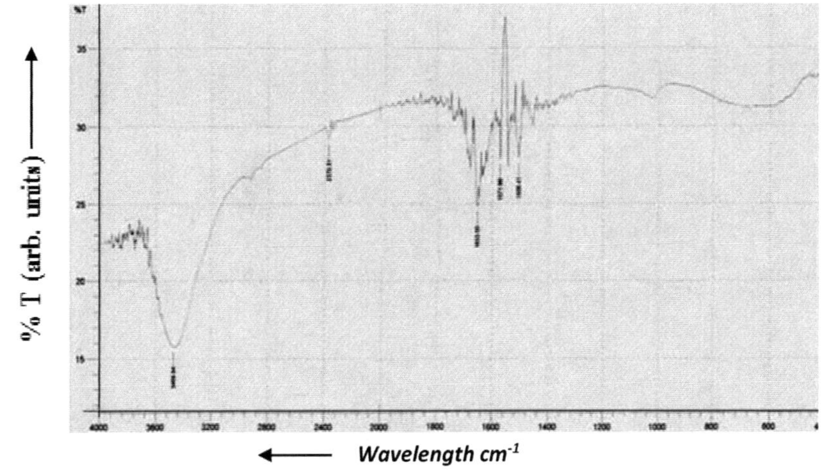

Figure 5. FTIR spectra of Silver Nanoparticles from 4000 to 400cm^{-1} wavelength.

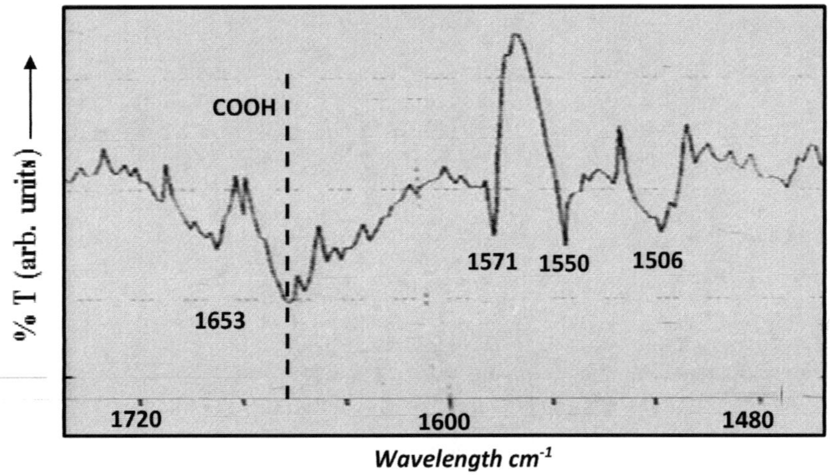

Figure 6. Enlarged FTIR spectra of Silver Nanoparticles capsuled with Carboxylic acid in wavelength range of 1760-1440cm^{-1}.

Table 3. Shows the values observed in XRD spectrum

Observed 2Θ values	d Value Angstrom	Intensity Count	Intensity Percentage
38.331	2.34634	1239	100
44.488	2.03484	500	40.3
64.644	1.44068	372	30
77.56	1.22975	371	29.9
81.751	1.17707	176	14.2

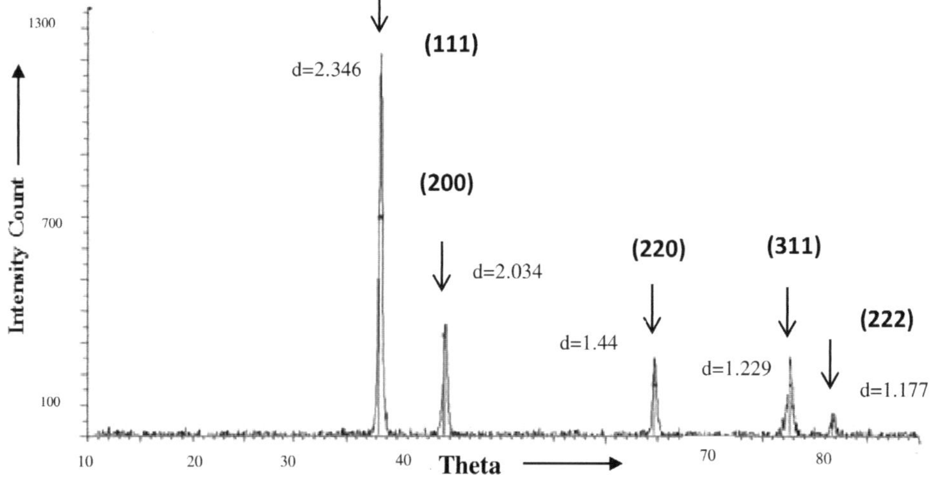

Figure 7. XRD Result showing various peaks.

The Miller indices are calculated by using the observed 2Θ values corresponding to each peak and the details are summarized in *Table 4*. The powdered silver nanoparticles show a cubic structure showing peaks at 2Θ: 38.331 (111), 44.488 (200), 64.644 (220), 77.568 (311) and 81.751 (222). The high intense peak for FFC material is (111), which is observed in the powdered sample. [8]

Table 4. Calculation Procedure for hkl Evaluation

Peak No.	2Θ	$\sin^2\Theta$	1 x $\sin^2\Theta$/ $\sin^2\Theta_{min}$	2 x $\sin^2\Theta$/ $\sin^2\Theta_{min}$	3 x $\sin^2\Theta$/ $\sin^2\Theta_{min}$	$h^2+k^2+l^2$	hkl
1	38.331	0.107	1.00	2.00	3.00	3	111
2	44.488	0.143	1.33	2.67	4.00	4	200
3	64.644	0.285	2.66	5.32	8.00	8	220
4	77.568	0.392	3.66	7.32	11.00	11	311
5	81.751	0.428	3.97	7.947	12.00	12	222

4. Conclusion

The Silver nanoparticles have been synthesized by wet chemical route and are characterized by using SEM, LASER Particle analyzer, FTIR spectrometry and XRD techniques. The mean particle size is found to be 120-325 nm. The concentration of DEA plays a vital role in controlling the size of particles. This work also reveals a quick route to synthesize Silver nanoparticles economically on large scale volumes using industrial grade chemicals and maintaining the purity and characteristics. In future, these Silver Nano Particles can be further analyzed for its heat transfer properties and application in nanofluids and conductive gels.

5. Acknowledgement

The authors are indebted to *Dr. Muhammed Tufail*, Dean faculty of Mechanical & Manufacturing and Chemical & Processing Engineering, *Dr. Mubashir Ali Siddiqui*, Chairman Mechanical Engineering Dept., *Dr. Ashraf Ali*, Chairman Materials Engineering Dept., NED UET, *Mr. Mudassir Ahmed* for technical discussion, *Mr. Sajid Ali Asghar* for SEM work and *Dr. Uzma* Ashiq for FTIR analysis.

6. References

[1] Vijay K Varadan, A Sivathanu Pillai, Debashish Mukherji *Nanoscience and Nanotechnology in Engineering* 2009 (Scientific Publishing Co. Pte Ltd., Singapore).

[2] H. Bönnemann; K. S. Nagabhushana, Chemical Synthesis of Nanoparticles, Max-Planck-InstitutfürKohlenforschung, Mülheim, Germany, 2004, *Encyclopedia of Nanoscience and Nanotechnology*, Electronic Publishing Solutions, (Copyright American Scientific Publishers).

[3] Guozhong Cao and Ying Wang. *Nanostructures and Nanomaterials Synthesis, Properties, and Applications* 2004 Edition 2nd, University of Washington, USA, Louisiana State University, USA (Imperial College Press).

[4] Robert Edward Lee *Scanning Electron Microscopy and x-ray Microanalysis* 1993 (P T R Prentice Hall, Englewood Cliffs, New Jersey 07632).

[5] Nath S. S., Chakdar D. and Gope G.; Synthesis of CdS and ZnS quantum dots and their applications in electronics, (2007), *Nanotrends- A Journal of Nanotechnology and its Application,* 02(03).

[6] Nath S. S., Chakdar D., Gope G., and Avasthi D. K.; Effect of 100 Mev Nickel Ions on Silica Coated ZnS Quantum Dot, 2008, *Journal of Nanoelectronics and Optoelectronics*, 3, 1-4.

[7] John Coates in Encyclopedia of Analytical Chemistry R.A. Meyers (Ed.) 2000 *Interpretation of Infrared Spectra, a Practical Approach.* (John Wiley & Sons Ltd, Chichester). pp. 10815–10837.

[8] Revathi Janardhanan, Murugan Karuppaiah, Neha Hebalkar, Tata Narsinga Rao, Synthesis and surface chemistry of nano silver particles, 2009, *Polyhedron*, Volume 28, issue 12, pp. 2522-2530. ISSN: 0277-5387, *Elsevier Science.*

[9] T. Theivasanthi and M. Alagar Electrolytic Synthesis and Characterizations of Silver Nanopowder, 2012, *Nano Biomedicine and Engineering*, Volume 4, No. 2, Pg. 58-65 ISSN 2150-5578.

[10] Sang Hyun Lee, Kiwhan Sung, Taek-Mo Chung, Sung-Goo Lee, Kyung Dae Min,Sangman Koo, and Chang Gyoun Kim, 2008, Preparation of Silver Nanoparticles and Antibiotic Test of Its Polycarbonate Films Composite, *Journal of Nanoscience and Nanotechnology* Vol.8, 1–4.

[11] Pinaki Segupta, Pradip C Saikia and Prakash C Borthakur, SEM-EDX, 2008, characterization of iron rick kaolinite clay, *Journal of Scientific and Industrial Research*, Vol. 67, pp. 812-818.

[12] Barbara H. Stuart, *Infrared Spectroscopy Fundamentals and Applications* 2004 University of Technology, Sydney, Australia (John Wiley & Sons. Ltd) pp. 72-152.

Synthesis and characterization of Mg-based amorphous alloys and their use for decolorization of Azo dyes

M Iqbal[1] **and W H Wang**[2]
[1]MSG, Physics Division, ATCOP, Islamabad, Pakistan
[2]BMG Group, Institute of Physics, Chinese Academy of Sciences, Beijing, China

E-mail: miqbalchishti@gmail.com, miqbalchishti@yahoo.com

Abstract. Mg-based alloys are light weight and have wide range of applications in the automotive industry. These alloys are widely used because of their very attractive physical and mechanical properties and corrosion resistance. The properties and applications can be further improved by changing the nature of materials from crystalline to amorphous. In this study, melt spun ribbons (MSRs) of $Mg_{70}Zn_{25}Ca_5$ $Mg_{68}Zn_{27}Ca_5$ alloys were prepared by melt spinning technique by using 3-4N pure metals. Characterization of the samples was done by X-ray diffraction (XRD), differential scanning calorimetry (DSC) and energy dispersive x-ray analyzer (EDAX). Microstructural investigations were conducted by using scanning electron microscopy (SEM), atomic force microscopy (AFM) as well as optical and stereo scan microscopy techniques. DSC results showed multistage crystallization. Activation energy was found to be 225 kJ/mol by Kissinger method indicating good thermal stability against crystallization. XRD, DSC, SEM and EDS (energy dispersive spectroscopy) results are agreed very well. In order to study decolorization, the MSRs of $Mg_{70}Zn_{25}Ca_5$ $Mg_{68}Zn_{27}Ca_5$alloys were treated repeatedly with various azo dyes at room temperature. In order to compare the results, MSRs of amorphous Zr- and Ni-based metallic glasses were also treated. Reaction of MSRs with azo dyes results in their decolorization in a few hours. Decolorization of azo dyes takes place by introducing amorphous MSRs which results in breaking the -N=N- bonds that exist in dye contents. It is concluded that Mg-based alloys are useful for paint and dye industries and will be beneficial to control water pollution. Comparison of results showed that Mg-based alloys are more efficient than Zr- and Ni-based amorphous alloys for decolorization of azo dyes.

1. Introduction

Amorphous materials are new class of materials with promising physical and mechanical properties. Mg-based alloys are light weight and have wide range of applications in the automotive and aeronautic industries. Mg alloys have attracted increasing interest in automotive due to their low density, high strength to weight ratio, good damping capacity, diecastability and recycling potential. These alloys are also potential candidate for a hydrogen storage material [1-3]. Amorphous materials have better properties than their crystalline counterparts due to disordered and non-equilibrium structure of metallic glasses. The first report on Mg-based metallic glasses was reported in 1991 [4]. As Mg-based alloys have very attractive physical and thermal properties so these materials are very important and favorite for commercial and industrial applications. Wang et.al reported degradation of various azo dyes using MgZn-based metallic glass powders which favor and broaden the functional applications of metallic glasses (MGs). The efficiency of MgZn-based metallic glasses reported is about 20 times better than their crystalline counter parts [5]. The reason is the non-equilibrium structure of amorphous materials. Azo dyes are typical water pollutants. The degradation efficiency of azo dyes using amorphous materials is thousands times higher than ordinary commercial crystalline Fe powders. Fe-based amorphous alloys are also reported for decolorization of various azo dyes [6-9]. Usually metallic powders of brittle materials such as Mg- and Fe-based alloys are used for decolorization of azo dye experiments because powders have high surface to volume ratio. In the present study, melt spun ribbons (MSR) of $Mg_{70}Zn_{25}Ca_5$ and $Mg_{68}Zn_{27}Ca_5$ alloys are synthesized and used repeatedly. In this manuscript, preliminary results of our investigations on thermal and mechanical properties of $Mg_{70}Zn_{25}Ca_5$ alloy as well as decolorization of azo dyes by using melt spun ribbons (MSRs) are reported. The results reported are part of the joint collaborated projects on "Amorphous materials and their application" between ATCOP, Islamabad, Pakistan and IOP, CAS, Beijing, China.

2. Experimental

Ingots of $Mg_{70}Zn_{25}Ca_5$ alloy were prepared by melting precisely weighted 3-4 N pure elements in quartz tubes using induction furnace under inert gas (5N pure Ar) atmosphere to avoid oxidation. Ca content was kept constant. Melt spun ribbons (MSRs) of 40-50 μm thickness and 5-7 mm width were prepared by melt spinning technique by using single roller Cu wheel at 4000 revolution/minute (rpm). The alloy was produced in two states, firstly fully amorphous and secondly partially (minor) crystalline. Structural studies of the material produced were done by X-ray diffraction (XRD) RIGAKU Diffractometer using Cu $K_{\alpha 1}$ radiation (wavelength λ = 1.54056 Å). Thermal behavior was studied by high and low temperature by using Perkin Elmer 7/Pyris DSC and NETZSCH DSC 404 F3 systems. Mechanical behavior was studied by dynamical mechanical analysis (DMA). Vicker's micro hardness was measured by "Everone hardness tester" under an appropriate load of 100 gmf. In order to conduct decolorization experiments, aqueous solutions of azo dyes were prepared by adding 200 mg of "Direct blue/yellow 2B ($C_{16}H_{11}N_2NaO_4S$)" and Direct Blue 6 ($C_{32}H_{20}N_6Na_4O_{14}S_4$)" azo dye powders in one liter deionized water. In order to conduct decolorization of azo dyes experiments, selected pieces of 0.1g of MSRs were treated with one liter of blue and yellow azo dye solutions independently at room temperature. Used MSRs were retreated repeatedly with fresh azo dyes under same experimental conditions to test their further decolorization ability. Selected pieces of MSRs were examined by stereo scan/optical microscope, scanning electron microscope (SEM LEO 440i) and atomic force microscope (AFM). Samples were analyzed by energy dispersive x-ray analysis (EDAX) technique. The work reported here on amorphous $Mg_{70}Zn_{25}Ca_5$ alloy is presented first time by us at Pakistan.

3. Results and discussion

Figure 1 shows XRD patterns of as cast and MSRs after reaction with blue and yellow azo dyes. Broad bands show amorphous nature of materials. It is very important to note that amorphous nature of MSRs does not change after reaction even after 24h. It means used MSRs can be reused. Reused MSRs were retreated for at least four times but still their nature remains amorphous. Low and high temperature DSC scans for $Mg_{70}Zn_{25}Ca_5$ alloy at various heating rates "r" are shown Figure 2(a, b). Exothermic and endothermic reactions are clear in DSC scans.

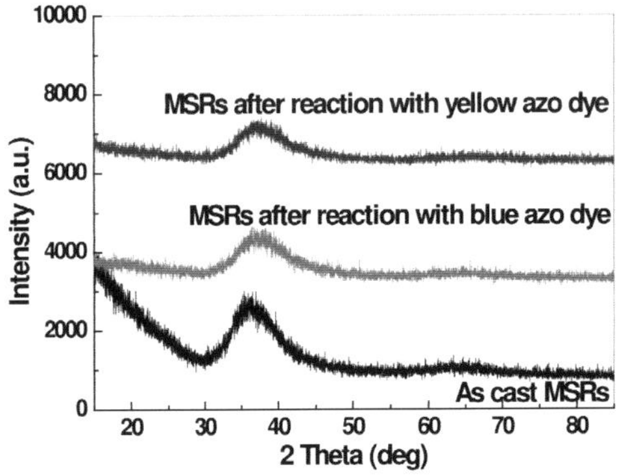

Figure 1 XRD patterns of as cast $Mg_{70}Zn_{25}Ca_5$ MSRs (a) after reaction with blue (b) and yellow (c) azo dyes

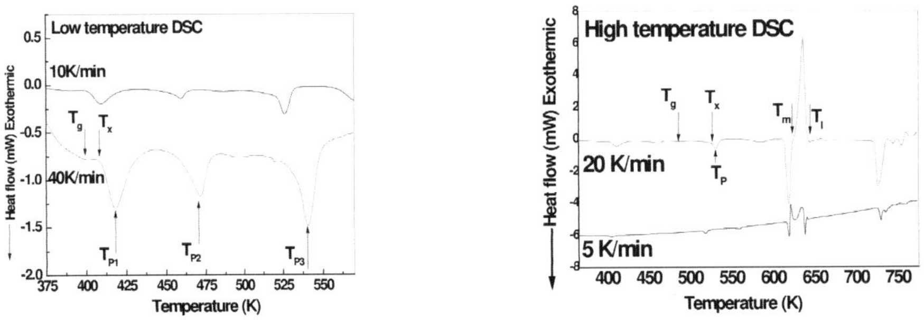

(a) **(b)**

Figure 2(a, b) Low temperature DSC scans of $Mg_{70}Zn_{25}Ca_5$ MSRs at 10 and 40 K/min (a) high temperature DSC at 5 and 20 K/min (b)

DSC scans shows multistage crystallization. Thermal parameters such as glass transition temperature T_g, crystallization temperature T_x, peak temperature T_p, melting temperature T_m and liquidus temperature T_l , a key parameter known as "reduced glass transition temperature" $T_{rg1} = T_g/T_m$ and $T_{rg2} = T_g/T_l$, $\gamma = T_x/(T_l+T_g)$ [10], $\delta = T_x/(T_l-T_g)$ [11], $\beta = T_g * T_x /(T_l-T_x)^2$ [12] and $\omega = T_g/T_x - 2T_g/(T_g+T_l)$ [13] are also deduced and the results are presented in table 1(a). In addition some more thermal parameters [14-18] such as Hruby parameter $K_H = \Delta T_x/(T_m-T_x)$ [13], thermal parameter introduced by Saad and Poulain $K_{SP} = (T_p-T_x)(\Delta T_x)/T_g)$ [15], $K_W = \Delta T_x/T_m$, parameter introduced by Lu and Liu $K_{LL} = T_x/(T_g+T_m)$ [16], $K_1 = T_m-T_g$, $K_2 = \Delta T_x$, $K_3 = T_x/T_m$ and $K_4 = (T_p-T_x)(\Delta T_x)/T_m)$ [17] were also calculated and the results are presented in table 1(b). The width of supercooled liquid region ΔT_x was found to be 40 K well agreed with the results reported by Park et al. [18].

Table 1(a). Thermal parameters (in K) for $Mg_{70}Zn_{25}Ca_5$ alloy

r	T_g	T_x	ΔT_x	T_p	T_m	T_l	T_{rg1}	T_{rg2}	γ	β	δ	ω
5	475	515	40	519	619	626	0.767	0.759	0.467	19.85	3.41	0.060
10	480	520	40	526	-	-	-	-	-	-	-	-
20	486	526	40	532	619	645	0.785	0.753	0.465	18.05	3.31	0.065
40	492	532	40	540	-	-	-	-	-	-	-	-

Table 1(b). Some more thermal parameters (in K) for $Mg_{70}Zn_{25}Ca_5$ alloy

r	K_H	K_W	K_{LL}	K_{SP}	K_1	K_2	K_3	K_4
5	0.385	0.065	0.471	0.34	144	40	0.832	0.29
20	0.430	0.065	0.476	0.50	133	40	0.850	0.39

All the above mentioned derived thermal parameters showed complete thermal behavior of the alloy studied. It shows good thermal stability of the alloy studied suitable for further R & D and applied nature of work. The activation energy "E_{ac}" required for crystallization of an alloy is an important kinetic parameter to determine thermal stability of the amorphous phase. The dependency of crystallization temperature on heating rate was used to determine the associated activation energy (E_{ac}) by means of Kissinger's equation $\ln (r/T_p^2) = -E_{ac}/RT_p$ + constant where "r" is the heating rate, T_p is the peak temperature in DSC scans, E_{ac} is the activation energy, R is the gas constant ~ 8.3145 J/mol K, with slope $-E_{ac}/R = B$, where B is a constant [19]. Kissinger plot is shown in figure 3. The E_{ac} was calculated to be 225 kJ/mol. It indicates high resistance against crystallization. XRD results confirmed that nature of MSRs still remains amorphous after use/reuse of MSRs for at least four times. It is due to high activation energy of the alloy studied.

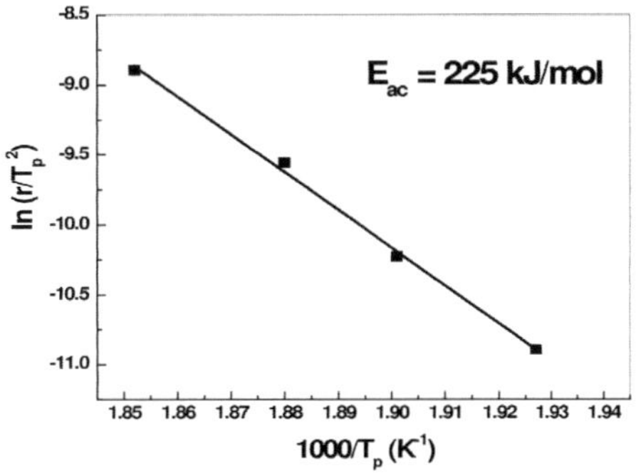

Figure 3 Kissinger plots for calculations of activation energy (E_{ac})

Selected broken pieces of MSRs were examined in AFM. Figure 4(a) shows precipitation and segregation that observed at the surface. EDAX analysis shows that precipitates and second phase particles are rich in Mg and Zn. Oxygen, carbon and some other organic elements are also detected. EDAX technique has the limitation to show actual concentration of the light elements. The surface was fond to be non-uniform and oxidized as clear in figure 4(b). Oxidation was confirmed by EDAX. Cut pieces of MSRs cut down by pair of scissors showed shear bands formations indicating brittle nature of materials. Figure 5(a) shows microstructure of blue dye powders. It looks like similar to flaks. At low magnifications, granular structure was observed. The EDS spectrum of liquid blue azo dye droplets dried over Cu stubs is shown in figure 5(b). Elements such as C, O, Na, S and Cl are identified by EDS. Light elements can't be detected by EDS. MSRs of the alloy were treated with blue and yellow azo dyes at room temperature for 24h. Decolorization achieved in about 6 hours. Figure 6(a) shows original blue and yellow azo dyes and their colorless solution achieved after 6h after treatment with MSRs. The decolorizaiton process was described by a pseudo-first-order kinetic model [6].

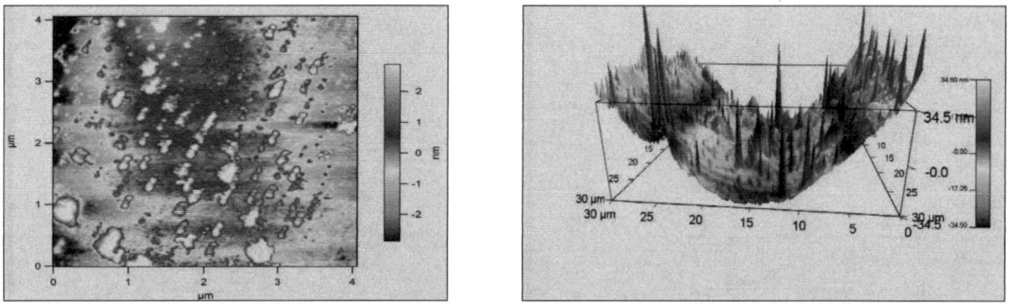

(a) **(b)**

Figure 4(a, b) AFM study of $Mg_{70}Zn_{25}Ca_5$ alloy showing precipitation (a) and non- smooth surface with second phase particles (b)

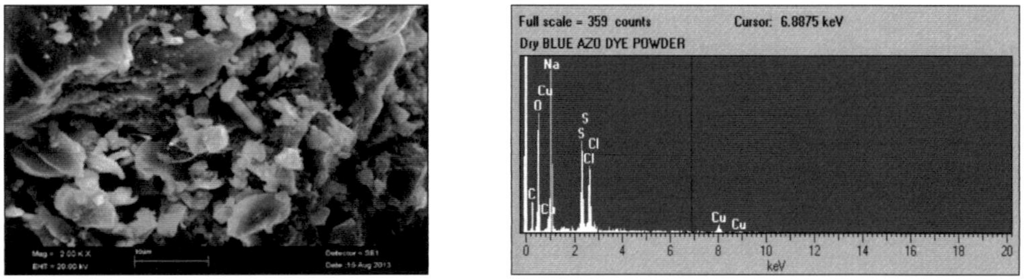

(a) **(b)**

Figure 5(a-c) SEM image of blue dye powders (a) EDS spectrum of blue dye powder (b)

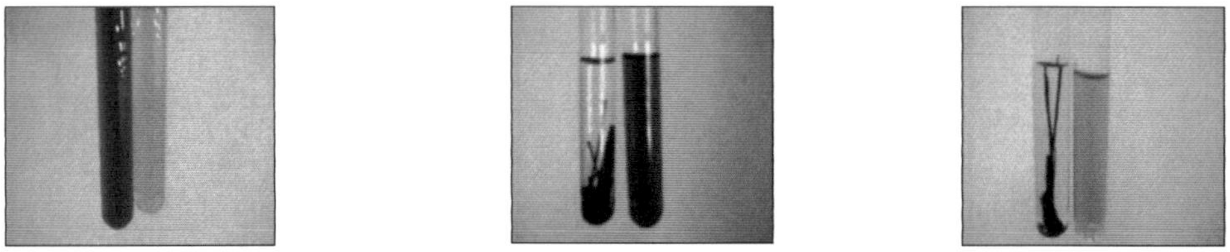

(a) **(b)** **(c)**

Figure 6(a-c) Original blue and yellow liquid dyes before reaction (a) reaction in progress of the liquid azo dyes with MSRs (b, c)

Used MSRs were re-treated repeatedly many times and similar results were obtained in ≤7 h. Precipitates settle down and mostly stuck over the surfaces of MSRs during reaction with blue and yellow azo dye solutions as shown in figure 6(b, c). MSRs change black and become dull due to corrosion. Both original azo dye solutions become colorless after completion of experiments. Comparison shows that reaction with blue solution is faster than yellow azo dye. Black precipitates settle down in the bottom of liquid solutions. After filtration, colorless liquids of azo dyes are obtained. It is believed that these experiments are useful for controlling colors of dyes in paint and textile industry as well as useful for controlling water pollution. The amorphous nature of MSRs after reactions with azo dyes remains unchanged. Therefore, the materials can be reused repeatedly with azo dyes for decolorization experiments. The conjugated structure formed by azo bonds of "-N=N-" type, becomes weaker (and broken down) with the reaction of azo dyes contents with MSRs used (due to their disorder and non-equilibrium structure). As a result decolorization of azo dyes was observed [5, 6]. These experiments open new windows for paint and dye industries and useful to control water pollution, filtration process, paint and textile industries. Figure 7(a, b) shows SEI images taken by SEM of precipitates (PPTs) that stuck over MSRs during reaction with azo dyes. Needle and spherical morphology was observed. After complete decolorization, precipitates settle down at the bottoms and stuck over the surfaces of MSRs. These PPTs are found to be rich in Mg and Zn. Their compositions correspond to $MgZn_2$ and Mg_2Zn. Flakes were also observed by SEM after reaction on dried MSRs as clear in figure 7(b). Carbon, O, Na and S were detected in yellow and blue azo dye powders by EDS.

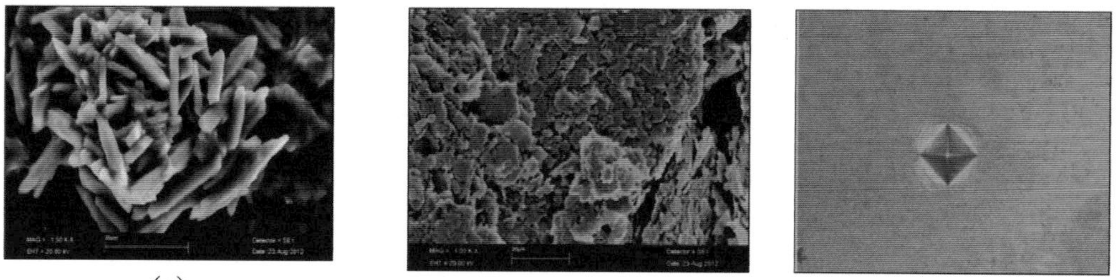

(a) (b) (c)

Figure 7(a-c) SEM images of second phase particles and intermetallic stuck over MSRs after reaction (a) image of dried flakes collected after reaction (b) optical microscopy (OM) of hardness tested MSRs (semicircular shear bands around the Vicker's indent)

Average Vicker's micro hardness of the MSRs of the alloy under study was measured and found to be 259 H_V for as cast samples. Treatment of MSRs with azo dyes results in the enhancement in hardness of alloy. Hardness increases a little after reaction with azo dyes. MSRs treated with blue and yellow azo dyes have hardness of 295 and 270 H_V respectively. Maximum variation in hardness is ± 10 H_V and error in measurement is not more than 2%. The increase in hardness MSRs is due to precipitation hardening by nucleation and depositions of precipitates (deposited in the bottom of transparent filtrate and dipped MSRs) over the surface of MSRs. PPTs are usually hard. MSRs become harder after drying. Majority of the precipitates are thin. Rod shaped PPTs are in minority. The concentration of PPTs is very low as compared to area under study of the MSRs (less than 2 %). Therefore, as a result, XRD scans don't show their contribution prominently (no diffraction peaks) and still shows broad bands in indicating their amorphous nature. Vicker's indents were examined by optical microscopy. Figure 7(c) shows semicircular shear bands over the as cast sample with featureless surface. Indents produce semicircular shear bands in used MSRs indicating a little ductile nature of materials. Similar shear bands were also observed by Cheng et al. [20]. The preset results on mechanical, thermal properties and activation energies are comparable or better than the results reported by Cheng for Mg-based amorphous alloys. Experiments on decolorization of azo dyes by using Zr- and Ni-based MSRs have been done by the authors (unpublished results). Results show that these systems are very resistive and not very efficient to color degradation of yellow and blue azo dyes. On the other hand Mg-based alloys are much faster for decolorization of azo dyes. The reason is fast reaction of alloy contents with atoms of azo dyes. The Zr- and Ni- based materials are mostly corrosion resistant and their reaction with liquid azo dyes is very slow as compared to Mg-based alloys. The results obtained by techniques such as SEM/EDS, DSC and XRD are agreed very well, authentic and confirmed.

4. Conclusions

$Mg_{70}Zn_{25}Ca_5$ alloy was found to be extremely reactive with azo dyes for decolorization of azo dyes. Used materials can be reused at least four times successfully for further treatments with azo dyes for color degradation. Complete transparent and colorless liquids solutions were achieved after a few hours' treatments at normal temperature and pressure. Fully amorphous Mg-based alloys are found to be more effective than partially crystalline materials and more efficient for decolorization than Ni- and Zr- based amorphous systems due to fast reaction of Mg-based alloys. Decolorization takes place due to breakage of the -N=N- bonds exists in-between azo dyes contents by introducing amorphous MgZnCa alloys. This work opens new windows for commercialization and broad based applications of amorphous materials for controlling of water pollution and beneficial for textile and paint/dye industries.

5. Acknowledgements

The first author gratefully acknowledges travel grant and the financial support provided by TWAS. The local hospitality provided by IOP, CAS, Beijing, China is highly appreciated. I am grateful to by Prof. D.Q. Zhao, P. Wen, D.W. Ding, D.P. Wang, Z.G. Zhu, all IOP friends such as Ma Jiang, Bo Huang, Gao, Cui, S.T. Liu and Jiao for providing technical support and best cooperation during my stay at IOP. Thanks to PAEC for permission and support to do this work at IOP, Beijing. Help provided by MSG officials is highly acknowledged. Special thanks to Prof. W.H. Wang, H.Y. Bai and D.Q. Zhao for excellent cooperation. The best cooperation of TWAS Authorities and all concerned official is highly appreciated. The authors are also grateful to Prof. H.F. Zhang and his co-workers for technical support.

6. References

[1] Cai J, Ma G C, Liu Z, Zhang H F and Hu Z Q 2006 *J. Alloy. Compd.* 422 92

[2] Xi X K, Zhao D Q, Pan M X and Wang W H 2004 *J. Non-cryst. Solids* 344 105

[3] Ma H, Ma E and Xu J 2003 *J. Mater. Res.* 18 2228

[4] Inoue A, Kato A, Zhang T, Kim S G and Masumoto T 1991 *Mater. Trans. JIM* 32 609

[5] Wang J Q, Liu Y H, Chen M W, Louzguine-Luzgin D V, Inoue A and Perepezko J H 2012 *Sci. Rep.* 418 1

[6] Zhang C Q, Zhang H F, Lv M Q and Hu Z Q 2010 *J. Non-Cryst. Solid.* 356 1703

[7] Zhang C Q, Zhu Z W, Zhang H F and Hu Z Q 2012 *J. Non-Cryst. Solid.* 358 61

[8] Wang J Q, Liu Y H, Chen M W, Xie G Q, Louzguine-Luzgin D V, Inoue A and Perepezko J H 2012 *Adv. Funct. Mater.* 1

[9] Zhang C Q, Zhu Z W, Zhang H F and Hu Z Q 2012 *J. Environ.Sci.* 24 1021

[10] Lu Z P, Bei H and Liu C T 2007 *Intermetallics* 15 618

[11] Chen Q J, Shen J, Zhang D L, Fan H B, Sun J F and McCartney D G 2006 *Mater. Sci. Eng. A* 433 155

[12] Yuan Z Z, Bao S L, Lu Y, Zhang D P and Yao L 2008 *J. Alloy. Compd.* 459 251

[13] Long Z L, Xie G, Wei H, Peng J, Zhang P and Inoue A 2009 *Mater. Sci. Eng.* A 509 23

[14] Hruby A 1972 *Czech. J. Phys. B* 22 1187

[15] Saad M and Poulain M 1987 *Mater. Sci. Forum* 19 11

[16] Li Y 1999 *J. Mater. Sci. Technol.* 15 97

[17] Nascimento M L F, Souza L A, Ferreira E B and Zanotto E D 2005 *J. Non-Cryst. Solids* 351 3296

[18] Park E S, Kyeong J S and Kim D H 2007 *Mater. Sci. Engg. A* 449-451 225

[19] Kissinger H E 1957 *Ana. Chem.* 29 1702

[20] Cheng Y T, Hung T H, Jang J S C, Tsao C Y A and Lee P Y *Intermetallics* 14 (2006) 866

Effect of interlayer configurations on joint formation in TLP bonding of Ti-6Al-4V to Mg-AZ31

A M Atieh[1] and T I Khan[1, 2]
[1]Department of Mechanical and Manufacturing Engineering, University of Calgary, Alberta, Canada
[2]Qatar Petroleum Chair, Department of Mechanical and Industrial Engineering, University of Qatar, Doha, Qatar

E-mail: anas.m.attieh@gmail.com

Abstract. In this research work, the transient liquid phase (TLP) bonding process was utilized to fabricate joints using thin (20μm) nickel and copper foils placed between two bonding surfaces to help facilitate joint formation. Two joint configurations were investigated, first, Ti-6Al-4V/CuNi/Mg-AZ31 and second, Ti-6Al-4V/NiCu/Mg-AZ31. The effect of bonding time on microstructural developments across the joint and the changes in mechanical properties were studied as a function of bonding temperature and pressure. The bonded specimens were examined by metallographic analysis, scanning electron microscopy (SEM), and X-ray diffraction (XRD). In both cases, intermetallic phase of $CuMg_2$ and Mg_3AlNi_2 was observed inside the joint region. The results show that joint shear strengths for the Ti-6Al-4V/CuNi/Mg-AZ31 setup produce joints with shear strength of 57 MPa compared to 27MPa for joints made using the Ti-6Al-4V/NiCu/Mg-AZ31 layer arrangement.

1. Introduction

The Ti-6Al-4V alloy has a high specific strength, good oxidation resistance, and high elastic modulus which makes it an attractive choice for aerospace, marine, and oil and gas industries [1–3]. The magnesium alloy Mg-AZ31 is a light structural alloy, with a high specific strength, good formability, and recently has been found to be suitable for use in the automotive industry [4–6]. However, differences in the physical properties of these alloys (e.g. melting point for Ti-6Al-4V is 1650° C and 680° C for AZ31) make the joining of these dissimilar alloys a great challenge. Transient Liquid Phase Bonding (TLP) process has been used to join various advanced metals and alloys [7–10]. In previous work [11,12] three eutectic forming interlayers based on nickel, copper and aluminum were tested to enhance joint formation between Ti-6Al-4V and Mg-AZ31. It was found that the nickel foil was the most suitable eutectic former and it gave the highest bond shear strength (39 MPa) compared to the copper and aluminum foils. However, the copper foil produced a eutectic liquid at a temperature below the eutectic temperature for Ni-Mg ($T_{E\ Mg\text{-}Cu}$= 485°C and $T_{E\ Mg\text{-}Ni}$= 512°C)[13,14]. Therefore, in order to take advantage of the lower eutectic temperature of Cu and the better joint quality produced using nickel foils a combination of both foils may produce fewer hard intermetallic phases inside the joint region and give better joint shear strengths. In this study, bonds were made between Ti-6Al-4V and Mg-AZ31 using a combination of nickel and copper as eutectic forming interlayers.

2. Materials and experimental work

Commercially available Ti-6Al-4V and Mg-AZ31 alloys were obtained from William Gregor Ltd (London, UK) and from Goodfellow (Cambridge, UK) respectively. The alloys were received in the form of extruded rods with10 mm diameter; the specimens were cut to 5 mm thick sections, ground and polished down to 1000 grit surface finish followed by rinsing and cleaning in an acetone bath. The chemical composition of these alloys obtained by WDS analysis is shown in Table 1. The Ni and Cu foils are cut into a dimension of 10x10 mm square pieces. Prior to diffusion bonding, the Ti-6Al-4V samples were pickled in diluted HCl (20%HCl-80%distilled water) for 3 minutes, the Ni and Cu foils were pickled for 1 minute to remove surface oxides and degreased in acetone solution. Figure 1 show the XRD spectrum for the as received Ni and Cu foils.

Table 1: Chemical composition (wt%) of Ti-6Al-4V and Mg-AZ31 using the WDS analysis.

Alloy	Al	V	Fe	Zn	Ni	Mn	Mg	Ti
Ti-6Al-4V	6.70	4.44	0.17	0.05	0.02	0.02	0.00	Rem.
Mg-AZ31	2.82	0.01	0.00	1.10	0.00	0.13	Rem.	0.00

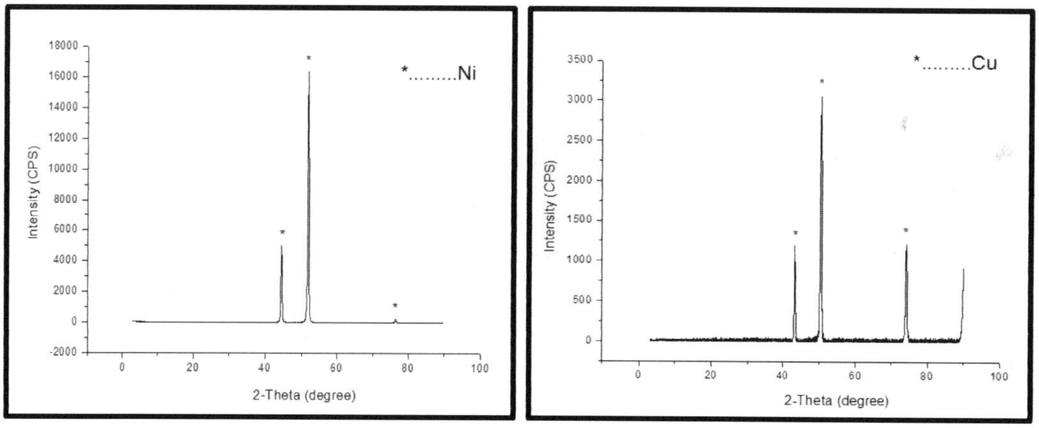

Figure 1. XRD spectra for Ni and Cu foils as received.

TLP bonding was carried out in a vacuum chamber with a pressure of $4X10^{-4}$ Torr (0.053 Pa). Induction heating was used with heating rate of 50 °C /min. A bonding pressure of 0.35 was applied to provide surface to surface contact and to hold the specimens together. For microstructural characterization the bonded samples were prepared by grinding and then polished using 0.25 micron diamond suspension. SEM micrographs were taken by JEOL JXA 8200 electron microprobe analyzer equipped with wavelength dispersive X-ray spectroscopy (WDS) and energy dispersive X-ray spectroscopy (EDS) for composition analysis. A Vickers micro-hardness test was carried out using a Shimadzu mini-load micro-hardness tester. A micro-hardness profile was established with 600 μm on each side and 50 g of load. The shear strength measurements were carried out using a 25kN load-cell tensile testing machine (Tinius Olsen H25KS) at a cross head-speed of 0.5mm/min. X-ray diffraction

(XRD) of fractured surface was performed to study the phase composition and intermetallic compounds formed inside joint region. A Rigaku Multiflex CuK α radiation source with a step size 0.1° and step time 5 s was used at 40 kW and 20 mA over a 2θ scanning range of 10-90° . Joints were made using interlayer configurations Ti-6Al-4V/CuNi/Mg-AZ31 and Ti-6Al-4V/NiCu/Mg-AZ31.

3. Results and discussion

3.1. Microstructural development

The SEM micrograph of joint interface for Ti-6Al-4V/NiCu/Mg-AZ31 made at 515°C bonding temperature, 0.35 MPa bonding pressure for10 min bonding times is shown in figure 2. At bonding times of less than 10 minutes joints could not be made and were seen to fail between Ni and Ti-6Al-4V interface. This was attributed to the absence of eutectic liquid formation at the Ni/Ti-6Al-4V interface although evidence of liquid formation was seen between Cu and Mg. At 10 minutes bonding time, the Cu-Mg eutectic formed and reacted with the Ni interlayer, then it started to dissociate and formed a small amount of liquid.

The SEM micrograph of joint interface for Ti-6Al-4V/CuNi/Mg-AZ31 made at 515°C bonding temperature, 0.35 MPa bonding pressure for 20 min bonding times is shown in figure 3. At bonding times less than 20 minutes all bonds failed to join, and failure was seen at the Cu foil on both sides between Ni-Cu and Cu/Ti-6Al-4Vinterface. This was attributed to the short bonding time preventing the Ni eutectic liquid from wetting and reacting with the Cu foil.

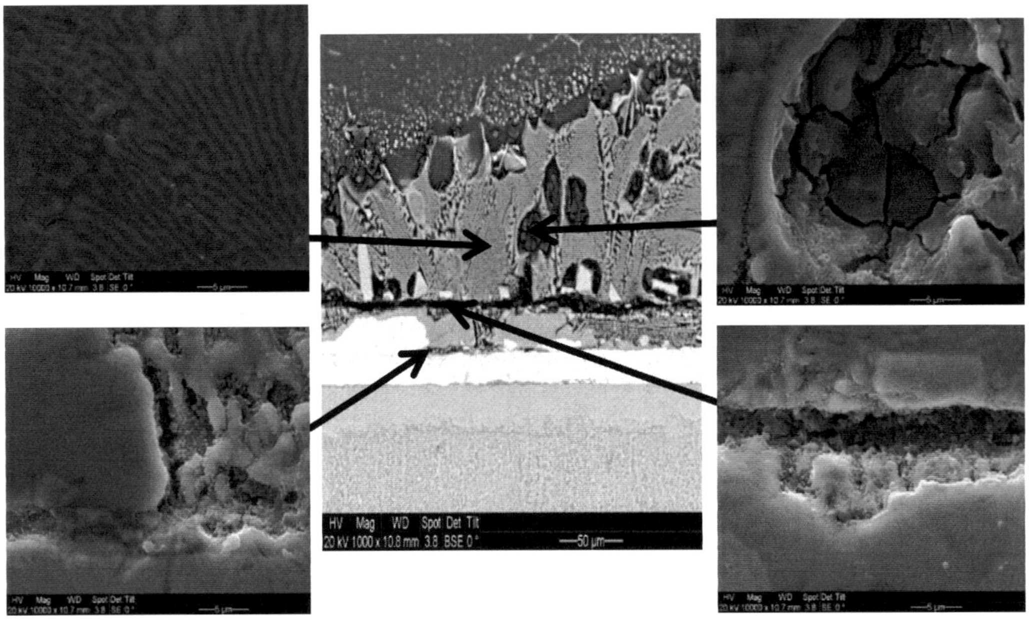

Figure 2. BSE-SEM micrographs of Ti-6Al-4V/NiCu/Mg-AZ31 joint interface after bonding at 515°C and 0.35 MPa for10 min [15].

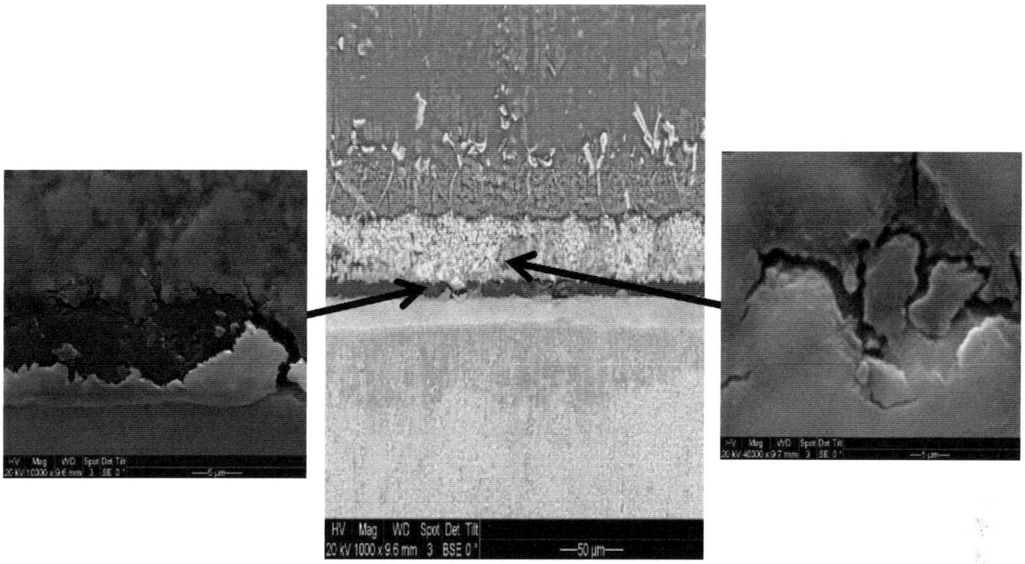

Figure 3. BSE-SEM micrographs of Ti-6Al-4V/CuNi/Mg-AZ31 joint interface after bonding at 515°C and 0.35 MPa for20 min [15].

3.2. Mechanical properties of the joint

Micro-hardness profiles for joint interface as a function of time are shown in figure 4. This profile can give a good indication on the nature of phases and intermetallics formed inside joint region. For the Ti-6Al-4V/NiCu/Mg-AZ31 configuration, as the bonding times increased from 10 to 20 minutes, at bonding time of 10 minutes a smooth transition between the two parent alloys was noticed from 250 VHN to 60 VHN. Comparing the two interlayer configurations at the same bonding time and pressure, it was observed that the Ti-6Al-4V/CuNi/Mg-AZ31 configuration resulted in higher hardness profile on both sides of joint interface towards Ti-6Al-4V and Mg-AZ31. The maximum hardness observed was at the joint center and a value of 308 VHN was recorded. After 200 μm inside Mg-AZ31 the micro-hardness profiles for the 10 minutes Ti-6Al-4V/NiCu/Mg-AZ31 and 20 minutes Ti-6Al-4V/CuNi/Mg-AZ31 measured the same value with an average of 56 VHN.

For joints produced using Ti-6Al-4V/NiCu/Mg-AZ31 setup; the bond shear strength was 27 MPa. However, increasing the bonding time to 20 minutes resulted in a decrease in shear strength to 18 MPa. This decrease in shear strength was attributed to the change in Cu-Mg eutectic concentration before complete reaction with Ni was possible. For a longer bonding time, a higher volume of Cu-Mg eutectic formed leaving less eutectic available to wet the Ni/Ti-6Al-4V interface. However, for Ti-6Al-4V/CuNi/Mg-AZ31 even the bond time less than 20 minutes failed to achieve a joint, the 20 minutes resulted in higher shear strength of 57 MPa which is twice the strength obtained for joints made using the Ti-6Al-4V/NiCu/Mg-AZ31 setup.

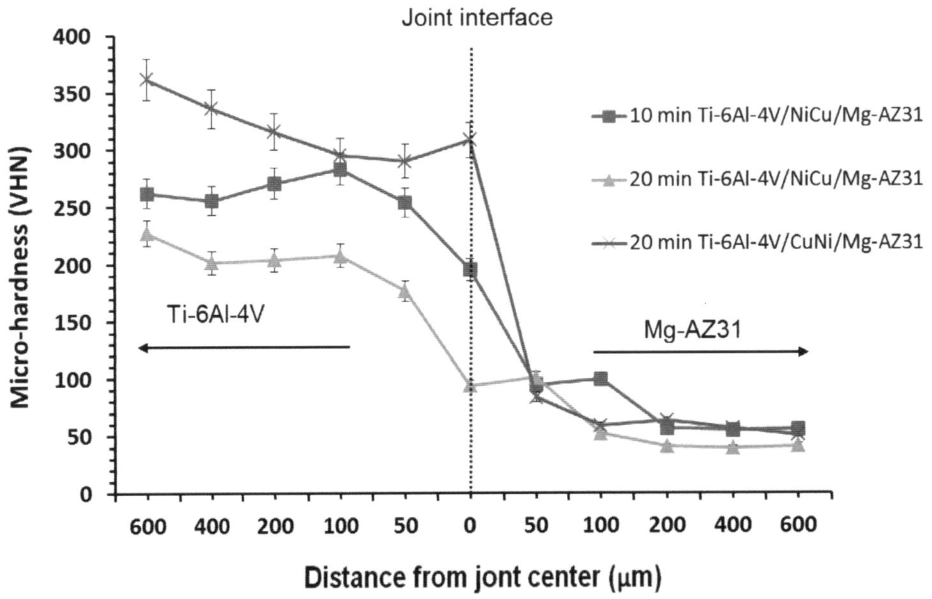

Figure 4. Micro-hardness profile for joint made at 515°C, 0.35 MPa for different bonding times.

3.3. Fracture surface and XRD spectra

The SEM micrograph and XRD spectra achieved from Mg-AZ31 fractured surface are shown in figures 5. The complexity of the system made difficult to identify the phases formed and the intermetallics inside joint region where 7 elements are reacting with each other to form the bonds. However, the application of XRD spectra with WDS and EDS allowed the prediction of phases formed. In the Ti-6Al-4V/NiCu/Mg-AZ31 configuration, at bonding time of 10 minutes 5 major compounds were detected on the fractured surface, only two was common with the fractured surface obtained for 20 minutes which are Mg and Mg_3AlNi_2. Mixed failure mode of brittle and ductile was observed. Those two observations suggested that the fracture path propagated between the reactions layers and contained both Cu and Ni. However, $CuMg_2$ was not detected at 20 minutes bonding time which suggested the decrease in Cu eutectic towards the formation of Ni eutectic as the bonding time progress. For the Ti-6Al-4V/CuNi/Mg-AZ31 setup, the formation of $CuMg_2$ and Mg_3AlNi_2 intermetallics was observed. Furthermore AlTi was detected at the Mg-AZ31 fracture surface, this suggested that the diffusion of Ti towards the joint interface and the fracture propagated close to Ti-6Al-4V interface. However, only ductile fracture mode was observed in the Ti-6Al-4V/CuNi/Mg-AZ31 set up.

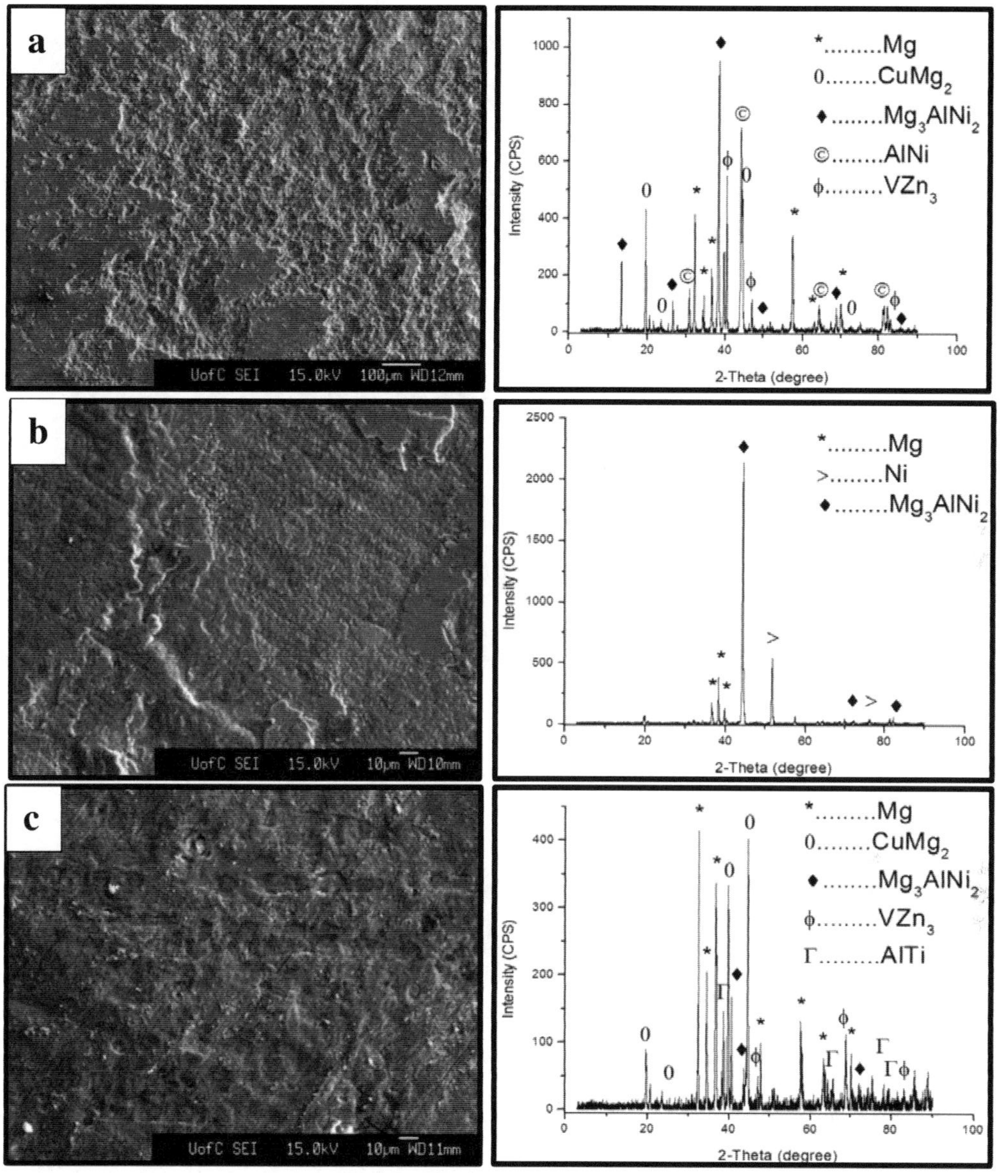

Figure 5. SEM micrograph and XRD spectra for Mg-AZ31 fractured surface for bond made 515°C, 0.35 MPa, and 20 minutes for a) Ti-6Al-4V/NiCu/Mg-AZ31 at 10 minutes b)Ti-6Al-4V/NiCu/Mg-AZ31 at 20 minutes c) Ti-6Al-4V/CuNi/Mg-AZ31 20 minutes.

4. Conclusion

In this study the TLP bonding process was used to join Ti-6Al-4V to Mg-AZ31 using double foil of Ni and Cu. The results showed that different configurations resulted in different minimum bonding times to achieve a joint. When the joint configuration was Ti-6Al-4V/NiCu/Mg-AZ31, a bonding time of 10 minutes was sufficient to achieve a joint. However for the Ti-6Al-4V/CuNi/Mg-AZ31configuration it required 20 minutes to achieve a suitable joint. The maximum shear strength obtained was 57 MPa for the Ti-6Al-4V/CuNi/Mg-AZ31 bonds made at 515°C, 0.35 MPa and 20 minutes. The formation of Mg and Mg_3AlNi_2 were detected at all joints and the fracture mode was mixed brittle-ductile for the Ti-6Al-4V/NiCu/Mg-AZ31 joints and ductile for the Ti-6Al-4V/CuNi/Mg-AZ31 joints.

5. Acknowledgment

The authors would like to acknowledge The German Jordanian University, NSERC Canada for the financial support for this research.

6. References

[1] Schutz R W and Watkins H B 1998 Recent developments in titanium alloy application in the energy industry *Materials Science and Engineering: A* 243 305–15

[2] Gurrapa I 2003 Characterization of titanium alloy Ti-6Al-4V for chemical, marine and industrial applications *Materials Characterization* 51 131– 139

[3] Lütjering G and Williams J C 2007 *Titanium* (Springer Berlin Heidelberg)

[4] Friedrich H E and Mordike B L 2006 *Magnesium Technology Metallurgy, Design Data, and Applications* (Springer Berlin Heidelberg)

[5] Mordike B L and Ebert T 2001 Magnesium Properties, applications, potential *Materials science and engineering A* 302 37–45

[6] Kulekci M K 2007 Magnesium and its alloys applications in automotive industry *The International Journal of Advanced Manufacturing Technology* 39 851–65

[7] Gale W and Butts D 2004 Overview Transient liquid phase bonding *Science and technology of welding and joining* 9 283–300

[8] Sun D, Liu W and Gu X 2004 Transient liquid phase bonding of magnesium alloy (Mg-3Al-1Zn) using copper interlayer *Materials Science and Technology* 20 1595–8

[9] Sun D, Gu X and Liu W 2005 Transient liquid phase bonding of magnesium alloy (Mg-3Al-1Zn) using aluminum interlayer *Materials science and engineering A* 391 29–33

[10] Cooke K O, Khan T I and Oliver G D 2012 Transient liquid phase diffusion bonding Al-6061 using nano-dispersed Ni coatings *Materials & Design* 33 469–75

[11] Atieh A M, Elbatanouny M and Khan T I 2012 Diffusion Bonding of Ti 6Al 4V and AZ31 Using Ni Foil *Proceeding of Materials Science & Technology 2012 Conference & Exhibition, (JASM XIV)* (Pittsburgh, Pennsylvania, USA: MS&T Partner Societies, 2012)

[12] Atieh A M and Khan T I 2013 Transient liquid phase bonding of Ti-6Al-4V and Mg-AZ31 using eutectic forming interlayers *Proceeding of the 24th Canadian Congress of Applied Mechanics (CANCAM)* (Saskatoon, Saskatchewan, Canada)

[13] Y. Z and A. C Y 1993 Thermodynamic Calculation of the Mg-Cu Phase Diagram *Z. Metallkd.* 84 662–7

[14] Okamoto H 2007 Mg-Ni (Magnesium-Nickel) Supplemental Literature Review: Section III *Journal of Phase Equilibria and Diffusion* 28 303–303

[15] Atieh A M and Khan T I 2013 Transient liquid phase (TLP) bonding of Mg-AZ31 and Ti-6Al-4V using Ni and Cu sandwich foils *Journal of Science and Technology of Welding and Joining, submitted paper*

IOP Conf. Series: Materials Science and Engineering **60** (2014) 012037 doi:10.1088/1757-899X/60/1/012037

Effect of V addition to Al and Al grain refined by Ti on chemical corrosion rates in NaOH solution with and without inhibitor at different temperatures

Adnan I O Zaid[1], Ahmad M Al-Haj-Ali[2] and G T A Allawi[3]
[1]Industrial Engineering Department, University of Jordan, Amman, Jordan
[2,3]Chemical Engineering Department, University of Jordan, Amman, Jordan

E-mail: a_kilani43@hotmail.com, ahmad.ay55@gmail.com, ghunaallawi@yahoo.com

Abstract. The paper presents the effect of vanadium addition to aluminum grain refined by titanium, in the range from 0.005 wt % to 0.236 wt %, on corrosion rate of these micro alloys in caustic soda (NaOH) solution at different temperatures. The corrosion rate decreased by the addition of any percent of vanadium at 25^{0}C compared to pure Al. The corrosion rate increased with increase of solution temperature from 25 to 40 and 60 ^{0}C at any percent of vanadium addition. However, addition of vanadium reduced the corrosion rate at these temperatures at most wt.% of V addition. The maximum achieved reduction in corrosion rate due to vanadium addition was 40 % at 0.148 wt % and 60^{0}C. The addition of 3 % potassium dichromate as corrosion inhibitor to NaOH solution reduced the average corrosion rates at any wt.% of vanadium. The inhibition efficiency ranged from 44- 58% at 25 ^{0}C and from 97-99% at 40 and 60^{0}C.

1. Introduction

Aluminum (Al), as well as its micro alloys, has been extensively applied in industry due to their particular properties such as high strength-to-weight ratio, good appearance and corrosion resistance [1-2]. Aluminum and its alloys solidify in a coarse columnar structure of large grain size [3], which tends to reduce their mechanical strength and surface quality, therefore they are grain refined by different rare earth elements such as titanium (Ti) or Ti plus boron (B) to produce fine and equiaxed homogeneous structure, to enhance their mechanical properties and surface quality [4]. The literature on the effect of addition of different grain refiners to Al and its alloys to improve their metallurgical structure and their mechanical properties, e.g. hardness, fracture toughness and fatigue strength is voluminous. However, little research work is reported on the effect of these grain refiners on their chemical corrosion resistance [5]. Vanadium (V) enhances the grain refining efficiency of the Al micro alloys when added at a weight percentage of 0.1% or more, i.e., above the peritectic limit on the Al-V phase diagram [4]. Vanadium addition was also found to improves hardness, mechanical strength and surface quality of Al and its alloys [6]. A wide spectrum of corrosion problems are encountered in industry as a result of combination of materials, environments and service conditions. Corrosion may not have a deleterious effect on a material immediately but it affects its mechanical strength, physical appearance and it may lead to serious operational problems in future [7]. Corrosion attack on Al surfaces is usually quite obvious since the products of corrosion are white and generally more voluminous than the original base metal taking the forms of general etching, pitting or roughness of metal surfaces [8,9]. In highly alkaline solutions, e.g. sodium hydroxide (NaOH), the dissolution process occurs easily and fast resulting in hydrogen gas (H_2) evolution [10]. First the reaction begins with the dissolution of the protective oxide film (hydration process at the film) and as a result hydroxide ions (OH^-) are formed:

$$2H_2O + 2e^- \rightarrow H_2 + 2OH^- \qquad (1)$$

Then aluminum reacts with hydroxide ions in alkaline solution as follows:

$$Al + 4OH^- \rightarrow Al(OH)^-_4 + 3e^- \qquad (2)$$

The overall main reaction taking place can be represented by the following equation:

$$2Al + 2OH^- + 6H_2O \rightarrow 2Al(OH)^-_4 + 3H_2 \qquad (3)$$

Potassium dichromate ($K_2Cr_2O_7$) is a common inorganic chemical reagent; it is a crystalline ionic solid with a very bright, red-orange color most commonly used as an oxidizing agent in various laboratory and industrial applications, [11]. As with all hexavalent chromium (Cr) compounds (chemical compounds that contain this element in the +6 oxidation state), dissolve as stable complexes in water, transport easily, and adsorb on oxide surfaces. The octahedral, d3, trivalent compounds of Cr form very stable inert oxides; $K_2Cr_2O_7$ is a mild oxidizer, the reaction between aluminum and $K_2Cr_2O_7$ in alkaline media is given in the following equation:

$$Al + Cr_2O_7^{-2} + 7H_2O + 3e^- \rightarrow Al(OH)^-_4 + 2Cr^{+3} + 10OH^- \qquad (4)$$

2. Materials and Experimental Procedures

2.1 Materials

Commercially pure Al of 99.8% having the chemical composition shown in Table1, high purity titanium and vanadium were used. Graphite crucibles were used for melting and graphite rods were used for stirring.

2.2 Preparation of the maser and binary alloys

The binary Al-Ti and Al-V master alloys were laboratory prepared and used for preparing the five different micro alloys which are used throughout this work are shown in table2.

The grain refinement process started by manufacturing the binary Al-Ti and Al-V master alloys by adding the calculated amount of pure Ti or V to the molten Al in the graphite crucible at 1100 C under cryolite flux to avoid oxidation. The temperature was kept constant for 30 minutes, then the crucible was brought out from the furnace, and stirred for two minutes before casting the alloy into a plate form of less than 10mm thickness by spreading it over a thick cast iron plate. The preparation of the different microalloys by melting the predetermined quantity of Al and the calculated amount of the binary master alloy was added to the molten Al bath, stirred and its temperature was raised to 820 C, then lowered to 750 C for 10 minutes and finally the crucible was brought out of the furnace, stirred for two minutes and left to solidify and cool inside the crucible, outside the furnace.

Table 1. Chemical composition (wt %) of commercially pure Aluminium [5]

Element	Fe	Si	Cu	Mg	Ti	V	Zn	Mn	Na	Al
Wt. %	0.09	0.05	0.005	0.004	0.004	0.008	0.005	0.001	0.005	Bal.

Table 2. Chemical composition of aluminum and its different microalloys (wt.%)

Specimen No.	Ti	V	Al.
1	0	0	Rem.
2	0.15	0.005	Rem.
3	0.15	0.053	Rem.
4	0.15	0.112	Rem.
5	0.15	0.148	Rem.
6	0.15	0.236	Rem.

2.3 Corrosion Testing Procedure

Cylindrical specimens having 10 mm diameter and 25 mm^2 working surface area were prepared from Al and its five micro alloy, immersed in 1 % HCl solution at room temperature for two minutes to remove any protective or oxide layer, rinsed in distilled water, kept in acetone for one minute, dried in hot air and then fixed in position in the PVC teflon holder, shown in figure1. The specimens were then immersed in 0.2 M NaOH solution of pH 13.3 at 25, 40 or 60 °C in a stand still condition, (without stirring). After a specified period of time (taken as 24 hr), the specimens were removed from the teflon holder, rinsed with acetone, dried and kept in the desiccator for two hours and weighted again from which the weight loss per mm^2 is calculated in each specimen. Each test was repeated twice to guarantee the repeatability of the results within an accepted difference of 3%. The average corrosion rate was then obtained.

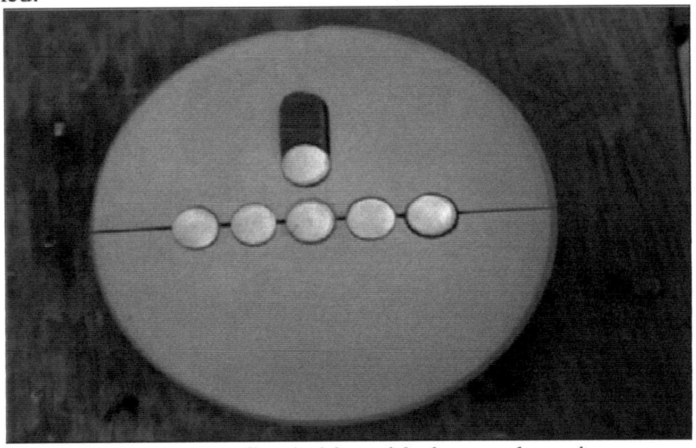

Figure 1. PVC teflon holder with the metal specimens

2.3.1 Corrosion Rate Mmeasurement.

The weight loss technique involves exposing a specimen of material to a process environment for a given duration, then removing the specimen for analysis. The weight loss taking place over the period of exposure in the corrosive medium [2]. is being expressed as a corrosion rate according to the following expression:

$$CR = \frac{\Delta W}{t \cdot A} \quad (5)$$

where CR is the corrosion rate, ΔW is weight loss (mg), t is the time (yr), A is the surface area exposed in the corrosive medium (mm^2).

2.4 Surface Examination Procedure

Microstructural examination of the surface of each specimen was examined by light microscope before and after corrosion test. Photographs of the etched specimens were taken using optical microscope. Detailed procedure of etching and surface examination is described elsewhere [2].

3. Results and Discussion

3.1 Effect of vanadium addition

Figures. 2, 3 and 4 show the effect of vanadium addition on the corrosion rate of commercially pure aluminum grain refined by Ti, (Al-0.15 % Ti) at solution temperatures of 25, 40 and 60 °C, respectively. It can be seen from Fig.2 that, except at infinitesimally small percent of V addition (less than 0.05%), vanadium addition has generally resulted in a decrease in corrosion rate comparable with the commercially pure Al specimen. The maximum decrease in corrosion rate was 38 % and the minimum was 11 % at 0.053 and 0.148 wt % vanadium addition respectively.This is attributed to the small grain size caused by the grain refinement caused by the addition of Ti or V either alone or together, which results in increasing the number of grain boundaries and redcing the surface roughness, both of which reduces corrosion rate, references, [3,6].

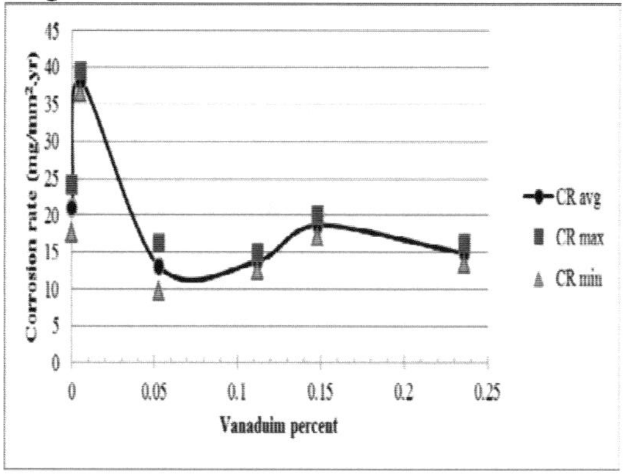

Figure 2. Corrosion rate vs. vanadium percent of specimens at solution temperature of 25°C.

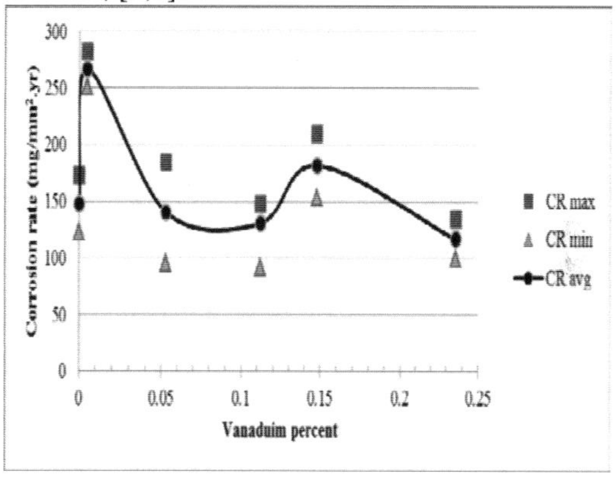

Figure 3. Corrosion rate vs. vanadium percent of specimens at solution temperature of 40°C.

Increasing the temperature of the NaOH solution to 40 and then to 60 °C resulted in higher rates of corrosion for a given micro alloy (fixed wt.% V). However, as the wt. % of vanadium is increased in the microalloy, the corrosion rate decreased significantly at the three solution temperatures . At the higher temperatures, the trend remained the same but addition of V became more effective as the temperature of the NaOH solution is increased. As can be deduced from Fig.3, the maximum reduction in the corrosion rate was approximately 28 % at 0.112 wt % V, and the minimum reduction was approximately 5% at 0.053 wt % V at 40 °C. Figure 4 indicates that the maximum reduction in the corrosion rate was approximately 40 % at 0.112 wt % V, compared to the minimum reduction of approximately 6 % at 0.148 wt % V at 60 °C. All reduction percentages are obtained from comparison with the performance of pure aluminum with zero wt % V.

IOP Conf. Series: Materials Science and Engineering 60 (2014) 012037 doi:10.1088/1757-899X/60/1/012037

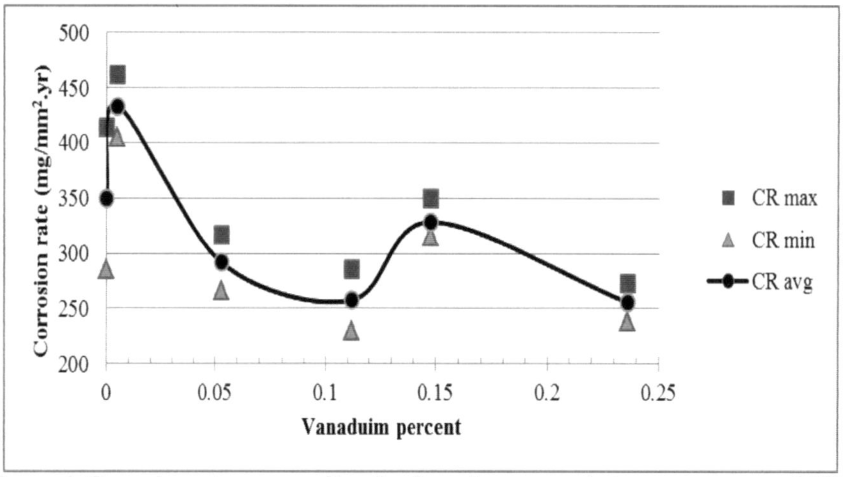

Figure 4. Corrosion rate vs. vanadium% of specimens at solution temperature of 60°C

3.2 Effect of Solution Temperature

In Figure 5, the average corrosion rates of micro alloys vs. wt% V at the three temperatures 25, 40 and 60 °C are compared. The Figure reveals that at for a given micro alloy, the average corrosion rates at 60 °C is approximately 280 mg/mm^2.yr compared to 140 at 40 °C and 15 mg/mm^2.yr at 25 °C. In making this comparison, the corrosion rate at 0.05 wt. % V is excluded due to the different behavior exhibited by this micro alloy. The corrosion rate rises sharply from zero wt.% to 0.053 wt. % V before declining to values virtually below those of vanadium-free micro alloy. The reduction in average corrosion rates reached a maximum of 40% at 0.112 wt% V at 40 °C and a minimum of 6% at 0.148 wt.% V at 60 °C.

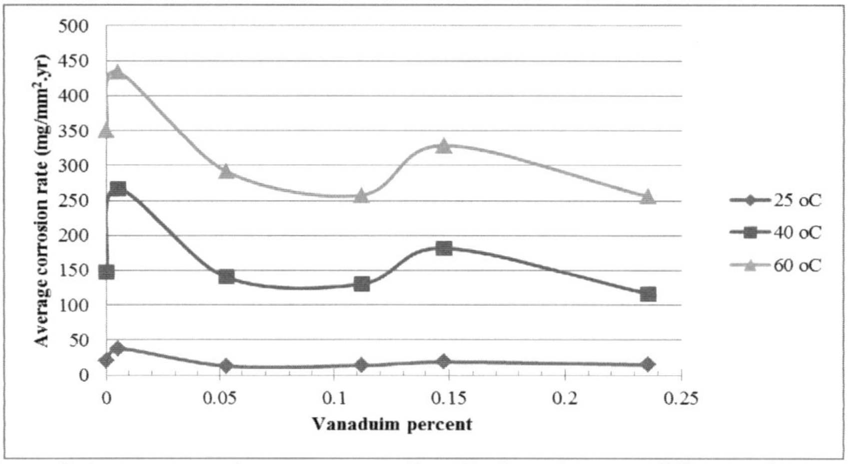

Figure 5. Average corrosion rate vs. vanadium % of specimens at different solute temp.

3.3 Surface Micrographs

Figure 6a shows the original microstructure of the commercially pure Al at magnification of X400 while figure6B is a photomicrograph showing the microstructure of commercially pure Al after an exposure of 24 hrs to 0.2 M NaOH solution at 60°C at the same magnification. The difference between the two graphs indicates clearly the corrosion action in the forms of pitting as well as wiping off the grains and their boundaries.

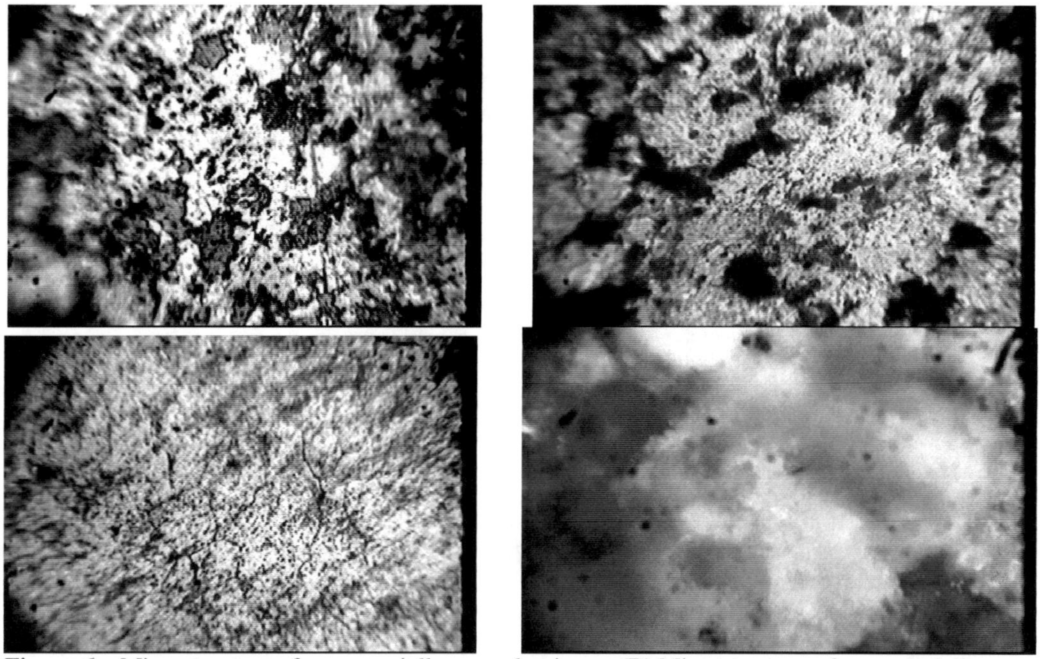

Figure 6a. Microstructure of commercially pure aluminum. (B) Microstructure of commercially pure aluminum after exposure of 24 hrs to 0.2 M NaOH solution at 60°C. (C) Original microstructure of Al-0.15 % Ti-0.112 % V micro alloy. (D) Microstructure of Al-0.15 % Ti-0.112 % V micro alloy after exposure of 24 hrs to 0.2 M NaOH solution at 60°C (all magnifications at X400).

Figure 6c is a photomicrograph of the original microstructure of Al-0.15 % Ti-0.112 % V micro alloy at magnification of X400. It shows the grain refinement effect of 0.112 wt % vanadium addition where the grains are much finer than those of commercially pure Al depicted by Figure 6A. A photomicrograph showing the microstructure of Al-0.15 % Ti-0.112 % V micro alloy after an exposure of 24 hrs to 0.2 M NaOH solutions at 60°C at magnification of X400 is shown in figure 6D. This Figure shows that the grains and their boundaries are wiped off. This agrees with the results obtained in the previous Figures for the corrosion rate of Al and its micro alloys in the NaOH solution.

3.4 Effect of Inhibitor and Inhibition Efficiency

Figure 7 shows the effect of V addition at different wt.% to commercially pure Al grain refined by Ti on the corrosion rate in NaOH solution at different temperatures in the presence of 3 wt% potassium dichromate ($K_2Cr_2O_7$) as inhibitor. It can be seen from this Figure that the corrosion rate increases as the V percent increases, but remained much less than the corrosion rate in the absence of the $K_2Cr_2O_7$ inhibitor. A comparison between Fig.7 and Fig.5 reveals that the highest corrosion rate in presence of inhibitor is only 8 mg/mm^2.yr whereas that in absence of inhibitor is 330 mg/mm^2.yr. In making this comparison, the corrosion rate data at 0.05 wt. % V is excluded as indicated earlier.

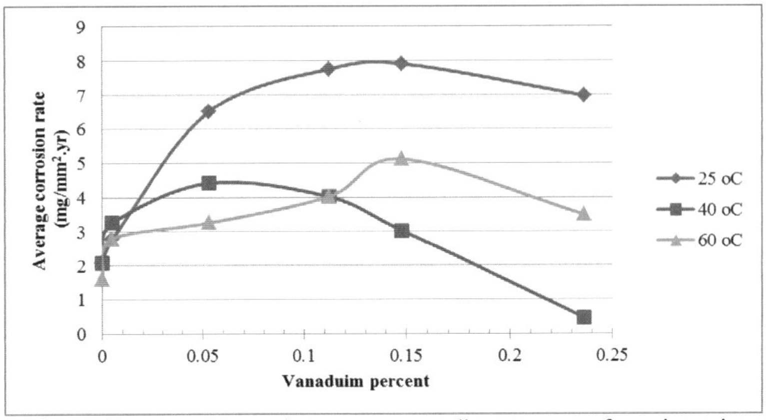

Figure 7. Average corrosion rate vs. vanadium percent of specimen in presence of $K_2Cr_2O_7$ inhibitor at different solution temperatures.

The presence of $K_2Cr_2O_7$ as an inhibitor caused significant reduction in the corrosion rate (that is, promotes corrosion resistance) at all rates of V addition as compared to the uninhibited NaOH solution at the three temperatures considered. This effect is due to adsorption of $K_2Cr_2O_7$ molecules on the Al surfaces thus forming a barrier layer that minimizes metal dissolution and release of its ions into solution. In Figures 8, 9 and 10 show comparison among the average corrosion rate vs. V addition percent at temperatures of 25, 40 and 60 °C, respectively in both blanks in the(uninhibited and inhibited NaOH solutions. These Figures indicate that addition of $K_2Cr_2O_7$ at 3 wt.% to the strong alkaline solution of NaOH was effective in corrosion inhibition at any these temperature for all of the micro alloys at any wt.% vanadium. The inhibition effect is obviously an increasing function of temperature as shown from the corrosion rate curves at 25 and 40 °C. After 40 °C, the temperature effect on inhibition effect diminishes as indicated by comparing the corrosion rate curves at 40 and 60 °C. This can be explained by the enhancing effect of temperature on the adsorption of inhibitor molecules on Al surface. Based on these observations, it is suggested that below 40 °C, corrosion rate is highly temperature dependent as metal dissolution requires high energy of activation. Over 40 °C, the amount of activation energy becomes much less, thus further temperature increase will not have significant effect on corrosion rate.

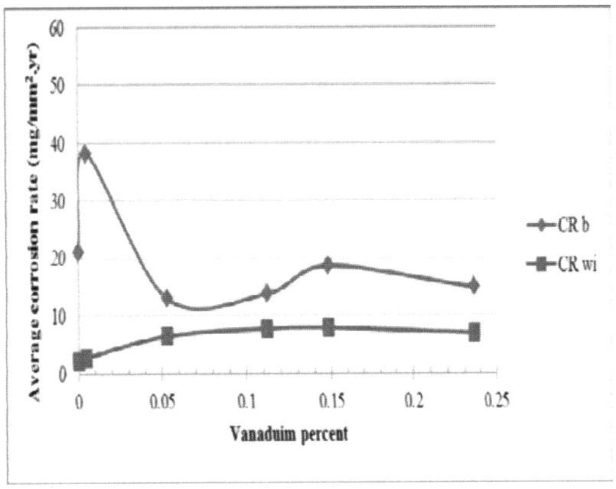

Figure 8. Average corrosion rate vs. vanadium percent of specimens in blank (uninhibited) and inhibited NaOH solution at 25°C.

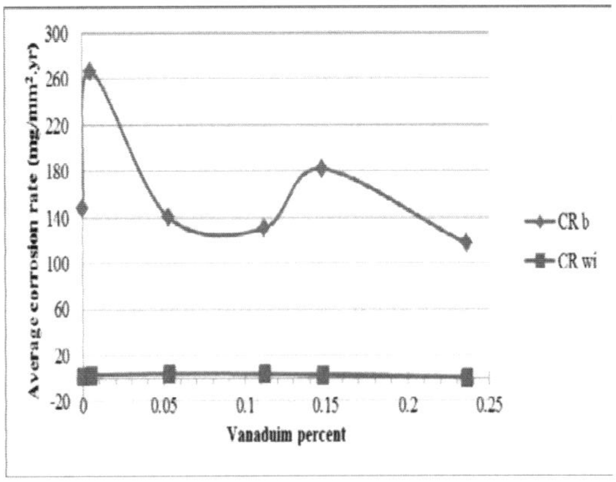

Figure 9. Average corrosion rate Vs. % of specimens In blank (uninhibited) and inhibited NaOH Solution at 40 C.

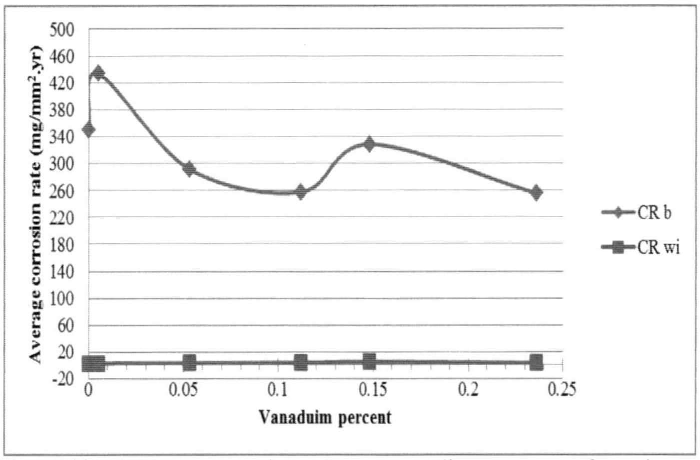

Figure10. Average corrosion rate vs. vanadium percent of specimens in blank (uninhibited) and inhibited NaOH solution at 60°C.

The corrosion inhibition efficiency, IE, is defined as [13]:

$$IE\ \% = \frac{CR_b - CR_{wi}}{CR_b} \times 100\% \tag{6}$$

Where CR_b is the corrosion rate in blank (uninhibited) solution and CR_{wi} is the corrosion rate in solution with inhibitor, both in ($mg/mm^2.yr$) units. The effect of V addition on the inhibition efficiency at different temperatures is shown in Table3 It can be seen from the result in this table that the presence of $K_2Cr_2O_7$ as an inhibitor in NaOH solution improved the inhibition efficiency of the solution by 90-99.5% for commercially pure Al and 97- 99 % in case of addition of Al-vanadium micro alloys. IE is obviously a function of temperature.

Table 3. Effect of V Addition on the Inhibition Efficiency in NaOH-$K_2Cr_2O_7$ Solution at Different Temperatures

wt. % Vanadium	Inhibition Efficiency, IE %		
	@ 25 °C	@ 40 °C	@ 60 °C
0	90	98.58	99.53
0.005	92.68	98.77	99.36
0.053	50	96.85	98.88
0.112	43.75	96.92	98.44
0.148	57.5	98.34	98.44
0.236	53.13	99.6	98.66

4. Conclusions

The addition of vanadium in small weight percentages (0.053-0.236) to micro alloys of aluminum and Al grain refined by Ti is beneficial in improving their corrosion resistance in highly alkaline (pH=13.3) NaOH solution, both blank (uninhibited) and in presence of potassium dichromate at 3%. The corrosion rates of these micro alloys increased with temperature increase in the range 25-60°C in absence of the inhibitor, however, in the presence of inhibitor, the corrosion rates decreased with temperature resulting in inhibition efficiency much higher at 40-60 C than at 25 C.

5. References

[1] Kalpakjian S and Schmid S R 2002 *Manufacturing Processes for Engineering Materials* (4th ed.), New Jersey: Pearson Education In.

[2] Fontana M G 1986 *Corrosion Engineering*, New York, McGraw-Hill international editions, 3rd Edition.

[3] Abdel-Hamid A A 1985 On the Mechanism of Grain Refinement of Aluminum by Small Addition of Ti and B *The Second Arab Aluminum Conference*, Arabal, Cairo.

[4] Abdel-Hamid A A 1989 Effect of Other Elements on the Grain Refinement of Al by Ti or Ti and B Part1: A critical Review Egypt.

[5] Zaid A I O, Allawi G T Aand Al-Haj-Ali A 2012 *Key Engineering Materials*, 510-511 481-486.

[6] Zaid A I O and Abdel-Hamid A A1999 Effect of Vanadium on Mechanical Behavior, Mechanability and Wear Resistance of Aluminum Grain Refined by Ti & B *International Symposium on Advanced Materials* ISAM-99, Islamabad.

[7] Zaki Ahmad 2006 Principles of Corrosion Engineering and Corrosion Control Elsevier, Amsterdam.

[8] Pierre R Roberge 2008 Corrosion Engineering: Principles and Practice, McGraw-Hill.

[9] Kaufman J G 2005 Corrosion of aluminum and aluminum alloys *ASM Handbook* 13B, Corrosion Materials ASM International Materials Park, OH 95 – 124.

[10] Emregul K C and Abbas A Aksut 2000 The Behavior of Aluminum in Alkaline Media, *Corrosion Science* 42 2051-2067.

[11] Cotton F A, Wilkinson G and GausP L 1995 *Basic Inorganic Chemistry*, 3rd Edition, Wiley.

[12] Allawi G T 2011 Effect of Vanadium Addition on the Chemical Corrosion Resistance of Commercially Pure Aluminum Grain Refined by Titanium at Different Temperatures, M.Sc. Thesis, Dept. of Chemical Engineering University of Jordan, Amman 120.

[13] El Maghraby A A 2009 *The Open Corrosion Journal* 2 189-196.

Molecular dynamics simulations of thermal conductivity in 2D complex Yukawa liquids

Aamir Shahzad[1,2], Mariam Sultana[1], Arffa Aslam[1] and Mao-Gang He[2]
[1]Department of Physics, Government College University Faisalabad, Allama Iqbal Road, Faisalabad 38000, Pakistan
[2]Key Laboratory of Thermo-Fluid Science and Engineering of Ministry of Education (MOE), Xi'an Jiaotong University, Xi'an 710049, P. R. China

E-mail: aamirshahzad_8@hotmail.com, aamir.awan@gcuf.edu.pk

Abstract. The thermal conductivity in strongly coupled complex dusty plasma liquids (SCCDPLs) has been investigated through an improved homogenous nonequilibrium molecular simulation (HNEMS) method, for the first time. The HNEMS method has been employed for two-dimensional (2D) Yukawa systems in a canonical ensemble. The thermal conductivities with suitable normalizations (plasma and Einstein frequencies), in the value of low force field strength, have been computed for a wide range of plasma state points of Coulomb coupling (Γ) and screening strength (κ). The new simulation results are found to obey the simple analytical temperature scaling law. The present HNEMS results are in generally with parts of earlier HNEMS, equilibrium molecular dynamics (EMD) and experimental data in the literature for the 2D and there-dimensional (3D) SCCDPLs. It is shown that the HNEMS method can be used to estimate the thermal conductivity very effectively and to understand the fundamental behaviours in 2D Yukawa systems.

1. Introduction

In recent applications, the transport problems in non-ideal systems of interacting particles are of significant interest in various fields of science and engineering (such are plasma physics, physics of polymers, medical, semiconductor and chemical industries, environmental safety and space plasmas, etc.) [1]. In practical applications, including those intended at developing novel thermal interface materials, micro and nano-level heat transport processes, and nano-particulate systems have been proposed to improve the efficient behaviour of the systems. Nanotechnology development in the latest decades revealed new outcomes in semiconductor nanowires (NWs) investigate and their interaction with fluids and transport properties of NWs, including the thermal conductivities are extensively amended from the macroscopic property [2].

The development features of heart transfer in materials at micro- and nano-levels have moved into the areas of technology issues, however, there are some areas, such as phonon thermal transports in semiconductor superlattices have much attention for researchers. Different practical novel applications and technologies such as thermoelectronic and photoelectronic devices, solar cells, and NWs-based thermoelectric systems have been proposed. Nano-particles themselves have low conductivity, however, their interactions along with larger particles are considered to increase the percolation threshold leading to better materials overall [3-5].

In addition to fundamental features, the study of thermal properties of complex (non-ideal) systems is of practical interest for the developments of micro- and nano-systems. A four components plasma (nonideal complex dusty plasma system) plays extremely significant part in a number of new technological advances (such as heat technology tools for process/ chemical industries, plasma processing for semiconductor industries, medical tool treatments and surface cleaning, and future energy production and storage devices, etc) and this strongly coupled complex dusty plasma (SCCDP) is containing grains of dust particles (solid matter) which found everywhere in nature [6-8]. These systems require a deep understanding of the interaction of complex (nonideal) systems with nano- and microstructuring of surfaces and of the behaviour of particles (micro and nano-levels) and chemically active particulates. One important challenge in using micro and nanotechnologies is the lack of knowledge regarding their thermal conductivity. A novel homogenous nonequilibrium molecular simulation (HNEMS) approach is to be needed to compute the thermal conductivity of complex (non-ideal) systems of much of the thermophysical property research in the fields of science and technology [9-11].

The purpose of the current study is to investigate all possible dependences of the heat conductivity (λ) on the plasma parameters (κ, Γ) using the nonequilibrium molecular dynamics (HNEMD) technique as reported in Shahzad and He [12], for the first time in two-dimensional (2D) SCCDP liquids (SCCDPLs). In addition, the effects of the screening length (κ) on the dust particle interaction for λ with different system sizes and temperatures are also computed by using extensive computer simulations for 2D SCCDPLs. The design of this paper is as follows: in Section 2, the improved form of mathematical model for the λ coefficient for 2D Yukawa liquids (SCCDPLs) using HNEMS algorithm is given, and this section also presents the molecular mechanics dynamics (MMD) simulation method. In Section 3, the HNEMS results are illustrated and discussed with previous reported experimental and computer simulation data by other author's. Finally, in Section 4, the work is summarized.

2. Theory and Nonequilibrium MMD Algorithm

Classical mechanics (CM) provide a fundamental description of any system (simple, dense and complex). In theory, CM calculations depend on parameters used and it is desirable to use the most sophisticated approach to CM for any calculation.

Currently, a well known model in laboratory plasma of gas discharge is Yukawa interaction model (screened Coulomb interaction), which is used for the account of the interactions between dust particles (Yukawa particulates) computationally for our case and this model has been used in may physical and chemical systems (for instances, ionic systems, medicine and biological sciences, physics of polymers, environment, etc.)

$$\phi(|\mathbf{r}|) = \frac{Q^2}{4\pi\varepsilon_0}\frac{e^{-|\mathbf{r}|/\lambda_D}}{|\mathbf{r}|}, \qquad (1)$$

where λ_D is the screening strength (Debye length), Q is the charge on a dust particles and r is the interparticle distance.

The general microscopic expression for thermal conductivity computationally is regarded as Green-Kubo relation (GKR) for thermal coefficients of uncharged particles [13]. This relation has been employed for one component Coulomb plasmas (OCCPs) [14] and SCCDPLs [2,5,9-11], using different numerical methods.

$$\lambda = \frac{1}{2k_B AT^2}\int_0^\infty \langle \mathbf{J}_Q(t).\mathbf{J}_Q(0)\rangle dt \qquad (2)$$

where T is system temperature, k_B is Boltzmann's constant and V is volume of the system under HNEMS method. The heat energy flux vector $\mathbf{J}_Q$ in terms of microscopic form can be written as

$$\mathbf{J}_Q A = \sum_{i=1}^{N} E_i \frac{\mathbf{p}_i}{m} - \frac{1}{2}\sum_{i\neq j} \mathbf{r}_{ij}(\frac{\mathbf{p}_i}{m}.\mathbf{F}_{ij}) \text{, where} \qquad E_i = \frac{p_i^2}{2m} + \frac{1}{2}\sum_{i\neq j}\phi_{ij} \tag{3}$$

where energy E_i of particle i with ϕ_{ij} is the Yukawa pair interaction between particle i and j, and the position vector and force are $\mathbf{r}_{ij}$=$\mathbf{r}_i$-$\mathbf{r}_j$ and $\mathbf{F}_{ij}$, respectively, on particle i due to j and $\mathbf{p}_i$ is the momentum vector of the ith particle, and . The computationally microscopic heat energy flux vector $\mathbf{J}_Q$ of the Yukawa liquids is given in Refs. 2 and 5. The Evans [15] proposed the non-Hamiltonian linear response theory and has been used for a system moving representing equations of motion

$$\dot{\mathbf{r}}_i = \frac{\mathbf{p}_i}{m}, \tag{4}$$

$$\dot{\mathbf{p}}_i = \sum_{j=1}^{N} \mathbf{F}_i + \mathbf{D}_i(\mathbf{r}_i,\mathbf{p}_i).\mathbf{F}_e(t) - \alpha\mathbf{p}_i. \tag{5}$$

In Eq. (5), $\mathbf{D}_i$= $\mathbf{D}_i(\mathbf{r}_i, \mathbf{p}_i)$ is the tensor phase variable that describes the coupling of Yukawa system to the external force field $\mathbf{F}_e(t)$, the total interparticle force acting on particle i is $\mathbf{F}_i$=$(-\partial\phi_{ij}/\partial r_i)$, and α is the Gaussian thermostat multiplier. The linear response theory can be used in a straightforward fashion to a Yukawa system [2,5]. A thermostat (α) has been employed to retain constant temperature in order to remove heat problem due to work done and it is given as [5]

$$\alpha = \frac{\sum_{i=1}^{N}[\mathbf{F}_i + \mathbf{D}_i(\mathbf{r}_i,\mathbf{p}_i).\mathbf{F}_e(t)].\mathbf{p}_i}{\sum_{i=1}^{N} p_i^2 / m_i} \tag{6}$$

We can used external force field of $\mathbf{F}_e$(t) = (0, F_z), which is parallel to the z-axis, then the final expression for thermal conductivity is given as:

$$\lambda = \frac{1}{2k_B AT^2}\int_0^{\infty}\langle \mathbf{J}_{Q_z}(t)\mathbf{J}_{Q_z}(0)\rangle dt = \lim_{F_z\to 0}\lim_{t\to\infty}\frac{-\langle \mathbf{J}_{Q_z}(t)\rangle}{TF_z} \tag{7}$$

where $\mathbf{J}_{Qz}$ is the z-component of the heat energy flux vector for SCCDPLs. The thermal conductivity is normalized by plasma frequency λ_0= $\lambda/nm\omega_p a^2$. All time series data are recorded during HNEMS method by using Eq. (7) to compute λ_0.

2.1. *HNEMD Simulation Method*

In our simulation, three normalized parameters (in reduced units) are used for complete characterizing of the Yukawa model, given in Eq. (1), for interaction between particles [2,5,12-15]: the external force field strength F^*=F_z a_{ws}, the plasma coupling (Coulomb) parameter Γ = $(Q^2/4\pi\varepsilon_0)(1/a_{ws}k_BT)$ and the screening strength parameter $\kappa \equiv a_{ws}/\lambda_D$. It is mentioned that the k_B is Boltzmann's constant, T is system temperature and a_{ws} = $(n\pi)^{-1/2}$ is the Wigner-Seitz radius [2,16] with n is the equilibrium particle number density. Time scales of interest are taken as inverse of the plasma frequency ω_p = $(Q^2/2\pi\varepsilon_0 ma^3)^{1/2}$, where m is the Yukawa dust particle mass. In order to implement our HNEMS method, a system of N=1024-4096 particles is chosen and placed in a computational cell of Yukawa system with edge lengths of Lx/Ly=$2/\sqrt{3}$. For simulations of an infinite Yukawa system using HNEMS technique, we need to employ the periodic boundary conditions (PBCs) in all directions with the minimum image conventions of the dust particles. The Newton equations of motions are used for computer integration of number particles to obtain the particle trajectories, and in solving Eqs. (4) and (5) we use the predictor-corrector algorithm [17]. In the HNEMS method, the particle number N was taken large enough to confirm the system size effects with efficient accuracy. Therefore, in order to save simulation time and computational power the present particle number N is acceptable for numerical simulations. The interaction force acting on the *i*th particle $F_i = (-\partial\phi_{ij}/\partial r_i)$ is computed by the Yukawa pairwise potential between particle i (at $\mathbf{r}_i$) and the particle j (at $\mathbf{r}_j$) and with its periodic images conventions of the Yukawa particles. A canonical ensemble that the Gaussian thermostat method was used for our HNEMS computations in order to maintain the temperature of Yukawa systems [2,13].

For the Ewald summation method detail has been reported by Shahzad and He [5,16], for Yukawa pairwise κ-dependent cutoff radius. Our simulations are carried out between $3.0\times10^5/\omega_p$ and $1.5\times10^5/\omega_p$ time units in the series of data recording of λ_0. The HNEMD numerical time step $dt = 0.001/\omega_p$ is chosen to allow calculating the heat transport important data [2,5]. In this paper, the calculations are accounted for thermal conductivity of 2D SCCDPLs over wide range of plasma parameters of $(10 \leq \Gamma \leq 100)$ and $(1 \leq \kappa \leq 3)$.

3. Simulation Results and Discussion

In this section, thermal conductivity measurements with appropriate normalizations, in the limit of low equilibrium external force field strengths F^* $(= F_z\, a_{ws})$, are presented over a wide range of plasma coupling ($\Gamma\geq10$) and screening ($\kappa\geq1$) parameters. A different sequence of the thermal conductivity corresponding to a higher and lower sequence of applied external field strength is calculated to determine the linear regime of the HNEMD Yukawa system. The HNEMS data from at a reduced force field strength of (F^* = 0.2) have a usually good agreement with the previous equilibrium molecular dynamics (EMD) investigations in 2D dissipative Yukawa systems by Khrustalyov and Vaulina [1], nonequilibrium EMD (NEMD) simulations of Hou and Piel [11], and experimental results of Nosenko *et al* [3], Nunomura *et al* [4], and Fortov *et al* [6]. A simple analytical temperature representation of Yukawa thermal conductivity with normalized Einstein frequency (ω_E) has also been performed. The present calculations with appropriate normalizations show that the position of minimum of 2D thermal conductivity lies at higher Γ with increasing κ same as in 2D Yukawa systems [1,11], and earlier experimental data [3,4,6].

The chief results obtained from the HNEMD simulations are given in Figs. 1-3 for κ = 1, 2 and 3, respectively, for the whole range plasma of the 2D Yukawa systems. The Figs. 1-3 display the present results as well as compared the previous results obtained from the earlier three-dimensional (3D) HNEMD simulations of Shahzad and He [2,5], Green-Kubo EMD of Salin and Caillol [9], and theoretical predictions of Faussurier and Murillo [10], and in addition the calculations taken from 2D means of Green-Kubo EMD of dissipative Yukawa systems of Khrustalyov and Vaulina [1]. Moreover, we have been compared our simulation results with the earlier 2D NEMD results of Hou and Piel [11], and experimental data of Nosenko *et al* [3], Nunomura *et al* [4], and Fortov *et al* [6], nearly at the same plasma state points.

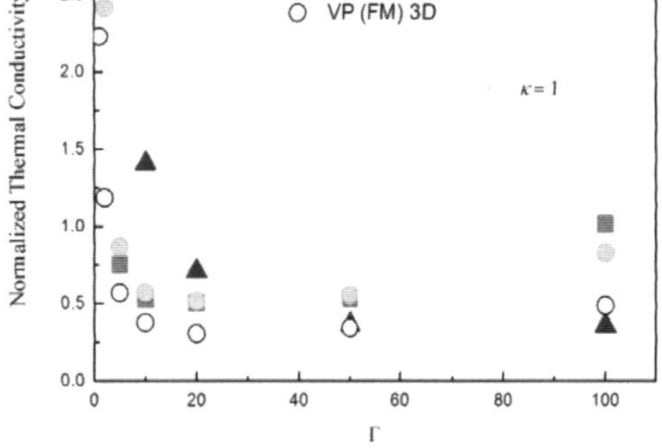

Figure 1. Comparison of results of thermal conductivity as a function of plasma coupling Γ (system temperature) for Yukawa liquid, obtained from HNEMS defined in Eq. (7), at κ=1. HNEMS of Shahzad and He: SH 2D HNEMS: present results (for N = 1024 particles), HNEMD of Shahzad and He [5]: SH 3D HNEMD, EMD computations results of Salin and Caillol [9]: SC 3D EMD, theoretical variational procedure calculations of Faussurier and Murillo [10]: FM 3D VP

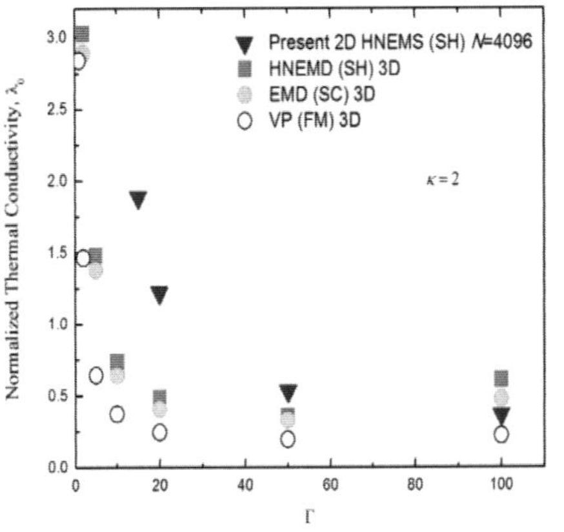

Figure 2. Comparison of results of thermal conductivity as a function of plasma coupling Γ (system temperature) for Yukawa liquid, obtained from HNEMS defined in Eq. (7), at κ=2 (for N = 4096 particles). For details, see the caption of Fig.

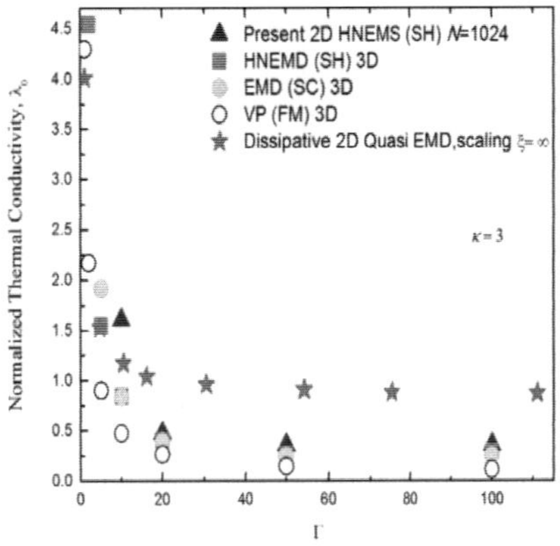

Figure 3. Comparison of results of thermal conductivity as a function of plasma coupling Γ (system temperature) for Yukawa liquid, obtained from HNEMS defined in Eq. (7), at κ=3. For details, see the caption of Fig. 1, and Green-Kubo EMD of dissipative Yukawa systems of Khrustalyov and Vaulina [1]: 2D Quasi EMD.

The present thermal conductivity obtained from HNEMD simulations is slightly higher than those previously 3D numerical simulation results [2,5,9,10] and lie close to the earlier 2D simulation results obtained from different simulation methods [1,11]. The minimum values of frequency decreases with increasing of κ. The thermal conductivity obtained from HNEMD simulations is found to be in very good agreement with that obtained through the previously known 2D numerical simulations and experimental results for 2D Yukawa liquids [1,3,4,6,11].

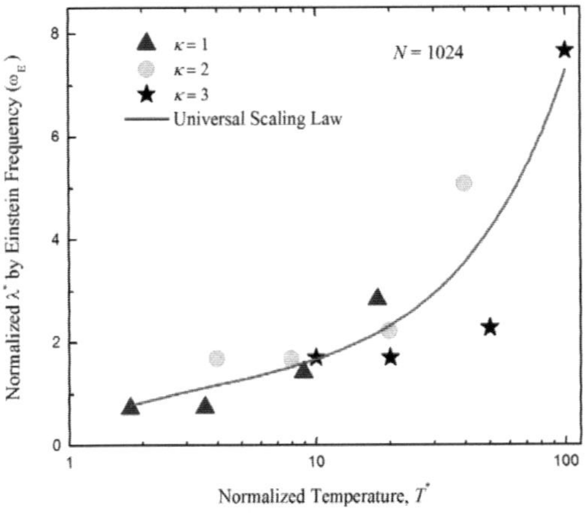

Figure 4. Variation of normalized thermal conductivity by Einstein frequency λ^* (T^*) with normalized temperature T^* for SCCDP Yukawa system at different κ=1, 2 and 3, where $T^* \equiv T/T_m = \Gamma_m/\Gamma$ and Γ_m corresponding to temperature at melting temperature. The solid line is obtained by using linear fitting for the simple Scaling Law: $\lambda^* = AT^* + B/T^* + C$, showing the temperature scaling law for 2D SCCDPLs.

Figure 4 shows the variation of normalized thermal conductivity λ^* (normalized by the Einstein frequency) at different normalized temperatures T^* (ratio of the temperature to melting temperature) for screening strength of κ=1, 2 and 3. It is evident from Fig. 4 that the λ^* is dependent on the normalized temperature in the 2D SCCDPLs, confirming earlier simulation and experimental data [3,4,6]. The simulation results obtained for λ^* using HNEMS method provide the best fitting and excellent trend for all the plasma state points (κ, Γ). The solid line given in figure 4 is obtained using the simple linear fitting of form, $\lambda^* = AT^* + B/T^* + C$, with coefficients: A= 0.06143, B= -0.8409, and C= 1.13778, showing the confirmation of temperature scaling law behaviours same as in 3D SCCDPLs Refs. 2 and 5. This scaling law is found to reproduce appropriately the shape of $\lambda^*(T^*)$ curves at three κ values and it gives suitable relationship between λ^* and T^* for the whole plasma state points of (κ, Γ), confirming the earlier numerical results [2,5].

4. Conclusions

HNE MD simulation method has been used to calculate the thermal conductivity of 2D SCCDPLs over a wide range of plasma parameters (κ, Γ). New HNEMD simulations give more consistent results for 2D Yukawa conductivity than previous known numerical and experimental results, over all plasma state points, and illustrate a fair agreement with earlier HNEMD, EMD simulations and experimental data. Our results show that the λ_0 dependent on the temperature (plasma coupling) in 2D Yukawa liquids. The present results obtained indicate that the value of λ_0 at intermediate and higher Coulomb coupling nearly the same as κ parameter increases, confirming earlier experimental results. The new HNEMD algorithm has strong advantages that easy to code, fast computing on small system sizes and showing efficient computing. In future work, it will also be of highly interest for analysis to see how quantum effects influence λ_0 in 2D and 3D SCCDPLs and it is suggested a need for an experiment or different simulations.

5. Acknowledgments

This work was sponsored by the China Postdoctoral Science Foundation (CPSF No. 2013M532042). The authors thank Z. Donkó (Hungarian Academy of Sciences) for providing his 3D thermal conductivity data of Yukawa Liquids for the comparisons of our simulation results, and useful discussions. We are grateful to the National High Performance Computing Centre of Xian Jiaotong University for allocating of computer time to test and run our MD code.

6. References

[1] Khrustalyov and Vaulina 2012 *Phy. Rev. E* 85 046405

[2] Shahzad A and He M-G 2012 *Physics of Plasmas* 19 083707

[3] Nosenko V and J. Goree 2004 *Phys. Rev. Lett.* 93 155004

[4] Nunomura S Samsonov D Zhdanov S and Morfill G 2005 *Phys. Rev. Lett.* 95 025003

[5] Shahzad A and He M-G 2012 *Contrib. Plasma Phys.* 52 667

[6] Shahzad A and He M-G 2014 *Computer Simulation of Complex Plasmas: Molecular Modeling and Elementary Processes in Complex Plasmas* Scholar's Press (Germany)

[7] Shahzad A and He M-G 2012 *Plasma Science and Technol.* 14 771

[8] Shahzad A, He M-G and He K He 2013 *Phys. Scr.* 87 035501

[9] Salin G and Caillol J M 2003 *Phys. Plasmas* 10 1220

[10] Faussurier G and Murillo M S 2003 *Phys. Rev.* E 67, 046404

[11] Hou L J and Piel A 2009 J. Phys. A: Math. Theor. 42 214025

[12] Shahzad A and He M-G 2013 *Interaction Contributions in Thermal Conductivity of Three-Dimensional Complex Liquids* AIP Conf. Proc. 1547, 173

[13] Evans D J and Morriss G P 1990 *Statistical Mechanics of Non-equilibrium Liquids* Academic (London)

[14] Pierleoni C Ciccotti G and Bernu B 1987 *Europhys. Lett.* 4 1115

[15] Evans D J 1982 *J. Phys. Lett. A* 91 457

[16] Shahzad A and He M-G 2012 *Phys. Scr.* 86 015502

[17] Rapaport D C 2004 *The Art of Molecular dynamics Simulation* Cambridge University Press (New York)

Correcting aberration in aspheric surfaces

K Ahmed, A N Khan, A Rauf and A Gul
Institute of Industrial Control System, G. P. O. Box - 1398, Islamabad

Email: drkamalahmed@yahoo.com

Abstract. New technique eases aspheric lens fabrication and overcome traditional limitation. An aspheric lens has been designed by using optical designing software to replace the achromat (Doublet) lens of eyepiece assembly of telescope. The devised physical parameters of aspheric lens have been incorporated into the CNC Aspheric machine to fabricate the lens. The antireflection coating for visible region has been carried out on lens by employing PVD technique. In this report diminished aberration effects due to non-spherical surface profile and comparison of optical parameters of achromat (doublet) and aspheric lens is presented.

1. Introduction

The advantages of utilizing aspheric surfaces in optical design have long been known. However, advances in optical manufacturing technology have opened up a new world of possibilities to the optical designer.

An aspheric lens is a rotationally symmetric optics with one or more surfaces having a non-spherical surface outline. This unique feature allows aspheric lenses to deliver improved optical performance, eliminating spherical aberration and greatly reduce other aberrations when compared to a simple spherical lens [1].

The software can easily and quickly produce a result for an aspheric that will improve a lens design. The designer can find only solutions that are manufacturable. Manufacturing technologies for aspherics have improved significantly over the last decades. Very accurate optical surfaces can be produced with exceptional quality. New methods like high speed computer controlled disc grinding have enabled machining methods to produce aspheric surfaces in brittle optical materials.

One more aspect is very important for optical assembly; utilizing aspheric surfaces in the design of an optical system can reduce the number of optical surfaces to process while achieving a more highly corrected imaging performance. To the designer, this mean an optical system with better performance, lighter weight, less fabrication time, and reduced tolerance restriction is possible now.

However, Interferometry has been the standard method of testing surface accuracy of conventional spherical optics since Newton's time. Still noncontact metrology of aspheric surface accuracy is not fully realized and is coming close to fruition [2]. So, the surface profilometer is the only choice to measure aspheric accurate profile without special tooling.

In this report, development of aspheric with the help of designing software and correction of spherical and chromatic aberrations due to aspheric surface contour is considered. The detailed comparison of optical parameters of achromat (doublet) and aspheric lens and their theoretical and experimental results are discussed.

2. Designing and manufacturing of aspheric lens

The standard way of representing a rotationally symmetric aspheric surface is given by [3-7]:

$$Z = \frac{CX^2}{1+\sqrt{1-(K+1)C^2X^2}} \quad (1)$$

Where C =1/radius of curvature, $K = -e^2$ and e is the eccentricity of the conic surface. The nature of conic surface depends upon the value of K: e.g., sphere: $K = 0$, ellipsoid: $K > -1$, paraboloid: $K = -1$, hyperboloid: $K < -1$.

Optical designing software allows designers significant design freedom to use aspheric terms on every surface in an attempt to coverage the optimal design solution. The software can easily and quickly produce a result for a manufacturable asphere that will improve a lens design.

The aspheric surface is designed using the equation 1 in the optical designing software. The design is optimized indicating the shape of aspheric as hyperboloid. The design parameters of the aspheric plano-convex lens are explained in the table 1:

Table 1. Design parameters of Aspheric Lens.

Parameter	**Dimension**
Radius of curvature (Side-1)	50.23 mm
Radius of curvature (Side-2)	∝
Conic constant (K)	-2.3
Diameter of lens	41.55mm
Centre thickness	17.55 mm
Optical material	K9

The ray tracing of the aspheric lens is shown in figure 1.

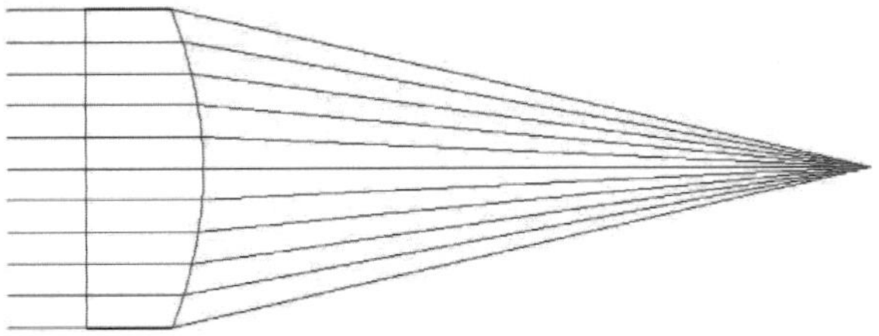

Figure 1. Ray tracing of aspheric lens

The design parameters like the radius of curvature, eccentricity etc. were fed into CNC grinding machine. The machine generated the desired aspheric surface and later it was polished on CNC polishing machine by using polyreuthine polishing pad. The size of the blank used for aspheric lens was having diameter: 43mm and thickness: 18mm.

3. Results and discussions

Figure 2 shows the comparison of defocusing effect on point spread functions for a peak-to-valley OPD (wave-front deformation) of a quarter-wave for doublet and aspheric lenses. It can be seen that the defocusing effect for aspheric lens (figure 2. (b)) with Strehl ratio equal to 0.51 is reduced as it obeys the Rayleigh criterion which is the OPD equal to a quarter-wave is identical to the Marechal criterion (Strehl ratio equal to 0.80) while for doublet lens (figure 2. (a)) the Strehl ratio is 0.37 which is away from Marechal criterion indicating the sharpness and height of the peak is low as compared to the aspheric lens.

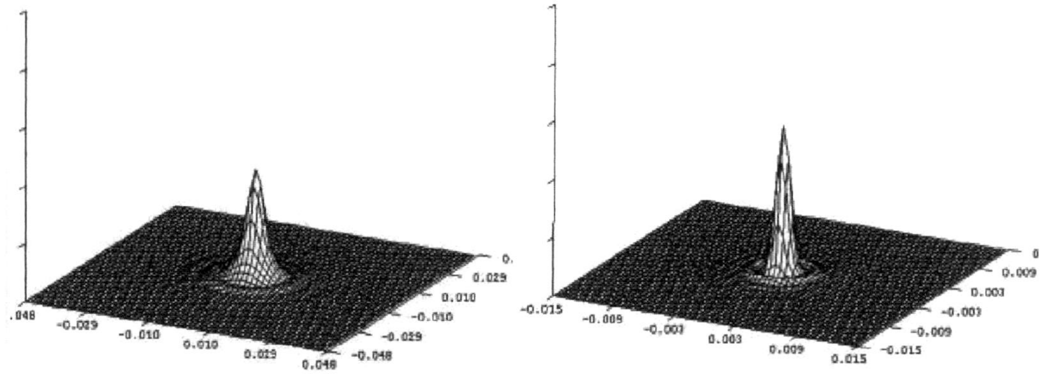

(a) (b)

Figure 2. Comparison of point spread function for doublet and aspheric lenses.

Figure 3 shows the focused spot size for doublet and aspheric lenses. It can be seen that in case of aspheric lens (figure 3. (b)) the spot size is focused up to a value of 0.34 mm where as for doublet lens (figure 3. (a)) its value is 0.55 mm. It can be concluded that the focused spot in case of aspheric lens is very sharp where as it is defused in case of doublet lens. The sharpness of the focused spot size for aspheric lens (figure 3(b)) is attributed to reduction in the aberrations while for doublet (figure 3(b)); the defused spot size is due to the presence of the aberrations.

(a) (b)

Figure 3. Comparison of focused spot size for doublet and aspheric lenses.

Figure 4 shows the comparison of transverse aberration fan for doublet and aspheric lenses. It can be seen that the transverse aberrations for doublet (figure 4 (a)) are much more than the aspheric lens (figure 4 (b)). As explained in [8] the transverse measure of an aberration is directly related to the size of the image blur. By comparing the figure 3 and figure 4 it is clear that the image quality in case of aspheric lens is much better than the image quality in case of doublet.

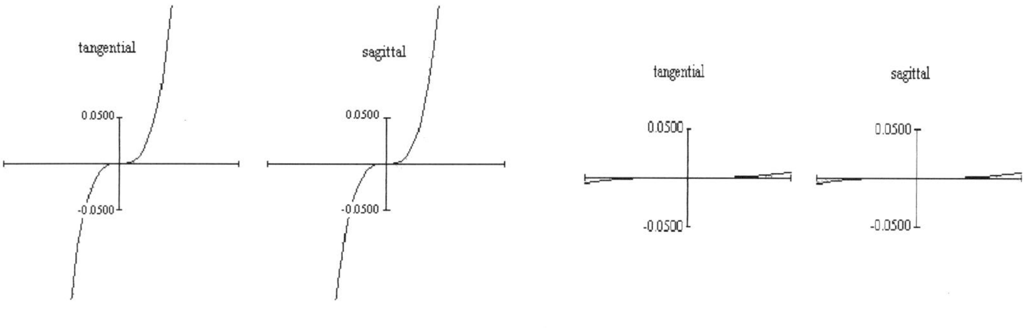

(a) (b)

Figure 4. Comparison of transverse aberration fan for doublet and aspheric lenses.

The manufactured aspheric lens was replaced with the doublet lens in the eyepiece assembly and measured optical parameters are listed in Table 2.

Table 2. Comparison of Optical parameters of doublet and aspheric lens fitted in eyepiece.

Parameter	**Doublet Lens**	**Aspheric Lens**
Exit Pupil Diameter	5.2 mm	5.2 mm
Resolution	4.12"	4.12"
Focal Length of eyepiece	28.07± 0.5 mm	28.06± 0.2 mm
Field of View	6.1°	6.1°
Exit Pupil Distance	22 mm	22 mm
Magnification	11.76	11.8
Focal Length of Individual Lens	98.12 ±0.6 mm	97.99 ± 0.2 mm

By comparing the optical parameters of both lenses given in table 2 it can be seen that almost all the parameters are same except the minor differences in the focal lengths of individual components as well as the eyepiece assembly. It can also be concluded that the deviation in focal lengths is minimum for aspheric lens than the doublet lens in both cases.

4. Conclusions

The development of aspheric lens provides the advantages over the achromat (doublet) lens such as reduction in optical components, manufacturing time as well as weight, more compact, improved in image quality (minimize aberrations), cost effective in the eyepiece assembly. However the short coming in the measurement of surface accuracy method is still in developing phase.

5. Acknowledgement

We are thankful to Mr. Kashif Dad Khattak and Mr. A. G. Molano for their valuable contributions in designing of aspheric lens and testing of optical parameters of eyepiece and lens assembly.

6. References

[1] Antonin M, Pavel N , Jiri N 2013 *Optics & Laser Technology* 45 708.

[2] Wang X, Zheng L, Zhang B, Li R, Fan D, Deng W, Wang X, Zhang F, Zhang Z, Zhang X 2009 *Proc. of SPIE* 72832J-1 7283.

[3] Juan C M, Pablo B, Wang L, José I, Fernando M, Asunción S 2009 *J. Opt. Soc. Am.* 17(26) 24036.

[4] Malacara D, "Optical Shop Testing" 2007 *John Wiley & Sons New York.*

[5] Jurek B "Optical Surfaces: Aspherical Optical Systems-X-Ray Optics-Reflecting Microscopes-Reflectors-Measurements" 1977 *Elsevier Scientific Pub. Co. Amsterdam.*

[6] Braunecker B, Hentschel R, Tiziani H. "Advanced Optics using Aspherical Elements" 2008 *SPIE Press Mc Graw Hill New York.*

[7] Rodgers J M "Nonstandard Representations of Aspheric Surfaces in Optical Design" 1984 *Ph.D. Theses University of Arizona.*

[8] Werran J S "Modern Optical Engineering" 2000 Third Edition *SPIE Press Mc Graw Hill New York.*

Synthesis of copper coated carbon nanotubes for aluminium matrix composites

Adnan Maqbool, F Ahmad Khalid, M Asif Hussain and Nabi Bakhsh
Faculty of Materials Science and Engineering, GIK Institute of Engineering Sciences and Technology, Topi 23640, KP, Pakistan

E-mail: adnanmaqbool247@gmail.com

Abstract. In this investigation copper coated carbon nanotubes (CNTs) were prepared to enhance the interfacial bonding between CNTs and aluminum matrix by the molecular-level mixing process. In optimized plating bath of (1:1) by wt. CNT with Cu, thickness of coated CNTs is reduced to 100 nm to promote uniform distribution of Cu nanoparticle on the surface of pretreated CNTs. The mixing of CNTs was accomplished by ultrasonication and ball milling. Scanning electron microscope analysis revealed the homogenous dispersion of Cu-coated CNTs in nanocomposites samples compared to the uncoated CNTs. The samples were pressureless sintered under vacuum. The densification increased with the increase in the CNTs content and is more pronounced in Cu-coated CNT nanocomposites.

1. Introduction

During past few decades, carbon nanotubes (CNTs) have attracted much attention. Especially for composite materials, it is predicted as an ideal reinforcement used in manufacturing of aircraft and sports goods, due to high aspect ratio, strength, elastic modulus and low density [1, 2]. Many researchers have encouraged the use of CNTs as reinforcement in metal matrices. For the fabrication of high strength and elastic modulus, homogeneous dispersion of CNTs is necessary and the interfacial bonding between metal and CNTs should be strong enough to transfer load efficiently. However, no major breakthrough is achieved with CNT/Al composites, because of obstacles related to the interfacial bonding between Al powder and CNTs, and the unavailability of suitable synthesis techniques for homogenous mixing at molecular level between Al and CNTs [3]. In addition, CNTs poor wetting property on Al due to entanglement and difference in density have been great hindrances in its uniform distribution for manufacturing the composites [4]. In case of Co-CNT, Ni-CNT and Cu-CNT composite powders, novel fabrication techniques to obtain outstanding composite powders are possible as the metal oxides (CuO and CoO) are easily reduced to metal Cu and Co [5, 6]. However, the fabrication of CNT/Al composite powder is irreversible once Al gets oxidized.

The researchers who have fabricated CNT/Al composites by conventional powder metallurgy and spray coating processes, have suggested CNT/Al composites as a new advanced structural material having high strength and modulus with low density for possible application in the next generation materials for automobile, sports and aerospace industries [7-9]. This consists of mixing CNTs with Al followed by spark plasma sintering or hot-extrusion. Kwon et al. fabricated CNT/Al composites by dispersion in natural rubber and a combination of processes of SPS and hot extrusion to get highly densified CNT/Al composites [10, 11]. Deng et al. performed isostatic pressing, followed by hot extrusion process. They distributed CNTs on the surface of 2024-Al powder by mechanical mixing assisted with ultrasonic shaker and fabricated composites which showed increased tensile strength, hardness and young's modulus [12-14]. Liao et al. fabricated CNT/Al composites by mixing the composite powder in the roller mill followed by SPS sintering and hot extrusion [15-18].

In this study, Cu-coated CNTs were prepared to enhance the interfacial bonding between CNTs and Al matrix by the molecular-level mixing process. To improve interfacial bonding, the enhancement in wettability of CNTs is necessary. For this purpose, CNTs were coated with Cu in the optimized plating bath. Therefore the influence of CNTs dispersion in the Al matrix for both, the Cu-coated and uncoated CNTs was ascertained.

2. Experimental procedures

For the fabrication of Al-matrix composites, Cu-coated and uncoated CNTs were used as the reinforcement in the Al matrix. Two processes: electroless plating and ultrasonic assisted mixing with N_2 purging were utilized for the uniform distribution of CNTs. For the enhancement in interfacial bonding of Al matrix and CNTs, a Cu/CNT precursor was prepared by using an electroless plating process and then mixed by planetary ball milling with Al powder.

For the fabrication of the composites, gas atomized Al powder (99.84% pure, with 10 μm average particle size, Angang Group Co. Ltd., China) and multiwalled carbon nanotubes (Sun Nanotech Co. Ltd., China, purity >90%, diameter 10–30 nm) were used in the present study. To get rid of the metallic particles and other impurities, purification and functionalization of CNTs carried out by acid treatment. Purification of CNTs was done in conc. HCl (37%), sonicated at 30 °C for 5 h and were then soaked for 12 h. Functionalization of purified CNTs was done by mixing conc. HNO_3 (65%) in conc. H_2SO_4 (98%) in the ratio of (1:3). After the functionalization process, CNTs were added in 1M solution of $SnCl_2.2H_2O$ along with 1 ml of HCl in 100 ml to adjust pH below 1 and sonicated for 30 min. Sensitized CNTs were washed with distilled water followed by filtration. For activation, CNTs were added in .0015 M solution of $PdCl_2$ along with 1.5 ml of HCl to adjust pH below 1 and sonicated for 30 min. Finally, activated CNTs were washed with distilled water followed by centrifugation and filtration. After two-step sensitization and activation, CNTs were exposed to catalytic treatment in $CuSO_4.5H_2O$ bath for 2 h at room temperature to suspend the CNTs in the solution. Formaldehyde was added in order to reduce the Cu in the bath for 20 min. Initially, 40 vol% CNT/Cu were used in the plating bath i.e. (1:9) wt ratio of CNTs with Cu. To optimize the Cu coating, bath composition was adjusted for the fixed Cu-CNT (1:1) wt. ratio i.e. for 50 mg CNTs bath composition was adjusted according to Cu content in the bath. The plating solution composition was $CuSO_4.5H_2O$ 196 g, $Na_3C_6H_5O_7.2H_2O$ 532 g and NaOH was used to adjust the pH equal to 12 and 1 ml of formaldehyde was added drop wise in the solution. The composition (1:1) wt ratio of CNTs with Cu showed satisfactory results and whole solution changed from black to red brown, showing the uniform coating of Cu.

Ultrasonic assisted mechanical mixing of the Al and different compositions of uncoated and Cu-coated CNTs named by CNTs additions (0.25, 0.50, 0.75 and 1.0 wt%) were performed in ethanol. To facilitate uniform distribution and mixing of CNTs; mixture was sonicated at 60 °C. Nitrogen gas was bubbled from bottom of the beaker by a glass rod that further enhanced the dispersion of Al powder with the CNTs by avoiding powder sedimentation. The process was continued until the thick slurry of the CNT/Al mixture was obtained. 12 g of mixed powder was placed in 250 ml stainless steel mixing jars together with stainless steel balls of 8 mm (Ø). The ball to weight ratio was also adjusted to (10:1). The jars were filled with argon gas and planetary ball mill was used for the mixing of powders, at 200 rpm for 1 h. Ethanol was used as the process control agent (PCA), for the purpose of minimizing cold welding of particles and to avoid sticking of powder with balls and walls of the jar. Uniaxial cold compaction of ball milled powder at 600 MPa was done by hydraulic press in 13 mm high-speed steel die. Pressureless sintering was performed in the tube furnace at 550 °C for 3 h in the evacuated quartz tube at 10^{-4} mbar.

Archimedes principle was used for the calculation of relative density of the sintered compact, with respect to the water as an immersing medium. IR spectrum was obtained by Fourier Transform Infrared Spectroscopy (FTIR) spectrum 2000 (Perkin Elmer). Al was preferentially etched to reveal the CNTs and grain boundaries. Etching was performed at room temperature in a solution of 2 g NaOH, 4 g

Na_2CO_3 and 94 ml of H_2O [19]. Scanning electron microscopy (SEM) was done to examine the distribution of CNTs with in the CNT/Al composites. Microhardness was evaluated with Vickers Hardness testing machine. To measure the microhardness, Vickers microindenter was used at load of 100 g_f. Each hardness value datum was the average of 3 samples.

3. Results and discussion

In this experimental study, Al is in the form of fine spherical powder with average particle size of 10 μm. CNTs are in the form of agglomeration and have the diameter 10–30 nm. Raw CNTs are highly tangled together and in the form of agglomerates. Impurities get attached with the walls and ends of CNTs during the production process. So, impurities are removed after functionalization and it improved the dispersion of CNTs in the matrix. This resulted in the better mechanical properties of composites.

3.1 CNTs functionalization analysis

The infrared spectrum of functionalized CNTs after acid treatment is shown in the Fig. 1. This curve indicates that few organic functional groups formed after the acid treatment on the surface of CNTs. In IR spectrum, there exist three valleys at wave number 3418, 1633 and 1110 cm^{-1} representing stretching vibration of hydroxyl group (-OH), aldehydic group (>C=O) and C-O groups respectively. The coexistence of hydroxyl and aldehydic groups may yield the carboxyl group (-COOH). This confirms that functionalization of the CNTs is done successfully. These results are very similar to the previous work and in close agreement with the literature [14, 20].

3.2 Electroless CNT/Cu plating

A fabrication approach such as the molecular-level mixing is necessary for CNT/Al composites. Homogeneous dispersion and strong interfacial bonding of CNTs with metal can be achieved by electroless coating [21]. By chemical reaction between functional groups on the surface of CNTs with metal ions, strong interfacial bonding between CNTs and a metal is achieved.

For the evaluation of coating thickness, initially (1:9) by wt CNTs with Cu were treated in the plating solution. Cu particles are decorated and distributed on the surface of CNTs, as shown in the Fig. 2a. It

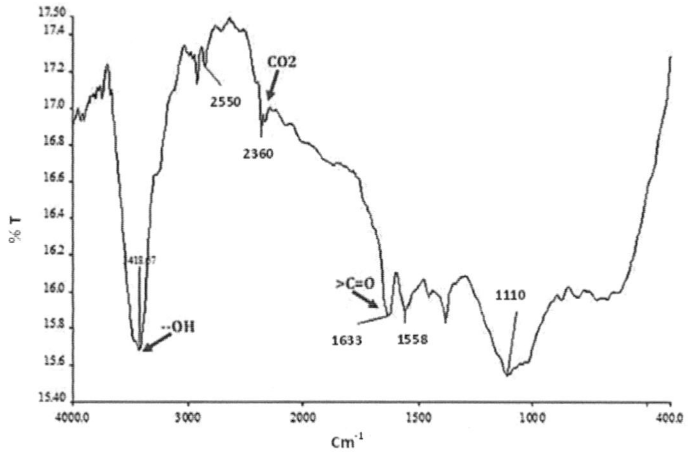

Figure 1. FTIR spectrum of CNTs after functionalization

Figure 2. SEM micrographs of (a) (1:9) by wt CNT with Cu coating, inset is showing that CNTs are completely embedded in the Cu nanoparticles and (b) (1:1) by wt CNTs with Cu coating, inset is showing Cu nanoparticles are the homogenously distributed on CNTs.

is observed that thickness of Cu-coated CNTs is about 300-500 nm. The size of Cu nanoparticle is less than 100 nm and densely coated on the CNTs, similar results are reported by many research studies [22, 23]. In addition, open end tips of CNTs are found and surface of CNTs is completely embedded in the Cu particles, as shown in the inset of Fig. 2a. In electroless process, some massive crystallite growths are also perceived at isolated spaces on Cu-coated CNTs, which is in accordance to the previous results [24, 25].

Optimizing the plating bath to (1:1) by wt CNT with Cu, thickness of Cu-coated CNTs is reduced to around 100 nm, as shown in Fig. 2b. By comparing the both inset micrographs in fig 2, it can be seen that thickness of Cu coating on the CNTs is successfully reduced. The Cu nanoparticles of few nm are homogenously distributed on the surface of pretreated (Sensitized and activated) CNTs. By the combination of molecular level growth of Cu on the CNTs (Bottom up approach) and controlling the coating thickness to very small and uniform level, excellent quality of Cu-coated CNTs has been achieved [26].

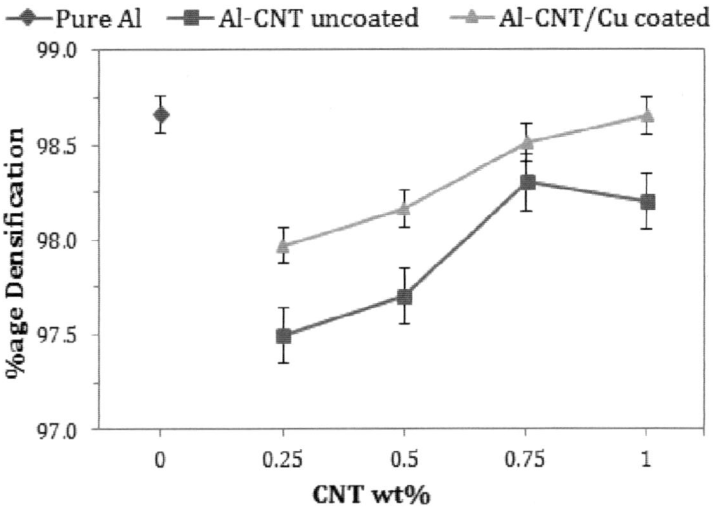

Figure 3. Effect of CNTs on densification behavior of uncoated and Cu-coated CNT/Al composites.

3.3 Densification

By mixing Al powder with Cu-coated and uncoated CNTs, a proper mechanical interaction between Al powder and CNTs can be achieved. As the CNTs content increased, more and more powder is covered with the CNTs, which is also suggested by previous research studies [16, 17]. The Cu-coated CNTs have provided sufficient wettability with the Al and hence a good bonding is achieved.

For the density measurement of Al and CNT/Al composites, Archimedes principle was used and distilled water as an immersion medium. The percentage relative densities of the samples prepared are shown in Fig. 3. The reference density of pure aluminum sample was close to intrinsic Al density (2.70 g/cm^3). The composites were fully densified near to their theoretical density. Densification of composite samples increased with increasing wt% of CNTs.

3.4 Microstructure

In sintered composite samples, morphology of Al grains did not change much relative to that of the raw powders, as shown in the Fig. 4a. It is due to the equal pressure being applied around the compacts, which is beneficial for the plastic forming of the powders in order to achieve the full density. The grain boundaries are more visible after repeated etching.

In composite samples, CNTs are homogeneously dispersed at grain boundaries and within the Al particles, as shown in the Fig. 4b. Equiaxed grains are observed in all the composites. This occurs because particles growth is restrained by the pinning effect of the CNTs. Here it is significant to note that because of Cu-coated CNTs, more interfacial bonding with Al matrix is achieved. Hence, small grain size resulted in the Cu-coated CNT/Al composites compared to uncoated CNT/Al composites, where a little grain growth is occurred and relatively bigger grains are observed. Owing to poor wettability in uncoated CNTs, more Al powder are cold welded during ball milling and pinning effect of CNTs in Al particles, is less pronounced compared to Cu-coated CNT/Al composites.

4. Summary

Molecular level mixed Cu-coated CNTs showed uniform dispersion in the Al matrix. Optimizing the plating bath to (1:1) by wt CNT with Cu, thickness of Cu-coated CNTs has been reduced to 100 nm. Wettability of Cu-coated CNTs was increased in Al matrix. The densification increased with the increase in the CNTs content, which is more pronounced in Cu-coated CNT/Al composites due to higher affinity

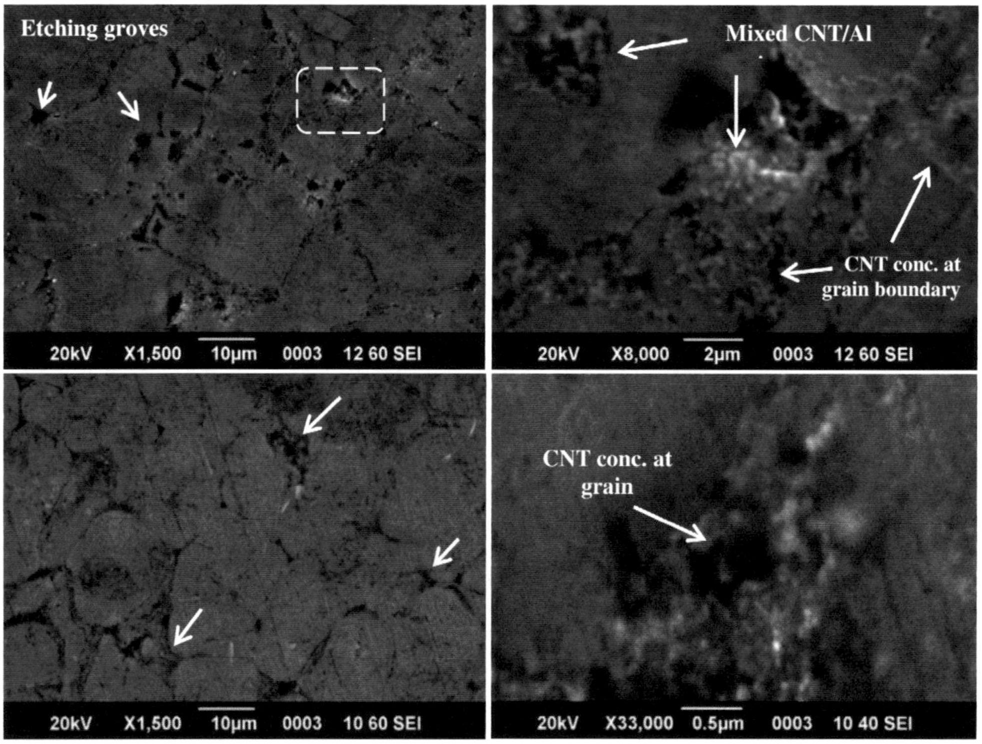

Figure 4. SEM micrographs of etched surfaces (a & b) 1.0 wt.% uncoated CNT/Al composites is showing the conc. of CNTs at grain boundaries and (c & d) 1.0 wt.% Cu-coated CNT/Al composites is showing the presence of CNTs within the etching grove at grain boundary

and strong interfacial bonding with Al matrix. Scanning electron microscope analysis revealed the homogenous dispersion of Cu-coated CNTs in nanocomposites samples compared to the uncoated CNTs. The densification increased with the increase in the CNTs content and is more pronounced in Cu-coated CNT nanocomposites.

5. Acknowledgments

The authors are thankful to GIK Institute for financial assistance and providing research facilities. They are sincerely thankful to Dr. A. Saeed Hakeem, Dr. Basit Yameen and Dr. Hamid Zaigham of the GIK Institute for their technical support and valuable suggestions.

6. References

[1] P.M. Ajayan, J.M. Tour, Materials science: nanotube composites, Nature 447 (2007) 1066-1068.

[2] P. Harris, Carbon nanotube composites, Int. Mater. Rev. 49 (2004) 31-43.

[3] T. Kuzumaki, K. Miyazawa, H. Ichinose, K. Ito, Processing of carbon nanotube reinforced aluminum composite, J. Mater. Res. 13 (1998) 2445-2449.

[4] R. George, K. Kashyap, R. Rahul, S. Yamdagni, Strengthening in carbon nanotube/aluminium (CNT/Al) composites, Scr. Mater. 53 (2005) 1159-1163.

[5] S.I. Cha, K.T. Kim, S.N. Arshad, C.B. Mo, S.H. Hong, Extraordinary Strengthening Effect of Carbon Nanotubes in Metal Matrix Nanocomposites Processed by Molecular Level Mixing, Adv. Mater. 17 (2005) 1377-1381.

[6] S.I. Cha, K.T. Kim, S.N. Arshad, C.B. Mo, K.H. Lee, S.H. Hong, Field Emission Behavior of a Carbon Nanotube Implanted Co Nanocomposite Fabricated from Pearl Necklace Structured Carbon Nanotube/Co Powders, Adv. Mater. 18 (2006) 553-558.

[7] C. Xu, B. Wei, R. Ma, J. Liang, X. Ma, D. Wu, Fabrication of aluminum-carbon nanotube composites and their electrical properties, Carbon 37 (1999) 855-858.

[8] T. Laha, S. Kuchibhatla, S. Seal, W. Li, A. Agarwal, Interfacial phenomena in thermally sprayed multiwalled carbon nanotube reinforced aluminum nanocomposite, Acta Mater. 55 (2007) 1059-1066.

[9] T. Laha, A. Agarwal, T. McKechnie, S. Seal, Synthesis and characterization of plasma spray formed carbon nanotube reinforced aluminum composite, Mater. Sci. Eng. A 381 (2004) 249-258.

[10] H. Kwon, M. Estili, K. Takagi, T. Miyazaki, A. Kawasaki, Combination of hot extrusion and spark plasma sintering for producing carbon nanotube reinforced aluminum matrix composites, Carbon 47 (2009) 570-577.

[11] H. Kwon, A. Kawasaki, Extrusion of Spark Plasma Sintered Aluminum-Carbon Nanotube Composites at Various Sintering Temperatures, J Nanosci Nanotechnol 9 (2009) 6542-6548.

[12] C. Deng, D. Wang, X. Zhang, A. Li, Processing and properties of carbon nanotubes reinforced aluminum composites, Mater. Sci. Eng. A 444 (2007) 138-145.

[13] C. Deng, X. Zhang, Y. Ma, D. Wang, Fabrication of aluminum matrix composite reinforced with carbon nanotubes, Rare Met 26 (2007) 450-455.

[14] C. Deng, P. Zhang, Y. Ma, X. Zhang, D. Wang, Dispersion of multiwalled carbon nanotubes in aluminum powders, Rare Met 28 (2009) 175-180.

[15] I. Sridhar, K.R. Narayanan, Processing and characterization of MWCNT reinforced aluminum matrix composites, J Mater Sci 44 (2009) 1750-1756.

[16] J.-z. Liao, M.-J. Tan, I. Sridhar, Spark plasma sintered multi-wall carbon nanotube reinforced aluminum matrix composites, Mater Des 31 (2010) S96-S100.

[17] J. Liao, M.-J. Tan, Mixing of carbon nanotubes (CNTs) and aluminum powder for powder metallurgy use, Powder Technol. 208 (2011) 42-48.

[18] J.Z. Liao, M.J. Tan, R.V. Ramanujan, S. Shukla, Carbon Nanotube Evolution in Aluminum Matrix during Composite Fabrication Process, Mater. Sci. Forum 690 (2011) 294-297.

[19] K. Morsi, A.M.K. Esawi, S. Lanka, A. Sayed, M. Taher, Spark plasma extrusion (SPE) of ball-milled aluminum and carbon nanotube reinforced aluminum composite powders, Compos. A 41 (2010) 322-326.

[20] C. Xu, G. Wu, Z. Liu, D. Wu, T.T. Meek, Q. Han, Preparation of copper nanoparticles on carbon nanotubes by electroless plating method, Mater. Res. Bull. 39 (2004) 1499-1505.

[21] S.-H. Joo, S.C. Yoon, C.S. Lee, D.H. Nam, S.H. Hong, H.S. Kim, Microstructure and tensile behavior of Al and Al-matrix carbon nanotube composites processed by high pressure torsion of the powders, J Mater Sci 45 (2010) 4652-4658.

[22] W.M. Daoush, Processing and characterization of CNT/Cu nanocomposites by powder technology, Powder Metall. Met. Ceram. 47 (2008) 531-537.

[23] J. Silvain, C. Vincent, J. Heintz, N. Chandra, Novel processing and characterization of Cu/CNF nanocomposite for high thermal conductivity applications, Compos. Sci. Technol. 69 (2009) 2474-2484.

[24] L. Ang, T.S.A. Hor, G. Xu, C. Tung, S. Zhao, J. Wang, Decoration of activated carbon nanotubes with copper and nickel, Carbon 38 (2000) 363-372.

[25] W.M. Daoush, B.K. Lim, C.B. Mo, D.H. Nam, S.H. Hong, Electrical and mechanical properties of carbon nanotube reinforced copper nanocomposites fabricated by electroless deposition process, Mater. Sci. Eng. A 513-514 (2009) 247-253.

[26] Y. Peng, Q. Chen, Ultrasonic-assisted fabrication of highly dispersed copper/multi-walled carbon nanotube nanowires, Colloids Surf. Physicochem. Eng. Aspects 342 (2009) 132-135.

Investigation of temperature dependent barrier height of Au/ZnO/Si schottky diodes

M Asghar[1], K Mahmood[1], S Rabia[1], Samaa BM[1], M Y Shahid[1] and M A Hasan[2]
[1]Department of Physics, The Islamia University of Bahawalpur, Pakistan
[2]Department of Electrical and Computer Engineering, UNC-Charlotte, NC, USA

E-mail: mhashmi@iub.edu.pk

Abstract. In this study, temperature dependent current-voltage (I-V) measurements have been performed to investigate the inhomogeneity in the temperature dependent barrier heights of Au/ZnO/Si Schottky barrier diode in the temperature range 150 – 400K. The room temperature values for ideality factor and barrier height were found to be 2.9 and 0.60 eV respectively indicating the inhomogenity in the barrier heights of grown samples. The Richardson plot and ideality factor verses barrier height graph were also drawn to verified the discontinuity between Au and ZnO. This barrier height inhomogenity was explained by applying Gaussian distribution model. The extrapolation of the linear Φ_{ap} (n) plot to n = 1 has given a homogeneous barrier height of approximately 1.1 eV. Φ_{ap} versus 1/T plot was drawn to obtain the values of mean barrier height for Au/ZnO/Si Schottky diode (1.1 eV) and standard deviation(δ_s) (0.02 V) at zero bais.

1. Introduction

ZnO is a material of interest among the research community due to its wide band gap of 3.37 eV and large exciton binding energy (60 meV) at room temperature [1, 2]. In order to utilized ZnO in practicle appications, metal contacts are neccessory. Therefore the study of characteristics parameters such as barrier height and ideality factor of metal-semiconductor Schottky diodes has fudamental importance. The performance and relaibility of these devices especially depend on the formation of insulator layer between metal and semiconductor interface, inhomogenities of Schottky barrier contacts and series resistance of diode [3]. Generally Schottky contacts show temperature dependent behavior as evidenced by the parameters such as ideality factor, barrier height and series resistance [4]. It is observed in the literature that I-V characteristics normally deviate from the ideal thermionic model because both, ideality factor and barrier height showed strong temperature dependence behavior [5-7]. This abnormal behavior of Schottky diodes was attributed to series resistance and inhomogenites between metal and semiconductor [8, 9]. Different models have been employed to understand the abnormal behavior but the subject is still under hot debate. Therefore the study of Schottky contacts of different metals on ZnO is an active area of research and needs to be further investigated.

In this study temperature dependence I-V measurements have been performed on Au/ZnO/Si Schottky diodes in the temperature range of 150 to 400 K to investigate the discontuniuty present between Au and ZnO. The ideality factor, series resistance and barrier height show strong dependence of temperature. There was also correlation between ideality factor and barrier height which showed the inhomogenity in the barrier height. The inhomogenity in barrier height was explained using Gaussian distribution model of barrier heights.

2. Experimental

ZnO layers were grown on 3 inch diameter p-type silicon (001) wafer (N_A ~ 1.2 x 10^{15} cm^{-3}) by means of Molecular Beam Epitaxy (MBE). The ZnO was grown under following conditions: the substrate temperature was set at 703 K while oxygen was supplied from an RF atomic source and Zn was evaporated from -Zn-Knudsen cell maintained at 562 K filled with Zn beads having a purity of 99.9999%. The pressure of the chamber during the growth was ~ 1× 10^{-4} mbar mainly due to the pressure required to operate the oxygen RF atomic source and growth time was 7 hours. The oxygen RF atomic source was operated at 300W. For electrical measurements, gold Schottky contacts with diameter 78 mm^2 were fabricated by electron beam evaporation. I-V measurements of Schottky diodes were carried out by the following equipment (manufacturer): DLTS (DLS-83D Hungry)

3. Results and Discussion

Fig. 1 shows the semi-logarithemic forwar biased I-V characteristic of the Au/ZnO/Si Schottky barrier diodes in the temperature range 150 to 400 K. Current in the Schottky barrier diode due to the thermionic emission, neglecting series resistance is given as [10],

$$I = I_S\left[\exp\left(\frac{qv}{nkT}\right) - 1\right] \quad \textbf{(1)}$$

Where n is the ideality factor and I_S is the reverse saturation current given by the relation,

$$I_S = AA^*T^2\exp\left[\frac{-q\phi_B}{kT}\right] \quad \textbf{(2)}$$

Where A is the contact area, A^* is the Richardson constant and its value is 32 AK^{-2} cm^{-2} for ZnO [11]. T is temperature in Kelvin, k is the Boltzmanns constant (k=1.38×10^{-23} J/K or 8.617×10^{-5}eV/K), q is the electric charge and ϕ_B is the barrier height

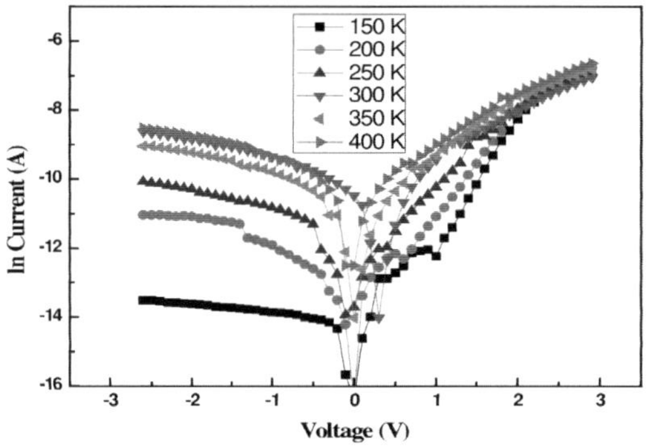

Figure 1. Semi-logarithm I-V plot of Au/ZnO/Si Schottky diodes at different temperatures.

The value of barrier height and ideality factor was determined from the slope of semi-logarithm forward biased I-V characteristics (shown in fig. 1) using following equations.

$$n = \frac{q}{kT \times slope} \quad \textbf{(3)}$$

$$\phi_{B(1-v)} = \frac{kT}{q}\ln\left(\frac{AA^*T^2}{I_S}\right) \quad \textbf{(4)}$$

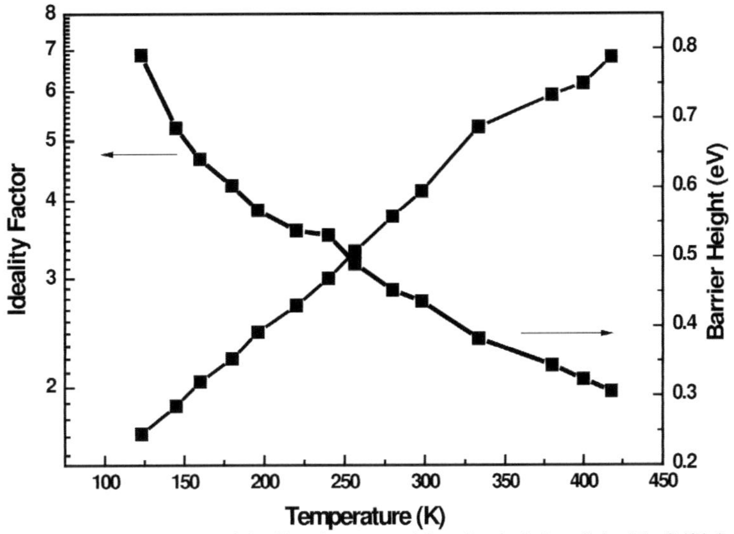

Figure 2. Effect of temperature on ideality factor and barrier height of Au/ZnO/Si Schottky diodes

The ideality factor and barrier height shows a strong dependence of temperature as shown in fig.2. The ideality factor decreases and barrier height increases with temperature. Room temperature n and ϕ_B values are found to be 2.9 and 0.60 eV respectively. The value of ideality factor attributed to interface states between gold and ZnO, image force lowering and barrier height inhomogenity [12]. Such results have been already published by many researchers [13]. Such temperature dependence is an obvious disagreement with the reported negative temperature coefficient of the Schottky barrier height. The reason of this disagreement may be the barrier height inhomogenity.

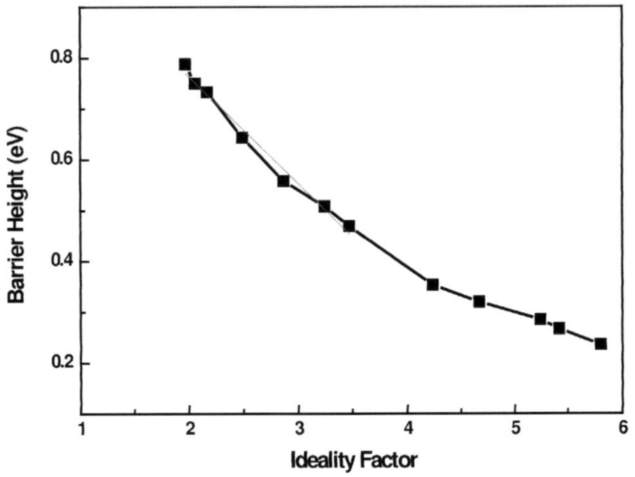

Figure 3. The plot between barrier height and ideality factor. The linear relationship demonstrated the presence of inhomogenity in barrier height

The barrier height inhomogenity can be also determined by plotting the graph between ideality factor and barrier height. Fig. 3 shows a plot of experimental BH and ideality factor. Fig. 3 demonstrated a linear relationship between experimental BH and ideality factor and is explained by lateral inhomogenities of BH in the schottky diode. The extrapolation of experimental BH verses ideality factor plot of n=1 has given homogenious barrier height of 1.1 eV. Therefore decreases of ideality factor and increases of barrier height at high temperature clearly demonstrated that there is discontinuity at Au/ZnO interface.

This barrier height inhomogeneity can be explained by using Gaussian distribution model of barrier height. According to this model the expression of barrier height can be written as [14- 16]

$$\Phi_{ap}=\Phi_{bo}-q\delta_s^2/2kT \quad \textbf{(5)}$$

where Φ_{ap} is the apparent BH which can be measured experimentally, Φ_{bo} is the mean BH and δ_s is the standard deviation of the BH distribution. The standard deviation is the measure of barrier height inhomogenity.

The temperature dependence of δ_s is usually small and can be neglected. In this model the observed variation of ideality factor with temperature is given by [17]

$$(1/n_{ap} -1) = - \rho 1 + q\, \rho 2/2kT \quad \textbf{(6)}$$

where n_{ap} is the apparent ideality factor and ρ1 and ρ2 are voltage coefficients which may depend on temperature and they quantify the voltage deformation of the BH distribution

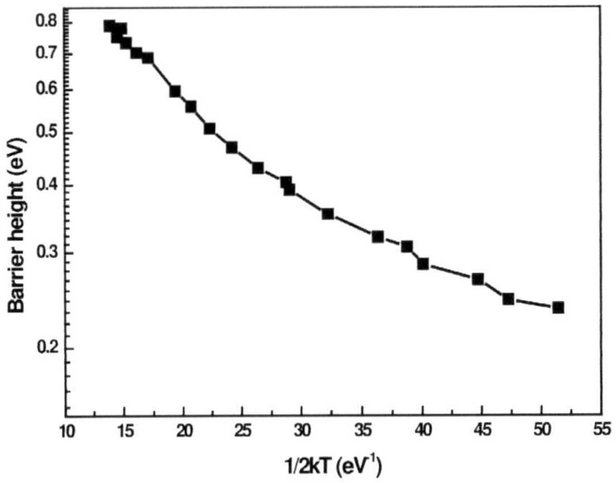

Figure 4. Zero bias apparent barrier height verses1/2kT curves of Au/ZnO/Si Schottky diodes according to the Gaussian barrier height

The linear relationship between barrier height and 1/2kT shows an agreement with current model that confirmed the discontinuity at metal and semiconductor interface. The fig shows a plot between Φ_{ap} and 1/2kT for Au/ZnO Schottky diode [18]. The intercept and slope of this plot gives the mean barrier height and the zero-bais standard deviation with values 1.1 eV and 0.02 eV respectively.

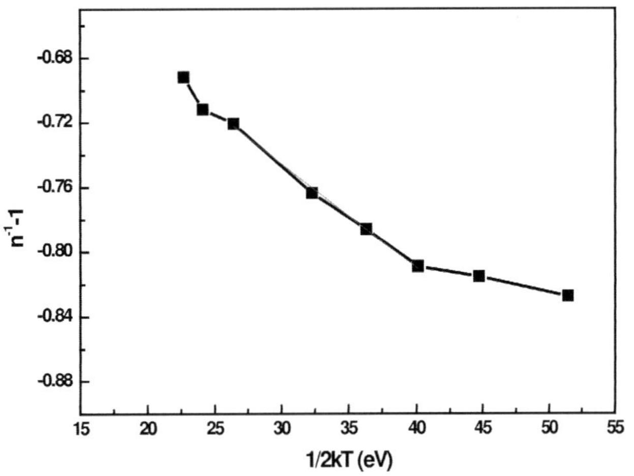

Figure 5. Zero bias ideality factor verses1/2kT curves of Au/ZnO/Si Schottky diodes according to the Gaussian barrier height

Again, the plot of n_{ap} versus 1/2kT is a straight line that gives voltage coefficients ρ1 and ρ2 from the intercept and slope, respectively (shown in fig. 5). The values of ρ1 = 0.55 and ρ2 = -0.006 V were obtained from the experimental n_{ap} versus 1/2kT plot (Fig. 5). As the inhomogenity of the interface is depend upon the value of δ_s. The lower value of δ_s corresponds to more homogeneous barrier heights.

4. Conclusion

In this study, temperature dependent current-voltage (I-V) measurements have been performed to investigate the inhomogeneity in the temperature dependent barrier heights of Au/ZnO/Si Schottky barrier diode in the range 150–400K. The room temperature values for ideality factor and barrier height found to be 2.9 and 0.60 eV respectively indicating the inhomogenity in the barrier heights of grown samples. This barrier height inhomogenity was explained by applying Gaussian distribution model.

5. References

[1] Asghar M, Mahmood K, Ali A, Hasan M-A, Raja M Y A, Hussain I, Willander M 2011 *ECS transactions* 35 (6) 149-154

[2] Asghar M, Mahmood K, Ferguson I.T, Yasin R, Tsu R, Xie Y-H 2013 *Semicond. Sci. Technol.* 28 105019

[3] Asghar M, Mahmood K, Faisal F, Hasan M-A 2013 *J. Phys: Conf. Ser.*439 012030

[4] Durmus A A, Ali K, Ahmet F Ö 2012 *Microelectronic Eng. B.* 98 6-11

[5] Jiang Y.L, Ru G.P, Lu F, Qu X.P, Li B. Z, Li W, Li A. Z 2002 *Chin. Phys. Lett.* 19 553–556.

[6] Chand S, Kumar J1997 *J. Appl. Phys.* 82 5005–5010

[7] Tung R. T, Levi A.F.J, Sullivan J. P, Schrey F 1991 *Phys. Rev. Lett.* 66 72–75

[8] Tung R. T 1992 *Phys. Rev. B* 45 13509–13523

[9] Potzger K, Shengqiang Z.J, Grenzer M.H, Fassbender 2008 *J Appl. Phys. Lett.* 92 182504

[10] Tneh S.S, Hassan Z, Saw K.G, Yam F.K, Hassan H.A 2010 *Physica B* 405 2045

[11] Asghar M, Mahmood K, and Hasan M A 2012 Key Eng. Mat. 511-512 132-136

[12] Wenksten H V, Kaidashev E M, Lorentz M, Hochmuth H, Biehne G, Lenzer J, Gottscalch V, Pickenhain R and Grundmann M 2004 Appl. Phys.Lett. 84 79

[13] Shawn C, Bernard J R and Kristin M P 2008 Appl. Phys. Lett. 92 012103

[14] Aydo gan S, Cinar K, Asıl H, Coskun C and Türüt A 2009 J. Alloy. Comp. 476 913

[15] Coskun C, Gedik N and Balc E 2006 Semicond. Sci. Technol. 21 1656–1660

[16] Dhananjaya, Nagarajua J and Krupanidhi S B 2007 Physics B 391 344

[17] Padovani F A and Summer G 1965 Appl. Phys. A 36 3744

[18] Altindal S, Karadeniz S, Tugluogiu N and Tataroglu A 2003 Solid-state Electronics 471847 1854

CO_2 laser welding of AISI 321stainless steel

A Hussain, A H Hamdani and R Akhter
[1]Pakistan Institute of Lasers & Optics, P.O. Box 505, Rawalpindi, Pakistan

E-mail: pilo786@yahoo.com

Abstract. CO_2 laser welding of AISI 321austenitic stainless steel has been carried out. Bead on plate welds on 2mm thick steel were performed with 450W CO_2 laser at speeds ranging from 200 to 900 mm/min. It was observed that weld depth and width was decreased with increasing the speed at constant laser power. Butt welds on different sheet thickness of 1, 2 and 2.5 mm were performed with laser power of 450 W and at speed 750, 275 and 175 mm/min, respectively. The microstructures of the welded joints and the heat affected zones (HAZ) were examined by optical microscopy and SEM. The austenite/delta ferrite microstructure was reported in the welded zone. The microhardness and tensile strength of the welded joints were measured and found almost similar to base metal due to austenitic nature of steel.

1. Introduction

Laser welding of austenitic stainless steel has established a significant attention in industry due to its wide applications in high pressure tanks for transportation of liquids and compressed gases, in power plants, petroleum refinement stations, pharmaceutical industry, households etc. Laser welding produces good metallurgical properties, high production rate and ease of automation [1]. Low contents of carbon in austenitic stainless steels prevent the intergranular corrosion at welds. Extensive work on weldability of stainless steel has been reported [2-4], which gives the essential conditions for ensuring good quality welds.

Laser welding is a high energy density and low heat input process resulting in a small heat-affected zone (HAZ), very little distortion, and a high depth-to-width ratio for the fusion zone. The flow of heat and molten metal (in the weld pool) can significantly influence the temperature gradients, the cooling rates and the solidification rates of the weld zone which ultimately controls the penetration and shape of the fusion zone [5].

Laser welding involves a number of variables such as laser power, welding speed, focused spot size, type of shielding gas, etc. These parameters considerably influence the microstructure of the fusion zone [6-8].

High power lasers are normally used, in kilowatt range, for welding of steels [9-11]. In this paper bead-on-plate and butt welding of 1, 2 and 2.5mm thick austenitic stainless steel (AISI 321) was carried out using a 450W CO_2 laser at different speeds. Microstructure, microhardness and tensile strength of the welds were analyzed and discussed.

2. Experimental setup

Figure 1 shows the experimental set up for laser welding of steel. A 500W CO_2 laser operated at 450W was used to carry out the welding experiments for bead-on-plate and butt joints on AISI 321 steels. Laser beam was focused on steel plates using a 100mm focal length ZnSe lens.

The sample was fixed in a jig on the CNC table, which moved beneath the focused laser at different speeds. Argon was used as shielding gas to protect the laser treated zone from oxidation. Beads-on-plate welds were formed on 2 mm thick plate. Weld samples with thickness of 1, 2, 2.5mm were prepared for butt joints with dimensions of 30x70 mm and were held firmly using fixture to prevent distortion. After welding, the samples were visually inspected and sectioned transversely. Samples were mounted for grinding and polishing and then etched in 2% ferric chloride to observe the microstructure. The shape and microstructure of the fusion zone and base metal were examined using optical microscope. SEM studies of microstructure of the weld zones were also carried out. Mechanical tests including tensile strength and hardness measurements of butt welds were performed.

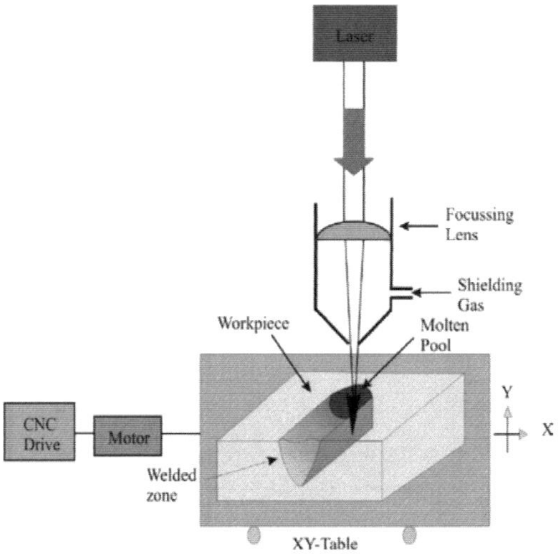

Figure 1. Experimental set up for laser welding process.

3. Results and discussion

3.1. Effect of traverse speed on weld depth and width

The welding of the samples was performed with focused laser beam having a spot size ~ 0.3 mm, which produced a continuous blue flash during the welding process. Figure 2 shows a graph for depth and width of the laser treated zone versus working speed for bead-on-plate weld. It can be seen that the depth and width of the welds decreases with increase of speed for fixed laser power. The interaction time of laser beam is decreased with increase of speed which ultimately reduces the heat input in the sample.

3.2. Microhardness and Microstructures

Figure 3 shows the microhard-ness of the laser weld samples along the depth at various speeds. It is clear from the figure that the microhardness of fusion zone remains almost the same as that of the base metal which is about 170±20Hv. It was observed that the hardness of weldment was almost same with increasing the speed of laser beam due to austenitic behavior of welded zone.

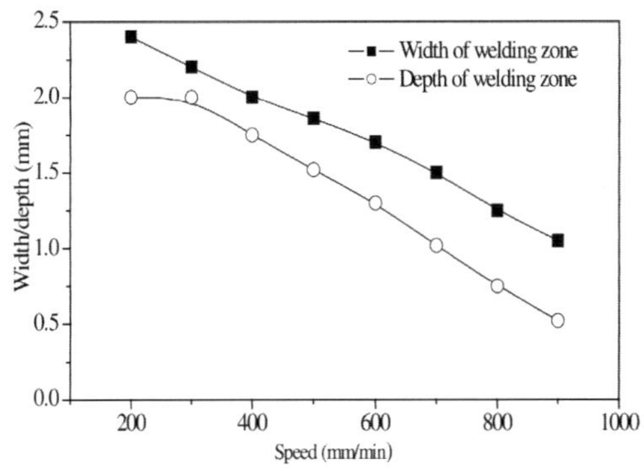

Figure 2. Variation of weld depth/width at different working speeds at laser power of 450W.

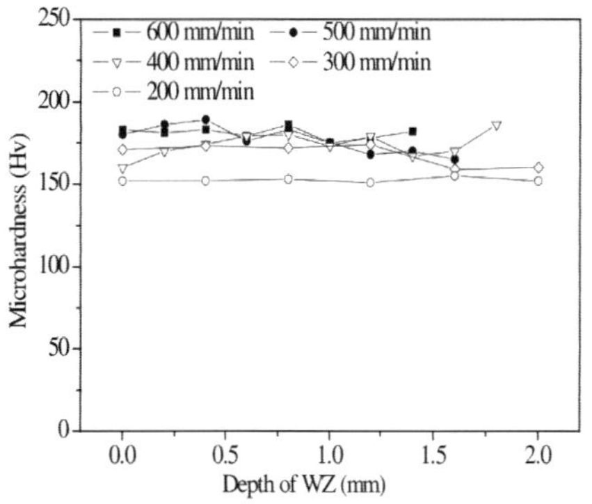

Figure 3. Microhardness Vs depth of weld zone at different speeds at constant laser power of 450W

Figure 4 shows a micrograph of the bead-on-plate (a) blind weld treated at 400mm/min using laser power of 450W. The micrograph shows three zones at higher magnification i.e., (b) heat affected zone (HAZ) (c) laser melted zone and (d) base metal. It can be seen that the fine dendritic structure is observed in fully laser melted zone which is attributed to high solidification rate. In this case the white region represents the austenite phase whereas the black lines show the δ-ferrite with vermicular morphology. In heat affected zone, the heat is transferred to base metal at very high cooling rate. Therefore heat energy finds insufficient time to melt the material which results in the partially melted zone as shown in Figure 4-b. HAZ is very small as compared to welded zone and is about 60-70 µm in width. Figure 4-d represents the microstructure of the base metal. It was also observed that the shape of the fusion zone was transformed from an elliptical shape to almost vertical and parallel sided for full penetration at low speed. Increase of laser power and decrease of working speed leads to increase in the heat affected zone (HAZ).

3.3. Butt welding

Three thicknesses 1, 2, 2.5mm of the austenite stainless steel AISI-321 sheets were welded with CO_2 laser at different speeds of 750, 275 and 175mm/min respectively at constant laser power of 450W. Complete penetrations of welded zones were obtained at parameters given in Table-1. The welded zones were symmetrical about the axis of the laser beam which shows that the welding was carried out at optimum laser parameters. No welding cracks or porosities were found in any of the welds by virtue of good crack resistance of the base metal and provision of suitable welding conditions.

Visual examinations of all butt joints showed that the surface of the joint's faces and roots were flat, smooth and with no undercuts. The weld heights were minimum and thermal distortions were negligible. It was found that the average width of butt welds of austenitic steels having sheet thickness 1, 2 and 2.5mm were 1.6, 1.8 and 2.2 mm, respectively which were less than the weld width reported elsewhere [8].

Table-1: Optimum welding parameters for various sheet thicknesses

S.No	Sheet thickness (mm)	Laser power (Watts)	Speed (mm/min)	Shielding gas
1.	1	450	750	Argon
2.	2.0	450	275	Argon
3.	2.5	450	175	Argon

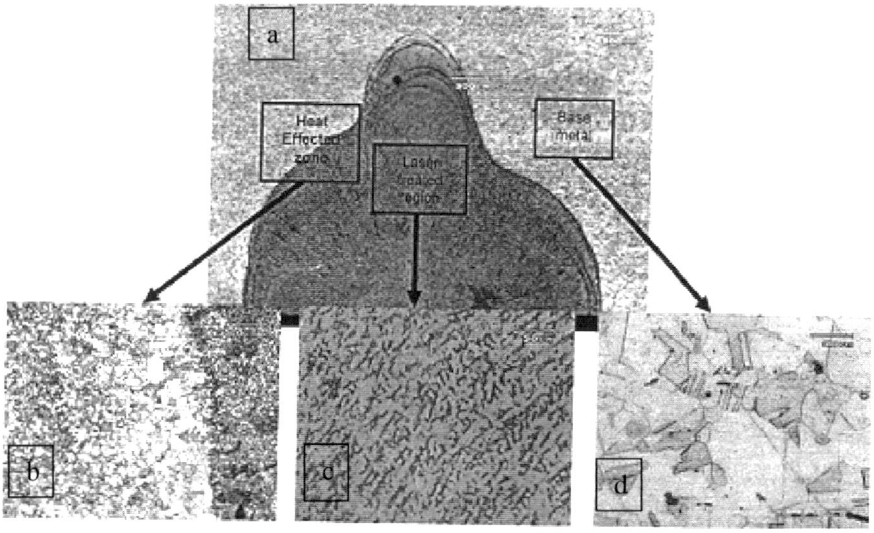

Figure 4. The microstructure of the laser treated sample (a) full bead, (b) heat affected zone, (c) melted zone and (d) base metal.

3.4. Microhardness and Microstructure

Measurements of microhardness profiles on the cross-section of the weld for each thickness showed that the fusion zone hardness was similar to base metal as shown in Figure 5. The steels of three thicknesses were welded at different speeds which were responsible for different solidification rates. However, it is was observed that the hardness of weld metal, HAZ and base metal remained in the range of 185±15Hv due to austenitic nature of stainless steel.

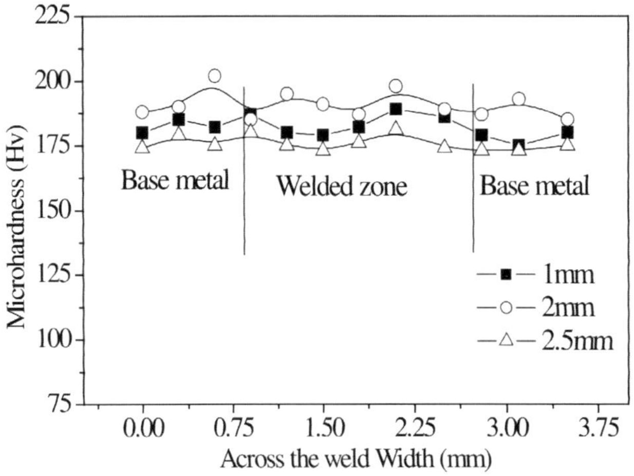

Figure 5. Microhardness across the welds width for three thicknesses

The analysis of macro and microstructure of butt welded steels at optimum laser parameters showed that the shape of the fusion zone was regular and symmetric and also the width of the HAZ is very narrow and is about 60-70μm.

In this paper, the microstructure of 2mm thick steel is being discussed because the similar behaviour of structural changes in the welding zones of 1 and 2.5mm steels was observed. Figure 6 represents the optical microstructures of base metal and fusion zone of 2mm austenitic steels which was carried out at speeds 275 mm/min with constant laser power. Figure 6-a shows the base metal, HAZ and fusion zone at low magnification. The microstructure of Figure 6-b represents the microstructure of annealed austenitic stainless steel of base metal. It is clear from the Figure 6-c that the laser melted zone has dark δ-Fe dendritic structure with vermicular morphology in austenite matrix which is attributed to high solidification and cooling rates due to low heat input. The width of heat effected zone is about 60-80 μm. The primary and secondary dendrite arm length in the structure is about 12-20 μm and 4-8μm, respectively. The average dendrite arm spacing of δ-Fe in laser welded zone is about 10-20μm.

Figure 7 shows the microstructure of welded zone using SEM; a) near the welding edge; b) center of the welding zone and, c) enlarge view of microstructure of "b". It can be seen that δ-ferrite exists as vermicular morphology near the edge of welding zone whereas skeletal morphology of delta ferrite in austenite structure near the center of the welding zone. This is due to the non-equilibrium rapid solidification condition in welding process. In laser welding, the peak temperature in fusion zone is much higher than the upper limit of phase balance between δ-Fe and γ-Fe phase. The austenite begins to precipitate at the ferrite grain boundaries during the cooling of welds in the temperature range 1573-1073K. Since the δ→γ transformation is a diffusion controlled process, the fast cooling in the welding process does not offer sufficient time to complete the phase transformation, which results in a large portion of δ-Fe is retained in weld. Further more, the incomplete transformation results in the retention of skeletal δ-Fe dendrite within the austenite matrix [9].

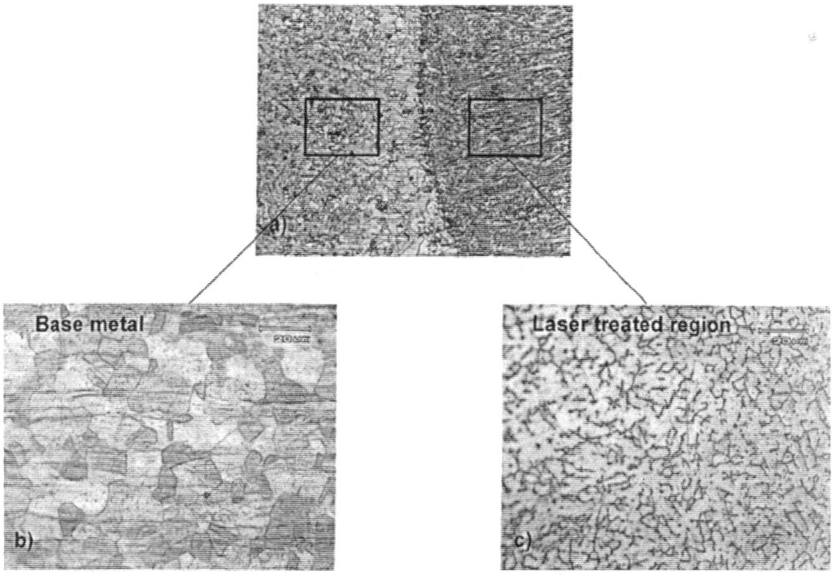

Figure 6. Optical microstructure of; a) weld zone b) base metal and c) fusion zone of 2mm thick AISI 321 steel.

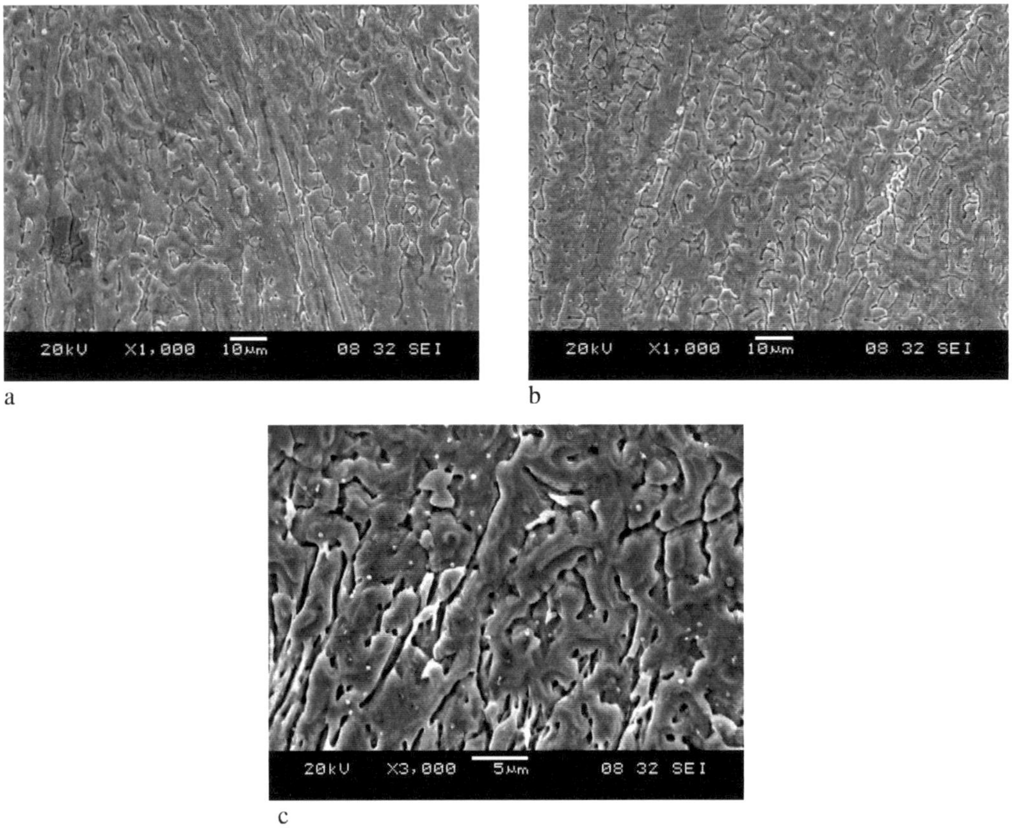

Figure 7. SEM microstructure "a, b and c" of laser welded zone for 2 mm thick AISI 321 steel.

3.5. Tensile test

Tensile test of four samples for each thickness was carried out. The average value of yield strength, ultimate tensile strength and percentage elongation of these samples are shown in Table-2. The results of all welds showed that failure had taken place near the heat affected zone of the welds. The tensile strength of these welds is close to the base metal which was about 650MPa [9]. The percentage elongation of 2mm steel is lower than 1 and 2.5mm thick austenitic stainless steel as given in the Table-II. The percentage elongation depends on grain size and grain boundary area. In smaller grains there exists large grain boundary area per unit volume which lowers percentage elongation. Low percentage elongation of 2mm steel is attributed to smaller grain size of the base metal as compared to other two steels. However the results show that the yield strength, UTS and % age elongation are close to the reported data [12].

Table 2. Tensile test result of the samples

Sheet thickness (mm)	Yield strength (MPa)	UTS (MPa)	% Elong (GL 25mm)
1	300	640	51.6
2	312	632	35.5
2.5	283	646	56.2

4. Conclusions

Bead-on-plate and butt welding of austenitic steel with varying thickness have been carried out at low power using a 450W CO_2 laser.

- The depth and width of weld zone decreases with increase in speed for constant laser power due to relatively lower thermal input at higher speeds.
- The δ-ferrite was found as vermicular morphology near the edge of welding zone whereas skeletal morphology of delta ferrite in austenite structure was found near the center of the weld zone.
- The mechanical properties of laser welded butt joints of austenitic steel sheets are almost similar to the base metal.

5. References

[1] Wollin P and Gooch T G 1995 *Welding World Journal* Vol. 36 75

[2] Bachman F 2003 *Appl. Surf. Sci.* Vol. 4 45

[3] Dilthery U, Risch A 2001 *Welding in the World* Vol. 36 67

[4] Curcio F 2006 *J. of Mater. Proc. and Techn.* Vol. 175 83

[5] Benyounis, Khaled and Olabi, Abdul-Ghani and Hashmi Saleem 2008 *Opt. & Laser Tech.* Vol. 40 (1) 76

[6] Zacharia T, David S A, Vitek J M & Debroy T 1989 *Metall. Trans.* Vol. 20 A 1125

[7] Suutala N 1983 *Metall. Trans.* 14 A 191

[8] Klimpel A and Lisiecki A 2007 *J. Achiev. Mater. and Manuf. Eng.* Vol. 25 1

[9] Yan J, GAO M and Zeng X 2010 *Opt. and Laser in Engin.* Vol. 48 512

[10] Quiroz V, Gumenyuk A and Rethmeive M 2012 *Welding in the world* Vol.56(11-12) 129

[11] Kumar H, Somileddy D A and Gureeraj K 2012 *Int. J. Mater. Sci. and Appl.* Vol. 1(1) 7

[12] http://www.ulbrich.com/files/austeniticstainlesssteel.pdf

Ferroelectric and impedance response of lead-free $(B_{0.5}N_{0.5})$ TiO_3-$BaZrO_3$ piezoelectric ceramics

Jamil ur Rahman[1], Ali Hussain[1], Adnan Maqbool[1], Jin Soo Kim[1], Tae Kwon Song[1], Je Hyun Lee[1], Won Jeong Kim[2] and Myong Ho Kim[1,]
[1]School of Advanced Material Engineering, Changwon National University, Gyeongnam 641-773, Republic of Korea
[2]Department of Physics, Changwon National University, Gyeongnam 641-773, Republic of Korea

Email: mhkim@changwon.ac.kr

Abstract. Lead-free piezoelectric $(0.96B_{0.5}N_{0.5}TiO_3)$-$0.04BaZrO_3$ (BNT-BZ4) was synthesized by using a solid-state reaction method. SEM micrograph shows dense microstructure. X-ray diffraction (XRD) indicated the formation of a BNB-BZ4 single phase having pseudocubic symmetry. A maximum value of remnant polarization (30μC/cm^2) and piezoelectric constant (112 pC/N) was observed for BNT-BZ4 ceramic. The temperature dependences of the dielectric properties of BNT-BZ4 were investigated in the temperature range of 25-600°C at various frequencies (0.1 Hz-1 MHz). The maximum dielectric constant value (ε_r) reaches a highest value of 4046 (at 10 kHz). The electrical properties were investigated by using complex impedance spectroscopy and provided better understanding of relaxation process.

1. Introduction

In the last few decades, there has been much interest in finding lead-fee piezoelectric ceramics. Among lead-free candidate materials, Bismuth sodium titanate, $Bi_{0.5}Na_{0.5}TiO_3$ (BNT), is one of the superior candidate material having excellent ferroelectric properties with a large remanent polarization (P_r = 38μC/cm^2) and a relatively high Curie temperature ($T_c \approx 320$ °C) [1]. At room temperature BNT has an ABO_3 perovskite structure with a ferroelectric rhombohedral phase. As temperature increases, BNT transform from rhombohedral to tetragonal and finally into cubic phase. The transition temperatures of rhombohedral-tetragonal phase ($T_{R\text{-}T}$) and tetragonal-cubic phase ($T_{T\text{-}C}$) are 300 and 540 °C, respectively [2]. On the other hand, BNT has drawbacks, such as a large coercive field (E_c) of approximately 7.3 kV/mm and a high conductivity, which causes problems in the poling process.

To improve the piezoelectric properties of BNT ceramics it is modified with other perovskite such such as $BaTiO_3$ (BT) [3, 4], $BaZrO_3$ (BZ) [5], $SrTiO_3$ (ST) [6, 7], $(Bi_{0.5}K_{0.5})TiO_3$ (BKT) [8, 9], $NaNbO_3$ (NN) [10, 11], $KNaNbO_3$ (KNN) [12]. These modifications reveal a significant enhancement in its piezoelectric and ferroelectric properties. However, previous reports are limited to the piezoelectric and the electromechanical properties, thus, the effects of doping and several solid solutions on the dielectric and the electrical properties are not fully understood.

In this work, $0.94(Bi_{0.5}Na_{0.5})TiO_3$-$0.04BaZrO_3$ (BNT-BZ4) attain a maximum ramanant polarization (P_r =30 μC/cm^2) with a maximum static piezoelectric coefficient of (d_{33} = 112pC/N). However, little attention has been paid to impedance and conductivity studies of this material. In this study, we prepared

BNT-BZ4 ceramics through a conventional solid-state reaction method and investigated the dielectric, electrical conductivity and impedance relaxations in the BNT-BZ4 ceramic.

2. Experimental

The $0.96(B_{0.5}N_{0.5})TiO_3$-$0.04BaZrO_3$ (BNT-BZ4) ceramics was fabricated by using conventional, solid-state reaction method with $BaCO_3$ (99%), Na_2CO_3 (99.95%), Bi_2O_3 (99.90%), ZrO_2 (99%), and TiO_2 (99.90%) from Sigma Aldrich Co, were used as starting raw materials. Prior to measuring the weight, the powders were dried in an oven at 100 °C for 24 hrs. The dried powders were weighed according to the stoichiometric formula and ball-milled for 24 h in ethanol. The dried slurries were calcined at 850 °C for 2 h and ball milled again to dissociate the agglomerates. After drying, the powders were pulverized, passed through a sieve of 150 μm mesh and mixed with an aqueous polyvinyl alcohol (PVA) solution as a binder for granulation. The granulated powders were subsequently pressed into green disks with diameter of 10 mm and thicknesses of 2 mm. The green disks were embedded in the same compositional powder to minimize the evaporation of the volatile elements (Bi, Na) and sintered at 1160 °C for 2 h in air.

The crystal structure was characterized using an X-ray diffractometer (XRD, X'pert MPD 3040, Philips) while microstructure was examined through scanning electron microscope (SEM) JP/JSM5200 (Japan). To measure the electrical properties, silver paste was coated on both polished faces of the sintered samples and fired at 650 °C for 0.5 h to form electrodes. Dielectric constant and complex impedance data for the BNT-BZ4 ceramic were measured using an automatic acquisition system using an impedance analyzer (SI1260, UK) in 25-600°C temperature range with frequency range of 0.1 Hz - 1MHz.

3. Results and discussions

3.1. Phase and microstructure analysis

SEM micrograph of the polished and thermally etched surface of BZ-modified BNT ceramic is presented in Fig. 1. The composition is well sintered and have compact microstructures with an average grain size of 5 ~ 6 μm.

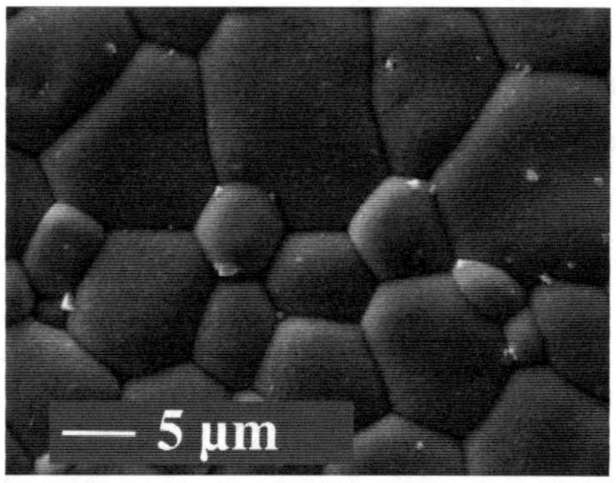

Figure 1. SEM micrographs of BNT-BZ4 ceramic.

Fig. 2 shows the room temperature XRD pattern of BNT-BZ4 ceramic. The X-ray pattern depicts the formation of a single phase without any trace of secondary phase. This indicates that BZ is successfully diffused into BNT to form a complete solid solution. The intensity of the peaks exhibit a pseudo-cubic symmetry having no characteristic peak splitting in the 2θ range of 39^0 - 41^0 and 45^0 - 48^0.

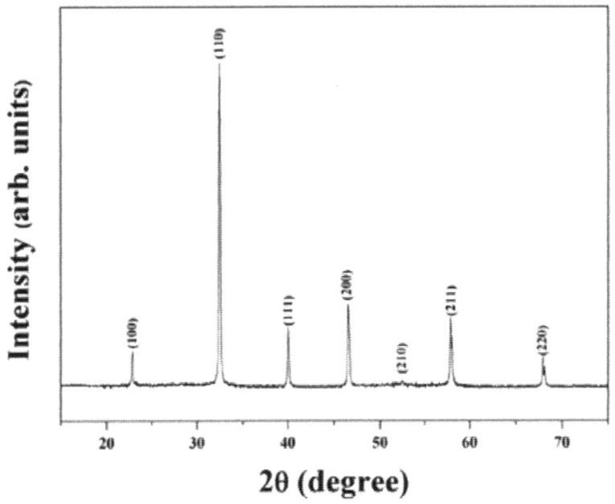

Figure 2. X-ray diffraction patterns of BNT-BZ4 ceramic.

3.2. P–E ferroelectric hysteresis

The ferroelectric response at room temperature of BNT-BZ4 is shown in Fig. 3. The hysteresis curve showed that ceramic exhibits typical ferroelectric behavior with a maximum remanent polarization (P_r) and coercive field (E_c) values of 30 μC/cm^2 and 3.6 kV/mm, respectively. In addition, a maximum static piezoelectric coefficient (d_{33}) of 112 pC/N is obtained for BNT-BZ4 ceramic. The significant improvement observed in d_{33} is attributed to a large P_r and a lower E_c. This is because a lower E_c enables the ceramics to be more easily poled, while a large P_r favors piezoelectricity.

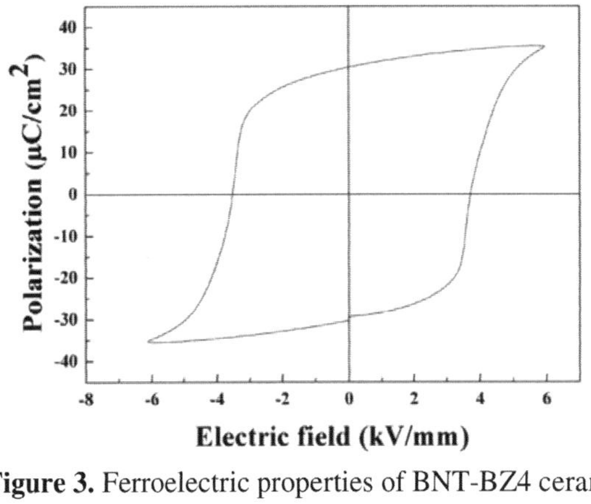

Figure 3. Ferroelectric properties of BNT-BZ4 ceramic.

3.3. Dielectric constant and impedance response

Fig. 4 shows the temperature dependence of dielectric constant (ε_r) of BNT-BZ4 ceramics at different frequencies (0.1 Hz ~ 1MHz) from 25-600°C.It has been observed that dielectric constant increases with increase in temperature up to a maximum value (ε_r) and then it decreases. The maximum dielectric constant (T_m) reaches a highest value of 4046 (at 10 kHz). In addition, the dielectric curve revealed that with increasing in measuring frequency, the observed dielectric constant decrease similar to other BNT-based relaxor ferroelectric materials [6]. The decreasing trend of (ε_r) from low frequency to high frequencies may be due to the active presence of all types of polarization at low frequency i.e., electronic, ionic, orientational and space charge polarization; however, at high frequencies, some of these contributions lagged behind and did not follow the field variation and became inactive. At high frequencies, all other modes of polarization fade away and electronic polarization becomes the dominant mode of polarization; therefore, ε_r decreased at high frequencies.

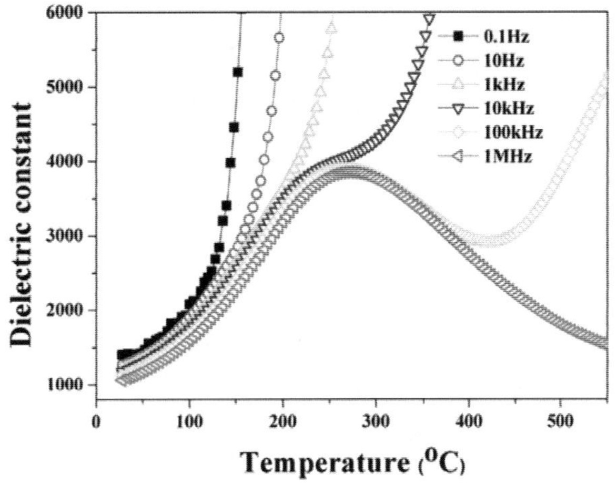

Figure 4. Temperature dependence of the dielectric constant and loss of BNT-BZ4 ceramics with different frequencies range.

Fig. 5 shows the frequency dependence of the real part Z' (i.e., resistance) of the complex impedance for temperatures ranging from 50 °C to 600 °C which calculated by using the basic formula given below

$$\text{Complex impedance} \quad Z^* = Z' - jZ'' \quad \ldots\ldots\ldots\ldots\ldots\ldots (1)$$

It is observed that the magnitude of Z' decreases with increase in both frequency as well as temperature, which indicates the possibility of an increase in the ac conductivity [13]. However, the Z' values for all temperatures combine at high frequency, which can be attributed to the release of space charge as a result of reduction in barrier properties of the material with rise in temperature[14] and may be a responsible factor for the enhancement of ac conductivity of the material with temperature at high frequencies. Also at low frequencies, the Z' values decrease with increasing temperature, and the compound exhibits a negative temperature coefficient of resistance (NTCR)-type behavior like that of semiconductors [15].

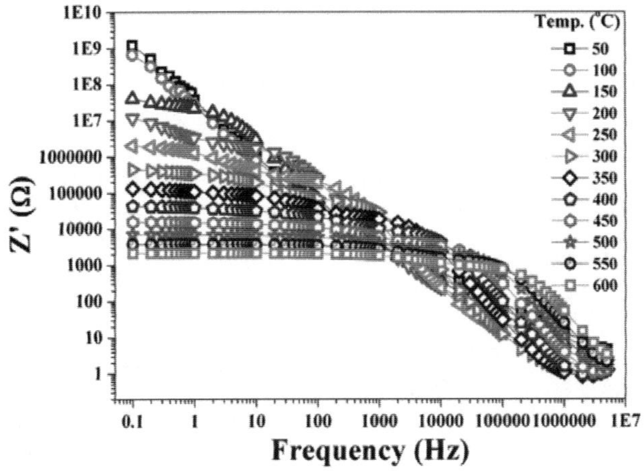

Figure 5. Variation of the real part of impedance of the BNT-BZ4 with frequency at different temperatures.

Fig. 6 shows the imaginary part of the impedance (Z'') as a function of frequency with temperatures ranging from 50 ^{0}C to 600 ^{0}C. At low temperature the values of Z'' monotonically decrease on increasing frequency without getting any peak. However with increasing temperature, Z''-frequency plot show peaks at certain frequency (f_{max}) which shifts toward higher frequency as well as the broadening of the peak. The magnitude of Z''at the peak (Z''_{max}) decreases with increase in temperature and the corresponding f_{max} shifts towards higher frequency. The behavior may be considered due to the temperature dependence of electrical relaxation phenomena on the material [16].

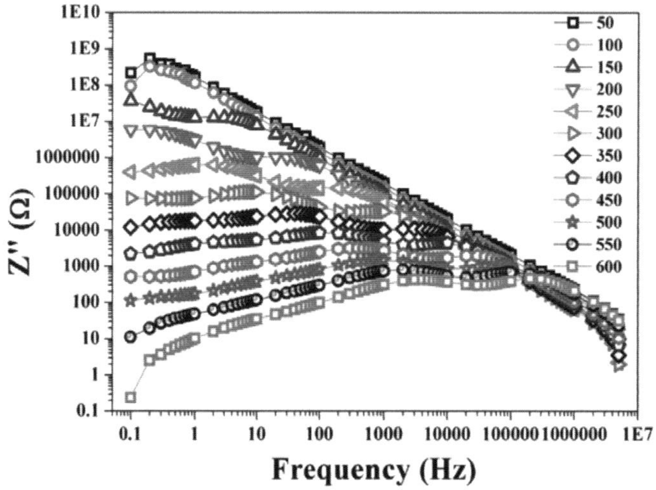

Figure 6. Variation of the imaginary part of the impedance of BNT-BZ4 with frequency at different temperatures.

4. Conclusion

A lead-free piezoelectric $0.96(B_{0.5}N_{0.5})TiO_3$-$0.04BaZrO_3$ (BNT-BZ4) was synthesized by using a solid-state reaction method. SEM morphology confirmed that the composition is well sintered and have compact microstructures with an average grain size of 5 ~ 6 μm. X-ray diffraction pattern reveals single phase with an ABO_3 perovskite structure. A maximum value of dielectric constant (T_m) 4046 (at 10 kHz), remnant polarization (P_r) 30μC/cm^2 and a piezoelectric constant (d_{33}) 112 pC/N were obtained for BNT-BZ4 ceramic. The ac conductivity data showed negative temperature coefficient of resistance character and of relaxation process. The resistances decrease with a rise in temperature.

5. Acknowledgements

This work was supported by National Research Foundation of Korea (NRF) grant funded by the Korean government (MOE) (2013R1A1A2055345) and Basic Science Research Program through the National Research Foundation of Korea (NRF) funded by Ministry of Education, Science and Technology (MEST) (2011-0030058).

6. References

[1] Smolenskii G A, Isupov V A, Agranovskaya A I and Krainik N N 1961 New ferroelectrics of complex composition *Sov. Phys. Solid State* **2** 2651-4

[2] Zvirgzds J A, Kapostis P and Zvirgzde J 1982 *Ferroelectrics 40* vol 40: Taylor & Francis)

[3] Takenaka T, Maruyama K I and Sakata K 1991 $(Bi_{1/2}Na_{1/2})TiO_3-BaTiO_3$ system for lead-free piezoelectric ceramics *Jpn. J. App. Phys.* 30 2236-9

[4] Bai Y, Zheng G-P and Shi S-Q 2011 Abnormal electrocaloric effect of $Na_{0.5}Bi_{0.5}TiO_3$-$BaTiO_3$ lead-free ferroelectric ceramics above room temperature *Mater. Res. Bull.* 46 1866-9

[5] Rahman J U, Hussain A, Maqbool A, Song T K, Kim W J, Kim S S and Kim M H 2014 Dielectric, ferroelectric and field-induced strain response of lead-free $BaZrO_3$-modified $Bi_{0.5}Na_{0.5}TiO_3$ ceramics *Curr. Appl Phys.* 14 331-6

[6] Hiruma Y, Imai Y, Watanabe Y, Nagata H and Takenaka T 2008 Large electrostrain near the phase transition temperature of $(Bi_{0.5}Na_{0.5})TiO_3$-$SrTiO_3$ ferroelectric ceramics *Appl. Phys. Lett.* 92 262904

[7] Krauss W, Schütz D, Mautner F A, Feteira A and Reichmann K 2010 Piezoelectric properties and phase transition temperatures of the solid solution of $(1\text{-}x)(Bi_{0.5}Na_{0.5})TiO_3\text{-}xSrTiO_3$ *J. Eur. Ceram. Soc.* 30 1827-32

[8] Yoshii K, Hiruma Y, Nagata H and Takenaka T 2006 Electrical properties and depolarization temperature of $(Bi_{1/2}Na_{1/2})TiO_3$-$(Bi_{1/2}K_{1/2})TiO_3$ Lead-free piezoelectric ceramics *Jpn. J. App. Phys.* 45 4493-6

[9] Sasaki A, Chiba T, Mamiya Y and Otsuki E 1999 Dielectric and piezoelectric properties of $(Bi_{0.5}Na_{0.5})TiO_3$-$(Bi_{0.5}K_{0.5})TiO_3$ systems *Jpn. J. App. Phys.* 38 5564

[10] Li Y, Chen W, Zhou J, Xu Q, Sun H and Xu R 2004 Dielectric and piezoelecrtic properties of lead-free $(Na_{0.5}Bi_{0.5})TiO_3$-$NaNbO_3$ ceramics *Mater. Sci. Eng. B* 112 5-9

[11] Hiruma Y, Nagata H and Takenaka T 2008 Phase diagrams and electrical properties of $(Bi_{1/2}Na_{1/2})TiO_3$-based solid solutions *J. App. Phys.* 104 124106-13

[12] Kounga A B, Zhang S-T, Jo W, Granzow T and Rödel J 2008 Morphotropic phase boundary in $(1\text{-}x)Bi_{0.5}Na_{0.5}TiO_3\text{-}xK_{0.5}Na_{0.5}NbO_3$ lead-free piezoceramics *Appl. Phys. Lett.* 92 222902-3

[13] Selvasekarapandian S and Vijayakumar M 2003 The ac impedance spectroscopy studies on $LiDyO_2$ *Mater. Chem. Phys.* 80 29-33

[14] Lily, Kumari K, Prasad K and Choudhary R N P 2008 Impedance spectroscopy of $(Na_{0.5}Bi_{0.5})(Zr_{0.25}Ti_{0.75})O_3$ lead-free ceramic *J. Alloys Compd.* 453 325-31

[15] Suman C K, Prasad K and Choudhary R N P 2006 Complex impedance studies on tungsten-bronze electroceramic: $Pb_2Bi_3LaTi_5O_{18}$ *J.Mater. Sci.* 41 369-75

[16] Pradhan D K, Choudhary R N P, Rinaldi C and Katiyar R S 2009 Effect of Mn substitution on electrical and magnetic properties of $Bi_{0.9}La_{0.1}FeO_3$ *J. App. Phys.* 106 024102

Synthesis and characterization of ternary Pt-Ni-M/C (M=Cu, Fe, Ce, Mo, W) nano-catalysts for low temperature fuel cells

Riaz Ahmed[1,2], Rabia Jamil[1] and Muhammad Shahid Ansari[1]
[1]Department of Chemistry, Quaid-i-Azam University, Islamabad, Pakistan
[2]COMSATS, Shahrah-e-Jamhuriat, G.5/2, Islamabad, Pakistan

E-mail: riaz_ecfc@yahoo.com

Abstract. Ternary metal catalysts were synthesized by impregnation method. The mixture of metal solutions was reduced slowly under inert atmosphere and the reduced metals were deposited on the Vulcan Carbon(VC). Tungsten, molybdenum, cerium, iron and copper were added to specified amounts of platinum and nickel. Addition of nickel generally improves catalytic activity of platinum. The XRD of the catalysts was done and the crystallite size and other parameters were calculated. Crystallite sizes were in the range of 5 to 16 nm. Electrochemical surface areas of the catalysts were determined by cyclic voltammetry (CV) in acidic media and are compared. Electro oxidation of methanol on the catalysts was done and peak potential, peak current, mass activity of the catalysts were calculated and are compared. These parameters were determined in acidic and basic media. It was found that mass activity increased significantly in basic media. Rate constants for the electro oxidation of methanol were also calculated in acidic and basic media and are compared and discussed. Rate constants were generally higher in basic media. Ternary catalysts showed improved catalytic activity than the binary catalyst. Nano alloying improved the catalytic activity and stability of the ternary catalysts.

1. Introduction

Demand for clean energy sources has become a strong driving force for sustainable economic development. To avoid irreparable damage to the environment as a consequence of burning fossil fuels, some viable alternates must be harvested. Fuel cells are poised for a breakthrough into the mainstream, and offer an attractive combination of highly efficient fuel utilization and environment friendly operation. A fuel cell is an electrochemical device that continuously and directly converts the chemical energy of externally supplied fuel and oxidant to electrical energy [1]. Electro catalysts are type of heterogeneous catalysts which facilitate electrochemical reactions either functioning at the electrode surface or the electrode itself. Due to a wide range of possible combinations of metals and their composition [2], choice of appropriate electro catalysts has become difficult in fuel cell technology [3]. Like platinum catalyst, particle size and shape influence the catalytic activity of binary and ternary catalysts [4-5]. Addition of two metals shows higher catalytic activity than single platinum metal [6-7] and even higher than binary catalysts [8]. Particle size, synergistic effects on the activity due to the difference in the redox potentials of the metals and atomic scale structural and chemical ordering in Nano alloy catalysts influence the catalytic activity [9-10]. Alloy catalysts are better catalysts than platinum alone [11-12]. Ternary alloys have a potential for the fine-tuning of the electronic structures to improve the catalytic activities [13-15]. Nano and high surface area catalysts have been thoroughly investigated but the catalytic activities of platinum ternary catalysts have not been fully investigated.

In this study Pt-ternary catalysts have been synthesized and characterized using XRD, electrochemical surface area, specific activity and kinetics of methanol oxidation on the catalysts.

2. Materials and Methods

2.1. Chemicals and equipments:

Platinum(IV) chloride PtCl4, Nickel(II) chloride $NiCl_2.6H_2O$, Sodium tungstate dihydrate $Na_2WO_4.2H_2O$, Copper(II) chloride $CuCl_2.2H_2O$, Ammonium Cerium(IV) sulphated hydrate $(NH_4)_4Ce(SO_4)4.2H_2O$, Ferric chloride $FeCl_3.6H^2O$, Moblydenum chloride $MoCl_5$, Sodium borohydride NaBH4 ,Vulcan Carbon (VC) Trisodium citrate $Na_3C_6H_5O_7.2H_2O$, Methanol, Ethanol, 2-Propanol C_3H_8O, Nafion solution, Hydrogen peroxide H_2O_2 35% , Hydrochloric acid, Sulphuric acid, Nitric acid, Potassium hydroxide, of analytical grades were used.

X-ray diffraction pattern of all samples were obtained at room temperature by using X-ray Diffractrometer (3040/60 X'pert PRO) using Ni-filtered cu-Kα radiations (40 kV,30mA). Cyclic voltammetric measurements were carried out using Eco Chemie Autolab PGSTAT 12 potentiostat/galvanostat Netherlands, with the electrochemical software package GPES 4.9.

2.2. Synthesis of catalysts:

Vulcan carbon was purified by treating with 6M HCl. The contents were refluxed and ultra sonicated for 4 hours at 60°C. Treated VC was filtered off and washed several times with de-ionized water till neutral pH of the filtrate. Finally the purified VC was vacuum-dried for 24 hours at 110°C. Functionalization of the purified VC was done in 4:1 v/v mixture of 8M HNO_3 and 35% H_2O_2 by ultra sonication, filtered, washed and vacuum-dried for 24 hours at 110°C.

Pt-Ni/C and Pt-Ni-M/C (M= Cu, Fe, Ce, Mo, W) catalysts were synthesized using $NaBH_4$ as reducing agent by conventional co-reduction method under argon atmosphere at room temperature. Aqueous solutions of $PtCl_4$, $NiCl_2.6H_2O$ and solutions of different metals (M= Cu, Fe, Ce, Mo, W) were added and stirred under argon atmosphere along with VC; the mixture became colorless from golden indicating adsorption of metal ions on the VC. Freshly prepared solution of $NaBH_4$ was added drop wise into the above dispersion. Four hours stirring resulted formation of black suspension which was filtered and washed several times with de-ionized water till neutral pH of the filtrate. The residue was vacuum-dried for 24 hours at 110 °C. Pt-Ni-M/C (M= Cu, Fe, Ce, Mo, W) catalysts were synthesized using $NaBH_4$ as reducing agent by conventional co-reduction method under argon atmosphere at room temperature.

2.3. Electrochemical measurements:

Graphite electrode (GE) was polished with 1.0 μm alumina (CHI Inc., USA) washed in ethanol and ultra-pure water ultrasonically to remove organic and inorganic impurities from the surface of electrode. Then, 20 mg catalyst and 1 mL of solvent (water + 2-propanol) were mixed ultrasonically for 30 min in a test tube. 50 μL of the suspension was pipetted out and pasted over the tip of polished graphite electrode, dried in air. After the solvent evaporation, 10 μL binder (5% nafion solution) was spread over the tip of GE. The air dried, working electrode was ready to use.

3. Results and discussions

3.1. XRD analysis:

XRD was done to find out the crystallite size of the particles, alloy formation and other parameters. Standard XRD data for platinum (00-001-1190) was used for comparison purposes. Fig.1 shows the comparison of XRD patterns of Pt–Ni–M/C (M= W, Cu, Fe, Mo, Ce) of all the five ternary catalysts. The broad peaks located at 2 theta (θ) position 26.19°, 44.14°, 54.39°and 59.54° are associated with the (002), (100), (004) and (101) planes of the graphite-like structure of the VC. Platinum face centered cubic peak at 2 theta position (39.67^0) is prominent in all the catalysts and used as reference peak. Shift in platinum 2 theta position for different ternary catalysts shows the alloy formation.

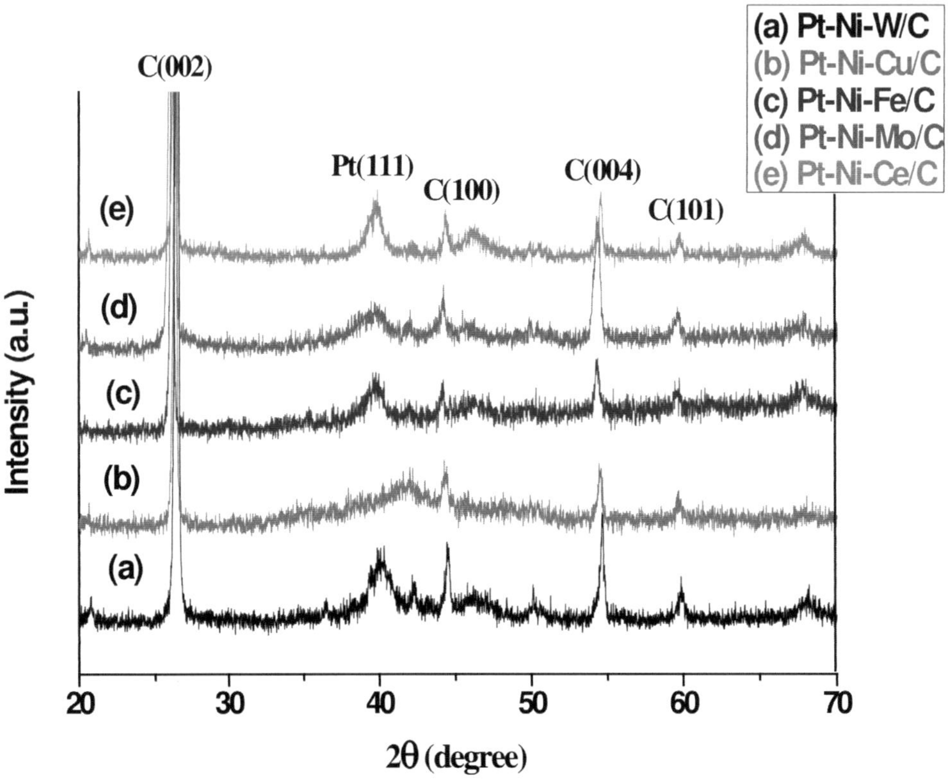

Figure 1. XRD patterns of Pt-Ni-W/C, Pt-Ni-Cu/C, Pt-Ni-Fe/C, Pt-Ni-Mo/C and Pt-Ni-Ce/C

The peaks observed at 2θ position 40.10°, 39.98°, 39.82^0 and 39.89° are due to Pt(111) plane of fcc crystalline structure in Pt–Ni–W/C , Pt–Ni–Ce/C , Pt–Ni–Fe/C and Pt–Ni–Mo/C catalysts respectively. Maximum shift in 2θ position is observed for Pt–Ni–W/C. The respective additions of Fe, Ce and Mo to Pt-Ni have caused some shift in platinum peak which indicates its alloy formation and more shift shows more alloy formation and crystalline structure of platinum has also not changed. Different parameters evaluated from XRD patterns of Pt–Ni–M/C (M= W, Cu, Fe, Mo, Ce) respectively are displayed in Table.1. There is a significant decrease in d-spacing and density. Decrease in density is due to the alloy formation of platinum with other low molecular weight metals. There is a change in lattice parameter and cell volume for all the additives. The rough estimated average crystallite sizes are 4–16 nm and is 4.5 nm for Pt-Ni-Mo/C, 8.0 nm for Pt-Ni-Fe/C, 11.9 nm for Pt-Ni-W/C and 16.0 nm for Pt-Ni-Ce/C. Shift in 2θ position shows alloy formation. This alloy formation has significant effect on catalytic activity.

Table 1. XRD parameters evaluated for Pt-Ni-W/C, Pt-Ni-Ce/C, Pt-Ni-Fe/C and Pt-Ni-Mo/C

Catalysts	Struct -ures	Position 2θ (degree)	*d*-spacing Å	Lattice param-eter a / Å	Cell volume V / Å^3	Density $\rho_{X\text{-ray}}$ /g·cm^{-3}	Avg. Crystall-ite size D/nm
Pt-Ni-W/C	cubic	40.10	2.248	3.894	59.04	18.5	11.9
Pt-Ni-Ce/C	cubic	39.98	2.255	3.906	59.59	16.9	16.0
Pt-Ni-Fe /C	cubic	39.82	2.264	3.921	60.30	13.9	8.0
Pt-Ni-Mo/C	cubic	39.89	2.260	3.914	59.95	15.3	4.5
Pt 00-001-1190	cubic	39.67	3.917	-	59.87	21.51	-

3.2. Electrochemical Surface area (S_{ESA})

Various transition metals have been incorporated to Pt-Ni in a view to further enhance the activity of bi-metallic catalysts. Table.2. shows the comparison of various tri-metallic catalysts, among them Pt-Ni-W/C shows the highest S_{ESA} followed by Pt-Ni-Fe/C, Pt-Ni-Cu/C, Pt-Ni-Mo/C and Pt-Ni-Ce/C. Roughness factor for the catalysts can be calculated by dividing the real surface area (electrochemical surface area) by the apparent area of the electrode.

Table 2. Electrochemical surface area from CVs in 0.5M H_2SO_4 on various catalysts

Catalysts	Pt Loading %	Mass of Pt/mg	Apparent Surface Area/(cm^2)	Qcat /μC	S_{ESA}/ (cm^2.mg^{-1})
Pt-Ni-W/C	10	0.1	0.785	40070	1908.1
Pt-Ni-Cu/C	10	0.1	0.785	12010	571.9
Pt-Ni-Ce/C	10	0.1	0.785	10440	497.1
Pt-Ni-Fe/C	10	0.1	0.785	14230	677.6
Pt-Ni-Mo/C	10	0.1	0.785	11510	548.1

3.3. Evaluation of catalytic activity from cyclic voltammograms

Cyclic voltammograms of the catalysts were taken both in acidic media and basic media. Cyclic voltammograms in acidic media are shown in Fig.2. For all the catalysts forward and backward anodic peaks for methanol oxidation were observed and peak heights were different for different catalysts. The height of the peaks shows the catalytic activity of the catalysts. Peak current, mass activity and current density of the catalysts were calculated from voltammograms and are given in table-3.

Table 3. Activity parameters evaluated from CVs in 1M CH_3OH in acidic and basic media

Catalysts	Peak Potential E_p/V Acidic	Peak Potential E_p/V Basic	Peak current /mA Acidic	Peak current /mA Basic	Mass activity /mA.mg^{-1} Acidic	Mass activity /mA.mg^{-1} Basic
Pt-Ni/C	0.69	-0.18	1.7	21.8	17	218.1
Pt-Ni-W/C	0.71	-0.11	7.8	37.4	78	373.6
Pt-Ni-Cu/C	-	-0.21	-	3.04	-	30.4
Pt-Ni-Ce/C	0.67	-0.20	1.8	15.5	18	154.8
Pt-Ni-Fe/C	0.70	-0.29	1.7	0.8	17	8.0
Pt-Ni-Mo/C	0.68	-0.20	3.0	24.2	30	242.0

It can be seen that the Pt-Ni-W/C has the maximum mass activity followed by Pt-Ni-Mo/C, Pt-Ni-Ce/C and Pt-Ni-Fe/C catalysts. Catalytic activity of the ternary catalysts is considerably higher than the binary catalyst.

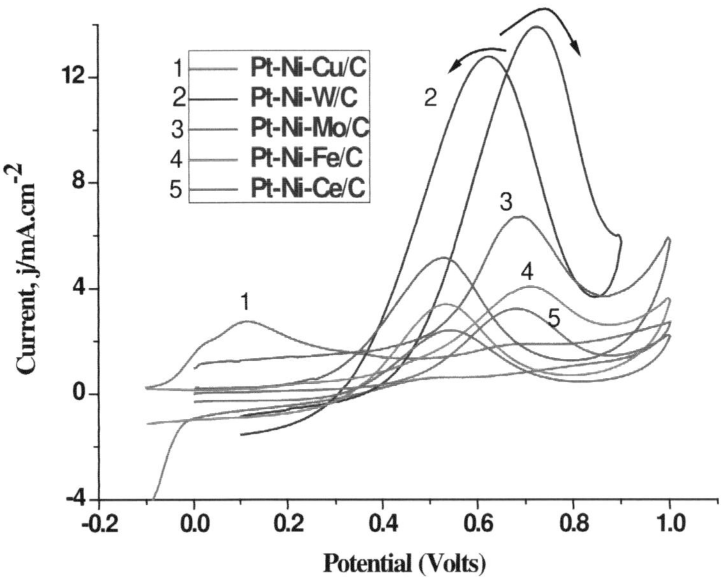

Figure 2. CVs measured in 1M CH_3OH + 1M H_2SO_4 on Pt-Ni-M/C (M= Cu, W, Mo, Fe, Ce)

In Fig.3 is given voltammogram for methanol oxidation in basic media for all the catalysts. Here we also get a forward anodic peak and a backward anodic peak. In basic media the backward anodic peak is much smaller than the forward anodic peak. Specific activities of the catalysts in basic media are given in table-3. In basic media the specific activity of the Pt-Ni-W/C catalyst is highest followed by Pt-Ni-Mo/C , Pt-Ni-Ce/C, Pt-Ni-Cu/C and Pt-Ni-Fe/C catalysts. Mass activities of the catalysts have nearly the same trend in both the media but are much higher in basic media. Generally the catalytic activity of the ternary catalysts in basic media also is better than the binary catalyst.

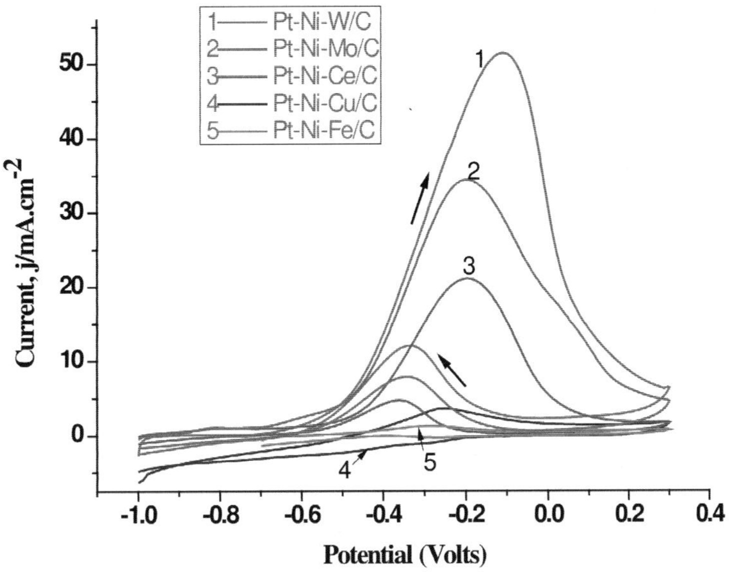

Figure 3. CVs measured in 1M CH_3OH + 1M KOH on Pt-Ni-M/C (M= Cu, W, Mo, Fe, Ce)

3.4. Kinetics of methanol oxidation on catalysts

Better catalysts will show fast kinetics and rates of reaction will be higher. Rate of an electrochemical process can be calculated by the Klinger and Kochi or Nicholson and Shain methods. In this work Nicholson equation was used [16-17] which can be written as:

$$i_p = 0.227\ nFAC_ok_s{\cdot}\exp[(\alpha n_\alpha{\cdot}F/RT)(E_p-E_i)]$$

where i_p is the peak current, n is the number of electrons transferred, A/cm^2 is the area of the electrode, C_o/mol·cm^{-3} is the bulk concentration of the reactant and ks /cm·s^{-1} is the standard heterogeneous rate constant. F and R are the Faraday and gas constants. E_i is the potential at the root of the wave. A graph is plotted between ln(I_p) vs. E_p-E_i to get a straight line and from the value of the slope, k_{het} was calculated. The values of heterogeneous rate constants for the MO on the catalysts in acidic and basic media have been calculated and are listed in Table.4. It can be seen that the rate constants are better on ternary catalysts as compared to binary catalyst and generally rate constants are higher in basic media.

Table 4. Rate constants of CH_3OH oxidation

Catalysts	Rate constant k_{het} /cm.s^{-1} *10^{-5} Acidic	Rate constant k_{het} /cm.s^{-1} *10^{-5} Basic
Pt-Ni/C	1.7	4.1
Pt-Ni-W/C	4.3	15.3
Pt-Ni-Cu/C	-	0.8
Pt-Ni-Ce/C	1.7	3.7
Pt-Ni-Fe/C	1.6	0.2
Pt-Ni-Mo/C	2.8	4.1

3.5. Comparison of catalytic activities:
In Fig.4 are compared the catalytic activities of different ternary catalysts both in acidic and basic media. Pt-Ni-W/C is the best catalyst followed by Pt-Ni-Mo/C and others. Catalytic activity is much better in basic media for all the catalysts except for the Pt-Ni-Fe/C catalyst. Basic binary catalyst used here Pt-Ni/C has also better catalytic activity in basic media (Table-3). Decrease in catalytic activity in basic media due to iron may be its tendency for hydrolysis in basic media.

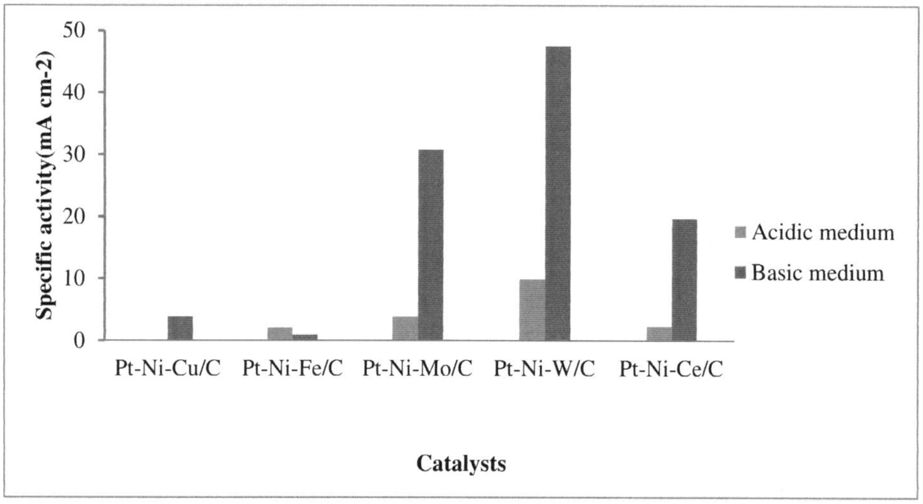

Figure 4. Specific activity comparison of tri-metallic catalysts in acidic and basic medium

4. Conclusions

Ternary catalysts were prepared by adding different metals to the binary catalyst Pt-Ni. The addition of non noble metals to the binary catalyst has improved the catalytic activity of the ternary catalyst. Elecrochemical surface area is higher for the ternary catalysts. Catalytic activity both in the acidic and basic media has improved in the ternary catalysts. Similarly rate of reaction both in acidic and basic media are also better in ternary catalysts. Alloying of the metals have improved the catalytic activity of the catalysts. Nano alloys through better structural and chemical ordering has improved the catalytic activity of the ternary catalysts. It can be concluded that the ternary catalysts are better than the binary and single platinum catalysts and addition of non noble metals will also reduce the costs of the catalysts.

5. References

[1] Zang J.(Ed.) 2008, PEM fuel cell Electrocatalysts and catalysts layer fundamentals and applications, Springer-Verlag, London

[2] Ferrando, R.; Jellinek, J. Johnstons R.L.C. 2008, Revs. Lett. 3, 108-119.

[3] Perry, M. L.; Fuller, T. F.; 2002, J. Electrochem. Soc149.59–67.

[4]. Stamenkovic, V. R.; Mun, B. S.; Mayrhofer, K. J. J.; Ross, P. N.; Markovic, N. M. 2006, J. Am. Chem. Soc., 128 (27), 8813.

[5] Fang, B.; Luo, J.; Chen, Y.; Wanjala, B. N.; Loukrakpam, R.; Hong, J.; Yin, J.; Hu, X.; Hu, P.; Zhong, C. J. 2011, ChemCatChem, 3, 583.

[6] Wanjala, B. N.; Fang, B.; Luo, J.; Chen, Y.; Yin, J.; Engelhard, M.; Loukrakpam, R.; Zhong, C. J. 2011, J. Am. Chem. Soc., 133, 12714.

[7] Loukrakpam, R.; Luo, J.; He, T.; Chen, Y.; Xu, Z.; Njoki, P. N.; Wanjala, B. N.; Fang, B.; Mott, D.; Yin, J.; Klar, J.; Powell, B.; Zhong, C. J. 2011, J. Phys. Chem. C, 115, 1682.

[8] Koh, S.; Leisch, J.; Toney, M. F.; Strasser, P. J. Phys. Chem. C 2007, 111, 3744.

[9] Greco, G.; Witkowska, A.; Minicucci, M.; Olivi, L.; Principi, E.; Dsoke, S.; Moretti, A.; Marassi, R.; Cicco, A. D. 2012, J. Phys. Chem. C,116, 12791.

[10] Wanjala B. N., Fang B., Shan S., Petkov V., Zhu P., Loukrakpam R., Chen Y., Luo J., Yin J., Yang L., Shao M., Zhong C.-J, Chem. Mater. 2012, 24, 4283−4293

[11] Wang C., Li D., Chi M., Pearson J., Rankin, R. B. Greeley J., Duan Z., Wang, G. van der Vliet D., More K. L., Markovic N. M., Vojislav R. Stamenkovic, J. 2012, Phys. Chem. Lett., 3, 1668−1673.

[12]. Stamenkovic, V. R.; Mun, B. S.; Arenz, M.; Mayrhofer, K. J. J.; Lucas, C. A.; Wang, G. F.; Ross, P. N.; Markovic, N. M. 2007, Nat. Mater., 6, 241−247.

[13] Antolini, E. 2007, Appl. Catal., B, 74, 337−350.

[14] Mani, P.; Srivastava, R.; Strasser, P. 2011, J. Power Sources, 196, 666−673.

[15] Wanjala, B. N.; Loukrakpam, R.; Luo, J.; Njoki, P. N.; Mott, D.; Zhong, C. J.; Shao, M. H.; Protsailo, L.; Kawamura, T. 2010, J. Phys. Chem. C, 114, 17580−17590.

[16] Nicholson, R.S., and Shain. I. Anal. Chem; 1964, 36, 70,

[17] Reinmuth, W.H., Anal. Chem. 1960, 32, 1891.

Indexing the structural parameters and investigating the magnetic properties of lanthanum doped strontium hexaferrites

M Azim, S Atiq, S Riaz and S Naseem
Centre of Excellence in Solid State Physics, University of the Punjab, Lahore-54590, Pakistan

E-mail: azimnadeem@gmail.com

Abstract. $Sr_{1-x}La_xFe_{12}O_{19}$ (x= 0, 0.1, 0.15, 0.2) nano-crystallites were synthesized using a sol-gel based combustion technique. Sintering of the samples at 800 °C helped to develop the stable hexagonal phase. Data received from the X-ray diffraction was used to confirm and determine the crystal structure of the samples by manipulating the lattice parameters, volume of the unit cell, crystallite size, X-ray density and porosity of the samples. Room temperature magnetic measurements were performed using a vibrating sample magnetometer with an applied field of ± 10 kOe. The magnetic hysteresis loops reveal that lanthanum doping at Sr-sites has enhanced the coercivity of the samples in the series with some compromise on magnetization while remanence remains almost constant, enabling the lanthanum doped samples appropriate for hard magnetic applications.

1. Introduction

Strontium hexaferrites are important magnetic materials due to their significant applications in diversity of fields. Owing to their easy, low cast manufacturing and excellent chemical stability, these materials are being synthesized by various methods [1]. In addition to co-precipitation, microemulsion [2] and conventional ceramic methods [3], these ferrites are also being synthesized by sol-gel method [4]. The dynamic properties of these ferrites can be tuned according to the requirement with the assistance of conventional doping process. Doping of some appropriate element in a particular proportion can play a vital role in altering the astonishing properties of the materials. For example, doping of La in $SrFe_{12}O_{19}$ lowers its coercivity and remanence due to which general behavior of these ferrites is changed [1]. Crystal structure of the material controls its various properties [5]. Ferrites are classified with respect to their properties as well as crystal structure. Different types of ferrites like spinel, garnet and hexaferrites have different structures, varying from very easy to most complex [6]. Many scientists preceded their research work on investigation of crystal structure of $SrFe_{12}O_{19}$ and reported their detailed results. For instance, Lechevallier et al. [7] figured out the magnetic and structural properties of hexagonal ferrites $Sr_{1-x}Sm_xFe_{12}O_{19}$. The scientists have determined that the crystal structure of $SrFe_{12}O_{19}$ is hexagonal. Choit et al. [8] carried out experiment to synthesize La-Co doped Sr-ferrite $(La\text{-}Co)_xSr_{1-x}Fe_{12-x}O_{19}$ by sol-gel method in order to investigate its magnetic and structural properties. Researchers concluded that crystal phase of $(La\text{-}Co)_xSr_{1-x}Fe_{12-x}O_{19}$ is single M-type hexagonal phase. To investigate the material properties, it is necessary to know about its structure. Without attaining the knowledge of crystal structure of the materials, it becomes very hard to investigate its properties [5]. In the present paper, X-ray diffraction analysis of the compositions $Sr_{1-x}La_xFe_{12}O_{19}$ (x = 0, 0.1, 0.15, 0.2) has been discussed in details. A complete procedure of indexing and calculating various related structural parameters are explained in a simple way. This knowledge will provide a way to make a precise analysis of diffraction patterns of other materials.

2. Experimental

Sol-gel auto-combustion technique was used for the preparation of various compositions of La doped $SrFe_{12}O_{19}$. To prepare the series of samples with general formula, $Sr_{1-x}La_xFe_{12}O_{19}$ (x = 0, 0.1, 0.15, 0.2), stoichiometric amounts of analytical grade reagents, i.e., strontium nitrate (99%), iron (III) nitrate (98%), lanthanum (III) nitrate (99%) and citric acid (99.5%) were weighed with the help of a precise digital balance. These metal nitrates and citric acid (used as a fuel) were dissolved in de-ionized water separately, and then mixed to make a total volume of 50 ml. pH-meter measured the pH of the solution as 2.5. The magnetic capsule (stirrer) was put in the solution and beaker was placed on a hot plate. The setup was housed in an ESCO fume hood. Stirring of solution was started and temperature of the hotplate was increased gradually up to 150 °C. The solution was heated and stirred for about 45 minutes, till the gel was formed. As the gel was formed, the temperature of the sample was further increased up to 250 °C. The gel was heated at this temperature for about 30 minutes when its combustion was started. It was allowed to combust normally. The sample took about 30 minutes for its complete combustion. During this interval, various gases were evolved from the beaker. The end product was a homogenous dry powder.

After complete combustion of the sample, the temperature of the hotplate was decreased gradually. Within 10 minutes, the hotplate was switched off and the sample was cooled down in air at room temperature. Although, the material was already in powder form but it became very fine after grinding, using an Agate mortar and pestle. Fine powder was put in the sample bottle. All the compositions were prepared in the same way. The samples were sintered at 800 °C for two hours in a muffle furnace. The samples were pelletized using an Apex hydraulic press by applying a pressure of 4.5 ton. The prepared samples were analyzed for the crystal structure, using the diffraction pattern, obtained from X-ray diffraction (XRD). Magnetic properties were determined using a Lakeshore - 7404 vibrating sample magnetometer (VSM).

3. Results and Discussion

Data obtained from XRD for the first sample (x = 0) has been plotted in figure 1. The first task is to determine the crystal structure of the sample using this data. It is very tough job to determine the crystal structure of some unknown material, than that of a known one. In a diffraction pattern, the pattern of diffracted lines tells about the crystal structure, the positions of lines have information about the unit cell and the positions of the atoms can be judged by the intensities of the lines [5]. From the XRD pattern shown in figure 1, it is expected that it is the diffraction pattern of hexagonal $SrFe_{12}O_{19}$. The first step is to index this pattern that will tell about the crystal system, to which the material belongs to. For this, the values of sin2θ are calculated for all major diffraction lines, as tabulated in Table 1. These values provide a base to solve the pattern. If there are some diffraction lines due to impurities in material or due to other reasons, then it creates problems and needs extra skills to tackle with. A relation is developed by combing plane-spacing equation and Bragg's law equation, which determines the Miller indices of some specific crystal system. For instance, for a cubic system, this expression can be written as [5].

$$\sin^2\theta/(h^2+k^2+l^2) = \sin^2\theta/s = \lambda^2/4a^2 \qquad (1)$$

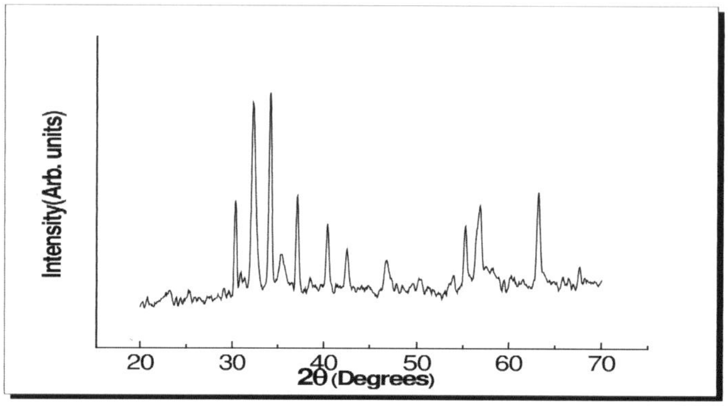

Figure 1. Plot of data received from X-ray diffractometer

Here $\lambda^2/4a^2$ is a constant and $s = h^2 + k^2 + l^2$, which is determined by $sin^2\theta$ values. For an unknown material, it has to be found that the values of sin2θ are integral multiple of which *hkl* values? First, we check the *hkl* values of a cubic system. If these do not satisfy, then we try for the other systems like tetragonal, hexagonal, etc. one by one. As the pattern under present study is expected to be of hexagonal crystal structure, hence it should obey the modified form of Eq. (1), given as [5].

$$\sin^2\theta = A(h^2 + hk + k^2) + Cl^2 \tag{2}$$

Table 1. Diffraction lines and their corresponding angles taken from diffraction plot

Line	2θ	θ	Sin θ	sin²θ	2θ	θ
1	23.11	11.55	0.2002	0.0401	23.11	11.55
2	30.36	15.18	0.2618	0.0685	30.36	15.18
3	32.35	16.18	0.2786	0.0776	32.35	16.18
4	34.15	17.07	0.2935	0.0862	34.15	17.07
5	35.19	17.59	0.3022	0.0913	35.19	17.59
6	35.4	17.70	0.3040	0.0924	35.4	17.70
7	37.1	18.55	0.3181	0.1012	37.1	18.55
8	40.35	20.18	0.3449	0.1189	40.35	20.18
9	42.55	21.27	0.3627	0.1316	42.55	21.27

Table 2. Values of sin²θ and its multiple (by 1/3, 1/4, 1/7) to find out *hkl*

Line	Sin²θ	Sin²θ/3	Sin²θ/4	Sin²θ/7	hkl
1	0.0401	0.0133	0.0100	0.0057	
2	0.0685	0.2285	0.0171	0.0098	
3	0.0776	0.0259	0.0194	0.0111	
4	0.0862	0.0287	0.0215	0.0123	
5	0.0914	0.0305	0.0229	0.0131	200
6	0.0924	0.0308	0.0231	0.0132	
7	0.1012	0.0337	0.0253	0.0144	
8	0.1189	0.0396	0.0297	0.0169	
9	0.1316	0.0439	0.3291	0.0188	

In this relation, $A = \lambda^2/3a^2$) and $C = (\lambda^2/4c^2)$. The value of A can be obtained at $hk0$ ($l = 0$). Hence, the permissible values of $(h^2 + hk + k^2)$ are 1, 3, 4, 7, 9, etc. Now, divide the values of $\sin^2\theta$ tabulated in Table 1, by 1, 3, 4, 7, etc. to get *hkl* values for hexagonal system, as shown in Table 2. Lattice parameter '*a*' is calculated for all values of sin2*θ*. There is some order in particular values of '*a*', being integral multiple of each other. Least value from them is the value of '*a*' for 100 plane. Value *a/*2 is value of '*a*' for 200 plane, and *a*/3 for 300 plane and so on. Generally, smallest repeating value in all columns is found and is considered to be '*A*'. Lattice parameter is calculated for this value and crystal structure is determined. But sometimes, '*A*' cannot be found for 100 plane, hence lattice parameter '*a*' is calculated at *hk0* for all sin2*θ* values, as has been proceeded in the present case. It is found that the diffraction line having sin2*θ* value of 0.0914, gives the most nearest value of '*a*' for 200 plane. This repeating value was also found in the next column of Table 2, hence it was considered as '*A*' value. '*C*' was determined by the following relation [5].

$$\sin^2\theta - A(h^2 + hk + k^2) = Cl^2 \qquad (3)$$

The values of *A, 3A, 4A,* etc. were subtracted from corresponding sin2*θ* values to find the remainders 1, 4, 9, 16, etc., as shown in Table 3. The value of '*C*' was found to be 0.0011. The corresponding value of *sin2θ* is combination of *2A* and *C*, hence *hkl* assigned to this value are 201. Such combinations (*A+C, 2A+2C, 3A+C, 3A+2C,* etc.) of sin2*θ* values were found to assign *hkl* values to various diffraction lines. These *hkl* values are listed in the most right column of Table 3 and are shown in figure 2. These *hkl* values confirm that the diffraction pattern belongs to a hexagonal crystal structure. The lattice parameter '*c*' was calculated with the help of Eq. (4) [5].

$$c = \lambda^2 / 4C^2 \qquad (4)$$

Table 3. Values of sin2θ and subtracted ones to find out values of *C*

Line	$Sin^2\theta$	$Sin^2\theta$-*A*	$Sin^2\theta$-3*A*	*hkl*	Line
1	0.0401			006	1
2	0.0685			110	2
3	0.0776			107	3
4	0.0862			114	4
5	0.0914	0.0000		200	5
6	0.0924	0.0011		201	6
7	0.1012	0.0098		203	7
8	0.1189	0.0275		205	8
9	0.1316	0.0402		206	9

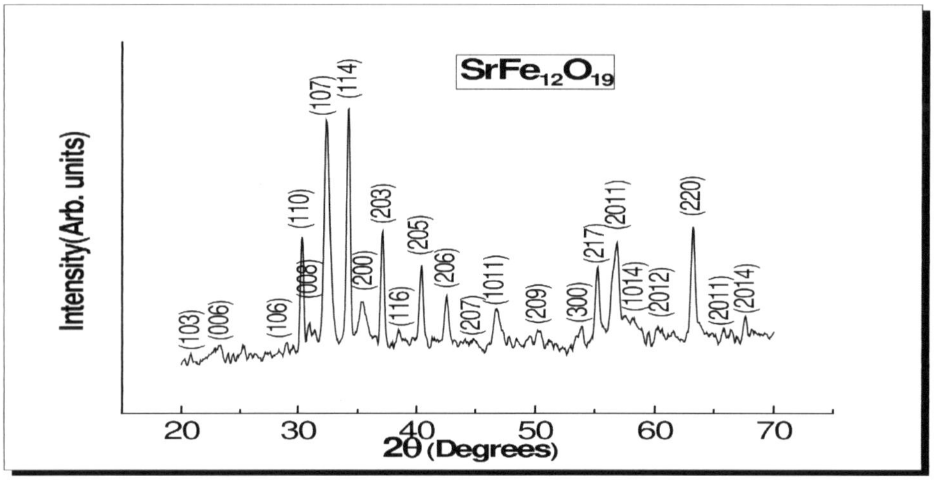

Figure 2. Indexing of X-ray diffraction pattern

The same procedure was repeated for indexing the patterns of all the compositions as shown in figure 3 and their lattice parameters were determined in the same way. These values are listed in Table 4. The results show that the lattice parameters '*a*' and '*c*' are increased with increase in La concentration, as shown in figure 4. These changes may be due to larger ionic radius of La^{3+} (117.2 pm) than that of Sr^{2+} (112 pm) [9].

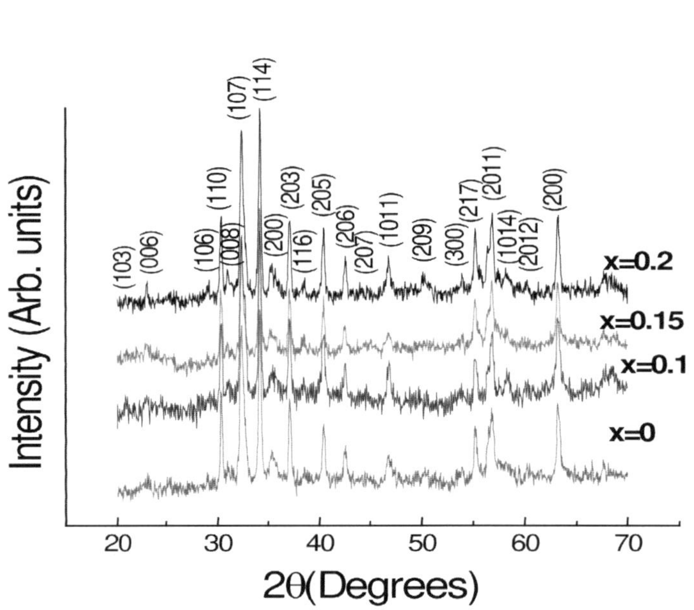

Figure 3. X-ray diffraction patterns of all the samples

The *c*/*a* ratio was also calculated for all the samples. The volume of the unit cell was calculated by the relation [10].

$$V = 3^{1/2} xa^2 xc/2 = 0.866a^2c \quad (5)$$

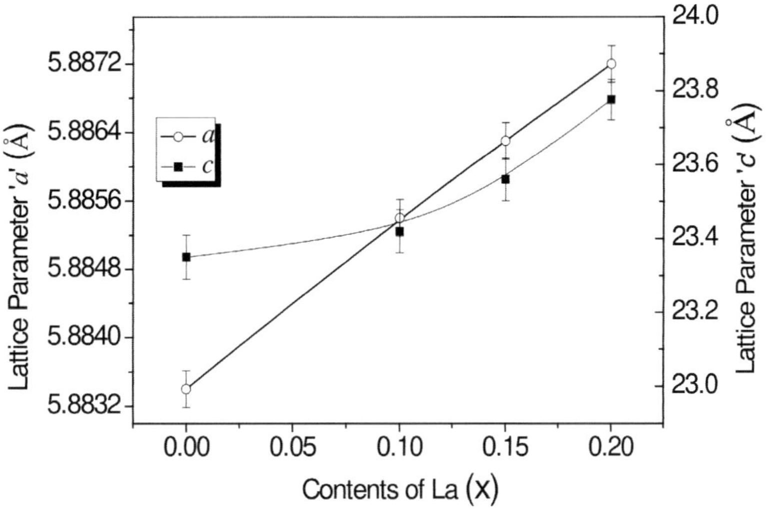

Figure 4. Trend of lattice parameters with increasing La-concentration

Unit cell volumes of all the samples are also given in Table 4. Due to increasing trend of lattice parameters '*a*' and '*c*', the volume of the unit cell is also increased with increasing La contents, as reported previously [9]. The crystallite size was measured for peak (114) using Scherrer's formula, given as [5].

$$D = k\lambda / \beta \cos\theta \quad (6)$$

Here, '*D*' is crystallite size, 'λ' is wavelength of incident X-ray, 'β' represents full width half maximum (FWHM) in radians, and 'θ' is Bragg's angle. '*k*' is a shape factor, having a value of 0.94 in the recent case. Crystallite size calculated for various compositions is given in Table 4. It is clear that crystallite size is increased with increase in doping ratio, which might be due to larger ionic size of La^{3+} [9]. Although, there was no distinct peak of lanthanum in diffraction patterns of La-doped compositions, but the increase in lattice parameters, unit cell volume and the crystallite size proves the presence of Lanthanum. The bulk density of the material was calculated [10].

$$\rho_b = m / \pi r^2 h \quad (7)$$

In this relation, '*m*' is mass, '*r*' is radius of the pellet in disk form and '*h*' is height or thickness of the sample. Bulk density for all the materials has been calculated and it is observed that its value is increased with increase of lanthanum concentration which is according to the previous results [9]. The X-ray density was determined with the help of following expression [10].

$$\rho_x = 2M / N_A V \quad (8)$$

Where '*M*' is molecular weight of the relevant composition and it is multiplied by '2' because one unit cell contains two molecules of the material. 'N_A' is Avogadro's number and '*V*' is volume of the unit cell, determined by Eq. 1.5. The calculated values of X-ray density are tabulated in Table 4. Now, porosity of the material can easily be determined by the equation [10].

$$P = 1 - \rho_b / \rho_x \tag{9}$$

The porosity of the material decreases with the increasing concentration of La as reported previously [8]. The calculated values of porosity for all the samples are given in Table 4.

Table 4. Calculated values of lattice parameters '*a*' and '*c*', *c/a* ratio, volume of the cell (*V*), crystallite size (*D*), bulk density of the material (ρ_b), X-ray density (ρ_x), porosity (*P*), Magnetization (*M*), remanence (M_r), Coercivity (H_C) and ratio (M_r/M)

Parameter	x =0.00	x =0.1	x =0.15	x =0.2
a(Å)	5.8834	5.8854	5.8863	5.8872
c(Å)	23.35	23.42	23.56	23.776
c/*a*	3.9687	3.9793	4.0025	4.0385
V(Å)^3	699.941	702.516	706.933	713.63
D(nm)	29.86	31.654	33.483	34.576
ρ_b(g/cm^3)	3.19	3.71	4.51	3.72
ρ_x(g/cm^3)	5.28	5.045	5.025	4.991
Porosity(%)	39.5	26.4	10.2	25.4
M (emu/g)	45.80	40.96	38.03	37.86
M_r(emu/g)	22.55	21.20	22.82	22.032
H_C(Oe)	2360	2093	3437	6547
M_r/M	0.492	0.517	0.600	0.581

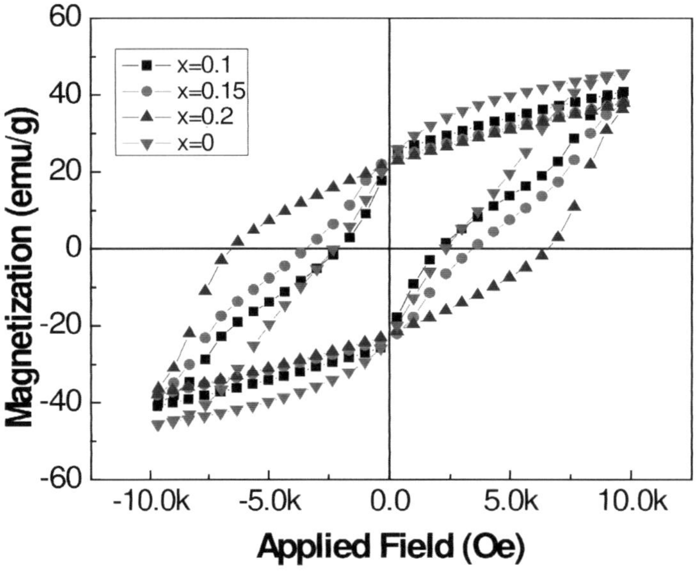

Figure 5. Combine M-H loop for all 'x'

Room temperature magnetic measurements were performed in an applied magnetic field of ±10 kOe, as shown in figure 5. As obvious from the plotted magnetic hysteresis (*M-H*) loops, magnetization has not been completely saturated depicting that ±10 kOe as an applied field is not sufficient to saturate the samples in the present case. Anyhow, specific magnetic parameters of $La_XSr_{1-X}Fe_{12}O_{19}$ samples (x = 0, 0.1, 0.15, 0.2), such as value of magnetization (*M*) at ±10 kOe, coercivity (H_c), remanence (M_r) and ratio of remanence to magnetization (M_r/M) were determined and summarized in Table 4. The value of M was found to decrease from 45.8 emu/g at x = 0, to a value of 37.86 emu/g at x = 0.2. Hard ferrites usually display large coercive fields, and similar type of behavior has been observed in the present series with the values varying in the range of 2360-6547 Oe. The overall trend of H_c was increasing with increasing La contents. It has been figured out previously that H_c increases due to increase in magneto crystalline anisotropy constant [8]. On the other hand, M_r does not vary significantly with the variation in 'x' and it almost remains constant. The M_r obtained from the M-H loops is observed as 22.03-22.82 emu/g. The observed trends are also in good agreement with previously reported results [11, 12]. The intrinsic magnetic properties, such as M, H_c and anisotropy field depends on the atomic magnetism. The lanthanum addition to the strontium hexagonal ferrites induces major refinement in these properties. The La substitution might result in a decrease of the rectangularity of the demagnetizing curves [12]. Nevertheless, the subsequent tuning of the magnetic parameters results in exploiting these compositions for device applications where hard magnetic features are required.

4. Conclusion

La doped Sr-hexaferrites were successfully synthesized using sol-gel auto-ignition technique. A complete procedure has been adopted to determine the crystal structure from X-ray diffraction data. It was observed that lattice parameters and crystallite size increased with the increase of La content attributed to the larger ionic size of 'La' than that of 'Sr'. This was due to the larger ionic size of La than that of Sr. Crystallite size had an increasing trend from 29.86 to 34.576 nm. Bulk density, X-ray density and porosity were also calculated for all compositions. The increase in bulk density and slight decrease in X-ray density reported with the increasing concentration of La which decreases the porosity of the material from 39.5 to 10.2. Saturation magnetization was observed to decrease and coercivity increased with the increase in concentration of La content. The increase in coercivity could mainly be attributed to the increase in magneto crystalline anisotropy constant.

5. References

[1] Hussain S, Maqsood A, Ali A and Naeem M 2011 *J. Supercond. Nov. Magn.* 24 1245

[2] Drmota A, Drofenik M and Znidarsic A 2012 *Ceram. Int.* 38 973

[3] Onreabroy W, Papato K, Rujijanagul G, Pengpat K and Tunkasiri T 2012 *Ceram. Int.* 38S S415

[4] Martinez Garcia R, Bilovol V and Socolovsky L M 2011 *Physica B* 407 3109

[5] B. D. Cullity, 1978 *Elements of X-rays Diffraction, 2nd Ed.*, (Addison-Wesley)

[6] Carter C B and Norton M G 2007 *Ceramic Materials/Science and Engineering* (New York, Springer)

[7] Lechevallier L, Le Breton J M, Morel A and Teillet J 2003 *J. Alloy Compd.* 359, 310

[8] Choi D H, Lee S W, Shim In-Bo and Kim C S 2006 *J. Magn. Magn. Mater.* 204 e243

[9] Onreabroy W, Papato K, Rujijanagul G, Penpat K and Tunkasiri 2012 *Ceram. Int.* 38S S415

[10] Hussain S and Maqsood A 2007 *J. Magn. Magn. Mater.* 316 73

[11] Tenaud P, Morel A, Kools F, Le Breton J M and Lechevallier Dong L 2004 *J. Alloys Compd.* 370 331

[12] Van Diepen A M and Lotgering F K 1974 *J. Phys. Chem. Solids.* 35 1641

Synthesis and characterization of multi-responsive poly (NIPAm-co-AAc) microgels

Irfan Haider[1], Mohammad Siddiq[2], Syed Mujtaba Shah[2], and Saif ur Rehman[2]
[1]Institute of Industrial Control System, P.O.Box 1398 Rawalpindi, Pakistan
[2]Department of Chemistry, Quaid-i-Azam University, Islamabad 45320, Pakistan

E-mail: niceiffi@gmail.com, m_sidiq12@yahoo.com

Abstract. Two samples of multi-responsive copolymer microgels, poly(N-isopropylacrylamide-co-acrylic acid) [P(NIPAm-co-AAc)], were synthesized by emulsion polymerization using different amounts of emulsifier. The unreacted monomers and surfactant were removed from the microgels by dialysis. Dynamic Laser Light Scattering (DLS) was used to determine the diffusion coefficient and hydrodynamic radius (Rh) of the microgels at different pH values and temperatures. At pH 7 the microgel showed rapid response to temperature and underwent volume phase transition (VPT) as the temperature was raised. At lower temperature the microgels were in the hydrated or swollen state but beyond 38oC, the size was abruptly decreased. Under different conditions of pH and NaCl concentrations, variations in VPTT values were noticed. By increasing pH the VPTT was increased but the addition of salt had an inverse effect. The weight average molar mass (Mw) and radius of gyration (Rg) were measured by Static Laser Light Scattering (SLS). FTIR spectroscopy was used to confirm the chemical structures of microgels.

1. Introduction

A polymer gel consists of an elastic cross-linked polymer network with a fluid filling the interstitial space of the network. Recently microgels have gained tremendous importance in theoretical studies on soft matter [1] and applied fields [2, 3]. In particular, they have proved very useful in the domain of material science owing to their potential applications in drug delivery,[4-6] sensing,[7,8], fabrication of photonic crystals,[9-11] template-based synthesis of inorganic nanoparticles,[10,12], separation and purification technologies[13]. The vast array of applications that microgels are suitable for arises from their stimulus-responsive nature, that is, their ability to undergo reversible volume phase transitions in response to external stimuli such as a change in pH,[14-16] temperature,[17,18] ionic strength of the surrounding medium,[19,20] nature of the solvent,[21,22] and the effect of an external electromagnetic field[23,24]. The swelling and deswelling transitions of stimulus-responsive microgels are governed by the imbalance between the repulsive and attractive forces acting within the molecules.

Several methods have been developed to enhance the response rate of microgels to the surrounding environment. Incorporation of hydrophilic polymers into the P(NIPAm) microgel network by interpenetrating polymer network (IPN) technologies has been reported to be an effective approach [25-27]. Zhang et al. synthesized a thermo-sensitive hydrogel with macroporous structures using a pore-forming strategy and expanded network structures [28]. Okano et al. achieved rapid deswelling volume changes in a comb-type grafted P(NIPAm) hydrogel with freely mobile chains grafted on a backbone network [29,30]. To increase the response rate of dual thermo- and pH-responsive microgel, several distinctive methods have also been reported. For instance, Lee et al. generated a series of rapidly responsive comb-type hydrogels by grafting thermo-responsive P(NIPAm) chains onto the backbone of alginate networks and chitosan chains [31]. Asoh et al. developed rapidly responsive porous semi-IPNs composed of P(NIPAm) networks and PAAc linear chains [32]. Kishi et al. prepared fast responsive and pH-/thermo-responsive copolymer microgel by creating micro-porous structures with γ-ray irradiation treatment [33]. However, because of their appeal, dual thermo- and pH-sensitive microgels with rapid responsivity continue to increasingly attract attention. Thus, in order to promote new applications, it remains essential to develop multi-responsive microgel with new synthetic strategies.

2. Experimental

Materials: N-Isopropylacrylamide (NIPAm) and Acrylic acid (AAc) were purchased from Sigma-Aldrich. N,N-methylenebis(acrylamide) (BIS) was obtained from ACROS. NIPAm was recrystallized from hexane–toluene (1:1 volume ratio) mixture and dried in an oven. Acrylic acid (AAc) was purified by distillation under reduced pressure to remove hydroquinone inhibitor whereas N, N-methylenebis (acrylamide) (BIS) was used without further purification.

Co-Polymer Microgel Synthesis: Poly(NIPAm-Co-AAc) microgel was prepared by free radical emulsion polymerization. SDS was used as emulsifying agent.1.4 g NIPAm, 0.033 g BIS and 0.095 mL of AAc were dissolved in 100 mL of deionized water in a three-necked round bottom flask equipped with a condenser and nitrogen inlet. The reaction mixture was heated at 70oC under a gentle stream of nitrogen. After the passage of one hour polymerization was started with 5 mL Ammonium per sulphate (APS) (0.0108g dissolved in 5mL of aqueous solution) at 70oC under constant stirring and continued supply of nitrogen for 5 hours. The resulting polymer microgel was refined by dialysis for 14-21 days (Spectra/Por molecular porous membrane tubing, cutoff, 12000-14000) with very frequently changed water at room temperature to remove unreacted monomers followed by centrifugation. Finally the microgel was dispersed in 100 mL of deionized water and labelled as G-1. Another microgel sample was synthesized using the same procedure but different amount of the emulsifier. This sample was labeled as G-2. The Polmer microgel synthesis is shown by the equation in Figure 1 and chemical composition of both samples synthesized is shown in Table 1.

Table 1. The composition of the both microgel samples

Sample Code	NIPAm	AAc	BIS	SDS	APS (0.06M)	H2O
G-1	1.4 g	0.095 mL	0.033 g	0.057 g	5 mL	100 mL
G-2	1.4 g	0.095 mL	0.033 g	0.114 g	5 mL	100 mL

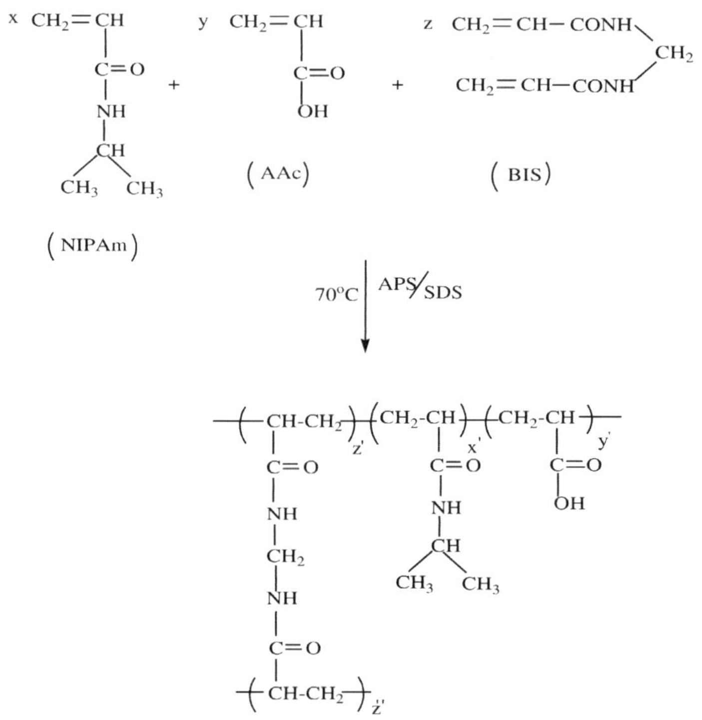

Figure 1. Equation showing the synthesis of crosslinked P(NIPAAm-co-AAc) microgel, where NIPAm and AAc are monomers, BIS is crosslinker, and APS is initiator

Characterization: FTIR spectroscopy was used to identify different functional groups present in the microgels. By interpreting the infrared absorption spectrum, the chemical bonds in a molecule can be determined. FTIR spectroscopy confirmed the chemical structure of product microgel P(NIPAm-co-AAc) and helped to estimate the completeness of polymerization reaction. Light scattering was used to study the solution behavior of microgels. The static and dynamic light scattering were performed using a standard laser light scattering spectrometer BI-200SM motor-driven goniometer equipped with BI-9000AT digital autocorrelator and a cylindrical 22mW uniphase He–Ne laser (wave length = 637nm) and BI-ISTW software. Dynamic light scattering experiments were carried out at different temperatures ranging from 15-70 °C and at different pH values ranging frm 2-9 at angle 90o. DLS of the microgel was also conducted in the presence of different concentrations of NaCl in order to study the effect of electrolyte on the polyelectrolyte i.e. P(NIPAm-co-AAc) microgel. The experiment duration was 5 min. Each experiment was repeated two or more times. In DLLS, Γ is related to the translational diffusion coefficient D by (Γ/q2)C→0,q→0 = D, where q = (4πn/λ) sin(θ/2) with n, λ, and θ being the solvent refractive index, the wavelength of the incident light in vacuum, and the scattering angle, respectively. D was converted to a hydrodynamic radius (Rh) distribution by using the Stokes-Einstein equation, Rh = (kBT/6πη)D-1, where T, kB, and η are the absolute temperature, the Boltzmann constant, and the solvent viscosity, respectively. The Static Laser Light Scattering experiment was carried out at 25°C. Dilute solution (concentration approaches to zero) of microgel was used to find the molar mass and radius of gyration according to Zimm equation.

3. Results and discussions

Multi-responsive copolymer microgels, poly(N-isopropylacrylamide-co-acrylic acid) [P(NIPAm-co-AAc)], were synthesized by using different amounts of emulsifier through emulsion polymerization. The FTIR spectra of crosslinked copolymer microgels G-1 and G-2 are shown in Figures 2 and 3 respectively. All the absorption peaks observed in the spectra of the microgel samples are given in table 2. The FTIR spectra, depicts bands for –CH2 (b) near1550 cm-1 and for C=O near 1620 cm-1 but no peak for C=C. The details of absorption ranges are also shown in table 2, which shows that polymerized microgels contain all the main functional groups required to confirm the chemical structure. The observed absorption bands are in good agreement with the reported range shown in Table 2. Both the spectra show similarity in absorption bands i.e. close values are observed for peaks of the same functional groups.

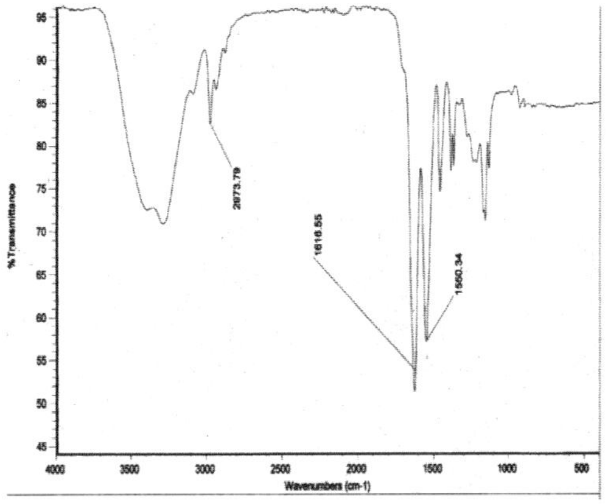

Figure:2; FTIR spectrum of P(NIPAm-co-AAc) microgel (G-1).

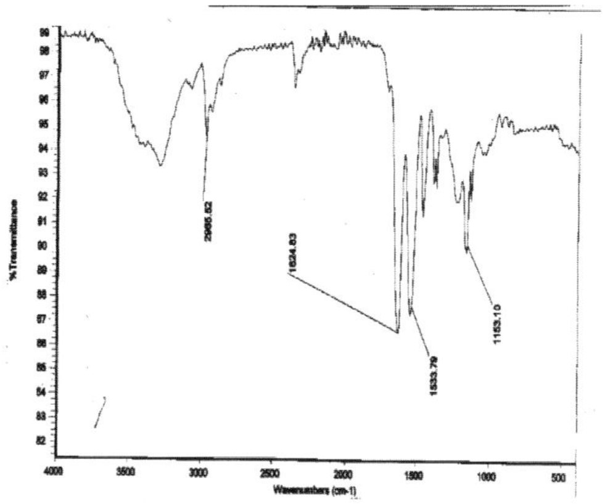

Figure:3; FTIR spectrum of P(NIPAm-co-AAc) microgel (G-2).

Table 2. Summary of FT-IR absorption bands of different chemical bonds of P(NIPAm-co-AAc) microgels G-1 and G-2.

Chemical Bond /Functional Group	Observed Peaks for G-1 (cm-1)	Observed Peaks for G-2 (cm-1)	Range of Peaks (cm-1)
C-N	1150.00	1153.10	1100-1200
CH2 (b)	1550.34	1533.79	1460-1550
C=O (amide)	1616.55	1624.83	1600-1750
C-H (s) sp3	2973.79	2965.52	2900-3000
N-H (s)	3305.00	3310.50	3300-3500

4. Thermo-responsive behavior of poly(NIPAm-co-AAc) microgels

Figure 4 shows the thermo-responsive behavior of microgel. The polt shows the decrease in hydrodynamic radius of poly(NIPAm-co-AAc) microgel with temperature (15-60oC) at constant pH. At lower temperature, it is observed that the size of the microgel is large but as the temperature is increased, the size of the microgel is slightly decreased up to 38oC. After 38oC there occurs a dramatic decrease in the size of the microgel and this transition in size is called volume phase transition while the temperature at which this transition takes place is named as volume phase transition temperature (VPTT). Actually at low temperatures, microgel is highly solvated due to hydrogen bonding between the amide residues on the polymer chain and the water molecules. Hence at low temperature, the polymer-solvent interactions are stronger than the polymer-polymer interactions and poly(NIPAm-co-

AAc) exists in a swollen state. At elevated temperature, the hydrogen bonds between the polymer and the water molecules are broken, leading to an entropically favored expulsion of water from the polymer network. Consequently the polymer-polymer interactions become stronger than the polymer-solvent interactions, resulting in phase separation as the polymer assumes a collapsed (shrunken) state. Hence, at T < VPTT the microgels are individually swollen with water and at T > VPTT the microgels deswell due to expulsion of water from the microgel interior. It is reported that VPTT of homopolymer microgel P(NIPAm) occurs at 32oC in water [34]. But our copolymer microgel is composed of N-isopropylacrylamide and Acrylic acid for which VPTT was observed at 38oC at pH 7. It shows the VPTT of our poly(NIPAm-co-Aac) microgel system is higher than the VPTT of poly(NIPAm) microgel. This increase in transition temperature is observed because of the increase in hydrophilic nature of the microgel particles because of presence of AAc.

5. Thermo-responsive behavior of the microgels at different pH values

The swelling and deswelling behavior of the microgels G1 and G2 as a function of temperature at different pH values ranging from 3 to 9 is shown in figures 6 and 7. There is a continuous decrease in the VPTT and hydrodynamic radius (Rh) as we go on decreasing the pH. In swollen state maximum size of gel is observed at pH 9, but the transition is less pronounced and after transition there is less reduction in hydrodynamic radius (Rh) of microgel. This is because in basic media, the microgel is in hydrophilic state and the pH effect is dominant over the thermosensitivity of microgel. Minimum size is observed at pH 3 in swollen state i.e.; at low temperature (below VPTT) but thereafter the VPTT instead of going into collapsed state by showing reduction in size, Rh increases abruptly indicating the aggregation of colloidal particles at elevated temperature. A reasonable explanation for this different phenomenon of microgel is inter-particle hydrogen bonding interactions between the carboxyl group of the acrylic acid side chain and the amide group of NIPAm which seems to favor larger aggregates. The decrease in size and VPTT of microgel with decrease in pH means that our microgel is more temperature sensitive at lower pH because it undergoes transition at lower temperature. The observed VPTT is higher at high pH because P(NIPAm-co-AAc) becomes more pH sensitive at high pH as acrylic acid deprotonates its carboxyl moieties in alkaline solution, due to which repulsion between these groups increases.

6. Effect of Emulsifier (Surfactant)

Figure 5 shows that both samples of microgel (G-1 and G-2) underwent phase transition on changing temperature from 15oC to 60oC at pH 7. But difference in Rh size was observed between two synthesized microgel samples. Different amounts of SDS were used as emulsifier during emulsion polymerization. The microgel with lower SDS concentration (G-1) has larger size than the microgel with higher SDS concentration (G-2). Surfactants play a major role during the particle nucleation and growth, with direct impact on latex particle size, size distribution, polymerization rate, molecular weight and particle morphology. Emulsion polymerization is propagated in micelles which make more stable structure by surrounding the monomer, usually in continuous phase (water) [35]. When surfactant concentration is high, more micelles are formed with smaller size. Due to this small size, micelles can accommodate small amount of monomers which react to produce polymer particles with smaller size. The smaller size of the hydrodynamic radius of G-2 could be ascribed to the higher SDS concentration in G-2 than G-1.

7. Effect of Medium pH

The pH dependence of the particle size of the poly(NIPAm-co-AAc) microgel is shown in Figure 8. Small size is observed at lower pH values, but size increases with the rising pH. This is due to deprotonation of the acrylic acid and the resultant electrostatic repulsion between the negatively charged carboxylate residues. The ionized structure provide increased electrostatic repulsion between chains and hence the hydrophilicity of network. Under these conditions, microgels are capable of uptaking large amounts of solvent and forming very loose structure consequently the hydrodynamic radius of microgel increases. It is evident from Figures 8 and 9 that the size of gel particle and swelling critically depend on the pH of medium.

In Figure 9 hydrodynamic radius is higher at 15oC and 25oC because both temperatures are below VPTT and microgel is in swollen state. As 35oC is just near the collapsing temperature that's why the size of G-1 starts decreasing as compared to 15 and 25oC at lower and higher pH values. But G-2 shows larger size at lower pH at 35oC because of greater concentration of surfactant, the aggregates formation takes place earlier as shown in Figure 10. Similar behavior is observed at 45oC for both the microgel samples at low and high pH values. It is observed that at lower pH values (2, 3) the size is surprisingly large. This is due to the formation of aggregates at higher temperature and lower pH. Because at lower pH the carboxylic acid groups are largely protonated and have little effects on phase transition.

There is a small change in size at pH 9 at different temperatures (15-45oC). This is because at higher pH the microgel is less temperature responsive and the pH effect is dominant to increase its size.

8. Influence of salt concentration on the swelling behavior of the microgels

The hydrodynamic radius in the presence and absence of salt (NaCl) is given in figure 11. It is observed that the presence of salt decreases the size of microgel particles and VPTT also shifts to lower value. In the absence of salt at pH 9 the VPTT is 42oC but when 0.05M NaCl is added into the microgel the VPTT decreases to 34oC which is a large shift. In general, an increase in ionic strength decreases the Debye screening length, and hence will diminish the repulsive electrostatic forces between the dissociated acrylic acid groups at pH 9. Therefore, it is expected that addition of salt will have a strong effect on the swelling behavior. Effect of salt on the microgel size is investigated by changing the NaCl concentrations up to 0.5M. It is noted that the size of the swollen particles at low temperature decreases with increasing salt concentration. This effect could be attributed to reduced osmotic swelling pressure, because at higher salt concentration the concentration gradient between the NaCl concentrations inside and outside the microgel particles decreases.

At low pH, the effect of salt on the microgel size is shown in Figure 12. At lower temperature the size was decreased but after VPTT, due to formation of aggregates size was increased further by the addition of salt. Figure 13 shows the salt effect on the microgels comparing both samples (G-1 and G-2). There is more reduction in size of G-2 on adding salt because of higher surfactant concentration. We have observed that multi-responsive microgels show transition in their size when their external environment is changed. Therefore, the VPTT can be shifted by changing the stimuli. The summary of VPPT of P(NIPAM-co-AAc) microgels is given in the tables 3 and 4 under different external conditions.

Table 3. Summary of the VPTT of the microgels at different pH values

pH	VPTT of G-1 (oC)	VPTT of G-2 (oC)
3	34	32
5	36	36
7	38	38
9	42	40

Table 4. Summary of VPTT of microgels in different NaCl conc. at pH 3 and pH 9

NaCl (M)	VPTT at pH 3 (oC)		VPTT at pH 9 (oC)	
	G-1	G-2	G-1	G-2
0.00	34	32	42	40
0.05	30	30	34	34
0.1	28	28	30	30
0.5	26	26	28	28

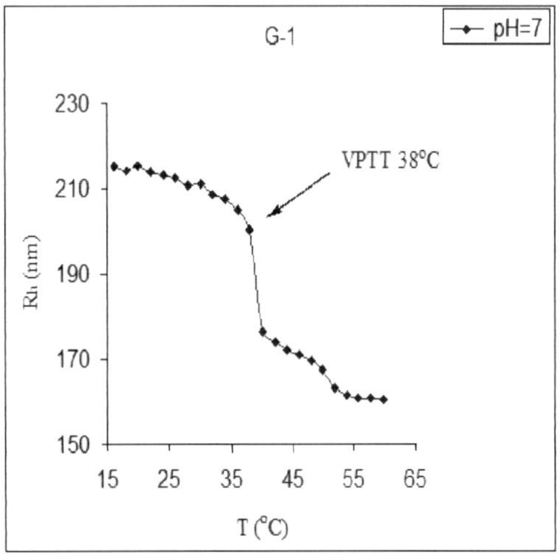

Figure 4. R_h of G-1 as a function of temperature at pH 7

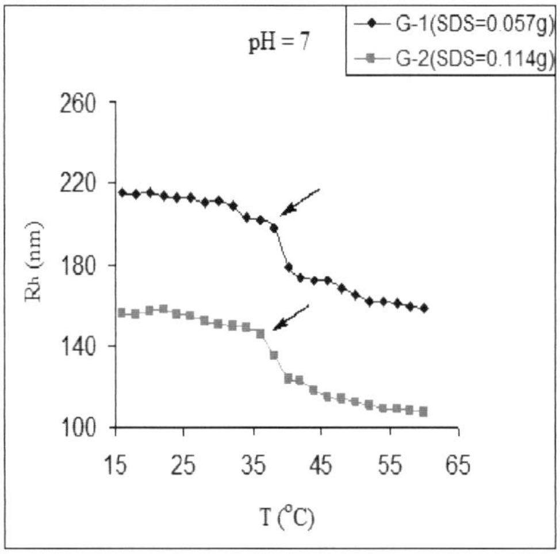

Figure 5. R_h of G-1 and G-2 as a function of temperature at pH 7

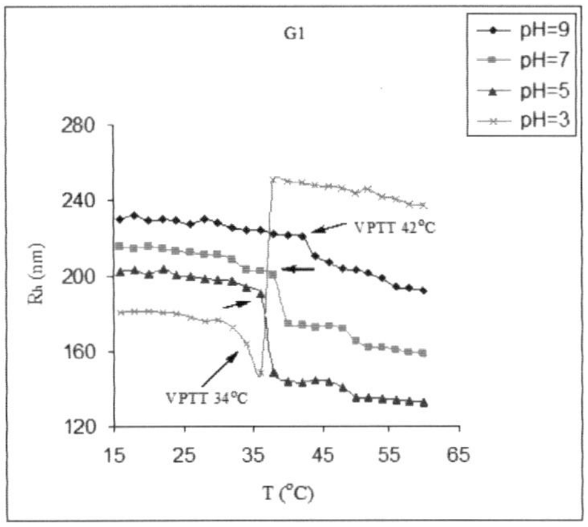

Figure 6. R_h of G-1 vs Temperature at different pH

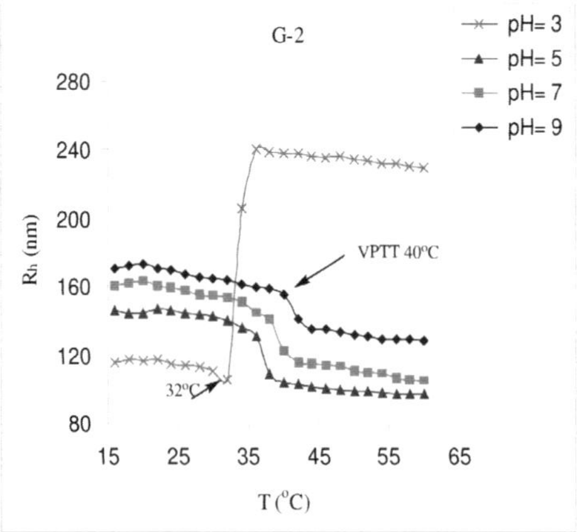

Figure 7. R_h of G-2 vs Temperature at different pH

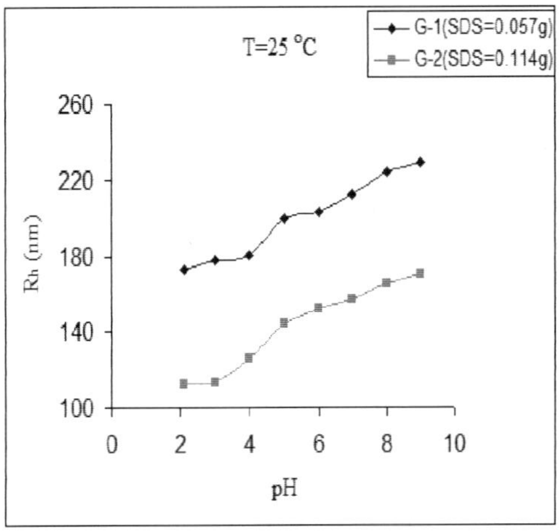

Figure 8. Hydrodynamic radius vs pH at 25°C

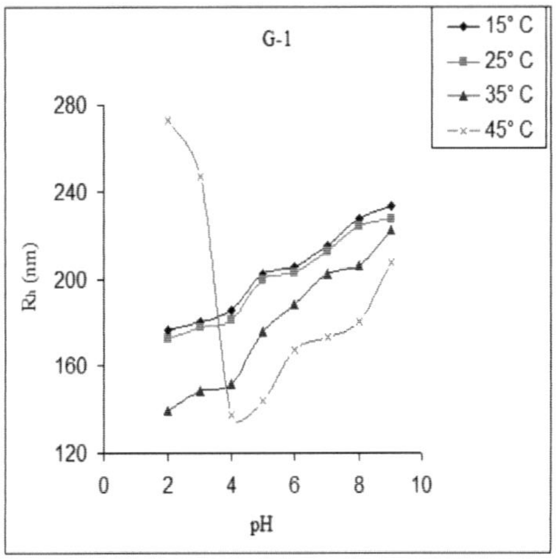

Figure 9. Rh of G-1 vs pH at different temperatures

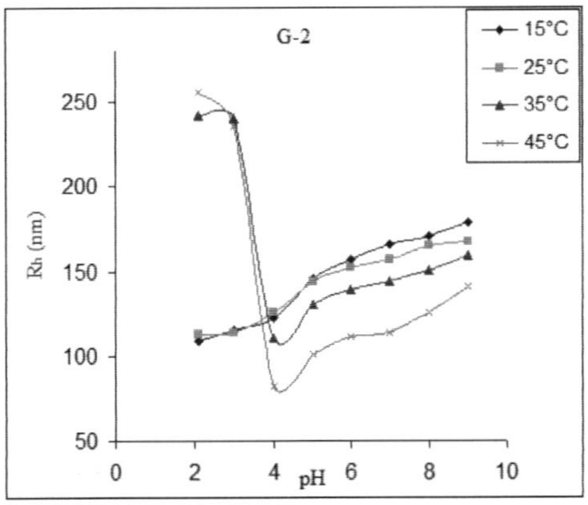

Figure 10. R_h of G-2 vs pH at different temperatures

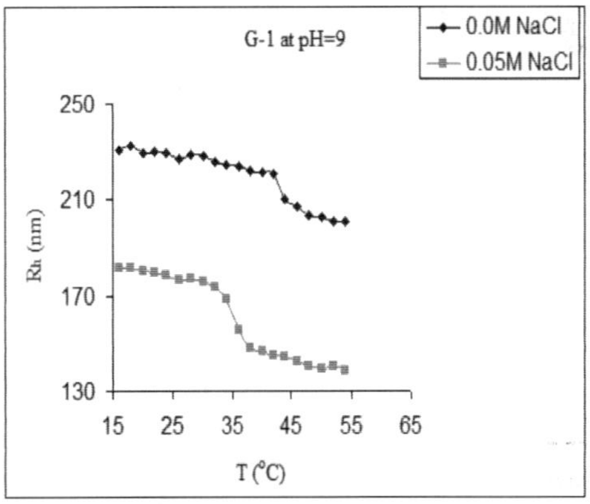

Figure 11. R_h vs T of G-1 in the presence of 0.0M and 0.05M NaCl at pH 9.

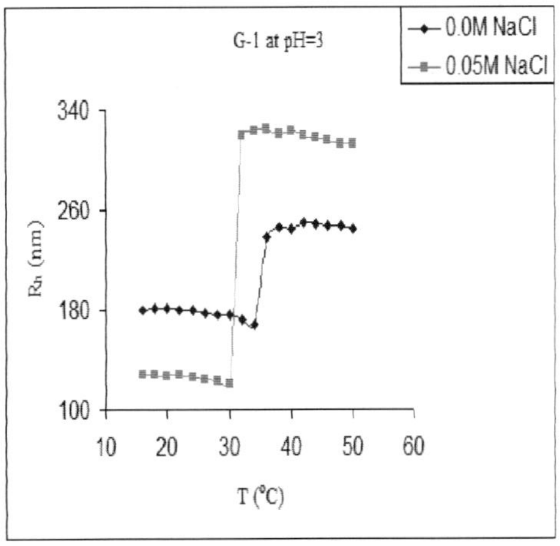

Figure 12. R_h of G-1 vs T at 0.0M and0.05M NaCl at pH 3

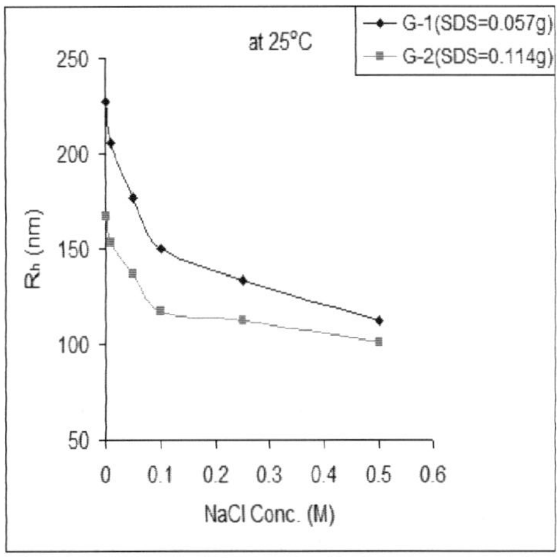

Figure 13. R_h vs NaCl concentration at 25°C and pH 9

9. Static Laser Light Scattering (SLLS)

The Static Laser Light Scattering experiment was carried out at 25°C to find the Molar mass (Mw) and radius of gyration (Rg) of both samples as shown in the table 4.4. As clear from the table, both the samples of P(NIPAm-co-AAc) microgel have the same molar mass but different values of radii of gyration. We used different amount of surfactant during polymerization keeping the monomers and crosslinker concentrations the same in both samples. After synthesis the surfactant was removed by dialysis. There is a small difference in the values of radii of gyration of both the microgels because static light scattering measurements were performed at 25oC which is below VPTT and microgel is in swollen state.

Table 5. Static laser light scattering results of micogels

Microgel sample	Mw (g mol-1)	Rg (nm)
G-1	5 x105	137.4
G-2	5 x105	134.8

10. Conclusion

Our results indicate that P(NIPAm-co-AAc) microgel is multi-responsive. It is in hydrated state at low temperature and show response upon exposure to various environmental stimuli such as change in pH, temperature and ionic strength. Emulsifier concentration has an important role during emulsion polymerization. The size of microgel decreases with increasing the concentration of surfactants and vice versa. Presence of salt also suppresses its electrolyte behavior by decreasing the VPTT. Increasing the pH up to basic value, VPTT increases and increasing the salt concentration VPTT decreases. At pH 7 and very low salt concentration, VPTT is near human body temperature, so it can be used for drug delivery. The responsive properties of these microgels make them ideal candidates for biological and optical applications as smart materials.

11. References

[1] Senff H.; Richtering W.; Norhausen C.; Weiss A.; Ballauff M.; Langmuir 1999, 15, 102-106.

[2] Jeong B.; Gutowska A.; Trends in Biotechnology 2002, 20, 360-360.

[3] Ogawa Y.; Ogawa K.;, Wang B. L.; Kokufuta E.; Langmuir 2001, 17, 2670- 2674.

[4] Murthy N.; Xu M. C.; Schuck S.; Kunisawa J.; Shastri N.; Frechet J. M. J.;Proceedings of the National Academy of Sci. USA 2003, 100, 4995-5000.

[5] Bromberg L.; Temchenko M.; and Hatton T. A.; Langmuir 2002, 18, 4944-4952.

[6] Standley S. M.; Mende I.; Goh S. L.; Kwon Y. J.; Beaudette T. T.; Engleman E. G.; Frechet J. M. J.; Bioconjugate Chemistry 2007, 18, 77-83.

[7] Lapeyre V.; Gosse I.; Chevreux S.;Ravaine V.; Biomacromolecules 2006, 7, 3356- 3363.

[8] Qu J. B.; Chu L. Y.; Yang M.; Xie R.; Hu L. and Chen W. M., Advanced Functional Materials 2006, 16, 1865-1872.

[9] Jones C. D.; and Lyon L. A.; Macromolecules 2000, 33, 8301-8306.

[10] Xu S. Q.; Zhang J. G.; Kumacheva E.; Composite Interfaces 2003, 10, 405-421.

[11] Lyon L. A.; Debord J. D.; Debord S. B.; Jones C. D.; McGrath J. G.; Serpe M. J.; J. of Phy. Chem. 2004, 108, 19099-19108.

[12] Zhang J. G.,. Xu S. Q and Kumacheva E., Journal of the Amer. Chem. Soci. 2004, 126, 7908-7914.

[13] Nilsson P.; Hansson P.; J. of Phy. Chem. 2005, 109, 23843-23856.

[14] Kim K. S.; Vincent B.; Polym. J. 2005, 37, 565-570.

[15] Mohan Y. M.; Lee K.; Premkumar T.;Geckeler K. E.; Polymer 2007, 48, 158-164.

[16] Nisato G., Munch J. P.; Candau S. J.; Langmuir 1999, 15, 4236-4244.

[17] Dowding P. J.; Vincent B.; Williams E., J. of Colloid and Interface Sci. 2000, 221, 268-272.

[18] Andersson M.; Maunu S. L.; Colloid and Polymer Science 2006, 285, 293- 303.

[19] Lopez-Leon T., Elaissari A., Ortega-Vinuesa J. L and D. Bastos-Gonzalez, Chemphyschem 2007, 8, 148-156.

[20] Snowden M. J., Thomas D.and Vincent B., Analyst 1993, 118, 1367-1369.

[21] Molina M. J., Gomez-Anton M. R. and Pierola I. F., J. of Phy. Chem.2007, 111, 12066-12074.

[22] Dragan S.; Ghimici L.; Wandrey C.; Macromolecular Symposia 2004, 211, 107-119.

[23] Bhattacharya S.; Moss R. A.; Ringsdorf H.; Simon J.; Langmuir 1997, 13, 1869-1872.

[24] Brugger B.; Richtering W.; Advanced Materials 2007, 19, 2973.

[25] Zhang J.T.; Cheng S.X.; Zhuo R.X.; Colloid Polym Sci. 2003, 281, 580-584.

[26] Zhang G.Q.; Zha L.S.; Zhou M.H.; Ma J.H.; Liang B.R.; Colloid Polym Sci. 2005, 283, 431-436.

[27] Owens Donald E, Yicun Jian, Fang Justin E, et al. Macromolecules 2007 ,40, 7306-7310 .

[28] Zhang X.Z.; Yang Y.Y.; Chung T.S.; Langmuir 2001, 18, 2013.

[29] Yoshida R.; Uchida K., Kaneko Y.; Sakai k.; Kikuchi A.; Sakurai Y.; Nature 1995, 374:240.

[30] Kaneko Y.; Sakai K.; Kikuchi A.; Yoshida R.;Sakuri Y.; Okano T.; Macromolecules 1995, 28, 7717.

[31] Ju H.K.; Kim S.Y.; Lee Y.M.; Polymer 2001, 42, 6851-6855.

[32] Asoh T.; Kaneko T.; Matsusaki M.; Akashi M.J.; Controlled Release 2006, 110, 387-393.

[33] Kishi R.; Miura T.; Kihara H.; Asano T.; Shibata M.; Yosomiya R.; J. Appl Polym Sci. 2003, 89:75.

[34] Huang J. and Wu X. Y., Journal of Polymer Science Part a-Polymer Chemistry 1999, 37, 2667-2676.

[35] Rosen, S.L.; "Fundamental Principles of Polymeric Materials" 2nd Ed., John Wiley and Sons, Inc., New York, 1993.

Structure and temperature dependent electrical properties of lead-free $Bi_{0.5}Na_{0.5}TiO_3$- $SrZrO_3$ ceramics

Adnan Maqbool[1], Jamil ur Rahman[1], Ali Hussain[1], Jong Kyu Park[1], Tae Gone Park[1], Jae Sung Song[2] and Myong Ho Kim[1]
[1]Engineering Research Center for Integrated Mechatronics Materials and Components, Changwon National University, Gyeongnam 641-773, Republic of Korea
[2]Korea Electrotechnology Research Institute, Gyeongnam 641-773, Republic of Korea

E-mail: mhkim@changwon.ac.kr

Abstract. Lead-free $SrZrO_3$-modified $Bi_{0.5}Na_{0.5}TiO_3$ (BNT-SZ) ceramics were fabricated by a conventional solid state reaction method. X-ray diffraction analysis reveals a pure perovskite phase without any traces of secondary phases. Scanning electron microscopy images depicts dense grain morphology. The temperature dependences of the dielectric behavior was measured in the temperature range of 50–500 °C at 100 kHz. With the increase in SZ content, the dielectric constant (ε_r) constantly decreased and the maximum dielectric constant temperature (T_m) shifted towards lower temperatures. In addition to this, ferroelectric hysteresis loops indicated a disruption of ferroelectric order and increase in the relaxor character of BNT ceramics with increase in SZ concentration. A maximum values of remnant polarization (32 μC/cm^2) and piezoelectric constant (100 pC/N) were observed at SZ5 and SZ4, respectively.

1. Introduction

Lead-based ceramics with perovskite structure, such as $Pb(Zr,Ti)O_3$ (PZT) and PZT-based multicomponent materials are widely used for actuators, sensors and micro-electromechanical devices owing to their superior piezoelectric properties [1]. However, the use of lead-based materials produce serious lead pollution and environmental problems due to high toxicity of lead oxide.

In lead-free materials, $Bi_{0.5}Na_{0.5}TiO_3$ (BNT) is considered one of the most promising candidate materials for the piezoelectric applications. It has ABO_3 perovskite structure with a ferroelectric rhombohedral symmetry (R3C) at room temperature [2, 3]. As temperature increases, the phase of BNT will transform from rhombohedral to tetragonal and finally into cubic phase [4]. The transition temperatures of rhombohedral-tetragonal phase (T_{R-T}) and tetragonal-cubic phase (T_{T-C}) are 300 and 540 °C, respectively. BNT shows strong ferroelectricity (P_r = 38 μC/cm^2), but also has drawbacks such as a large coercive field (E_c) of approximately 7.3 kV/mm and a high conductivity. To decrease its conductivity and improve its piezoelectric properties, it has been modified with various types of perovskite materials, such as $CaTiO_3$ (CT) [5], $BaTiO_3$ (BT) [6-8], $BaZrO_3$ [9], $SrTiO_3$ (ST) [4, 10, 11], $BiAlO_3$ (BA) [12, 13], $(Bi_{0.5}K_{0.5})TiO_3$ (BKT) [14, 15], $NaNbO_3$ (NN) [16], KNbO3 (KN) [17, 18] and $BaSrTiO_3$ (BST) [19, 20]. Furthermore, it has been also modified with divalent (Ca, Sr, Ba) [21] and trivalent (La, Nd, Ho, Yb) ions [22].

Hiruma *et al.* [18] have fabricated BNT-based solid solutions, such as BNT-NN, BNT-KN and BNT-BKT and demonstrated relationship between the phase diagrams and their electrical properties. They reported that piezoelectric properties and strain response increased at the phase boundary between the ferroelectric rhombohedral phase and the ferroelectric relaxor phase with the mixed pseudocubic tetragonal phase. In addition, Krauss *et al.* [10] studied the piezoelectric properties and phase transition temperatures of the solid solution of (BNT-ST) ceramic system. From their X-ray diffraction (XRD) measurements a broad maximum of the lattice parameter was observed around $x = 0.5$, but no structural evidence for a morphotropic phase boundary was found. Above studies [4-8, 10-24] revealed that formation of a new solid solution is an effective way to improve the piezoelectric properties of the BNT ceramics.

Strontium zirconate ($SrZrO_3$) belongs to the perovskite family and has an orthorhombic structure at room temperature, with space group P*bnm* [25]. Due to its high dielectric constant, high breakdown strength and low leakage current density, $SrZrO_3$ has been extensively investigated as a possible candidate material for high-k gate dielectrics (E_g = 5.6 eV) [26, 27]. In the present work, new SZ-modified BNT ceramics were produced by continues solid solution reaction and its ferroelectric and piezoelectric properties were investigated systemically.

2. Experimental

A conventional mixed oxide method was utilized to prepare $SrZrO_3$-modified $Bi_{0.5}Na_{0.5}TiO_3$ (BNT-SZ100x with x = 0-0.15) ceramics. The commercially available reagent grade oxide or carbonate powders of Bi_2O_3 (99.90%), Na_2CO_3 (99.95%), TiO_2 (99.90%), $SrCO_3$ (99.90%) and ZrO_2 (99.0%) of Sigma Aldrich Co. St. Louis, MO, were used as starting raw materials. Prior to measuring the weighs, the powders were dried in an oven at 100 °C for 24 hrs. For each composition, the starting materials were weighed according to the stoichiometric formula and ball-milled for 24 hrs in ethanol. The dried slurries were calcined at 850 °C for 2 hrs and then ball-milled again. After calcinations, the mixture was ball milled for 24 h, dried. The powders were pulverized, mixed with an aqueous polyvinyl alcohol (PVA) solution as a binder for granulation and passed through a sieve of 250 mesh. The granulated powders were subsequently pressed into green discs of diameter 10 mm. The compacts were sintered at 1150 °C for 2 hrs in in a covered alumina crucible. To minimize the evaporation of the volatile elements Bi and Na, the disks were embedded in the powder for same composition.

Silver paste was coated on both faces of the sintered samples and fired at 650 °C for 0.5 hrs to form electrodes. The specimens for measurement of piezoelectric properties were poled in silicone oil bath with a dc field of 4 kV/mm for 15 min. All the electrical measurements were performed after aging for at least 24 h. The crystal structure of sintered samples was characterized using X-ray diffractometry (XRD, X'pert MPD 3040, Phlips, Netherland). Surface morphology was checked through scanning electron microscope (SEM, JP/JSM 5200, Japan). Dielectric constant and loss of the specimens were measured using an automatic acquisition system using an impedance analyzer (Agilent HP4292A, USA) in 25-500 °C temperature range at different frequencies. The piezoelectric properties were measured using a Berlincourt d_{33} meter (IACAS, ZJ-6B). The dependence of the electric polarization (P) under an external electric field (E) were measured in a silicon oil bath by using Precision Material Analyzer (Radiant Technologies, Inc. Albuquerque, NM).

3. Results and Discussion

3.1 Phase and microstructure analysis

The XRD patterns of sintered SZ-modified BNT ceramic samples are shown in Fig. 1. Patterns of all samples revealed a pure perovskite phase, indicating that SZ successfully diffused into the lattice structure of the BNT ceramics to form complete solid solutions. Pure BNT ceramics revealed the

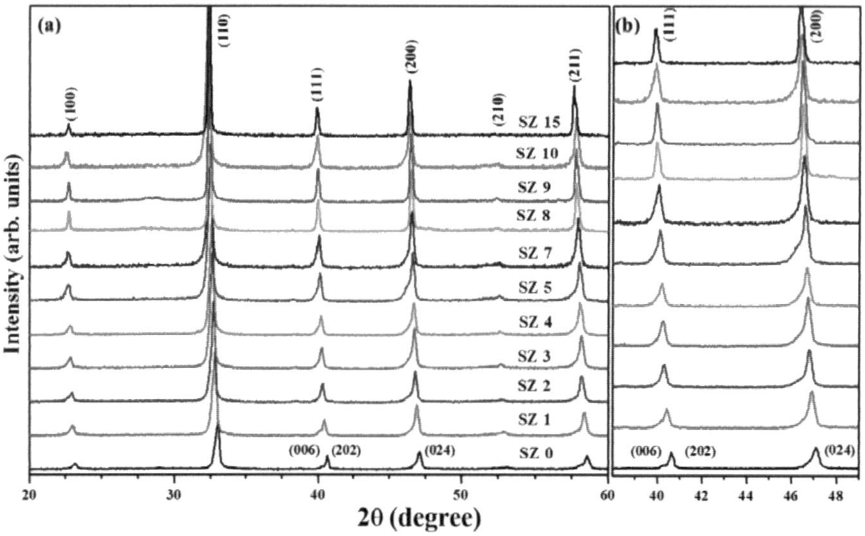

Figure 1. X-ray diffraction patterns of the BNT–SZ100x ceramics in the 2θ range of (a) 20–60° and (b) 38–48°.

characteristic rhombohedral structure. With increasing SZ content, the rhombohedral structure gradually transforms to the pseudo-cubic symmetry. The overall effect of SZ substitution on the XRD patterns of BNT ceramics is the slight shifting of intensity peaks towards lower angle, this peak shifting behavior increase with increasing SZ concentration. This may be due to replacement of small ions with the larger ionic radii Sr^{2+} (1.44 Å) on A site (for Bi^{3+} = 1.36 Å and Na^{+} = 1.39 Å) and Zr^{4+} (0.72 Å) on B site (for Ti^{3+} = 0.605 Å) in BNT ceramics. Similar peaks shifting behavior was also observed in Zr and BZ-modified BNT ceramics [9, 28] and in BST-modified BNT ceramics [19, 20].

Figure 2 presents the SEM micrographs of the polished and thermally etched surfaces of SZ-modified BNT ceramics with x = 0, 0.02, 0.04, 0.05, 0.07 and 0.09. All ceramics were uniformly distributed and tightly bound with homogeneous macrostructures of similar grain morphology. SZ had small influence on the grain morphology of the BNT ceramics. The grain size decreased with an increase in the SZ concentration. Using a linear intercept method, the average grain size was found to decrease from 25.2 μm for x = 0 to 2.4 μm for x = 0.10.

3.2 Dielectric constant and loss

The temperature dependent dielectric constant (ε_r) and loss ($tan\delta$) for the BNT-SZ ceramics measured at 100 KHz are presented in Fig. 3. It can be observed that with increase in measuring frequency, the observed dielectric constant decrease for all the samples similar to other BNT-based relaxor ferroelectric materials [10, 11, 18]. For pure BNT ceramics, the maximum dielectric constant (T_m) peak is observed at highest value 3600 at 10 KHz. The curves of temperature dependence of relative dielectric constant for different samples look similar, as indicated in Fig. 3. By the substitution of SZ, T_m gradually decreases and peaks become much broader. The decrease of T_m can be associated with the structural transition point from the rhombohedral to pseudo-cubic phase, as discussed in XRD analysis. The dielectric loss slightly increased with increasing SZ content in the BNT. The dielectric loss has not change much at room temperature because of the less distortion in the crystalline structure and then increase around T_m due to increase in the conductivity of the conductivity of SZ-modified ceramics. The dielectric loss slightly decreased with increasing SZ content and varied from of SZ7-10.

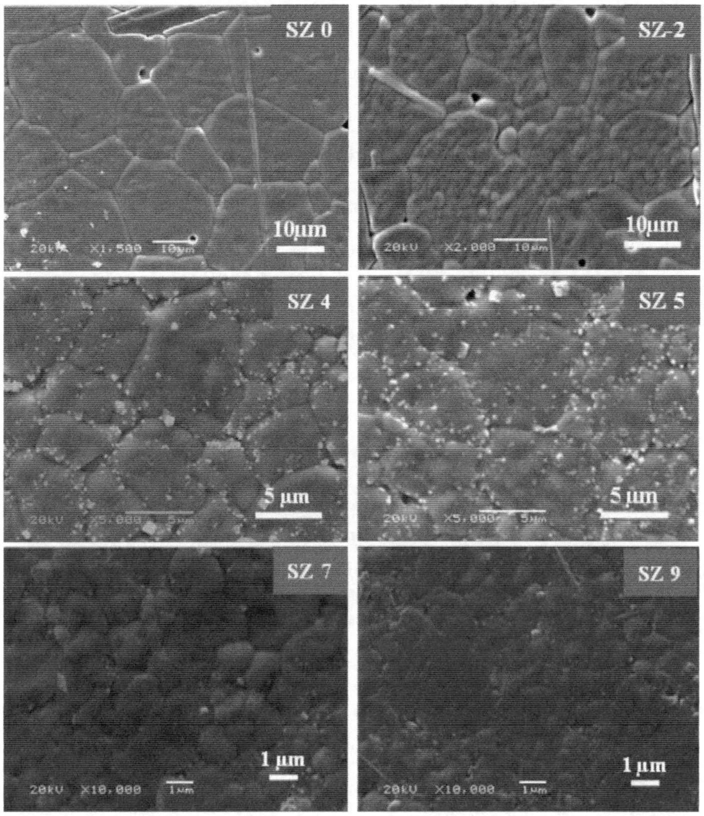

Figure 2. SEM micrographs of thermally etched samples of BNT-SZ ceramics with different SZ content.

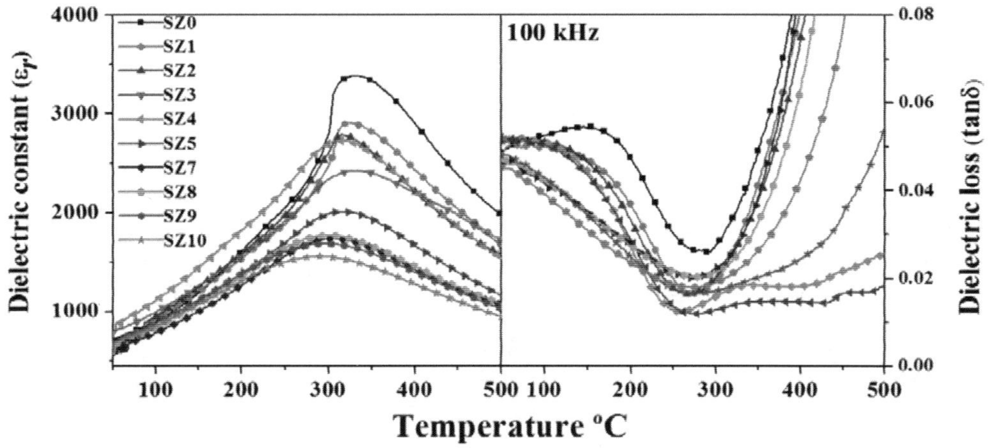

Figure 3. Direct comparisons of the dielectric constants and losses of all the samples of BNT-SZ ceramics at 100 kHz

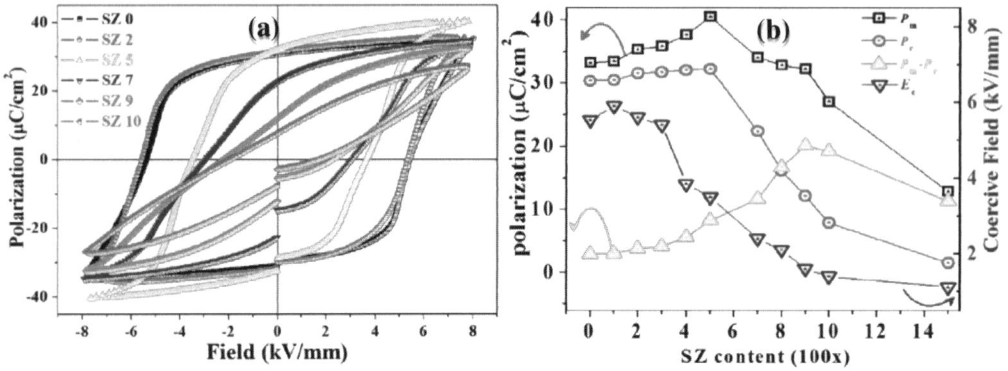

Figure 4. Effect of SZ-modification on the (a) *P-E* hysteresis loops. (b) Characteristic values of P_{max}, P_r, their difference (P_{max}-P_r) and E_c as function of SZ content.

3.3 P–E ferroelectric hysteresis

The measurement of polarization versus electric field (*P–E*) hysteresis loops was conducted to investigate the ferroelectric properties of BNT-SZ ceramics. Fig. 4(a) shows the room temperature *P–E* hysteresis loops of the BNT–SZ ceramics with different SZ content measured at 20 Hz. Pure BNT ceramics without SZ content show a typical saturated FE behavior, which is characterized by definite squareness in the *P–E* hysteresis loop with a remanent polarization (P_r) of 30 μC/cm² and a coercive field (E_c) of 5.5 kV/mm. It can be clearly seen in Fig. 4(a), SZ exerts significant influence on the loop shape and polarization values, similar behavior is also perceived in ST and BA-modified BNT ceramics [11, 13]. A significant reduction in the E_c is observed at SZ5 along with the clear increase in the P_r and P_m. However, at SZ7 the P_r drastically decreased from 32 to 22 μC/cm², and the E_c decreased from 3.5 to 2.4 kV/mm. At higher SZ concentrations i.e. for SZ10, both P_r and E_c drastically decreased, indicating the material became electrostructive without any apparent switching.

The characteristic values of P_r, P_m, their difference (P_m - P_r) and E_c are shown in Fig. 4(b). At higher SZ concentrations P_r and E_c significantly decreased with increasing SZ content which is characterized by a slim *P-E* hysteresis loop. The large difference between the P_m and P_r at SZ9 indicates the large filed induced strain response which is similar to other reports on BNT-based ceramics [7, 10, 11]. This significant decrease in P_r and E_c along with concurrent minor decreases in P_m implies that the long-range FE order dominant in BNT is disrupted and FER type behavior become dominant with the addition of SZ.

3.4 Temperature dependent P-E loops

To better understand the temperature stability and ferroelectric behavior of BNT-SZ ceramics, the temperature dependent polarization hysteresis loops were illustrated in detail in Fig. 5. For pure BNT ceramics, P_r decreases slightly however sharp reduction in E_c is observed with the increase in the temperature. For SZ-modified ceramics, with increase in temperature a sharp reduction in P_r and E_c is observed, indicating the disappearance of FE order and appearance of a relaxor phase [7, 29, 30]. For SZ4 and SZ5, a pinched-type character is also observed at elevated temperatures. In case of pure BNT, no pinched-type loop is revealed even at 175 °C. However, pinched-type characteristic can be easily seen around 100 °C for SZ4 and 75 °C for SZ5 samples. In SZ4 and SZ5, at elevated temperature around 150 °C, *P-E* loops become very slim and narrow but still pinched-type character is not completely vanished. However, SZ9 ceramics exhibited like the linear dielectric materials at elevated temperature. Similar results of the temperature dependent *P–E* loops have been reported for other BNT-based ceramics [7]. This change of the *P-E* loops are well consistent with the temperature dependence of dielectric properties. As suggested by literature, it can be concluded that pinched type anomalies in *P–*

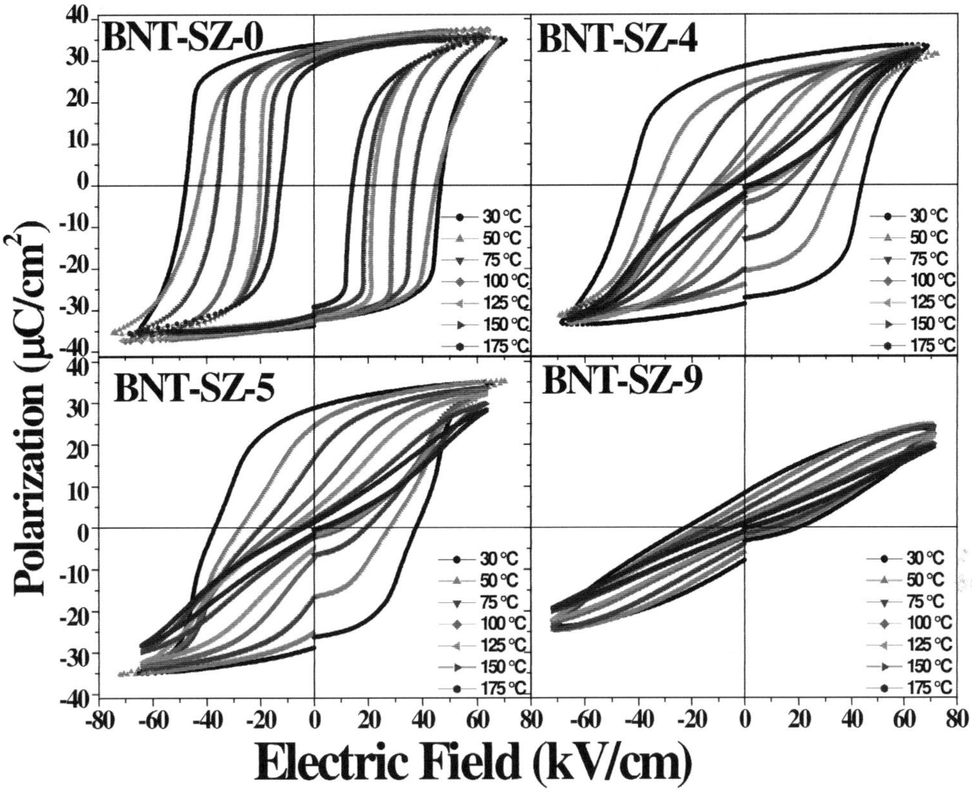

Figure 5. The *P-E* hysteresis loops of BNT-SZ ceramics with different SZ content as a function of temperature

E loops resulted from the electromechanical interaction between the polar and non-polar regions which coexisted in the ceramics at high temperatures.

3.5 Piezoelectric constant

Figure 6 shows the piezoelectric constant d_{33} and K_p of the BNT ceramics as a function of SZ content. The d_{33} and K_p both enhance with an increase in the SZ content, reaches a maximum value of $d_{33} = 100$ pC/N and $K_p = 0.19$ at SZ4. Further increase in SZ concentration resulted in a significant reduction in d_{33} and K_p values. The observed trends is in good agreement with polarization hysteresis loops in Fig. 4(a), similar behavior is also in other studies [12, 13]. The substantial increase observed in d_{33} and K_p at SZ4 is attributed to a large P_r and a lower E_c. This is because a lower E_c enables the ceramics to be more easily poled, while a large P_r and P_m favors piezoelectricity.

4. **Conclusions**

The influence of SZ-substitution on crystal structure, microstructure, dielectric, ferroelectric piezoelectric behavior of BNT ceramics was investigated. XRD revealed that rhombohedral structure gradually transforms to the pseudo-cubic symmetry with increasing SZ. SEM-analysis revealed a decrease of the grain size for increasing SZ-content. The maximum dielectric constant temperature T_m gradually decreased and shift toward lower temperature with increasing SZ content. Deformed hysteresis loops at small SZ concentrations were observed, suggesting the coexistence of ferroelectric and relaxor phases for modified BNT ceramics. A significant reduction in the E_c is observed at SZ5 along with the clear increase in the P_r and P_m. At SZ5, the slightly deformed hysteresis curve with high remnant polarization (32 μC/cm^2) and at SZ4 high piezoelectric constant (100 pC/N) was observed. In temperature dependent P–E loops, pinched type behavior were revealed which indicated the coexistence of polar and non-polar regions in the ceramics at high temperatures.

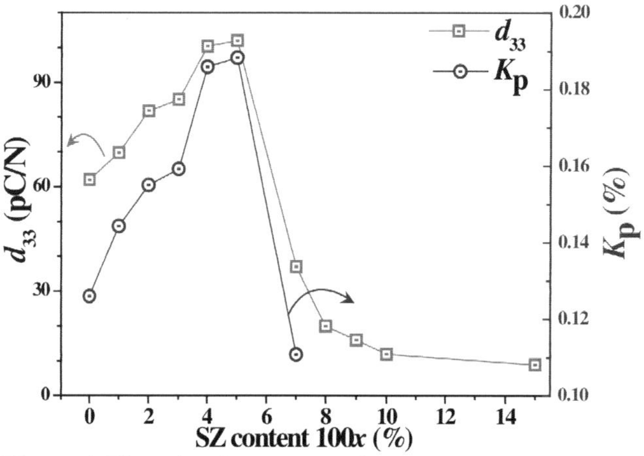

Figure 6. Piezoelectric constant (d_{33}) and coupling coefficient (K_p) of BNT-SZ100x ceramics.

5. Acknowledgements

This work is supported by the Basic Research program through the National Research Foundation of Korea (NRF) funded b Ministry, Science and Technology (MEST) (2012-0009458).

6. References

[1] Jaffe B, Cook W R and Jaffe H 1971 *Piezoelectric ceramics* vol 1 (New York: Academic Press)

[2] Smolenskii G A, Isupov V A, Agranovskaya A I and Krainik N N 1961 New ferroelectrics of complex composition *Sov. Phys. Solid State* 2 2651-4

[3] Rödel J, Jo W, Seifert K T, Anton E M, Granzow T and Damjanovic D 2009 Perspective on the Development of Lead-free Piezoceramics *J. Am. Ceram. Soc.* 92 1153-77

[4] Sakata K and Masuda Y 1974 Ferroelectric and antiferroelectric properties of $(Na_{0.5}Bi_{0.5})TiO_3$-$SrTiO_3$ solid solution ceramics *Ferroelectrics* 7 347-9

[5] Takenaka T, Sakata K O and Toda K O 1990 Piezoelectric properties of $(Bi_{1/2}Na_{1/2})TiO_3$-based ceramics *Ferroelectrics* 106 375-80

[6] Takenaka T, Maruyama K I and Sakata K 1991 $(Bi_{1/2}Na_{1/2})TiO_3$−$BaTiO_3$ system for lead-free piezoelectric ceramics *Jpn. J. App. Phys.* 30 2236-9

[7] Zhang S-T, Kounga A B, Aulbach E and Deng Y 2008 Temperature-Dependent Electrical Properties of $0.94Bi_{0.5}Na_{0.5}TiO_3$-$0.06BaTiO_3$ Ceramics *J. Am. Ceram. Soc.* 91 3950-4

[8] Xu C, Lin D and Kwok K W 2008 Structure, electrical properties and depolarization temperature of $(Bi_{0.5}Na_{0.5})TiO_3$–$BaTiO_3$ lead-free piezoelectric ceramics *Solid State Sci.* 10 934-40

[9] Rahman J U, Hussain A, Maqbool A, Song T K, Kim W J, Kim S S and Kim M H 2014 Dielectric, ferroelectric and field-induced strain response of lead-free $BaZrO_3$-modified $Bi_{0.5}Na_{0.5}TiO_3$ ceramics *Curr. Appl. Phys.* 14 331-6

[10] Krauss W, Schütz D, Mautner F A, Feteira A and Reichmann K 2010 Piezoelectric properties and phase transition temperatures of the solid solution of $(1-x)(Bi_{0.5}Na_{0.5})TiO_3-xSrTiO_3$ *J. Eur. Ceram. Soc.* 30 1827-32

[11] Hiruma Y, Imai Y, Watanabe Y, Nagata H and Takenaka T 2008 Large electrostrain near the phase transition temperature of $(Bi_{0.5}$Na0.5)$TiO_3-SrTiO_3$ ferroelectric ceramics *Appl. Phys. Lett.* 92 262904

[12] Watanabe Y, Hiruma Y, Nagata H and Takenaka T 2009 Fabrication and electrical properties of $(Bi_{1/2}Na_{1/2})TiO_3$-$BiAlO_3$ ferroelectric ceramics *Key Eng. Mater.* 388 229-32

[13] Ullah A, Ahn C W, Jang K B, Hussain A and Kim I W 2010 Phase Transition and Electrical Properties of $BiAlO_3$-Modified $(Bi_{0.5}Na_{0.5})TiO_3$Piezoelectric Ceramics *Ferroelectrics* 404 167-72

[14] Sasaki A, Chiba T, Mamiya Y and Otsuki E 1999 Dielectric and Piezoelectric Properties of $(Bi_{0.5}Na_{0.5})TiO_3$-$(Bi_{0.5}K_{0.5})TiO_3$ Systems *Jpn. J. App. Phys.* 38 5564

[15] Yoshii K, Hiruma Y, Nagata H and Takenaka T 2006 Electrical Properties and Depolarization Temperature of $(Bi_{1/2}Na_{1/2})$TiO3–$(Bi_{1/2}K_{1/2})TiO_3$ Lead-free Piezoelectric Ceramics *Jpn. J. App. Phys.* 45 4493-6

[16] Wada T, Toyoike K, Imanaka Y and Matsuo Y 2001 Dielectric and piezoelectric properties of $(A_{0.5}Bi_{0.5})TiO_3$-$ANbO_3$ (A= Na, K) systems *Jpn. J. App. Phys.* 40 5703

[17] Kounga A B, Zhang S-T, Jo W, Granzow T and Rödel J 2008 Morphotropic phase boundary in (1-x) $Bi_{0.5}Na_{0.5}TiO_3-xK_{0.5}Na_{0.5}NbO_3$ lead-free piezoceramics *Appl. Phys. Lett.* 92 222902-3

[18] Hiruma Y, Nagata H and Takenaka T 2008 Phase diagrams and electrical properties of $(Bi_{1/2}Na_{1/2})TiO_3$-based solid solutions *J. App. Phys.* 104 124106

[19] Lee W-C, Huang C-Y, Tsao L-K and Wu Y-C 2009 Chemical composition and tolerance factor at the morphotropic phase boundary in $(Bi_{0.5}Na_{0.5})TiO_3$-based piezoelectric ceramics *J. Eur. Ceram. Soc.* 29 1443-8

[20] Lee W-C, Huang C-Y, Tsao L-K and Wu Y-C 2010 Crystal Structure, dielectric and ferroelectric properties of $(Bi_{0.5}Na_{0.5})TiO_3$–(Ba,Sr)TiO_3 lead-free piezoelectric ceramics *J. Alloys Compd.* 492 307-12

[21] Watanabe Y, Hiruma Y, Nagata H and Takenaka T 2008 Phase transition temperatures and electrical properties of divalent ions (Ca^{2+}, Sr^{2+} and Ba^{2+}) substituted $(Bi_{1/2}Na_{1/2})TiO_3$ ceramics *Ceram. Int.* 34 761-4

[22] Hiruma Y, Watanabe Y, Nagata H and Takenaka T 2007 Phase transition temperatures of divalent and trivalent ions substituted $(Bi_{1/2}Na_{1/2})TiO_3$ ceramics *Key Eng. Mater.* 350 93-6

[23] Yu H and Ye Z-G 2008 Dielectric, ferroelectric, and piezoelectric properties of the lead-free (1-x) $Na_{0.5}Bi_{0.5}TiO_3$-$x$$BiAlO_3$ solid solution *Appl. Phys. Lett.* 93 112902-3

[24] Parija B, Rout S K, Cavalcante L S, Simões A Z, Panigrahi S, Longo E and Batista N C 2012 Structure, microstructure and dielectric properties of 100−x(Bi0.5Na0.5)TiO3−x[SrTiO3] composites ceramics *Appl. Phys. A* 109 715-23

[25] Ahtee A, Ahtee M, Glazer A and Hewat A 1976 The structure of orthorhombic $SrZrO_3$ by neutron powder diffraction *Acta. Crystallogr. B* 32 3243-6

[26] Popescu B, Enache S, Ghica C and Valeanu M 2011 Solid-state synthesis and spark plasma sintering of $SrZrO_3$ ceramics *J. Alloys Compd.* 509 6395-9

[27] Chen C, Zhu W, Yu T, Chen X and Yao X 2003 Preparation of metal–organic decomposition-derived strontium zirconate dielectric thin films *Appl. Surf. Sci.* 211 244-9

[28] Rachakom A, Jaiban P, Jiansirisomboon S and Watcharapasorn A 2012 Crystal structure and electrical properties of bismuth sodium titanate zirconate ceramics *Nanoscale Res. Lett.* 7 57

[29] Hussain A, Ahn C W, Lee J S, Ullah A and Kim I W 2010 Large electric-field-induced strain in Zr-modified lead-free $Bi_{0.5}(Na_{0.78}K_{0.22})_{0.5}TiO_3$ piezoelectric ceramics *Sens. Actuator A-Phys.* 158 84-9

[30] Hussain A, Ahn C W, Ullah A, Lee J S and Kim I W 2012 Dielectric, ferroelectric and field-induced strain behavior of $K_{0.5}Na_{0.5}NbO_3$-modified $Bi_{0.5}(Na_{0.78}K_{0.22})_{0.5}TiO_3$ lead-free ceramics *Ceram. Int.* 38 4143-9

IOP Conf. Series: Materials Science and Engineering **60** (2014) 012048 doi:10.1088/1757-899X/60/1/012048

Synthesis of Mn_{1-x} Zn_x Fe_2O_4 ferrite powder by co-precipitation method

Salma Irfan[1], M Ajaz-un-Nabi[1], Yasir Jamil[2] and Nasir Amin[1]
[1]Department of Physics, GC University, Faisalabad, Pakistan
[2]Department of Physics, University of Agriculture, Faisalabad, Pakistan

Abstract. Ferrimagnetic substances referred to as ferrites are ionic crystals whose chemical composition is of the form XFe_2 O_4 where X signifies a divalent metal. Magnetic Nano sized ferrites have found a significant potential in many applications, such as magnetic recording media, Ferro fluids and radar absorbing coating. Ferrites are widely used in many industrial applications due to their spontaneous magnetization. Soft ferrites of Mn-Zn, Ni-Zn and Mg-Mn are well known for their high magnetic permeability. In the present research work we have prepared fine $Mn_{1-x}Zn_xFe_2O_4$ ferrite powder with varying x concentrations (0.25-0.75) by metal chloride precursors through a co-precipitation technique by pipette drop method using aqueous NaOH solution for comparing their spontaneous magnetization and particle size. The co-precipitation technique is a high way to produce chemically homogeneous powder with fine particle size in nanometers (22.5nm-74.5nm).The effect of x-concentration on the particle size of the $Mn_{(1-x)}$ $Zn_{(x)}$ ferrite has been discussed on the basis of XRD. The crystalline phases have been identified by X-ray diffraction with Cu-Kα radiations. The XRD patterns have verified that the specimen has spinal type structure. The observable peaks are broad since the size of the particles is small. We have concluded that at constants temperature particle size increases with increasing x-concentrations. Effect of different concentrations of x (Zn^{+2}) on the spontaneous magnetization of different $Mn_{(1-x)}$ $Zn_{(x)}$ Fe_2O_4 sample is determined. We have reached the conclusion that all the samples of $Mn_{(1-x)}$ $Zn_{(x)}$ Fe_2 O_4 ferrites were magnetic either of low or high magnetization. The maximum spontaneous magnetization and minimum particle size is obtained at x=0.25 (at digestion temperature=65°C).

1. Introduction

The study of spinel ferrites in the nanoregime gets significance from a fundamental point of view as well as an applied perspective. Mn-Zn ferrites, among all the ferrites, are commercially the most important class of materials. Soft ferrites such as Mn-Zn, Ni-Zn and Mg-Mn are well known for their high magnetic permeability [1]. Current research efforts have been concentrated on synthesizing nanometre sized (22.5-90 nm) ferrite particles to minimize energy losses associated with bulk systems. Mn and Mn-Zn ferrites both have been proved as essential materials for higher frequency (up to MHz) applications such as inductors and transformers [2].

The structural and magnetic properties of ferrites are found to be sensitive to their composition and microstructure, which depends upon the processing conditions. Chemical co-precipitation was selected as a best method to synthesize nanoparticles. Co-precipitation is the least expensive and the simplest approach for making nanoparticles. This method produces nanoparticles in large quantities (order of grams) in a relatively short interval of time and utilizes inexpensive and readily available chemicals as precursors. Inexpensive iron salts may be mixed with a precipitating agent such as NaOH to form nanoparticles of iron oxides [3]. In Co-precipitation, the nucleation and growth steps overlap, but the particles are usually harvested in the time period before Ostwald ripening fully takes into effect.

X-ray diffraction technique is used to study the effect of variation in the Zn concentration and its effect on particle size and spontaneous magnetization [4]. In the present work Mn-Zn ferrites are prepared by Co-precipitation technique and subsequent digestion process (below 100°C). Digestion, which is an Ostwald ripening process, did not change the crystalline structure but increased the particle size and crystallite size. The particle size is characterized by using XRD [5].

2. Experimental Details

Manganese zinc ferrites belonging to the series of $Mn_{1-x}Zn_xFe_2O_4$(x = 0.25, 0.50, 0.75) were synthesized by wet chemical co-precipitation technique and digestion process. The samples with varying x concentration were prepared. The value of x was 0.25, 0.50, and 0.75.

For each sample the concentration of both NaOH and $FeCl_3$ were kept constant. For each sample the ratio Fe: Na was taken 1:4. All the salts $MnCl_2\,4H_2O$, $ZnCl_2$ and $FeCl_3$ with different quantities for different samples were dissolved in distilled and de-ionized water [6].

The beaker containing 100 ml solution of NaOH was placed on magnetic stirrer of moderate speed (50rpm) at room temperature. The solution of beaker containing metallic ions was added drop by drop into the beaker containing NaOH solution. During the addition dark grey precipitates were obtained. The beaker containing the dark grey precipitates was placed into pre-heated water bath containing water and ethylene glycol. The temperature of water bath was varied from sample to sample. Digestion was performed for 85 minutes. The particles were settled down at the bottom of the beaker after 85 minutes. The beaker was cooled to a moderate temperature after taken out from water bath. The particles were obtained through filtration. The filtered particles were dried at 50°C in an oven for 18 hours. Finally the fine materials were yielded and powder form was obtained [7-8].The spontaneous magnetization of the particles was checked with the help of permanent magnet. These samples were characterized by x-ray powder diffraction technique using Cu Kα radiation (λ = 1.5405 Å). Lattice parameter of these samples of different compositions was calculated. The average crystallite size of each sample was estimated by employing the Debye–Scherrer formula.

3. Results and Discussion

The properties of ceramics are greatly affected by the characteristics of the powder, such as particle size, morphology, purity and chemical composition. Using chemical methods, e.g. co-precipitation, sol-gel, hydrothermal and colloid emulsion technique, it has been confirmed that they efficiently control morphology and chemical composition of the prepared powder [9]. Among these wet chemical technique, sol-gel using alkoxides, hydrothermal and colloid emulsions are time consuming and involve highly unstable alkoxides and difficult to maintain reaction condition. Co-precipitation is one of the more successful techniques for synthesizing ultrafine powders having narrow particles size distribution [10-11]. This process can avoid complex step such as refluxing of alkoxides, resulting in less time consumption compared to other techniques. [12].

The x-ray diffraction patterns of mixed ferrites belonging to the series $Mn_{1-x}Zn_xFe_2O_4$ is shown in fig. a, Fig. b and Fig. c They are found to be characteristic of a spinel structure [13].

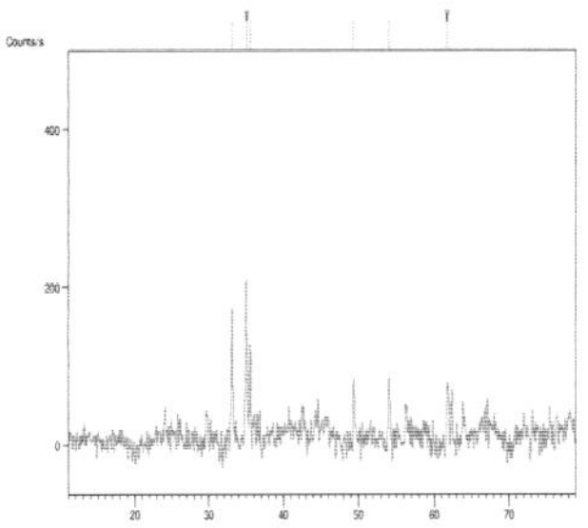

Fig.1

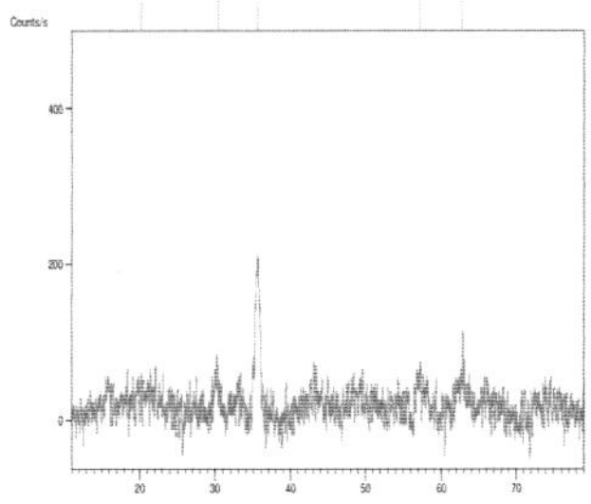

Fig. 2

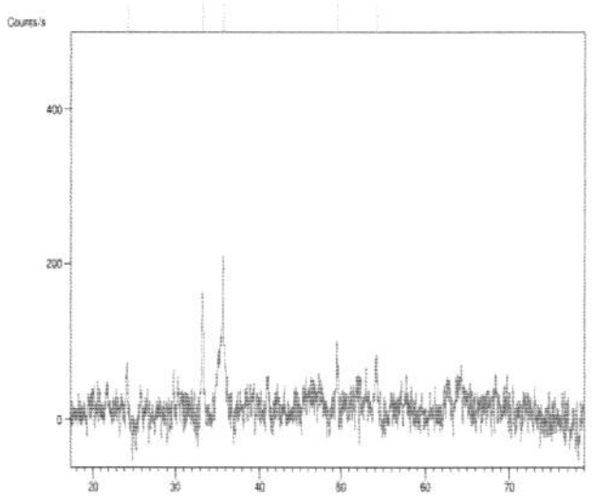

Fig. 3

Figure 1. Sample 1, prepared at x=0.25 and at digestion temperature 65^{0}C. **Figure 2.** Sample 4, prepared at x=0.50 and at digestion temperature 65^{0}C. **Figure 3.** Sample 7, prepared at x=0.75 and at digestion temperature 65^{0}C

Spontaneous magnetization was checked with the help of permanent magnet. All the samples of $Mn_{1-x}Zn_x$ Fe_2O_4 showed spontaneous magnetization whether of low or high magnetization.

3.1. Discussion on the results of XRD patterns

3.1.1. Comparison between 0.25g concentration samples

In this study MnZn Fe_2O_4 was prepared by varying the digestion temperature. Temperature of the water bath was kept at 65°C, 75°C and 95°C. Each sample was hosted in water bath for 85 minutes. According to the calculated particles sizes, the particle size of sample 1 which was 22.35 nm prepared at 65°C was found better as compared to others samples (e.g. sample 2 = 36.18nm, sample 3 = 45.29nm) of 0.25g concentrations.

3.1.2. Comparison between 0.50g concentration samples

In this study MnZn Fe_2O_4 was prepared by varying the digestion temperature. Temperature of the water bath was kept at 65°C, 75°C and 95°C. Each sample was hosted in water bath for 85 minutes. Spinel peak of sample 4 prepared at 65°C of digestion temperature was more prominent as compared to sample 5 and sample 6. According to the calculated particles sizes sample 6 = 57.25nm (prepared at 95°C digestion temperature) was better as compared to sample 4 and sample 5. The characteristic peak was obtained at an angle of 33, 35.86 and 35.66 degree 2 theta with Cu Kα x-rays.

3.1.3. Comparison between 0.75g concentration samples

In this study MnZn Fe_2O_4 was prepared by varying the digestion temperature. Temperature of the water bath was kept at 65°C, 75°C and 95°C. Each sample was hosted in water bath for 85 minutes. Spinel peak of sample 8 prepared at 75°C of digestion temperature was more prominent as compared to sample 7 and sample 9. According to the calculated particles sizes sample 7 = 57.25nm (prepared at 65°C digestion temperature) was better as compared to sample 8 and sample 9. The characteristic peak was obtained at an angle of 35.66, 35.228 and 35.36 degree 2 theta with Cu Kα x-rays.

3.1.4. Comparison between 0.25g, 0.50g and 0.75g concentration samples

It was observed that variation in concentration from 0.25g to 0.75g changed the size of the particle(as shown in table 1) but peak intensity of the concentration 0.50g(as shown in Fig. 2) showed that it had better magnetic qualities. So concentrations 0.50g had better crystal structure and had spinel-type structure. [8]

3.2. Particle Sizes

Particle sizes of the prepared manganese zinc ferrites were measured with the help of Debye–Scherer formula using the peak of maximum intensity for each pattern [14]. The peak of maximum intensity for each sample was obtained at angle 34.60, 35, 35, 33, 35.86, 35.66, 35.66, 35.228 and 35.36 respectively. The particle size so calculated is given in Table 1.

Sample No.	X	Particle Size (nm)
1	0.25	22.35
2	0.25	36.18
3	0.25	45.29
4	0.50	72.94
5	0.50	74.51
6	0.50	57.25
7	0.75	57.25
8	0.75	74.51
9	0.75	90.58

Table 1. particle sizes of the sample

The particles sizes calculated for all the nine samples at different digestion temperatures and with varying Zn concentrations (x) 0.25, 0.50 and 0.75 are 22.35, 36.18, 45.29, 72.94, 74.51, 57.25, 57.25, 90.58 and 74.51 (nm) respectively, using Debye–Scherer formula. This showed that the particle size of MnZn ferrites increased with increase in the concentration of (x). The minimum particle size obtained was 22.35 nm which is comparable to particle size obtained in the previous work of Dunlap et al. (2003).

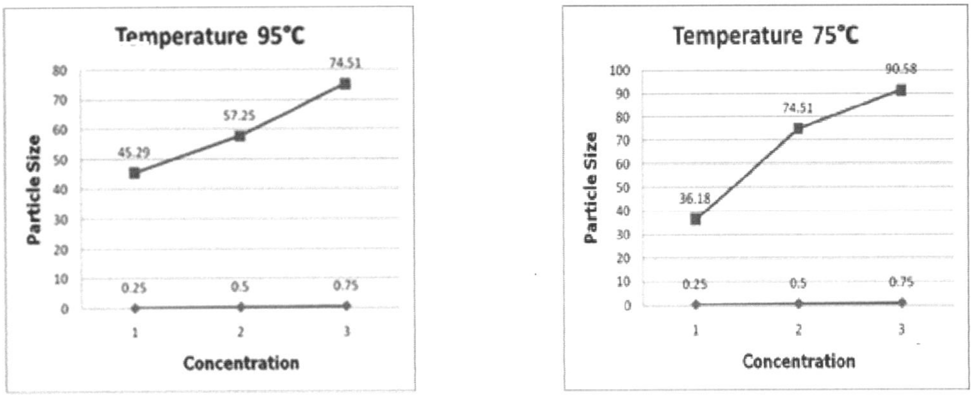

Figure 4. Variation in particle size with concentration at different digestion temperatures

4. Conclusions

At constant temperature, particle size increases with increasing x concentration. The maximum spontaneous magnetization and minimum particle size was obtained at x = 0.25 at digestion temperature 65°C.

5. References

[1] Singh M A., 2003. Ceramics International; 29:505.

[2] Ahmed M A., Okasha N and El-Sayed M M, 2007. Enhancement of the physical properties of rare-earth-substituted Mn–Zn ferrites prepared by flash method. Ceramics International, 33 (1): 49-58.

[3] Shokrollahi H, 2008. Magnetic properties and densification of Manganese – Zinc Soft ferrites ($Mn_{1-x}Zn_xFe_2O_4$) doped with low melting point oxides. Journal of Magnetism and Magnetic Materials, 320 (3-4): 463-474.

[4] Parvatheeswara Rao B, Caltun O, Cho W S, Chong-Oh Kim, CheolGi Ki, journal of magnetism and magnetic materials, Volume 2, Part 3 pp 50-59, March 2007.

[5] Xuan Y, Qiang Li, Gang Yang Journal of Magnetism and Magnetic Materials 312 (2007) 464–46

[6] Arulmurugana R, Jeyadevanb B, Vaidyanathana G, Sendhilnathanc S. Journal of Magnetism and Magnetic Materials Volume 288, March 2005, Pages 470–477.

[7] Wahab, M A, 2005. Solid State Physics: Structure and Properties of Materials Narosa Publishing House, India.

[8] Manzoor A, K. Hasanain S,Mumtaz A, F. Bertino M and Franzel L,Journal of Nanoparticle Research November 2012, 14:1310

[9] Musat V, Potecasu O, Belea R, Alexandru P, Materials science and engineering, Volume 167, Issue 2, 2010.

[10] Dhag S R., Ravi V. and Date S K, 2003. Bull. Material Science, 26:215.

[11] Gaikwad S P, Dhesphande S B, Khollam Y, Samuel V, and Ravi V, 2004. Materials Letters, 58:3474

[12] Amirabadizadeh A, Farsi H, Dehghani, M, Arabia H, Effect of Substitutions of Zn for Mn on Size and Magnetic Properties of Mn–Zn Ferrite Nanoparticles Journal of superconductivity and novel magnetism, December 2012, Volume 25, Issue 8.

[13] Puri R K. and Babbar V K, 2001. 2nd Edition. S. Chand & Company Ltd., Ram Nagar, New Dehli.

[14] Jamil Y, Ahmad MR, Hafeez A, Zia-ul-Haq, Amin N. Microwave Assisted Synthesis of Fine Magnetic Manganese Ferrite Particles Using Co-Precipitation Technique. Pak J Agri Sci 2008; 45(3).

Synthesis, characterization and electrochemical studies of Pt-W/C catalyst for polymer electrolyte membrane fuel cells

Riaz Ahmed[1,2], Saliha Shahid[1] and Muhammad Shahid Ansari[1]
[1]Department of Chemistry, Quaid-i-Azam University Islamabad-45320, Pakistan
[2]COMSATS, Shahrah-e-Jamhuriat, G.5/2, Islamabad, Pakistan

E-mail: riaz_ecfc@yahoo.com

Abstract. Pt-W/C catalyst was synthesized by slow reduction of platinum and tungsten solutions in the desired ratio with subsequent deposition on the Vulcan carbon already added to the solution. Crystallite size of catalyst was about 9nm and its density, cell volume, d-spacing and lattice parameter were also calculated. EDX analysis of the catalyst was also done. Electrochemical surface area of the catalyst was determined by cyclic voltammetry (CV). CV of the catalyst was done both in acidic and basic media to find out the peak potential, peak current, specific activity and mass activity of the catalyst. Peak potential versus scan rate plots showed that the electro oxidation of methanol is an irreversible process. Tafel equation was used to plot polarization curves to find out the exchange current density. Higher values of exchange current indicate better catalysts. Specific activities of the catalyst were determined in acidic and basic media and it was found that the specific activity in basic media increased substantially as compared to acidic media. The specific activity in acidic media was 83 mA/mg pt whereas in basic media it was 137mA/mg pt which is a substantial increase. Heterogeneous rate constant in acidic media was 6.15 x 10^{-6} cm/ s and in basic media it was 4.92 x 10^{-5} cm/s which is much higher in basic media. In this binary catalyst addition of tungsten has increased the catalytic activity but it is non-noble metal thus will decrease the cost. Stability studies of the catalyst were done upto fifty cycles both in acidic and basic media and was found quite stable in both the media.

1. Introduction

Increase in energy demand and environmental concerns require the development of clean energy sources for sustainable development. Fuel cells may meet all the above requirements and are being developed as one of the primary energy technologies of the future [1-2]. Low temperature fuel cells particularly polymer electrolyte membrane fuel cells are important for automotive industry. Important technical parameters required for the automotive industry have been met including membrane electrode assembly (MEA) and diffusion media optimization and development of platinum alloy cathode catalysts. Vulcan carbon (XC-72R, 30–60 nm) is a commonly used support that fulfills the requirements of having better dispersion ability, conductivity and adequate strength [3-4] besides being cheaper. Work on carbon supported bimetallic Pt-nanoalloys with different metals have been reported for methanol oxidation which exhibited better efficiency but durability did not improve much [5-6]. Catalyst durability is still a challenge. Corrosion of carbon supports is the major reason for catalyst degradation [7]. Considerable work has been done on the development of ceramic materials as supports for low temperature fuel cells [8]. However, there are other important factors for catalyst supports such as surface area, porosity, electrical conductivity, electrochemical stability and surface functional groups [9]. It is reported that stability of WC is higher than carbon [10] and addition of platinum to WC increases the activity of the catalyst significantly [11]. The main drawback of WC supports is their low surface area as compared to carbon [12]

In this study carbon tungsten composite was used as a support and platinum was deposited to prepare a nanocomposite catalyst. The synthesized catalyst was characterized using XRD, EDX and cyclic voltammetry.

2. Materials and Methods

2.1. Chemicals and equipment

Platinum(IV) chloride $PtCl_4$, from Johnson Mathey, UK, Sodium tungstate $Na_2WO_4.2H_2O$, Aldrich, Hydrazine hydrate $N_2H_4.2H_20$, Aldrich, Sodium borohydride $NaBH_4$ from Merck, Vulcan carbon , (XC-72) Aldrich, Methanol Sigma-Aldrich, 2-Propanol Merck, Nafion® 117 solution Sigma-Aldrich, Hydrogen peroxide 35% Applied Chem, Hydrochloric acid Sigma-Aldrich, Sulphuric acid Fluka, Nitric acid 65% Merck, Potassium hydroxide from Fluka were used.

The XRD spectra were recorded on X-ray diffractometer model 3040/60 X'Pert PRO. The XRD analysis of the catalyst was carried out in powder form with scan rate of 0.04 and 2θ values ranged from 10-90°.

Cyclic voltammetric measurements were carried out using Eco Chemie Autolab PGSTAT 12 potentiostat/galvanostat Netherlands, with the electrochemical software package GPES 4.9.

2.2. Synthesis of catalyst

Vulcan carbon employed had the diameter in a range of 30-60 nm (45 nm from XRD).Vulcan carbon was taken in a round bottom flask to which 6 M HCl was added. The contents were refluxed and ultrasonicated for 4 hours at 60 °C using E-30H (240 watts) ultrasonicator from Elmasonic, Germany. The acid treated Vulcan carbon was filtered off and was vacuum-dried for 24 hours at 110 °C. Functionalization of the purified vulcan carbon was also done by ultrasonication method in 4:1 v/v mixture of 8 M HNO_3 and 35% H_2O_2 .

Aqueous solutions of $Na_2WO_4.2H_2O$ and $PtCl_4$ were added to Vulcan carbon and stirred for 30 minutes. the mixture became colourless from light brown indicating adsorption of metal ions on carbon. Freshly prepared solution of $NaBH_4$ in the solvent mixture was drop wise added into the above dispersion. 4 hours stirring resulted formation of black suspension which was filtered vacuum dried for 24 hours at 110 °C. Syntheses were carried out under argon atmosphere to avoid formation of metal oxides.

2.3. Electrochemical measurements

For the CV measurements, three-electrode system was employed throughout the electrochemical measurements. Graphite electrode of surface area 0.785 cm^2 was used as a working electrode

Graphite electrode (GE) polished with 1.0 μm alumina washed in ethanol and ultra-pure water ultrasonically to remove organic and inorganic impurities from the surface of electrode. Then, 10 mg catalyst and 0.5 mL of water were mixed ultrasonically for 30 min. in a test tube. 40 μL of the suspension was pipetted out and pasted over the tip of polished graphite electrode, dried in air. After the solvent evaporation, 10 μL binder (5% nafion solution) was spread over the tip of surface of GE. Again dried it in air.

3. Results and Discussions

3.1. X-ray diffraction analysis

Fig.1 reveals the XRD patterns of Vulcan carbon, Pt/C, Pt-W/C, The diffraction peak at 24–26° and 54.01° observed in all the XRD patterns of the carbon-supported catalysts is due to the (002)

Table 1. XRD parameters evaluated for bi-metallic catalyst

Catalyst	Position 2θ°	*d*-spacing /Å	Lattice parameter a / Å	Cell volume V / Å³	Density $\rho_{X\text{-}ray}$/g.cm⁻³	Ave. crystallite size D/nm
$Pt_{0.10}$-$W_{0.05}$/C	39.57	2.2773	3.9443	61.36	20.68	8.93
PSP*	39.67	2.27	3.917	59.87	21.51	-----

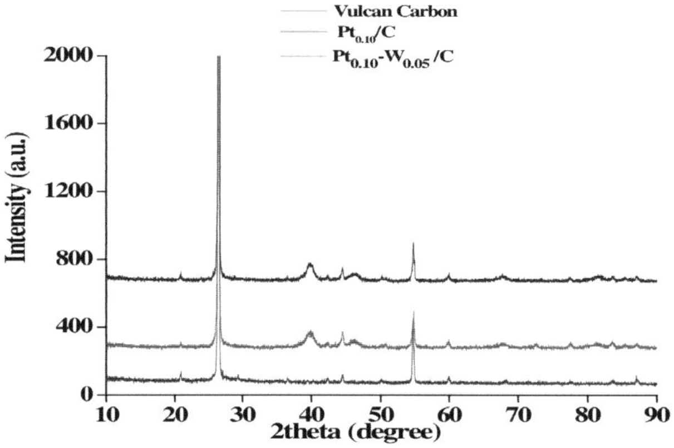

Figure 1. XRD spectrum for the catalyst and support.

and (004) planes of the hexagonal graphitic structure of Vulcan XC-72 carbon [13]. Four characteristic peaks corresponding to (111), (200), (220), and (311) planes of the fcc crystalline Pt are observed in XRD patterns [14-15]. Pt standard pattern *(00-001-1190)* is used as a reference in order to confirm nanoparticles (NPs) formation. The standard peak observed at 2θ position 39.67° was due to Pt (111) plane of fcc crystalline structure. A significant shifting of peaks was observed from 39.67° to 39.57° in Pt-W/C which is due to alloy formation of Pt with W. No peak for W metals and their oxides were observed. W metal forms amorphous oxide which was not indicated by XRD [9-11].It indicates that secondary metal forms single phase crystalline structure with Pt.

Table.1 shows different parameter evaluated from the XRD patterns of mixed structures of Pt-W/C. There is small decrease in X-ray density. However, lattice parameter and cell volume increased.

3.2. EDX analysis

Compositional analysis of the synthesized catalyst was done by EDX. Initially 10% platinum, 5% Tungsten and 85 % functionalized Vulcan carbon were added. EDX results are given in Table-2. It shows that in the synthesized catalysts the values are nearly the same with some small losses.

Table 2. EDX Analysis of catalyst

Catalyst	**Pt**	**Co**	**W**	**Mo**	**C**	**O**
	±0.001	**±0.001**	**±0.001**	**±0.001**	**±0.001**	**±0.001**
Pt-W/C	9.51	------	4.38	------	83.69	2.42

3.3. Electrochemical Studies

3.3.1. Electrochemical surface area

Electrochemical surface area (S_{ESA}) is usually proportional to catalytic activity. Electrochemical surface area (S_{ESA}) can be calculated from the amount of charge transfer (Q_{cat}) in hydrogen adsorption region by applying equation

$$S_{ESA} = \frac{Q_{cat}}{Q_{Pt}.L} \quad (1)$$

Where, Q_{cat} is charge transfer from catalyst in hydrogen adsorption region (µC), Q_{Pt} is the charge transfer for monolayer hydrogen adsorption (210 µC.cm^{-2}) and L is the mass of Pt loading (mg).

Table 3. Electrochemical surface area from CV in 1M H_2SO_4

Catalyst	**Mass of Pt /mg**	**$Q_{pt\text{-}H}$ /µC**	**S_{ESA} /(cm^2.mg^{-1})**	**S_{CSA} /(cm^2.mg^{-1})**	**Roughness factor(cm^2)**
$Pt_{0.10}$-W $_{0.05}$/C	0.10	3270	156	32.49	19.83

Values of Q_{cat} and S_{ESA} are directly proportional to electro catalytic activity and are given in Table.3 for Pt-W/C catalyst. Higher value of roughness factor shows more efficient catalyst which show high activity for methanol oxidation [16]. Table-3 clearly shows that the introduction of non-noble metals such as W to Pt catalyst may play an important role in proficiently improving the S_{ESA}, enhancing surface accessibility for hydrogen adsorption/desorption and affecting the electrochemical activity of bi- metallic catalysts [17].

3.3.2. Cyclic Voltammetry for methanol electro-oxidation

CV was performed for the study of methanol oxidation on the graphite electrode modified with synthesized catalyst in acidic medium. Amount of methanol oxidized and the peak heights

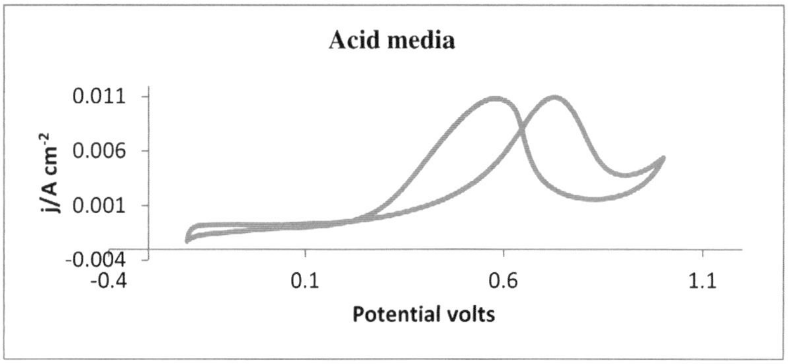

Figure 2. Cyclic voltammogram in acidic media

Table 4. Activity parameters evaluated from CVs measured at scan rate 50 mVs^{-1}

Catalyst	Peak potential E_p/V	Peak current $I_p/(mA)$	Specific activity $j\ /mA.cm^{-2}$	Mass activity mA/mg-Pt
$Pt_{0.10}$-$W_{0.05}$ /C, Acid media	0.723	8.29	10.56	82.9
$Pt_{0.10}$-$W_{0.05}$ /C, Basic media	-0.241	13.71	17.45	137

(current) obtained is proportional to the catalytic activity of the catalyst. Fig.2. shows CVs measured in 1.0 M CH_3OH +1.0 M H_2SO_4 at scan rate 50 $mV.s^{-1}$. In anodic sweep, the electro-oxidation of methanol gives a well-defined forward anodic peak, while in cathodic sweep the electro-oxidation of methanol begins again and gives a reverse anodic peak. The presence of only anodic peaks in both anodic and cathodic sweeps indicates that electro-oxidation of methanol is an irreversible process [18]. Electro catalytic activity of various synthesized catalysts for methanol oxidation was also determined in basic medium using 1.0 M KOH solution by evaluating peak current (I_p) through

CVs. Fig.3 shows CVs measured in 1.0 M CH_3OH + 1.0 M KOH at scan rate 50mVs^{-1} from -0.7 to 0.4V on catalysts. In anodic sweep, a well-defined and sharp forward anodic peak was observed near at -0.3V, while in cathodic sweep very small reverse anodic peak was observed of the methanol oxidation.

Table.4 compares the peak potential, peak current, specific activity and mass activity for the oxidation of methanol in acidic and basic media. Catalyst activity is better in both the media but it is much better in basic media.

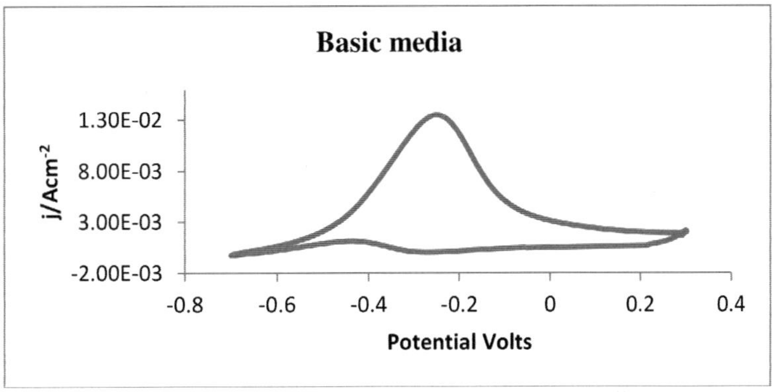

Figure 3. Cyclic voltammogram in basic media

3.3.3. Exchange current densities from the Tafel plots

Tafel equation was used for the determination of exchange current densities. Tafel equation can be written as [19].

$$\Delta V_{act} = a + b \log i \qquad (2)$$

Where $a = -2.3\ (RT/\alpha F) \log i_0$ $\quad b = 2.3\ RT/\alpha F$

The value of i_0 can be evaluated from the intercept of the ΔV_{act} vs log i plots

Table-5 shows the values of exchange current density as measured both in acidic and basic media. Values of exchange current are nearly similar in both the media.

Table 5. Polarization data evaluated from Tafel plots

Catalyst	**Tafel's slope "b" /(V decade^{-1})**	**αn_α**	**Intercept of E vs. Log í**	**i^0 mA·cm^{-2}**
$Pt_{0.10}$-W $_{0.05}$ /C, acid media	0.218	0.27	1.064	15.47
$Pt_{0.10}$-W $_{0.05}$ /C, Basic media	0.2279	0.26	0.1079	16.21

3.3.4. Rate constant of methanol oxidation

Rate of an electrochemical process can be calculated using different methods. Heterogeneous rate constant (k_s) was determined using the Nicholson-Shain equation [20]:

$$i_p = 0.227\, nFAC_o k_s \exp[(\alpha n_\alpha F/RT)(E_p - E^o)] \qquad (3)$$

where i_p is the peak current, n is the number of electrons transferred, A/cm^2 is the area of the electrode, C_o/mol·cm^{-3} is the bulk concentration of the reactant and k_S / cm·s^{-1} is the standard heterogeneous rate constant, F and R are the Faraday and gas constants.

Using equation a graph is plotted between $\ln(I_p)$ vs. E_p-E^o to get a straight line and from the value of slope, k_{het} was calculated . Rate constants were measured both in acidic and basic media and are given in Table-6. It can be seen that rate constants are better in basic media than in acidic media and are better by a factor of ten. Performance of the catalyst is better in basic media.

Table 6. Rate constants of CH_3OH oxidation

Catalyst	Peak current i_p /mA	Rate constant k_{het}/cm.s^{-1}
$Pt_{0.10}$-W $_{0.05}$ /C, acidic media	8.29	6.15x10^{-6}
$Pt_{0.10}$-W $_{0.05}$ /C, basic media	13.71	4.92x10^{-5}

3.3.5. Durability of the catalysts

Great focus has been paid to the durability of catalyst in order to commercialize it for fuel cells. Bifunctional mechanism, comprising the incorporation of non-noble metal to Pt-catalyst, improved the durability of catalysts. Fig.4. shows durability of the catalyst in terms of specific

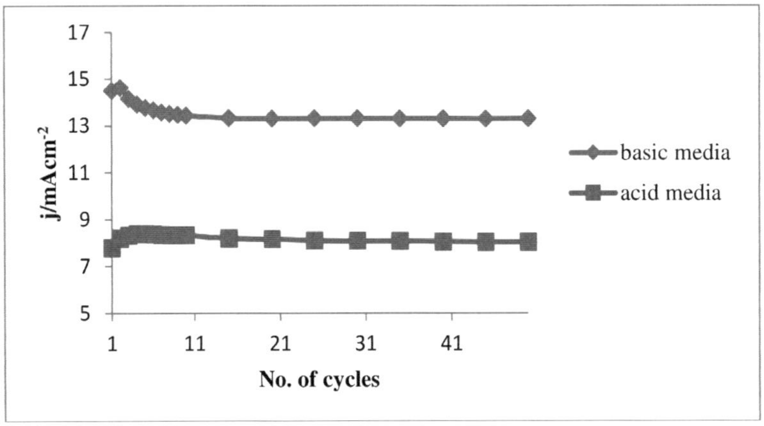

Figure 4. Durability study of catalysts

activity obtained at scan rate 50mV.s^{-1} for fifty numbers of cycles in a single run both in acidic and basic media. After each cycle, the value of current decreased in acidic medium. CO formation at the catalyst surface is a big reason which decreased the catalytically active sites of catalyst. In basic media after each cycle, the value of SA increased because catalyst becomes more and more active and less poisoning occurs in basic medium.

4. Conclusions

Addition of tungsten increased catalytic activity of platinum. XRD showed some alloy formation between platinum and tungsten. Catalytic activity of the catalyst was evaluated both in acidic and basic media. Catalytic activity was evaluated using Peak current, specific activity and mass activity and it was found that catalyst is efficient in both the media and comparatively more efficient in basic media. Exchange current and rate constants for the oxidation of methanol were also calculated and it was found that the performance of the catalyst was better in basic media. Durability studies were also done of the catalyst for oxidation of methanol repeating for fifty cycles and again the performance of the catalyst was better in basic media but it was generally stable in both the media. It can be concluded that rate of corrosion of the catalyst decreased with the addition of tungsten in both the media but it was better in basic media.

5. References

[1] Zhao, S.; Kreuer, K-D.; Nguyen, T. V. 2007, *Advances in Fuel Cells,* 1st ed., Elsevier,UK.

[2] Munuswamy, S.; Nakamura, K.; Katta, A. 2011, Renew. Energy 36 2978–2983.

[3] Xu, C.; Shen, P. K.; Liu, Y. 2007, J. Power Sources164*,* 527–531.

[4] Kim, C.; Kwon, H-H.; Song, I. K.; Sung, Y-E.; Chung, W. S.; Lee, H-I. 2007, J. Power Sources 171, 404–411.

[5] Jiang, Q.; Jiang, L.; Hou, H.; Qi, J.; Wang, S.; Sun, G.2010, J. Phys. Chem. C 114 19714–19722.

[6] Lima, F. H. B.; Lizcano-Valbuena, W. H.; Teixeira-Neto, E.; Nart, F. C.; Gonzalez, E. R.; Ticianelli, E. A. 2006, Electrochim. Acta 52*,* 385–393.

[7] Baumgartner, W.R.; Parz, P.; Fraser S.D.; Wallnöfer, E.; Hacker, V; 2008, J. Power Sources. 182, 413.

[8] Anatolini, E.; Gonzalez, E.R.; 2009, Solid State Ionics, 746-763.

[9] Sharma, S.; Pollet, B.G.; 2012, J. Power Sources. 208, 96.

[10] Chhina, H.; Campbell, S.; Kesler, O.; 2007, J. Electrochem. Soc. 154, B533-B539.

[11] Meng, H.; Shen, P.K.; Electrochem. Commun.2006, 8 , 588-594.

[12] Perchthaler , M.; Ossiander , T.; Juhart, V.; Mitzel , J.; Heinzl , C. ; Scheu , C.; Hacker, V.; 2013, Journal of Power Sources 243, 472-480

[13] Li, W.; Liang, C.; Zhou, W.; Qiu, J.; Zhou, Z.; Sun, G.; Xin, Q. 2003, J. Phys. Chem. B 107,6292-6299.

[14] Leontyev, I. N.; Chernyshov, D. Y.; Guterman, V. E.; Pakhomova, E. V.; Guterman, A. V. 2009, *Appl. Catal. A: General 357,* 1–4.

[15] Huang, Q.; Yang, H.; Tang, Y.; Lu, T.; Akins, D. L.2006; Electrochem. Commun. 8, 1220–1224.

[16] Khan,A. S. A.; Ahmed, R.; Mirza,M.L, 2009. Portugaliae Electrochim. Acta *27,*429-441.

[17] Hsieh, C. T.; Lin, J. Y.; Wei, J. L, 2009. Int. J. Hydrogen Energy 34, 685–693.

[18] Pattabiraman, R. 1997, Appl. Catal. A 153, 9–20.

[19] Khan, A.S.A., Ahmed, R., Mirza, M.L., Turk. J. Chem. 2008, 32, 743–753.

[20] Nicholson, R.S., and Shain. I. 1964, Anal. Chem;,36, 70.

Development and characterization of fatigue resistant Aramid reinforced aluminium laminates (ARALL) for fatigue Critical aircraft components

M H Qaiser, S Umar and S Nauman
Department of Material Science & Engineering, Institute of Space Technology, Islamabad, Pakistan

E-mail: haroon.qaiser@yahoo.com, suniya.butt@yahoo.com, aadnauman@gmail.com

Abstract. The structural weight of an aircraft has always been a controlling parameter that governs its fuel efficiency and transport capacity. In pursuit of achieving light-weight aircraft structures, high design stress levels have to be adopted and materials with high specific strength such as Aluminum etc. are to be deployed. However, an extensive spectrum of fatigue load exists at the aircraft wings and other aerodynamic components that may cause initiation and propagation of fatigue cracks and concludes in a catastrophic rupture. Fatigue is therefore the limiting design parameter in such cases and materials with high fatigue resistance are then required. A major improvement in the fatigue behavior was observed by laminating Kevlar fibers with Aluminum using epoxy. ARALL (Aramid Reinforced ALuminum Laminates) is a fatigue resistant hybrid composite that consists of layers of thin high strength aluminum alloy sheets surface bonded with aramid fibers. The intact aramid fibers tie up the fatigue cracks, thus reducing the stress intensity factor at the crack tip as a result of which the fatigue properties of can be enhanced with orders of magnitude as compared to monolithic high strength Aluminum alloy sheets. Significant amount of weight savings can be achieved in fatigue critical components in comparison with the traditional materials used in aircraft.

1. Introduction

The structural components of an aircraft endure severe cyclic loads during their service life. Some of the stress critical zones may develop a growing crack that propagates with each stress reversal cycle. In such cases, even a minor crack has a very sharp radius at tip and the corresponding stress intensity factor is very high. This results in a rapidly expanding crack that terminates in a catastrophic rupture and hence fatigue resistant materials are highly looked-for in order to prolong the service life of the component. The desired features of the materials for aircraft structures include its low weight, damage tolerance and resilience against the rigorous environmental factors. High strength aluminum alloys are considered as the eligible material for the aircraft structures as they meet the demanding set of properties required in aerospace materials [1].

A relatively new class of composite materials is Fiber Metal Laminates (FML) that are light in weight as well as damage tolerant against fatigue loads. Alternate layers of fibers and metal sheets bonded together via resin impregnation produce a component with comparatively low density, high specific strength and the desired Fatigue tolerance [2].

In 1983, the Delft University of Technology developed a new composite material called ARALL (Aramid Reinforced Aluminum Laminate) [2] which is a laminated structure containing alternating of thin sheets of Aluminum based alloys and Kevlar impregnated with epoxy [2]. The presence of Kevlar significantly reduced the weight and imparted the fatigue tolerance in the structure [3]. Further studies were made in 1984 the durability of ARALL was figured out against the environmental factors [4]. Gunnink et al. analyzed the aerospace structural designs made out of these laminates in 1987 [5] while the damage tolerance of ARALL in fuselage skin was examined in 1989 [6]. The fatigue behavior was reported along with the different laminate orientation with respect to the applied loading direction in 1995 [7]. In 2011, Sham Rasad et. al. tested ARALL for the interlaminar & intralaminar strength of the interfaces [8]. These interfacial strengths affects directly on the performance of the material and are desired to be controlled.

The current research work is related with the fabrication of these Aramid reinforced Aluminum laminates by sandwiching Kevlar fiber between two Aluminum sheets. Anodizing the aluminum sheet prior to VARTM process is a must in order to improve the adhesive bonding between two materials that belong to dissimilar classes.

2. Experimental Work

2.1. Fabrication of ARALL

For the fabrication of these laminates, three different materials are utilized:

- Aluminum 2024 T3 alclad
- Kevlar 49 Fibre Cloth
- Araldite LY 5052 Epoxy

The major concern in the fabrication of these laminates is the adhesion of Kevlar-49 and Aluminum-2024. Interfaces are the weak links in such kind of materials, and it is desired to alter the surface properties of the aluminum sheets for greater surface area and a stronger adhesion [9]. One way is to create micro scale pores on the surface of mechanically cleaned Aluminum via a process called Anodizing. Such metal surface modification process includes mechanical cleaning, chemical cleaning and electrochemical anodizing.

Although solvent degreasing is carried out to degrease the surface and to remove the contaminants but alone it is not sufficient to improve the bond durability for longer period [10]. Mechanical methods of abrasion are used [11] in order to remove Al_2O_3 layer and to create micro roughness on the metal surface, leaving it more adherent to the other surface. The surface topology can be altered through abrasion of the surface using sand paper and hence the wet-ability of surface can be increased [12]. The processes of alkaline cleaning and acid etching are then carried out to remove the debris and to dissolve the oxide layer. Anodizing is performed on these surfaces which creates a porous oxide layer on the surface of the metal. This layer is highly rough and contains micro pores which ensure the permanent bond between metal and the fiber [10].

Figure 1. Setup for Aluminum Anodizing Process

Multiple sets of Aluminum 2024 sheet of 0.5mm thickness were cut in the size of 150mm x 80mm and were grinded using 1200C fine emery paper to attain a uniform surface. The sheets were alkaline cleaned using the NaOH solution (12g/liter) and acid etched in a solution containing 10wt% sodium dichromate, 30wt% sulphuric acid and 60wt% distilled water held at 80°C using hotplate and magnetic stirrer. To validate whether the degreasing is achieved fully or not, water break test was done at each stage of cleaning.

Anodizing was then carried out on the mechanically cleaned acid etched sheets in 12wt% phosphoric acid (H_3PO_4) bath held at ambient temperature. Stainless steel sheet was used as cathode and the Aluminum sheet was made anode while a voltage of 15V±0.1V was maintained for 25 min (figure 1). Optical imagery of the anodized surface of Aluminum at a magnification of 600x is shown in figure 3.

Following Anodizing, the Aluminum sheets were laminated with Kevlar via the process of Vacuum Assisted Resin Transfer Molding (VARTM). To serve as a mold base, acetone cleaned glass slab was used with mold release wax applied on its surface. Aluminum and Kevlar were arranged in 2:1 configuration and the auxiliaries of the process such as peel plys and resin transfer mesh were placed in accordance with the standard VARTM procedure. The VARTM setup is illustrated in figure 2. Vacuum was created by applying pressure bags and vacuum suction pump and resin (Araldite LY 5052) was infused into the mold which spread due to the capillary action. After primary gelation time, samples was extracted from the fabrication panel and post cured at 100°C for 4 hrs.

Figure 2. Setup for VARTM Process

2.2. Fatigue Testing

The post cured samples were cut in dimensions of 135mm x 25mm using MetaCut M250. Bending Fatigue testing was performed on the samples keeping a displacement rate of 50mm per minute and a stress level of 80% of the yield stress of ARALL. A stress reversal of 50 cycles was applied and the corresponding behavior of flexural stress-strain was plotted.

3. Results and Discussion

A marked difference in surface morphology of Anodized Aluminum alloy was observed when examined under Metkon Metallurgical Microscope. The observed microstructure taken at 600x magnification is depicted in figure 3. The presence of porous oxide layer improves its adherence with Kevlar since the infiltration of epoxy becomes easy.

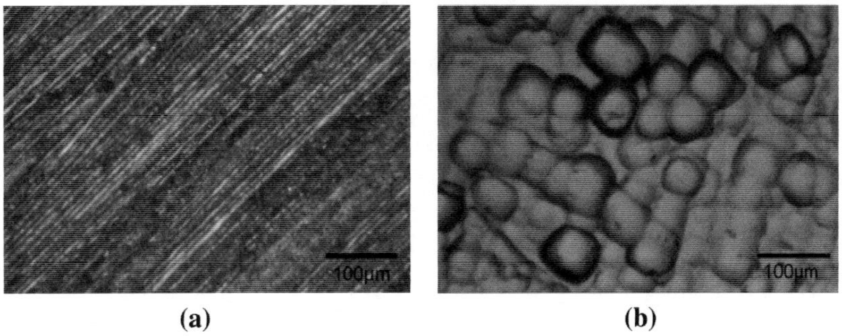

Figure 3. Optical micrograph of **a)** mechanically cleaned Aluminum **b)** anodized Aluminum surface showing porous oxide layer

Figure 4 illustrates the bending fatigue cycles of ARALL given at a displacement rate of 50mm per minutes. The sample was tested at stress levels ranging from a maximum value of 205 MPa to a minimum of 40 MPa. Even at such level of stress range and the applied strain rate, no visible drop in stress value was observed after 50 cycles. No visible signs of interfacial disbondment were observed on the sample after testing.

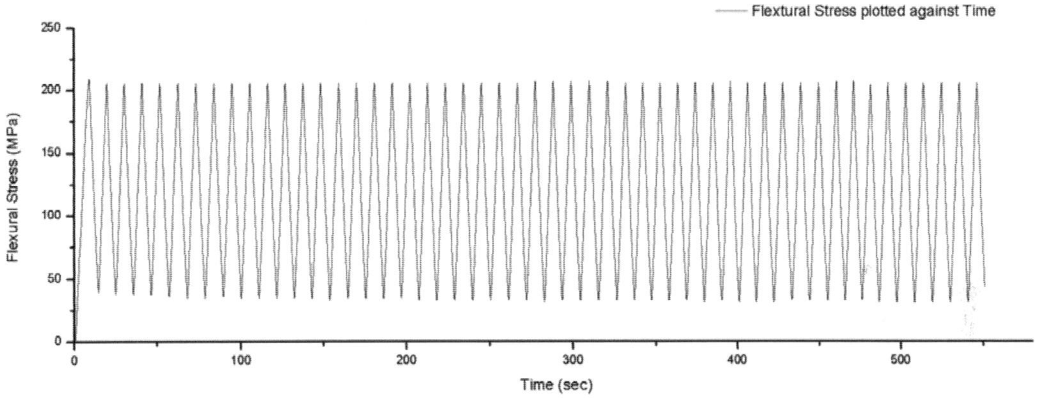

Figure 4. Flexural stress of ARALL plotted against time. (50 cycles, 50mm/min)

When this data was plotted for flextural stress vs. flextural strain, the corresponding stress reversal cycles were attained as shown in figure 5. Initially, before the first cycle, the stright line proportionality was observed till the maximum applied stress (80% of the yield stress) was reached. When the load decreased to its lower value, the curve didn't adopted the same track, rather it formed in a closed loop for the given number of cycles. The material kept its elastic modulus to a constant value throughout the stress reversal cycles as elucidated from slope of the curve.

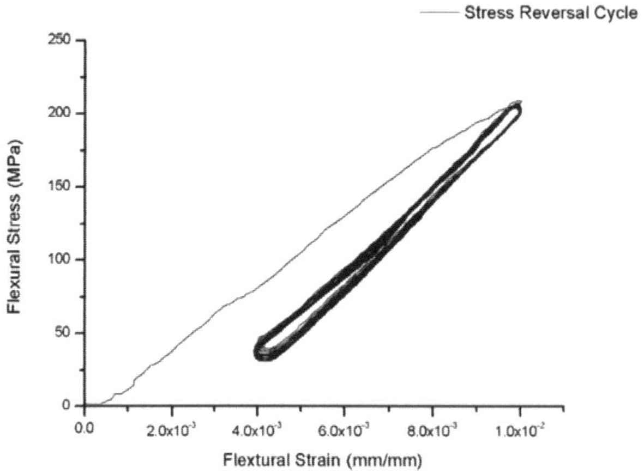

Figure 5. Stress Reversal cycle of ARALL.

4. Conclusion and Future Work

For sufficient weight savings that is a main concern in the case of aerospace industry, aramid reinforced Aluminum laminates can serve as a structural material with high damage tolerance. If the process of Aluminum anodization is well optimized, a strong interface can be achieved via infiltration of epoxy into the porous Al_2O_3 layer that has been developed intentionally. A strong interface of ARALL promoted resistance to stress reversal cycles and bending fatigue loads in comparison with monolithic materials out of which ARALL is fabricated. Moreover, between the two of these anodized Aluminum layers, woven Kevlar 49 fibers hinder any crack propagation due to 'crack arrest' mechanism. Tests to measure the Interlaminar Shear Strength and Tee Peel resistance [13] can assess the bonding strength and interfacial adhesion and will be conducted in future.

5. References

[1] Marissen R. , "Mechanical aspects related to fiber fracture in ARALL®2 laminates," Volume 159, 1989, pp 697-707

[2] Vogelesang L.B., and Gunnink J.W., "ARALL, a material for the next generation of aircraft. A state of the art," Delft University of Technology, Department of Aeronautics, Netherlands, Aug 1983.

[3] Marissen R., DFVLR, D-5000 Koln FRG. Institute for Materials Research Fatigue Department, 6160 MD Geleen Netherlands Material Technology, Department Composites.

[4] Verbruggen M.L.C.E., "Durability of ARALL," department of Aerospace Engineering, Delft University of Technology, March 1984.

[5] Gunnink Ir.J.W., "Design studies of primary aircraft structures in ARALL laminates," presented at ICCM 6/ECCM 2, Imperial College London, United Kingdom, July 1987.

[6] Chen D., "Damage tolerance aspects of ARALL fuselage skin structure," faculty of Aerospace Engineering, Delft University, Netherlands, Feb 1989.

[7] Keming Li., "Fatigue behavior and damage analysis in engineering composites," ProQuest Dissertations and Theises, 1995.

[8] Sham Rasad M.S., Venkatesha C.S., and Jayaraju T., "Experimental Methods of Determining Fracture Toughness of Fiber Reinforced Polymer Composites under Various Loading

Conditions," Journal of Minerals & Materials Characterization & Engineering, Vol. 10, No.13, pp.1263-1275, 2011 jmmce.org Printed in the USA.

[9] Sinmazçelik Tamer, Avcu Egemen, Özgür Bora Mustafa, Çoban Onur, "A review: Fibre metal laminates, background, bonding types and applied test methods," Received 28 January 2011,Accepted 4 March 2011, Materials and Design 32 (2011) 3671–3685.

[10] Davis M, Bond D, "Principles and practices of adhesive bonded structural joints and repairs," Int J Adhes Adhes 1999;19:91–105.

[11] SY Park, WJ Choi, HS Choi, H Kwon, SH Kim, "Recent trends in surface treatment technologies for airframe adhesive bonding processing: a review," (1995–2008). J Adhes 2010;86:192–221.

[12] AF Harris, A Beevers, "The effects of grit-blasting on surface properties for adhesion," Int J Adhes Adhes 1999;19:445–52.

[13] ASTM D1876 – Standard Test method of Peel resistance of adhesives, 1995

International Symposium on Advanced Materials (ISAM 2013) IOP Publishing
IOP Conf. Series: Materials Science and Engineering **60** (2014) 012051 doi:10.1088/1757-899X/60/1/012051

The effect of metallic oxide deposition on the electrochemical behaviour of Al-Zn-Mg-Sn alloy in natural tropical seawater

M S Din Yati[1], Mohd Nazree Derman[1], M C Isa[2], M Y Ahmad[3], N H N Yusoff[2], M M Muhammad[2] and H Nain2

[1]School of Materials Engineering, Universiti Malaysia Perlis (UniMAP), 02600 Jejawi, Perlis, Malaysia

[2]Maritime Technology Division, Science & Technology Research Institute for Defence (STRIDE), RMN Naval Base, 32100, Lumut, Perak, Malaysia

[3]Mechanical & Aerospace Technology Division, Science & Technology Research Institute for Defence (STRIDE), 43000, Kajang, Selangor, Malaysia

E-mail: mohdsubhidinyati@yahoo.com

Abstract. The potential of aluminium alloys as anode materials in cathodic protection system has been explored and a significant improvement has been achieved. However, for marine application, it is quite difficult to maintain continuous activation process due to passivation behavior of aluminum alloys. Therefore, to choose the best activation mechanism for aluminium alloy in marine environment, it has to be considered from various points such as alloy composition and surface treatment. This paper report the effect of metallic ruthenium oxide (RuO_2) deposition on the surface of as-cast Al-Zn-Mg-Sn alloy and to study the effect of its presence on the electrochemical behavior using direct current (DC) electrochemical polarization and current capacity measurement. The morphology and topography of corroded surface were studied by the aid of scanning electron microscope (SEM) and confocal laser scanning microscope (CLSM) respectively. Results from this study showed that the presence of intermetallic compound (Mg_2Sn) and also mixed metal oxide compound (Al_2O_3 and RuO_2) on the alloy surface has been very useful in improving electrochemical reaction and charge transfer activities in chloride containing solution. This study also showed that RuO_2 catalytic coating applied on the surface of Al-Zn-Mg-Sn alloy has slightly increased the corrosion current density compared to Al-Zn-Mg-Sn without RuO_2. The corrosion morphology and topography of corroded surface of Al-Zn-Mg-Sn alloy deposited with RuO_2 was found more uniform corrosion attack with the formation of porous and fibrous mud-like crack on outer layer. Based on surface morphology and 3D topographic studies, these features were believed to facilitate ionic species adsorption and diffusion through corrosion product layer at solution-alloy interface. Deposited RuO_2 films also was found to increase of current efficiency by more than 10%.

1. Introduction

Cast aluminium was first highlighted in the public on year 1885 in Paris [1]. However, due to the technology limitation at that moment and also the unknown characteristic and potential of aluminium, industry has refused to further explore it capability. A few decade later, aluminum has vastly used in multidiscipline area and one of it was in the corrosion control by cathodic protection system. It all started when marine engineers in that time started to realize that stern drives and outboard motors require a little more care than zinc anode can give[2]. This component was the most expensive and very important to all boat and ship. The application of aluminium sacrificial anodes for the cathodic protection of steel structures has been practiced since early 1960s and has become a popular choice nowadays due to it abundance and excellent electrochemical properties [3-5]. For example, aluminium alloy generally used in ports to protect structures (e.g., docks, pontoon, etc) because of its higher theoretical efficiency and other merits, such as a low density, vast availability, reasonable cost and its capability in providing higher driving potential throughout service life [6-10]. Sacrificial anodes have a difficult task, since they have to protect what is already a very active aluminium assembly. Installing cheap or sub-standard anodes will undoubtedly increase corrosion risk and potentially very expensive corrosion problems mitigation cost [11]

Domination of pitting dissolution, surface passivation and unstable corrosion potential, were among problems that has been reported associated with aluminium sacrificial anode [12]. Previous researchers have reported that the presence of passive oxide layer on aluminum alloy can be modified by the addition of alloying element and also by applying catalytic coating [13-18]. Both surface activation methods were meant to generate surface modification such as micro-structural electronic defects that may lead to passive oxide film rupture. The alloying element and catalytic coating will disturb Al_2O_3 bonding strength by sit placing impurities in between [12]. The addition of alloying element such as Hg, Ga, In, Sn Bi may weaken the oxide layer by ease of ion penetration from electrolyte through the defected oxide layer thus generating continuous localized corrosion [16,18]. In order to improve electrochemical activity of aluminum alloy, IrO_2 and RuO_2 has been extensively used as catalytic coating for chlor-alkali production and dimensionally stable anode due to it good catalytic activity [16,18,20]. In this study, the potential of RuO_2 as a surface activator in improving electrochemical properties of Al-Zn-Mg-Sn alloy in tropical seawater will further be explored.

2. Materials and Methods

2.1. Specimen preparation and alloy composition

Cast aluminium alloy was prepared in a graphite crucible under the protection of argon gas (inert atmosphere). Pure metals were melted in the crucible at the temperature of 850°C using Al bead (purity 99.90%) as primary component. Samples were made into rod shape using preheated mild steel mould. As for Al-Zn-Mg-Sn alloy fabrication, 1.5%wt Sn was added into aluminium melted solution when approaching 650°C to minimized Sn oxidation. In the same time, Argon gas flowed into the chamber surrounding was increased from 5 to 8 litres/minute. Sample was then cooled down at room temperature before proceeds to the next process. The as-cast aluminium was then cut by low speed cutter at speed of 250 rpm. Sample then polished by using 400, 600, 800, 1000 grades emery paper for the final surface preparation. Before dipped in $RuCl_3$ precursor solution, sample was first degreased and washed using acetone and distilled water. To obtain an aluminium alloy deposited with RuO_2, samples were then dipped in 1000 ppm $RuCl_3$ solution, dried and heated at 400°C for 30 minutes. Prepared aluminum alloy samples compositions were shown in Table 1.

2.2. Electrochemical test & corroded surface evaluation

Polarization test was conducted by using three-electrochemical cell GAMRY Instruments Reference 3000 potentiostat/galvanostat/ZRA controlled by GAMRY Framework V5.65 and using Echem Analyst software for data analysis. Tafel fit was used in determining the corrosion rate (μm/y) as well as the corrosion current density. Saturated calomel electrode (SCE) is used as the reference electrode

Table 1. Aluminium alloy chemical composition analysed by wavelength dispersive x-ray fluorescent spectrometer (WDXRF)

	Composition (wt. %)					
Sample	**Zn**	**Mg**	**Sn**	**Fe**	**Pb**	**Al**
Pure Al	-	-	-	0.035	-	Bal.
Al-Zn-Mg-Sn (substrate)	5.38	1.87	1.42	0.013	0.0007	Bal.
Substrate deposited RuO_2	5.42	1.86	1.42	0.044	0.006	Bal.

(RE) and high density graphite as counter electrode (CE). Sample was cut from the as-cast cylindrical alloy and drilled. 5 mm thickness of sample then polished up to grit 1000 followed by ultrasonic cleaning and finally dried in warm flowing air prior to each experiment. Potentiostatic polarization test started from cathodic to anodic potential by the range of ±250 mV with scanning rate set at 0.5mA/sec. Specimen exposed area was 1 cm^2, embedded in specimen water tight Teflon holder from Princeton Applied Research Company. Corrosion current density, i_{corr} can be calculated from stern-Geary equation [21].

$$i_{corr} = B/R_p, \tag{1}$$

$$B = b_a b_c / 2.303\,(b_a + b_c), \tag{2}$$

Where B is so-called the "Stern-Geary constant", b_a and b_c are the Tafel slopes for the anodic and cathodic reactions, respectively. The fitted value of R_p represents the overall resistance of the surface corrosion, but does not identify the passive or active corrosion areas even though the latter has a major contribution to the measured R_p. Derived from Faraday's law, the following equation is applied to calculate the corrosion rate (CR).

$$\text{CR } (\mu\text{m/y}) = 3.27 \times i_{corr} \times \text{E.W.}/d \tag{3}$$

Where E.W. is the equivalent weight of alloy (g) and d is the density of alloy in gm/cm^3. By substituting Eq. (1) into Eq. (3) and setting the values of E.W. and d, Eq. (4) is obtained.

$$\text{CR } (\mu\text{m/y}) = 304.2/R_p \tag{4}$$

Where the units of polarization resistance was Ω cm^2. Experiments will be held in tropical seawater solution at 27°C of pH 8.1. The solution was taken from Teluk Batik, Perak which was located at the west coast of peninsular Malaysia. Surface morphology of the samples after RuO_2 deposition and polarization test was observed by Scanning Electron Microscope (SEM) model Zeiss VP 1430 equipped with Inca Oxford energy dispersive x-ray (EDX). Confocal laser scanning microscope (CLSM) Model Zeiss LSM 5 Pascal will be used for surface topography analysis. Data and images reconstruction will be examined by LSM Zen software to extract the important information related to surface roughness parameters.

2.3. Current capacity and efficiency

Both anode (working electrode) and cathode (steel) specimens were polished to 1000 grit, washed and finally rinsed with acetone. Individual dimension for disc shape anode was 0.6 cm in thickness and 2.10 cm in diameter. A cathode of rectangular shape had a dimension of 0.8 x 4.1 x 16.0 cm. The separation or distance between anode and cathode was 10 cm and anode weight was recorded before the test started. Both anode and cathode were immersed in a lightproof perspex tank containing 40 liter of filtered tropical seawater medium for about 72 hours. A current density was supplied to the anode at amount of 0.5 mA/cm^2 and charge transfer reading recorded at coulometer after 72 hours would be taken for anode capacity determination. Anode specimen was then washed with water and soft brushed and dried. For weight loss determination, anode specimen was immersed in ammonium hydroxide solution for several minutes, dried and then dipped in a mixture of 50 g/l chromic acid and 10 g/l argentum nitrate for 30 seconds. Sample was then washed with water, rinsed with acetone and dried to determine their weight loss. Anode capacity measurement were determined by ASTM G97 (2007) as below [22].

$$\text{Current capacity} = \frac{A.h}{M_1 - M_2} \text{ Ah/kg} \tag{5}$$

where Ah = recorded charge transfer (Ampere-hour).
M_1 = sample weight before polarized (g).
M_2 = sample weight after polarized (g).

While for the current efficiency measurement were determined by:

$$\text{Current efficiency} = \frac{\text{Measured current capacity}}{\text{Theoritical current capacity}} \text{ x } 100\ \% \tag{6}$$

where

$$\text{Theoritical current capacity} = \frac{(96\,480\text{ C} / 3\,600\text{ S})}{\text{Sample equivalent weight}} \text{ Ah/kg} \tag{7}$$

3. Results and Discussion

3.1. Physical evaluation

As cast Al-Zn-Mg-Sn alloy was cut, polish and prepared for physical surface characterisation by the aid of SEM. As-cast alloy showed a dendrite structure with the existence of bright irregular shape spot spread all over the alloy surface (Figure 1a). The bright spot was believed to be an intermetallic compound located at the grain boundary of aluminum matrix. EDX analysis has been carried out to determine chemical composition at two different locations marked as spot 1 and spot 2 in Figure 1b. EDAX spectra showed the presence of high intensity peaks for elements such Mg, and Sn from the bright spot area (Figure 1c), meanwhile Al, Zn and Mg peaks were detected from the dark spot area (Figure 1d). Based on the EDX analysis, the bright spot was believed to be an intermetallic compound known as Mg_2Sn. Based on XRD spectrum data in previous study, Mg_2Sn formation can be correlated with the low solid solubility characteristic of Sn in aluminum matrix [23]. When Sn is added more then it solid solubility, it has a tendency to react with Mg and rejected to interdendrite structure, forming a stable Mg_2Sn intermetallic compound or secondary phases. This compound also has been reported to play an important role in modifying the oxide layer and electrical properties of Al-Zn-Mg-Sn alloy [24]. The surface morphology of as-cast Al-Zn-Mg-Sn alloy deposited with 1000 ppm RuO_2 and heated at 400°C for 30 minutes were shown in Figure 2. SEM micrograph showed the alloy surface has changed with a distinctive surface layer morphology including cracked surface layer ranging from micro to macro-crack, that similar to dry mud-crack feature reported by Roger *et al.* [25]. High intensity elemental peaks arises from EDAX spectra (Figure 2c & 2d), showed the presence of mixture or bimodal oxide layer such as Al_2O_3 and RuO_2 on the specimen surface. This mixed oxide particles were similar to the TiO_2/RuO_2 system in dimensionally stable anode (DSA) as reported by Shibli et. al [17, 18].

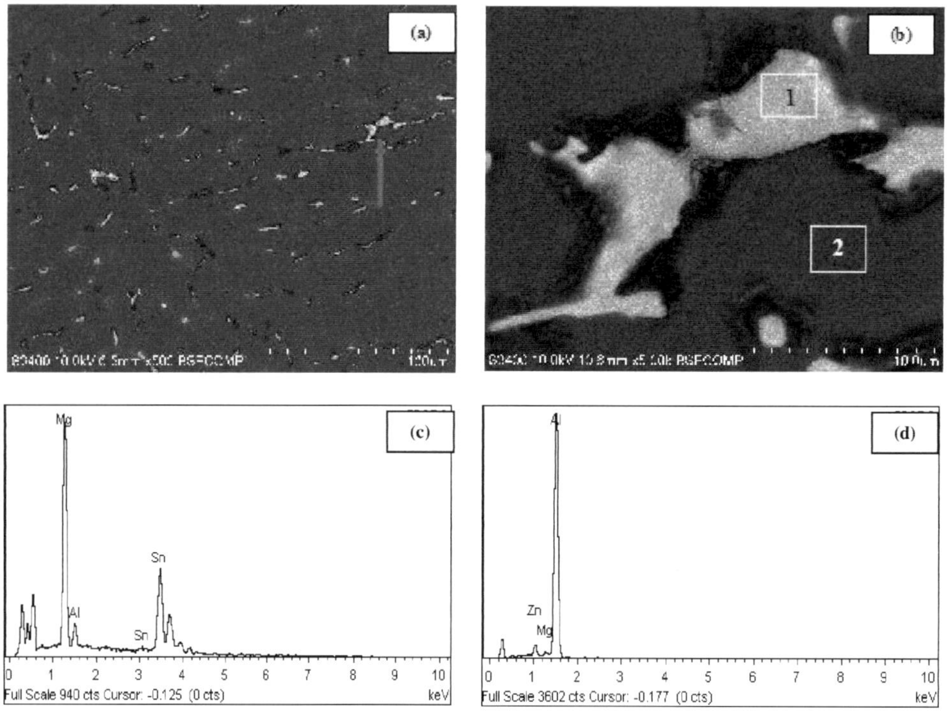

Figure 1. SEM images for (a) Polished surface of Al-Zn-Mg-Sn alloy (b) Magnified area of interest (c) EDX spectra at spot 1 and (d) EDX spectra at spot 2

3.2. Electrochemical evaluation of deposited and undeposited alloy samples

3.2.1. Polarization test & current capacity measurement

All samples has been polarized in order to investigate the effect of surface chemistry on activation of pure aluminium, deposited and undeposited aluminium alloy corrosion in seawater. Figure 5 showed polarization curves obtained for the samples after being polarized in tropical seawater medium. In this scope of study, parameters from Tafel plot has been analysed and presented in Table 2. The curved line obtained as Figure 3 is the total current (the sum of the anodic and cathodic currents). The sharp point in the curve is due to the logarithmic scale and it is the point where the current changes from anodic to cathodic or viceversa. This intersection point has shows both the i_{corr} and the E_{corr} value. The key parameter is the i_{corr} as it is directly proportional to the corrosion rate of the aluminium alloys. RuO_2 deposited sample showed an enhanced corrosion rate with 336 μm/year and pure Al with the smallest (0.55 μm/year).

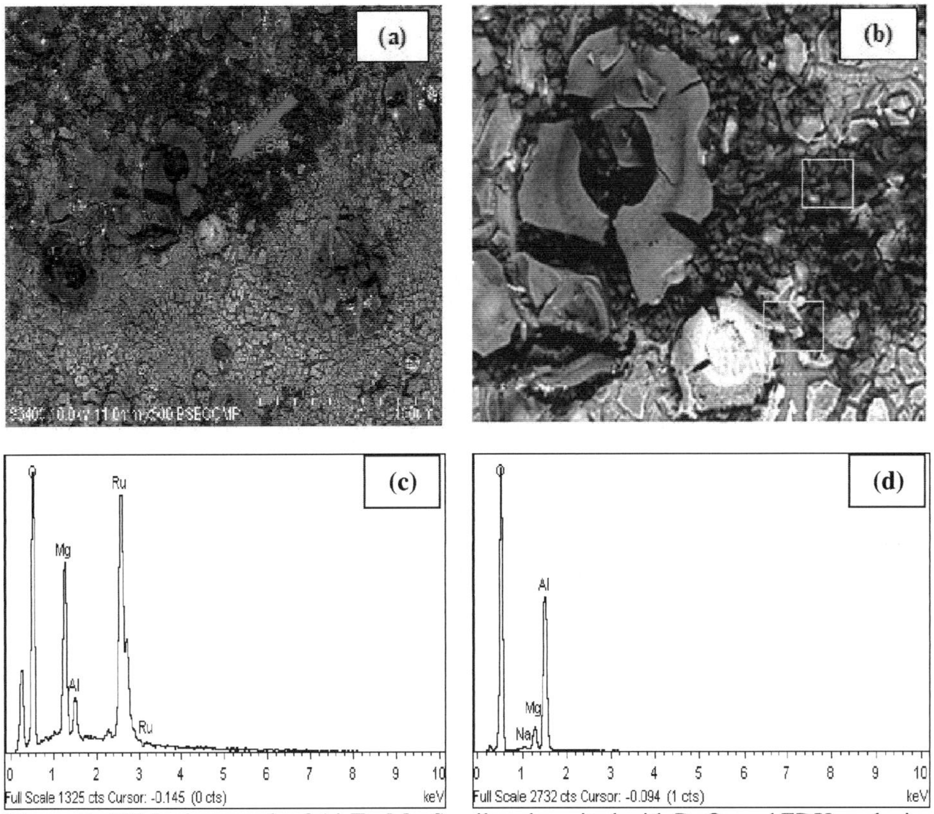

Figure 2. SEM micrograph of Al-Zn-Mg-Sn alloy deposited with RuO_2 and EDX analysis at spot 1 and 2 (Fig. 2c= spot 1, Fig. 2d= spot 2).

The increasing of corrosion rate for deposited sample is believed contributed from the presence of metallic RuO_2 coating on Al-Zn-Mg-Sn alloy surface which has synergistic effect in mobilizing the ionic species or enhanced ion-electron movement and catalyzing the the electrochemical process. Higher corrosion current density indicates active dissolution process occurs in the system and usually dominated by uniform corrosion. The uniformity of corrosion attack morphology for Al-Zn-Mg-Sn has been reported by Isa et al. which found that Al-Zn-Mg-Sn alloy tend to form secondary phase or intermetallic particle known as Mg_2Sn [14]. The presence of Mg_2Sn particle the in alloy will generates defect to the oxide layer by forming a thin and weak oxides properties on the surface of the alloy. When expose to aggressive chloride solution, this weak spot will became more vulnerable to the chloride attack. This attack usually started at the outer passive layer on top of Mg_2Sn intermetallic particles [12].

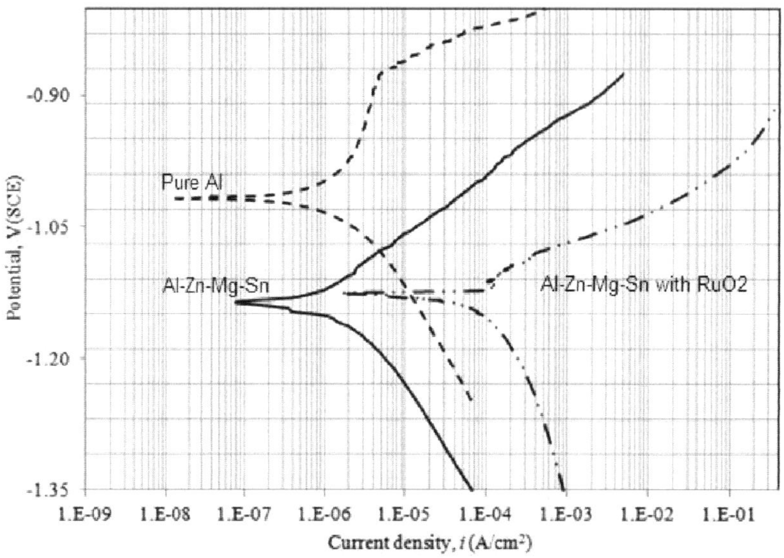

Figure 3. Polarisation (Tafel) plots of pure aluminium, Al-Zn-Mg-Sn and deposited RuO_2 aluminium alloy in tropical seawater

Table 2. Parameters from Tafel analysis based on polarization plot in Figure 3.

Specimen	βa (mV/dec)	βc (mV/dec)	i_{corr} (μA/cm²)	Corr. Rate (μm/year)	E_{corr} (mV,SCE)
Pure Al	215.1	241	0.014	0.55	-980
Al-Zn-Mg-Sn	85.7	131.5	1.6	101.9	-1130
Al-Zn-Mg-Sn deposited RuO_2	31.1	125.5	227	336	-1120

Figure 4a, c and e showed surface morphology of pure Al, bare and Al-Zn-Mg-Sn alloys deposited RuO_2 after polarization test in tropical seawater solution. At this stage, it can be observed that corrosion phenomena visible on all samples with the severity of corrosion attack patterns were different. On pure Al sample, corrosion is initiated and occurs but slow and may be not propagated, hence the appearance of corrosion product layer is seen intact. A dense adherent layer of the corrosion product formed on the surface can block or reduce the current flow (I_{corrs},) by increasing the electronic and ionic resistance during the test. On Al-Zn-Mg-Sn sample corrosion attack is initiated at grain boundary and propagated leading to the appearance of much widespread generalized pattern on its surface. Morphological aspect of corroded surface showed that both Al-Zn-Mg-Sn alloy surface (4c& 4e) and was heavily disintegrated compare to pure aluminium sample (4a). Deposited RuO_2 aluminium alloy surface showed mud-crack surface morphology and porous surface layer compared to undeposited Al-Zn-Mg-Sn alloy this type of corrosion product layer normally can be correlated with aggressive corrosion attack on the alloy surface [26].

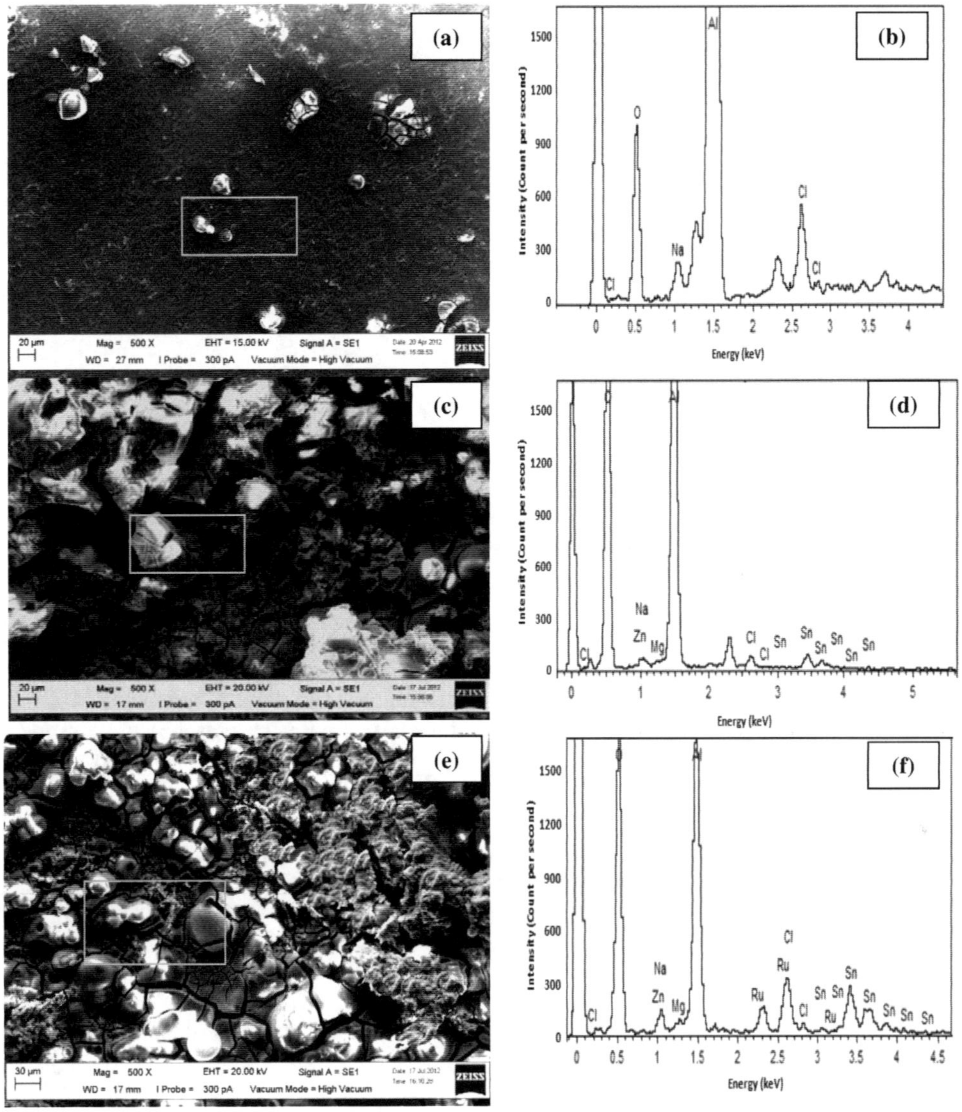

Figure 4. Surface morphology and EDX spectra of samples after polarization test in tropical sea water. (a&b) Pure Al, (c&d) Al-Zn-Mg-Sn alloy, (e&f) Al-Zn-Mg-Sn alloy deposited with RuO_2

3.2.2 Current Capacity measurement

Tables 3 showed that current capacity and efficiency differs according to the chemical composition. Without the presence of alloying elements, pure aluminium gave a significant decrease in current capacity. It is clearly shown that both alloying elements and metallic oxide (RuO_2 deposited on the sample) play a very important role on the current efficiency. For example, sample pure aluminium give 1756 Ah/kg current capacity while sample Al-Zn-Mg-Sn alloy gave only 2201 Ah/ kg. Highest current capacity recorded for Al-Zn-Mg-Sn deposited with 1000 ppm RuO_2 and it is believed that the increase in the efficiency (88.78%) for this sample is caused by the presence of alloying elements (Mg,

Sn) and metal oxide deposited on the surface which can attributed to synergistic activation and enhanced electrochemical activities at alloy-solution interface.

Table 3. Current capacity and efficiency in tropical seawater for 72 hours

Specimen	Weight loss (g)	Current capacity Ah/kg	Efficiency (%)
Pure Al	0.1522	1756	59.00
Al-Zn-Mg-Sn	0.1317	2201	79.71
Al-Zn-Mg-Sn deposited RuO_2	0.1048	2452	88.78

3.3. Surface topography study

Surface roughness parameter such as average surface roughness (Ss_a), surface skewness (Ss_k) and surface kurtosis (Sk_u) were used to translate the specific topographical surface entity because single parameter was unable to describe the surface quality. Table 4 showed surface roughness parameters after polarization test and all the data were extracted from 3D scanning process and images in Figure-5. The average surface roughness for deposited RuO_2 sample slightly decrease after polarisation test and this result of lower R_{sa} value after polarisation test is consistent with the nature uniform corrosion on the specimen surface. This decrease agrees with the widespread nature of corrosion attack that affects a large number of the scanning pixels in an image as shown in Figure 5. The negative value of surface skewness for undeposited Al-Zn-Mg-Sn sample (-0.07) representing predominance of valley structures on the alloy surface. The positive skewness value (0.102) for RuO_2 deposited Al-Zn-Mg-Sn alloy reflects unsymmetrical height distribution on the surface, with the presence of high peaks or filled valleys, and less deep scratch or a lack of peaks. From this observation, it was strongly believed that the presence of ruthenium oxide, RuO_2 in particles form on the alloy surface was believed to contribute to the better alloy dissolution process by creating a more uniform galvanic current distribution at the alloy-electrolyte interface and hence promoting non-localised and non-pitting corrosion attack. A positive R_{ssk} value indicative of flat surface comprised of disproportioned number of peaks structure meanwhile surface with predominantly deep valleys, scratches or pores will tend to have a negative skewness value [27]. In this case, negative skewness values after polarisation test can be ascribed from the existence of deep pore structures, grain falling or dissolution process of an alloy when the sample was polarised in natural seawater. Kurtosis was a measure of peakness or sharpness of surface height distribution or sometimes qualifies the flatness of height distribution. Based on Table 4, kurtosis values in the range of 1.055 – 1.110 ($S_{Ku} < 3$) were recorded for samples after polarization test which indicate a corroded surface (Fig. 5c) with less peaks and valley structures. [28,29].

Table 4. Surface roughness parameter after polarization test

Specimen	Ss_a (μm)	Ss_k (μm)	Sk_u (μm)
Pure Al	98.031	0.062	1.110
Al-Zn-Mg-Sn	57.357	-0.073	1.128
Al-Zn-Mg-Sn deposited RuO_2	48.358	0.102	1.055

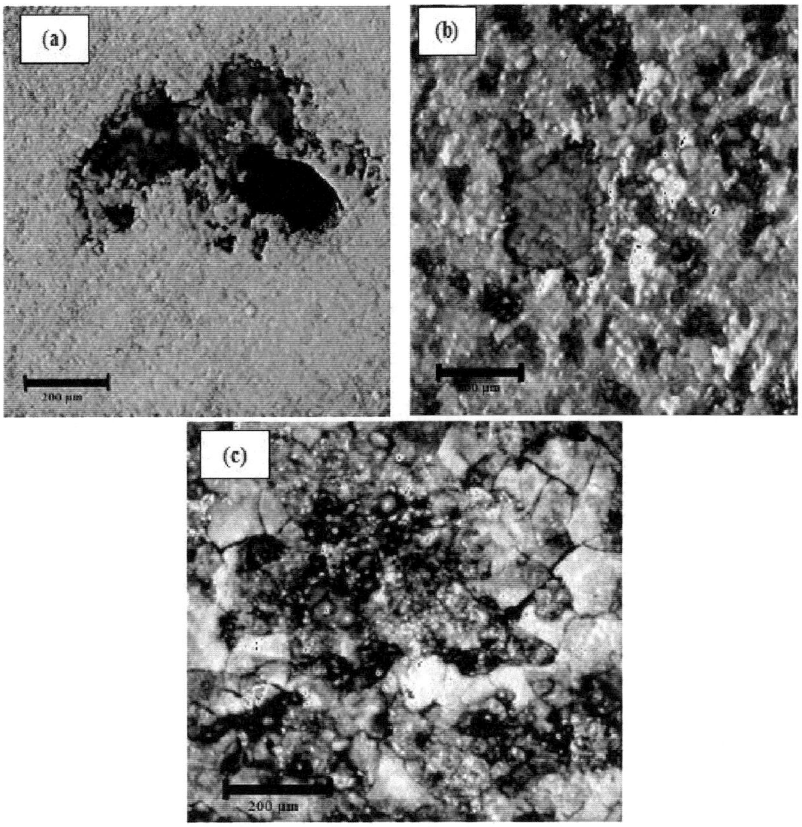

Figure 5. 3D CLSM constructed images of corroded surface after polarization test (a) Pure Al (b) Al-Zn-Mg-Sn alloy and (c) Al-Zn-Mg-Sn alloy deposited with RuO_2

4. Conclusion

In this study, both alloying addition (Mg, Sn) and metal oxide deposition (RuO_2) have proved their role in activating aluminium alloys in natural sea water. The presence of intermetallic compound (Mg_2Sn) in the Al-Zn-Mg alloy produced higher corrosion current density during polarisation process due to the cathodic nature of Sn as compared to Al matrix and thus initiated corrosion attack by uniform dissolution. As for the RuO_2 deposited Al-Zn-Mg-Sn alloy, it has successfully overcame the tendency of localised attack or pitting corrosion on the alloy surface by forming as catalytic activator layer for producing uniform and continuous dissolution process. The current capacity study showed that the presence of RuO_2 on the alloy surface can be very useful in reducing polarisation resistance of Al-Zn-Mg-Sn alloy at the early stage of immersion and maintaining ionic movement even after 72 hours of immersion. The RuO_2 metallic oxide has shown its usefulness by modifying the electrical properties of Al-Zn-Mg-Sn oxide layer by forming a porous and loose outer layer at the surface of the alloy. Surface topographic parameters for deposited Al-Zn-Mg-Sn alloy produce a lowest S_{sa} values after polarisation process which was in a good agreement with current capacity of the alloy. This observation is a manifestation of better dissolution process occurs at the alloy Al-Zn-Mg-Sn deposited with RuO_2 in natural seawater.

5. Acknowledgement

The authors would like to thank the Science and Technology Research Institute for Defence (STRIDE) and University Malaysia Perlis (UniMAP) for providing research facilities and technical assistance in this study. We also acknowledge academic and technical staff from School of Materials Engineering, UniMAP for a series of fruitful discussions and technical assistance in improving this manuscript.

6. References

[1] Ashby J 1999 *Construction History* 15 79-90

[2] Bohnes H and Richter B 1997. Cathodic Protection of Ships. Handbook of Cathodic Corrosion Protection (Third Edition) 391-414

[3] Barbucci A, Bruzzone G and Sacconec A 1997 *Electrochimica Acta* 42(15) 2369

[4] Andrea Mao M L M, Pineau S, Barillier D and Caplat C 2011 *Marine Pollution Bulletin* 62 2707

[5] Shibli S M A and Gireesh V S 2005 *Corr. Sci.* 47 2091

[6] Qingfeng L and Bjerrum N J 2002 *J. of Power Sources* 110 1

[7] Jingling M and Jiuba W 2010 *J. of Alloys and Compounds* 496 110

[8] Jiuba W, Junguang H and Xianwen L 2011 *Corr. Sci.* 53 3861

[9] Junguang H, Jiuba W and Xudong L 2011 *Corr. Sci.* 53 1948

[10] Shibli S M A, Archana S R and Muhamed Ashraf P 2008 *Corr. Sci.* 50 2232

[11] Thompson N G, Yunovich M and Dunmire D 2007 *Corrosion Reviews* 25 247

[12] Jafarzadeh K, Shahrabi T and Hosseini M G 2008 *J. Mater. Sci. Technol.* 24 215

[13] Ige O O and Umoru L E 2007 *J. of Minerals & Materials Charac. & Eng.* 7 105

[14] Isa M C, Din Yati M S, Yusoff N H N, Mohd Fauzy M N, Muhammad M M, Ismail O and Mohd Nor I 2011 *Defence S&T Technical Bulletin* 4 119

[15] Isa M C, Ahmad M Y, Daud A R and Daud M 2010 *Key Eng. Material* 442 322

[16] Shibli S M A and Gireesh V S 2003 *Applied Surface Science* 219 203

[17] Shibli S M A Gireesh V S and George S 2004 *Corr. Sci.* 46 819

[18] Shibli S M A Dilimon V S and Saji V S 2007 *J. Solid State Electrochem.* 11 201

[19] Bessone J B 2006 *Corr. Sci.* 48 4243

[20] Likun X, Yonglei X and Juntao W 2009 *Electrochimica Acta* 54 1820

[21] Ohtsu M I M 2006 *Construction and Building Materials* 20 458

[22] ASTM G97 2007 Standard Test Method for Laboratory Evaluation of Magnesium Sacrificial Anode Test Specimens for Underground Applications *American Society for Testing Materials* West Conshohocken PA USA

[23] Isa M C 2007 *Fabrikasi dan pencirian aloi Al-Zn-Mg-Sn mengandungi unsur Sn dan Cu bagi perlindungan kakisan* PhD Thesis Universiti Kebangsaan Malaysia

[24] Isa M C, Yusoff N H N, Din Yati M S, Muhammad M M and Mohd Nor I 2011 *Defence S&T Technical Bulletin* 4 1

[25] Roger M H, Gatepally K V and Patangia D N 1991 *Proc. Arkansas Academy of Science* 45 33

[26] Muazu A and Yaro S A 2011 *J. of Minerals & Mat. Charac. & Eng.*10 185

[27] Amal Abd El-Samad Sakrane O B A, Matsumura H, Koizumi H and Tanoue N 2004 *Int. Chin J. Dent* 4 85

[28] Sedlaček M, Vilhena L M S, Podgornik B and Vižintin J 2011 *J. of Mechanical Eng.* 57 674

[29] Sedlaček M, Podgornik B and Vižintin J 2012 *Tribology Int.* 48 102

Effect of molybdenum addition on aluminium grain refined by titanium on its metallurgical and mechanical characteristics in the as cast condition and after pressing by the equal angular channel process

A I O Zaid and A M Atieh
Department of Industrial Engineering, University of Jordan, Amman, Jordan

E-mail: a.kilani@ju.edu.jo, anas.m.attieh@gmail.com

Abstract. Aluminium and its alloys are versatile materials which are widely used in industrial and engineering applications due to their attractive characteristics. However, they solidify in columnar structure which tends to reduce their surface quality and mechanical strength. It is therefore, grain refined by grain refiners i.e. titanium or titanium+boron. The equal angular channel pressing, ECAP, process is a recent method for producing severe plastic deformation in materials. In this research work, the effect of addition of molybdenum either alone or in the presence of titanium to commercially pure aluminium on microstructure and mechanical behaviour is investigated in two conditions; first, in the as cast condition, and second after pressing by the ECAP process at room temperature. It was found that addition of Ti alone at a rate of 0.15% weight to commercially pure Al resulted in grain refining of microstructure and a grain size of 91μm was obtained. However, after pressing by the ECAP process further refinement was achieved and the grain size was reduced to 18μm. Addition of Mo alone to aluminium at a rate of 0.1% resulted in grain size of 76μm in the as cast condition and 32μm after pressing by the ECAP process. The combination of the two elements Ti and Mo together resulted in 48μm grain size in the as cast condition, compared to 40μm after pressing by the ECAP process. Furthermore, it was found that in the as cast condition: addition of Ti alone to Al resulted in enhancement of its mechanical behaviour by an increase of 5.2% increase in its flow stress at 20% true strain, whereas addition of Mo either alone or in the presence of Ti resulted in decrease of its flow stress at 20% by 9% and 5.6% respectively. However, after pressing by ECAP: it was found that addition of Ti or Mo either alone or together to Al resulted in increase of its flow stress at 20 % strain by the following percentages 5.49, 4.74 and 10.3% respectively.

1. Introduction

Aluminium and its alloys are versatile metals. They widely used in industrial and engineering applications due to their attractive properties. These include: high strength-to-weight ratio, corrosion resistance and high electrical and thermal conductivities. Against these preferable characteristics, Al and its alloys solidify in a grain columnar structure which tends to affect their mechanical strength and surface quality. Therefore, it became customary in aluminium foundry to grain refine aluminium and its alloys by Ti or Ti+B, as well, different master alloys were developed, which are now commercially available. The grain refinement of aluminium and its alloys by the binary Al-Ti and ternary Al-5Ti-1B master alloys is well established and currently in use in the aluminium industry. However, the mechanisms of grain refinement and the enhancement of grain refining efficiency by the presence of some elements and the deteriorating effect by the presence of others are still a controversial matter, although many attempts have been made and reported in the literature [1- 4]. In earlier work [3] the effect of addition of Mo, up to 0.3% to commercially pure aluminum was investigated. In addition, the grain refined Al by 0.15% Ti was also investigated. The tests was carried on its hardness, mechanical strength and machinability and found that improvement up to 16% in hardness and more than 30% in flow stress at 20% strain were achieved by the addition of 0.3% Mo. Furthermore, the addition of Mo resulted in enhancement of the surface quality. Examination of the available literature review revealed that the effect of addition of grain refiner's on the formability of aluminium and its alloys is very rare. The equal channel angular pressing which is referred to as ECAP is relatively recent processes for producing sever plastic deformation (SPD). This technique can effectively improve strength of various metallic alloys by producing ultrafine grain. ECAP is considered as a source of grain refinement mechanism, and it's also used to produce ultra-fine grained materials. So far ECAP method was unsuccessful at attempts of obtaining nano-metric materials, i.e. materials with grain size under 0.1 μm [5-7]. Sivaramen and chkkingal have investigated the workability of commercial pure aluminium processed by ECAP and found an improvement in strength, grain refinement, and reduction in workability [5]. The effect of addition of grain refiners and ECAP on the formability of aluminium and its alloys has not previously reported except recently by the authors [8].

2. Materials and experimental work

2.1. Materials
Commercially pure aluminium wires supplied by the Jordanian electrical authority with the chemical composition shown in Table 1 were used in preparing the master and microalloys. Mo and Ti powders of 95.95% and 99.98% purity respectively were used as grain refiners.

Table 1. Chemical composition (wt %) of commercially pure aluminium.

Element	Fe	Si	Cu	Mg	Ti	V	Zn	Mn	Na	Al
wt. %	0.09	0.05	0.005	0.004	0.004	0.008	0.005	0.001	0.005	Bal.

2.2. Equipment
The master and the microalloys were all prepared in graphite crucibles and stirring was carried out by graphite rods. The die which was designed and manufactured for carrying out ECAP tests consists of two parts which accommodate the punch and die in position is made of D2steel. All tests were carried out on a universal testing machine at the maximum compressive load of 100 KN and cross head speed of 5 mm/min. All specimens and inner surface of the ECAP die were lubricated using a semi-solid lubricant, MoS_2, to reduce friction. The autographic was obtained for each test after pressing. A schematic diagram and photograph of the ECAP die is shown in figure 1. The design which consists of two parts was introduced and manufactured by Electric discharge machining (EDM).

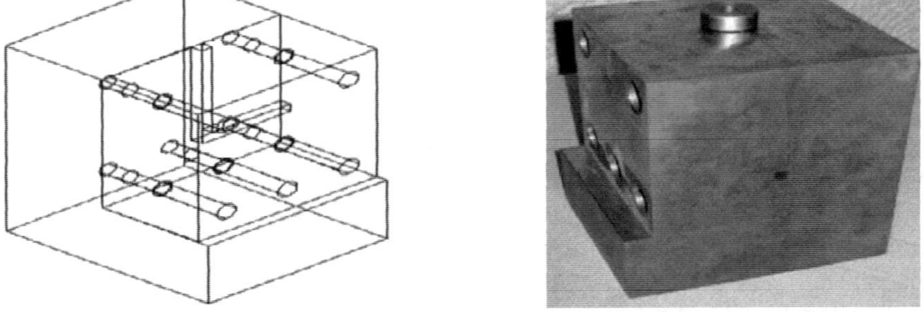

Figure 1. Isometric and photograph of the ECAP apparatus used.

2.3. Experimental procedures and Preparation of the different microalloys

The experimental procedures started with preparing the Al-Mo and Al-Ti master alloys followed by determining their chemical composition, then the microalloys were prepared. To prepare the Al-Mo master alloy; fifty grams of commercially pure aluminium were introduced into the graphite crucible under a layer of cryloite flux, charged into an electrical furnace at 1100 C for 15 min, after which 3 grams of pure molybdenum powder are added to the aluminium melt and stirred with graphite rod for one minute and brought back to the furnace for 30 min. finally the crucible is taken out of the furnace, stirred for one minute and left to solidify and cool down in air to room temperature. SEM was used to determine Mo weight % and found to be 6.01%. the same procedure was followed for preparing the Al-Ti master alloy using only 2 grams of pure Ti powder with 50 grams of commercially pure aluminium. The Ti weight% as determined by SEM was found to be 2.02 %.

In addition to commercially pure aluminium four microalloys were prepared Al-0.1%Mo, Al-0.15%Mo, Al-0.15%Ti-0.1%Mo. The calculated amount of each master alloy was added to predetermined amount of aluminium at 1100°C in case of Al-Mo and at 800°C in case of Al-Ti master alloys. The mixture was agitated for two minutes in the graphite crucible using graphite rod and finally poured to solidify in hollow brass cylinders. The square cross section of brass die was 10 x50x100 mm. Mo was added to Al grain refined by Ti at a rate of 0.1 % by weight, which corresponds to the peritictic limit on the Al-Mo phase diagram. The chemical analyses of different prepared microalloys are shown in Table 2.

Table 2. Chemical composition by wt.% of microalloys

Alloy	Mo wt.%	Ti wt.%	Al wt.%
Al	0.00	0.00	Rem.
Al-Mo	0.05%	0.00	Rem.
Al-Mo	0.10%	0.00	Rem.
Al-Mo	0.15%	0.00	Rem.
Al-Ti	0.00	0.15%	Rem.
Al-Ti-Mo	0.1%	0.15%	Rem.

Specimens from the casted microalloys were cut to the required dimensions using a CNC milling machine at 1000 rpm rotational speed (37.7 mm/min), 0.8 mm depth of cut and 600 mm/min using 12 mm diameter cutter with 4 cutting teeth to be pressed in the ECAP die set. Finally, two sets of each microalloy one in the as cast condition and another one after pressing were prepared for metallurgical examination by mounting in Bakelite, ground with different grades of silicon carbide abrasive paper and polished with 1 micron diamond paste and finally etched in a solution of 0.5 % HF acid + 2.5% HNO3+1.5%HCL+95.5% distilled water for a period ranging from 15 to 30 seconds for the metallurgical examination, which was carried out using Vickers optical microscope M17. The grain size was determined using the linear intercept method. The Vickers micro-hardness was determined using a load of 600 grf.

3. Results and discussion

3.1. Effect of Mo addition on the metallurgical aspects of Al grain refined by Ti

The effect of addition of Mo or Ti alone or both together on the grain size of Al is shown in the histogram of figure 2. It can be seen from this histogram that addition of 0.15% Ti to Al resulted in pronounced reduction in its grain size as 34.5% reduction was achieved in the as cast condition. However, after pressing by ECAP die, it produced better refinement as 87% reduction in grain size was achieved when compared to Al in the as cast condition and 43.8% when compared with Al after pressing in the ECAP die (see figure 2). This is also explicitly demonstrated by the photomicrographs of figure 3 (a), (b), (c), and (d) for Al and Al-Mo micro alloys as cast, and figure 3 (a'), (b'), (c'), and (d') for Al and Al-Mo micro alloys after pressing, respectively. Addition of Mo at 0.1% weight to Al grain refined by Ti resulted in refinement of its structure in the as cast condition where 47.3% reduction was achieved in its grain size, but resulted in poisoning its grain size after pressing in the

ECAP die as 122.2% in grain size was obtained as indicated by figure 2. This means although their effect is positive when added together in the as cast condition, the effect is negative after pressing in the ECAP die. This may be explained due to the existence of the intermetallic phases of Al-Mo which do not respond to the refinement by pressing as the Al-Ti in the as cast condition .This however needs further verification by studying and analyzing these phases.

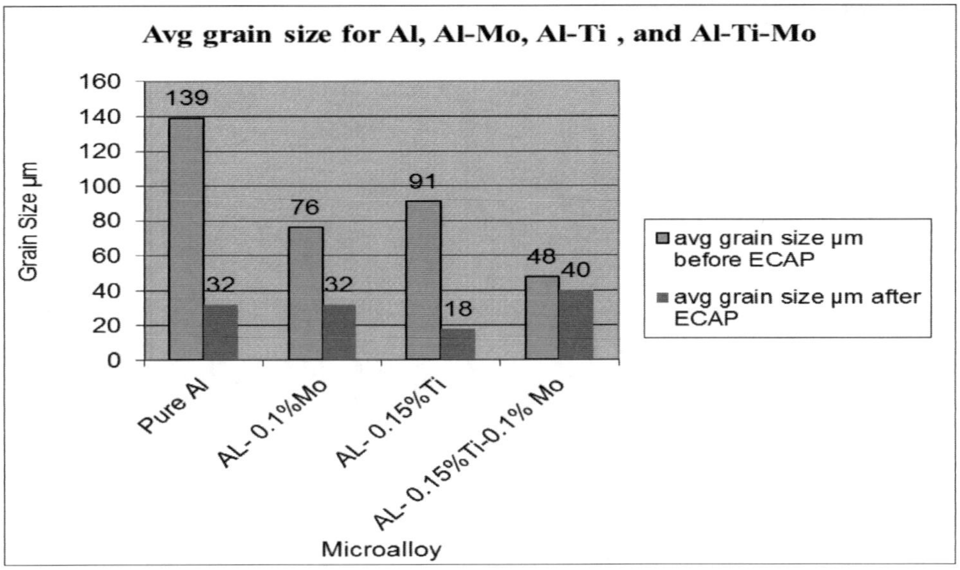

Figure 2. Effect of Mo addition on the grain size of Al grain refined by Ti in the cast and after pressing by the ECAP process.

It is worth noting that although addition of Ti or Mo either alone or together to Al resulted in grain refinement of grain size in the as cast condition; however, addition of both Ti and Mo together (Ti+Mo in the Al-0.15%Ti-0.1%Mo) resulted in poisoning of the grain size after ECAP process. This means that the effect of addition of both of them together is not additive. This agrees with the previous findings for addition of other grain refiners to Al [8].

3.2. Effect of Mo addition on the mechanical characteristics and hardness of Al grain refined by Ti

The effect of addition of Ti or Mo either alone or both together to Al on its mechanical behaviour is presented by the true stress, σ, true strain, ε, in the as cast condition in figure 4 .It can be seen from this figure that addition of Ti alone resulted in enhancement of Al mechanical behaviour whereas addition of Mo alone resulted in deterioration of its mechanical behaviour. Similarly the addition of both Ti and Mo together resulted in deterioration of its mechanical behaviour in the as cast condition as illustrated in table 3, but at less amount than when added alone. For example addition of Ti alone resulted by 5.2% enhancement in the flow stress at 20% strain and resulted in decrease of the flow stress at 20% strain by 9% and by 5.6% in case of Mo and Ti+Mo respectively. However, the work hardening index, n, was reduced by 3.2%, 11% and 6% in case of Mo, Ti and Mo+Ti respectively.

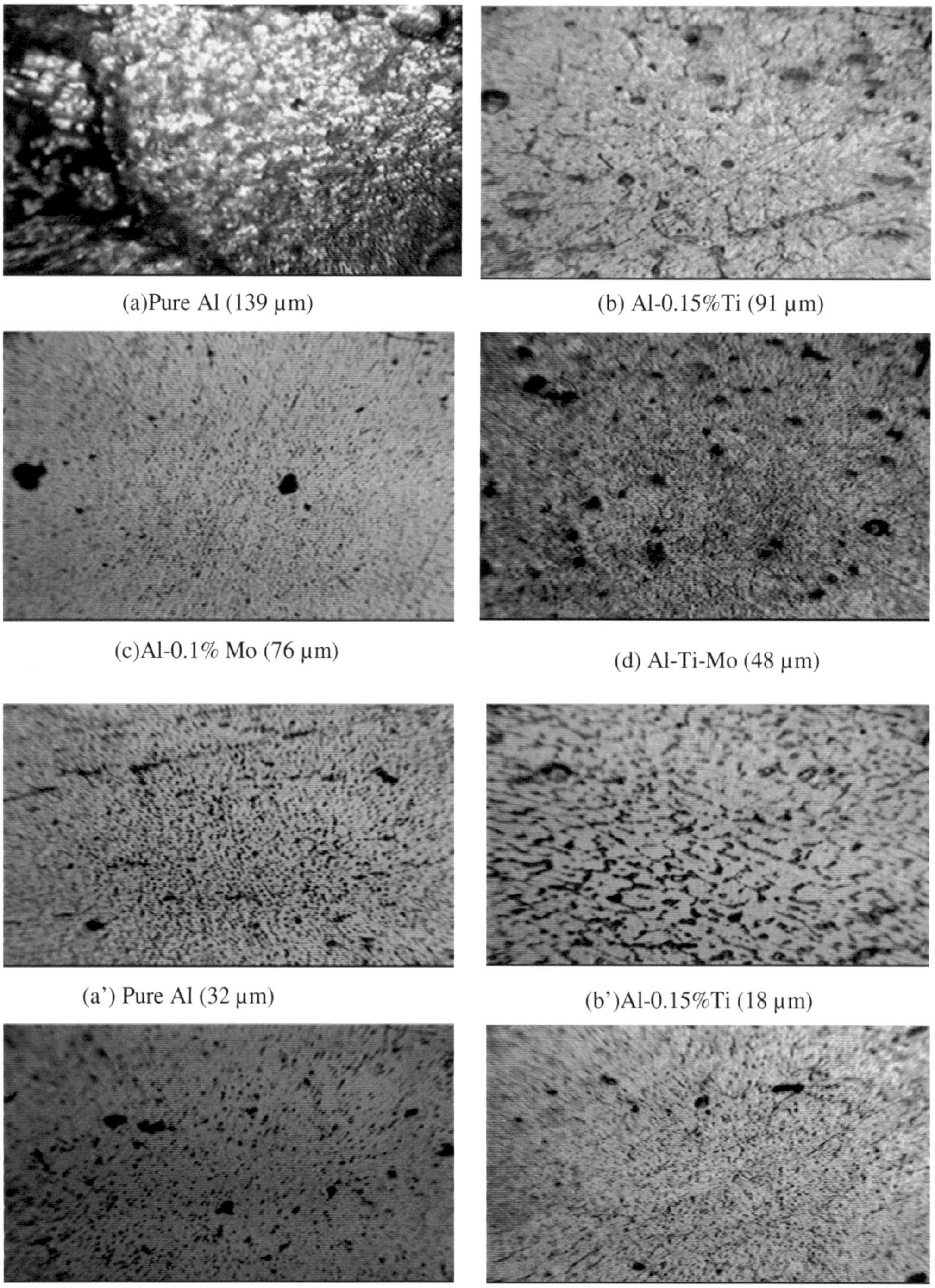

(a)Pure Al (139 μm) (b) Al-0.15%Ti (91 μm)

(c)Al-0.1% Mo (76 μm) (d) Al-Ti-Mo (48 μm)

(a') Pure Al (32 μm) (b')Al-0.15%Ti (18 μm)

(c')Al-0.1%Mo (32 μm) (d') Al-Ti-Mo (41 μm)

Figure 3. Photomicrographs of Al and Mo micro alloys in two conditions (a to d) in the cast condition and (a' to d') after pressing by ECAP process conditions, X250

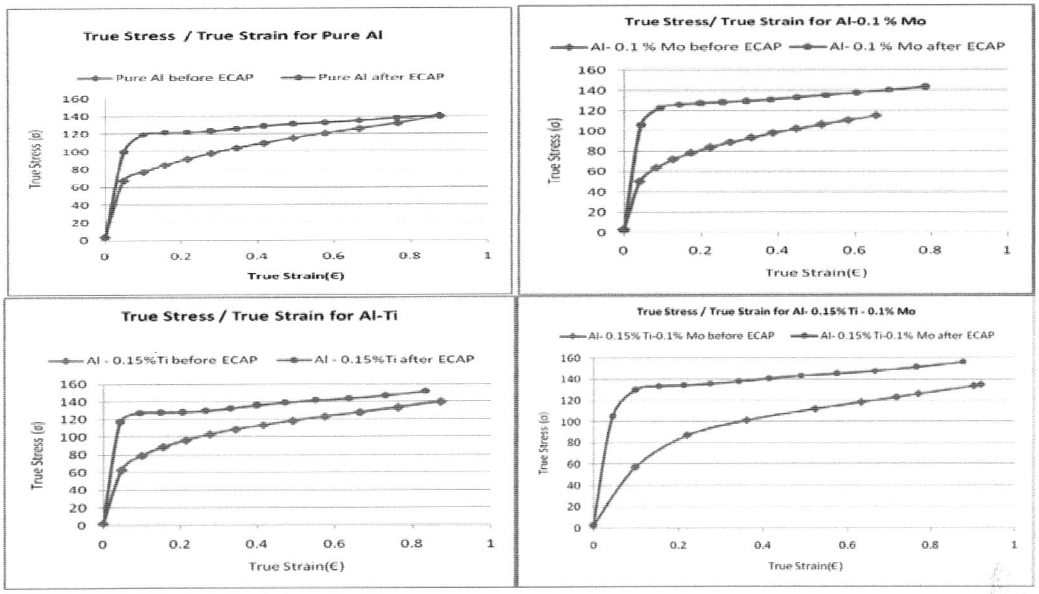

Figure 4. The true stress- true strain of Al and its alloys in the as cast condition and after pressing by ECAP process conditions.

Table 3. The mechanical characteristics of Mo addition to Al and Al grain refined by Ti in the cast condition.

Micro Alloy	General equation representing the mechanical behaviour in the plastic region MPa	Flow stress (MPa)at strain=20%	Flow stress (MPa)at strain=50%
Pure Al before ECAP	$\sigma = 143\ \varepsilon^{0.3092}$	89.51	115.59
Al- 0.1%Mo before ECAP	$\sigma = 130\ \varepsilon^{0.2994}$	81.45	105.76
Al- 0.15%Ti before ECAP	$\sigma = 143\ \varepsilon^{0.2753}$	94.18	118.51
Al- 0.15%Ti-0.1% Mo before ECAP	$\sigma = 136\ \varepsilon^{0.2908}$	84.50	110.89

After pressing by the ECAP process, the obtained results (see figure 4), shows that the behaviour is opposite to those obtained in the as cast condition. It can be seen from figure 4, that the mechanical behaviour is improved by the addition of Ti or Mo alone or both together to Al, being 5.5%, 4.8%, and 10.3%, respectively. The best improvement was achieved in the addition of both together followed by the addition of Ti alone and finally by addition of Mo alone. This improvement is explicitly demonstrated in table 4, which summarizes the mechanical characteristics of Al and its Mo and Ti microalloys and the Al-Ti-Mo ternary alloy. However, for 0.1% Mo alone, 0.15% Ti alone and in case of addition of both together respectively, the enhancement in flow stress at 20% strain was 4.75%, 5.5% and 10.3%. The enhancement in strength coefficient, K, was 4.25%, 7.8%, and 10.64% and an increase in the work hardening index by 12.55%, 10.2% and 9.4% as compared to Al after pressing by the ECAP process respectively.

Table 4. Mechanical characteristics of Mo addition to Al and Al grain refined by Ti after pressing by the ECAP process

Micro Alloy	General equation representing the mechanical behaviour in the plastic region MPa	Flow stress (MPa)at strain=20%	Flow stress (MPa)at strain=50%
Pure Al after ECAP	$\sigma = 141\ \varepsilon^{0.1139}$	121.84	130.95
Al- 0.1%Mo after ECAP	$\sigma = 147\ \varepsilon^{0.1282}$	127.61	134.50
Al- 0.15%Ti after ECAP	$\sigma = 152\ \varepsilon^{0.1255}$	128.53	139.80
Al- 0.15%Ti-0.1% Mo after ECAP	$\sigma = 156\ \varepsilon^{0.1246}$	134.38	143.65

3.3. Effect of Mo addition on the hardness of Al grain refined by Ti

The histogram of figure 5 shows that addition of Ti or Mo either alone or together to Al resulted in increase of its hardness in the as cast condition. The maximum increase is in the case of addition of Ti alone, being 22.2% followed by the addition of Ti and Mo together, 14.8% and only 3.7% in case of Mo addition. However after pressing in ECAP process, addition of Ti alone to Al resulted in very small increase in its hardness, only 2.27%, whereas resulted in decrease of its hardness in case of either addition of Mo alone or by the addition of Mo and Ti together being 9.1% and 13.6% respectively. This may be explained by the different response of the existing inter compounds of Ti and Mo in the Al matrix.

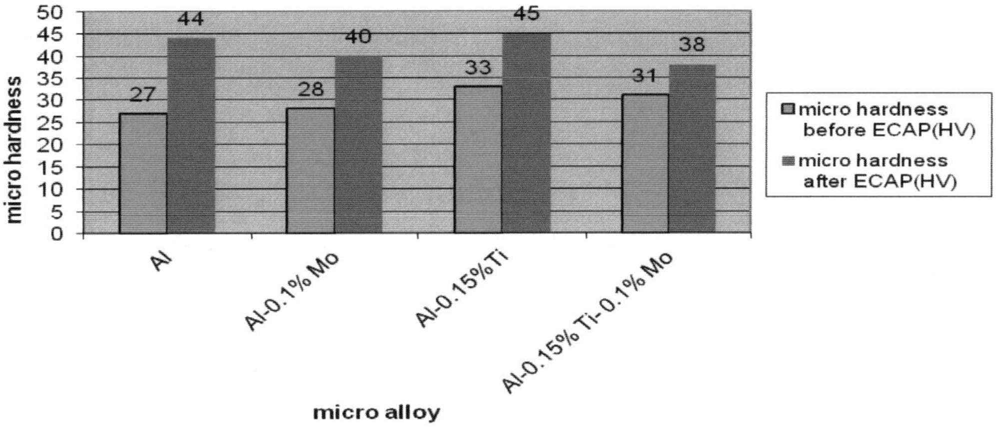

Figure 5. Effect of addition of Ti and Mo to Al and Al grain refined by Ti on hardness.

4. Conclusion

- Addition of Mo or Ti alone or together at any rate to Al within the experimental limitations, resulted in grain refinement, However, after ECAP it resulted in further refining in when Ti or Mo added alone, and poisoning effect, i.e. grain coarsening when added together.
- The addition of Mo to Al within the experimental limitations resulted in deterioration of its mechanical behaviour and reduction in its work hardening index, n, in the as cast condition, However Ti addition resulted in increase of the mechanical behaviour and also reduction in its work hardening index, n, in the as cast condition.
- Pressing Al and Al-Mo, Al-Ti, and Al-Mo-Ti micro alloys by the ECAP process resulted in enhancement of their mechanical behaviours.
- Addition of Ti or Mo either alone or together to Al resulted in increase of its hardness in the as cast condition, where after pressing in ECAP process, addition of Ti alone to Al resulted in very small increase in its hardness, whereas resulted in decrease of its hardness in case of either addition of Mo alone or by the addition of Mo and Ti together.

5. References

[1] Abdel-Hamid A A and Zaid A I O 2000 *proc. Internat. MDP-6 Conf.* Cairo Egypt

[2] Jones G P and Person J 1976 *Metal. Trans. B* 223-234

[3] Davies I G, Dennis J M and Hellawell A 1970 *Metal. Trans. A* 1 275-279

[4] Abdel-Hamid A A 1989 Z. Metall Kd 80 463-466

[5] Sivaramen A and Chakkingal U 2007 *Mat. Sci. Eng. A* 487

[6] Liu M, Roven H J, Yingda Y and Werenskiold C J 2006 *Mat. Sci. Eng. A* 483-484

[7] Greger M, Vodarek V, Kander L, Cizek L 2009 *Arch. of Mater. Sci. Eng.*37 13-20

[8] Zaid A I O and Atieh A M 2013 *proc.Canad. conf appl. mech.* Saskatoon Canada

Deposition and characterization of plasma sprayed Ni-5Al/ magnesia stabilized zirconia based functionally graded thermal barrier coating

M N Baig[1] and F A Khalid
Faculty of Materials Science & Engineering, GIK Institute of Engineering Sciences & Technology, Pakistan

E-mail: mnbaig8167@gmail.com

Abstract. Thermal barrier coatings (TBCs) are employed to protect hot section components in industrial and aerospace gas turbine engines. Conventional TBCs frequently fail due to high residual stresses and difference between coefficient of thermal expansion (CTE) of the substrate & coatings. Functionally graded thermal barrier coatings (FG-TBCs) with gradual variation in composition have been proposed to minimize the problem. In this work, a five layered functionally graded thermal barrier coating system was deposited by atmospheric plasma spray (APS) technique on Nimonic 90 substrates using Ni-5Al as bond coat (BC) and magnesia stabilized zirconia as top coat (TC). The coatings were characterized by SEM, EDS, XRD & optical profilometer. Microhardness and coefficient of thermal expansion of the five layers deposited as individual coatings were also measured. The deposited coating system was oxidized at 800°C. SEM analysis showed that five layers were successfully deposited by APS to produce a FG-TBC. The results also showed that roughness (Ra) of the individual layers decreased with an increase in TC content in the coatings. It was found that microhardness and CTE values gradually changed from bond coat to cermet layers to top coat. The oxidized coated sample revealed parabolic behavior and changes in the surface morphology and composition of coating.

1. Introduction

Thermal barrier coatings (TBCs) are widely used to protect internally cooled hot section components of gas turbine engines for both power generation and propulsion. These coatings are applied to provide a temperature drop (~170-200 °C) between the coating surface and the load-bearing alloy component underneath thus increasing the Turbine Entry Temperature (TET) and consequently turbine engine's efficiency. The eventual target of TBCs application is to achieve an enhancement in fuel efficiency by increasing the operating temperature (~100-300 °C) thereby increasing the engine thrust and efficiency up to 5 % and 1 %, respectively. TBCs are also meant to extend the life of turbine components and hence to improve engine reliability [1-8].

A TBC system is a multilayered coating structure having a complex geometry and usually consists of four layers, (1) a superalloy substrate, (2) an oxidation & corrosion resistant bond coat (BC), (3) a thermally grown oxide (TGO) and (4) a low thermal conductivity ceramic top coat (TC) [9]. It has been reported that most superalloys oxidize rapidly at temperatures above 900 °C as they are unable to form a protective oxygen diffusion barrier during early stages of oxidation [10]. The use of an approximately 100-150 μm thick layer of a metallic BC has now been a norm to protect the superalloy substrate from high temperature oxidation and corrosion [11]. The typical BC materials used in TBCs are NiCrAlY/CoCrAlY/NiCoCrAlY alloys, Ni-Cr-Al, Ni-Cr and Ni-Al with or without Pt modification [8, 11-13]. The Ni-Al type bond coats such as Ni-20Al and Ni-5Al are recognized for their enhanced bond strengths attributed to an exothermic reaction apparently resulting from an extra heat of solidification associated with the partial oxidation of Ni and Al during spraying [14-16]. It has also been reported that Ni-5Al coatings develop a protective and adherent oxide scale when cyclically oxidized at 900 °C resulting in a reduction in weight gains of the superalloy substrates thus lowering the rate of oxidation [16].

The TC is characterized by a high melting point, excellent thermal stability, good erosion resistance, comparatively high coefficient of thermal expansion (CTE), and low thermal conductivity and is usually 200-250 μm thick layer attached directly to the BC [8,10-11]. Partially or fully stabilized zirconia by adding various oxides like Y_2O_3, CeO_2, CaO, Al_2O_3, Gd_2O_3 and MgO as stabilizers is used as TC material because of its low thermal conductivity and relatively good mechanical properties [8, 17-18]. Among these, ZrO_2-Y_2O_3 (7-8 % YSZ) has been documented as an outstanding TC material for last few decades due to its high thermal shock resistance, low thermal conductivity, and a higher melting point than most oxides [10]. However, it is more expensive as compared to CaO and MgO-stabilized zirconia [8, 19-20]. Zirconia stabilized with 15-24 % MgO (magnesia stabilized zirconia) is comparatively cheaper and is utilized in those regions where the temperature intensity is relatively low such as in the exhaust nozzle of the jet engines [20-21]. Recently, magnesia stabilized zirconia has been investigated for its hot corrosion behavior in LiCl-Li_2O molten salt owing to its excellent chemical stability [21]. Moreover, magnesia stabilized zirconia has low thermal conductivity (1.0-1.5 W/mK), fairly comparable to YSZ, which suggests that it has a huge potential to be used as TC material for suitable applications.

The main problem in using a conventional TBC system is its premature failure because of the delamination or spallation of the ceramic TC from the metallic BC due to large residual stresses generated during cooling from the spraying temperature at or near the interface and poor bond strength between the coating and the substrate. These residual stresses are induced due to mismatch in CTEs of the substrate and BC & TC and can have a considerable effect on the properties of the deposited coating system [22-25].

In order to reduce the thermal expansion mismatches and to enhance the service life of the TBC system, the introduction of one or more intermediate layers with compositional gradient between the bond and top coats has been proposed. These thermal barrier systems are usually referred to as functionally graded-thermal barrier coatings (FG-TBCs). FG-TBCs may consist of several intermediate layers whose composition changes gradually from substrate metal to ceramic coating [24-29].

In this study, a five layer FG-TBC system and five individual layers comprising of the coating system were prepared by APS to study their characteristic properties. The coatings were characterized by SEM, EDS, XRD and optical profilometer. CTEs and microhardness of the individual coatings were also measured. Oxidation of the coating system was also performed at 800 °C.

2. Materials and methods

The substrate used in this study were discs of a nickel-based alloy, Nimonic 90, 12.8 mm in diameter and approximately 3 mm in thickness from Goodfellow Cambridge Limited, UK [(nominal composition Ni-20Cr-18Co-2.5Ti-1.5Al-1.5Fe & 3.5-others (in weight %)]. Two types of commercial feedstock powders from Sulzer Metco, USA, namely Metco 450NS (Ni-5Al, will be referred as Ni) and Metco 210NS-1 (ZrO_2-24MgO, will be referred as MSZ) were used as BC and TC materials, respectively. The two powders were also pre-mixed with each other in three different ratios (by weight) to form three powder blends, i.e., 75 Ni/25 MSZ, 50 Ni/50 MSZ, and 25 Ni/75 MSZ. These powders were then used to prepare five layer FG-TBC system. The coatings were deposited by APS using a Metco 9MB plasma spray gun (Sulzer Metco, USA). Prior to deposition, the substrates were grit blasted with Corundum 500 particles to a surface roughness of ~4.0 μm followed by ultrasonic cleaning. Deposition parameters are summarized in table 1 and the coating architecture is shown in figure 1.

Table 1. Plasma spraying parameters

Parameter	Limits used
Primary gas type/pressure/flow rate	Ar/100± 5Psi/100± 5 ft^3/min
Auxiliary gas type/pressure/flow rate	H_2/56± 5Psi/15± 5 ft^3/min
Powder carrier gas/flow rate	Ar/37 ft^3/min
Powder feed rate	51, 60, 68 g/min
Arc current	500 Amps
Arc voltage	70 ± 5 V
Spray distance	75, 100, 125 mm
Gun to work angle	90 degree
Atmosphere	Ambient

The powders and coatings were characterized by an analytical low vacuum scanning electron microscope (JEOL-Japan, Model JSM 6490 LA) operated at 20 kV and coupled with an energy dispersive spectrometer (EDS). Roughness profiles of the individual coatings were recorded by a 2D optical profilometer (Nanovea-USA, Model PS 50). The distance scanned was 1 mm and a scan velocity of 0.1 mm/sec with a scan step of 0.1 µm was used. The coating in as sprayed and oxidized conditions was analyzed by using X-ray diffractometer (STOE-Germany, Model Theta/Theta) with CuKα radiation at 40 kV and 40 mA. Microhardness profiles of the polished surfaces of individual coatings were obtained utilizing a Vickers Indenter (Wilson Instruments-USA, Model 401MVD) with a 300gf load and 15 sec. dwell time (loading time). The CTEs of the two individual freestanding coatings (bond and top coats) were measured as a function of temperature using a dilatometer (Netzsch Germany, Model 402E). CTE of the remaining three types of coatings representing composite layers was calculated using Vegard's rule of mixture. Approximately 03 mm thick individual coatings were deposited onto a 05 mm thick steel plate (SS 304L). The specimens for CTE measurements were then removed from the SS substrate by an automatic diamond cutter. The approximate sample size was 03x03x25 mm. The coating system was oxidized at 800 °C in a tube furnace in still laboratory air for 100 hours. For measuring weight gain during oxidation, the heating cycle consisted of 11 hours in furnace at 800 °C followed by cooling in still laboratory air to room temperature for 30 minutes. The coated discs were then weighed by a precise weight balance.

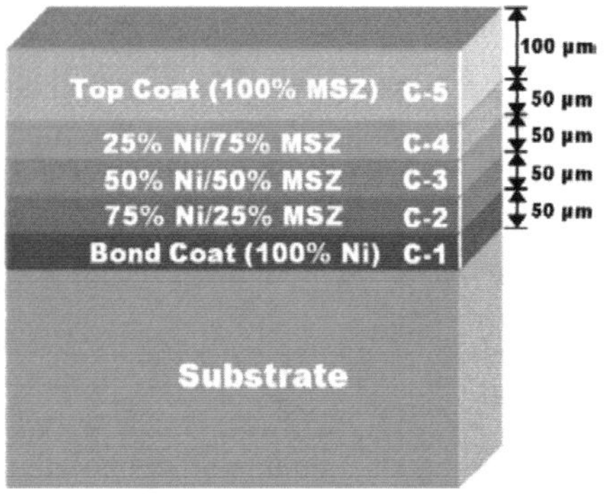

Figure 1. Schematic diagram showing architecture of the five layer FG-TBC system.

3. Results and discussions

3.1. Characterization of powders and coating morphology

SEM micrographs showing the morphology and particle size distribution of the two powders used as BC and TC materials are shown in figure 2 (a) & (b), respectively. Both the powders showed a spheroidal morphology however MSZ powder particles showed a perfectly round spherical morphology as compared to Ni powder particles. Surface morphologies of the single particles of both the powders at higher magnification showed that Ni particle has an irregular and rough surface whereas MSZ particle has a very smooth surface. Furthermore, SEM micrographs of both the powders showed that MSZ has a higher range of finer size particles as compared to Ni powder.

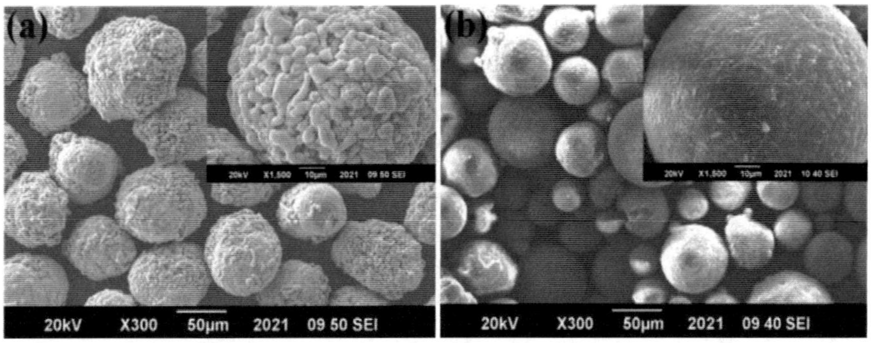

Figure 2. SEM micrographs of the two feedstock powders showing surface morphology and particle size distribution, (a) Ni powder and (b) MSZ powder.

SEM micrograph of polished cross section of the five layered TBC system is shown in figure 3 (a). The microstructure showed that there were no sharp boundaries formed between the layers and the structure was continuous. EDS line scan analysis for each element comprising the TBC system is shown in figure 3 (b). It can be seen from the figure that there were no clear boundaries formed between the deposited layers resulting in a relatively continuous microstructure and the distribution of elements comprising the TBC was relatively unbroken. Three graded layers of different compositions as shown in figure 1 were deposited between the bond and top coats thus giving rise to a microstructure having a gradient in composition. The introduction of such graded layers between the bond and top coats in a TBC system is believed to be an important factor in ensuring a sound bonding between them and thought to offer a great potential for increased service life due to microstructural and compositional stability at the interface [28].

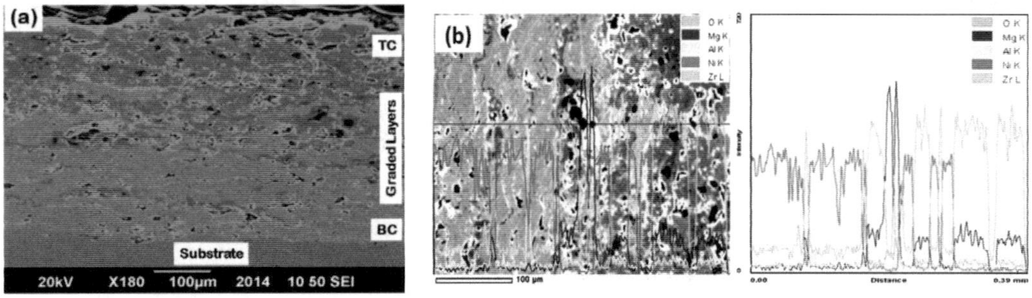

Figure 3. (a) SEM micrograph of the polished cross section of the five layered coating system showing typical plasma sprayed microstructure and (b) EDS line scan analysis of the elements comprising the five layered FG-TBC system.

3.2. Surface Roughness

Surface roughness profiles of the five individual coatings were recorded by a 2D optical profilometer and are shown in figure 4. The results showed that coating C-1 (100% Ni) has the maximum Ra value whereas coating C-5 (100% MSZ) has the minimum Ra value. Furthermore, Ra values of the intermediate layers were found to be decreasing with an increasing amount of MSZ content in the cermets, respectively. The surface morphology of the sprayed coatings is thought to be strongly dependent on the size and shape of the starting powder particles. The decrease in surface roughness of the coatings with an increasing amount of MSZ content is due to the fact that MSZ powder particles have a perfectly round morphology as compared to BC powder. Furthermore, MSZ powder has a particle size smaller than Ni powder and has a larger range of finer particles as compared to Ni powder [29].

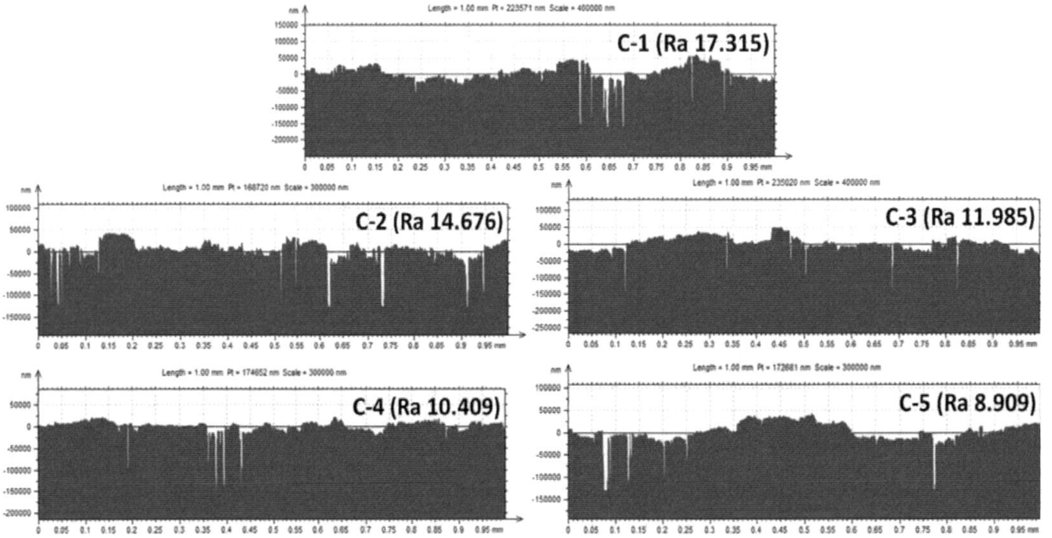

Figure 4. Surface roughness profiles of the as sprayed five individual coatings.

3.3. Microhardness measurements

Vickers microhardness (HV0.3) of the polished surfaces of five individual coatings were measured and drawn in order to simulate the coating system as shown in figure 5. It was observed that the microhardness changed gradually through the five layers of the coating in such a way that a sharp mismatch could be avoided. Therefore, it is expected that the absence of a sharp mismatch in microhardness due to the presence of compositionally gradient layers might cause suppression in the development of otherwise high residual stresses at the coating layers interfaces thus increasing the expected life of the TBC system [30-31].

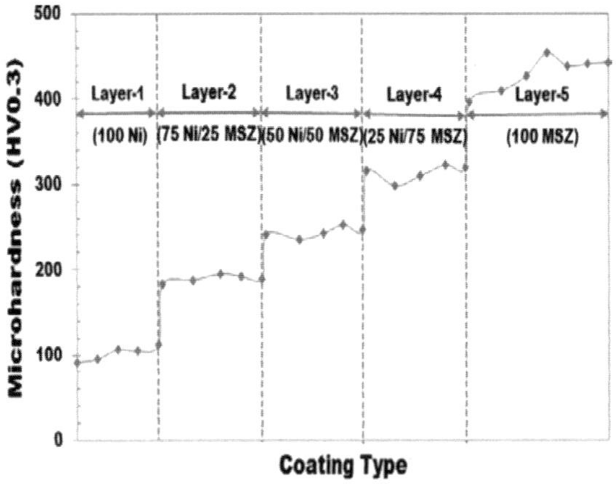

Figure 5. Microhardness distributions of the five individual layers comprising the TBC system

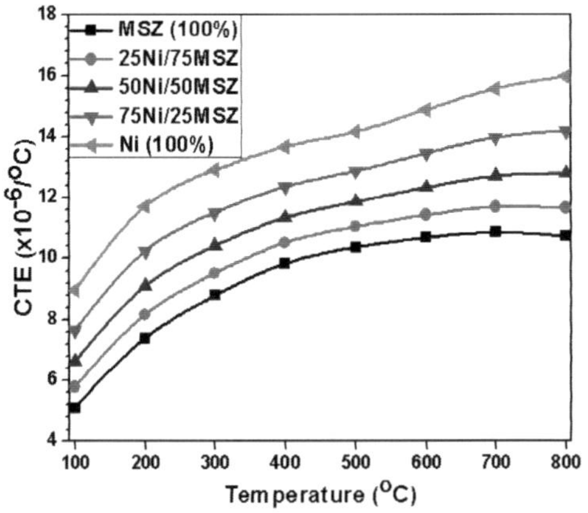

Figure 6. CTE values plotted as a function of temperature for five layered TBC system.

3.4. Measurement of CTE

Linear CTEs of the five individual freestanding coatings as a function of temperature from room temperature to 800°C on heating were plotted against temperature in order to simulate the conditions of the coating system as shown in figure 6. It was observed that coating C-5 exhibited the lowest linear expansion rates whereas coating C-1 exhibited the largest linear expansion rates. While for the cermet coatings, there was a gradual increase in the linear expansion rates with the reduction of MSZ content in the coatings (C-4, C-3, & C-2, respectively). It can also be seen from figure 6 that the CTE values of the five coated layers gradually increased thus steadily reducing the difference between their thermal expansion coefficients. This gradual variation in thermal expansion coefficients between the five deposited layers is expected to further reduce the level of residual stresses within the FG-TBC system.

3.5. Oxidation Behavior

SEM micrographs in as sprayed and oxidized conditions of the coating system after 100 hours at 800 °C are shown in figure 7. As can be seen from figure 7 (b), a network of continuous cracks was formed on the surface of the coating after oxidation. These multiple cracks were formed because the top zirconia layer became thinner as the graded layers were introduced and this top layer started behaving like a thin single-layered zirconia coating. Such behavior has been reported for single layered zirconia coatings [32]. During oxidation, sintering, healing or bridging of these microcracks as marked by arrows can be seen as shown in figure 7 (c) [33]. Cross sectional SEM micrograph of the coating system within the top zirconia layer is shown in figure 7 (d). Both horizontal and vertical cracks were seen within the top zirconia layer.

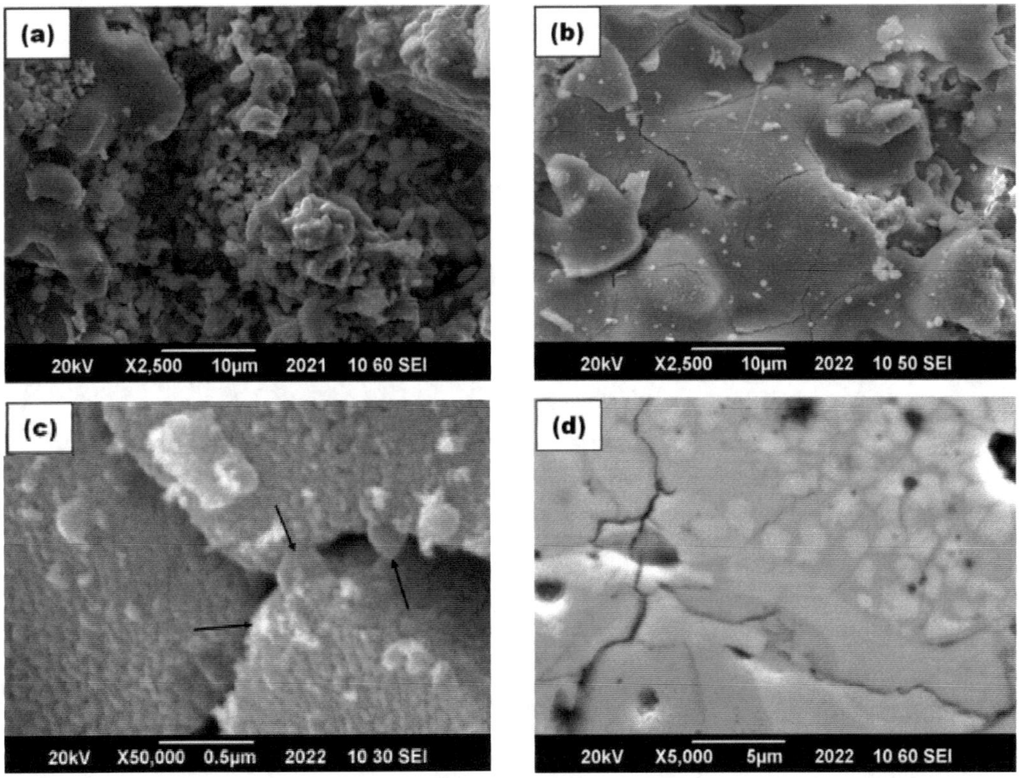

Figure 7. SEM micrographs of the coating system, (a) in as-sprayed condition, (b) after oxidation showing multiple surface cracks, (c) sintering of inter-splat microcracks and (d) cross section showing vertical and horizontal cracks within the top MSZ layer.

Weight gain per unit area (mg/cm^2) as a function of oxidation time of the TBC system is shown in figure 8 (a). The oxidation kinetic curve showed a sharp increase in weight gain for the initial 11 hours followed by an extensive period of slower weight gain rate throughout the oxidation period. The oxidation rate was calculated to be 0.0676 mg/cm^2/hour. Oxidation kinetics was further examined to find out what oxidation rate law is more relevant to be followed by the coating. The relationship between the specific weight gain and the square root of time is shown in figure 8 (b). It was observed that the data almost fell along a straight line indicating that the oxidation kinetics obeyed a parabolic rate law according to the equation [34]:

$$\Delta m/A = (k_p t)^{1/2} + C$$

where $\Delta m/A$ is the weight gain per unit surface area (mg/cm^2) at anytime t (h), k_p is the parabolic rate constant ($mg^2cm^{-4}h^{-1}$) and C is a constant.

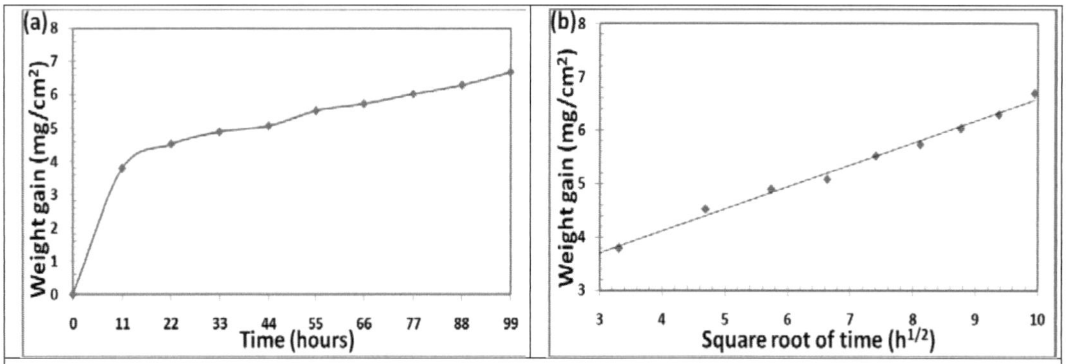

Figure 8. Oxidation kinetics after 99 hours at 800 °C, (a) weight gain as a function of time and (b) weight gain as a function of square root of time.

XRD patterns of the coating system oxidized at 800 °C for 99 hours and in as sprayed condition are shown in figure 9. Both the as sprayed coating and after oxidation treatment consisted of cubic zirconia (c-ZrO_2), monoclinic zirconia (m-ZrO_2) and cubic magnesia (c-MgO) phases. However, the intensity of the c-ZrO_2 peaks in the oxidized coating was found to be higher than those in the as sprayed coating. This may be due to the fact that at high temperature, through diffusion of the stabilizer (MgO), the metastable c-ZrO_2 phase transformed to the stable c-ZrO_2 phase [35].

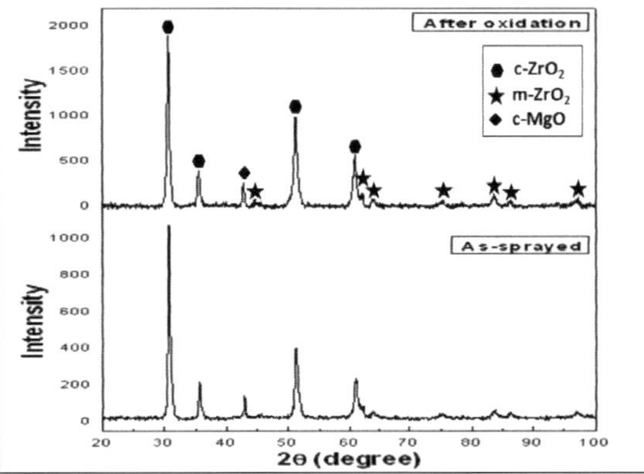

Figure 9. XRD patterns of the five layer coating system in as sprayed condition and after oxidation at 800 °C for 99 hours.

4. Conclusions

Microhardness changed gradually across the coating layers indicating no distinctive boundaries across the coating. There was a gradual increase in the CTE of the five coated layers thus steadily reducing the difference between their thermal expansion coefficients. Multiple surface cracks were formed in oxidized coating. Oxidation kinetics was found to follow a parabolic rate law. No phase change was observed after oxidation however intensity of the c-ZrO_2 peaks was found to be increasing after oxidation as compared to as-sprayed condition.

5. Acknowledgements
M. N. Baig would like to thank the Higher Education Commission of Pakistan (HEC) for providing financial assistance and PIEAS, ATCOP and NUST for use of facilities.

6. References

[1] Zhang Y Y, Deng H X, Shi H J, Yu H C and Zhong B 2012 *Surf. Coat. Technol.* 206 2977-85

[2] Okazaki M, Yamagishi S, Yamazaki Y, Ogawa K, Waki H and Arai M 2013 *Int. J. Fatigue* 53 33-39

[3] Wang X, Tint S, Chiu M and Atkinson A 2012 *Acta Mater.* 60 3247-58

[4] Bargraser C, Mohan P, Lee K, Yang B, Suk J, Choe S and Sohn Y H 2012 *Mater. Sci. Eng.* A 549 76-81

[5] Ramachandran C S, Balasubramanian V, Ananthapadmanabhan P Vand Viswabaskaran V 2012 *Ceram. Int.* 38 4081-96

[6] Yugeswaran S, Kobayashi A and Ananthapadmanabhan P V 2012 *J. Eur. Ceram. Soc.* 32 823-34

[7] Kumar S and Cocks A C F 2012 *J. Mech. Phys. Solids.* 60 723-49

[8] Bhatty M B, Khalid F A and Khan A N 2012 *J. Therm. Spray Technol.* 21 121-31

[9] Liang J J, Matsumoto K, Kawagishi K and Harada H 2012 Surf. Coat. Technol. 207 413-20

[10] Hardwicke C U and Lau Y-C 2013 *J. Therm. Spray Technol.* 22 (5) 564-76

[11] Song S-H, Xiao P and Weng L-Q 2005 *J. Eur. Ceram. Soc.* 25 1167-73

[12] Pan C and Xu X 2003 *Surf. Coat. Technol.* 162 194-201

[13] Darzens S and Karlsson A M 2004 *Surf. Coat. Technol.* 177-178 108-12

[14] Khor K A, Chia C T and Gu Y W 2000 *Mater. Sci. Eng.* A 279 166-71

[15] Sampath S, Jiang X Y, Matejicek J, Prchlik L, Kulkarni A and Vaidya A 2004 *Mater. Sci. Eng.* A 364 216-31

[16] Mahesh R A, Jayaganthan R and Prakash S 2008 *Mater. Sci. Eng.* A 475 327-35

[17] Choi H, Yoon B, Kim H, Lee C 2002 *Surf. Coat. Technol.* 150 297-308

[18] Portinha A, Teixeira V, Carneiro J, Martins J, Costa M F, Vassen R and Stoever D 2005 *Surf. Coat. Technol.* 195 245-51

[19] Khan A N and Qureshi I N 2009 *J. mater. Process. technol.* 209 488-96

[20] Khan A N, Khan S H, Ali F and Iqbal M A 2009 *Comput. Mater. Sci.* 44 1007-12

[21] Cho S H, Park S B, Lee J H, Hur J M and Lee H S 2012 *Mater. Chem. Phys* 131 743-51

[22] Begley M R and Wadley H N G2012 Acta. Mater. 60 2497-2508

[23] Khoddami A M, Sabour A and Hadavi S M 2007 *Surf. Coat. Technol.* 201 6019- 24

[24] Kawasaki A and Watanabe R 2002 *Eng. Frac. Mech.* 69 1713-28

[25] Khor K A and Gu Y W 2000*Mater. Sci. Eng.* A 277 64-76

[26] Schulz U, Peters M, Bach Fr-W and Tegeder G 2003 *Mater. Sci. Eng.* A 362 61-80

[27] Khor K A, Dong Z L and Gu Y W 1999 *Mater. Lett.* 38 437-44

[28] Kwon J-Y, Jung S-I, Lee S Y, Lee P H, Lee J H, Jung Y G, Paik U, Cho H and Chang J C 2010 *Prog. Org. Coat.* 67 95-101

[29] Demirkıran A S and E. Avcı 1999 *Surf. Coat. Technol.* 116–119 292-95

[30] Khor K A, Dong Z L and Gu Y W 2000 *Thin Solid Films* 368 86-92

[31] Khor K A and Gu Y W 2000 *Mater. Sci. Eng.* A 277 64-76

[32] Kokini K, Dejonge J, Rangaraj S and Beardsley B 2002 *Surf. Coat. Technol.*154 223-31

[33] Eriksson R, Brodin H k, Johansson S, Östergren L and Li X-H 2012 *Surf. Coat. Technol.* xxx

[34] Khor K A and Gu Y W 2000 *Thin Solid Films* 372 104-13

[35] Prawara B, Yara H, Miyagi Y and Fukushima T 2003 *Surf. Coat. Technol.* 162 234-41

IOP Conf. Series: Materials Science and Engineering **60** (2014) 012054 doi:10.1088/1757-899X/60/1/012054

Optimization and characterization of adhesion properties of DLC coatings on different substrates

B Waseem, S Alam, M Irfan, M Shahid, B D Soomro, S Hashim and R Iqbal
Pakistan Institute of Technology for Minerals & Advanced Engineering Materials (PITMAEM) PCSIR Labs. Complex, Lahore, Pakistan

E-mail: pitmaem.lhr@gmail.com

Abstract. The Diamond Like Carbon coatings (DLC) are gaining prime importance in the field of surface engineering especially cutting tools technology. The self lubricating property of these coatings makes them unique among other coatings like TiN, TiAlN, CrN etc. Unlike other coatings, DLC coatings give better surface finish and their self lubrication reduces the wear of a part to large extent. In present work, different substrates were selected to study the wear and adhesion behavior of DLC coatings. The coating was produced by physical Vapor Deposition (PVD) technique and the adhesive properties of DLC coatings were analyzed under ambient conditions using nano Scratch testing. Scanning electron microscope (SEM) was used to observe the scratches and their mechanisms.

1. Introduction

The great versatility of carbon materials arises from the strong dependence of their properties on the ratio of (graphite-like) sp^2 to (diamond-like) sp^3 bonds in the films whereas amorphous carbon with a high fraction of sp^3 bonds is named as diamond-like carbon (DLC) [1, 2]. DLC coatings have been the subject of extensive study since years because of their potential applications in science and technology. In science, the diamond-like materials preparation and their expanded applications is a subject yet to be explored. In technology, the diamond-like carbon (DLC) films are currently used in the production of hard coatings because of their extreme hardness, chemical inertness and excellent tribological, corrosion and adhesive properties [3]. The most important industrial applications of DLC films are wear-resistant coatings for computer hard-disks and optical coatings for the infrared and for the visible light range [4]. Moreover DLC coatings are being applied upon eye glass frames, pen and cell phone casings and chains because of their scratch resistance and pure black color [5].

Physical vapor deposition (PVD) is fundamentally a vaporization coating technique, involving transfer of material up to atomic level. It is an alternative process to electroplating technique. The process is similar to chemical vapor deposition (CVD) except that the raw materials or precursors, i.e. the material that is going to be deposited starts out in solid form, whereas in CVD, the precursors are introduced to the reaction chamber in the gaseous state [6].

Wear is the deformation and loss of a material in moving contact with another material. Wear and erosion of a film can be measured by: weight loss, material transfer and wear scars. It is extremely sensitive to the application, temperature, material etc [7-8]. The mechanical properties of films are important in their response during subsequent processing and to mechanical stresses [9-10].

The general method for the measurement of adhesion is the scratch test in which a diamond indenter is drawn across the coated surface and cohesive or adhesive behavior of the coating is assessed. The load on the coated surface during the scratch test can be increased progressively and the load, at which the adhesive failure of the coating appears, is known as the critical load (Lc). Usually three critical loads are defined during a scratch test. First critical load (Lc1) defines the start of initial cracking in the coating surface. Second and third critical loads (Lc2 & Lc3) define the start of buckling by merging the cracks and complete failure of the coating respectively.

In the present study, steel substrates of different types were selected and DLC (diamond like coating) coating was deposited by the process of physical vapor deposition (PVD). Scratch tests were carried out at each coated surface to evaluate the critical loads in order to check which substrate gives the best adhesion with DLC coating. The scratch tracks were also analyzed under the optical microscope to check the adhesion mechanism.

2. Experimental Work

2.1 Sample Preparation

Die steel (D2), high strength steel (HSS), stainless steel (SS), high carbon steel, mild steel (MS), hard steel (H13) were chosen as substrates. Degreasing of the substrates was carried out through ultrasonic cleaning in which samples were dipped in CCl_4 for 5 minutes. After that, samples were thoroughly washed with distilled water to remove the traces of CCl_4 . The samples were then placed in cleano gel solution at a temperature of 60°C for about 10 minutes. Finally the samples were cleaned in ultrasonic bath with acetone followed by careful transfer to coating chamber avoiding any contamination.

2.2 Chemical composition of substrates

The chemical composition analysis of the substrates ware carried out by optical emission spectrometer. The following table shows the chemical composition of the substrates used.

Table. 1 Chemical composition of the substrates used for coating

	%C	%Mn	%P	%S	%Si	%Fe
Medium carbon steel	0.403	1.388	0.066	0.022	0.312	97.680
Mild Steel	0.193	0.664	0.084	0.027	0.258	98.470
Die steel	2.358	0.563	0.061	0.021	0.643	91.415
H13	0.951	0.280	0.004	0.016	0.233	97.184

2.3 Deposition of Coating

PLATIT π^{80} coating system was used to deposit DLC coatings upon the above described substrates. The system is fully automated and is controlled by the process computer. The coating deposition was conducted in a flowing pure nitrogen atmosphere. The cleaned coupons were mounted on carrousel substrate holders which rotate continuously around the vertical central axis at a speed of 12 rpm. During deposition, a negative DC bias of -70 V was applied to the substrates. The parameters are shown in Table 2.

Table 2. Technological Parameters of PVD Process

Parameter	Coat	Substrate Temperature (°C)	Nitrogen Pressure, (Pa)	Accelerating Voltage, (Volts)	Coat Period (min.)
Set Value	DLC	450	1.5	70	120

2.4 Characterization of Coatings

The adhesion of the coating was assessed by micro scratch testing. Scratch testing was carried out by a Rockwell diamond indenter of radius 0.1 mm. Three scratches were placed at each load to ensure the accuracy of the results. Loads of 20 and 25N were used to place the scratches at each coated substrate. Multiple scratches of incremental loads were placed in which an area of the coated surface was scratched three times at the same incremental load. Pre-scan and post scans were also carried out that describe the surface profile before and after the scratch testing respectively. The initial load was set at 0.03N at which, pre-scan and post-scan were carried out. The loading rate was adjusted at the half of final load in each test. The length of all scratches was 3mm and the acquistic emission sensitivity was set at 7. All scratches were run at a speed of 1.5 mm/min. The contact force was kept same as the scanning load i.e: 0.03N.

These scratches were then analyzed through an optical microscope to define the critical loads and observe the mechanism of failure as well.

3. Results and Discussions

Figure.1 shows the optical micrographs of a complete multi scratch on DLC coated die steel at 20N load. The first micrograph depicts the region of first critical load where initial cracking in the coated surface appears. Second and third micrographs depict the appearance of first adhesive failure and the end of scratch respectively. It can be observed that there is no appreciable wear on the coated surface, which is the indication of good adhesion of the coating with the substrate.

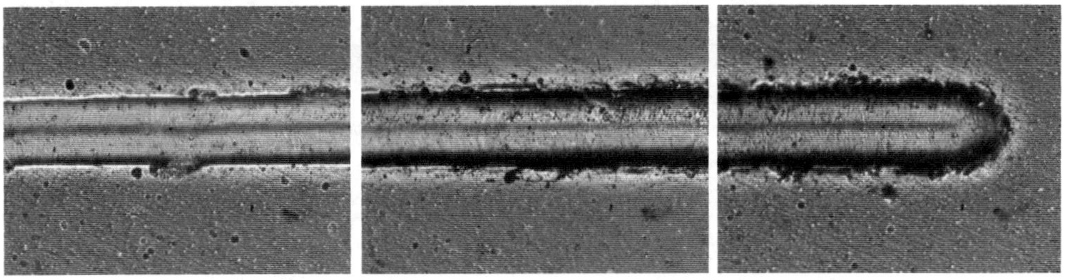

Figure 1. Optical micrographs of die steel at 20N (20x)

Figure 2 shows the optical micrographs of die steel at a load of 25N and same magnification. It is clear from the micrographs that the wear is much more obvious in this case. But it is evident from the micrographs that there is still no appreciable wear on the coated surface and the traces of coating are still there even at the highest load. The dense cracking within the scratch is the result of excessive amount of load and also because the same area was scratched three times.

IOP Conf. Series: Materials Science and Engineering **60** (2014) 012054 doi:10.1088/1757-899X/60/1/012054

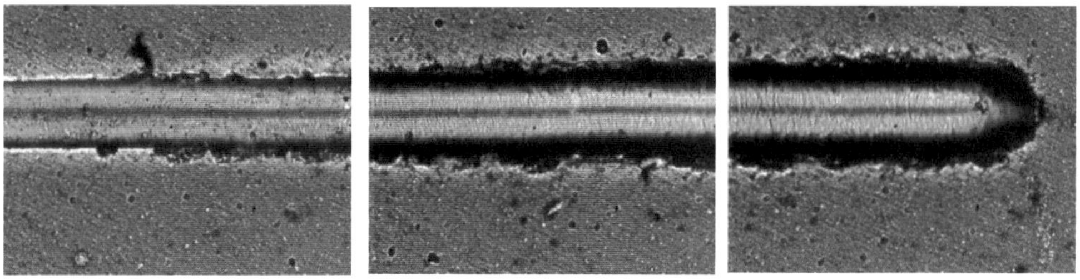

Figure 2. Optical micrographs of scratch at Die steel at 25N 20X

The figure-3 shows the comparison of the penetration depth curves for different substrates. It is obvious that mild steel shows the largest penetration depth. This is because the mild steel is the softest among all the substrates and the indenter easily penetrated the surface. Least penetration is shown by high strength steel (HSS) and die steel. This fact is associated to their high hardness and hence showed least penetration depth.

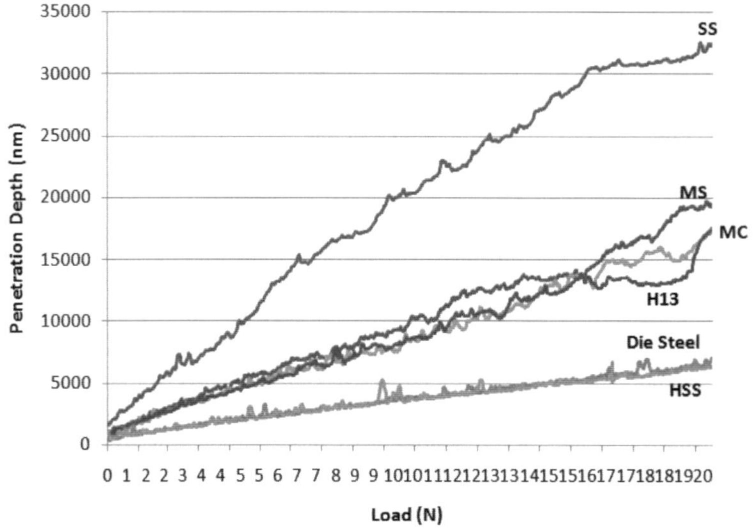

Figure 3. Load versus penetration depth curves of different substrates at 20N load

The figure 4 shows the acoustic emission curves for different substrates. It can be noticed that die steel and high strength steel show maximum acoustic emission, while H13 steel shows least acoustic emission. This might be because of the brittleness of hard steels, due to which coating produces the heavy noise when it delaminates.

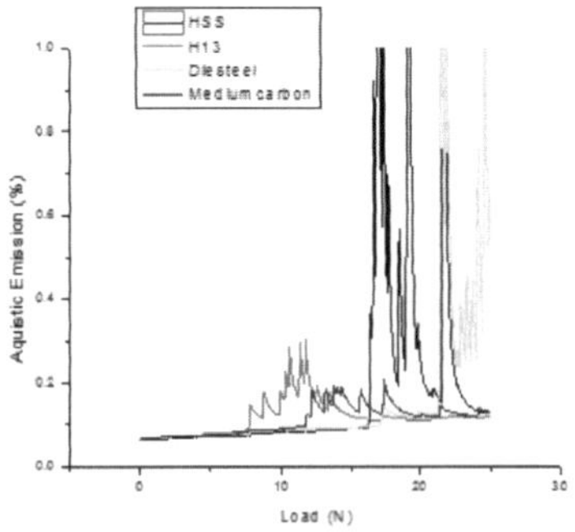

Figure 4. Load versus acoustic emission curves of different substrates at 25N load

4. Conclusions

The comparisons of penetration depth and acoustic emission curves shows that high strength steel and die steel show the best adhesion with DLC coating. Least adhesion is shown by the mild steel substrate. It can be concluded that harder substrates show better adhesion properties with DLC coatings than the softer ones. The softness of the substrate is also associated to lesser acoustic emission intensity and vice versa.

5. References

[1] Robertson, Journal of Material Sciences & Engineering, 3(2002)129.

[2] Jacob W, Moller W. Appl. Phys. 63(1993)1771.

[3] H. O. Pierson, *"Hand Book of Carbon, Graphite, Diamond and Fullerenes"* Noyes, Park Ridge, New Jersy. (1993).

[4] S. S. Tinchev et al., J. Opto. *Adv. Mater.* 8 (2006) 308.

[5] Kenneth Holmberg and Allan Matthews, "*Coatings Tribology: Properties, Mechanisms,* Techniques and Applications in Surface Engineering". Elsevier, (2009)

[6] http://www.azom.com/article.aspx?ArticleID=1558

[7] Bunshah, R.F, *Selection and use of wear Tests for coatings,* ASTM STP 769, p. 3 (1982)

[8] Kato, K. "*Micro wear mechanisms of coatings*" Surf. Coat. Technol., 76/77: 469 (1995)

[9] Pulker, H.K., "*Mechanical Properties of Optical Thin Films,*" Thin Solid Films, 89:191 (1982)

[10] Campbell, D.S., "*Mechanical Properties of Thin Films,*" Handbook of Thin Film Technology, (L.I Maisell and R. Glang, eds.), Ch. 12, Mc-Graw Hill (1970).

[11] R Chattopadhyay, "*Surface Wear- Analysis, Treatment, & Prevention*", SM International, Materials Park, OH, USA, 2001, ISBN 0-87170-702-0.

Influence of Nanostructure Geometry on Electronic Properties

A Tavkhelidze
Ilia State University, Cholokashvili Ave. 3-5, Tbilisi 0162, Georgia

E-mail: avtotav@gmail.com

Abstract. Recently, new quantum features have been studied in the area of nanostructured layers. It emerges that properties of nanostructures depend not only on their size but also on their geometry. Particularly, a nanograting (NG) on the surface of the thin layer imposes additional boundary conditions on electron wave function and forbids some quantum states. Density of quantum states reduces. Unlike conventional quantum well, state density per volume, is reduced in the case of NG layer. This leads to changes in electronic properties. Electrons, rejected from forbidden quantum states, have to occupy states with higher energy. In the case of semiconductor layers, electrons rejected from the valence band have to occupy empty quantum states in the conduction band. Such increase in conduction band electron concentration can be termed as geometry-induced doping or G-doping. G-doping is equivalent to donor doping from the point of view of the increase in electron concentration. However, there are no ionized impurities. This preserves charge carrier scattering to the intrinsic semiconductor level and increases carrier mobility with respect to the donor-doped layer. As rejected electrons occupy quantum states with the higher energy, the chemical potential of NG layer increases and becomes NG size dependent. We regard a system composed of NG layer and an additional layer on the top of the NG forming periodic series of p–n junctions. In such system, charge depletion region develops inside the top of NG and its effective height reduces, becoming a rather strong function of temperature T. Consequently, T-dependence of chemical potential magnifies and Seebeck coefficient S increases. Calculations show one order of magnitude increase in the thermoelectric figure of merit ZT relative to bulk material. In the case of metal layers, electrons rejected from forbidden quantum states below Fermi energy, occupy quantum states above Fermi energy. Fermi energy moves up on energy scale and work function (WF) reduces. WF reduction was observed in thin amorphous Au films grown on thermally oxidized Si substrates. WF was measured using Kelvin Probe and PEEM microscope. WF reduction depended on film internal structure and NG sizes. Maximum reduction in WF of 0.56 eV has been obtained.

1. Introduction

Developments in nanotechnology allow fabrication of densely packed, periodic structures [1-4]. Recently, ultra-short period nanopore arrays and nanogratings have been obtained by block copolymer lithography [1, 2]. Another method for nanograting (NG) fabrication is a multi-beam interference lithography. Using these techniques, gratings with 10-nm [3] and even sub-10 nm pitch [4] were fabricated. At the same time, it has been shown that NG dramatically improves thermoelectric [5] and electron emission properties [6] when grating pitch becomes comparable with electron de Broglie wavelength. This is owing to special boundary conditions imposed by NG on electron wave function. Supplementary boundary conditions forbid some quantum states, and density of quantum states (DOS) reduces (in all bands). Electrons rejected from NG-forbidden quantum states have to occupy empty states with a higher energy. Fermi energy increases and electronic properties of NG layer change.

Qualitatively, DOS reduction can be described by the de Broglie wave reflection. Let us regard a solid/vacuum interface with NG on the surface (figure 1).

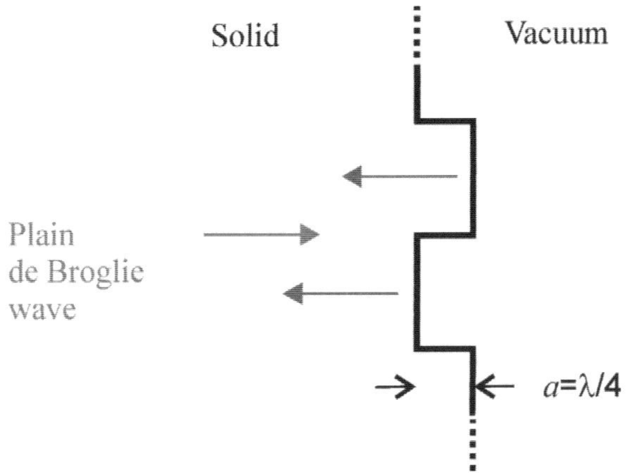

Figure 1. Reflection of plane de Broglie wave from the NG surface.

Let us consider an electron with energy $E < E_F$ (where E_F is Fermi energy), moving towards the border of the solid as a planar de Broglie wave. It will reflect back from the interface as the electron does not have enough energy to leave the solid. There will be two reflected waves. One reflected from the top of the indent and another from the bottom of the indent. If the indentation depth $a = \lambda / 4$, where λ is electron de Broglie wavelength, two reflected waves will interfere destructively resulting in no reflected wave. Consequently, an electron of certain energy cannot reflect back from the surface. On the other hand, the electron cannot leave solid to vacuum as it does not have enough energy to overcome the potential barrier ($E < E_F$). It cannot reflect under arbitrary angle as it is in the Fermi gas in which all quantum states are already occupied. Therefore, all possible final quantum states for that particular electron are forbidden. As all the final states are forbidden, the initial state with wave vector $k = \pi / 2a$ is also forbidden. Obviously, all states with wave vector component $k_Z = (\pi / a)(p + 1/2)$, where p is integer are also forbidden. As a result, DOS near the interface is reduced with respect to the plain solid.

Qualitative description of DOS reduction is demanding. The calculation requires solving of the time-independent Schrödinger equation in the NG geometry. Mathematically, there is no difference between DOS reduction and the electromagnetic (TM) mode depression [7, 8]. Helmholtz equation and Dirichlet boundary conditions are used in both cases. Unfortunately, there is no exact analytical solution for NG geometry. The approximate analytical expression known as Weyl's formula [9, 10] allows calculation of DOS using a ratio of layer surface and volume. Perturbation method [11] has been used to find an approximate analytical expression for *G* in NG geometry. DOS for electromagnetic modes in resembling geometries has been numerically calculated in the literature related to the Casimir effect [12].

Influence of periodic structures on electronic properties has been studied in related geometries such as periodic curved surfaces [13, 14], nanotubes [15], cylindrical surfaces with non-constant diameter [16] and strain-driven nanostructures [17].

2. Semiconductor nanograting layers

In the case of semiconductor materials, electrons rejected from the valence band (VB) occupy empty quantum states in the conduction band (CB). Electron concentration n in the CB increases which can be termed as geometry-induced electron doping or G-doping. G-doping is equivalent to donor doping from the point of view of increase in n and Fermi energy. However, there are no ionized impurities. This maintains charge carrier scattering to intrinsic semiconductor level and increases carrier mobility with respect to the donor-doped layer of same electron concentration. G-doping is temperature independent as it originates from layer geometry and no ionized impurities are involved.

Other methods of doping without impurities are well known modulation doping and recently introduced polarization doping [18]. Both are 2D in their nature. However, a 3D approach of modulation doping was introduced in [19] to improve thermoelelectric characteristics of nanocomposites [20].

Electron confinement to the NG layer is needed to obtain G-doping. The layer can be made free standing but in practice it is usually sandwiched between wide band gap layers. NG layer thickness is fundamentally limited by the requirement of having quantum properties. However, thin layers have low optical absorption. Layer thickness also limits lateral charge and heat transport.

Figure 2 shows a cross section of a single NG layer. The grating has depth a and period $2w$. To make a comparison, we choose reference layer as plain layer with thickness H such that it has the same cross section area.

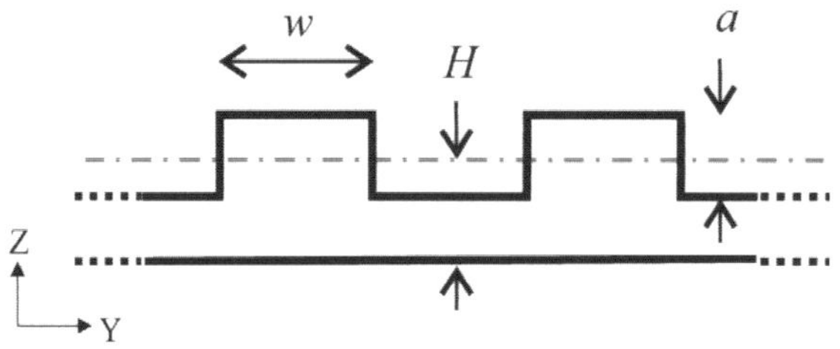

Figure 2. Cross section of nanograting layer.

Nanograting imposes additional boundary conditions on electron wave function and forbids some quantum states. The DOS in energy $\rho(E)$ reduces [21] with respect to the reference well

$$\rho(E) = \rho_0(E)/G. \qquad (1)$$

Where $\rho_0(E)$ is the DOS (number of quantum states within the unit energy region and in the unit volume) in a reference well and $G = G(H, w, a) > 1$ is the geometry factor. With increasing NG depth DOS reduces.

Nanograting reduces DOS in all bands. Electrons rejected from the NG-forbidden quantum states have to occupy empty quantum states with higher E. In a semiconductor, electrons rejected from the VB have to occupy empty (and not forbidden by NG) energy levels in CB (figure 3). Electrons reject from low energy levels and occupy high energy ones. During this process, Fermi energy increases from $E_F^{(0)}$ to E_F. To simplify the presentation, in figure 3, we presume that T=0 (T is absolute temperature) and energy levels are equidistant on the energy scale (geometry-induced energy level shift is also ignored).

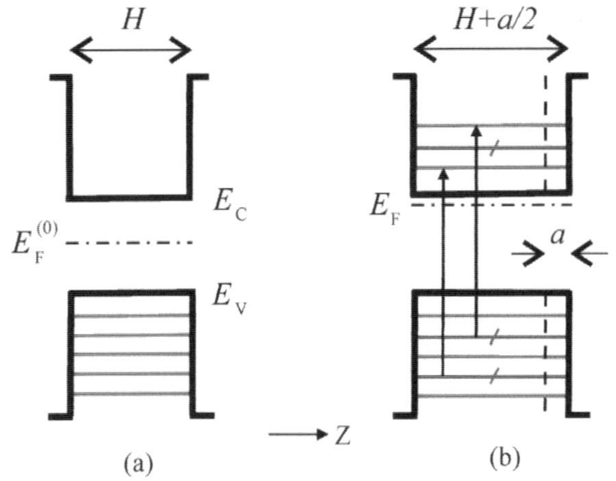

Figure 3. Energy diagrams of a) reference semiconductor quantum well and b) NG layer. Green lines depict occupied energy levels. Red lines depict NG-forbidden energy levels.

We will employ (1) to calculate the DOS and investigate G-dependence of n and E_F. The density of the NG-forbidden quantum states is

$$\rho_F(E) = \rho_0(E) - \rho_0(E)/G = \rho_0(E)(1-1/G). \qquad (2)$$

To determine the number of rejected electrons n_r (per unit volume), (2) should be integrated over the electron confinement energy region.

$$n_r = \int_{con} dE\, \rho_F(E) = (1-1/G) \int_{con} dE\, \rho_0(E) = (1-1/G)\, n_{con}. \qquad (3)$$

Where, $n_{con} = \int_{con} dE\, \rho_0(E)$. n_r Here, we assume that electron confinement takes place only in narrow energy intervals inside which G is energy-independent. In most cases, thin layer is grown on a semiconductor substrate and confinement energy regions are band offsets.

There are distinctions and similarities between the NG forbidden quantum state and a hole. The state is forbidden by the boundary conditions and cannot be occupied. However, it is not forbidden in an irreversible way. If the boundary conditions change (e.g., owing to charge depletion), then it can recombine with the electron (like the hole recombines with the electron). As the NG forbidden state is confined to the boundary conditions (macroscopic geometry), it is not localized in the lattice and cannot move like a hole.

3. Electronic properties of multiple homojunction nanograting layer

To obtain electron confinement in multiple NG layers transitional barrier layers are required. We regard cases of wide band gap material forming homojunction. Confinement regions originate from barrier layer donor doping.

Here, we investigate G-doping in multiple NG layers. Such layers are quasi-3D and have improved optoelectronic and thermoelectric characteristics with respect to a single layer. Multiple NG layers consist of a replicated structure including main and barrier layers forming a series of hetero or homo junctions. Main layer is thicker than a barrier layer and plays a leading role in carrier transport and optical absorption. Barrier layer forms electron confinement energy regions. It also contributes to carrier transport and optical absorption. Both main and barrier layers have NG geometry.

One of the objectives of this work is to calculate n and in multiple NG layers and find out how they depend on layer thicknesses and material properties. First, we introduce G-doping in a single NG layer. Next, we calculate n and in a homojunction multilayer structure. Subsequently, we calculate the same for a heterojunction multilayer structure. Finally, the possibility of realization of such structures and their advantages for optoelectronic and thermo-electric devices are discussed. An analysis was made within the limits of parabolic band, wide quantum well and degenerate electron gas approximations.

We begin from calculating n in a homojunction NG layer, as it is relatively straightforward. Homojunction is formed between intrinsic main layer and donor-doped barrier layer. Let us consider a case when all layers have a plain geometry (NG is not formed yet). Figure 4a shows an energy diagram of such structure (typically obtained by homoepitaxal growth) [22]. Charge depletion regions form inside the both i-type main layer and n-type barrier layer. Band edges curve, shaping confinement regions in the main layer VB and in the barrier layer CB. Electrons with energies $E_V - \Delta E_{con}^{(0)} < E < E_V$ are confined to main layers and ones with energies $E_C - \Delta E_{con}^{(0)} < E < E_C$ are confined to barrier layers. Here, we assume that both main and barrier layers are relatively thick so that electron wave functions do not overlap and we can ignore the mini band formation.

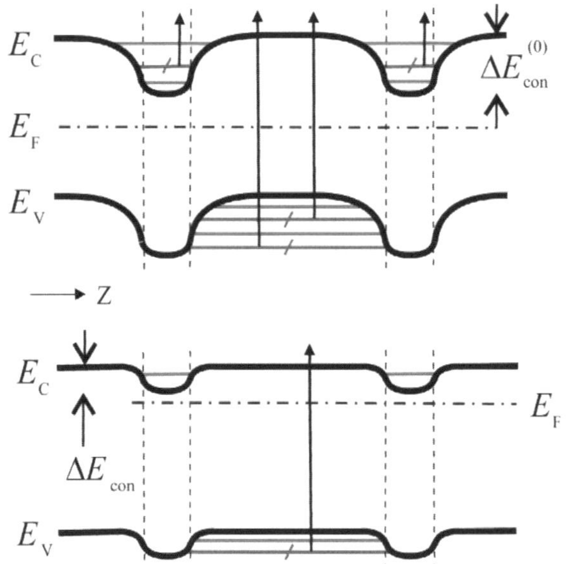

Figure 4. Homojunction structure energy diagrams: a) before NG formation, b) after NG formation. Green lines depict occupied energy levels. Red lines depict NG-forbidden energy levels.

3.1. Influence of Nanograting

Next, we fabricate NG on the surface of all layers simultaneously (fabrication of such structure will be discussed later in this work). Nanograting forbids some energy levels (red lines) and induces G-doping depending on geometry factor value. We introduce two geometry factors G_m and G_b for the main and barrier layers separately (layers have different geometry factors owing to a different thickness).

If $G_m > 1$ some electrons reject from the VB of main layer (figure 4). If $G_b > 1$ some electrons reject from the CB of barrier layer and occupy higher energy levels in the same band. In the beginning, let us regard the case $G_m > 1$ and $G_b = 1$. As $G_m > 1$, the Fermi level increases in main layer, which is equivalent to shifting band edges down on the energy scale. As $G_b = 1$, the bang edges remain intact in barrier layer. With the downward shift in main layer, VB confinement region will shrink and fewer electrons will reject to the CB. This will continue until some equilibrium value of rejected electrons is achieved. In other words, band edge downward shift provides negative feedback to the G-doping process. Next, let us regard the case $G_m > 1$ and $G_b > 1$. Here, barrier layer band edges also shift down. Owing to the downward shift in barrier layer, confinement energy regions widen. This increases the number of rejected electrons in main layer. In other words, barrier layer G-doping provides positive feedback to main layer G-doping. Finally, we have a combination of negative feedback originating from main layer and positive feedback origination from barrier layer. In the case of overall negative feedback G-doping of two layers will saturate and equilibrium value of ΔE_{con} will be attained (figure 4b). Last depends on G_m, G_b and semiconductor material parameters.

3.2. Electron confinement energy region

To find ΔE_{con} we have to calculate electron concentrations in both main and barrier layers first. Applying (2) to the main layer we find the number of rejected electrons

$$n_r = (1 - 1/G_m) \int_{E_V}^{E_V - \Delta E_{con}} dE\, \rho_{0V}(E) \quad (4)$$

where, $\rho_{0V}(E)$ is the initial (to NG) DOS in the main layer VB. Let us regard the case when the rejected electron number is much more than the initial electron number in CB $n_r >> n_0$. Then, the CB electron concentration will be merely n_r. Further, we use the following well known expression for valence band DOS to calculate n_r [23]

$$\rho_{0V}(E) = 2\pi^{-1/2} N_V (k_B T)^{-3/2} (E - E_V)^{1/2} \quad (5)$$

Here, N_V is the effective VB density of states. Inserting (5) in (4) and integrating, choosing zero of energy scale as $E_V = 0$ we find

$$n_r = (4/3\sqrt{\pi}) N_V (1 - 1/G_m)(\Delta E_{con} / k_b T)^{3/2} \quad (6)$$

Further, we find Fermi levels in both layers using corresponding electron concentrations. According to the well known expression for degenerate semiconductor [24] Fermi level in main layer can be found from

$$E_F - E_C^{(m)} = k_B T \left[\ln (n_r / N_C) + 2^{-3/2} (n_r / N_C) \right] \quad (7)$$

where, N_C is the effective CB density of states. CB electron concentration of barrier layer does not change during NG formation, as there are no electrons transferred from the VB to the CB. However, Fermi level in the barrier layer moves up on an energy scale owing to NG-induced reduction in N_C

$$E_F - E_C^{(b)} = k_B T\left[\ln(G_b n_d / N_C) + 2^{-3/2}(G_b n_d / N_C)\right] \tag{8}$$

Here, n_d is barrier layer initial electron concentration (obtained by donor doping). Energy diagram (figure 4) shows that $\Delta E_{con} = E_C^{(m)} - E_C^{(b)}$. Subtracting (7) from (8) and inserting in the last expression we find

$$(\Delta E_{con}/k_B T) = \ln(G_b n_d / n_r) + 2^{-3/2}\left[(G_b n_d - n_r)/N_C\right]. \tag{9}$$

Inserting (6) in (9) and depicting $\eta \equiv \Delta E_{con}/k_B T$ we get following nonlinear equation for η

$$\eta + \frac{3}{2}\ln\eta + 2^{-3/2}\frac{4}{3\sqrt{\pi}}\frac{N_V}{N_C}\frac{G_m - 1}{G_m}\eta^{3/2} - \ln\frac{3\sqrt{\pi}G_m G_b n_d}{4N_V(G_m - 1)} - 2^{-3/2}\frac{G_b n_d}{N_C} = 0 \tag{10}$$

Equation (10) was solved numerically to find ΔE_{con} using two geometry factor and n_d, N_C, N_V values for Si and GaAs materials.

3.3. Electron concentration and Fermi level

Numerical solutions of (10) were inserted in (6) to find n_r. Next, we found the Fermi level of the main layer by inserting values of n_r in (7). Formula (8) was used to find a Fermi level position in barrier layer. Table 1 shows the results for geometry factors G_m=1. 02, $G_b = 1.1$ and T=300 K.

Table 1. Electron concentration and Fermi levels in main and barrier layers for Si and GaAs materials. Energy was measured from corresponding layer CB edge.

Material	n_d [cm^{-3}]	ΔE_{con} [meV]	n_r [cm^{-3}]	$E_F^{(m)}$ [meV]	$E_F^{(b)}$ [meV]
	3 x 10^{18}	45	6 x 10^{17}	103	58
Si	1 x 10^{19}	64	1 x 10^{18}	89	25
	3 x 10^{19}	87	1 x 10^{18}	77	-10
	1 x 10^{20}	126	2.8 x 10^{18}	62	-64
	1 x 10^{18}	47	3.2 x 10^{17}	3.4	-44
GaAs	3 x 10^{18}	86	8 x 10^{17}	-29	-115
	5 x 10^{18}	119	1.3 x 10^{18}	-52	-172
	1 x 10^{19}	197	2.8 x 10^{18}	-110	-297

For barrier donor doping levels 10^{18}-10^{20} cm^{-3} and G values rather close to unity (G_m=1.02, $G_b = 1.1$) main layer G-doping levels of 10^{17}-10^{18} cm^{-3} were obtained. We chose this range of G-dropping levels, as they are frequently used in applications. Higher and lower G-doping levels can be obtained as well. Table 1 shows that one order higher donor doping n_d (10^{20} cm^{-3}) is needed in Si with respect to GaAs (10^{19} cm^{-3}) to obtain a G-doping level 2.8 10^{18} cm^{-3}. It is convenient to give an interpretation of this result using terms of negative and positive feedback. Negative feedback in main layer is weaker for Si material as it has higher n_0 value with respect to GaAs. Owing to the high value of n_0, band edges shift down less rapidly with increasing n_r, following well known logarithmic dependence

$\Delta E_F = k_B T \ln(n_r / n_0)$. As ΔE_F is positive in our case, higher n_0 results in weaker negative feedback. Weaker negative feedback in the main layer requires weaker positive feedback in the barrier layer to obtain equilibrium. Positive feedback is weaker in highly donor-doped barrier layers (as shifting band edges require more rejected electrons). This explains why higher n_d is required in Si barrier layer with respect to GaAs.

As figure 4 shows there are potential barriers for electrons in the CB and holes in the VB of multilayer NG. They are affecting carrier transport in Z direction. However, barrier height is ΔE_{con} which is of the order of a few $k_B T$ as table 1 indicates (except for very high level of donor doping. $k_B T$ =26 meV for T=300 K). Charge carriers easily overcome such obstacles (thermionic emission) and their influence can be ignored. In the case of high values of ΔE_{con} barriers may block carrier transport in Z direction. To avoid this, high level of G-doping should be obtained not by increasing donor doping of barrier layers but by increasing of G_m and G_b values.

During numerical calculations, we kept $E_F^{(m)}$ and $E_F^{(b)}$ in proximity of corresponding CB edges to stay within the limits of degenerate electron gas approximation (so that (7) and (8) are valid). For lower doping levels, other expressions should be used instead of (7) and (8).

3.4. G-doping for high electron mobility applications

Realization of G-doping is attractive from the point of view of high mobility applications. Let us use a multi junction solar cell [25] as an example to estimate G-doping improvement of its characteristics. In this device, window, emitter and tunnel junction layer doping levels are roughly 10^{18} cm^{-3}. At this level of donor doping, ionized impurities reduce electron mobility by a factor of 4 in GaAs [26] (at T=300 K) and by the factor of 10 in Si [27]. Most types of solar cells use transparent conductive oxides with doping levels of 10^{20}-10^{21} cm^{-3}. At this donor doping level ionized impurities reduce electron mobility by the factor of 30-50 in GaAs. Thus, using G-doping in these layers can dramatically improve characteristics of solar cells.

Multiple NG layers can be realized by epitaxial growth on the top of lead grating, previously formed on the substrate. Such structure has been modelled [28] and fabricated [29] on GaAs substrate using epitaxial growth technique. Layers, grown using this technique, have diverse geometry compared to single NG layer. They have gratings from both sides and interfaces resemble sine instead of square. With increasing number of layers sine amplitude reduces and finally one gets plain layers. Geometry factor and G-doping level reduce with increasing of layer number. It is challenging to obtain high G values or considerable DOS reduction in both NG and alike geometries. However, as table 1 shows, G-doping required values are quite close to unity (especially for the main layer). The obtaining of such values in multiple NG systems seems to be straightforward.

The above analysis was made using the assumption of G energy independence (3). Usually, electron confinement energy regions are small (tens or hundreds of meV) and this assumption is valid. However, the NG layer can be made free standing or grown on very wide band gap substrate. In these cases, confinement energy intervals are much wider and care should be taken when assuming G energy independence.

4. Thermoelectric properties of nanograting layers

Thermoelectric materials are characterized in terms of the dimensionless figure of merit *ZT*. Here, *T* is the temperature and *Z* is given by $Z = \sigma S^2 / (\kappa_e + \kappa_l)$, where S is the Seebeck coefficient, σ is electrical conductivity, κ_e is an electron gas thermal conductivity, and κ_l is lattice thermal conductivity. The difficulty in increasing *ZT* is that materials having high S usually have low σ. When σ is increased, it leads to an increase in κ_e, following Wiedemann–Franz law, and *ZT* does not improve much. Another approach is to eliminate the lattice thermal conductivity by introducing vacuum nanogap between the hot and cold electrodes [30–32] and using electron tunnelling. Cooling in such designs was observed in [33] and theoretically studied in [34]. However, vacuum nanogap devices appear extremely difficult to fabricate. In this work, we present a solution that allows large enhancement of S without changes in σ, κ_e, and κ_l. It is based on having a series of p–n junctions on the top of the NG.

It is another objective of this work is to calculate *ZT* of NG with series of p–n junctions and compare it with Z_0T of reference nanograting (RNG), in which $G \neq G(T)$, and bulk material. The reduced figure of merit ZT/Z_0T is calculated and μ dependences are presented for such traditional thermoelectric materials as Si and Ge. Analysis was made within the parabolic bands approximation and the abrupt junction's approximation. Since only heavy G-doping was considered, we neglected the hole contribution in transport.

4.1 Charge and heat transport in the NG layer with p-n junctions

Here, we propose a system composed of NG layer and an additional layer on the top of the ridges forming periodic series of p+–n+ junctions (Figure 5). In such system, charge depletion region

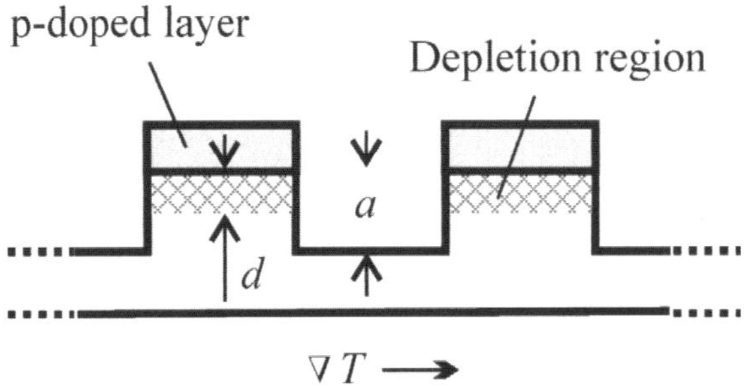

Figure 5. NG with series of periodic junctions grown at the top of the ridges.

develops inside the ridges and effective ridge height reduces, becoming a rather strong function of temperature T. Consequently, T-dependence of chemical potential μ magnifies and Seebeck coefficient S increases. Depletion region width $d(T)$ depends rather strongly on temperature. The ridge effective height $a_{eff}(T) = a - d(T)$ and consequently the geometry factor G becomes temperature-dependent, $G = G(T)$. All parameters of NG layer including μ become stronger functions of *T* than it will be in an NG layer without the junctions. Seebeck coefficient and thermoelectric figure of merit increase.

Analysis was made using the Boltzmann transport equations. First, we calculate $\nabla\mu$ for the system of NG layer and junctions and express it as

$$\nabla\mu = \nabla\mu_0 + \nabla\mu_J \qquad (11)$$

Here, $\nabla\mu_0$ is the chemical potential gradient in an RNG (NG layer without p-n junctions) and $\nabla\mu_J$ is introduced by the junctions. Next, we insert $\nabla\mu$ in Boltzmann transport equations and calculate S for the system of NG layer and junctions, expressing it as $S = S_0 + S_J$, where S_0 is Seebeck coefficient of RNG and S_J is introduced by the junctions.

Cross-section of the system of NG and periodic junction is shown in figure 5. We assume that there is a temperature gradient ∇T in the Y-dimension. Consequently, depletion depth depends on the Y-coordinate, and geometry factor gradient ∇G appears in the Y-direction. The presence of ∇G and ∇T modifies the electron distribution function and causes electron motion from the hot side to the cold side. This motion is compensated by thermoelectric voltage. Let us write Boltzmann transport equations [35] for the system of NG layer and periodic series of junctions

$$J = \mathbf{L}^{11}(\mathcal{E} + \nabla\mu/e) - \mathbf{L}^{12}\nabla T \text{ and } J^Q = \mathbf{L}^{21}(\mathcal{E} + \nabla\mu/e) - \mathbf{L}^{22}\nabla T \qquad (12)$$

Here, J is the electric current density, J^Q is the heat current density, $\mathbf{L}^{ij}$ are coefficients, $\mathcal{E}$ is the electric field, and e is the electron charge. Within the parabolic band approximation, $\mathbf{L}^{ij}$ are the functions of integrals of type [36].

$$\Omega^{(\alpha)}(E) = \int_{-\infty}^{+\infty} dE\left(-\partial f_0/\partial E\right)\rho(E)\tau(E)\mathrm{v_y}^2(E-\mu)^\alpha \qquad (13)$$

where f_0 is the electron distribution function, $\tau(E)$ is the electron lifetime, $\mathrm{v_y}$ is the electron velocity in the Y-direction, and α = 0, 1, 2. Let us find how NG affects $\Omega^{(\alpha)}(E)$. The NG does not change dispersion relation and consequently $\mathrm{v_y}$. It reduces density of states (3D density) G times, i.e. $\rho(E) = \rho_0(E)/G$ and increases the transport lifetime G times [21], i.e., $\tau(E) = G\tau_0(E)$ (the later follows from Fermi's golden rule). Here, $\rho_0(E)$ and $\tau_0(E)$ are the density of states and carrier lifetime, respectively, in the case $a = 0$. Consequently, product $\rho(E)\tau(E)\mathrm{v_y}^2$ in the NG layer is the same as in the conventional wide quantum well of the same width (3D case) and in the bulk material. This product does not depend on G and consequently it becomes independent of depletion depth. The NG changes the distribution function f_0, since it increases μ. The NG influences integrals $\Omega^{(\alpha)}(E)$ by changing μ alone. Therefore, the integrals are the same as in the bulk material having the corresponding chemical potential

$$\mathbf{L}^{ij} = \mathbf{L}_0^{ij}. \qquad (14)$$

where $\mathbf{L}_0^{ij}$ are the coefficients of bulk material having the value of μ obtained by the conventional doping (for instance).

Further, we have to find μ for the system of NG layer and junctions and insert it in (12) together with (14). The chemical potential of degenerated semiconductor ($-2 < \mu^* < 2$ where $\mu^* = \mu/k_BT$) can be written as [37] (note that frequently Fermi energy is used instead of chemical potential in semiconductor literature)

$$\mu = k_BT\left[\ln(n/N_{NG}) + 2^{-3/2}(n/N_{NG})\right] \qquad (15)$$

where n is the electron concentration and N_{NG} is the effective conduction band density of states of NG layer. In the case of heavy G-doping ($n >> n_i$, where n_i is intrinsic concentration), we can use (3) for n. Density of states reduces G times in NG layer, i.e. $N_{NG} = N_C / G$, where N_C is the CB effective density of states for bulk semiconductor. Inserting this and (3) in (15), we get

$$\mu = k_B T \left\langle \ln\left[n_{con}(G-1)/N_C\right] + 2^{-3/2}\left[n_{con}(G-1)/N_C\right] \right\rangle \tag{16}$$

Further, we rewrite (16) as

$$\mu = \mu_0 + k_B T \left\langle \ln\left[(G-1)/(G_0-1)\right] + 2^{-3/2}\left(G - G_0\right) n_{con}/N_C \right\rangle \tag{17}$$

where G_0 is constant. Introduction of $\mu_0 \equiv \mu(G_0)$ defines reference material as n+-type semiconductor with electron concentration of $N_D = n_{CON}(G-1)$, or RNG layer having constant geometry factor $(\partial G/\partial T) = 0$ and $G = G_0$. Next, we calculate the gradient of (10) taking into account that $N_C \propto T^{3/2}$. The result is

$$\nabla\mu = \nabla\mu_0 + \theta\nabla T \tag{18}$$

$$\text{where } \theta \equiv k_B\left[\ln\frac{G-1}{G_0-1} - \frac{1}{2}\xi\left(G-G_0\right) + T\left(\frac{1}{G-1} + \xi\right)\frac{\partial G}{\partial T}\right] \tag{19}$$

and $\xi \equiv 2^{-3/2} n_{con}/N_C$. Inserting (18) and (14) in Boltzmann equations, we find charge and heat currents in the system of NG layer + Junctions

$$J = \mathbf{L}_0^{11}\left(\mathcal{E} + \nabla\mu_0/e\right) - \left(\mathbf{L}_0^{12} - \mathbf{L}_0^{11}\theta/e\right)\nabla T \text{ and} \tag{20a}$$

$$J^Q = \mathbf{L}_0^{21}\left(\mathcal{E} + \nabla\mu_0/e\right) - \left(\mathbf{L}_0^{22} - \mathbf{L}_0^{21}\theta/e\right)\nabla T. \tag{20b}$$

Further, Seebeck coefficient, electrical conductivity, and electron gas heat conductivity can be found from (20a) and (20b) in a conventional way.

$$S = \left(\mathbf{L}_0^{12} - \mathbf{L}_0^{11}\theta/e\right)/\mathbf{L}_0^{11} = S_0 - \left(\theta/e\right) \tag{21}$$

$$\kappa_e = \mathbf{L}_0^{22} - \mathbf{L}_0^{21}(\theta/e) - \left(\mathbf{L}_0^{12} - \mathbf{L}_0^{11}\theta/e\right)\mathbf{L}_0^{21}/\mathbf{L}_0^{11} = \kappa_{e0} \tag{22}$$

$$\sigma = \mathbf{L}_0^{11} = \sigma_0 \tag{23}$$

Electrical and thermal conductivity in the system of NG layer and junctions remain unaffected (with respect to RNG layer or bulk material having the same μ value) by a series of junctions, and S change according to (21). To calculate θ and then S, we have to find $\partial G/\partial T$ first (19).

Such performance of transport coefficients has following physical interpretation. The geometry factor G is a rather strong function of temperature. Consequently, T-dependence of chemical potential magnifies and Seebeck coefficient, having approximate value $S \approx -(1/e)(\partial\mu/\partial T)$, increases. Electrical conductivity $\sigma = \Omega^{(0)}$ is G independent since, the reduction of state density leads to a proportional increase of lifetime, according to Fermi's golden rule (resulting in (14)). In a degenerate electron gas (n+-type semiconductor) Wiedemann–Franz law works well. Thus, the thermal conductivity κ_e follows σ and is also G independent.

4.2. Seebeck coefficient of NG layer with p–n junctions

It is reasonable to calculate the Seebeck coefficient of the NG layer with junctions relative to RNG layer, in which geometry factor is constant $(\partial G/\partial T) = 0$. This will allow comparison of dimensionless figure of merit *ZT* with Z_0T, using similarity of electric (23) and heat conductivities (22). In (17)–(19) G_0 is arbitrary. To simplify the comparison, we choose G_0 so that NG layer with junctions and reference RNG layer have the same μ value. Equation of chemical potentials leads to $G_0 = G$ (16). Inserting this in (19) and further inserting the obtained result in (21), we obtain for the Seebeck coefficient of the NG layer with junctions

$$S = S_0 - \frac{k_B T}{e}\left(\frac{1}{G-1} + \xi\right)\left(\frac{\partial G}{\partial T}\right) \tag{24}$$

3D Seebeck coefficient of reference RNL does not differ from the Seebeck coefficient of the bulk material, and within the parabolic band approximation it is equal to [38]

$$S_0 = \frac{k_B}{e}\left[\frac{r+5/2}{r+3/2}\frac{F_{r+3/2}(\mu^*)}{F_{r+1/2}(\mu^*)} - \mu^*\right] \tag{25}$$

Here, r refers to scattering parameter and it is assumed that electron lifetime $\tau(E) \propto E^r$, $\mu^* = \mu/k_B T$ is the reduced chemical potential, and $F(\mu^*)$ are Fermi integrals. In the case of no impurities (G-doping) and low energies, acoustic phonons are responsible for electron scattering and $r = 0$. It should be noted here that NG layer has G times more width [21] with respect to the conventional layer. As 3D $\rho(E)$ in wide quantum wells tends to $\rho(E)$ of the bulk material of the same width, using (28) for RNG is a good approximation. However, in the case of thin layers oscillatory behaviour of transport coefficients should be considered [39].

To find the ratio Z/Z_0, we use relations (22) and (23) and obvious relation between lattice thermal conductivities $\kappa_l = \kappa_{l0}$, all together leading to

$$Z/Z_0 = (S/S_0)^2 \tag{26}$$

Figure 6 shows the dependence $S_0(\mu)$ according to (25). The $S(\mu)$ in the same figure is determined by first, calculating S (G) by inserting $\partial G/\partial T$ (which was calculated in [5]) in (24), and then (16)

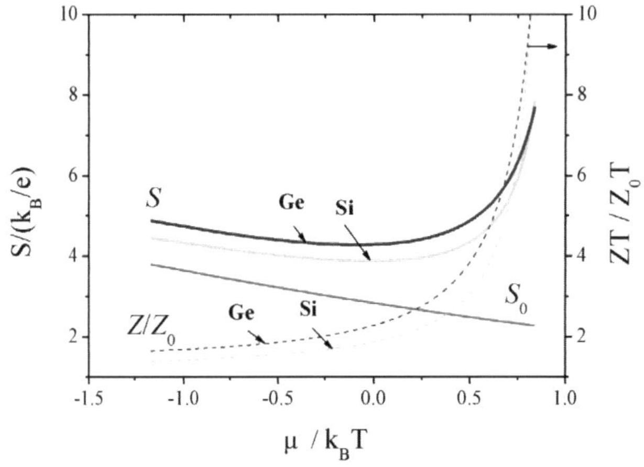

Figure 6. $S(\mu)$ – solid lines, $S_0(\mu)$ – dotted line, and $Z(\mu)/Z_0(\mu)$ – dashed lines (belong to right Y-axis). Dependences $S(\mu)$ and $Z(\mu)/Z_0(\mu)$ are calculated for Si and Ge materials.

was used to convert X-axis so that $S(\mu)$ was obtained from $S(G)$. The ratio $Z(\mu)/Z_0(\mu)$ in the same figure is calculated according to (26). We present μ dependences as they allow the understanding of possible μ ranges, within which real devices can operate without changing the sign of $(\partial G/\partial T)$ (see reference 5) and consequently S sign. Traditional thermoelectric materials Si and Ge are chosen as examples. Dependences are plotted for the following parameters: Material Si, n_{CON} = 4.5×10^{18} cm^{-3}, a = 20 nm, $\gamma = 3.3$ (γ is reduced chemical potential of p-type layer), $\beta = 0.1$ ($\beta = a_{eff}/d$), $T = 300$ K; Material Ge, n_{CON} = 1.4×10^{18} cm^{-3}, a = 35 nm, $\gamma = -1$, $\beta = 0.25$, $T = 300$ K. For both Si and Ge, we choose n_{CON} so that, for value G=10, μ was close to the optimum value $\mu_{OPT}/k_BT = r + 1/2$ [38]. It should be noted here that NG layer exhibits quantum properties at G times more widths with respect to the conventional quantum well.

Figure 6 shows large enhancement of figure of merit in p+–n+ junction NG layer both for Si and Ge materials. The dependences are quite identical for materials having rather different band gaps (1.12 eV for Si and 0.66 eV for Ge). This shows that despite the band gap value entering expressions for depletion depth (Reference 5) and $(\partial G/\partial T)$, almost similar $S(\mu)$ and $Z(\mu)/Z_0(\mu)$ dependencies can be obtained by matching such parameters as β and γ.

5. Metal nanograting layers

In the case of metal layers, electrons rejected from forbidden quantum states below Fermi energy, occupy quantum states above Fermi energy. Fermi energy moves up on energy scale and work function reduces.

Low work function (WF) electrodes [40] are essential for cold emission and room temperature operation of thermionic [41, 42] and thermotunnel [43-45] energy converters and coolers. Such devices consist of plain emitter and collector electrodes, separated by the thin vacuum gap. Electrons absorb heat energy inside the emitter electrode, overcome the vacuum barrier (or tunnel through it) and release heat energy inside the collector electrode. Electrodes require materials with work function $e\phi = 0.2 - 0.4$ eV (here, e is electron charge and ϕ potential). For most metals $e\phi$>4 eV, and only some compounds show $e\phi$=2-3 eV. This is one order of magnitude higher than required. WF values of about 1 eV were obtained in sophisticated systems like Mo-Cs and Ag-O-Cs. However, these types of electrodes have a short lifetime even in good vacuum conditions. To overcome difficulties, quantum mechanical tunnelling was utilized. Tunnelling through vacuum nanogap allows sufficiently large currents from the electrodes, having relatively high $e\phi$ values. It was found that image force reduces the potential barrier and increases tunnelling current, giving a cooling power of 100 W/cm^2 for $e\phi \approx 1$eV [43] in mixed thermotunnel and thermionic regime. Electrons were filtered by collector coating, to increase the cooling coefficient [44] and confor mal electrode growth technology was developed [45]. However, the vacuum nanogap device appears extremely difficult to fabricate as it requires an electrode spacing of 5-10 nm. Earlier, WF reduction in Au, Nb and Cr metal NG layers was observed experimentally [46]. If an $e\phi < 1$eV electrode could be obtained, poor thermionic regime can be realized at increased electrode spacing. Fabrication of wide vacuum gap, using conformal electrode technology, is much more straightforward. Low WF energy converters are essential for heat to electricity direct conversion in the temperature range 400-1000 K in which waste heat is available from the combustion sources. Low WF coolers are essential for integrated circuit cooling and other applications where low weight and ecological purity are major requirements.

Here, we calculate $e\phi$ in the metal NG layer, grown on semiconductor and metal substrates, and find out how it depends on NG dimensions and electron confinement. In the beginning we calculate $e\phi$ values in the metal NG layer forming a Schottky barrier or ohmic contact with a semiconductor substrate. Next, we calculate $e\phi$ in the metal NG layer grown on a metal substrate. Finally, the possibility of realization of such electrodes using conventional thermionic and other materials is discussed. An analysis was made within the limits of parabolic band, wide quantum well and degenerate electron gas approximations.

5.1. Work function of metal NG layer grown on a semiconductor substrate

To maintain the uniform vacuum nanogap over the whole area, electrodes should have plane geometry and smooth surface. The simplest solution is to use semiconductor substrate as an electrode base [47] and grow a thin metal NG layer on it.

The Energy diagram of the metal film grown on n+ type semiconductor substrate is shown in Figure 7a. We begin from the case when the difference between metal initial work function $e\phi_0$ and

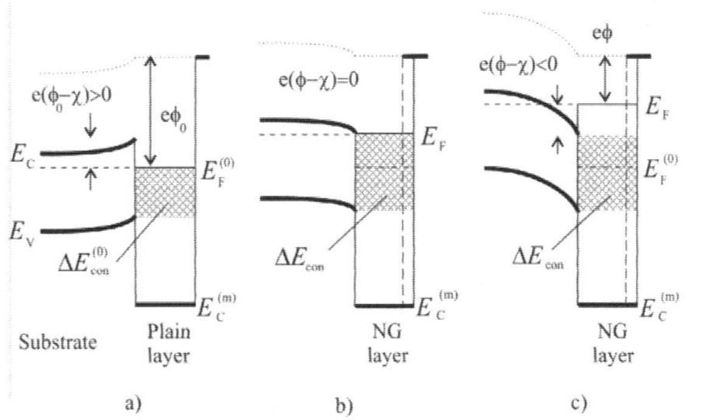

Figure 7. Energy diagram of metal-semiconductor contact in the case $e(\phi_0 - \chi) > 0$, a) without NG or G=1, b) with NG at $G = G_0$, c) with NG at $G > G_0$. Hatch depicts a confinement energy region.

semiconductor electron affinity $e\chi$ is positive, i. e. $e(\phi_0 - \chi) > 0$. We presume that electron gas in metal layer is degenerate, so that all quantum states are occupied below Fermi energy E_F and are empty above E_F. Further, suppose that the semiconductor band gap $E_g = E_C - E_V$ is wide enough, so that $e(\phi_0 - \chi) < E_g$ is satisfied. Then, electrons having energies within the region

$$\Delta E_{con}^{(0)} = E_g - e(\phi_0 - \chi) \qquad (27)$$

are confined to the metal film. Within $\Delta E_{con}^{(0)}$, quantum states for electrons are filled in the metal film and forbidden in the semiconductor substrate. Next, we fabricate ridges on the metal film surface (Fig. 7b). Additional vertical lines near the vacuum boundary in figures 7b and 7c depict extra boundary conditions, imposed by the ridges. Owing to quantum state depression, supplementary boundary conditions reduce the quantum state density G times, within the energy region $e(\chi_m - \chi) - E_g < E < e(\chi_m - \chi)$, having a width $\Delta E_{con} = E_g$. Here, we measure energies from the metal conduction band bottom $E_C^{(m)}$ and depict $e\chi_m \equiv E_F^{(0)} + e\phi_0$ as a metal conduction band width. Electrons, rejected from the forbidden quantum states within $\Delta E_{con}^{(0)}$, occupy the empty states above $E_F^{(0)}$. The Fermi level and the semiconductor band edges E_C and E_V move up on the energy scale. At the same time, $e\phi$ decreases and the energy region $\Delta E_{con} = E_g - e(\phi - \chi)$ (Eq. (5)) gets extended. As $e\phi$ decreases, the semiconductor band edge curving follows the $e(\phi - \chi)$ value and reverses its curving direction at $e(\phi - \chi) = E_C - E_F^{(0)}$.

The $e\phi$ value was calculated using electron number conservation in the NG layer conduction band. The number of electrons rejected from the forbidden quantum states was equal to the number of electrons accommodated above the initial Fermi level n_{ACC}. The rejected electron number, according to Eq. (3) was

$$n_r = (1 - 1/G) \int_{con} dE\, \rho_0(E) = (1 - 1/G) \int_{e(\chi_m - \chi) - E_g}^{e(\chi_m - \varphi_0)} dE\, \rho_0(E). \qquad (28)$$

In Eq. (28), we take into account that electrons were rejected from the energy interval $e(\chi_m - \chi) - E_g < E < e(\chi_m - \phi_0)$. They were accommodated in the interval $E_F^{(0)} < E < E_F$ or $e(\chi_m - \phi_0) < E < e(\chi_m - \phi)$. If n_r was low, the last interval fit within the ΔE_{con} (Fig. 7b) where the density of states is reduced Eq. (1). The accommodate electron number was equal to the number of empty quantum states between $E_F^{(0)}$ and E_F

$$n_{INS} = \int_{E_F^{(0)}}^{E_F} dE\, \rho(E) = \int_{e(\chi_m - \varphi_0)}^{e(\chi_m - \varphi)} dE\, \rho(E) = (1/G) \times \int_{e(\chi_m - \phi_0)}^{e(\chi_m - \phi)} dE\, \rho_0(E) \qquad (29)$$

Using condition $n_r = n_{ACC}$ and putting in $\rho_0(E) \propto E^{1/2}$, which is true within the limit of parabolic band approximation, we found after integration and simplification that

$$\underset{e\phi \geq e\chi}{e\phi} = e\chi_{\rm m} - e\left[G(\chi_{\rm m} - \phi_0)^{3/2} - (G-1)(\chi_{\rm m} - \chi - E_{\rm g}/e)^{3/2}\right]^{2/3} \tag{30}$$

Here, we use 3D density of states $\rho_0(E) \propto E^{1/2}$ (wide quantum well). Further, if $n_{\rm REJ}$ was high enough, $e(\phi - \chi)$ changed sign as shown in figure 7c. Here, $n_{\rm r}$ was calculated using Eq. (28) again. However, in this case, electrons were accommodated in quantum states from two different energy intervals. The primary interval $e(\chi_{\rm m} - \phi_0) < E < e(\chi_{\rm m} - \chi)$ was above $E_{\rm F}^{(0)}$ and within the energy region $\Delta E_{\rm con}$ where the DOS was reduced, and the secondary interval $e(\chi_{\rm m} - \chi) < E < e(\chi_{\rm m} - \phi)$ was above $E_{\rm F}^{(0)}$ and out of $\Delta E_{\rm con}$. The two intervals differ by the DOS. The DOS was equal to Eq. (1) in the primary interval and $\rho_0(E)$ in the secondary one. Consequently, $n_{\rm ACC}$ was

$$\int_{E_{\rm F}^{(0)}}^{E_{\rm F}} dE\,\rho(E) = \int_{e(\chi_m-\phi_0)}^{e(\chi_m-\phi)} dE\,\rho(E) = (1/G)\int_{e(\chi_m-\phi_0)}^{e(\chi_m-\chi)} dE\,\rho_0(E) + \int_{e(\chi_m-\chi)}^{e(\chi_m-\phi)} dE\,\rho_0(E) \tag{31}$$

Further, equalizing Eq. (28) and Eq. (31) and repeating the above described steps we found

$$\underset{e\phi<e\chi}{e\phi} = e\chi_{\rm m} - e\left\langle(\chi_{\rm m} - \phi_0)^{3/2} + (1 - G^{-1})\left[(\chi_{\rm m} - \chi)^{3/2} - (\chi_m - \chi - E_{\rm g}/e)^{3/2}\right]\right\rangle^{2/3} \tag{32}$$

Next, we consider the case when the difference between $e\phi_0$ and $e\chi$ was negative, i.e, $e(\phi_0 - \chi) < 0$ from the beginning (Fig. 7a). Here, $\Delta E_{\rm con}^{(0)} = E_{\rm g}$. When ridges were fabricated (Fig. 3b), the rejected electrons reduced $e\phi$. As $e\phi$ reduced $e(\phi - \chi)$ also reduced and got even more negative. Semiconductor bands curve in the direction of $E_{\rm C}^{(\rm m)}$. At the same time, $\Delta E_{\rm con}$ remains constant. Electrons were rejected from the interval $e(\chi_{\rm m} - \chi) - E_{\rm g} < E < e(\chi_{\rm m} - \chi)$, and their number was calculated by applying this interval to Eq. (4). Electrons were accommodated in the interval $e(\chi_{\rm m} - \phi_0) < E < e(\chi_{\rm m} - \phi)$ and their number was

$$\int_{e(\chi_m-\phi_0)}^{e(\chi_m-\phi)} dE\,\rho_0(E). \tag{33}$$

Using $n_{\rm r} = n_{\rm ACC}$, and repeating the above steps, we found that Eq. (32) is true for the case $e(\phi_0 - \chi) < 0$ as well.

Finally, Eq. (32) allows the calculation of geometry factor value G_0, at which an ideal ohmic contact was obtained or $e(\phi - \chi) = 0$. Inserting this in Eq. (8), we found

$$G_0 = \left[(\chi_{\rm m} - \chi)^{3/2} - (\chi_m - \chi - E_{\rm g}/e)^{3/2}\right]\left[(\chi_{\rm m} - \phi_0)^{3/2} - (\chi_m - \chi - E_{\rm g}/e)^{3/2}\right]^{-1} \tag{34}$$

Analysis of the metal NG layer/semiconductor contact shows that in the case $e(\phi_0 - \chi) \geq 0$, metal NG layer $e\phi$ was calculated using Eq. (30) if $G \leq G_0$ (or $e\phi \geq e\chi$) and Eq. (32) if $G > G_0$ (or $e\phi < e\chi$).

For the opposite case, $e(\phi_0 - \chi) < 0$, $e\phi$ was calculated using Eq. (32) for all G values (note that G>1 by definition).

6. Material pairs for nanograting layer coated electrode

The thickness of a conventional metal quantum well is only 1-5 nm [48]. NG fabrication on such thin layer surface seems complicated. However, owing to reduced quantum state density, metal NG layer retains quantum properties at G times more thickness (as found in Ref. 21) with respect to a conventional metal quantum well. This makes NG fabrication straightforward. In some cases, the layer thickness can be increased even more than G times. As figure 7 shows, confined electrons have energies $E < E_F^{(0)} - K_B T$ (here, K_B is Boltzmann constant and $K_B T \approx 25$ meV << $E_F^{(0)}$ for T=300K). For these energies, Fermi-Dirac distribution function $f(E, E_F, T) \approx 1$. Consequently, electron scattering on phonons and lattice defects is limited. Such scattering requires the exchange of a small portion of energy with the environment, which is quantum mechanically restricted as all nearby quantum states are occupied. The mean free path of such electrons is very large and is limited only by structural defects, such as grain boundaries. In epitaxial films, such electrons allow phase coherence at large distances (in transverse direction) and metal NG thickness can be increased. In this work, we do not concentrate on thermal processes in the metal NG layer by following reasons. Transport coefficients in the NG have the same values as in a conventional NG layer (having the same value of E_F) [5]. In vacuum device, the temperature gradient is applied mainly to the vacuum gap and only significant thermal process inside the electrode is Joule heating [49]. Such heating inside the metal NG layer is negligible. We regard only n+ semiconductor substrates in which, the difference $E_C - E_F$ is low and, heat absorption or release at the metal/semiconductor interface is insignificant.

Regular semiconductor materials can be used as base substrate. Thermionic and thermotunnel converters are high current low voltage devices. Consequently, the main limitation is electrical conductivity. Equations (30) and (32) show that the wide band gap material allows more reduction in $e\phi$. Unfortunately, it is problematic to achieve low resistivity in such materials. The most promising seems GaN in which relatively low resistivity was obtained. Other possible substrate materials are GaAs and Si. Table 2 shows possible donor concentrations N_D for GaN [50] , Si, GaAs [23] and corresponding

Table 2. Parameters of electrode base materials. The P_d and ∇T are given for 1 mm thick substrate

Substrate material	E_g (eV)	$e\chi$ (eV)	N_D (cm^{-3})	r (Ωcm)	P_d (mW) at 10 Acm^{-2}	κ (Wcm^{-1}K^{-1})	∇T (K) at 10 Wcm^{-2}
Si	1.12	4.05	8 10^{20}	2 10^{-4}	2	1.6	0.6
GaAs	1.42	4.07	8 10^{19}	1 10^{-4}	1	0.5	2.0
GaN	3.20	4.10	1 10^{19}	7 10^{-3}	70	5.0	0.2
Mo				5.3 10^{-6}	5.3 10^{-2}	1.4	0.7
Ni				6.2 10^{-6}	6.2 10^{-2}	0.9	1.1

electrical resistivity r and heat conductivity κ . Metals, Ni and Mo, are also included since they are frequently used in thermionic converters. Table 2 also shows the dissipated power per cm^2 area P_d, and temperature gradient ∇T over 1 mm thick substrate, calculated for typical current density (10 A/cm^2) and heat flux (10 W/cm^2). Power dissipation in GaN substrate is considerable (note that ∇T is low for GaN). On the other hand, GaN has a wide band gap and allows lower $e\phi$ values. A possible solution is to grow thin GaN epitaxial layer on the GaAs or Si substrates. Such bi-layer substrate ensures low power loss together with low $e\phi$ in the metal NG layer. Fortunately, GaN can be grown epitaxially on both GaAs and Si substrates [51].

Most technological metals have $e\phi_0 > 4\,\text{eV}$ and form Shottky barriers with above semiconductor materials i.e. $e(\phi_0 - \chi) > 0$. Table 3 shows the parameters of some metals frequently used as electrode materials (other interesting metals are included as well), collected from the literature: $e\phi_0$ from Ref. 52 and $E_F^{(0)}$ from Ref. 53.

Table 3. Characteristic energies for some metals and values of $e\phi$, calculated for $G=10$

NG Material	$e\phi_0$ [eV]	$E_F^{(0)}$ [eV]	E_F^{Band} [eV]	$e\phi$ (eV) on Si	GaAs	GaN
La B_6	2-3.2		10.0[54]	1.15	0.95	0
Ag	4.26	5.48	7.5[54]	3.32	3.10	2.02
Nb	4.30	5.32	5.5[55]	3.36	3.14	2.06
W	4.55		6.7[56]	3.58	3.34	2.14
Cu	4.65	7.0	9.1[54]	3.65	3.43	2.20
Mo	4.60		5.0[58]	3.63	3.41	2.32

It also contains data on E_F^{Band}, which is the Fermi energy obtained from *ab initio* calculations and experiments [56- 61]. Parameter E_F^{Band} differs more or less from $E_F^{(0)}$. We used E_F^{Band} instead of $E_F^{(0)}$ for Mo, W, Ni and Pt since no $E_F^{(0)}$ data was found. This substitution is acceptable as far as a parabolic band is good approximation for these metals. The $e\phi$ values presented in the table were calculated using Eq. (32). WF values were reduced by $\approx 1\,\text{eV}$ on Si, by $\approx 1.2\,\text{eV}$ on GaAs and by $\approx 2.3\,\text{eV}$ on GaN substrates. Reduction is limited by the ΔE_{con} scale. The last can only be increased by increasing band gap. However, wide band gap substrates have high resistivity and were not acceptable for thermotunnel and thermionic devices. Obviously 1-2.3 eV reduction in $e\phi$ was not enough for cold emission, but it was still interesting since the plain Mo and Ni electrodes, coated with an ultra thin layer of Cs atoms, show very low WF (1-1.5 eV). WF reduction by less than mono layer of Cs atoms is a surface effect. At the same time, quantum state depression is not a surface effect. Most probably, the two mechanisms of WF reduction will sum up and result in an $e\phi$ considerably less than 1 eV. The same is true for Ag-Ba and W-Li electrodes.

Borides have $e\phi_0 < 4\,\text{eV}$ and for them $e(\phi_0 - \chi) < 0$ (Fig. 7). The most frequently used is Lanthanum hexaboride LaB_6, which shows $e\phi$=2-3.2 eV. Fermi energy for LaB_6 is E_F^{Band}=10 eV [54]. Inserting these values in Eq. (32) gave $e\phi$ = 0-0.85 eV for ridged LaB_6 layer on GaN substrate, $e\phi$ = 0.94-2.05 eV on GaAs substrate and $e\phi$ = 1.15-2.28 eV on Si substrate ($G = 10$ was used in all cases). These values were low enough for thermotunnel and thermionic devices operating at room temperatures.

Geometry factors, at which ideal ohmic contact was obtained, were calculated using Eq. (34). They were in the range $G_0 = 1-5$ for most material pairs. Such G_0 can easily be obtained in practice. Exceptions were Ni/Si ($G_0 = 59$) and Au/Si ($G_0 = 16$) pairs.

7. Experimental investigation of metal nanograting layers

Thin gold films, having NG on both sides, were prepared to observe the WF reduction (Figure 8). Gold was the material of choice because it does not form a natural oxide on the surface and allows

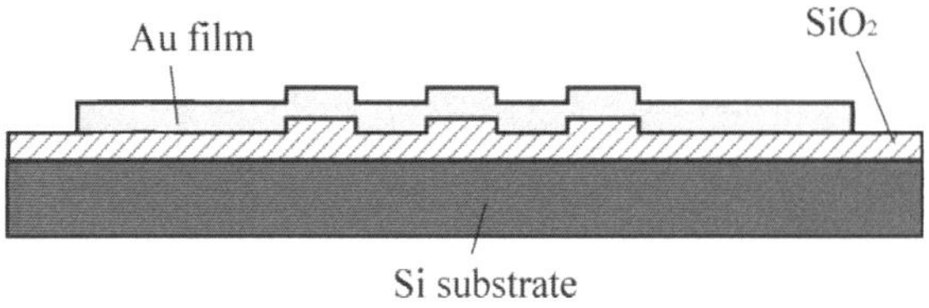

Figure 8. Cross section of Au film grown on Si substrate.

exposure of the films to the atmosphere. Au film was vacuum deposited on a Si/SiO_2 (dry thermal oxide) substrate in which NG was fabricated using both e-beam lithography and photolithography. Au was grown on a variable temperature NG substrate (T=80-300 K) using the quench deposition technique. Previously, the substrate was cooled to temperatures between -16 C and -22 C, and a thin film of 2-3 nm thickness, evaporated from a mixture of Au and Cr, was deposited on SiO2 to form an adhesive layer (not shown on figure 8 for simplicity). Following the deposition, the wafer was moved rapidly (maximum 5 s) to another location, where an Au film of thickness of 60 nm was deposited using the rapid thermal evaporation of Au wire (99.999% purity). In other cases NG in Au was made only on the surface of the film using etching techniques, while film base remained plain. Finally, large structures such as NG and plain circles and chess desk like areas containing both NG and plain squares were fabricated using photolithography.

Measurements of the work function were made using Kelvin Probe (KP) method and Photo Electron Emission Microscope (PEMM). All measurements were comparative to exclude absolute inaccuracies: KP reading from the NG area was compared to the reading from the plane area of the film. PEEM images were collected using high-resolution photoelectron microscope [62]. The internal structure of the films was analyzed using X-ray diffraction.

For all samples measured, the indented regions showed a reduced work function (WF) compared to the plain regions. The magnitude of reduction depended on the structure of the film and the dimensions of NG. The internal structure of the film was varied by changing substrate temperature. Amorphous films show a much higher reduction in WF than films with polycrystalline structure. For amorphous films the reduction of WF was in the range of 0.2-0.5 eV. All polycrystalline films show a WF reduction less than 0.1 eV. The reduction in WF was more pronounced for samples that were deposited in a cleaner environment (by plasma cleaning the deposition chamber prior to deposition). Residual gas pressure and composition has considerable influence on the results. One order difference in WF reduction shows that the internal structure of the film has principal importance.

It was observed that the strength of the effect depends on the depth of the indents. Amorphous Au films with thickness of 60 nm, and an indent depth of 50, 20 and 10 nm, show WF reduction of 0.16 eV, 0.25 eV and 0.56 eV respectively.

A WF reduction of 0.1 eV was observed in NG films grown on plain substrate. Low value of WF reduction could be explained by the fact that Au films were deposited on room temperature substrate. Another explanation is the difference in geometry. One of the samples was prepared so that it contained five areas each having a different surface geometry. The first area was plain and the other four areas had various indent widths of 800, 600, 400, 200 nm, but the same depth and length. Comparative measurements made between 1-2, 1-3, 1-4, 1-5 show that WF difference increases with reduction of indent width.

Reduction of WF of 0.2 eV was also observed in the indented areas of SiO_2. The measurement was made on several samples and yielded almost identical results.

PEEM measurements show that NG areas in all microscope images appeared brighter than adjacent plain areas (two of them were placed as on chess desk). Brighter areas corresponded to more electron emission from NG areas. This was in quantitative agreement with KP measurements. Unfortunately

PEEM microscope does not allow WF measurement on the base of image intensity. More information about experiments can be found in Reference 47.

8. Conclusions

Geometry-induced electron doping (G-doping) was investigated in multiple NG layers. They were composed of main and barrier layers forming a series of isotype homo junctions. Barrier layers were used to form electron confinement energy regions. Both main and barrier layers were n-type. Barrier layers were donor-doped to obtain electron confinement and induce G-doping in main layers. Such parameters as electron concentration, Fermi level and confinement region width were calculated for this system. Main layer G-doping level of 10^{17} -10^{18} cm^{-3} was obtained at barrier layer donor doping of 10^{18} -10^{20} cm^{-3} and geometry factor values G_m=1.02, $G_b = 1.1$. One order higher donor doping was required in Si with respect to GaAs to obtain the same G-doping level. For high electron mobility it was preferable to have high values of geometry factors. Electron mobility was higher in G-doped layers with respect to donor-doped layers of same doping level. G-doping opens prospects for new quasi-3D optoelectronic and thermoelectric systems. At the same time G-doping is temperature independent and can be used to extend working temperature ranges of cryogenic and power electronics.

Thermoelectric transport coefficients were investigated in the system of NG layers and periodic series of p–n junctions at the top of the ridges. An analysis was made on the basis of Boltzmann transport equations. It was shown that the Seebeck coefficient increases considerably. At the same time, electrical and thermal conductivities remain unaffected by the series of junctions. This allows large enhancement of thermoelectric figure of merit. Dependence of Seebeck coefficient on the geometry factor *G* and junction parameters was investigated and the analytical expression was obtained. Seebeck coefficient changes sign for some value of *G*. Dependences of S and ZT on chemical potential were presented for p–n junction NG layers. Calculations show one order of magnitude increase in thermoelectric figure of merit with respect to the bulk material.

Low work function was obtained in the metal NG layer having a reduced quantum state density. Metal LG layers, grown on semiconductor and metal substrates, were analyzed. Electron confinement to the NG layer was essential in both cases. When using semiconductor substrate, the wide band gap material allows more electron confinement and lower values of the resulting $e\phi$. Dependence of $e\phi$ on the band gap was analyzed for a number of cases and the corresponding formulae derived.

Thin Au films, having NG on both sides, were prepared using vacuum deposition, e-beam lithography and photolithography. WF reduction was measured using Kelvin Probe and verified using PEEM microscopy. Film internal structure was monitored using X-ray structural analysis. WF reduction dependence on film internal structure and NG dimensions were recorded. WF reduction of 0.56 eV was observed in Au NG films with amorphous internal structure.

9. Acknowledgements

Author thanks Ismat S. Shah and Juejun Hu for useful discussions. Apparatus received from EU TEMPUS project 530278-TEMPUS-1-2012-1-DE-TEMPUS-JPHES was used for preparation of this work.

10. References

[1] Andreozzi A, Lamagna L, Seguini G, Fanciulli M, Schamm-Chardon S, Castro C and M Perego M 2011 Nanotechnology 22 335303

[2] Xia G, Jeong S-J,Kim J E, Kimand B H, Koo C-M and Kim S O 2009 Nanotechnology 20 25301

[3] Paivanranta B, Langner A, Kirk E, David C and Ekinci Y 2011 Nanotechnology 22 375302

[4] Bergmann K, Danylyuk S V and Juschkin L 2009 J. Appl. Phys. 106 073309

[5] Tavkhelidze A 2009 Nanotechnology 20 405401

[6] Tavkhelidze A, 2010 J. Appl. Phys. 108 044313

[7] Kim J H, Barth M, Kuhl U, Stockmann H-J and Bird J P 2003 Phys. Rev. B 68 045315

[8] Berggren K-F,Yakimenko I I and Hakanen J 2010 New J. Phys. 12 073005

[9] Baltes H P and Hilf E R 1976 Spectra of Finite Systems (Wissenschaftsverlag, Mannheim)

[10] Eckhardt B 1988 Phys. Rep. 163 205-297

[11] Tavkhelidze A, Svanidze V and Noselidze I 2007 J. Vac. Sci. Technol. B 25 1270

[12] Emig T 2010 Casimir Forces and Geometry in Nanosystems, in Nonlinear Dynamics of Nanosystems, ed G. Radons, B. Rumpf, H. G. Schuster (Willey-VCH) p. 165

[13] Ono S, Shima H 2010 Physica E, 42 1224-1227

[14] Kartashov Y V, Szameit A, Keil R, Vysloukh V A, and Torner L 2011 Optics Letters 36 3470

[15] Gupta S and Saxena A 2011 J. Appl. Phys. 109 074316

[16] Fujita N 2004 J. Phys. Soc. Jpn. 73 3115-3120

[17] Ortix C, S. Kiravittaya S, Schmidt O G, and van den Brink J 2011 Phys. Rev. B. 84 045438

[18] Simon J, Protasenko V, Lian C, Xing H and Jena D 2010 Science 327 60-64

[19] Yu B, Zebarjadi M, Wang H, Lukas K, Wang H, Wang D, Opeil C, Dresselhaus M,Chen G, and Ren Z 2012 Nano Lett. 12, 20772082

[20] Goldsmid H J 2010 Introduction to Thermoelectricity (Springer)

[21] Tavkhelidze A and Svanidze V 2008 Int. J. Nanosci. 7 333

[22] Sanders A et al 2011 Nanotechnology 22 465703

[23] Sze S M and Ng K K 2007 Physics of Semiconductor Devices (New Jersey: Wiley-Interscience)

[24] Joyce W B and Dixon R W 1977 Appl. Phys. Lett. 31 354

[25] Kirk A P 2010 Solar Energy Materials & Solar Cells 94 2442

[26] Casey H C Jr and Panish M B 1978 Hetrostructure lasers (Academic press, New York)

[27] Bulucea C 1993 Solid-State Electron. 36 489

[28] Ohtsuka M and Suzuki A 1993 J. Appl. Phys. 73 7358

[29] Pickrell G W, Xu C F, Louderback D A, Lin H C, Fish M A, Hindi J J, Simpson M C, Guilfoyle P S, Zhang Z H and Hsieh K C 2004 J. Appl. Phys. 96 4050

[30] Despesse G and Jager T 2004 J. Appl. Phys. 96 5026

[31] Wachutka G and Gerstenmaier Y C 2006 proc. Int. Conference on Mixed Design (MIXDES Poland) p 48

[32] Goldsmid, H J 2003 in Thermoelectrics, 2003 Twenty-Second International Conference on ICT pp. 433- 438

[33] Hishinuma Y, Geballe T H, and Moyzhes B Y 2003 J. Appl. Phys. 94 4690

[34] Zeng T 2006 Appl. Phys. Lett. 88 153104

[35] Ashkroft N E and Martin N D 1976 Solid State Physics, (Saunders College Publishing, NY)

[36] Mahan J D and Sofo J O, 1996 Proc. Natl. Acad. Sci. USA 93 7436

[37] Joyce W B and Dixon R W 1977 Appl. Phys. Lett. 31 354

[38] Bhandari C M, Rowe D M 1995 *CRC Handbook of Thermoelectrics,* ed Rowe D M (CRC Press) p 43

[39] Rogacheva E I, Nashchekina O N, Grigorov S N, Us M A, Dresselhaus M D and Cronin S B 2002 Nanotechnology 13 1

[40] Yamamoto S 2006 Rep. Prog. Phys. 69 pp 181–232

[41] Wachutka G and Gerstenmaier Y. C. 2007 AIP Conf. Proc. 890 - 349

[42] Koh W S and Ang L K 2008 Nanotechnology 19 235402

[43] Hishinuma Y, Geballe T H, Moyzhes B Y and Kenny T W 2001 Appl. Phys. Lett. 78 2572

[44] Tavkhelidze A, Svanidze V. and Tsakadze L 2008 J. Vac. Sci.Technol. A 26 5

[45] Tavkhelidze A , Skhiladze G, Bibilashvili A, Tsakadze L, Jangidze L, Taliashvili Z, Cox I and Berishvili Z, Electron tunnelling through large area vacuum gap - preliminary results in *International Conference on Themoelectrics (ICT2002)* pp 435-439 (2002)

[46] Tavkhelidze A, Bibilashvili A, Jangidze L, Shimkunas A, Mauger P , Rempfer G F, Almaraz L, Dixon T and Kordesch M E 2006 J. Vac. Sci. Technol. B 24 1413

[47] Jangidze L, Tavkhelidze A, Blagidze Y and Taliashvili Z 2012 J. Electrochem. Soc., 159 D413-D417

[48] Su W B, Chang C S and Tsong T T 2010 J. Phys. D: Appl. Phys. 43 013001

[49] Hishinuma Y, Moyzhes B Y, and Geballe T H 2002 Appl. Phys. Lett. 81 4242

[50] Eiting C J, Grudowski P A and Dupuis R D 1998 Journal of Electronic Materials 27 206

[51] Yu J W, Lin H C, Feng Z C, Wang L S, Tripathy B S and Chua S J 2006 Thin Solid Films 498 108 – 112

[52] Linde D R 1991 *CRC handbook of Chemistry and Physics*, (71-st edition) p. 12-84

[53] Ashkroft N E and Martin N D 1976 *Solid State Physics* (NY: Saunders)

[54] Westover T L and Fisher T S 2008 Phys. Rev. B 77 115426

[55] Mizurani U 2001 *Electron theory of metals*, (Cambridge University Press) p. 27

[56] Lekka Ch E, Mehl, M J, Bernstein N and Papaconstantopoulos D A 2003 Phys. Rev. B 68 035422

[57] Lassner E and Schubert W-D 1999 *Tungsten: properties, chemistry, technology of the element, alloys, and chemical compounds* (Kluwer academic/plenum publishers NY)

[58] McAvoy T, Zhang J, C. Waldfried , D. N. McIlroy, P. A. Dowben, O. Zeybek, T. Bertrams, and B. S. Barrett 2000 The European Physics Journal B 14 747–755

[59] U. Mizurani, *Electr* J 2001 O*n theory of metals*, (Cambridge University Press) p. 27

[60] Yamasaki A and Fujiwara T 2003 J. Phys. Soc. Jpn. 72 607-610

[61] Baud S, Ramseyer C, Bihlmayer G, Blügel S, Barreteau C ,Desjonquères M C, Spanjaard D and Bernstein N 2004 Phys. Rev. B 70 235423

[62] Rempfer G F, Skoczylas W P, and Hayes Griffith O 1991 Ultra-Microscopy 36 196

A process for the development of strontium hydroxyapatite

N Zahra, M Fayyaz, W Iqbal, M Irfan and S Alam
Pakistan Institute of Technology for Minerals and Advanced Engineering Materials
PCSIR Laboratories Complex Lahore, Pakistan

E-mail: naseem.zahra1981@gmail.com

Abstract. A procedure for the preparation of Strontium Hydroxyapatite is adapted to produce high purity and better homogeneity ceramic with good Crystallinity. The strontium substituted bone cement has potential for use in orthopedic surgeries. Ionic Strontium (Sr) in humans shares the same physiological pathway as calcium and can be deposited in the mineral structure of the bone. In the present study, a novel concept of preparing Sr-contained Hydroxyapatite bone cement by using a precipitation method is proposed to get an ideal biomaterial that possesses potential degradability and more excellent pharmacological effect. Chemical analysis, Fourier Transform Infra Red analysis and Thermogravimetric/ Differential Scanning Calorimetric studies were conducted on prepared Strontium Hydroxyapatite sample to characterize the incorporation of 15% Sr^{2+} into the crystal lattice of Hydroxyapatite. Strontium was quantitatively incorporated into Hydroxyapatite where its substitution for calcium provoked a linear shift of the infrared absorption bands of the hydroxyl and phosphate groups. Thus, the formation of Sr–HAp was confirmed by Chemical Analysis, FT–IR and TGA/DSC results.

1. Introduction

Bone grafting is widely used in orthopaedic practices to deal with problems associated with bone loss, bone defects and reconstruction. Numerous researches have been dedicated to the development and evaluation of a synthetic substitute because of the significant problems associated with the limitation of suitable autogenous bone graft. Several materials have been tried and found to be inappropriate but the most promising materials to date are those based on calcium Hydroxyapatite (HA) [1]. Various HA synthesis techniques have been developed such as precipitation method [2], hydrothermal synthesis [3], solid state reaction [4] and sol-gel procedure [5]. HA is one of the advanced ceramic materials that have been universally accepted in biomedical applications due to its excellent biocompatibility and tissue bioactivity properties. HA is also a bioactive ceramic which actively takes part in physiological processes to enhance bone tissue formation in our body [6]. Due to its bioactive properties, it can integrate into bone structures and supporting bone growth without breaking down or dissolving [7]. The synthetic HA cannot be used in load bearing application due to its brittleness and lack of strength. These poor mechanical properties are the reasons why pure HA is not ideal to be used as bone substitutes [8]. Other shortcomings include high degree of crystallinity which result in the un-degradability of pure HA when it is implanted into the organ. Various dopants metal such as magnesium [9- 11], manganese [12], zinc [13], titanium are believed to have significant role towards improving cell-material interactions of HA and strengthening its mechanical properties. As calcium and strontium share the properties of group 2A elements, strontium can replace calcium in Hydroxyapatite, and hence in bone, without much difficulty. It is reported that the stable strontium is non-toxic even when it is administered in large doses in our body for prolonged periods [14-15]. The minerals and traces of metal elements accelerate bone formation and resorption on bone cells or bone mineral in vivo and in vitro. The process of cell interaction on materials is very dynamic, thus encouraging an exceptional tissue response at the biomaterials surface. Strontium is one of the metallic elements found in bones and teeth. Strontium appears to be one of the most effective substances for the treatment of osteoporosis and other bone-related conditions [16]. Various works on strontium doped HA have been done by other researchers. The different routes for the synthesis of Sr-doped HA have been employed. Various morphology, stoichiometry and level of crystallinity have been achieved depending on the technique and methods used for the synthesis process. Strontium is used to improve the mechanical strength of bone cement. The precipitation method, hydrolysis method, the solution reaction method and hydrothermal methods have been used to prepare strontium substituted calcium phosphate ceramics [17-19].

2. Materials and Methods

Strontium doped Hydroxyapatite (Sr-doped HA) has been synthesized using a precipitation method. Calcium nitrate tetrahydrate [$Ca(NO_3)_2.4H_2O$] (Merck) and strontium nitrate [$Sr(NO_3)_2$] (Merck) were dissolved in distilled water in two sets of beakers and stirred with magnetic bar stirrer to make it homogeneous. These solutions were poured into the di-ammonium hydrogen phosphate [$(NH_4)_2HPO_4$] (Merck) solution. pH was maintained at 10 and suspension was kept for digestion. The precipitates were then filtered, washed and dried. The resultant solid was crushed into fine powder using pestle and mortar. The powder was then subjected to 900 °C calcination where white powder was obtained. The elemental analysis for lead, cadmium, mercury, arsenic and chromium was performed on Inductively Coupled Plasma (ICP) [20-21]. To confirm the Ca+Sr/P molar ratio of the synthesized Strontium-Hydroxyapatite powder, calcium was determined by Gravimetric method using Ammonium Oxalate [22] that was further confirmed by Atomic Absorption Spectrometer [ASTM-E 350]. Phosphorous was determined using spectrophotometer [ASTM E-325295]. The weight loss and thermal behavior of sample were also evaluated from the Thermo gravimetric analysis data by using DSC-SDT Q 600 at heating rate of 10 °C up to 1400 °C. [ASTM-E 794]. The chemical nature and molecular bond structure of the synthesized powder was determined using FTIR (Fourier Transformed Infrared Spectroscopy). These spectral studies were conducted using Thermonicolet FTIR-200 by KBr disks in the range of 4000-400 cm^{-1}.

3. Results and Discussions

The prepared Strontium Hydroxyapatite was analyzed for various elements i.e. Arsenic, Cadmium, Lead, Chromium and mercury. It was found that all the elements were not found in the prepared sample (Table. 1).

Table 1. Determination of Elements in Strontium Hydroxyapatite

Elements	Permissible limit (standard)	Sr-HAp prepared
As (Arsenic)	3.0 ppm	Nil
Cd (Cadmium)	5.0 ppm	Nil
Pb (Lead)	30.0 ppm	Nil
Cr (Chromium)	-	Nil
Mercury (Hg)	-	Nil

Pure HAp has Ca/P molar ratio of 1.67 and the measured Sr+Ca/P molar ratio for Prepared Strontium Hydroxyapatite powder was found to be 1.71 which indicates the formation of Strontium

Hydroxyapatite and which is close to the theoretical value (Ca/P=1.67) that found in the Hydroxyapatite of natural bone (Table 2).

Table 2. Determination of Ca+Sr/P ratio in Prepared Strontium Hydroxyapatite

Mol % Sr	**15%**
Ca+Sr/P ratio (Nominal)	1.67
Ca+Sr/P ratio (Measure)	1.71

Fourier Transform Infrared Spectroscopy (FTIR) spectra of prepared sample was used to characterize the chemical nature and molecular bond structure of the synthesized powder. FTIR analysis showed all typical absorption characteristics of Strontium Hydroxyapatite. A typical HA structure containing sharp PO_4 bands at 1000 cm^{-1} – 950 cm^{-1} was observed which match the vibration modes of the phosphate group at 1092 and 1040 cm^{-1}. The broad band observed at 1110.45 cm^{-1} may be due to a minor fraction of octacalcium phosphate (OCP), $Ca_8H_2(PO_4)_6 \cdot 5H_2O$, which has a similar structure to HA and is considered one of its precursors [23].

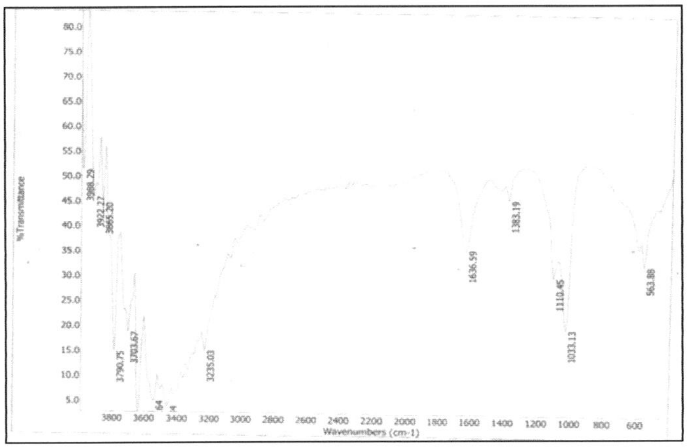

Figure 1. FTIR Analysis of Strontium Hydroxyapatite

15%-doped HA experienced weight loss (5.18%). It is expected from the analysis that 15%-doped HA should contain more secondary phase of TCP. From this figure, it is observed that the mass loss during heating could be classified into three different stages.

The first stage weight loss occurs at 30 to 200°C due to evaporation of water. Then, it experienced about 40% weight loss in the temperature range between 200-470°C which attributed to the crystallization of the powders from the amorphous phase. Finally, at temperature between 760-790°C, the third stage of weight loss took place and the powder experienced about 1.8-3.1% weight drop. This weight loss is attributed to the decomposition of pyrophosphates ($P_2O_7^{4-}$) to biphasic mixtures [24-25].

$$P_2O_7^{\ 4-} + 2OH^- \longrightarrow 2PO_4^{3-} + H_2O$$

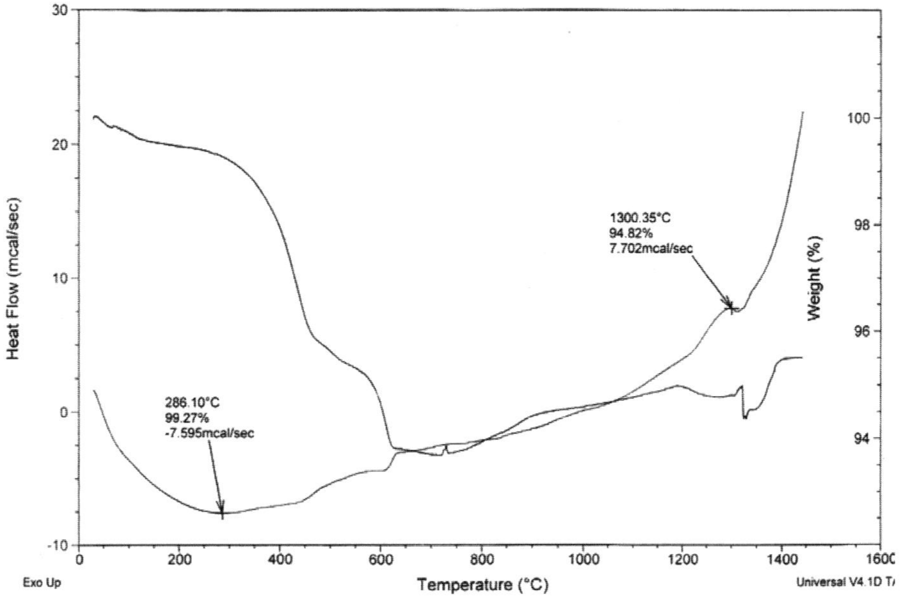

Figure 2.TGA-DSC Analysis of Strontium Hydroxyapatite

4. Conclusion

Pure and stiochiometric Strontium-Hydroxyapatite powder was successfully synthesized by precipitation method. Further studies will be carried out considering the effects of the sintering temperature and doped solution concentration. However, incorporation of cations like Sr^{+2} in Hydroxyapatite structure will play a vital role to enhance the bioactivity and physiochemical properties of the apatite [26]. DSC-TGA results showed the thermal stability of Sr-HAp on heating up to 1400 ^{0}C and FTIR analysis clearly indicated the presence of phosphorous and hydroxyl group in its structure. The Ca+Sr/P molar ratio was found to be 1.71 which falls in the range of compounds to be used for implants. The proposed route is most economic way of synthesizing Sr-HAp for biomedical applications.

5. References

[1] Black R J, Zardiackas L D, Teasdall R and Hughes J L, 1990 *Clinical Material* 251-264.

[2] Kannan S, Lemos I A F, Rocha J H G and Ferreira J M F, 2005 *Journals of Solid State Chemistry* 178 3190-3196.

[3] Komarneni S and Katsuki H 2002 *Pure Appl. Chem,* 74 1537-1543.

[4] Ryu H S, Hong K S, Lee J K and Kim D J 2006 *Materials Research Society* 21 428-436.

[5] Xu W, Hu W, Li M. and Wen C, 2006 *Materials Letters* 60 1575-1578.

[6] Paula W and Sharma C P, 2006 *American Journal of Biochemistry and Biotechnology* 2 41-48.

[7] Prokopiev O and Sevostianov I, 2006 *Materials Science and Engineering* 218-227.

[8] Fu Y F and Chen D M, 2005 *J Oral Tissue Engin* 2 76-78.

[9] Landi E , Logroscino G, Proietti L, Tampieri A, Sandri M and Sprio S, 2008 *J Mater Sci: Mater Med* 19 239-247.

[10] Landi E, Tampieri A, Mattioli-Belmonte M, Celotti G, Sandri M and Gigante A 2006 *Journal of the European Ceramic Society* 26 2593-2601.

[11] Gibson R and Bondfield W, 2002 *J Mater Sci: Mater Med* 13 685-693.

[12] Acchar W and Ramalho E G, 2008 *Materials Science and Engineering C* 28 248-252.

[13] LeGeros R Z and LeGeros J P, 1993 "An introduction to bioceramics: dense Hydroxyapatite", World Scientific Publishing Co.

[14] Li M, Xiao X, Liu R, Chen C and Huang L, 2008 *J Mater Sci: Mater Med* 19 797-803.

[15] Simon V, Muresan D, Popa C and Simon S, 2005 *Journal of Optoelectronics and Advanced Materials* 7 2823-2826.

[16] Kim H W, Koh Y H, Kong Y M, Kang J G and Kim H E, 2004 *Journal of Materials Science: Materials in Medicine* 15 1129-1134.

[17] Jegou S, Camire C L, Nevsten P, Hansen S and Ginebra M P, 2005 Journal *of Materials Science: Materials in Medicine* 16 993-1001.

[18] Bigi, Boanini E, Capuccini C and Gazzano M, 2007 *Inorganica Chimica Acta* 360 1009.

[19] Kim T G and Park B, 2005 *Inorganic Chemistry* 44 9895-9901.

[20] Boevski I and Daskalova N, 2007 *J Univ Chem Tech Metallurgy* 42 419-26.

[21] Ozturk M, Ozozen G, Minareci O and Minaeci E 2009 *Iran J Environ Health Sci Eng* 6 73-80.

[22] Vogel I., 1961 *A Text-Book Of Quantitative Inorganic Analysis*. 3rd Edition (Richard Clay and company, Ltd. Great Britain).

[23] Joice T, Erico R D, Jean-Guillaume E, Donald E E, Gabriela G and Alexander M R, 2009 *Phys. Chem. Chem. Phys.* DOI: 10.1039/b802841a

[24] Lazic S, Zec S, Miljevic N. and Milonjic S, 2001 *Thermochimica Acta* 374 13-22.

[25] Habib F, Alam S, Zahra N, Irfan M, Iqbal W 2012 *J Chem Soc Pak* 34 584-588.

[26] Ahmed S and Ahsan M, 2008 *Bangladesh Journal of Scientific & Industrial Research* 43 501.

Effect of lead ion concentration on the structural and optical properties of nano-crystalline PbS thin films

S Zaman[1], S K Mehmood[1], M Mansoor[2] and M M Asim[1]
[1]Institute of Industrial Control Systems, P.O Box 1398, Rawalpindi Pakistan
[2]School of Chemical & Materials Engineering, NUST, H-12, Islamabad, Pakistan

E-mail: superbs_1@yahoo.com

Abstract. PbS thin films have received considerable attention because of their potential applications in opto-electronics applications. Spontaneous reaction of lead acetate and thiourea in aqueous hydrazine hydrate has been used for depositing PbS thin films on glass substrates. Structural and optical properties of PbS thin films are greatly influenced by the morality of the reactants and crystal defects in the lattice. Our work focuses on the variation in lead ion concentration and its effect on the structural and optical properties of PbS thin films. The deposited films were analyzed using XRD, SEM, spectrophotometer and dark resistance measurement. XRD patterns indicated the formation of major phase of nano crystalline PbS with minor presence of lead oxide phase. We also noticed that peak intensity ratio of I_{111}/I_{200} varied by changing the Pb ion concentration. The film thickness and dark resistance increased whereas optical band gap decreased with the decreasing Pb ion concentration. SEM scans showed that the grain size is less than 100 nm and is not affected by varying Pb ion concentration.

1. Introduction

Thin films form the basis for today's electronic components. Even complicated device structures are constructed from thin films, deposited one by one. Nanocrystalline thin films make up a group of materials attracting a great deal of interest both from fundamental and applied point of view. This is due to the fact that these materials display a large surface to bulk ratio, which distinguishes them from bulk materials. A nano-crystalline material contains a high concentration of grain boundaries or in case of porous films, a high internal surface compared to the bulk material. These properties are of considerable importance for electrical transport, optical and mechanical properties [1-4]. Thin films of metal nanoparticles connected in a percolating network display high optical absorption in infrared wavelength range due to the small size of individual metal particle [5-6].

Lead sulphide (PbS) has been a rapidly growing area of importance to study the quantum size effect in recent years [7-9]. Structural, optical and electrical properties of PbS nanoparticles have been studied extensively [10-13]. Nanocrystalline PbS is a material of interest due to its potential application for infra-red detection. However, reduction of grain size extends the employability of the material over wide spectral range. Grain boundaries in compound semiconductor films are different from those in elemental semiconductors. Boundaries for larger crystallite films differ from those for small-grained ones. . In a polycrystalline semiconductor thin film, factors like film defects, surface scattering and grain boundaries complicate identification of electrical properties.

The physical, structural, electrical and optical characteristics of grain boundary regions are drastically altered by exposure to impurities, diffusion and field effects [14]. Considering the fact that stoichiometric excess of lead (Pb) in PbS (bulk) imparts n-type character whereas deficient lead makes it p-type [15], a similar change in carrier type can be expected in PbS films by manipulating the lead content. In chemical bath deposition technique, which has been used for the growth of PbS films for the present study, the stoichiometry is determined by the relative availability of the ions reaching the substrate to undergo reactions at the surface. A change, intentional or otherwise, in relative concentrations of the ions reaching the substrate surface can lead to non-stoichiometry in the resulting films. It is thus possible to grow films with excess or deficit lead by modifying its supply to the substrate during growth of films. We have reported that an appropriate choice of lead ion concentration could modify the carrier or conductivity type of PbS thin films. The availability of lead ions to the substrate during growth effect average grain size, film thickness and optical band gap of PbS films. An attempt to correlate the lead ion concentration with thickness, structural and optical properties has been made.

2. Experimental

The thin films of PbS were deposited onto chemically cleaned glass substrates using AR grade chemicals by CBD process. Thiourea was used as a source of sulphide ions and lead acetate as Pb^{+2} precursors in alkaline medium, using hydrazine hydrate as base instead of commonly used basis such as sodium hydroxide. Chemical bath was prepared by mixing aqueous solutions of thiourea and hydrazine hydrate in 1M: 7M respectively while lead acetate concentration was varied from 1.2M to 0.6M according to Table 1. The starting solution had pH 12. The substrates were mounted on the external walls of a beaker fitted with an immersive heater. The beaker was then filled with water and placed in the chemical bath solution. The temperature of substrates was kept close to boiling temperature of water throughout the experimentation. The deposition time was 30 minutes, as given in Table 1. After deposition, films were rinsed in de-ionized water and dried in oven at 60-80°C.

Table 1. Concentration of Lead ions used for different baths

Bath	Conc. Of Pb^{+2}	Bath	Conc. Of Pb^{+2}
A	1.2	G	0.8
B	1.1	H	0.75
C	1	I	0.7
D	0.95	J	0.65
E	0.90	K	0.60
F	0.85	L	0.55

The annealing of films was done at 100°C for 7 hours. The structure and crystallite sizes were measured by X-ray diffractometer. The films were well adherent to the substrates, as tested qualitatively by tape test. Film thickness was determined by gravimetric method. The surface morphology of films was studied by scanning electron microscope. The dark resistance of film was measured with a two probe method. The optical properties were studied on UV/VIS/NIR spectrophotometer.

3. XRD Studies

Figure 1 shows the diffractograms of PbS thin films prepared at different Pb^{+2} ions concentrations, from 0.6M to 1.2M for a constant deposition time of 30 mins. The diffraction peaks observed are at 2θ equal to 26, 30 and 43 degrees corresponding respectively to (111), (200), and (220) planes, confirmed cubic structure of PbS (JCPDS reference no. 78-1901) with slightly preferential (200) orientation of grains. In our study we reduced the concentration of lead ions while keeping the concentration of hydrazine hydrate constant. Hydrazine hydrate decreases the releasing rate of Pb^{+2} ions by making complex with these ions. It was observed that the growth of crystallographic orientation and microstructure of the films depend on Pb^{+2} ion concentration in the solution of the deposition process. The diffractogram 'a' shows that peak ratio of (111)/(200) is 80/100, when we lowered the concentration of lead acetate, the peak ratio of (111)/(200) also reduced almost linearly up to an optimum level and then became constant.

In the pattern 'e' this peak ratio of (111)/(200) is reduced to 25/100, that is an indicative that decreased Pb^{+2} ion concentration favors growth along (200) plane. This can be explained by the fact that when Pb^{+2} ion concentration decreases, the reaction rate slows down and chances of rapid precipitation of PbS salt decrease and hence nucleation growth increases.

One of the most important aspect revealed from these patterns is the growth of lead oxide phase Pb_3O_4 at 2θ equal to 26.15 and 32 degrees corresponding respectively to (211) and (310) planes according to JCPDS card # 76-1799 whereas at 2θ equal to 38 degree correspond to (020) plane of lead oxide phase PbO according to JCPDS card # 03-610, which appeared when Pb^{+2} ion concentration was decreased up to 0.95M as shown in pattern 'c'. The peak height of lead oxide phases increased with decreasing Pb content and become maximum at 0.8M Pb^{+2} ion concentration. On further reducing the concentration of lead ions, the growth of lead oxide phases decreased as well.

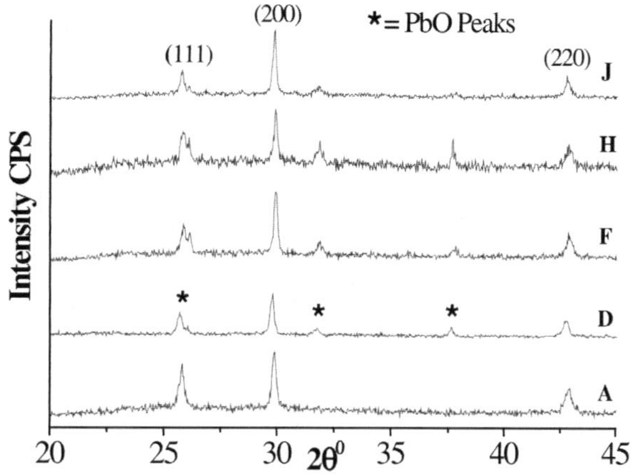

Figure 1. XRD comparison of PbS thin films deposited at different Pb^{+2} ion concentrations.

the explanation of oxide growth comes from the changes occurring in chemical bath as a result of gradual decrease in Pb^{+2} ion concentration. When we lowered lead acetate concentration, the influence of hydrazine hydrate became prominent. The reaction process slowed down and formation of lead oxide phases became more probable, and this process of oxide growth reached to maximum at an optimum level of Pb^{+2} ion concentration and after that became constant.

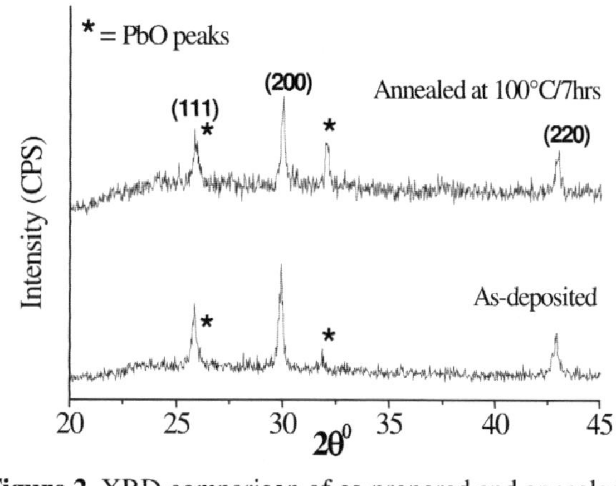

Figure 2. XRD comparison of as-prepared and annealed PbS thin films for Bath F

the deposited PbS thin films were thermally treated in air. The optimal temperature of treatment has been found to be 100°C in order to improve photoelectric properties. Figure 2 shows the comparison between diffraction patterns of as deposited and annealed PbS thin film for 0.85 M Pb^{+2} concentrations.

Thermal treatment in air favors oxidation and hence we observed increase in intensity of lead oxide peaks. However this effect of thermal oxidation in air decreased with decreasing Pb^{+2} ion concentration. When Pb^{+2} ions decreased from an optimum level in bath, the chances of free lead to incorporate in the PbS structure/lattice also decreased and we obtained more closer towards pure galena phase. Due to decreased free Pb in the deposited PbS thin films the thermal oxidation becomes less pronounced after 0.8M Pb^{+2} ion concentration.

4. Morphological Studies

Surface morphology and grain size of the deposited films was studied by using scanning electron microscope. It is obvious from figure 3, that all deposited PbS films are nanocrystalline and compact with no voids or pinholes. The grains are homogenously distributed and less than 100nm in size. As we decreased the concentration of Pb ions there is no significant change in surface morphology, the overall grain size of the base is nearly same. There is some increase in number of facets at surface that can be attributed to increasing film thickness and resulting non homogeneity.

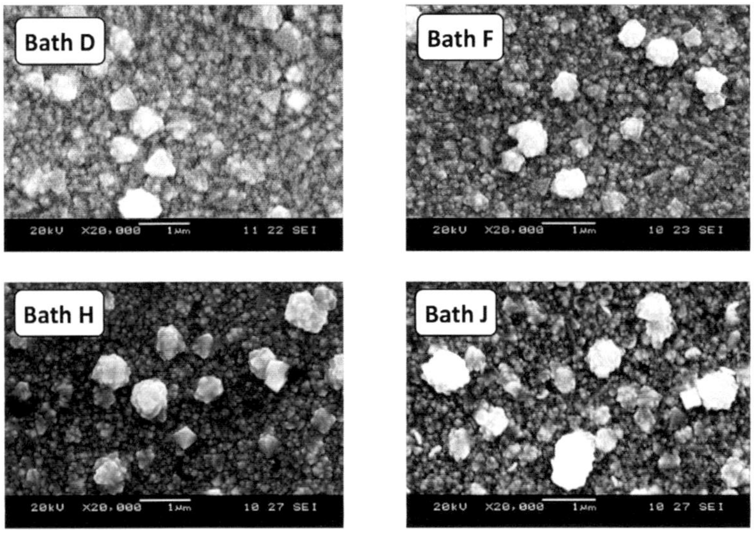

Figure 3. SEM micrographs of PbS thin films deposited at various Pb^{+2} concentrations.

5. Thickness measurement

The thickness of deposited PbS thin films was measured by gravimetric method. Figure 4 shows the effect of lead ion concentration on the thickness of PbS thin films. It was observed that lead ion concentration has marked effect on film thickness. Initially at 1.2M Pb^{+2} ion concentration the observed film thickness was 268nm and it increased up to 935nm when Pb^{+2} ion concentration was decreased to 0.6M. The observed results can be explained on the basis of ion-ion condensation. The formation of PbS is governed by the equation (1),

$$[Pb\,(HH)^{+2}] + NH_2CSNH_2 + H_2O \rightarrow PbS\downarrow + NHCNH + [HH.2H]^{+2} + H_2O \quad (1)$$

at a particular temperature PbS will be deposited if the ionic product of Pb^{+2} and S^{-2} exceeds the solubility product. The Pb^{+2} released by the hydrazine hydrate (HH) complex and S^{-2} released by thiourea combine at the nucleation centers on the substrate to produce PbS. The quantity of ions utilized for film formation depends on the rate of formation of PbS on the substrate surface, which in turn, depends not only on the ratio of the nucleation centers available at the surface of the substrate to those in the volume of the solution, but also on the rate of release of Pb^{+2} and S^{-2} ions. As we reduce the concentration of Pb^{+2} ions in chemical bath, the release of Pb^{+2} from its HH complex becomes slow and the whole deposition process slows down, as a result chances of rapid precipitation of PbS decrease and that of nucleation growth at the substrates increase. Due to increased nucleation growth, the growth along (200) plane dominated and as a result thickness of PbS film also increased.

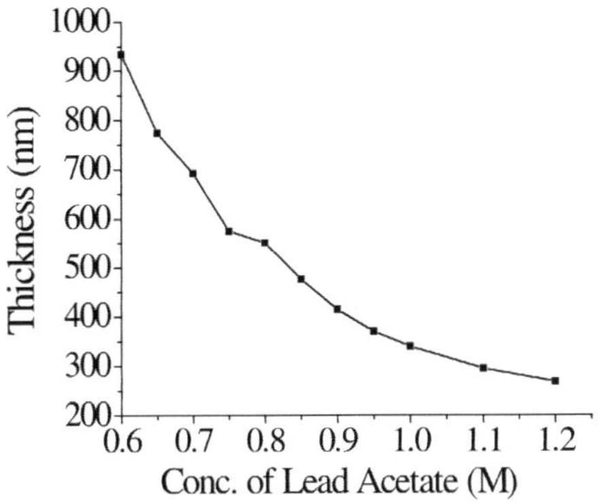

Figure 4. The effect of Pb^{+2} concentration on the thickness of PbS thin films.

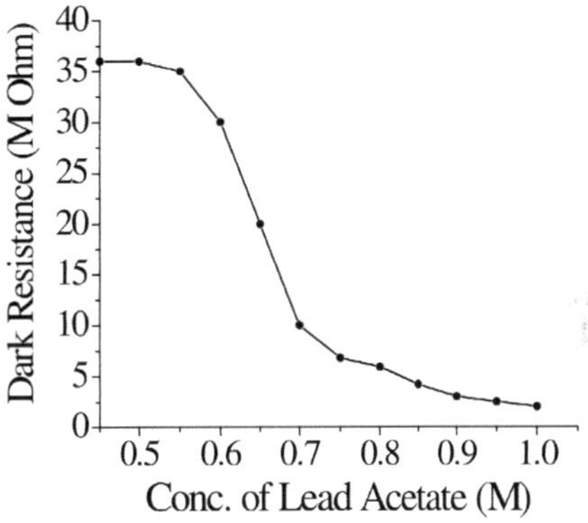

Figure 5. The effect of Pb^{+2} concentrations on the dark resistance of PbS thin films.

6. Dark Resistance

The dark resistance of the deposited films was measured at room temperature by using a multimeter. Figure 5 shows the effect of decrease in Pb ion concentration on the dark resistance of deposited films. There is a gradual increase in dark resistance up to 0.70M Pb ion concentration and then resistance increases rapidly up to 0.55M, after that it remained almost constant. This effect can be described by keeping in view that with the decreasing lead in chemical bath there is a stoichiometric decrease of Pb content in the deposited PbS thin films. The carrier type is shifted from n to p-type, as a result hole concentration decreases and the net dark resistance of PbS films increases.

7. Optical Studies

The optical reflection spectra of chemically deposited PbS thin films were recorded on a Shimadzu UV/VIS/NIR 3600 spectrophotometer in the range of 190-3100nm. Reflectance is less than 20% for almost all samples but increases slightly with annealing. With the help of this reflectance data, absorptivity was calculated by using Kubelka-Munk function, i.e.

$$F(R) = (1-R)^2 (2R)^{-1} = \alpha S^{-1}$$

where α is absorptivity and S is scattering factor. The energy band gap was found by plotting Kubelka-Munk function "$F(R)^2$" against energy, the intercept of the plot gives the value of band gap. Figure 6 shows the effect of decreasing lead ion concentration on lowering the energy band gap of deposited films. This effect can be contributed to increase in film thickness. As we decreased the lead ion concentration the film thickness increased.

Many references from literature show that annealing improves the photosensitivity of PbS films. The photosensitivity increases about four to ten times by optimized thermal treatment [16]. The effect of prolonged annealing could be explained by desorption of water and by the appearance of oxidation products [17].

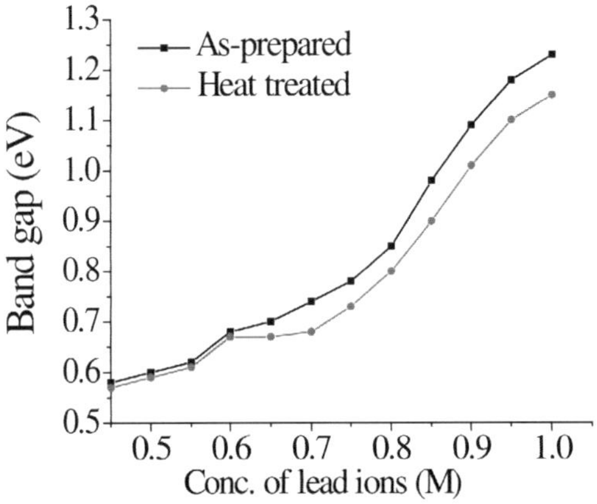

Figure 6. The effect of lead ion concentration on energy band gap of as-prepared and annealed PbS thin films.

Annealing process is expected to relieve stresses effectively and modify grain boundaries to some extent [16]. This results in photoconductivity enhancement. The deposited films were annealed at 100°C for 30 minutes. Annealing decreased the optical band gap of samples. This effect is marked at the start but less pronounced after 0.6M Pb ion concentration.

8. Conclusion

Structural and optical properties were determined and correlated with the different molarities of Pb ion concentrations. Several conclusions can be drawn on the basis of the experimental data.

- Thickness of the film increased as a result of decreasing Pb ion concentration, it is due to the fact the nucleation phenomena of PbS growth became more dominant and more time was available for the reaction to complete as a result thickness of PbS films enhanced.
- Dark resistance increased as a result of decreasing Pb ion concentration, it is due to fact that with the decreasing Pb content in the deposited PbS thin films, the carrier type is shifted from n to p-type, as a result the net dark resistance of PbS films increases.
- SEM scans indicated the formation of nanocrystalline PbS thin films having grain size less than 100 nm, it is also evident the grain size is not affected by the Pb ion concentration but the number of facets increases which is also responsible for the enhancement of thickness with decreasing Pb ion concentration.
- Optical studies revealed that by lowering the Pb ion concentration decreases the optical band gap of PbS thin films. The optical band gap decreases for annealed thin films which is responsible for the enhanced photosensitivity of annealed films.

9. References

[1] Tang F and Okuda S 1996 *Mater. Trans.* 37 1813

[2] Lang M J, Duarte-Dominguez M, Birringer R, Hampelman R, Natter H and Arnold W 1999 *Nanostruct. Mater.* 12 1811

[3] Tanimoto H, Sakai S and Mizubayashi H 1999 *Nanostruct. Mater.* 12 951

[4] Andrierskii R A and Glozer G M 2000 *Phys. Met. Metallogr.* 89 83

[5] Granqvist C G, Buhrman R A, Wyns , and Sievers A J 1976 *Phys. Rev. Lett.* 37 625

[6] Becker W, Fettig R and Ruppel W 1999 *Infrared Phys. Technol.* 40 1431

[7] Wang Y, Suna A, Mahler W, Kasowski R J 1987 *J. Chem. Phys.* 87 7315

[8] Gao F, Lu Q Y, Liu X Y, Yan Y S and Zhao D Y 2001 *Nano Lett.* 1 743

[9] Nanda K K, Kruis F E, Fissan H and Acet M 2002 *J. Appl. Phys.* 91 2315

[10] Nanda K K, Kruis F E, Fissan H 2002 *Phys. Rev. Lett.* 89 256103

[11] Joshi R K, Sehgal H K 2004 *Physica* E 23 168

[12] Kuang D B, Xu A W, Fang Y P, Liu H Q, Frommen C and Fenske D 2003 *Adv. Mat.* 15 1747

[13] Joshi R K, Kanjilal A and Sehgal H K 2004 *Appl. Surf. Sci.* 221 43

[14] Lawrence L. and Kazmerski 1980 *Polycrystalline and amorphous thin films and devices*, (U.K.: Academic press)

[15] Brown F C 1967 *Physics of Solids* (New York: Benjamin) p 320

[16] Hirata H and Higashiyama K 1971 *Bull. Chem. Soc. Jpn.* 44 2420

[17] Pop I, Ionescu V, Nascu C, Vomir V, Grecu R and Indrea E 1996 *Thin Solid Films* 283 119

Effect of substrate temperature on structure and optical properties of Co_3O_4 films prepared by spray pyrolysis technique

S Z Abbas[1], A A Aboud[2], M Irfan[1] and S Alam[1]
[1]Pakistan Institute of Technology for Minerals and Advanced Engineering Materials (PITMAEM), PCSIR Labs Complex, Lahore
[2]Department of Physics, Beni-Suif Faculty of science, Beni-Suif, Egypt

E-mail:pitmaem@yahoo.com

Abstract. Polycrystalline Co_3O_4 films of spinel cubic phase were prepared by spray pyrolysis on amorphous substrate. The deposition time was kept constant (30min) with variable ranges of substrate temperature (573-723 K). The structural and optical properties of films were studied. The average crystallite size calculated by Scherrer's equation was found to increase as substrate temperature increased. Very small particle size was achieved. Even at the highest deposition temperature, the size did not exceed 11.6nm. The average crystallite size obtained from AFM photos was ranged in from 182 to 248nm which reveals the presence of agglomerates. Optical data were recorded in the wavelength range300-2500nm. A considerable difference in transmittance was observed for films prepared at different substrate temperatures. The absorption coefficient was measured and then correlated with the photon energy to estimate the energy gap, which was the value of the spinel phase with a direct transition

1. Introduction

Transition metal oxides crystallizing in the spinel structure have been under attention for many decades due to their interesting electronic, magnetic and optical properties related to the multi-valence nature of the constituent transition metal ions. As one of those metal oxides, cobalt oxide is a vital material under consideration. Cobalt forms two stable oxides: - CoO (NaCl structure, a=4.26°A) and Co_3O_4 (normal spinel structure, a=8.08°A). Co_3O_4 is stable at lower temperature than CoO. Above 1073K CoO is thermodynamically a stable phase in air [1]. Co_3O_4 is known to have spinel structure in which the Co^{2+} ion in the formula (a mixed valence material that is formally [2]) unit occupies the tetrahedral site, while the two Co^{3+} ions occupy the octahedral sites [3]. Thin films of Co_3O_4 have been prepared on a variety of substrates by different methods. Such as RF sputtering [4], pulsed laser deposition (PLD) [5-7], solgel process [8-12], spray pyrolysis [12-15] and chemical vapor deposition (CVD) [16]. Thin films of these oxides have tremendous applications like gas sensor [17], solar selective absorber [18] and fuel-cell [19]. In the present work the spray pyrolysis technique is used for preparation of (Co_3O_4) films. The effect of substrate temperature on the structure and optical properties of sprayed films is then investigated.

2. Experimental

Co_3O_4 films were deposited by spraying 0.2M solution of Co $(NO_3)_2.6H_2O$ in doubly ionized water onto preheated glass substrate kept at different temperatures ranged from 573 to 723 K. The spray rate was kept constant at 0.2 ml min−1. Film thickness was measured by Tolansky's method of multiple beam Fizeau fringes [25] Structural characterization of the produced films has been carried out using JEOL JSDX-60PA diffractometer (Cu Kα radiation - Ni filtered λ = 1.54184°A – scanning speed is 1° min−1 - time constant is 1s) in the range 5 to 90°. The optical measurements were studied in the range from 300 to 2500 nm using a UV-310 Pc:UV-VISNIR; Shimadzu, double beam spectrophotometer with V-N specular reflection attachment. Orgin6.1 program, based on sloving the exact equations were used to calculate the complex optical constants, absorption coefficient and energy ban gap.

3. Results and Discussion

3.1 *Structural characterization*

Figure 1 shows XRD scans of films sprayed at the lowest and highest studied substrate temperatures, 573 and 723K respectively, in a sθ-range between 5 to 90°. The XRD investigations showed that polycrystalline cobalt oxide is formed in its spinel cubic phase at all the investigated temperatures. A Co_3O_4 phase with (311) major plane is formed at the lowest tested temperature, 573 K, and no detectable traces of any other cobalt oxide phases were observed within the investigated temperature range. Considering initial deposition or precursor layer, the (311) and (111) peaks are reported to be interchangeably the 100 percent peak for Co_3O_4 phase [12, 16].

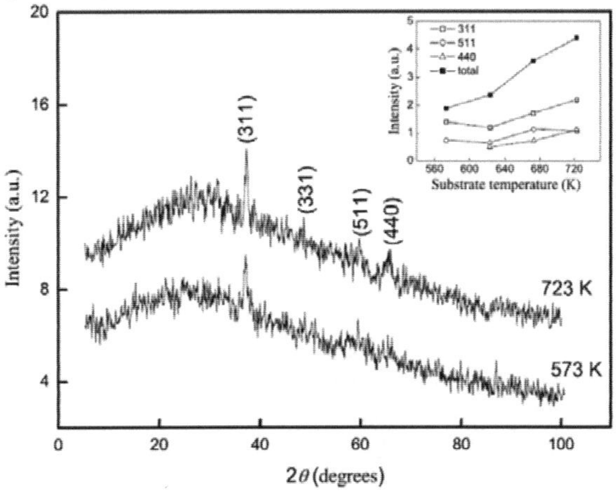

Figure 1. The XRD charts of Co_3O_4 films sprayed at two substrate temperatures. The inset shows the effect of the substrate temperature on intensity of the major peaks

The measured intensities of the film major peaks showed substrate temperature dependence as evidenced by the inset curve of Fig.1. A small change in the measured intensities of the film major peaks is observed as the temperature increased from 573 to 623 K. On the other hand, a larger increase in these intestines with temperature is observed for the higher temperature range, 623- 723 K. This observation is consistent with the important role played by the substrate temperature in the film growth mechanism. Hot substrate not only pyrolysis the sprayed solution but also provides nonvolatile species with the kinetic energy for relatively ordered film growth. Thus, a different effect of the substrate temperature on the sprayed films is expected based on the temperature range.

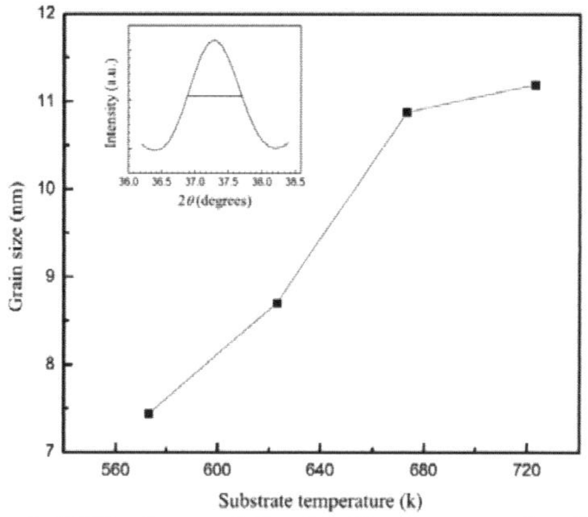

Figure 2. XRD charts of Co3O4 films prepared at 573 K.

For qualitative purposes, the intensity sum of the measured peaks within the XRD of the film can be taken to be directly related to the film crystallinity [20]. The inset of figure 1 indicates that the film crystallinity increases as the substrate temperature increase within the investigated temperature range. Crystalline enhancement can manifest itself in preferred orientation and particle size growth. XRD pattern in Fig.1 shows the presence of the main peaks of the Co_3O_4 phase as that reported for the powder sample in the ICDD cards [21]. This indicates that the Co_3O_4 is formed in a nearly random polycrystalline form. In order to estimate the size of the formed Co_3O_4 crystallites, the Scherrer method was applied to the (311) peak and plotted as a function of the substrate temperature as shown in figure 2. The inset shows a smoothed (311) peak for FWHM measurements. The FWHM is directly measured from the XRD chart using the regular scan rate without further measurements. The inset of figure 2 shows the smoother (311) peak for the FWHM measurements. The average crystalline was found to have a value between 7 nm to 12 nm, and increased as the substrate temperature increased.

Figure 3 presents the AFM morphological images of Co_3O_4 films deposited at substrate temperatures 573 K and 723 K. However, it is observed that even at the highest deposition temperature (723 K), the size of crystallites calculated by Scherrer's equation dose not exceed 12nm, which is much smaller than average sized obtained from AFM photos, 182nm and 248nm.

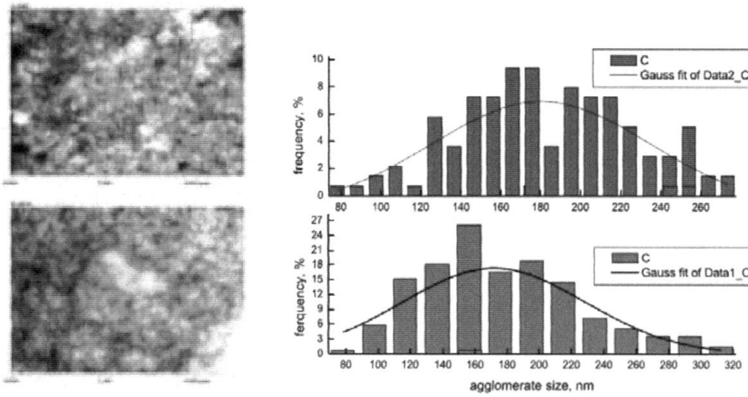

Figure 3. Estimation of particle size according to AFM analysis.at (A) 573 K and (B) 723 K

The variation of film thickness with substrate temperature is shown in figure 4. It is seen that the film deposited at 623K has the highest thickness as compared to the others.

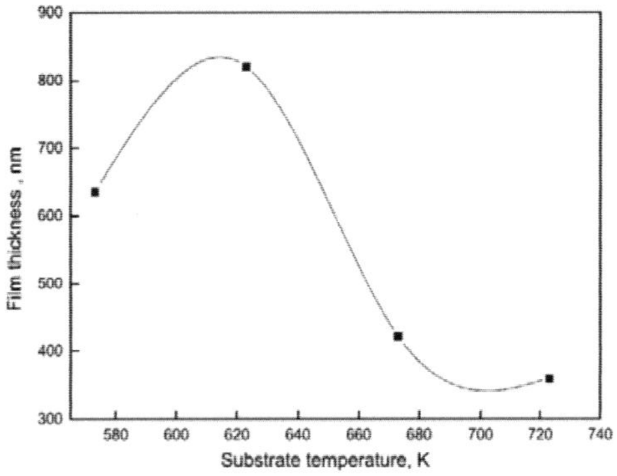

Figure 4. Variation of Co_3O_4 film thickness with substrate temperature.

The decrease in thickness for substrate temperature (lower than 623K) can be attributed to incomplete thermal decomposition of the spray droplets. For higher substrate temperatures, the thickness is found to decrease. The decrease in film thickness is attributed to an increase in the rate of re-evaporation of initial constituent with increasing substrate temperature.

3.2 *Optical characterization*

The optical transmission reflection of the Co_3O_4 films is shown in fig.6 for samples prepared at various substrate temperatures for a deposition time of 30min.

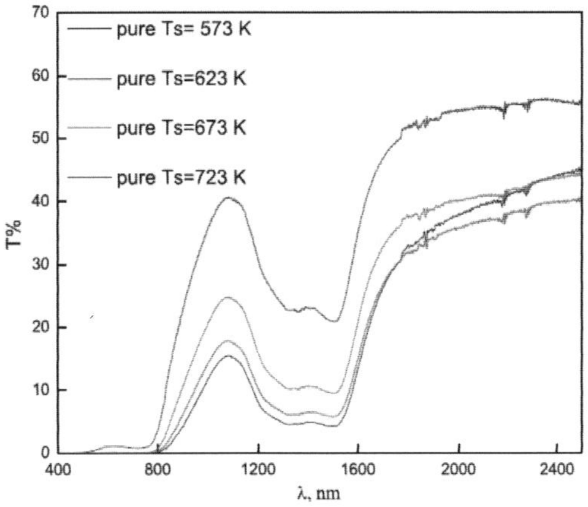

Figure 5. The transmittance and reflectance variation of Co_3O_4 thin films as a function of photon wavelength for different substrate temperatures at constant deposition time 30min

It can be observed that in particular spectral range, 800 nm $\leq \lambda \leq$ 1600 nm, an increase in substrate temperature improves the transmission. This improvement can be attributed to the improvement in stoichiometry of the films. The difference in thickness between the films prepared at 623 K and 673 K is larger than that between 673 K and 723 K as seen in figure 4. On the other hand, the difference in transmittance is larger between the films prepared at 673 K and 723 K. This indicates that the effect of thickness is less predominant than the effect of perfection and stoichiometry. The absorption coefficient α and extinction coefficient k were obtained from the transmittance and reflectance R curve, using the approximate formula [22].

$$T = \frac{(1-R)^2 \exp(-\alpha t)}{(1-R)^2 \exp(-2\alpha t)}$$

where 't' is the film thickness and 'α'' is related to the extinction coefficient k by

$$\alpha = \frac{4\pi k}{\lambda}$$

In this way the band gap is calculated with reasonable accuracy, as reported by curve [23].

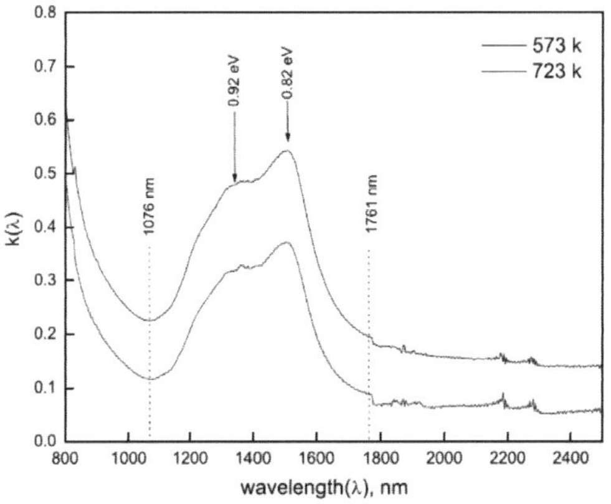

Figure 6. Variation of absorption index with wavelength for Co_3O_4 films prepared at various substrate temperatures

Figure 6 shows the variation in k as a function of wavelength. The behavior of k for the sprayed films in the covered range of λ is the same, therefore two representative curves, at substrate temperatures 573 K and 723 K, are present. Also two absorption bands are observed at 0.92 eV and 0.82 eV. The values obtained for k at higher wavelengths are in accordance with literature [24].

The optical energy gap of the sprayed films can be estimated from the optical data. The absorption coefficient of Co_3O_4 was found to follow the relation [22]:

$$\alpha = \frac{B\,(h\nu - Eg)}{h\nu}$$

The energy dependence of the absorption coefficient (Fig.8) indicates a direct allowed interband transition. Two straight lines portions are observed with intercepts at the energy gaps. Patil et al.[13-14] showed that there are two energy levels of the direct allowed transition at 1.3 eV and 2.0 eV.

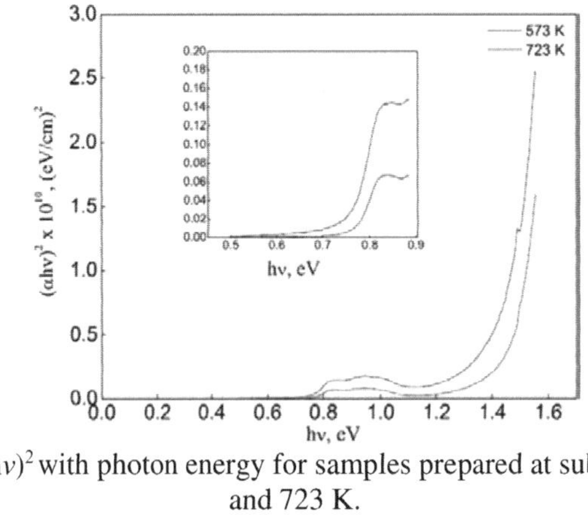

Figure 7. Variation of $(\alpha h\nu)^2$ with photon energy for samples prepared at substrate temperatures 573 K and 723 K.

Other band gaps were present at lower photon energies, about 0.7eV [20]. The energy equivalent to 0.7 eV and 1.47 eV is assigned to the charge transfer transition between the $Co^{2+}(\sigma^*e)$ to $Co^{2+}(\pi^*t_2)$ and $Co^{3+}(\pi^* t_2)$ to $Co^{2+}(\sigma^* t_2)$

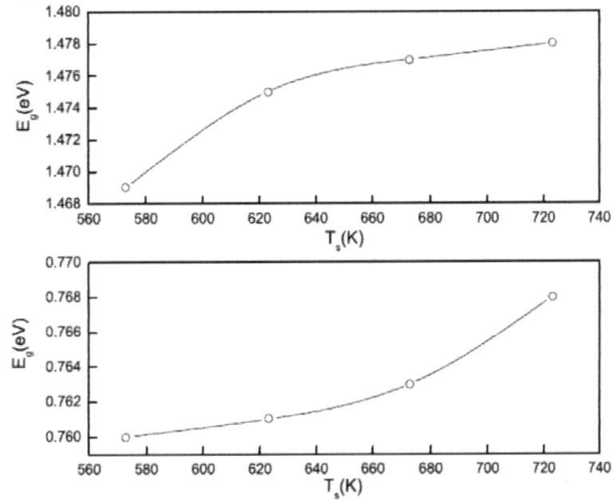

Figure 8. Variation of optical band gap of Co_3O_4 films as a function in substrate temperature

4. Conclusion

In present work, the effect of substrate temperature on the structural and opical properties of Co_3O_4 thin films was investigated. It is concluded that the crystallinity improved as substrate temperature increases. AFM images indicate the formation of agglomerates as well. T-λ curvs showed an increase in transmission as the substrate temperature rises. Also band gaps corresponding to Co^{2+} (σ^*e) to Co^{2+} $(\pi^* t_2)$ and Co^{3+} $(\pi^* t_2)$ to Co^{2+} $(\sigma^* t_2)$ recorded.

5. References

[1] NAS *phase diagram for ceramists* AM 1969 Ceramic Soc. 2 5

[2] Greenwood N N, Earnshaw A 1980 *Chemistry* 4*th* ed. Wiley Chicester p 767

[3] Spencer C D and Schoroeer D 1974 *J. Phy.Rev.* B9 3658

[4] Cook J G and Van Der Meer M P 1986 *J. Thin Solid Films* 144 165

[5] Kennedy R J 1995 *IEEE Trans. Magn.* 31 3829

[6] Hayes E, Bellingham B, Mark H, Galal A 1996 *J. Electrochem.* 41 337

[7] T Sasaki Q Li and Koshizaki N 1999 *J. Appl.*Phys. A69 115.

[8] Barrera E, Viveros T, Avila A, Quintana P, Morales M, Batina N 1999 *J. Thin Solid Films* 346 138.

[9] Singh N K, Singh J P, Singh R N 2002 *Int. J.of Hydrogen Energy* 27 895

[10] Tototzinata-Huitle T, Prokhove E, Mendoza- Galvain A, E.Urbina J, Gonzalez-Hevnandez J 2003 *J.Phy. and Chem. of solids* 64 975

[11] Morales U, Compero A, O.So ref12

[12] Kadam L D, Patil P S 2001 *J. Material Chemistry and Phy.* 68 225

[13] Patil P S, Patil P R, Kadam L D, Bawar S H 1999 *J. Electrochem* 15 307.

[14] Patil P S, Kadam L D, Lokhande C D 1996 *J. Thin Solid Films* 272 29

[15] Singh S P, Samuel S, Towari S K, Singh R N 1996 *Int. J. Hydrogen Energy* 21 171

[16] Bahlawane N, Ribera E F, Koshe-Hinghaus K, Brechling A, Kleineberg U 2004 *J. Appl. Cata. B.* 53 245

[17] Willenstein J, Burgmair M , Plescher G, Sulima T, Hildenbrand J, Buttner H, Eisde I 2003 *Senes.and Actu.* B 93 442

[18] Mc-Donald G E 1980 *J. Thin Solid Films* 72 83

[19] Mane A U, Shivashankar S A 2003 *J. Cry. Growth* 254 368

[20] Yamamoto E, Tanaka S 2003 *J.Appl. Phys.* 93 4158

[21] ICDD No. 9-418.

[22] Ashour A, Afify H H, Mohmoud S A 1994 *J. Thin solid Films* 248 253

[23] Couve S 1973 *J. Thin Solid Films* 15 72

[24] Athely P R, Urbanill F K, Tablet M F, Ma- Gahan W A 1996 *J. Vac. Sci. Technol.* A14 685

[25] Tolansky S 1970 *Multiple Beam Interferometry of Metal* Acdemic London p 55

Effect of sintering temperature on densification and mechanical properties of pressureless sintered CNT-alumina nanocomposites

N Bakhsh[1], F A Khalid[1] and Abbas S Hakeem[2]
[1]Faculty of Materials Science and Engineering, Ghulam Ishaq Khan Institute of Engineering Sciences and Technology, Topi, KPK, Pakistan
[2]Centre of Excellence in Nanotechnology, King Fahd University of Petroleum and Minerals, Dhahran, Kingdom of Saudi Arabia

E-mail: khoja72@gmail.com

Abstract. Carbon nanotubes (CNTs) exhibit excellent mechanical, electrical and thermal properties thus have been considered for applications in the production of nanocomposite materials. This work presents results of sintered CNTs-reinforced alumina nanocomposites that can be used for many structural and engineering applications. Gas purging sonication (GPS) was used to achieve homogeneous dispersion of CNTs in alumina powder. Nanocomposites were synthesized by conventional pressureless sintering technique using varying amounts of CNTs in alumina matrix. Densification, hardness and fracture toughness of the resulted nanocomposites were examined. It is found that considerable improvement in fracture toughness at 1 wt% CNT-alumina nanocomposite has occurred. Role of CNTs in improving the fracture toughness of nanocomposites is explained which is attributed to well known bridging and pullout mechanism of CNTs in the matrix.

1. Introduction

Carbon nanotubes (CNTs) have excellent properties and a distinct structure that have fascinated the researchers' interest worldwide. The high values of modulus (1 TPa), tensile strength (upto 100 GPa), large aspect ratio (1000-10,000) and low density of CNTs are being utilized to improve the toughness and strength of materials in various composite systems, including polymer, metal, cermets and ceramic matrix composites [1].

The higher degree of entanglement between CNTs; due to their high aspect ratio and strong Van der Waals forces, restrict them to get full benefits out of each individual CNT. Various chemical and physical methods are used for the exfoliation and dispersion of CNTs in the matrices. These methods include surface functionalization, shear mixing, high energy ball milling, ultrasonication and gas purging sonication etc. [2-9]. Ceramic materials possess high brittleness and low fracture toughness and low thermal stability which are undesirable for various engineering applications[10]. Alumina is extensively being used for many engineering utilities but; its toughness is too low to make it feasible for several advanced applications [11, 12]. The introduction of CNTs in alumina matrix enhances its mechanical properties by the suppression of grain growth during sintering and well known mechanism of CNTs pull out, crack bridging between grains during load applications. The improved fracture toughness and modulii of CNT-alumina nanocomposite makes it a strong candidate for an efficient structural material suitable for many advanced engineering applications [2, 13]. Several sintering techniques such as hot isostatic pressing, hot pressing, microwave sintering and spark plasma sintering (SPS) have been used to improve the mechanical properties of alumina and other ceramics [13-17]. However, all these processes impose many geometrical limitations to the sintered products and hence require simple shapes for the sintering purposes. Thus pressureless sintering technique is still widely being used due to its accommodation of intricate sintered parts [18]. The sintering of CNT-alumina nanocomposites was carried out by the same authors at 1600°C and the data had been published in reference [9]. Sintering at a further higher temperature is thought to be investigated to optimize the temperature range using pressureless sintering technique.

In the present work, pressureless sintering of CNT–Al_2O_3 nanocomposites is carried out at 1700° C in an inert atmosphere. The Vickers hardness and fracture toughness of the CNT–Al_2O_3 nanocomposites containing 1, 2 and 3 weight % CNTs are investigated. Synthesis and processing parameters are optimized and presented to explain the role of CNTs in CNT-alumina nanocomposites showing better fracture toughness at 1wt% CNT-alumina nanocomposites.

2. Experimental Procedure

Multiwalled Carbon nanotubes (diameter 10-40 nm) were purchased from Sun Nanotech, China. The reported purity level of CNTs was greater than 90%. α-Alumina powder, with an average particle size around 400 nm and purity >99%, was procured from Nabond Technologies, China.

The as-received CNTs and alumina powder were separately dispersed in de-ionized water through sonication and ball milling to get homogeneous and de-agglomerated suspensions. The suspensions were mixed together through a unique mixing and dispersion technique, which consisted of simultaneous sonicating and purging of nitrogen gas through the solution till no sonicating effects could be observed in the thick slurry. This novel technique of sonication and purging with N_2-gas for the de-agglomeration of CNTs and alumina resulted in a uniform mixture of the composite powder [9]. The slurry was dried at 120°C for 24 hours before sieving it through a 250 mesh to homogenize the particle size of the resultant composite powder. The powder was uniaxially pressed in a steel die to get pallets suitable for handling during the sintering process. The sintering was carried out at 1700° C for 15 minutes in flowing argon.

Particle size of alumina powder was analyzed with the help of Microtrac S3500 particle analyzer. Scanning electron microscope (Philips XL30) was used for microstructural investigation. Vickers hardness tester (DVK-2) was used for fracture toughness measurements with a load of 10 kg. A microhardness tester with Vickers' sensor was employed for the measurement of hardness at 200g load with 15 seconds of dwell time.

3. Results and Discussion

3.1. Morphology of CNTs and alumina powder

SEM image shown in figure 1(a) displays the surface features and morphology of as-received CNTs used in current research.

Figure 1. SEM images of (a) as-received CNTs, (b) alumina powder.

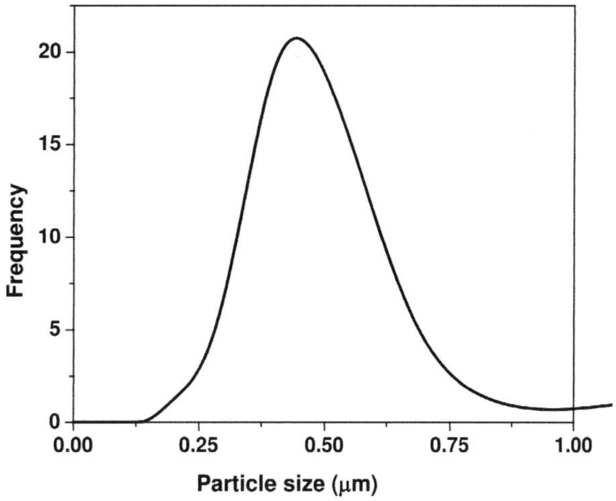

Figure 2. Particle size analysis of alumina powder [9].

These highly entangled and agglomerated CNTs have diameter in the range of 10-40 nm with a length of few microns. The alumina powder particle size ranges from less than 100 to 500 nm with irregular shape morphology as shown in figure 1(b). The approximate average particle size of the starting alumina powder is 0.4μm as shown in figure 2.

3.2. Sintering and Densification

The sintering process has produced a near full densification of around 99% for monolithic alumina. However, the addition of CNTs results in the reduction in final density of alumina nanocomposites as shown in figure 3. Density of the nanocomposite at 1 wt% CNT-alumina is achieved as 98.6% as compared to corresponding theoretical density, which is higher than the similar nanocomposites sintered at 1600°C [9]. The mass transportation caused due to bulk diffusion and pores elimination are most important reasons for enhanced densification of alumina [19].

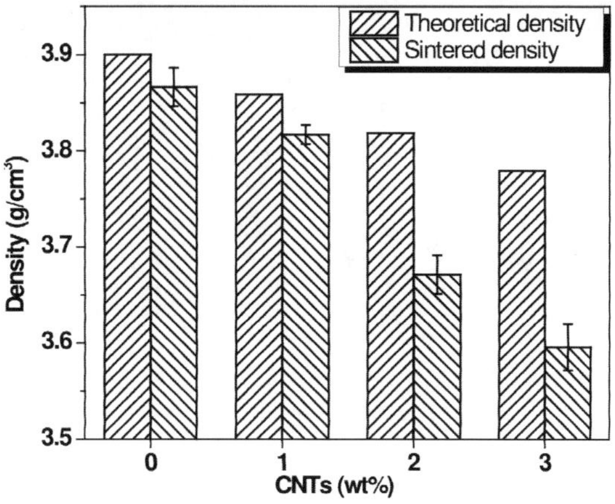

Figure 3. Comparison of sintered density of alumina and CNT-alumina nanocomposites.

Further addition of 2 and 3wt% CNTs in alumina result in reduced density values of 96.1 and 95.1 % respectively as compared to their corresponding theoretical densities. This reducing trend in density with increased CNT contents is due to CNTs stagnation at the grain boundaries which results in impeding the grain growth at the sintering temperature. The higher degree of CNT-agglomerations at a high CNT-concentration and the nano to submicron size range of alumina powder has the inherited porosity which also results in decreased densification of the nanocomposites [18-22].

3.4. Fracture Behavior and Phases

Figure 4 (a) reveals the SEM image of fractured surface morphology of alumina after sintering at 1700°C for 15 minutes. The finer alumina grains are surrounded by larger ones thus filling the gaps between the grains that help in enhancing the density of the sintered alumina. Average grain size of the sintered monlithic alumina lies between 6 ± 1μm. The fractured surface of pure alumina reveales mostly the intergranular fracture alongwith transgranular behaviour. The SEM image of fractured surface from 1 wt % CNT-alumina nanocomposite shows uniform and reasonably good dispersion of CNTs in the alumina matrix as shown in figure 4(b). The addition of 1wt% CNTs to alumina results in the decrease in average grain size to approximately 1.2 μm after sintering compared to monolithic alumina. The decreasing trend in grain size due to addition of CNTs is in accordance with the previous literature [9, 18].

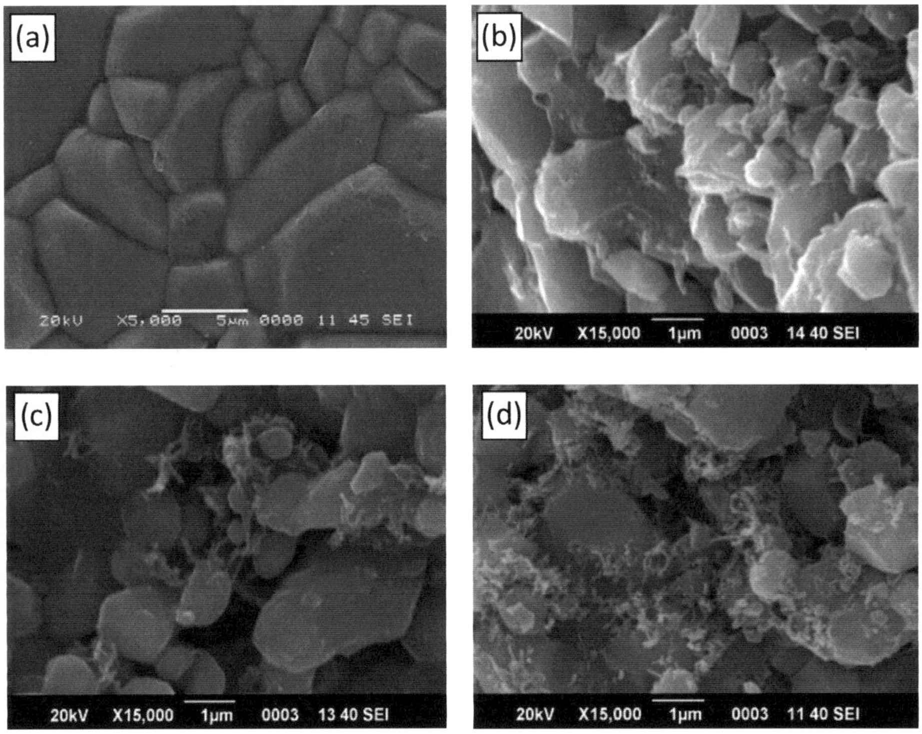

Figure 4. SEM images of the fractured surface of: (a) monolithic alumina, (b) 1wt%CNT-Al_2O_3, (c) 2wt%CNT-Al_2O_3 and (d) 3wt%CNT-Al_2O_3 nanocomposites.

The decrease in the grain size of CNT-alumina nanocomposite can also be attributed to the well dispersed nanosize carbonaceous impurities present in the as-received CNTs. This has also been confirmed that the addition of CNTs not only results in the pinning of grain boundaries but also retards the grain growth during sintering as described in reference [13]. Pull out pattern of CNTs indicates that these are embedded in the alumina grains and tied them up and hence resulted in the improved mechanical properties. At 2 and 3 wt% CNTs in alumina, the agglomeration can be observed as shown in figure 4 (c and d). The SEM images of CNT-alumina nanocomposites also depict the same intergranular and transgranular fracture modes [9]. The images show no evidence of CNTs degradation reflecting that the CNTs can withstand at 1700°C in an inert atmosphere using conventional sintering.

3.5. Mechanical Behavior

Figure 5(a) shows the comparison of hardness of sintered monolithic alumina and CNT-alumina nanocomposites. The monolithic alumina has a hardness of 27.5 GPa after sintering at 1700°C which is better than sample sintered at 1600°C as reported in the previous work [9]. At a concentration of 1wt% CNTs in alumina matrix, the hardness is reduced by 9% compared to monolithic alumina. On further increasing the concentration to 2 and 3wt% CNTs, the hardness values follow the same decreasing trends with 27% and 40% respectively compared to monolithic alumina.

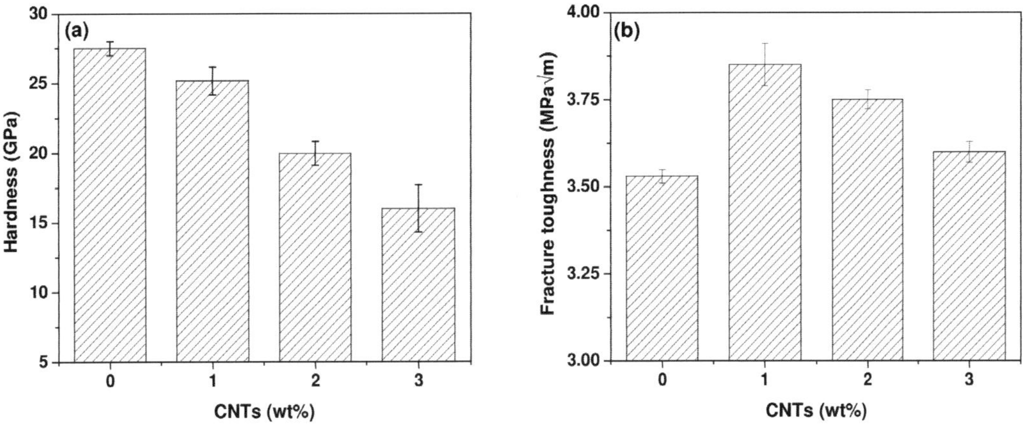

Figure 5. Comparison of mechanical properties of alumina and alumina-CNTs nanocomposites: (a) Hardness and (b) Fracture toughness.

Figure 5(b) shows the comparison of fracture toughness of CNT-alumina nanocomposites with monolithic alumina. Fracture toughness of monolithic alumina after sintering is determined as 3.50 MPa m $^{1/2}$. On addition of 1wt% CNTs, fracture toughness of the nanocomposite is improved by 10% which is attributed to better dispersion of CNTs using gas purging sonication, reduced grain size of alumina, and crack deflection and bridging due to CNTs [1, 18, 19]. On further increasing the CNTs concentration, fracture toughness results have shown a decreasing trend but still maintaining a higher value compared with monolithic alumina.

4. Conclusions

(1) The CNT-alumina nanocomposites were fabricated by a novel mixing technique of simultaneous sonication and gas purging prior to conventional sintering at 1700°C.
(2) Microstructure analysis presented a uniform distribution of 1wt% CNTs which were strongly bonded with the matrix and revealed a decrease in grain growth due to pinning effect showing an improvement in fracture toughness by 10% over monolithic alumina.
(3) Pressureless sintering at 1700°C provided better densification and hardness at 1wt% CNT-alumina nanocomposite but deteriorated the fracture toughness as compared to sintering at 1600°C.

5. Acknowledgement

Higher Education Commission (HEC) of Pakistan is acknowledged for the financial support and scholarship to Mr. Nabi Bakhsh.

6. References

[1] Zhang T, Kumari L, Du G, Li W, Wang Q, Balani K and Agarwal A 2009 *Composites Part A* 40 86-93.

[2] Yu M F 2004 *J. Eng. Mater. Technol.* 126 271.

[3] Srivastava D, Wei C and Cho K 2003 *Appl. Mech. Rev.* 56 215.

[4] Chen Y, Balani K and Agarwal A 2008 *Appl. Phys. Lett.* 92 011916.

[5] Wang J, Kou H, Liu X, Pan Y and Guo J 2007 *Ceram. Int.* 33 719-722.

[6] Peigney A, Laurent C, Flahaut E and Rousset A 2000 *Ceram. Int.*26 677-683.

[7] Balani K, Zhang T, Karakoti A, Li W, Seal S and Agarwal A 2008 *Acta Mater.* 56 571-579.

[8] Cho J, Boccaccini A R and Shaffer M S P 2009 *J. Mater. Sci.* 44 1934-1951.

[9] Bakhsh N, Khalid F A and Hakeem A S 2013 *Mater. Sci. Eng. A* 578 422-429.

[10] Zhan G D and Mukherjee A K 2004 *Int. J. App. Ceram. Technol.* 1 161-171.

[11] Jiang D, Thomson K, Kuntz J, Ager J and Mukherjee A 2007 *Scripta Mater.* 56 959-962.

[12] Thostenson E T, Ren Z and Chou T W 2001 *Compos. Sci. Technol.* 61 1899-1912.

[13] Inam F, Yan H, Peijs T and Reece M J 2010 *Compos. Sci. Technol.* 70 947-952.

[14] Kumari L, Zhang T, Du G, Li W, Wang Q, Datye A and Wu K 2009 *Ceram. Int.* 35 1775-1781.

[15] Langer J, Hoffmann M J and Guillon O 2009 *Acta Mater.* 57 5454-5465.

[16] Nygren M and Shen Z 2003 *Solid State Sci.* 5 125-131.

[17] Cheng J, Agrawal D, Zhang Y and Roy R 2002 *Mater. Lett.* 56 587-592.

[18] Zhang S C, Fahrenholtz W G, Hilmas G E and Yadlowsky E J 2010 *J. Eur. Ceram. Soc.* 30 1373-1380.

[19] Ahmad I, Kennedy A and Zhu Y 2010 *Wear* 269 71-78.

[20] Rahaman M N 2003 *Ceramic processing and sintering* (New York: CRC Press).

[21] Rice R W 2000 *Mechanical properties of ceramics and composites: grain and particle effects* (New York: CRC Press).

[22] Gupta A, Sharma S, Joshi M R, Agarwal P and Balani K 2010 *Mater. Sci. Forum* 653 87-130.

IOP Conf. Series: Materials Science and Engineering **60** (2014) 012060 doi:10.1088/1757-899X/60/1/012060

Preparation of zeolite from incinerator ash and its application for the remediation of selected inorganic pollutants: A greener approach

J Iqra, M Faryal, R Uzaira and T Noshaba
Department of Environmental Sciences, Fatima Jinnah Women University, The Mall Rawalpindi, 46000 Pakistan

E-mail: geminie1991@gmail.com

Abstract. Zeolites are potential materials and can conveniently be processed as adsorbents for the removal of environmental pollutants. A wide range of commercial zeolites have been marketed but due to high cost are of limited use. The present research offers a green approach for the synthesis of zeolite using Incinerator waste (ash) as precursor. The significance rests on the conversion of hazardous waste into a useful resource (adsorbent).Incinerator ash (IA) was converted hydrothermally under strongly alkaline conditions into zeolite (ZIA). The synthesized ZIA was applied as virgin adsorbent in batch mode for the removal of primary metal pollutants of Cadmium, Chromium and Lead. The residual concentration of each metal was analyzed on Flame Atomic Absorption Spectrophotometer. Each series of batch was conducted at four varying induced concentration of metal salts as a function of time. The synthesized adsorbent was characterized on FTIR spectrophotometer to assess the involvement of functional group in metal binding to adsorbent surface. The results depicted the performance of ZIA (1mg/Kg) in removing 99.5 %, 84% and 78% of Cadmium, Chromium and Lead, respectively, upon adsorbate-adsorbent contact for 30 minutes, at lower dose of ZIA. It was also found that higher is the induced concentration, less is the removal efficiency. It may be due to limiting factor of adsorbent dose. Correlation matrix suggests positive relationship of Pb and Cr, Pb and Cd, Cr and Cd. No negative correlation was found. The study recommends the reutilization of Incinerator ash as a potential adsorbent, which can greatly enhance the sustainability of useful resources.

1. Introduction

The discharge of produced water or pollution by textile and petroleum industries has become a significant issue of environmental concern due to importance of this natural resource in the energetic matrix. In most cases, petroleum related pollution is a chronic problem, due to port activities and the outflow of urban and industrial wastewater contaminated with petroleum and its derivatives. Industries discharge pollutants in liquid, solid and gaseous form in the environment. Heavy metal pollution is a problem associated with areas of intensive industry. The term heavy metal refers to any metallic chemical element that has a relatively high density and is toxic or poisonous at low concentrations. Examples of heavy metals include mercury (Hg), cadmium (Cd), arsenic (As), chromium (Cr), thallium (Tl) and lead (Pb) [1]. Heavy metals are dangerous because they tend to bioaccumulate. In order to treat these inorganic pollutants, various methods and technologies are used). Adsorption is one of the most feasible options available due to its high removal efficiency for inorganic pollutants.

Adsorption process may also provide an advantage over costly and energy-intensive alternative processes [2]. The adsorption process depends upon the type of adsorbent used [3].

Large amount of waste that is generated from the municipal activities can be converted to useful products. The low cost adsorbent is much preferable for the removal of pollutants [4]. The present environmental concerns over municipal solid waste ash disposal have sparked a resurgent interest in its conversion to value added product such as zeolites. Municipal solid waste ash has a tremendous potential for conversion to zeolite and its subsequent utilization as adsorbents for the removal of heavy metals in the industry [5]. In this investigation, hydrothermal processing was employed to recycle IA by converting it to zeolite-like materials through synthesis. More than 100 types of zeolite have been synthesized from various raw materials in the past. For example, coal-fired power plant fly ash has been successfully converted to many types of zeolite such as phillipsite, analcime, and zeolite P .Zeolite synthesis process using municipal solid waste ash as a source for silica and alumina has not been addressed extensively in the literature [6]. Zeolites are hydrated alumina silicates. Their structure consists of primary building blocks of inorganic tetrahedrons of silicon and alumina oxides. These atoms are strongly bonded together via oxygen bridges to form well-defined channels and cavities [7]. Zeolites are negatively charged as a result of the substitution of silica by alumina in the structure. Water molecules and cations, such as Na, K and Ca, are adsorbed on the pore surfaces [8] .These cations balance the negatively charged zeolite structure and are exchangeable species. Simply, any source that can provide a reactive form of silica and alumina can be used to synthesize zeolite under thermal high alkaline conditions [9] .Zeolite has been synthesized massively from pure silica and alumina chemical compounds, such as sodium silicate and sodium aluminates. Other sources including natural clay and fired coal ash have been also investigated [10].

The aim of this study is to find out the removal ability of ZIA using alkali reaction for harmful inorganic ions from wastewater and to find original ways to utilize waste IA.

2. Experimental

In this study the incinerator fly ash was obtained from national cleaner production center (NCPC) in Rawalpindi, Pakistan.

2.1. Samples preparation

Before the synthesis Zeolite, incinerator fly ash was characterized by FTIR. The total content of heavy metals was also determined on Flame atomic absorption spectrophotometer by US EPA 3050 method. The sample was pre treated and dried in oven at temperature 110 ^{o}C then milled until 80% of the sample passed through 40 micrometer sieve. The sieved sample was then subjected to treat hydrothermally.

2.2. Experimental setup

Zeolite synthesis process was performed by placing 5 grams of ash samples in a 50 ml conical flasks, then 50 ml of deionized water was added to the ash samples to obtain solid/ liquid ratio ,sodium hydroxide was added in different ratios to obtain 1.5N and 2.5N solutions. Then the flasks were sealed and placed in hot plate that was maintained either at 60 ^{0}C or at 100 ^{0}C for different periods of times including 6, 24 and 72 hours. Conditions that are mentioned above were used for another set of ash samples .After the hydrothermal process is completed the reacting solution was decanted and samples were washed three times with deionized water, the residue was separated from samples by using what man filter paper 60.[11].

2.3. Characterization of synthesized product

The synthesized products were characterized by the properties listed as follows: (1) Functional group identification, Spectra of adsorbent recorded from 4000cm-1 to 400 cm-1 on FTIR spectrophotometer (FTIR 8400, Shimadzu, Japan) ;(2) mineral species, by X-ray diffraction –XRD analysis using Cu anode material, 45 kV, 40 mA, and a scanning rate of one degree per second .The results of determined d-spacing and relative intensities for corresponding 2u values were analyzed by Xpert

High score PANalytical Malaysia ;(3) TGA and (4) SEM of sample to identify paricle size. So doing, each of the mineral species including zeolite like materials was identified according to the major and minor peaks. Due to the complexity of crystalline phases, the overlapping of peaks for different mineral species is a commonplace.

2.4. Preparation of Stock Solutions

2.4.1. Preparation of 1000 mg/L stock solution of Lead Nitrate
A salt of 0.1599 grams of Pb $(NO_3)_2$ was dissolved in 10 mL of deionized water and transferred to a 100 mL volumetric flask. The volume is made up to 100 mL with deionized water. The 1000 mg/L stock solution is prepared.

2.4.2. Preparation of 1000 mg/L stock solution of Cadmium Nitrate
A salt of 0.2744 grams of Cd $(NO_3)_2.4H_2O$ was dissolved in 10 mL of deionized water and transferred to a 100 mL volumetric flask. The volume is made up to 100 mL with deionized water. The 1000 mg/L stock solution is prepared.

2.4.3. Preparation of 1000 mg/L stock solution of Chromium Nitrate
A salt of 0.7695 grams of Cr $(NO_3)_2.9H_2O$ was dissolved in 10 mL of deionized water and transferred to a 100 mL volumetric flask. The volume is made up to 100 mL with deionized water. The 1000 mg/L stock solution is prepared.

2.5. Batch adsorption studies
The laboratory scale experimental studies were conducted using synthetic solutions prepared using deionized water with constant metal concentration for each of Cd, Cr and Pb separately. The uptake of heavy metals on the zeolite was carried out using the batch method.Batch adsorption experiments were conducted using 1mg dose of zeolite with 10 ml of solutions containing heavy metal ions of desired concentrations (50,100, 150 and 200mg/L) for each metals at ambient conditions. The solutions were placed for 60 min. The applied time intervals were 10, 20,30,40,50 and 60 min. and after that solutions containing heavy metals were filtered through Whatman 40 filter paper.

3. Results and Discussion

3.1. Characteristics of incinerator ash
After digestion of raw material, it was run on Flame atomic absorption spectrophotometer to check the metal concentration of Lead, Cadmium, Chromium, Copper, Nickel and Zinc. The metal concentrations of raw material are reported in Table 1.

Table 1: Metals concentration in incinerator ash

Metal	Concentration(mg/L)
Lead	2.4268
Cadmium	0.0423
Chromium	2.8387
Copper	12.1336
Nickel	2.2327
Zinc	7.887

3.2. Characteristics of synthesized product
The characterization of adsorbent material ZIA is done by using FTIR and XRD analysis.

3.2.1. FTIR characterization of adsorbent
The FTIR characterization of these adsorbent is showed in figure. In it the major peaks and band were taken and studied to find out the functional group present in Zeolite. Compare these spectrums to the

standardized FTIR spectrum table. In all type of Zeolite the following functional groups exist showed in Table 2 (FTIR 8400, Shimadzu, Japan).

Table 2: FTIR Characterization of adsorbents

Wave number (cm^{-1})	Assignment
3600 - 3650 cm^{-1}	Hydroxyl group
1640 cm^{-1}	Adsorbed H_2O
1438 -1452 cm^{-1}	Aluminum containing entities; $Al_xO_y^{n+}$
1250 - 920 cm^{-1}	Tetrahedral $SiO_4/_2$ / $AlO_4/_2$
650 - 500 cm^{-1}	Si-O/ Al-O bend; external linkages
200 - 50 cm^{-1}	Cationic vibration; far infrared region

Vibrations of the frameworks of zeolites in following figure give rise to typical bands in the mid and far infrared. A distinction is made between external and internal vibrations of the Si or Al $O_4/_2$ tetrahedral. The original assignments of the main IR bands in zeolite adsorbent ZIA is as follows: internal tetrahedral 1250 - 920 cm^{-1}, asymmetrical stretch (n_{asym}); 720 - 650 cm^{-1}, symmetrical stretch (n_{sym}; 500 - 420 cm^{-1}, Si/Al-O bend; external linkages: 650 - 500 cm^{-1}, double ring vibrations; 420 - 300 cm^{-1}, pore opening vibrations; 1150 - 1050 cm^{-1}, asymmetrical stretch; 820 - 750 cm^{-1}, symmetrical stretch.The positions of bands due to vibrations of external linkages are often very sensitive to structure [12].

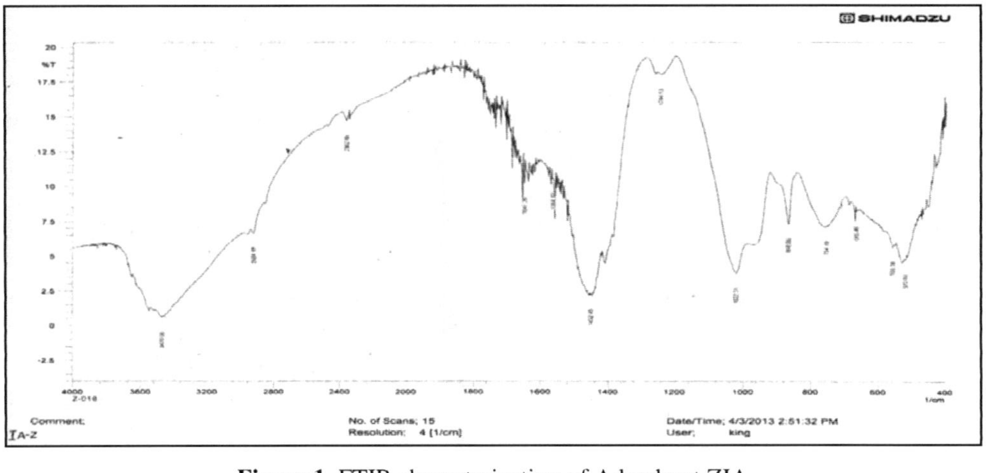

Figure 1. FTIR characterization of Adsorbent ZIA

3.2.2. *X-ray diffraction analysis*

Results of X-ray diffraction analysis have proved that hydrothermal processing is capable of converting IA into zeolite like materials. After synthesis several types of zeolite have been identified on basis of different conditions. XRD powder pattern for the synthesized product resulted from various IA mineralizer ratios are similar, but different in relative intensities for various mineral species [13].

Table 3. XRD identified pattern of compounds in hydrothermally treated Zeolite at 2.5N NaOH at 100°C

Visible	Ref. Code	Score	Compound Name	Displacement [°2Th.]	Scale Factor	Chemical Formula
*	01-086-2335	58	Magnesium Calcium Carbonate	0.000	0.820	(Mg.064 Ca.936) (CO_3)
*	01-083-2467	39		0.000	0.277	SiO_2

3.2.2. *X-ray diffraction analysis*

Results of X-ray diffraction analysis have proved that hydrothermal processing is capable of converting IA into zeolite like materials. After synthesis several types of zeolite have been identified on basis of different conditions. XRD powder pattern for the synthesized product resulted from various IA mineralizer ratios are similar, but different in relative intensities for various mineral species [13].

Table 3: XRD identified pattern of compounds in hydrothermally treated Zeolite at 2.5N NaOH at 100°C

Visible	Ref. Code	Score	Compound Name	Displacement [°2Th.]	Scale Factor	Chemical Formula
*	01-086-2335	58	Magnesium Calcium Carbonate	0.000	0.820	(Mg.064 Ca.936) (CO_3)
*	01-083-2467	39	Silicon Oxide	0.000	0.277	SiO_2
*	98-017-0486	27	Zeolite	0.000	0.133	O_2Si_1
*	01-078-1254	41	Silicon Oxide	0.000	0.238	SiO_2

On the other hand, different types of compounds were identified when 1.5 N NaOH was used; whereas magnesium calcium carbonate and silicon oxide present which shows the existence of zeolite shown in fig 3[14].

3.2.3. *TGA of synthesized product*

The number of water molecules attached with the hydrothermally synthesized zeolite and its thermal stability was investigated using TGA. Upon heating the sample from room temperature a continuous weight loss of 55% is clearly observed in TGA. This weight loss is may be due to the dehydration of physically adsorbed water. When the sample is further heated in the temperature range of 200 to 700°C the weight loss observed is attributed to desorption of remaining water enclosed in the material matrix. Reduced weight loss in this region was observed with increase in crystallization of the sample which is consistent with the fact that zeolite becomes more hydrophobic as the Al content decreases [14].

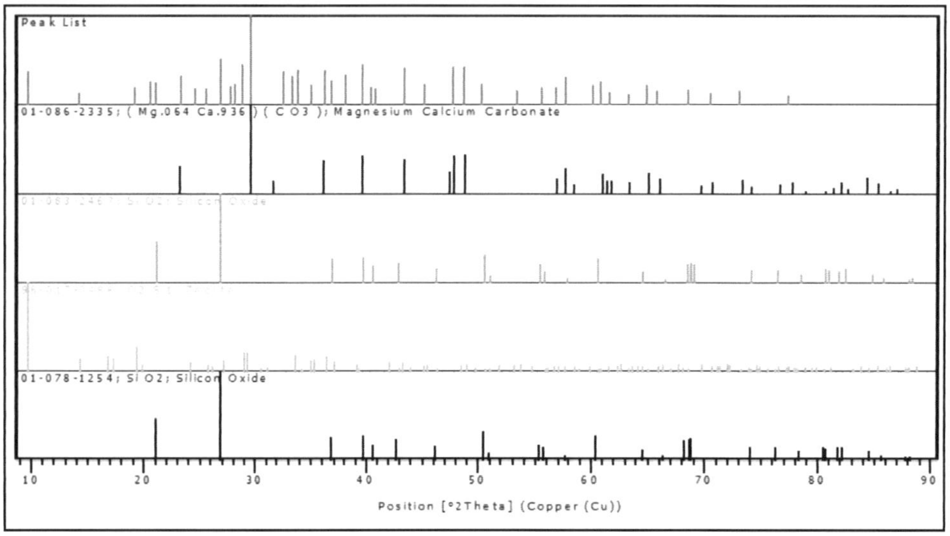

Figure 2. Pattern of X-ray diffraction of hydrothermally treated zeolite at 2.5N NaOH with at 100°C

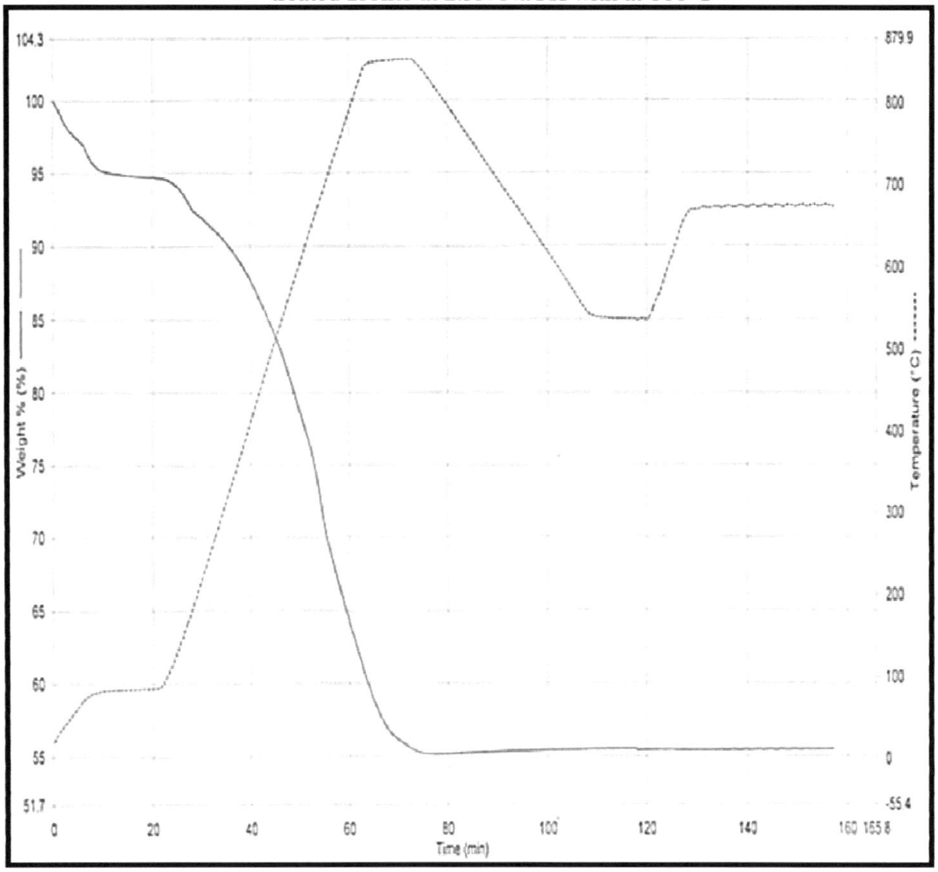

Figure 3: TGA of synthesized I.A

3.2.4. SEM of synthesized product

The synthesized product as examined under an electronic microscope. The micrograph is presented in Figure 4 by using different magnifications. The sample revealed the presence of some cubic crystal of zeolite. The micrograph indicated a plate like structure which is an indication that the silica and alumina are sliding over one another. Some partial damage of the plate like structure was observed in the synthesized sample. The SEM image revealed a uniform particle size of the sample with a regular shape. The synthesis SEM results give a narrow distribution of particle size with average crystal size of 1μm [14].

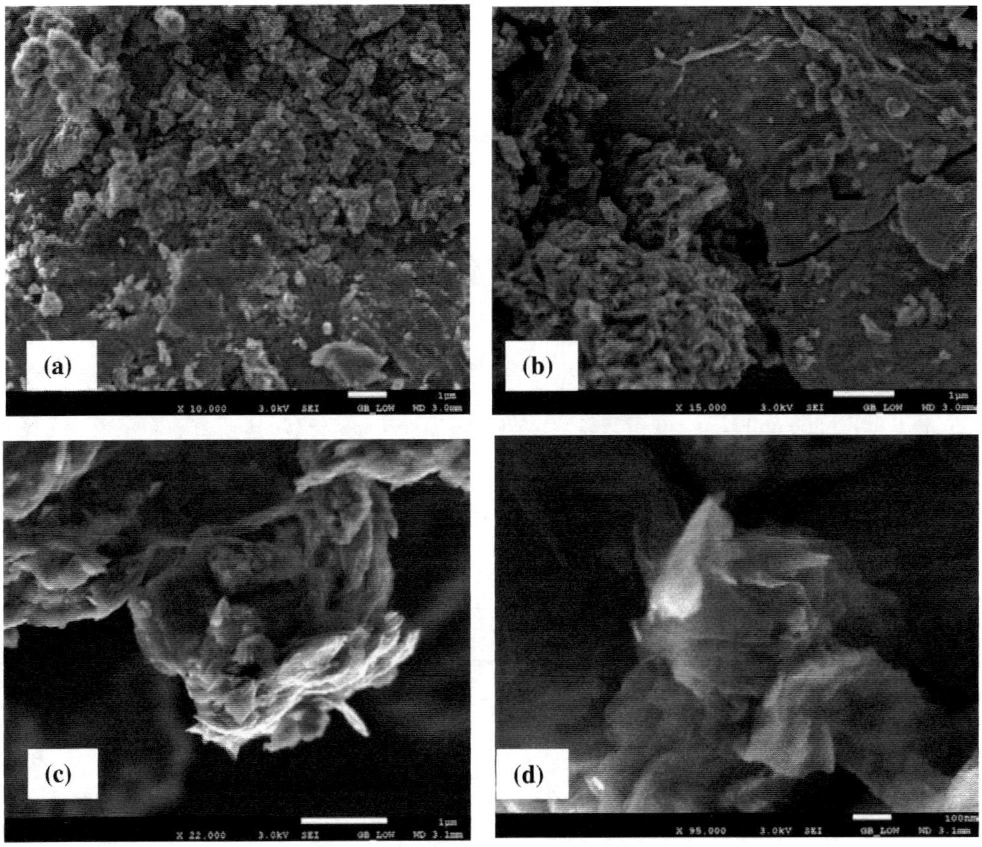

Figure 4: SEM of adsorbent at different magnifications **(a)** 10,000 X, **(b)** 15,000 X, **(c)** 22,000X and **(d)** 95,000X

3.3. Adsorption of heavy metals

3.3.1. Effect of contact time

The studies of different contact time helps in determine the removal of heavy metal from constant mass of adsorbent (1mg) at different time intervals of 10, 20,30,40,50 and 60 min. The results given in Figure showed that the adsorption of the studied heavy metals has gradual increase with contact time because ions has maximum time to hold on to the surfaces of zeolite by physical or chemical bonding until 50 min then the removal percent became constant i.e. equilibrium is attained[15]. From the calculated removal percentage, the efficiency of IA-Z is 99.5 %, 84% and 78% of Cadmium, Chromium and Lead, respectively at optimum contact time although. This may be attributed to the fact that both Zeolite can act as ion exchanger. Accordingly, the pore diameter is not effective but the operating capacity depends up on other parameters such as concentration of adsorbent, contact time and particle size [16].

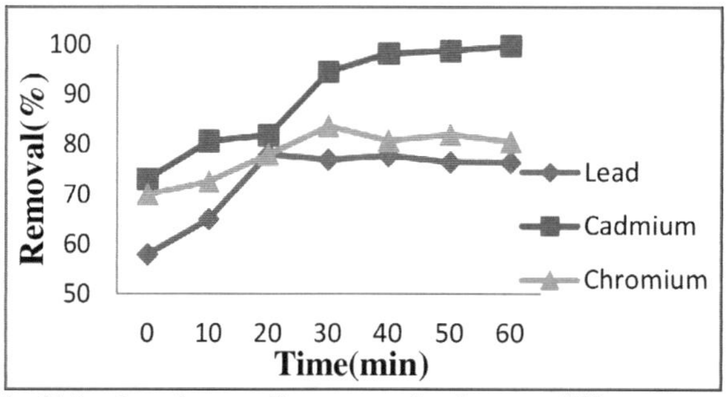

Figure 5: Removal of Metals using zeolite as an adsorbent at different contact time (optimum operating conditions).

3.3.2. Effect of induced concentrations

Four different concentrations (50,100,150,200mg/L) of Heavy metals solution were introduced on Zeolite adsorbent with different contact time.It is investigated from data that change in induced concentration has significant impact on adsorption process. It is deducted that adsorbent showed the decreasing trend of adsorption with increase in induced concentrations of metal solutions. The reason behind decrease in adsorption is attributed to the less number of available adsorption active sites as compared to heavy metal concentrations [17].From the result showed in Figure the percentage removal at initial concentrations of the heavy metals are high, particularly for Lead, Cadmium and Chromium whereby the ability to absorb Pb, Cd and Cr cations at higher concentrations is less because it involved energetically less favorable sites of zeolite in the uptake of heavy metals. The higher metal uptake at low concentration is due to the availability of greater surface area with active centers on the adsorbent for lesser amounts of adsorbate ions [18].

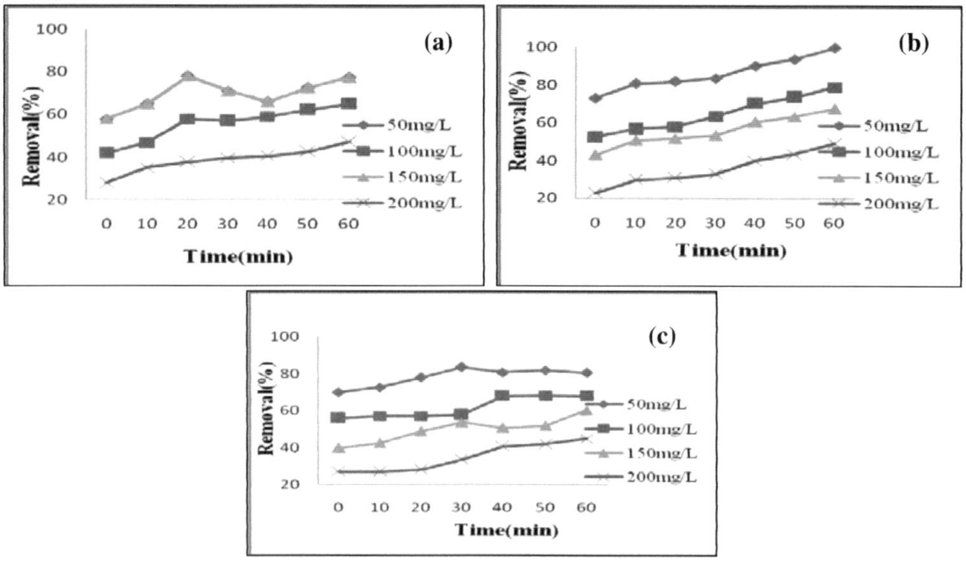

Figure 6: Removal of Metals (a) Lead, (b) Cadmium and (c) Chromium using ZIA adsorbent at different induced concentration

3.3.3. Batch Adsorption studies for Industrial Effluents

Industrial effluents were collected from point source of Plastic factory (PE) located on Millet road (Sargodha road), Textile I (TE-I) and Textile II (TE-II) located on Bawachuk, Manawala, Faisalabad. Water samples were also collected from D- separator(D-S) and C- separator (C-S) of Attock oil refinery limited located in Morgah, Rawalpindi The industrial effluents were subjected to batch removal for the removal of metals and azo dyes using zeolites synthesized from incinerator ash under varying experimental parameters. The results were analyzed on FAAS. The 1 mg dose of zeolite was placed in each effluent taken from different industries. The effluents were placed for 60 minutes for the removal of metals .The samples were then filtered and analyzed .The results are shown in following figure 7. The results are quite similar to the synthetic solution batch experiment performed in lab. The Sequence of removal efficiency of IA-Z is Cd > Cr >Pb >Cu >Ni >Zn. Along with the heavy metals Cu, Zn and Ni removal from effluents are also investigated.

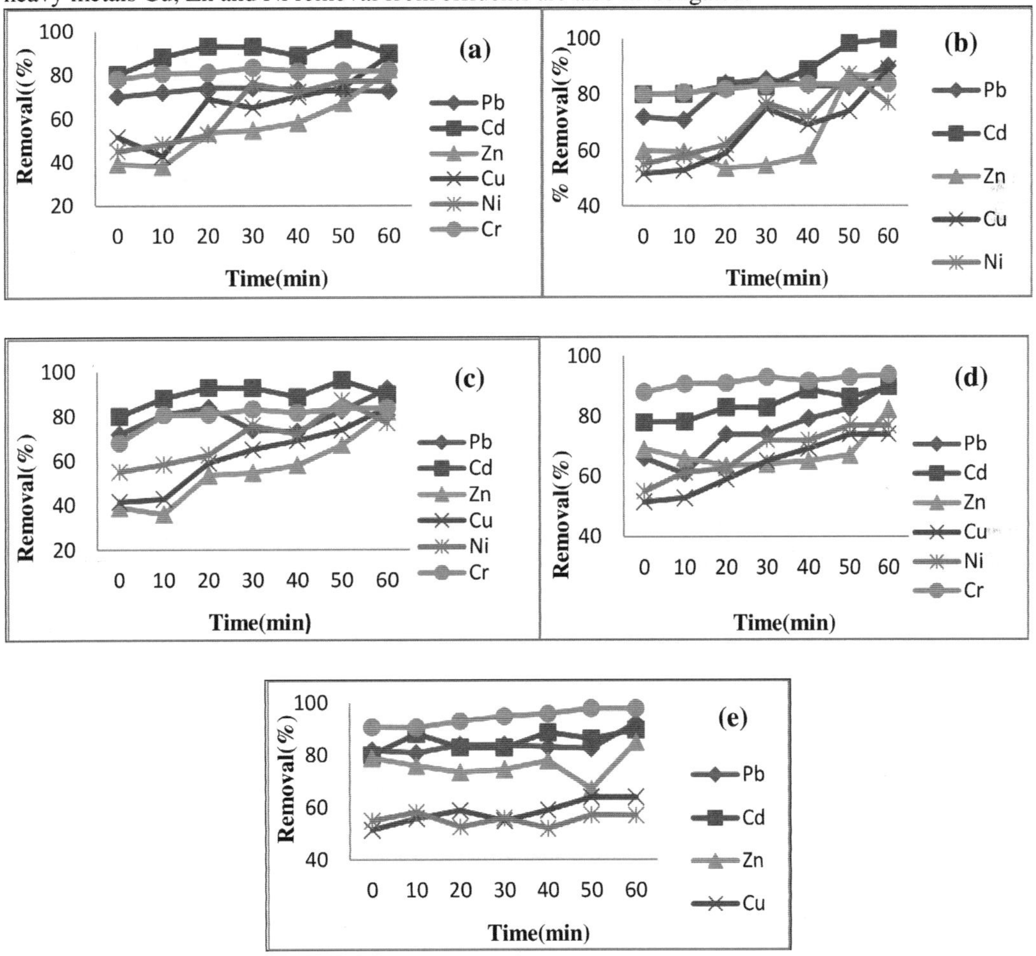

Figure 7: Removal of Metals by using CFA-Z adsorbent from industrial effluents (a) PE, (b) TE-I, (c) TE-II, (d) D-S and (e) C-S

4. Conclusion

The zeolitic type materials synthesized by hydrothermal conversion of incinerator ash demonstrated promising adsorption performance for the three heavy metal contaminants tested (Cd^{2+}, Cr^{2+} and Pb^{2+}). The key findings of this study are:

- Adsorption studies indicated higher affinity for Cd, Cr and Pb for ZIA.
- The adsorption performance of ZIA was generally superior to that of the natural zeolite for both synthetic heavy metal solutions and contaminated industrial effluents. ZIA achieved 99.5 %, 84% and 78% of Cadmium, Chromium and Lead respectively from 1 mg sorbent dosage.
- The Sequence of average removal of metals at optimum operating conditions on ZIA as Cd > Cr >Pb

Given the positive results found in this study, further work aims to optimize the performance of the ZIA. The affinity of this material for heavy metals, its flexibility and low-cost, also open the possibility for its application in other industrial processes, including chemical and catalyst synthesis. In any case, the reutilization of incinerator ash achieves the goal of reducing the burden of industry on its safe disposal and improves the sustainability of waste incineration technology.

5. References

[1] Tarek S. J, Hanan S. I, Maksoud A.E, Wakeel S.T. (2010). Application of Zeolite prepared from Egyptian kaolin for removal of heavy metals:I. Optimum conditions. *Desalination*, 34-40.

[2] Argun M.E, Dursum S. (2008). A new approach to modification of natural adsorbent for heavy metal adsorption. *Bio resource Technology*, 2516-2527.

[3] Cristian C, Rafael G, Rena A.J, Maria T. (2006). Cr (III) exchange on Zeolites obtained from kaolin and natural mordenite. *Microporous and Mesoporous Materials*, 220-231.

[4] Chojnacki A, Chojnacka K, Hoffmann J , Gorecki H.(2004) .The application of natural Zeolites for mercury removal:from laboratory tests to industrial scale. *Minerals Engineering*, 933-937.

[5] Ojha K, Narayan C, Amarnath S. (2004). Zeolite from fly ash: synthesis and characterization. *Material sciences* , 555-564.

[6] Dinu M.V, Dragan E.S. (2010). Evaluation of Cu^2, Co^2 and Ni^2 ions removal from aqueous solution using a novel chitosan/clinoptilolite composite: kinetics and isotherms. *Journal of Chemical Engineering,* 157-163.

[7] Erdem E, Karapinar N, Donat R. (2004). The removal of heavy metal cations by natural Zeolites. *Journal of Colloid and Interface Science*, 280-309.

[8] Auerback S.M, Carrado K.M, Dutta P.K. (2003). *Handbook of Zeolite Science and Technology*. Marcel Dekker, Inc., New York.

[9] Baerlocher C, Meier W.H, Olsan D.H. (2007). *Atlas of Zeolite framework types, 6th edition.* Elsevier.

[10] Maria T. A, Francisco C F. B, Giselle S. C. R, Cícero P.M, Ronaldo F. N. (2012). Metal Ions Removal From Synthetic Solutions and Produced Water Using Activated Zeolite. *International Journal of Civil & Environmental Engineering* , 20-25.

[11] Chiang Y.W, Ghyselbrecht K, Santos R.M, Meesschaert B, Martens J.A. (2012). Synthesis of zeolitic-type adsorbent material from municipal solid waste incinerator bottom ash and its application in heavy metal adsorption. Catalysis Today, 190, 23-30.

[12] Karge H. G. (1991). Characterization by IR spectroscopy. In Karge H. G., *Verified Syntheses of Zeolitic Materials,2nd Revised Edition.* Berlin, Germany: Institut der Max-Planck Gesellschaft.

[13] Li, G. (2005). FT-IR studies of zeolite materials: characterization and environmental applications. *dissertation*, University of Iowa

[14] Omisanya N.O, Folayan C.O, Aku S.Y, Adefila S.S. (2012). Synthesis and characterisation of zeolite a for adsorption refrigeration application. *Advances in Applied Science Research* , 3746-3754

[15] Hernandez M. V, Perez C.M.A, Mendoza C. D.I, Moreno V.M.R, Bonilla P.A. (2013).Competitive adsorption of dyes and heavy metals on zeolitic structures. *Journal of Environmental Management* , 213-221.

[16] Halimoon N, Goh R.S.Y. (2010). Removal of Heavy Metals from Textile Wastewater using Zeolite. *Environment Asia* , 124-130.

[17] Wang S, Mehdi S, Li L, Zhu Z.H. (2005). Coal ash conversion into effective adsorbents for removal of heavy metals and dyes from wastewater. *Elsevier,* 243-251

Preparation of Silver Nanoparticles from Synthetic and Natural Sources: Remediation Model for PAHs

Maryam Abbasi, Fatima Saeed and Uzaira Rafique
Department of Environmental Sciences, Fatima Jinnah Women University, The Mall Rawalpindi, 46000 Pakistan

E-mail: maria_zulfiqar13@yahoo.com

Abstract. The emergence of nanoscience and technology is gaining popularity with an increasing demand for metal nanoparticles applicability in various areas such as electronics, catalysis, chemistry, energy and medicine. Metallic nanoparticles are traditionally synthesized by wet chemical techniques, where the chemicals used are quite often toxic and flammable. In this work, an attempt is made to compare the efficiency of two different synthesis methods and application of each for the remediation of poly aromatic hydrocarbons (PAHs). In this regard, silver nanoparticles are prepared by green and wet chemical method using plant extract of garlic (Allium sativum). The extract is known to reduce the metal during synthesis and acts as stabilizing ligand. These synthesized silver nanoparticles (Agp) and (AgW) were applied as adsorbents in synthetic batch mode experiments at varying parameters of pH and temperature. A concentration of 0.01mg/L of Phenanthrene, Anthracene, and Pyrene were induced at fixed dosage of 1mg/Kg of adsorbent. Residual concentration of each PAH was analyzed on UV-Visible spectrophotometer. The results indicated that both adsorbents follow the sequence of Phenanthrene>Pyrene>Anthracene with optimal removal of higher than 85% in each case. A distinguishing privilege is attained by Agp adsorbent showing 3, 3 and 11 orders of magnitude higher efficiency than Agw. It may be attributed to more functional groups in the plant extract participating in binding of PAH to the surface. Each synthesized adsorbents was characterized by FTIR, SEM and EDX. The average particle size was determined to be of the order of 13-26 nm. The study concludes the use of alternate economical and green adsorbents for control of poly aromatic hydrocarbons (PAHs).

1. Introduction

Toxic organic pollutants are released to water with the rapid industrialization of human society. Polycyclic aromatic hydrocarbons (PAHs) are a class of diverse organic compounds present in the environment with adverse impacts on ecosystems and human health. PAHs are among the most carcinogenic, teratogenic, mutagenic and toxic contaminants [1] and are classified among persistent organic pollutants. PAHs are known to pose, even at very low concentrations, a threat to environment due to their benzene structures [2].

PAHs have received immense attention because of their omnipresence in industrial and municipal wastewater and can be generated from both natural processes and anthropogenic activities [3]. PAHs are by-products of incomplete combustion (pyrolysis) of wood, oil, coal, garbage or hydrolysis of organic materials [4].

Removal of PAHs is of considerable interest in environmental analysis. Several methods including immunoassay [5], gas chromatography [6] and high performance liquid chromatography (HPLC) with UV-Vis absorbance or fluorescence detection [7] and capillary electrophoresis (CE) equipped with laser-induced fluorescence [8] have been used for the determination of PAHs. These methods are expensive, time consuming, require large sample volumes as well as large amount of organic solvent with separation and extraction procedures [9].

Adsorption technology is regarded as the most promising one due to its low cost, high efficiency and simple to operate for removing trace levels of effluents. Traditional sorbents could remove effluents from wastewater but the low sorption capacities and efficiencies limit their application. Nanoparticles have exhibited much higher efficiency and faster rates in water treatment [10] due to high specific surface area and occupy a large number of unsaturated atoms on the surfaces [11].These specific advantages enhance adsorption capacity for the removal of organic and inorganic pollutants [12].

Recently metal based nanoparticles have been successfully applied for removal of various toxic materials due to high ordered structure, high mechanical and thermal strength. Sumesh et al., [13] applied synthesized silver nanoparticles for the removal of mercuric ions present in contaminated water. Removal of methyl orange (MO) from aqueous solutions was achieved by silver nanoparticles loaded on activated carbon [14]. Liu *et al.*, [15] removed trace Cr (VI) ions from aqueous solution by titanium oxide–Ag composite adsorbents.

The present investigation is step forward to prepare Silver based nanoparticles from plant extract (*Allium Sativum*) following slight modification of method reported by White et al., [16] and extend its application for the removal of persistent organic pollutants. Use of plants in synthesis of nanoparticles provide advancement over chemical and physical method as it is cost effective and environment friendly that do not require any special culture preparation and isolation techniques [17].

For comparison, Ag particles are also synthesized from wet chemical co-precipitation method to determine the efficiency of both as adsorbents in removing selected PAHs.

2. Experimental

2.1. Synthesis of Silver nanoparticles

Siver nanoparticles were synthesized separately from plant extract (*Allivum sativum*) and wet chemical method. The products obtained are coded as Ag_p and Ag_w, respectively.

A procedure adapted from White et al., [16] has been used for the preparation of garlic extract from *Allivum sativum*. The extract obtained was then subjected to silver particles synthesis using the following protocol.

Allivum sativum (2.5 ml) extract solution was added to 50ml of 0.98mM of $AgNO_3$ in distilled water. After standing of two hours, change in color to light orange was observed. The solution was allowed to age for one week to yield a deep orange-brown color and monitored on UV-Visible spectrophotometer (full scan range 190-1100 cm^{-1}) to indicate the formation of silver nanoparticles. 10 ml of 2M NaOH solution was added to facilitate the precipitate formation. Grayish silver nanoparticles were obtained and dried under vacuum.

Silver particles were also synthesized by wet chemical co-precipitation method adapted from Moghaddam et al, [18]. The beaker containing 0.9 M aqueous solution of sodium hydroxide was heated at 55°C followed by drop wise addition of 0.45 M aqueous solution of silver nitrate($Ag(NO_3)_2.4H_2O$) under speed stirring. The precipitates of Ag nanoparticles were collected after 2 hours standing, cleaned with distilled water, ethanol and finally air dried at 60 °C.

2.2. Characterization

Each synthesized silver nanoparticle was subjected to characterization by standardized spectroscopic tools of Fourier Transform Infrared (FTIR) and UV-Visible spectroscopy. FTIR spectra of pressed pellet were recorded from 4000 cm^{-1} to 400 cm^{-1} on FTIR-8400, Shimadzu, Japan. The dried massof garlic extract (reducing agent) was also pressed into pellet for observation under FTIR. UV-VIS spectra were fit with Gaussian curves for measurement of λ_{max} keeping distilled water for background correction. The samples surface morphology and size was determined with Scanning Electron Microscopy (SEM) coupled with EDX. Sample was sonicated and analyzed at different magnifications on 6490(LA) JEOL machine with gold sputtering at a potential of 20 kV prior to recording SEM. EDX analysis was done on JED-2300 analysis station.

2.3. Batch Experiment for Adsorption

The synthesized silver nanoparticles (Ag_P and Ag_w) were applied as adsorbents in time-dependent batch experiment to evaluate the removal of poly aromatic hydrocarbons. Each batch was investigated at variable temperature and pH to determine the optimal removal efficiency. The following general protocol is adopted for batch experiment.

Synthetic solution of known concentration (0.01 mg/L) of each poly aromatic hydrocarbon was prepared and a known volume (5ml) of it was added in test tubes containing 1mg of each adsorbent. The aliquot was drawn after 5 minutes, filtered and analyzed on UV-visible spectrophotometer at wavelengths of 250 nm, 260 nm and 241 nm for phenanthrene, anthracene, and pyrene, in respective order.

3. Results and Discussion

3.1. UV-Visible analysis

The process of synthesis of silver nanoparticles (Ag_P) was followed on UV-Vis spectrophotometer by recoding the spectra of dark brown colored mixture of garlic extract and silver nitrate. Reduction of silver ions can easily be observed by UV-vis spectra presented in Figure 1. It is clearly visible that optimum absorbance is obtained at 400 nm, depicts the characteristic of surface plasmon absorption band [19]. This band is attributed to silver nanoparticles free electrons [20]. Saha et al., [21] reported absorption band for synthesized silver particles at 420 nm due to colloidal silver.

A broadening of peak in the range from 400-410 nm is also observed indicating that the particles are polydispersed [22], further confirmed by SEM images.

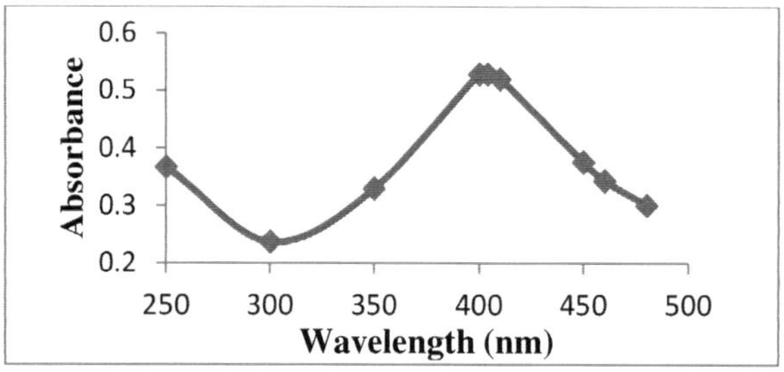

Figure 1. UV-visible spectra of silver nanoparticles synthesized from *Allium sativum* extract (Ag_P).

3.2. FTIR characterization of silver nanoparticles

The important frequencies (in cm^{-1}) recorded under FTIR for *Allium sativum* and two synthesized silver particles (Ag_P and Ag_w) are presented in Figure 2.

FTIR spectra of garlic extract consist of S=O (at 1026 $cm^{-1)}$ and S group (at 1261 $cm^{-1)}$ indicates the presence of allicin as major component [23]. Further elucidation of FTIR spectra of garlic extract alone shows large –OH and –CH stretches at 3373 cm^{-1} and 2931 cm^{-1}, characteristic of nonstructural sucrose and fructose [16]. On the other hand, the synthesized silver nanoparticles (Ag_P) represent a peak shift of 41 cm^{-1} in –OH and –CH stretches confirming its complete formation.

FTIR spectra of silver nanoparticles (Ag_w) synthesized by wet chemical method presents two peaks of absorption bands at 596 cm^{-1} and 848 cm^{-1} representing presence of Ag-O bond [24].

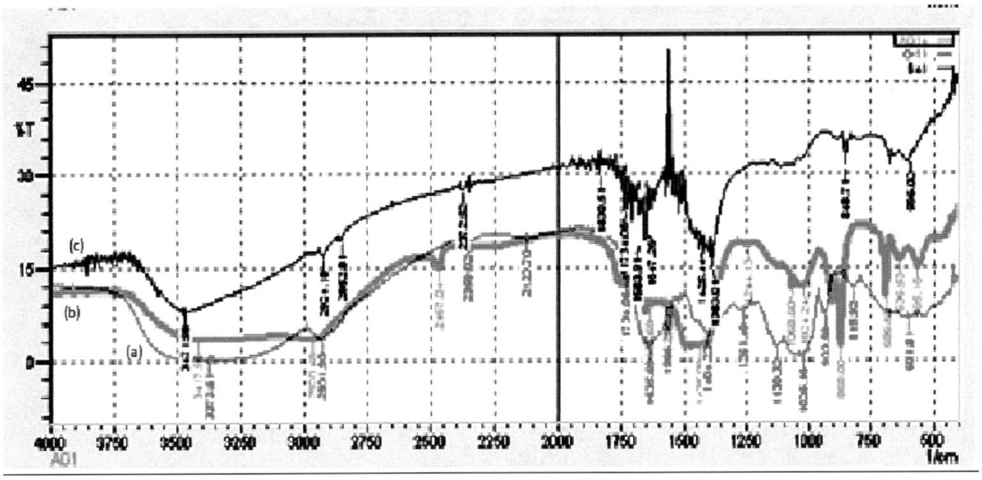

Figure 2. (a) FTIR spectra of dried powder of *Allium sativum* extract. (b) silver nanoparticles synthesized from *Allium sativum* extract (Ag_p). (c) silvernanoparticles synthesized by wet chemical method (Ag_w).

3.3. SEM & EDX analysis of silver nanoparticles

SEM images of synthesized silver nanoparticles Ag_p and Ag_w are shown in Figure 3 (a and b), respectively, whereas Figure 3(c) image presents the garlic extract loaded with silver particles.

It is evident that silver particles synthesized by two routes can safely be regarded as nanoparticles showing particle diameter range of 13 to 26 nm for Ag_p and 16 to 28 nm for Ag_w.

EDX spectra and quantitative data is reported in Figure 4(a and b), showing signals having 4.97 and 59.57 mass % of silver in Ag_p and Ag_w, respectively. Higher mass percentage of silver in Ag_w may be assigned to higher molar ratio of reagents used in the synthesis protocol in comparison to Ag_p. The presence of oxygen, carbon, sodium and calcium signals in Ag_p indicates their origin from biomolecules bound to the surface [25].

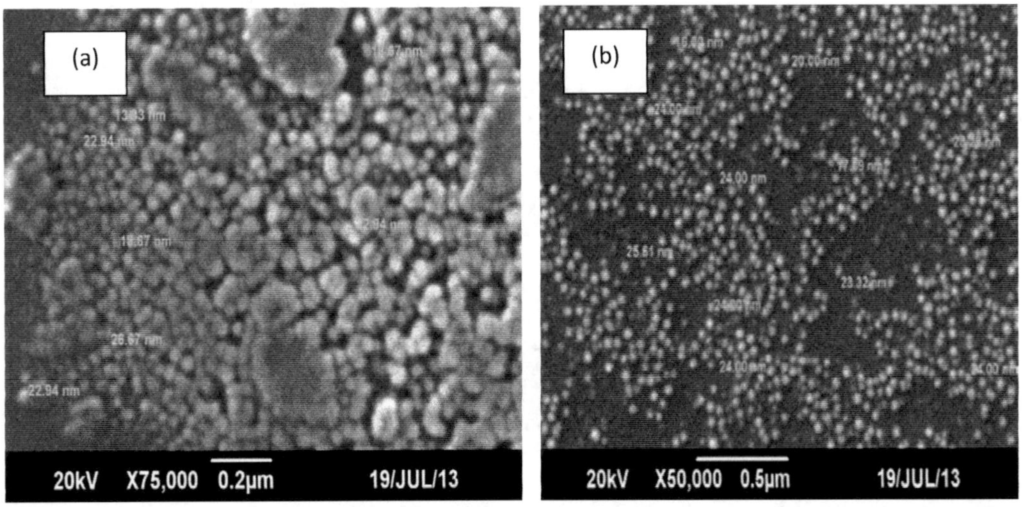

Figure 3. SEM images of synthesized silver nanoparticles (a) silver nanoparticles (Ag_p) at 75,000magnification. (b) silver nanoparticles (Ag_w) at 50,000 magnification. (c) *Allium sativum* extract loaded with silver nanoparticles (Ag_p).

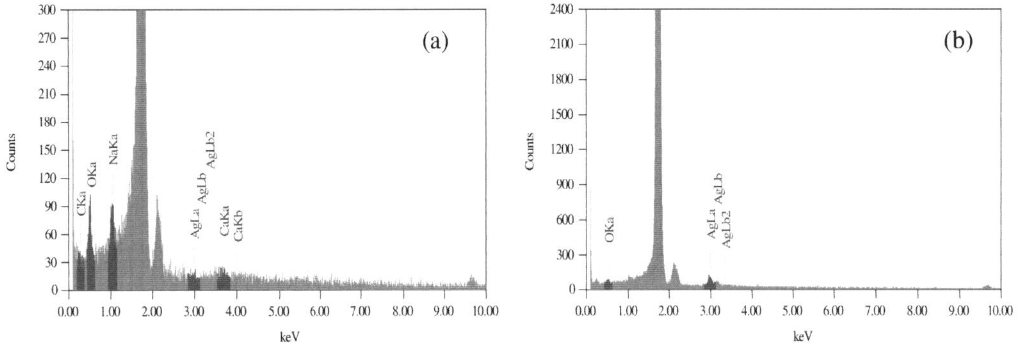

Figure 4. EDX image of synthesized silver nanoparticles. (a). Silver nanoparticles synthesized from *Allium sativum* extract (Ag_p). (b) Silver nanoparticles synthesized by wet chemical method (Ag_w).

3.4. Adsorption experiments

The synthesized silver nanoparticles (Ag_w and Ag_p) were applied as adsorbents to evaluate the removal of selected poly aromatic hydrocarbons.

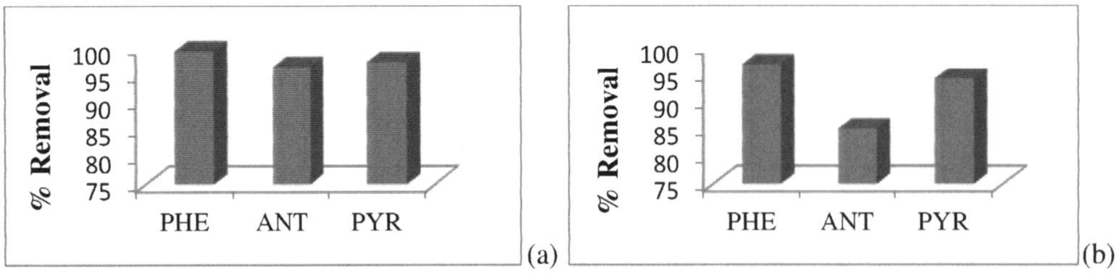

Figure 5: Removal (in %age) of phenanthrene, anthracene and pyrene on (a) Ag_p and (b) Ag_w adsorbents.

The results are graphically presented in Figure 4 (a and b), respectively. It is clearly evident that both silver particles are good adsorbent for the removal of PAHs. PAHs are optimally removed showing removal efficiency of more than *80% at ambient conditions.* The good sorptive behavior of PAH may be related to increase in hydrophobicity of the non-polar moiety of the aromatic rings [26]. Adsorption process in polyaromatic hydrocarbons results due to hydrophobic interactions [27].

Sequence depicted more efficient removal of phenanthrene followed by pyrene and anthracene. This order is also observed by Kumar et al., [28] in the batch experiments reporting lower removal percentages of pyrene to its recalcitrant toxicity and higher molecular weight. However, Ag_P particles stamp its preference by showing 3, 3 and 11 orders of magnitude more removal of selected PAHs in respective order.

3.4.1. Effect of pH.

The pH is one of the important parameters controlling the removal of poly aromatic hydrocarbons. In adsorption processes, the pH of the aqueous solution directly affects the surface
binding properties of the proposed adsorbents and the nature of the functional groups of both adsorbate and adsorbent molecules [12].

Batch was conducted at different pH to cover acidic, neutral and alkaline range. pH effect was determined on adsorption keeping other parameters constant.

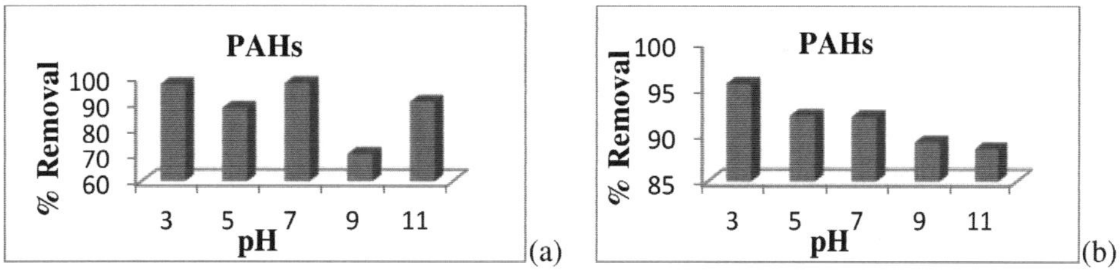

Figure 6. Effect of pH on the removal (in %age) of PAHs (a) Ag_P and (b) Ag_w adsorbents.

The results for the average removal of PAHs on silver particles are presented in Figure 5 (a) and (b). It is observed that generally acidic pH is more suitable for optimum removal of PAHs. This trend is also depicted in study conducted by Kumar et al., [28] reporting higher PAHs removal in acidic compared to alkaline conditions. Sponza et al., [29] relates higher removal under acidic conditions with the effect of protonation. It is concluded that Ag_P has greater removal efficiency at wide pH range through acidic to neutral (3-7 pH) in comparison to regular decreasing adsorption trend with increasing pH for Ag_w.

3.4.2. Effect of temperature

The temperature influence is an important controlling factor in the applications of removal process. In the present investigation, adsorption experiment was conducted at 5 different temperatures (25°C, 35°C, 45°C, 55°C, and 65°C). Results are shown in Figure 5 (a and b).

A general increase in percent removal of PAHs with increasing temperature is observed. The tendency to adsorb PAHs is more pronounced in moving from 25°C to 35°C followed by only incremental rise upto 65°C. However, Christensen et al., [30] observed that biological PAH removal was significantly enhanced by increasing the temperature from 35°C to 55°C

On the other hand, Ag_w as adsorbent depicts 45°C as optimum temperature for 97% removal of PAH. This suggests relatively wider applicability of Ag_P in terms of temperature like effect of pH. It is also supported by Belguith et al., [31] suggests that compounds present in *Allium sativum* are stable at high temperature (80°C).

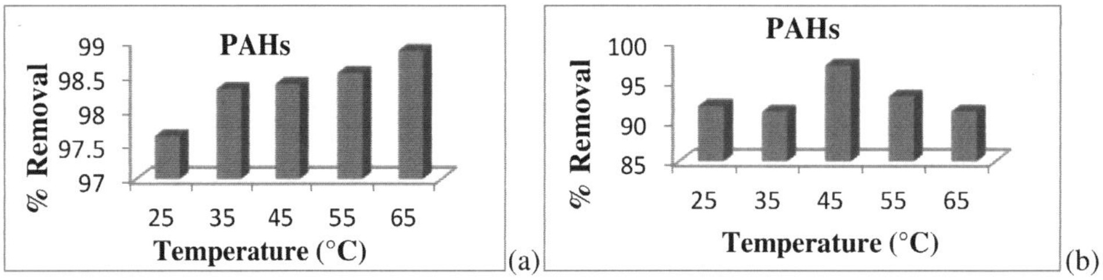

Figure 7. Effect of temperature on the removal (in %age) of PAHs (a) Ag_p and (b) Ag_w adsorbent

4. Conclusions

- The present study offers a simple and convenient method of synthesis of nano-sized particles by using a low cost and non-hazardous sodium hydroxide rather than toxic sodium borohydride as reducing agent.
- PAHs are removed successfully with optimal efficiency of more than 85% following the order of Phenanthrene>Pyrene>Anthracene. The adsorption of PAHs on silver nanoparticles is attributed to hydrophobic interactions.
- The natural and green source provides an adsorbent that has comparable efficiency to chemically synthesized expensive adsorbent under ambient conditions.
- Ag_p is relatively a better adsorbent than Ag_w in removing PAHs. It may be attributed to more functional groups in the plant extract participating in binding of PAH to the surface.
- Ag_p can be applied over a wide range of temperature, due to stability of compounds present in *Allium sativum* at high temperatures.
- The study recommends the use of alternate economical and green natural adsorbents for control of poly aromatic hydrocarbons (PAHs).

5. References

[1] Zhang W, Wei C, Yan B, Feng C, Zhao G, Lin C, Yuan M, Wu C, Ren Y and Hu Y 2013 Identification and removal of polycyclic aromatic hydrocarbons in wastewater treatment processes from coke production plants *Environ. Sci. Pollut. Res.* 20 6418-32

[2] AlSalka Y, Karabet F and Hashem M S 2011 Removal efficiency of polycyclic aromatic hydrocarbons from synthetic water samples by electrocoagulation processes, *J. for BASIC Sciences*27 65-80

[3] Omar N Y M J, Mon T C, Rahman N A and Bin Abas M R 2006 Distributions and health risks of polycyclic aromatic hydrocarbons (PAHs) in atmospheric aerosols of Kuala Lumpur, Malaysia *Sci. Total Environ.*369 76–81

[4] Gao Y, Ling W and Wong H M 2006 Plant-accelerated dissipation of phenanthrene and pyrene from water in the presence of a nonionic-surfactant *Chemosphere.* 63 1560–67

[5] Kim I S, Ritchie L, Setford S, Taylor J, Allen M, Wilson G, Heywood R, Pahlavanpour B and Saini S 2001 Quantitative immunoassay for determining polyaromatic hydrocarbons in electrical insulating oils *Analytica. Chimica.Acta.*450 13–25

[6] Krupadam R J, Bhagat B and Khan M S 2010 Highly sensitive determination of polycyclic aromatic hydrocarbons in ambient air dust by gas chromatography-mass spectrometry after molecularly imprinted polymer extraction*Anal. Bioanal. Chem.* 397 3097-06

[7] Eiroa A A, Blanco E V, Mahía P L, Lorenzo S M, Rodríguez D P and Fernández E F 2000 Determination of polycyclic aromatic hydrocarbons (pahs) in a complex mixture by second-derivative constant-energy synchronous spectrofluorimetry *Talanta* 51 677–84

[8] Kim Y M, Ahn C K, Woo S H, Jung G Y and Park J M 2009 Synergic degradation of phenanthrene by consortia of newly isolated bacterial strains *J. Biotechnol.*144 293-98

[9] Cai Z Q, Xian Y Z and Yong Z 2008 Simultaneous determination of dissolved anthracene and pyrene in aqueous solution by synchronous fluorimetry*Spectrochim. Acta. A.* 69 130-33

[10] Wang X, Guo Y, Yang Li, Han M, Zhao J and Cheng X 2012 Nanomaterials as sorbents to remove heavy metal ions in wastewater treatment *J. Environ. Anal.Toxicol.* 2 154

[11] Chen Y H 2011 Synthesis, characterization and dye adsorption of ilmenite nanoparticles *J. Non-Cryst. Solids*357 136–39

[12] Ghaedia M, Sadeghiana B, Pebdania A A, Sahraeib R, Daneshfarb A and Duranc C 2012 Kinetics, thermodynamics and equilibrium evaluation of direct yellow 12 removal by adsorption onto silver nanoparticles loaded activated carbon, *Chemical Engineering Journal*187 133– 41

[13] Sumesh E, Bootharaju M S, Anshup and Pradeep T 2011 A practical silver nanoparticle based adsorbent for the removal of Hg2+ from water*Journal of Hazardous Materials*189 450–57

[14] Karimi H, Mousavi S and Sadeghian B 2012 Silver nanoparticle loaded on activated carbon as efficient adsorbent for removal of methyl orange *IndianJournal of Science and Technology*5 2346

[15] Liu Si Si, Chen Y Z, Zhang Li De, Hua G M, Xu Wei, Li Nian and Zhang Ye 2011 Enhanced removal of trace Cr(VI) ions from aqueous solution by titanium oxide–Ag composite adsorbents *Journal of Hazardous Materials*190 723–28

[16] White G V, Kerscher P, Brown R M, Morella J D, Allister W M, Dean D and Kitchens C L 2012 Green synthesis of robust biocompatible silver nanoparticles using garlic extract Journal of Nanomaterials2012 1-12

[17] Benjamin G and Bharathwaj S 2011 Biological synthesis of silver nanoparticles from Allium Cepa (Onion) & estimating its antibacterial activity, *International Conference on Bioscience, Biochemistry and Bioinformatics* 5 35-38

[18] Moghaddam A B, Nazari T, Badraghi J and Kazemzad M 2009 Synthesis of ZnO nanoparticles and electrodeposition of polypyrrole/ZnOnanocomposite film *Int. J. Electrochem. Sci.*4 247 – 57

[19] Shankar S S, Rai A, Ankamwar B, Singh A, Ahmad A and Sastry M 2004 Biological synthesis of triangular gold nanoprisms*Nat. Mater.* 3 482

[20] Noginov M A, Zhu G, Bahoura M, Adegoke J, Small C, Ritzo B A, Drachev V P, Shalaev V M 2007 The effect of gain and absorption on surface plasmons in metal nanoparticles Appl. Phys. B. 86 455–60

[21] Saha S, Malik M M and Qureshi M S 2012 Characterisation and synthesis of silver nano particle using leaf extract of epipremnumaureum*International Journal of Nanomaterials and Biostructures*2 1-4

[22] Geethalakshmi R and Sarada D V L 2010 Synthesis of plant-mediated silver nanoparticles using Trianthemadecandra extract and evaluation of their anti microbial activities *International Journal of Engineering Science and Technology*2 970-75

[23] Rajam K, Rajendran S and Saranya R 2013 Allium sativum (Garlic) Extract as Nontoxic Corrosion Inhibitor *Journals of Chemistry*20131-4

[24] Bashkova S, Deoki D and Bandosz T J 2011 Effect of silver nanoparticles deposited on micro/mesoporous activated carbons on retention of NOx at room temperature *Journal of Colloid and Interface Science*354 331–40

[25] Rao Y S, Kotakadi V S, Prasad T N V K V, Reddy A V and Gopal D V R S 2013 Green synthesis and spectral characterization of silver nanoparticles from lakshmitulasi (Ocimum sanctum) leaf extract *Spectrochimica. Acta. A.*103 156–15

[26] Valderrama C, Gamisans X, Heras F X, Cortina J L and Farran A 2007 Kinetics of polycyclic aromatic hydrocarbons removal using hyper-cross-linked polymeric sorbents MacronetHypersol MN200 *Reactive & Functional Polymers* 67 1515–29

[27] Crisafully R, Milhome M A L, Cavalcante R M, Silveira E R, Keukeleire D D and Nascimento R F 2008 Removal of some polycyclic aromatic hydrocarbons from petrochemical wastewater using low-cost adsorbents of natural origin *Bioresource Technology*99 4515–19

[28] Kumar M, Wu P C, Tsai J C and Lin J G 2009 Biodegradation of soil-applied polycyclic aromatic hydrocarbons by sulfate-reducing bacterial consortium*Journal of Environmental Science and Health A.* 44 12–20

[29] Sponza D T and Oztekin R 2011 Removals of some hydrophobic poly aromatic hydrocarbons (PAHs) and Daphnia magna acute toxicity in a petrochemical industry wastewater with ultrasound in Izmir-Turkey *Separation and PurificationTechnology*77301–11

[30] Christensen N, Batstone D J, He Z, Angelidaki I and Schmidt J E 2004 Removal of polycyclic aromatic hydrocarbons (PAHs) from sewage sludge by anaerobic degradation*Water Science and Technology*50 237–44

[31] Belguith H, Kthiri F, Chati A, Sofah A A, Hamida J B and Landoulsi A 2010 Study of the effect of aqueous garlic extract (Allium sativum) on some Salmonella serovars isolates. *Emir. J. Food Agric.* 22 189-206

Boriding of high carbon high chromium cold work tool steel

W Muhammad
Institute of Industrial Control Systems, P.O. Box 1398, Rawalpindi, Pakistan

E-mail: valimuhammad@yahoo.com

Abstract. High-carbon high-chromium cold work tool steels are widely used for blanking and cold forming of punches and dies. It is always advantageous to obtain an increased wear resistant surface to improve life and performance of these steels. In this connection boriding of a high-carbon high-chromium cold work die steel, D3, was conducted in a mixture of 30% B_4C, 70% borax at 950 °C for two, four and six hours. Case depth of the borided layer obtained was between 40 to 80 μm. After boriding, the surface hardness achieved was between 1430 to 1544 HV depending upon the process time. X-ray diffraction studies confirmed the formation of a duplex compound layer consisting of FeB and Fe_2B. It is generally considered that FeB is undesirable because of its inherent brittleness. Post boriding treatment (homogenization) transformed the compound layer into single-phase layer of Fe_2B, while surface hardness decreased to 1345-1430 HV. Pin-on-disc wer test showed that wear resistance of the borided samples was superior as compared to non-borided material and increased with boriding time.

1. Introduction

The surfaces of tools, dies and majority of the machine parts are commonly subjected to higher stresses, wear and corrosive damages. In such working conditions, surface properties are often most important for reliable and long economic service life. In order to reduce this loss, properties of the surfaces should be improved. The diffusion of atoms of different elements into the surface of the materials is a commonly used technique to improve surface properties of tools and mechanical parts. Among these surface hardening treatments, boriding is highly effective method for increasing surface hardness, wear resistance, corrosion resistance and high temperature oxidation resistance [1-3].

Boriding is a thermochemical surface treatment that involves diffusion of boron into a substrate at elevated temperatures. Diffusion of boron into the surface of metals and alloys by means of gaseous, liquid or solid substances forms intermetallic borides [4]. Other techniques include plasma boriding and fluidized bed boriding [5]. Boriding can be applied to a wide range of steels including carbon steel, low alloy steel, tool steel and stainless steel. In addition, materials like nickel base alloys, cobalt base alloys, tungsten and niobium can also be borided to obtain very high hardness and wear resistance surface [1-8]. During boriding of ferrous alloys, generally, a boron-compound layer develops which consists of a surface-adjacent FeB layer and a Fe_2B underlying layer. If the concentration of boron id kept around 9%, the Fe_2B phase grows. On the other hand, if the concentration of boron rises to 16%, FeB forms and grows on the top of Fe_2B [2,4]. Other factors that influence the formation of single or duplex layers are potential of the medium, process temperature, time and chemical composition of the substrate material [4, 8-10]. The process temperature is in the range of 700-1000 °C [1-4, 8, 10].

In industry, the high carbon high-chromium cold work tool steels (D type) are the most commonly used cold-work tool steels. They offers high hardenability, low distortion after quenching, high resistance to softening and good wear resistance. They are commonly used for various types of cold work dies, rolls and punches. The use of appropriate hard coatings on these steels can further improve their wear resistance. Surface treatments provide improved tribological performance [3].

The object of present work is to investigate some physical and mechanical properties of borided high carbon high chromium cold work tool steel, D3. The boriding is performed by molten salt bath technique. Three different boriding durations are selected and a post boriding heat treatment is carried out in an attempt to obtain single phase layer. Metallography, X-ray diffraction (XRD), microhardness, Scanning Electron Microscopy (SEM) and wear test measurements were performed to describe the effect of boriding time on the boride layer morphology, phases, surface hardness and wear properties.

2. Experimental Techniques:

The material used in this study was tool steel (D3) bar. The chemical composition of the material is (wt. pct.) 2.18C, 0.39Mn, 0.21Si, 12.05Cr, 0.35 Ni, 0.55V and 0.008S. The hardness of the as received material was 243HV. Cylindrical samples, Ø10 mm and 15 mm length were machined from the bar and were cleaned ultrasonically.Thermoreactive diffusion boriding was conducted in molten bath at 950 °C for 2, 4 and 6 hours. The composition of the bath was 30%B_4C and 70% borax. After boriding treatment, the samples were quenched in hot water. Post boriding heat treatment was carried out at 800°C for 6 hours in inert gas atmosphere.

The borided samples were cut and the cross section was mounted, polished and etched in 2% nital to measure the depth of the borided layer with the help of optical microscope fitted with a digital camera. The hardness profiles were taken on the polished cross sections from surface to bulk with the help of a microhardness tester. X-ray diffraction studies were carried out to identify the phases in the layer. Pin-on-disc wear tests were performed to study the worn surface and weight loss in non-borided, borided and borided & homogenized samples. The wear machine consisted of two discs. One disc was fitted with 320 grit size SiC carbide paper and acted as a rotor. The second disc was used as sample holder and remained fixed. The sample holding disc was attached to a hydraulic system for applying a load normal to the contact surface. Two sets of samples of each condition were tested. Each sample was tested individually under the same conditions of load and time. A new abrasive paper was used each time. Worn surfaces were examined using SEM and weight loss was calculated to compare the relative wear resistance of non-borided, borided and borided & homogenized samples.

3. Results and Discussion:

The depth of the boride layer was dependent on boriding time as shown in figure1. The average case depth of boride layer after 2, 4 and 6 hours of boriding was about 40, 52 and 90 μm. The boride layer contained 10 to 35 μm deep layer of FeB.

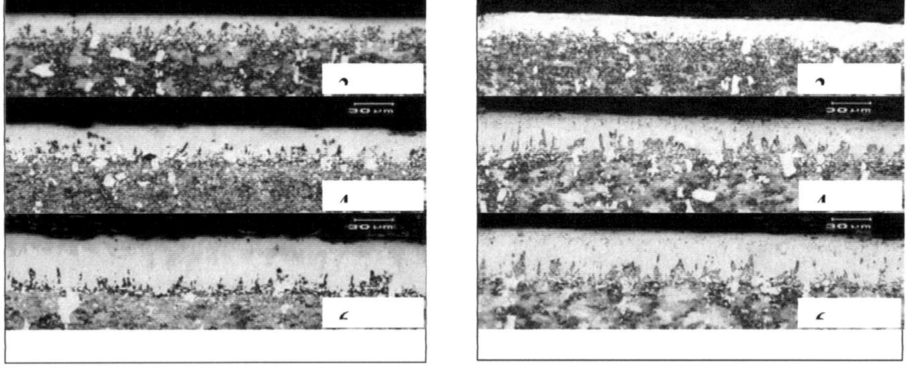

borided borided +homogenized

Figure 1. The boride layer after 2,4 and 6 hours of treatment

The hardness profile, figure 2, shows that plateaus of higher hardness bend down after some depth and finally reaches the uniform hardness of the matrix. The maximum surface hardness of boride layer after 2, 4 and 6 hours of boriding was 1430, 1505 and 1544 HV.

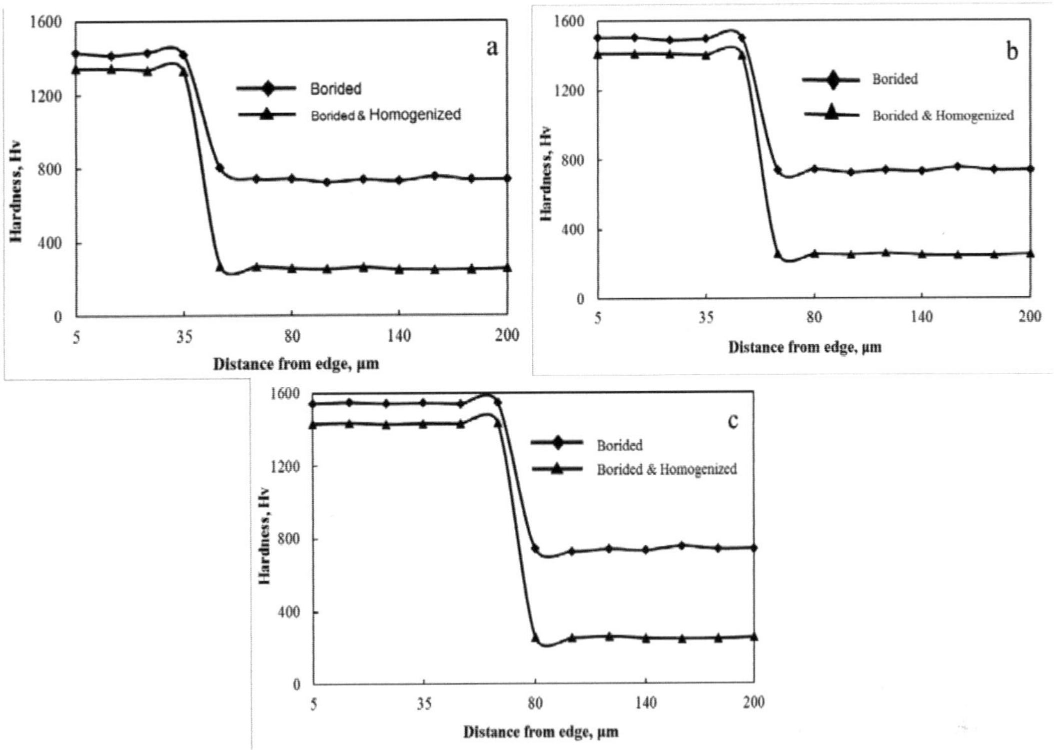

Figure 2. Hardness profile with depth of borided and borided&homogenized samples after a) 2; b) 4 and c) 6 hours

X-ray diffraction studies, figure 3, showed an almost single-phase layer of Fe_2B after homogenization, figure1b. However, small-scattered traces of FeB were found in outer most layer of the samples borided for six hours. The surface hardness was decreased from 1431, 1505 and 1544 to 1345, 1413 and 1430 HV after homogenization and the same is presented in figure 2. No significant change in case depth was observed after the post boriding treatment.

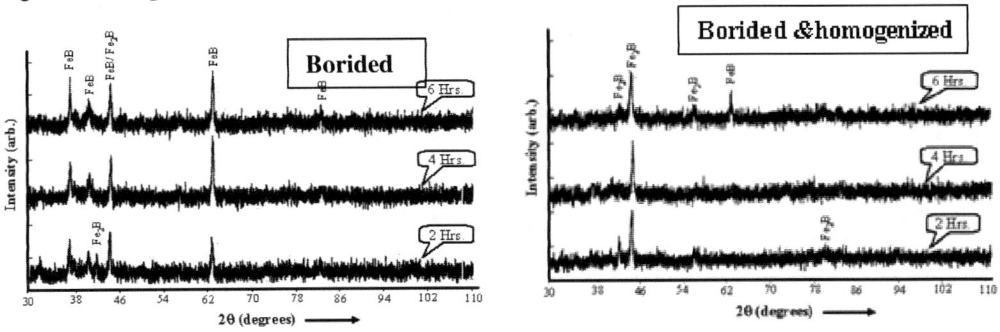

Figure 3. X-ray diffraction patterns of borided and borided&homogenized samples

The samples tested on pin-on-disc machine were analyzed under scanning electron microscope. SEM study showed grooved features on worn surfaces confirming an abrasive wear condition. SEM study of worn samples borided for 2 hours showed a complete removal of the boride layer along with the substrate. The reasons were less case depth and the presence of isolated patches of hard FeB phase. These patches after detachment from the layer remained trapped between the sample and the abrasive surface resulting in increased weight loss, figure 4(a). In the sample borided for 4 hours, the borided layer was partially removed because the hard FeB was present as a continuous layer very close to the outer edge. This layer resisted the wear for some time and when detached, enhanced the severity of

Figure 4. SEM micrographs of wear surfaces of as borided samples

wear mechanism to some extent, figure 4(b). In the samples borided for 6 hours the compound layer was completely removed along with the substrate. In this case, hard FeB layer was continuous but slightly deeper than in the samples treated for 4 hours. During the wear test this layer resisted the wear but once it worn off, the severity of the wear mechanism increased many times due to the trapped FeB particles, causing wear of the substrate, figure 4(c).

The 2 hours samples, which contained no FeB after homogenization, did not show this accelerated wear. The case depth of these samples were almost the same as that of the unhomogenized samples, but the resistance to wear was better, figure 5(a). Similarly, 4-hours borided & homogenized samples showed the same trend. Due to greater case depth, figure 5(b), the boride layer remained even after the test. The 6-hour borided & homogenized samples should have shown the same trend. But the layer was removed upto the substrate, figure 5(c). This was due to the presence of some retained FeB. The FeB particles after detachment increased the wear.

Figure 5. SEM micrographs of wear surfaces of borided and homogenized samples

This is supported by the XRD plot for the same sample. Wear resistance of the borided & homogenized samples was superior to the borided samples. The weight loss data is presented in figure 6. The plot further shows that best wear resistance was achieved in samples treated for 4 hours. The reason has already been discussed in preceding paragraphs. The wear mechanism observed was abrasive wear.

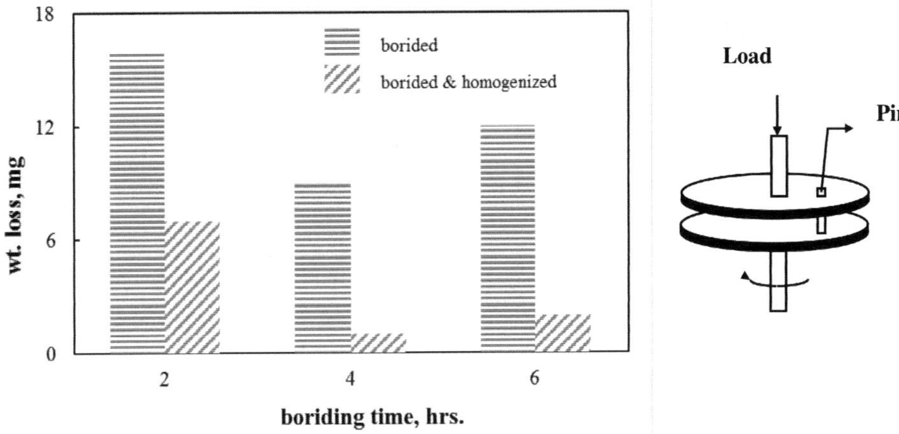

Figure 6. Comparison of wear test data of borided and borided &homogenized samples

4. Conclusions

- Surface hardness was improved from 243 to 1430 HV, after boriding and homogenization.
- Case depth was found sensitive to boriding time while the surface hardness was not significantly changed.
- Homogenized 4-hours samples showed the best wear properties in a pin-on disk rig.
- Borided samples showed better wear resistance than un-borided samples in a pin-on disk rig.

5. References

[1] Milinović A, Krumes D, Marković R 2012 *Technical Gazette 19* An investigation of boride layers growth kinetics on carbon steels 1 pp. 27-31

[2] Gunes I, Kayali Y and Ulu S 2012 *Indian journal of Engineering & materials Sciences* Investigation of surface properties and wear resistance of borided steels with different B_4C mixtures 19 397-402

[3] Oliveira C K N, Casteletti L C, Neto A L, Totten G E and Heck S C 2010 *Vacuum* Production and characterization of boride layers on AISI D2 tool steel 84 792–96

[4] Castro G R, Silva I C, Trinidad J M, López U F, Morales D M and Hernández J V 2009 *Advanced Materials Research* Effect of Boriding on the Mechanical Properties of AISI 1045 Steel, 65 63-68

[5] Celik O N, Gasan H, Ulutan M and Saygin M, 2009 *Journal of achievements in Materials and Manufacturing Engineering* An investigation on fatigue life of borided AISI 1010 steel 32 13-17

[6] Torun O 2008 *Surface and Coating technology* Boriding of nickel aluminide 202 3549-54

[7] Taktak S 2007 *Materials and Design* Some mechanical properties of borided AISI H13 and 304 steels 28 1836–43

[8] Genel K 2006 *Vacuum* Boriding kinetics of H13 steel, 80 451–57

[9] Campos I, Palomar M, Amador A, Ganem R and Martinez J 2006 Surface *& Coating technology* Evaluation of the corrosion resistance of iron boride coatings obtained by paste boriding process 201 2438–42.

[10] Ozdemir O, Usta M, Bindal C and Ucisik A H, 2006 *Vacuum* Hard iron boride (Fe_2B) on 99.97wt% pure iron 80 1391–95.

Ellipsometric study of GaN/AlN/Si(111) heterostructures grown by molecular beam epitaxy

M Ajaz-un-Nabi[1,2], A Ashfaq[1], M Imran Arshad[1], A Ali[1], K Mahmood[2], M A Hasan[3] and M Asghar[2]

[1]Department of Physics, GC University, Faisalabad, Pakistan
[2]Department of Physics, The Islamia University of Bahawalpur, Pakistan
[3]Department of Electrical and Computer Engineering, UNC, Charlotte, USA

E-mail: majazunnabi@gcuf.edu.pk

Abstract. GaN and related structures attracted a great interest in the recent years for electronic and optoelectronic applications due to their promising properties. GaN is grown popularly on foreign substrates like sapphire and SiC. However, silicon due to its favourable properties attended the great attention of material scientists and researchers to utilize as substrate for heteroepitaxy of GaN based structures and devices. Silicon substrates are low cost, available in large diameters and have well characterized thermal and electrical properties. In this study, GaN/AlN/Si(111) heterostructures were grown by molecular beam epitaxy. We performed x-ray diffraction spectroscopy and spectroscopic ellipsometry on these samples to study their structural and optical properties. XRD measurements performed on these samples revealed the presence of high quality GaN films as well as the presence of AlN buffer layer with the following miller indices: GaN (002), GaN (004), GaN (006) and GaN (110) along with Si peak of phase (111). The ellipsometric data obtained were used to characterize the GaN/Si samples as a function of film thickness. Refractive index, extinction coefficient and dielectric constant were calculated by the measured data.

1. Introduction

In the recent years, GaN based materials have been studied extensively and developments have been made in the areas of fabrication of UV/blue/green/amber/white light emitting diodes and violet laser diodes. The commercial fabrication of these devices now has become a reality. But most of GaN based devices are fabricated on sapphire and SiC substrates [1,2]. However, in the recent years, extensive research work has been started on the fabrication of GaN based devices on silicon substrate. Si has the advantages of low cost, availability in large size wafers and matured micro-technology. Despite of its advantages, the growth of high quality GaN films on Si is still a challenge. The high lattice constant mismatch (16.9 %) between GaN and Si incorporates high dislocation density at the GaN/Si interface which leads to the growth of low quality GaN films. Moreover, the large thermal (54 %) mismatch between GaN and silicon leads to the cracking of films formed which reduces the usable area of the film [3-5].The obvious solution to overcome these issues is the buffer layer growth of GaN films on silicon substrate. Various buffer layers i.e AlN, GaAs, SiC, SiN have been studied to enhance the quality of GaN layers. However, AlN buffer layers have advantages of small lattice mismatch and high density of nucleation centers. Moreover, AlN buffer layer promotes the lateral growth of GaN[6-8]. In this study, GaN layers were grown on Al/Si(111) substrate by molecular beam epitaxy. X-ray diffraction and spectroscopic ellipsometry were used to study the role of AlN buffer layer on the structural and optical properties of as-grown GaN samples.

2. Experimental Details

The detailed procedure and conditions of growth of GaN grown on Si(111) by molecular beam epitaxy is discussed here in this section. The MBE chamber was prepared for the growth of GaN epilayers on Si(111) substrate. The resident oxide on the polished surface of Si(111) substrate was removed by piranha procedure. In the first step of piranha procedure, the Si(111) wafer was dipped into a solution of H_2SO_4 (96%) and H_2O_2 (150:100 ml) for 20 minutes. Then the wafer was rinsed into flowing DI water for 10 minutes. In the next stage of procedure, the wafer was dipped into HF (49%) for 10 seconds. The wafer then again immersed into flowing DI water for 20 minutes followed by drying with nitrogen. Now as-cleaned Si(111) substrate was loaded into MBE chamber. During the 4 hours growth procedure, the chamber pressure, substrate temperature and gallium cell temperature were kept as 1.4×10^{-9} Torr, 1050 °C and 930 °C, respectively. The nitrogen plasma was generated by RF power supply maintained at 250 W. It is pertinent to mention here that an AlN layer to act as buffer layer was deposited for 10 minutes while maintaining the substrate temperature at 600°C. The structural study of as-grown samples was made by x-ray diffraction using PANALytical Xpert Pro and optical measurements were performed by using ultraviolet (UV) (SE-850 SENTECH) spectroscopic ellipsometer in the wavelength range of 380 – 800 nm. The elemental analysis was done by energy dispersive x-ray (EDX) spectroscopy.

3. Results and Discussion

Figure 1 displays the representative XRD spectrum of as-grown GaN layers on Si(111) substrate by molecular beam epitaxy. The spectrum reveals the presence of diffraction peaks at 2θ (degree) ~ 28 ± 0.41, 34 ± 0.57, 36 ± 0.01, 58 ± 0.83, 72 ± 0.93, 94 ± 0.93 and 126 ± 0.21. Peaks observed at 2θ ~ 28 ± 0.41, 57, 58 ± 0.83 and 94 ± 0.93 are due to the silicon substrate and 34 ± 0.57, 72 ± 0.93 and 126 ± 0.21 are related to the crystalline phase of GaN, whereas diffraction peak at 36 ± 0.01 corresponds to AlN buffer layer. Miller indices were calculated for these diffraction peaks which are (111), (222) and (333) for silicon, (002) for AlN and (0002), (0004) and (0006) for GaN, respectively. The XRD data confirm the growth of single crystalline GaN thin films on Si(111) substrate. The grain size (particle size) was calculated by using Scherrer's equation and was found to be 29 nm. It suggests the presence of good quality GaN films [9-18]. The elemental analysis by energy dispersive x-ray (EDX) spectroscopy, performed on the deposited films (not shown here), revealed that the Ga:N ratio was 37.45:18.03.

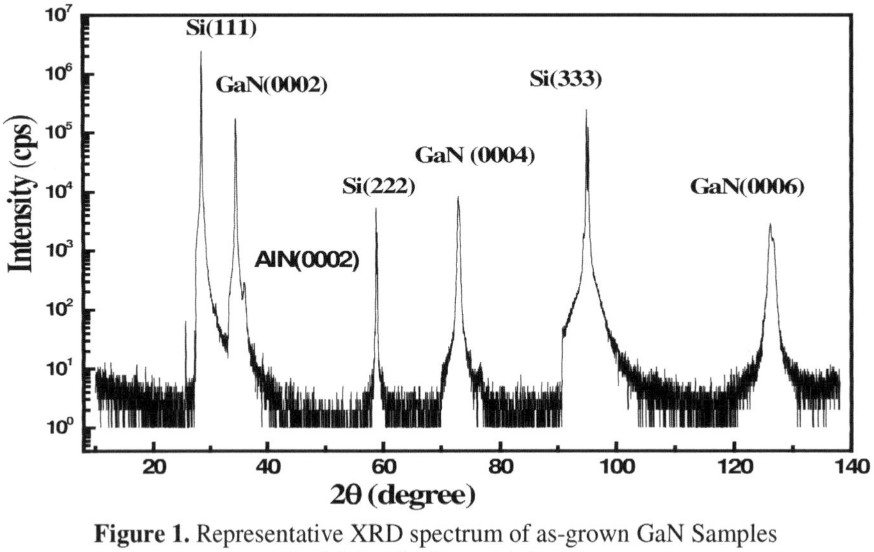

Figure 1. Representative XRD spectrum of as-grown GaN Samples by Molecular Beam Epitaxy

Spectroscopic ellipsometry was performed on as grown GaN samples to study the various optical parameters i.e. refractive index, extinction coefficient and dielectric constant as a function of film thickness. The ellipsometry measurements were carried out at various points on the as-grown GaN wafer. This article includes representative results at two points only. The thickness was calculated by employing Cauchy model on the measured $\psi(\lambda)$ and $\Delta(\lambda)$ data as shown in figure 2 and was found to be 358.86 nm and 496 nm. Moreover, figure 2 shows that the number of peaks increases with increasing the thickness of the films. This increase comes from the interference effects. Also, it is evidence of low refractive index of measured samples [19].

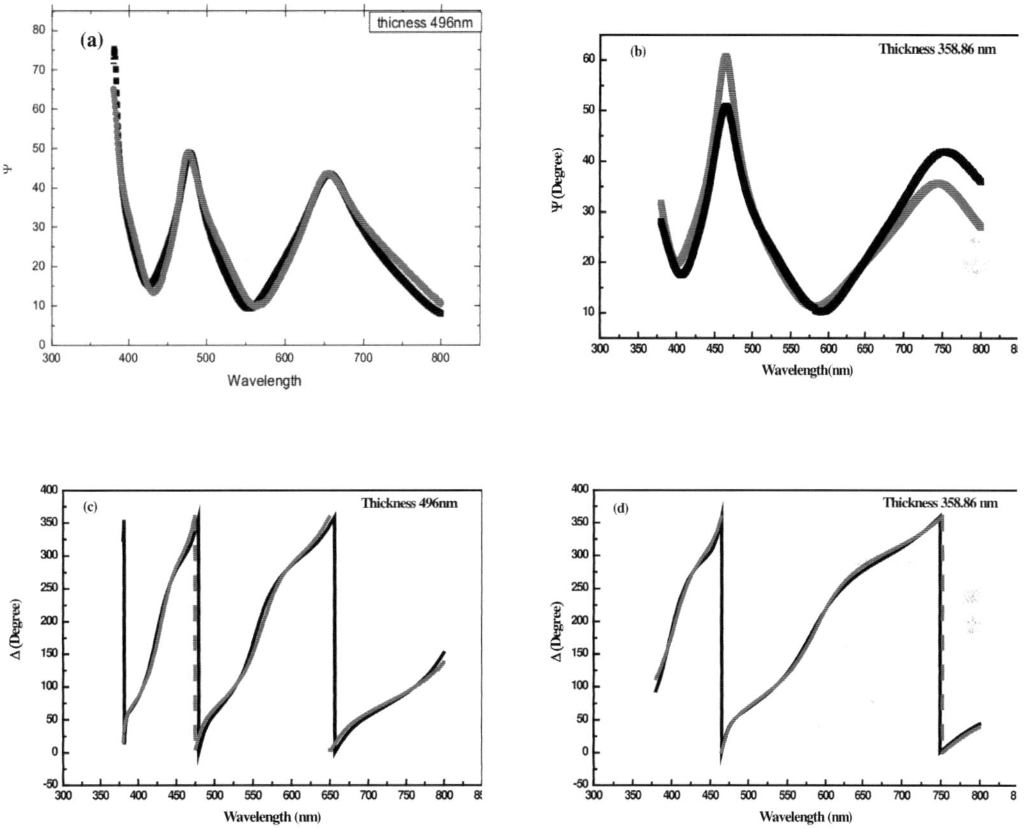

Figure 2. Ellipsometric parameters of as-grown GaN samples as a function of wavelength (solid lines/black) and results of the simulation (dotted lines/red).

The extinction coefficient and refractive index of as-grown GaN samples are displayed in the figure 3. The peaks appearing in the extinction coefficients may be attributed to the presence of defects in GaN films [20]. It also displays that extinction coefficient decreases with the thickness of the films. The refractive index of GaN samples with thickness 496 nm and 358 nm are found to be 1.88 - 1.95 and 1.84 - 1.90, respectively in the measured wavelength range. The figure 3 (b) clearly illustrates the dependence of refractive indices of GaN films on the films thickness and wavelength of incident light. The decrease in the refractive index with increasing film thickness may be attributed to the variation in film structure and/or to the surface roughness which increases with the increase in film thickness [21].

The refractive index values of our samples are found to be lower than the values reported in literature [22-23].

The dielectric function plays a prime role to study the optical properties of materials quantitatively. Figure 4 shows the dielectric constant of GaN samples as a function of wavelength. It can be clearly seen from figure 4 that dielectric constant of GaN epilayers decreases with increasing wavelength. Moreover, the dielectric constant of the film with thickness 496 nm is higher than the film with thickness 358.86 nm. The real dielectric constant of the GaN samples with thickness 496 nm and 358 nm are found to be 3.57 - 3.82 and 3.41 - 3.64 respectively, in the measured wavelength range that are compatible with the values reported in literature [24].

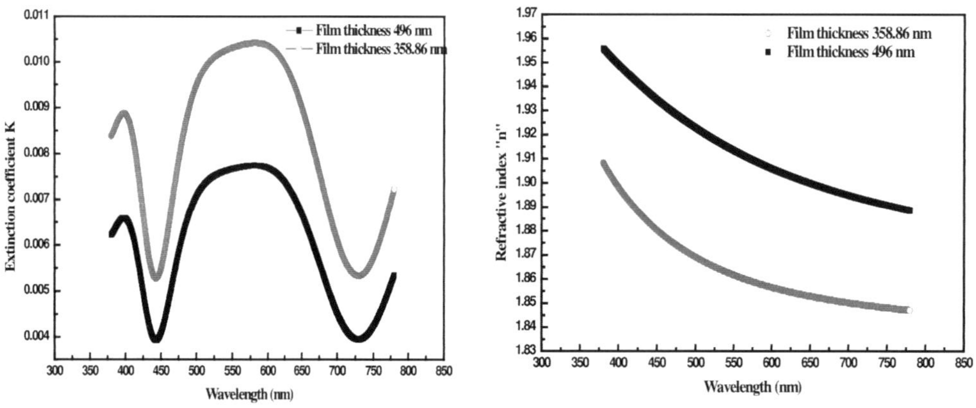

Figure 3. Extinction coefficient "K" and refractive index "n" of the as-grown GaN samples as a function of wavelength

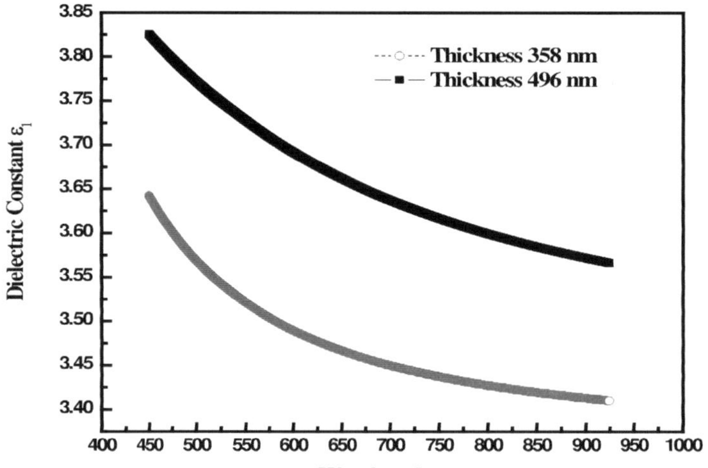

Figure 4. Plot of real dielectric constant ε_1 of as-grown GaN samples as a function of wavelength

4. Conclusion

In this study, GaN layers were deposited on Silicon substrate by molecular beam epitaxy. Prior to the growth of GaN layers, a thin layer of AlN was deposited to function as a buffer layer. The as-grown samples were investigated by using X-ray diffraction, energy dispersive x-ray spectroscopy and spectroscopic ellipsometry. The XRD spectrum confirms the presence of good quality single crystalline GaN films along (0002) plan with average particle size of about 29 nm. The optical parameters *i. e.* extinction coefficient, refractive index and dielectric function of GaN films were studied by spectroscopic ellipsometry. The measured data reveal that the as-grown GaN films exhibit low refractive index that can be attributed to the presence of defects in the crystal structure of GaN grown on Si substrate.

5. Acknowledgment

The authors are thankful to The Higher Education Commission, Pakistan and United States Educational Foundation for financial assistance.

6. References

[1] Tripathy S, Wang L S and Chua S J 2006 *Appl. Surf. Sci.* 253 236-40.

[2] Cao B, Xu K, Ishitani Y and Yoshikawa 2004 *Thin Solid Films* 455-6 661-4.

[3] Tungare M, Kamineni V K, Sandvik F -S and Diebold A C 2011 *Thin Solid Films* 519 2929-32.

[4] Ajaz-un-Nabi M, Arshad M I, Ali A, Asghar M and Hasan M-A 2011 *Adv. Mater. Res.* 295-7 777-80.

[5] Dadgar A, Schulze F, Wienecke M, Gadanecz A, Blasing J, Veit P, Hempel T, Diez A, Christen J and Krost A 2007 *New J. Phys.* 9 389.

[6] Ghosh B K, Saad I and Yamamoto A 2013 *Inter. J. Mater. Sci. Appl.* 2 43-6.

[7] Chen T, Wang Y, Xiang P, Luo R, Liu M, Yang W, Ren Y, He Z, Yang Y, Chen W, Zhang X, Wu Z, Liu Y and Zhang B 2012 *Appl. Phys. Lett.* 100 241112.

[8] Park S -E, Lim S -M, Lee C -R, Kim C S and Byungsung O 2003 *J. Cryst. Growth* 249 487-91.

[9] Saron K M A, Hashim M R and Allam N K 2013 *J. Appl. Phys.* 113 124304.

[10] Al Tahtamouni T M, Lin J Y and Jiang H X 2013 *J. Appl. Phys.* 113 123501.

[11] Chen D, Xu D, Wang J, Zhao B and Zhang Y 2008 *Thin Solid Films* 517 986-9.

[12] Chandolu A, Nikishin S, Holtz M and Temkin H 2007 *J. Appl. Phys.* 102 114909.

[13] Ni X, Zhu L, Ye Z, Zhao Z, Tang H, Hong W and Zhao B 2005 *Surf. Coat. Technol.* 198 350-3.

[14] Mele A, Giardini A, Palma T M D, Flamini C, Okabe H and Teghil R 2001 *Inter. J. Photoenergy* 3 111.

[15] Boo J -H, Lee S -B, Kim Y -S, Park J T, Yu K -S and Kim Y 1999 *Phys. Stat. Sol. (a)* 176 711.

[16] Al-Heuseen K, Hashim M R and Ali N K 2010 *Mater. Lett.* 64 1604.

[17] Schenk H P D, Kipshidze G D, Lebedev V B, Shokhovets S, Goldhahn R, Kraublich J, Fissel A and RIchter W 1999 *J. Cryst. Growth* 201-2 359.

[18] Lee J W, Jung S H, Shin H Y, Lee I-H, Yang C-W, Lee S H and Yoo J-B 2002 *J. Cryst. Growth* 237 1094.

[19] Joo H Y, Kim H J, Kim S J and Kim S Y 2000 *Thin Solid Films* 368 67.

[20] Massoud R A and Przewlocki H M 2002 *J. Appl. Phys.* 92 2202.

[21] Chung G -S, Kim K -S and Han K -B 2008 *Ceram. Int.* 34 841.

[22] Majumder S B, Jain M and Katiyar R S 2002 *Thin Solid Films* 402 90.

[23] El-Nahass M M and Farag A A 2012 *Opt. Laser Technol.* 44 497.

[24] Gila B P, Kim J, Luo B, Onstine A, Johnson W, Ren F, Abernathy C R and Pearton S J 2003 *Solid-State Electron.* 47 2139.

IOP Conf. Series: Materials Science and Engineering **60** (2014) 012064 doi:10.1088/1757-899X/60/1/012064

Effect of constituent-particles distribution on mechanical behavior of an AlMgSi alloy

Wali Muhammad, Noveed Ejaz and Iftikhar us Salam
Institute of Industrial Control Systems, P.O. Box 1398, Rawalpindi, Pakistan

E-mail: valimuhammad@yahoo.com, noveedejaz@yahoo.com, iftikhar.salam@gmail.com

Abstract. In the present work, the effect of constituent particles distribution on mechanical behavior of a high strength AlMgSi alloy is investigated. The study consists of the tensile and fatigue testing of the material in longitudinal and transverse directions. Monotonic tensile properties, in both the directions showed virtually identical yield and tensile strengths. However, higher values of elongation and reduction in area were observed in the longitudinal direction. SN curves produced after fatigue testing displayed a shorter fatigue life in transverse direction as compared to the longitudinal direction. Microstructural analysis exposed that the material had clusters of constituent particles. Post fracture analysis of tensile and fatigue samples, in SEM, revealed that the topographical features change with the sample orientation. The present study revealed that the distribution of constituent particles in longitudinal and transverse directions has a pronounced effect on the tensile and fatigue behavior of the alloy and also induce anisotropy in the material.

1. Introduction

Generally mechanical processing in the form of rolling, forging, drawing etc. is included in the manufacturing route of the wrought components. These processes are thought to be essential to attain the ultimate mechanical properties of such components. During these processes the physical dimensions of the material change and the effect can be seen on the microstructural scale. Grains as well as other constituent features undergo substantial changes which significantly influence the mechanical properties and often create anisotropy in the material. The anisotropy produced in the material results in dissimilar properties along different orientations of the product and is undesirable in several applications. Subsequent thermal processes are used to minimize the directionality effect and achieve an isotropic material. The microstructural features thus developed in the final product play a pivotal role to establish the mechanical properties.

The response of the microstructural features against these mechanical processes depends upon their mechanical behavior at the processing conditions. Metal in the form of grains, easily deform at high temperature and elongate in the allowable direction. Constituent particles in the material behave differently under loading conditions. For example, sulphides normally present in steels, show a ductile behavior and deform under the processing conditions. However, oxides in general exhibit a brittle behavior and fracture during processing. Debonding easily occurs where these particles show weak adherence with the matrix and deformation behavior of the material considerably changes. Chemistry of these constituent particles, their density, size, distribution, morphology and the adherence with the matrix are main controlling factors which ultimately affect the mechanical behavior of the material.

Aluminum alloys are of vital importance for many structural applications in automotive, marine and aircraft industries [1]. The principal design features of these alloy systems are their high strength with low density along with high corrosion resistance. In aluminum alloys, constituent particles are formed during the cooling process when some of the alloying elements solidify more rapidly than the aluminum. The mechanical properties of aluminum alloys are greatly influenced by the presence of these particles.

The AlMgSi alloys are frequently used due to relatively high strength, good corrosion resistance and high toughness in addition to their good formability and weldability [2-3]. In AlMgSi alloys ultimate mechanical properties are achieved in T6 temper. This includes solution heat treatment at 530 °C for adequate time to allow thorough heating and then water quenching. Aging precipitation heat treatment is done at 175 °C for 8 hours followed by air cooling [4].

Mechanical testing is frequently conducted to acquire the material's data required for modeling the mechanical systems. The monotonic tensile test is a simple and versatile test to evolve most of the mechanical properties under tension [5]. Fatigue testing is mandatory to explore the behavior of the material under cyclic loading.

Metal fatigue is significantly influenced by microstructure. Investigations of the aluminum alloys under fatigue loading have been extensively done [6-9] and the constituent particles are found to play an important role in the fatigue life of these alloys. Many researchers have made experimental observations of the types of features that become sites for nucleation of fatigue cracks in aluminum alloys. Generally, it was observed that the cracks were formed at debonded or broken constituent particle clusters or at particles that cracked during rolling. Minimizing discontinuities including the constituent particles in aluminum alloys is a key to good fatigue resistance. In a study on AlCuMg bars, it was observed that reducing the density of constituent particles significantly enhanced the fatigue resistance of the alloy [10]. The objective of the research work is to determine the effect of orientation on tensile and fatigue performance of high strength AlMg1SiCu alloy in T6 condition. The study consists of the tensile and fatigue testing of the samples taken from transverse (TR) and longitudinal (LD) directions. The results obtained from the monotonic tensile test of round samples and SN curves from the fatigue testing of middle tension (MT) samples are presented. Prior to mechanical testing, the microstructural analysis of the material was conducted with the help of optical and scanning electron microscopes. After the tests, the topographical features of the fractured surfaces were examined in Scanning Electron Microscope (SEM).

2. Experimental procedure

2.1. Material characterization

The material used for this study was AlMg1SiCu aluminum alloy in T6 condition. The chemical composition of the material was determined using Energy Dispersive spectrometer (EDS) in SEM. The microstructural features in two orientations were investigated with the help of optical microscope and SEM.

2.1.1. Chemical composition

The chemical composition of the material is given in table 1. The material confirms the specifications of aluminum alloy AlMg1SiCu. The major constituents of the alloy are Al, Mg and Si.

Table 1. Chemical composition of the material, wt%

Material	Mg	Si	Fe	Cu	Cr	Mn	Al
Rod	0.94	0.64	0.20	0.19	0.10	0.06	Bal.

2.1.2. Microstructural analysis
Metallographic sections were prepared using standard procedures and optical and scanning electron microscopes were used to examine the microstructural features of the alloy. Samples were studied in polished as well as etched conditions, in two orientations. Etching was done in 3% HF solution to reveal the grain size and their orientation. EDS analysis of the constituent particles was conducted in SEM.

2.2. Monotonic tensile testing

2.2.1 Test sample
Samples were taken in LD and TR directions as shown in figure 1. Round tensile samples were machined with final dimensions as given in figure 2. The gauge section of the samples was polished using aluminum oxide powder.

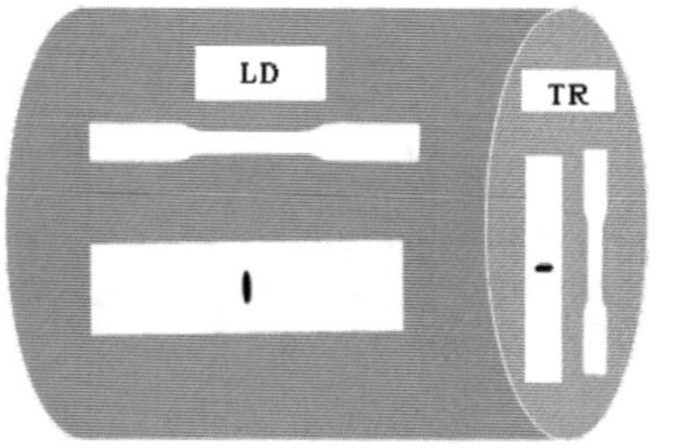

Figure 1. Orientations of tensile (T) and fatigue (F) specimens used in this study.

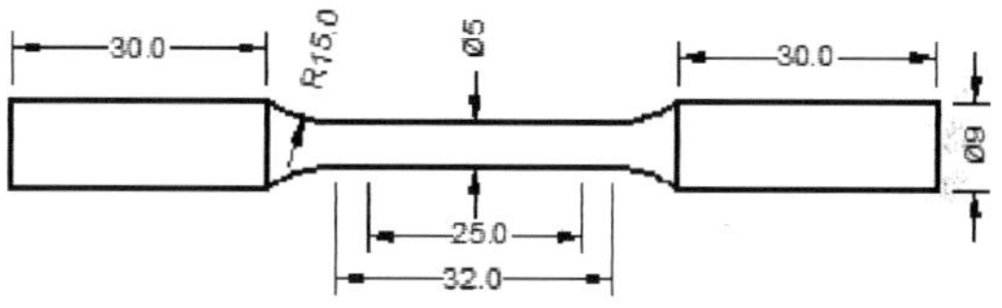

Figure 2. Dimensions of the tensile sample (in mm).

2.2.2 Testing procedure
Monotonic tensile tests were conducted to obtain stress-strain curves in two orientations. Tests were conducted at a strain rate of 3.3×10^{-5} per second. An extensometer of 25 mm gauge length was attached to the sample in the gauge section. The load and elongation data was recorded and subsequently stress strain curves were obtained. Five samples were tested for both orientations.

2.3. Fatigue testing

2.3.1 Sample preparation
Fatigue testing was performed on middle tension (MT) specimens prepared in two orientations. Samples were cut from the extruded cylinder and machined to final dimensions according to the standard ASTM E 647 [11]. Samples were subjected to mechanical grinding and subsequently fine polishing with 1 μm alumina powder to minimize the surface roughness effects. Dimensions of the sample in TR orientation are given in figure 3. The length, width and thickness of LD sample were 80, 20 and 2.5 mm, respectively.

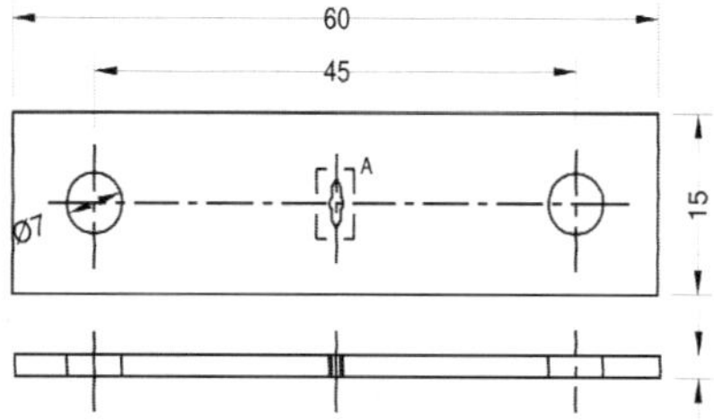

Figure 3.. Dimensions of the MT sample (in mm) - TR direction

2.3.2 Test parameters
Fatigue tests were performed on a servo-hydraulic testing machine. Tests were conducted in tension-tension mode under constant amplitude loading with R ratio 0.1. A sinusoidal waveform was applied at a loading frequency of 10 Hz. Tests were conducted at maximum stress levels representing 15 to 40% of the yield strength (YS) of the material. Seven samples were tested for each orientation and SN curves were

achieved. Crack length was measured with the help of traveling microscope at a magnification of 100x. Tests were interrupted for short times to measure the crack length. During testing, the number of cycles and crack extension data were recorded until failure. Testing was conducted in air at 20 °C and approximately 50% relative humidity.

2.4. Fractography
The macroscopic and microscopic examination of the failed samples was conducted after mechanical testing to understand the effect of orientation on fracture mechanism. The fractured surfaces of the specimens were cut, cleaned in acetone and examined in SEM. The images of the important features were recorded during examination.

3. Results and discussion

3.1. Microstructural features
The optical microstructure of the alloy is shown in figure 4(a,b), in two orientations. The etched microstructure revealed that it consisted of recrystallized grains. It seems that the grains elongated during processing and recrystallize into fine almost equiaxed grains during subsequent heat treatment. The morphology and the alignment of the constituent particles were not affected by heat treatment process.

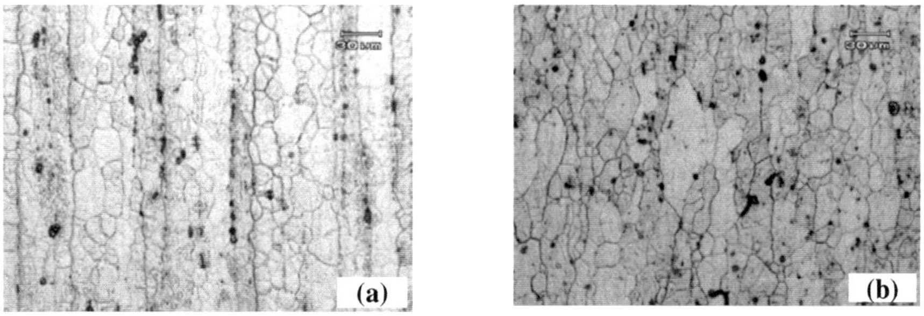

Figure 4(a, b). Optical micrographs revealing the microstructure of the alloy ina) LD and b) TR direction

Figure 5a shows the SEM micrograph of the polished sample revealing constituent particles in the LD direction. Further examination after etching exposed that the constituent particles were present preferably along the grain boundaries. The particles can be seen aligned along the extrusion direction. Inspection at high magnification revealed that clusters of particles were present at these locations; see figure 5b.

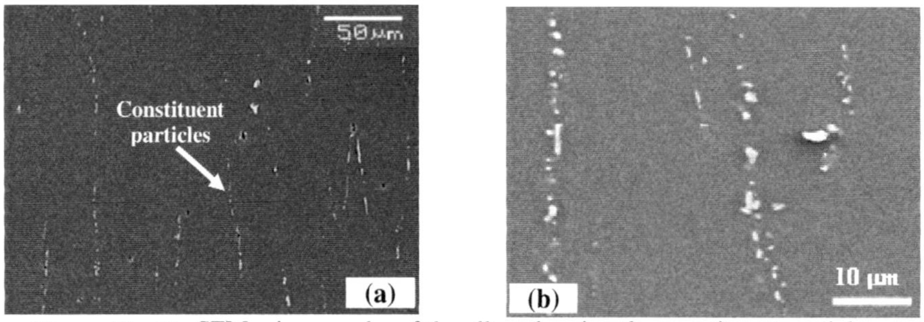

Figure 5(a, b). SEM micrographs of the alloy showing the constituent particles in longitudinal direction a) low mag. b) high mag. (clusters of particles)

It seems that these particles were formed during solidification and fractured in fragments and aligned along the direction of working. EDs analysis showed that these particles were rich in Al, Fe, Si, Cr and Mn. The other type of particles found in the material was rich in Al, Mg and Si.

3.2. Monotonic tensile test

The stress-strain diagrams obtained after tensile testing in two orientations are presented in figure 6. These graphs indicate that the behavior of the material, under tensile loading, remains essentially same up to the point of yield strength. This shows that the response of grain structure and the constituent particles is similar in both the samples up to this point. However, after this point, difference of the two curves is visible. A smaller strain value up to the point of failure in transverse direction indicates that the distribution of constituent particles in this orientation restricted the material to deform much before fracture.

The results obtained from the tensile tests of the alloy in two orientations are given in Table 2. The table also provides the hardness measured along two sections. The analysis of the above results shows that the yield strength and ultimate tensile strength of the material in the two orientations were not different. However, the difference in the deformation behavior was significant. In LD direction the material has higher ductility represented by the elongation and reduction in area. These properties indicate that the recrystallized grain structure has eliminated the effect of texture on the yield and tensile strengths of the material. However, the texture produced by the clusters of the constituent particles was not eliminated by subsequent heat treatment processes and that reduced the elongation and reduction of area in the transverse direction.

Table 2. Nominal tensile properties of the alloy

Properties	**Orientation**	
	Longitudinal	**Transverse**
0.2 % Yield strength, MPa	324	322
Ultimate tensile strength, MPa	354	353
Elongation, %	16	9
Reduction in area, %	50	24
Elastic modulus, GPa	71	71
Hardness, HV	115	115

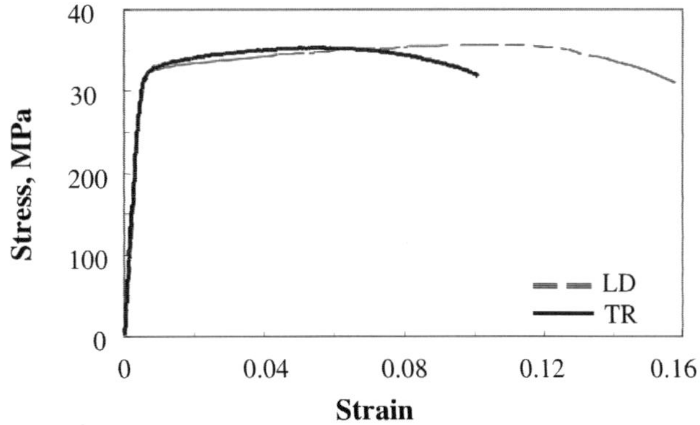

Figure 6. Stress-strain diagram of the material in monotonic testing

3.3. Fatigue test

3.3.1 Fatigue life analysis

The graphs of the stress range vs. the number of cycles to failure (SN curves) for TR and LD orientations are given in figure 7. Fatigue life increases as the stress range decreases for both the orientations. SN curves revealed a shorter fatigue life of the samples prepared in TR orientation; a possible explanation of which is the distribution of the constituent particles. Clusters of aligned particles in TR orientation were present in the direction perpendicular to the loading axis and more favorably oriented to assist the crack growth process. The difference in number of cycles to failure in two orientations at four stress levels is given on the graph. It can be seen that the difference is increasing with lowering stress level. This indicates that the microstructural variation in the two orientations is more effective at low stress levels.

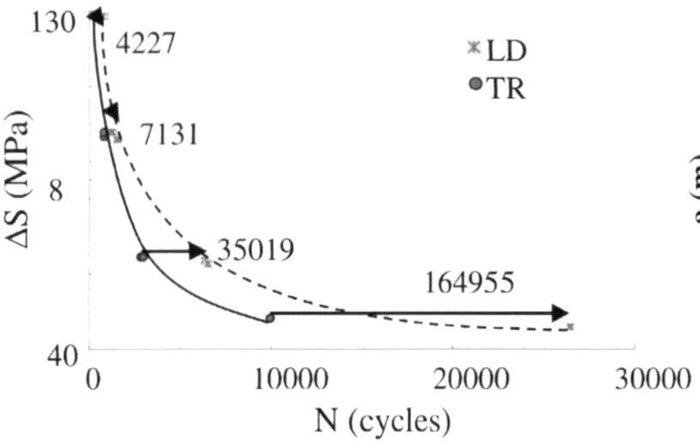

Figure 7. Stress range vs the number of cycles to failure for the two orientations

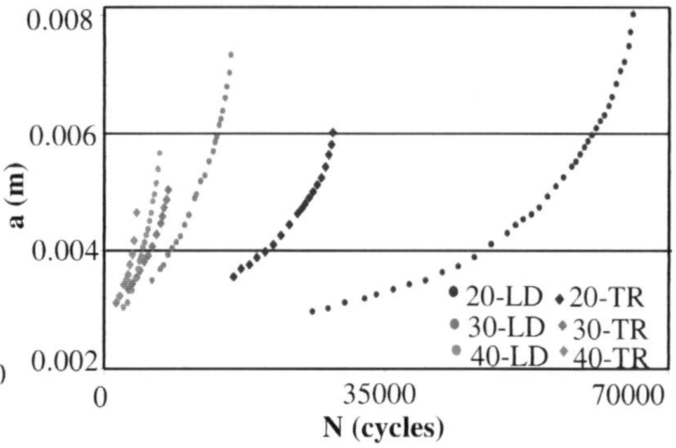

Figure 8. Plots of the crack length versus the number of cycles to failure for two orientations

3.3. Crack extension

Figure 8 shows the crack length versus the number of cycles to failure for two orientations at three stress levels. The data covers the range from start of the crack at the notch tip up to the specimen failure. In both orientations, the number of cycles to start the crack from the notch and the maximum value of the crack length increase with the decrease in the stress level. Comparing the curves for two orientations, at the same stress level, it is evident that in TR orientation the crack started earlier. Start of the crack at comparably low number of cycles indicates that the distribution of the constituent particles is favorable in TR orientation to nucleate the crack from the notch tip.

3.4. Post-fracture analysis

3.4.1 Tensile samples

SEM fractographs of the tensile samples in two orientations are shown in figure 9. In LD sample necking is visible indicating higher ductility in this orientation. The fracture is at an angle of about 45°. Figure 10 shows SEM fractographs at high magnification of the samples shown in figure 9. The micrographs in both the orientations showed typical dimple fracture indicating deformation before failure. The presence of constituent particles in the dimples showed that these were the crack initiation sites. Clusters of particles as were observed in the metallographic sections were present in the TR orientation.

3.4.2 Fatigue samples

The fracture surfaces of the fatigue samples tested at maximum stress of 20% yield strength in two orientations are shown in figure 11. The corresponding numbers of cycles to failure for LD and TR samples were 63672 and 28653, respectively. Starter notch in both the samples is marked on the micrographs. In LD sample, only the crack growth region is visible and the overload region is out of the micrograph. In TR sample, shown in figure 11b, both the crack growth and the overload failure

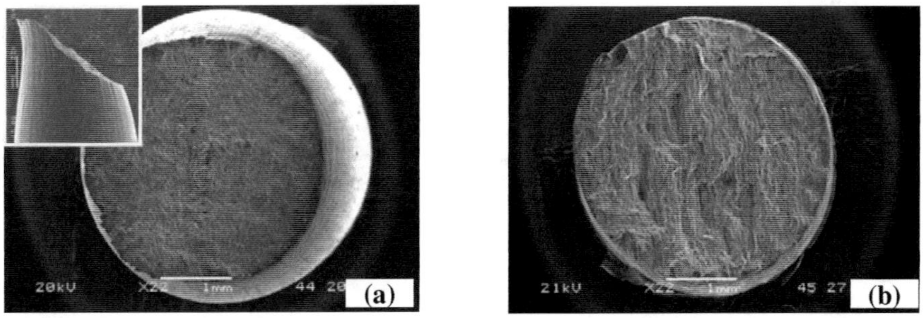

Figure 9(a, b). SEM micrographs of the fractured tensile samples a)LD b)TR orientation

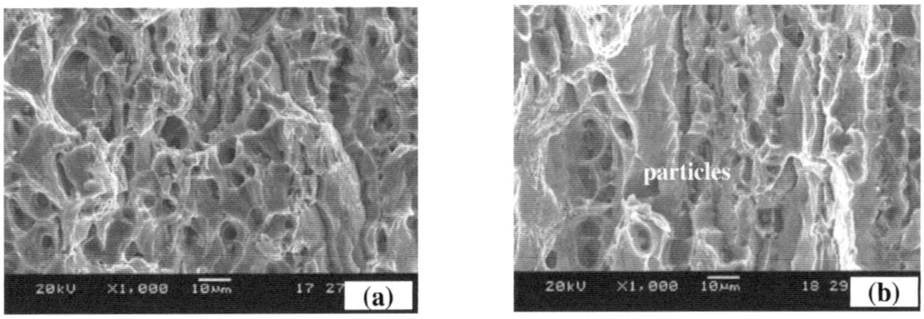

Figure 10(a, b). SEM micrographs of the fractured tensile samples at high mag. a)LD b)TR orientation

regions are shown. The macro appearance of the two fractured surfaces is markedly different and represents the corresponding microstructures observed during metallography. Figure 12 shows the higher magnification images of the fatigue crack growth region shown in figure 11. These microscopic views show fatigue facets created by propagation of the fatigue crack. Cluster of fragmented constituent particles can be seen in the center of figure 12b. The mode of failure locally changed due to the presence of these particles. Inspection of the fractured surfaces at higher magnification shows striations on the fatigue facets. This indicates that the general mechanism of crack growth was the formation of striations. In the final rapid fracture ductile dimples were observed in both the samples.

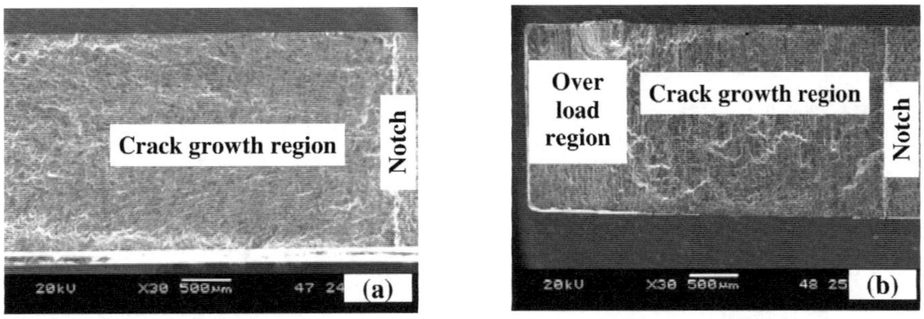

Figure 11(a, b).SEM micrographs of the fatigue fracture samples tested at maximum stress of 20 %YS a) LD b)TR orientation

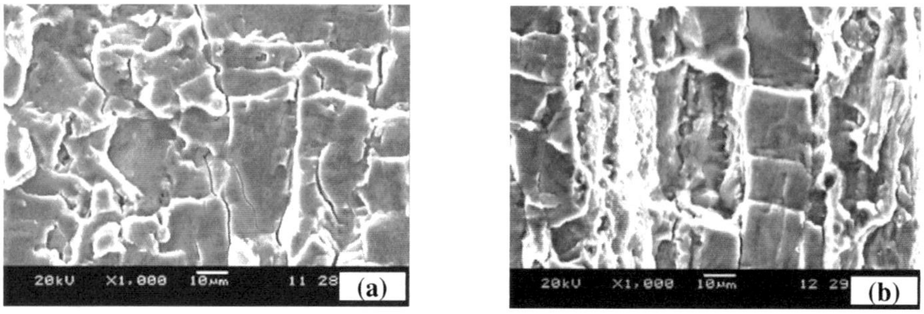

Figure 12(a, b). Higher magnification of figure 11; fatigue crack growth region
a) LD b) TR orientation

4. Conclusions

Following conclusions can be drawn on the basis of the present study;

- The material was composed of recrystallized grains. Clusters of constituent particles were present aligned along the direction of processing.
- Monotonic tensile properties of the material in two orientations revealed that the yield and tensile strengths were virtually identical. However, higher values of elongation and reduction in area were observed in the longitudinal direction.
- SN curves of the samples in two orientations showed a shorter fatigue life in TR direction. The difference was more visible at low stress levels.
- Post fracture analysis revealed that the topographical features change with the sample orientation. Tensile samples showed typical dimpled failure where constituent particles were the main crack nucleating sites.
- The present study revealed that the distribution of constituent particles after processing has a pronounced effect on the tensile and fatigue performance of the alloy and induced anisotropy in the material.

5. References

[1] Heinz A, Haszler A, Keidel C, Moldenhauer S and Benedictus R 2000 *Material Science and Engineering* Recent development in aluminum alloys for aerospace application A280 102–07.

[2] Borrego L P, Ferreira J M, Pinho da Cruz J M and Costa J M 2003 Engineering. Fracture Mechanics Evaluation of overload effects on fatigue crack growth and closure 70 1379–97.

[3] Oppenheim T, Tewfic S, Scheck T, Klee V, Lomeli S, Dahir W, Youngren P, Aizpuru N, Clark Jr R, Lee E W, Ogren J and Es-Said O S 2007 Engineering. Failure. Analysis On the correlation of mechanical and physical properties of 6061-T6 and 7249-T76 aluminum alloys 14 218–25.

[4] ASM Handbook 1989 Vol. 2 (New York: McGraw-Hill).

[5] Lee H T and Shaue G H 1999 Materials Science and Engineering. The thermomechanical behavior for aluminum alloy under uniaxial tensile loading., A268, 154-64

[6] Zabett A and Plumtree A 1995 Fatigue and Fracture of Engineering. Materials and Structures Microstructural effects on the small fatigue crack behavior of an aluminum alloy plate 18 801-09.

[7] Fonte M A, Stanzl-Tschegg S E, Holper B, Tschegg E K and Vasudevan A K 2001 International Journal of Fatigue The microstructure and environment influence on fatigue crack growth in 7049 aluminum alloy at different load ratios 23 S311–17.

IOP Conf. Series: Materials Science and Engineering **60** (2014) 012064 doi:10.1088/1757-899X/60/1/012064

[8] Al-Rubaie K S, Barroso E K L and Godefroid L B 2006 International Journal of Fatigue Fatigue crack growth analysis of pre-strained 7475–T7351 aluminum alloy 28 934–42.

[9] Stanzl-Tschegg S E and Mayer H 2001 International Journal of Fatigue Fatigue and fatigue crack growth of aluminum alloys at very high numbers of cycles 23 S231–S237.

[10] Li J X, Zhai T, Garratt M D, Bray G H 2005 Metallurgical and Materials Transactions A Four point-bend fatigue of AA 2026 aluminum alloys, 36A 2529–39.

[11] American Society for Testing and Materials 1996 *ASTM E 647. Annual Book of ASTM Standards* (Philadelphia).

IOP Conf. Series: Materials Science and Engineering **60** (2014) 012065 doi:10.1088/1757-899X/60/1/012065

Blanking and piercing theory, applications and recent experimental results

Adnan I O Zaid
Industrial Engineering Department, University of Jordan, Amman, Jordan

E-mail: a_kilani43@hotmail.com

Abstract. Blanking and piercing are manufacturing processes by which certain geometrical shapes are sheared off a sheet metal. If the sheared off part is the one required, the processes referred to as blanking and if the remaining part in the sheet is the one required, the process is referred to as piercing. In this paper, the theory and practice of these processes are reviewed and discussed The main parameters affecting these processes are presented and discussed. These include: the radial clearance percentage, punch and die geometrical parameters, for example punch and die profile radii. The abovementioned parameters on the force and energy required to effect blanking together with their effect on the quality of the products are also presented and discussed. Recent experimental results together with photomacrographs and photomicrographs are also included and discussed. Finally, the effect of punch and die wear on the quality of the blanks is alsogiven and discussed.

1. Introduction

Blanking and piercing are manufacturing processes by which engineering or industrial parts of certain geometrical shapes are sheared off sheet materials such that the produced parts do not need further or subsequent machining unless very high quality is required. If the sheared off part is the object the process is referred to as blanking and if the reining part is the required part the process is referred to as piercing. The literature on these processes is voluminous and goes back to the forties. Most of the work has been directed towards improving the quality of the product and the effect of punch speed on the quality of the produced blanks. They found that blanks produced at punch speed of 3m/s were of better quality than those produced at conventional speeds [1,2] They carried out experiments on cast iron, mild steel, brass, copper, zinc, aluminum and lead to investigate the effect of clearance, tool shear and the ratio of punch diameter, d, to thickness, t, in blanking circular discs. They found that as the ratio d/t increases the maximum blanking force per unit sheared area decreases and that the maximum blanking force is greatest at no clearance. They also concluded that similar influence of the d/t ratio on the energy required to effect blanking existed. The d/t ratio has been specifically studied by Maeda and Temura,[3],for mild steel, soft copper and soft aluminum. They found that for values of d/t >5 the maxmum blanking force per unit sheared area is almost independent of the ratio whereas for d/t ≤5 the blanking force is highly dependent on the d/t ratio. Following Zener and Hollomen findings, interest in the dynamic blanking and piercing has been stimulated in late sixties. An extensive amount of information on the comparison between dynamic and quasi-static blanking of ferrous and non-ferrous materials, regarding the quality of the produced blanks,the maximum blanking force and energy required to effect blanking, both in the cold and hot conditions, together with the effect of percentage radial clearance had been published in refs. [3-13]. In general it is concluded that improvement in the quality of the blank has been achieved in high speed blanking particularly for ferrous materials, and they are always higher than those of slow speed. However, the energy requirement was less for non-ferrous ones. As the quality of the produced parts is of prime importance, examination of the available literature reveals that most of the published work was directed towards studying the effect of the percentage radial clearance, strain rate and temperature on the blanking force and energy required to effect blanking aiming to improve the blank quality and tool (punch and die) lives. Little work is published on the effect of geometrical parameters of punch and die on the blank quality and tool life. Koura and Zaid [14] investigated the effect of out of roundness of the punch and die in determining the quality of blanks. Kazzutake Komori [15] investigated the blanking operation using the node operation method, giving attention to the effect of various kinds of ductile fraction criterion on crack initiation and propagation during the blanking process. He reported that the methods gave good agreement with the autographic record (punch force-punch displacement curve) as calculated by the suggested work method agrees fairly with the one obtained experimentally [15]. Thipprakmas reported that fine blanking method for improving the wear resistance method of sprocket parts. The sprocket is a profile wheel with teeth. Its periphery is widely used in the drive system which makes it subjected to wear. This requires that the sprocket should posses high strength and good wear resistance. The traditional method in manufacturing the sprocket is the hobbing process followed by heat treatment to increase its strength, hardness and wear resistance. Recently, the fine blacking process has seen increase in use for manufacturing the sprocket with the advantages of improving strength and better surface quality which in twin will improve the wear resistance, furthermore, the reduction of the process operations leading to saving of the production time and cost; as compared to its traditional manufacturing process by hobbing [16].He has also investigated the fine blanking process for manufacturing the sprocket regarding the improvements in its strength and wear resistance using the metallurgical examination of the fine blanked sprocket and found that the improvement in its hardness and wear resistance is attributed to the grain elongation in the plastically deformed zone as supported by the photomicrographs in this region. It is therefore, anticipated that using fine blanking fore producing the sprocket may select low carbon steel instead of the medium carbon in its production cost [16].

1.1 Accuracy of blanks

Although the blanking and piercing produces products which do not require subsequent machining, however in predict the products suffer small parentages of geometrical errors e.g., dishing, doming and edge taper as indicated and magnified for clarity purposes in Figure1. Also indicated on the figure1 the ideal blank without errors by the dotted line.

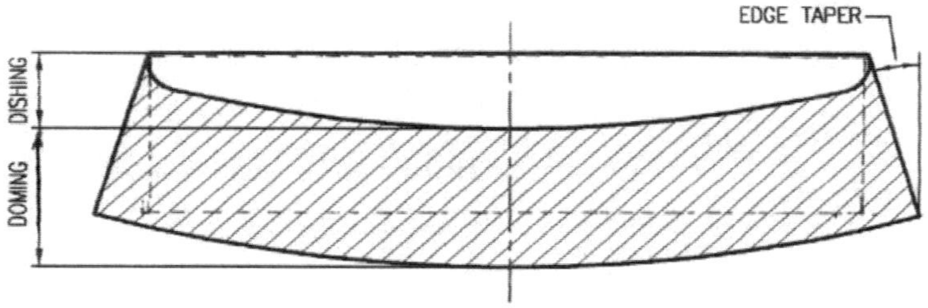

Figure1. Sectioned hatched blank showing the defects together with a projection of an ideal dotted blank

1.2 Main Parameters Affecting Blanking and Piercing

The main parameters affecting the blanking and piercing processes are summarized under four main elements namely;

1.2.1 Parameters related to the workpiece material

1.2.2 Parameters to the punch and die
1.2.3 Parameters related to the process.
1.2.4 Parameters related to the machines used for providing the force required to effect the process these are summarized in table 1.

Table 1. Paremeters Affecting Blanking and Piercing

Parameters related to workpiece	Dimensions, Geometrical shape, Physical properties Mechanical properties
Parameters related to the punch and die	Radial clearance %, c/t_o, punch and die profile radius, rp and rd, punch and die diameters, Dp, Dd., surface roughness, out of roundness.
Parameters related to the process	Strain rate, Working temperature (cold, warm or hot)
Parameters related to the machine	Presses (hydraulic, mechanical, electrical) Hammers (falling using hydraulic, high energy rate petroforge, dynaic)

2. Materials and experimental procedures

2.1 Materials

The specimens used for the blanking process were strips of sheet 50 mm wide and 2 mm thick, made of 0.84% carbon steel,Y8A, of the chemical composition shown in table 2. The punches and dies were all made of die steel X12M of the chemical composition shown in table 3.The die diameter was 25.001 mm and the group of punches wer calculated and manufactured to provide radial clearance percentages defined as, the radial clearance between the punch and die,(c) over sheet thickness,(to) i.e.(c//to) % to cover a range from 0.12% to 17.6 %.. All punches and dies were heat treated by quenchening in oil from 1040 C and tempered at 300 C to attain a Rockwell hardness RC54. All punches and dies diameters were measured using the universal measuring microscope to accuracy of 0.5 micron.

Table 2. Chemical Composition of Workpiece Material Y8A

Element	C%	Cr%	Mn%	Si%
Weight percentage	0.75-0.84	< 0.15	0.15-0.30	0.15-0.3

Table 3. Chmical Composition of Punches and Dies, Steel X12M

Element	C%	Mo%	Cr%	V%	Si%
Weight percentage	1.45-1.65	0.4-0.6	11.0-12.5	0.15-0.30	0.15-0.35

2.2 Experimental procedures

All the blanking tests were carried out using the designed and manufactured apparatus shown In Figure 2.

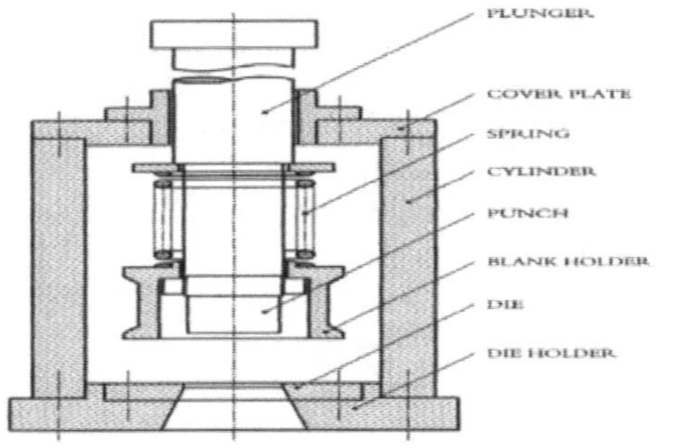

Figure 2. Blanking apparatus

The tests were carried out on a universal testing machine of 500KN capacity at a cross head speed of 5mm/min in two steps: the first step involved the variation of the punch profile radius, keeping the die profile sharp until blanking is completed and the second step involved the variation of the die profile radius keeping the punch profile sharp. The punch and die profile radii were measured using the profile measuring machine at a magnification of 30. the maximum blanking force was determined directly from the autographic record and the energy required to effect blanking is determined from the area under the autographic record, punch load-punch displacement curve, Figure 3a using a planimeter. The accuracy of the blanks was determined by measuring the amount of dishing, doming and edge taper on the sectioned and mounted blanks using a profile projector type PT300 at a magnification of X10. Finally , to study the mechanism of deformation and the metallurgical aspects of the process, the blanking tests were interrupted before, at and after the maximum blanking force was reached, at points a, b and c of Figure 3b. the partially and completely blanked specimens are then sectioned along their diameters, mounted in Bakelite, ground, polished with 1 micron diamond past and etched in 1% natal solution for 5 seconds, metallurgically examined for crack formation and photographed. Hardness and microhardness Vickers survey were carried out at the shear zone. The total number of performed tests were hundred and sixty.

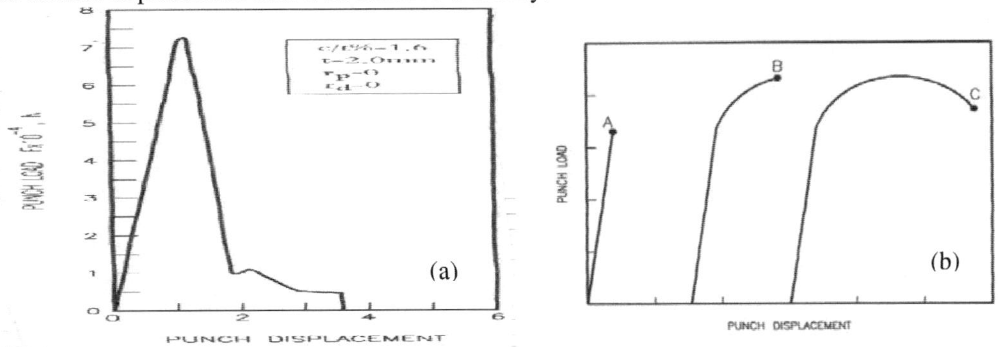

Figure 3. Punch displacement

3. Results and Discussions

3.1 Effect of radial clearance percentage

3.1.1 Effect on force

In all tests, the maximum blanking force is approximately constant at all values of radial clearance percentages, when the punch and die profile radii are kept sharp, Figure 4. The slight variations are due to small variations in the sheet thickness from the rolling process. This is due to the fact that the blanking force = the shear stress, T, x sheared periphery x sheared thickness. As it is obvious, all these parameters are not affected by the radial clearance percentages.

3.1.2 Effect on blanking energy

The blanking energy consumed to affect mechanical separation of the blank consists of four portions one the energy dissipated in friction between the punch and the penetrated part of the sheet plus energy dissipated in friction between the bottom part of the blank and the die and the main consumed in shearing off the blank from the sheet and finally the energy consumed in forming the dishing and doming. Examination of Figure 5, indicates an increase in energy due to the high friction between the punch and die within the clearance region being high at the beginning of the process forming the dishing. As the punch penetrates certain distance in the sheet during the punch travel this consumed energy is reduced in the clearance zone reaching its minimum value at the bottom of the curves shown in Fig.4, then after that the total blanking energy starts to increase because the amount of dishing and doming increase to their maximum at the end of the mechanical separation of the blank. The other portions of energy, namely the energy consumed in shearing off the blank from the sheet is constant. This is similar in all the curves. This agrees with the previous findings of other researchers, [4-8]. This is expected as the periphery and the thickness and the shear stress and the punch and die profile radii are all constants in the group of tests. This is of vital importance as it was found that the produced blanks at the optimum value of the radial clearance percentage has the best blanks quality i.e least dishing, doming and edge taper in addition it saves energy.

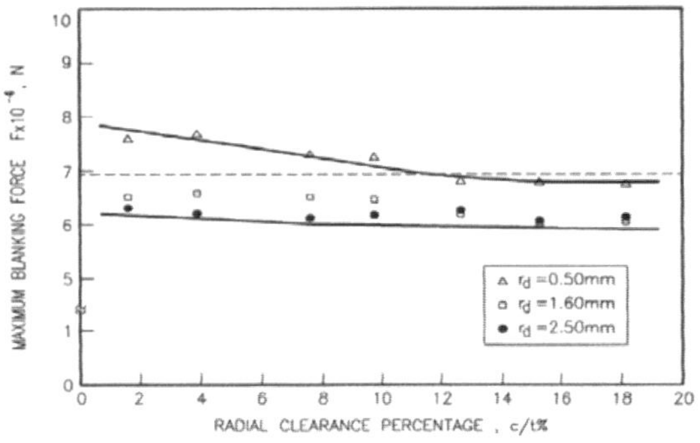

Figure4. Variation of maximum blanking force with the radial clearance percentage at sharp punch and die.Also shown the theoretical value of force,in dotted line

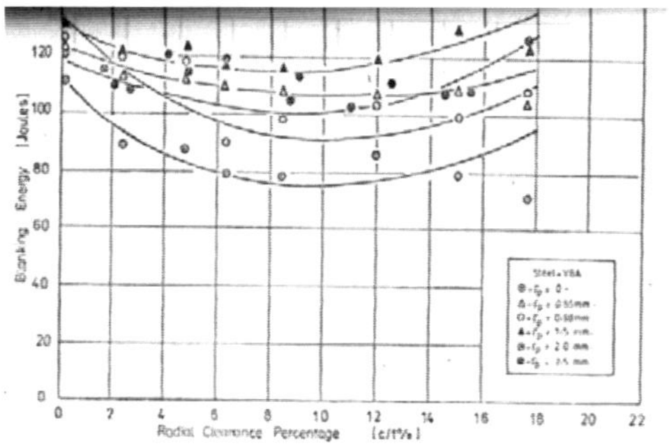

Figure5. Variation of the blanking energy with radial clearance percentages at different values of punch`profile radius, and square ended die.

3.2 Metallurgical examination

Microstrcutural examination of the partially blanked specimens (a total of 160 specimens) either with radiused punch or radiused die at different values of percentage radial clearance, first below the point of maximum force, at and beyond the maximum force revealed that the crack formation is dependent on all the aforementioned parameters in the following manner: The photomicrograph of Figs.6(a) and(b) indicate that at zero punch and die profile radii, the shear lines starts to form in the clearance zone before the maximum blanking force is reached. After it is reached, cracks started at the square ends of the punch and die and propagate until they meet and the blank is mechanically separated i.e. there is no shear stresses holding the blank to the parent sheet. Later, the blank is pushed through the die opening against the friction between punch and sheet, blank and sheet and blank and die opening.

Figure 6(a). Photomicrograph of the shear zone before maximum force is reached

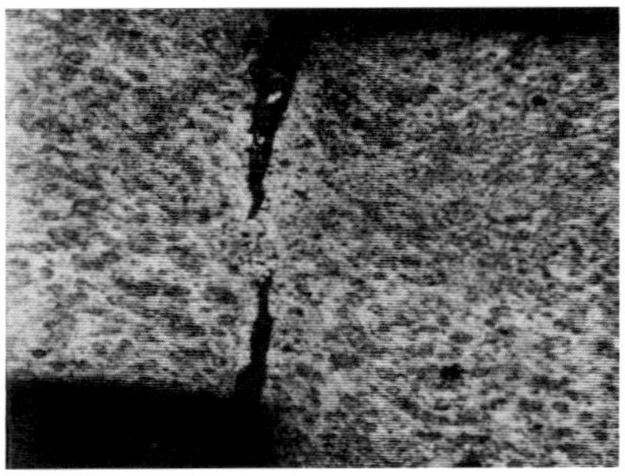

Figure 6(b). Photomicrograph of the shear zone after maximum force is reached

As the radial clearance percentages is increased keeping a square ended punch and die results in widening the shear zone and reducing the work hardening in the regions contiguous to the punch and die corners as observed from the microhardness measurements taken in this zone. This is clearly shown in Tables 4 and 5 at and beyond the maximum force. It can be seen from these tables that The hardness at the tip of the crack is found to be 328 HV and 303 HV near the corner of the die. No such cracks occurred at the point of maximum force (at 60% penetration of the sheet thickness), the metallurgical examination revealed that, for all values of percentage radial clearance two cracks occurred: one near the corner of the punch and the second near the corner of the die However the crack near the corner of the punch is longer. This suggests that crack has first occurred near the corner of the punch following the points of maximum hardness until the hardness falls to the value below that near the corner of the die. The crack then stops and second crack develops near the corner of the die and propagates towards the first one. Pera report,[17] shows that in the punching operation cracks occur in the region of maximum hardness gradient close to the corner of the punch and the die. In these tests it was found that the hardness of the material is less at the die edge than at the punch edge for all values of radial clearance percentages.

Radiusing either the punch or the die profile resulted in a decrease in both the level of hardness near the modified tool and depth of hardening; it also caused an increase in the width of the shear zone near the edge of the modified tool. The microhardness measurements indicated that hardness is less near the edge of the modified tool and that the hardness decreases as the profile radius increase. No appreciable change in the microhardness of the material at the corner of the square ended tool was noticed beyond the maximum force, table 3 on the other hand change in the order of the microhardness was noticed in the specimen material adjacent to both ends of the modified and non-modified tool at the maximum force. An example is given in Table 2. An interesting results is that the crack always initiated at the corner of the square ended tool, Figs. 6(a) and 6(b), and propagated towards the modified tool until the material is no longer able to carry the applied load, then shearing of the blank occurs. This can be explained using the plain strain deformation model for blanking suggested by Noble and Oxley [18] in which they used a simple stress analysis model to show that the crack formation can be related to the change of the shear flow stress across the shear zone. The greater the shear stress gradient the will be the chance of forming a crack. Thus cracking would be most expected in the regions where the workhardening is high, i.e. in the region adjacent to the square ended tool, and least expected in the regions where the workhardening is low, i.e. in the region adjacent to the modified tool. This agrees with the experimental results obtained in this work.

Table 4. Microhardness Vickers survey at the Maximum Blanking Force

Blanking conditions			Microhardness (HV160) (kp/mm^2)	
r_p (mm)	r_d (mm)	c/t_o %	Punch end	Die end
0	0	18.2	295	247
0	0.93	18.2	280	238
0	1.5	18.2	272	220
0	2	18.2	260	215

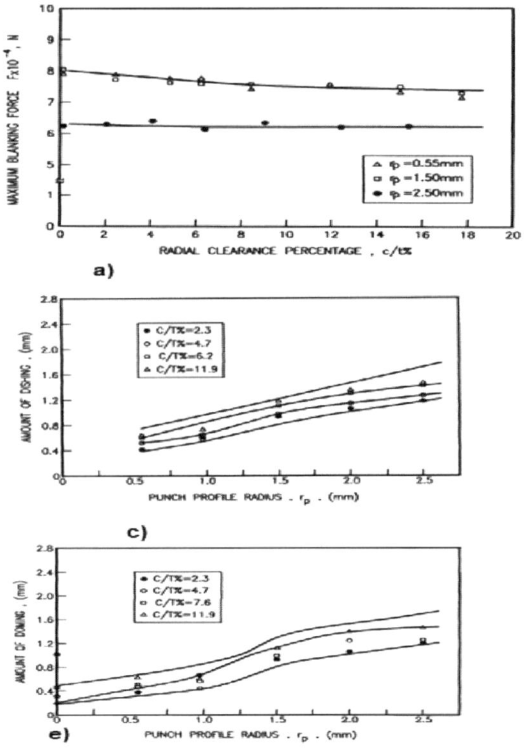

Figure 7. Effect of punch profile radius on blanking force, amount of dishing and amount of doming:a), c),and e) respectively.

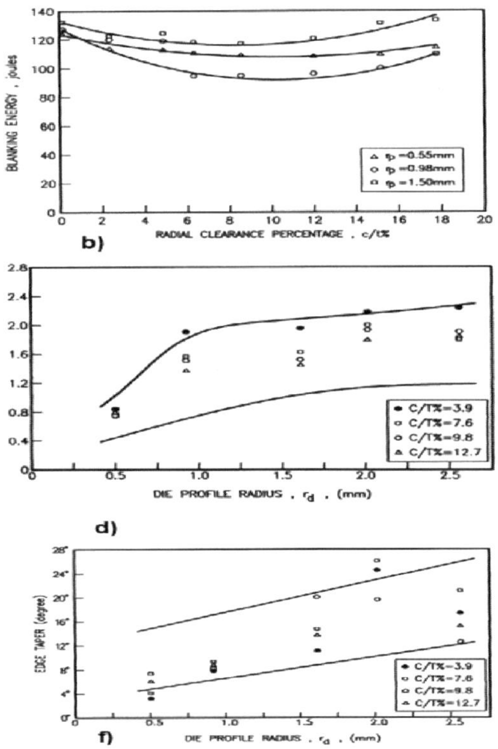

Figure 8. Effect of die profile radius on blanking energy,amount of dishing, doming and taper,b),d),and e)respectively

Tabble 5. Microhardness Vickers survey beyond the point of maximum blanking force

Blanking conditions			Microhardness (HV160) (kp/mm^2)	
r_p (mm)	r_d (mm)	c/t_o %	Punch end	Die end
0	0	18.2	295	295
0	0.93	18.2	295	265
0	1.5	18.2	290	240
0	2	18.2	288	228

Figures 7 show the effect of punch profile radius onblanking force and the accuracy of the produced blanks, from which it can be seen that increasing the punch profile radius results in deteriorating of the blanks quality.Similarly, Figure 8 indicates that increase of die profile radius radius also resulted in deterioration of blanks quality i.e.increase of the amounts of dishing, doming, and edge taper.

3.3 Effect of the wear and surface roughness on blanking and piercing

As it was found that the radial clearance percentage between punch and die is the most important parameters in blanking and piercing in determining the accuracy of the blank, therefore, wear of blanking and piercing tools will cause the optimum radial clearance percentageto change, therefore, wear should be considered in the repair and maintenance of their tools. Similarly, the wear causes the surface roughness of the punches and dies which will have to a lesser degree, the same effect as the change in the radial clearance percentage. Little work on the effect of the wear of tools in these processes is published therefore, this area is still open for research,[19].This lack of research may be attributed to the complexity of the process which still not fully understood despite the amount of theoretical and experimental research which was work carried out on these processes since mid forties till now. Analytical modeling, and finite element analysis have received much attention in recent years, [19,20]. Review of these current analyses and models are reviewed and discussed in reference [21}. A more detailed study concerning the mechanism of deformation and crack propagation is given by Noble and Oxley showed that the cracks occur in the region of maximum hardness i.e. where the metal has been subjected to the greatest amount of strain hardening, namely at the punch and die edges [17]. PERA report using a plain strain deformation model for blanking showed that crack formation can be related to the change of the shear flow stress a cross the shear zone, the greater the shear stress gradient the greater the chance for crack formation [18].

4. Conclusions

The following points are concluded:

- Although square ended punch and die produce blanks of better quality, it is essential to provide profile radiusing of punch and die to improve their lives. Increasing punch and die profile radii caused increase in both blanking force and energy particularly at small radial clearance percentage.
- In general, providing the profile radii tends to increase the energy and reduces the blanking force and reduces the quality of the blanks. Furthermore, they caused enlargement of the shear zone, being more affected by the die profile radius.

- Radiusing the punch and die profiles caused delay in crack formation at small values and non-occurrence at large values and resulted of lower levels of miccrohardness in the vicinity of the radiused end.

5. References

[1] Zener C and Hollomen J H 1944 *Journal of applied physics* 15 22.

[2] Chang I M 1950-51 *Journal of institute of metals* 78 393.

[3] Maeda T and Temura K 1960 *Bulletin Japanese society of mechanical engineer* 312.

[4] Daives R and Dhawan S M 1966 A preliminary investigation of high speed blanking and piercing of metals *Applied mechanics convention, Cambridge.*

[5] Daives R and Dhawan S M 1968 Comparison of some strain rate effects in slow and high speed blanking *proceeding of 9th international machine tool design research conference.*

[6] Jonson W and Slater R A C 1964-65 A Comparison of Energy Required for Slow Speed and Dynamic Blanking using an Improved Linear Motor (proceeding institute of mechanical engineers, 1964-65).

[7] Jonson W and Slater R A C 1965-66 Further Experiments in Quasistatic and Dynamic Blanking of Circular Discs from Various Materials *proceeding institute of mechanical engineers.*

[8] Slater R A C and Johnson W1967 *International journal of mechanical science* 9 271.

[9] Johnson W and Travis F W1966 High Speed Blanking of Copper *Applied Mechanics Convention*, London 1966.

[10] Johnson W and Travis F W 1965 High Speed Blanking of Steel Engineering plasticity edited by J. Heymon and F.A. Leckie 385.

[11] Dowling A R and Harding J 1970 *Journal of institute of metals* 96 215.

[12] Stock T A C and Wingrove A L 1970 *Journal of mechanical engineering science* 13(1971), p. 110.

[13] Balendra R and Travis F W1979 *Journal of machine tool design and research*, Vol. 10 . 249.

[14] Koura M M and Zaid A I O Parameters Affecting Blanking Operation second Cairo conference Cairo 1982.

[15] Kazzutake Komori 2013 Simulation of crack arrest in blanking using the node separation method *International Journal of Mechanical Sciences* MS2386.

[16] Thipprakmas S, Jin M and Murakawa M 1961 An Investigation of material flow analysis in fine blanking process *Journal of Materials Science Processing* 192-193 237-242.

[17] P.E.R.A.: Blanking and Piercing Report 93.

[18] Noble C F and Oxley P L B 1963 Crack Formation in Blanking and Piercing, *International Journal of Production Research*, Nov.

[19] Klingenberg W and de Boer T W 2008 Condition based maintenance in punching/ blanking of sheet metal *International Journal of Machine tools and Manufacture* 48 589-598.

[20] Klingenberg W and Singh U P 2005 Comparison of two Analytical Models of Blanking and Proposal of a newmodel *International Journal of machine tools and manufacture* 45 519-527.

[21] Kligenberg W and Singh U P 1980 Further Observations and Review of Numerical Solutions of Sheet Metal Punching *International Journal of Advanced Manufacturing Technology* 22 215- 231.

On the stress corrosion cracking of lean duplex steel in chloride environment

Qanita Tayyaba[1], Hina Farooq[1], Muhammad Shahid[1], Ammer Khan Jadoon[2], M Shahzad[3] and AH Qureshi[3]
[1]School of Chemical and Materials Engineering, National University of Sciences and Technology, Islamabad, Pakistan
[2]British Petroleum Ltd, UK
[3]Materials Division, ATCOP, Islamabad, Pakistan

E-mail: mshahid@scme.nust.edu.pk

Abstract. Duplex stainless steel having attractive combination of austenitic and ferritic properties is being used in industry such as petrochemical, pulp and paper mills. In this study, the corrosion and stress corrosion behavior of duplex stainless steel in 3.5% sodium chloride environment was investigated by weight loss measurements, electrochemical DC testing and slow strain rate test (SSRT). Weight loss data showed no significant corrosion after 1700 hours. Electrochemical polarization test in 3.5% NaCl solution exhibited a uniform corrosion rate of 0.008 mpy (calculated using Tafel analysis) showing passivity in the range of 735–950 mV. A comparison of the slow strain rate test in 3.5% NaCl solution with air shows almost a similar stress strain curve for duplex stainless steel. In comparison, the stress strain curves for 0.15% carbon steel show a loss of about 25% tensile elongation for the same comparison. The excellent corrosion and especially resistance to localized corrosion (pitting) is responsible for no loss of ductility in duplex stainless steel.

1. Introduction

Duplex stainless steels have vast applications in chemical industries, power plants and marine owing to their excellent mechanical properties and high corrosion resistance [1]. These steels consist of a ferrite/austenite volume ratio of about 1:1 placing them amongst good materials against aqueous corrosion along with high strength. However, existence of undesirable phases e.g. intermetallics, nitrides and carbides in the steels profoundly influence on these properties unless controlled during the manufacturing processes [2]. 'σ' phase with relatively rapid kinetics for formation is considered to be of particular importance causing a drastic decrease in toughness and corrosion resistance [3].

Nilsson and Wilson [4] investigated the influence of isothermal heat treatment of super duplex stainless steel on toughness and pitting corrosion resistance. They found it to be directly related to the existence of 'α' phase due to longer time–temperature regime. Secondary austenite was found to be present at the ferrite/austenite phase boundaries, which can be as detrimental to pitting corrosion as 'α' phase.

Pitting resistance of 25% Cr duplex UNS S32550 was investigated as a result of annealing temperature, by Garfias-Mesias et al [5]. Lower annealing temperatures were observed to be favourable to attain higher pitting potentials. Ferrite phase was found to be easy sites for pitting.

Pohl et al [6] studied the effect of intermetallics on corrosion properties of DSS. They had the opinion that the intermetallic precipitations influenced mechanical properties and the corrosive properties rather extensively. Rapid reduction of the toughness of the steel was observed due to precipitation of brittle phases the resultant tertiary austenite makes the alloy prone to corrosion. Sherif [7] studied the corrosion of DSS Alloy 2209 in acidic and neutral chloride solutions and its passivation by ruthenium as an alloying element and found that ruthenium increases its passivity against corrosion. Effect of annealing temperature on pitting corrosion resistance of DSS was investigated by Tan et al [8]. They found that variation in annealing temperature changes the amount of alloying elements in the ferrite and austenite phases and correspondingly affect the critical pitting temperature. In the present investigation, SCC behaviour of DSS has been explored in 3.5% NaCl solution and in air.

2. Experimental

2.1 Metallography and Characterization.

Microstructural characterisation was performed using optical microscope with associated image analysis equipment. This system was used to determine the phase percentages of ferrite and austenite in the duplex stainless carbon steel structures. The material surface was etched using a solution containing 200ml HCl, 200ml water, 200ml HNO_3 and 0.6g Sparbeize (pickling inhibitor) at 40°C to reveal the microstructure of the material. The etching was performed after mechanical grinding with silicon carbide papers (300-1200 grit size), and polished using 6 microns and 1 micron diamond pastes.

2.2 Mechanical Testing.

Hardness was performed using Vicker hardness tester. The minor load of 10kgf and major load of 150kgf (1470N) was applied on specimen surface for a dwell time of 5s. Micro-hardness was performed by a Vickers hardness tester with the load of 1000 gf with a dwell time of 10 seconds using diamond indenter in order to measure the hardness of austenite and ferrite phase. Tensile specimens having gauge diameter 2.5mm and gauge length 11.5 mm were tested under tensile load at strain rate of 0.69mm/min. The load displacement data from the machine was used to plot the stress strain curve for each tested samples.

2.3. General corrosion and pitting corrosion.

2.3.1. Immersion

Test. The 10.0 mm x 8.2 mm x 9.2 mm size and weight of 5.017 grams DSS was first cleaned and polished. A 3.5% NaCl solution (electrolyte) was prepared by taking 7g of NaCl in 200ml of distilled water and was stirred using electronic stirrer. pH of the solution after adding salt was 6.5. Sample of DSS was immersed in this electrolyte. After immersion for a specified time the sample was taken out, washed and cleaned, followed by weighing to determine the weight loss.

The corrosion rate was calculated as: [9]

Corrosion rate (mpy) = 534W/ DAT

Where

W= weight loss of the metal

D = density of the metal (in g/cm^3).

A = surface area of the metal exposed (in^2)

T = time in hours

2.3.2 Electrochemical Corrosion Test. Electrochemical studies were conducted using DC corrosion testing technique, in a 3-electrode electrochemical cell; a silver/silver chloride reference electrode in saturated potassium chloride (KCl), and a working electrode. The potentiodynamic polarization response of the DSS was recorded in 3.5% NaCl solution. Tafel Plots were determined at a scan rate of 1 mVs^{-1} starting from -0.30V to +0.60V with respect to open-circuit potential.

2.3.3 Slow Strain Rate Test. Stress corrosion cracking behaviour of DSS samples was conducted by slow strain rate test (SSRT). The gauge length of each tensile specimen was polished to 1000 grit paper finish and cleaned with acetone before using them for SSRT. These specimens were then uni-axially loaded in a modified slow strain rate testing machine with autoclave in an environment containing 3.5% NaCl solution at room temperature and at atmospheric pressure.

The specimens were then subjected to an initial strain rate of $1 \times 10^{-6} s^{-1}$ and loaded until fracture occurred. Samples were electrically isolated from the test rig by making use of ceramic washers. Stress-strain curves were obtained from each test. Fractured surfaces were examined inSEM to examine the mode of fracture while EDS helped to obtain chemical composition of the fractured surface and performing phase analysis.

3. Results

3.1 Metallography and Scanning Electron Microscopy.

The microstructural analysis shows approx. 70:30 (α: γ volume ratio) duplex microstructure with ferrite grains elongated in the rolling direction. The point count method was used to estimate the volume fractions of microstructural constituents.

The optical micrographs show dark phase (elongated phase) as ferrite and light phase (matrix) as austenite. The micrographs are shown in figure1.

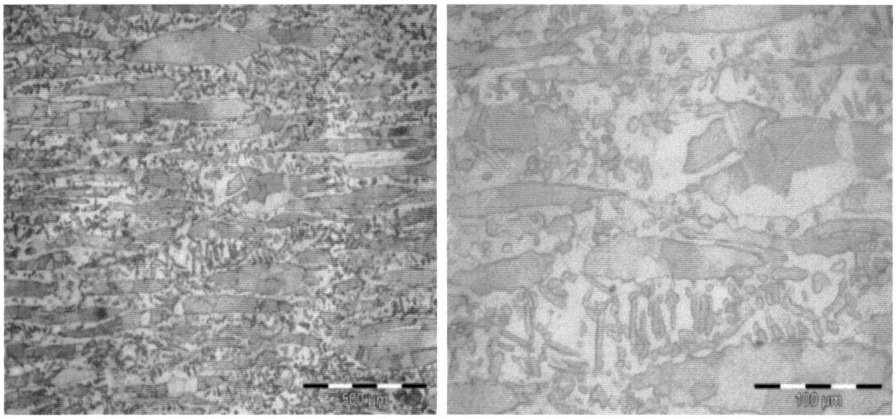

Figure 1. Optical micrographs showing Ferritic & Austenitic phases.

The ferrite contained chromium content in higher percentage compared with the matrix (austenite). The matrix contained nickel content in higher percentage than ferrite. The composition of the two phases is summarized in table1.

Table 1. EDX analysis of Ferrite phase and Austenite phase.

Elements (Ferrite)	Mass %	Elements (Austenite)	Mass %
Iron	62.56	Iron	63.20
Chromium	27.08	Chromium	24.45
Nickel	4.57	Nickel	8.13
Molybdenum	3.55	Molybdenum	1.67

3.2 Mechanical Testing

3.2.1 Hardness testing

Measurements of macro-hardness were performed by a Rockwell hardness tester using diamond indenter on scale C gave 21.0 HRC.Measurements of micro-hardness were performed by Vickers hardness tester showed that elongated phase (ferrite) was harder than austenite (matrix).The results showed that Ferrite phase is 37% harder than Austenite. Table-2 shows the hardness values of DSS.

Table 2. Hardness values of DSS.

Bulk hardness of DSS	290 HV
Micro hardness of ferrite (elongated) phase	361 HV
Micro hardness of austenite (matrix) phase	262 HV

.2.2 Tensile Testing

Stress-strain curve for DSS in air is given in figure 2. From the graph UTS was found to 750 N/mm^2, elongation 6.7% and reduction in area was observed as 3.4%.

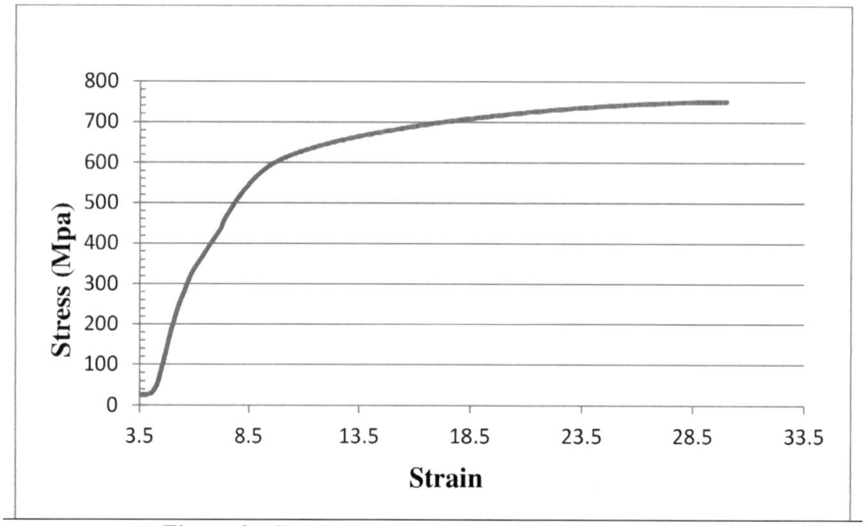

Figure 2. Graph showing stress-strain curve of DSS.

3.3 Testing for corrosion behavior

3.3.1Immersion Test

Immersion Corrosion Test in 3.5% NaCl solution showed no significant corrosion after 1700 hr. The graph in figure 3 shows no change in weight with the passage of time.

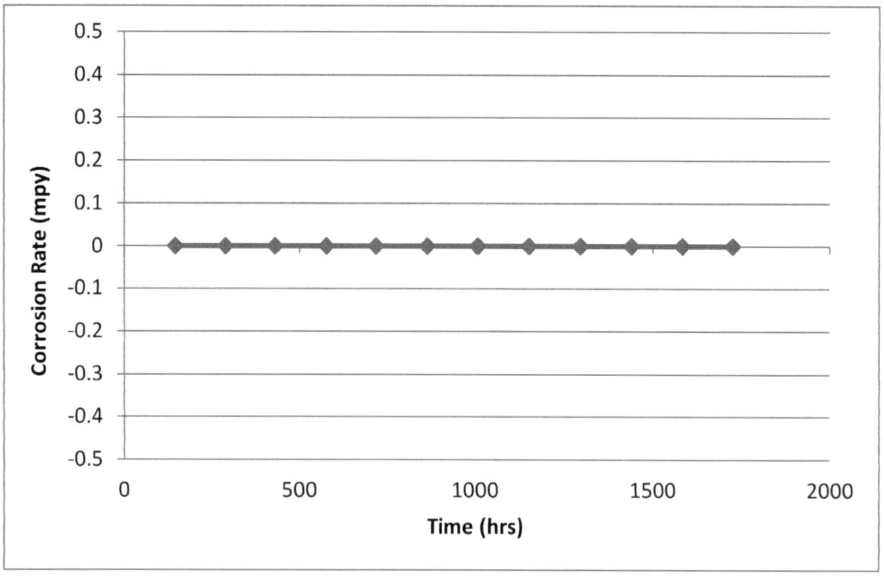

Figure 3. Corrosion rate vs. time of DSS in 3.5% NaCl solution.

3.3.2 Electrochemical Corrosion Testing

'E-log i' diagram of DSS in 3.5% NaCl solution is given in figure 4. Tafel analysis exhibited a corrosion rate of 0.008 mpy using a built-in software of Gamry® framework. The data obtained from the figure is summarized in table 3.

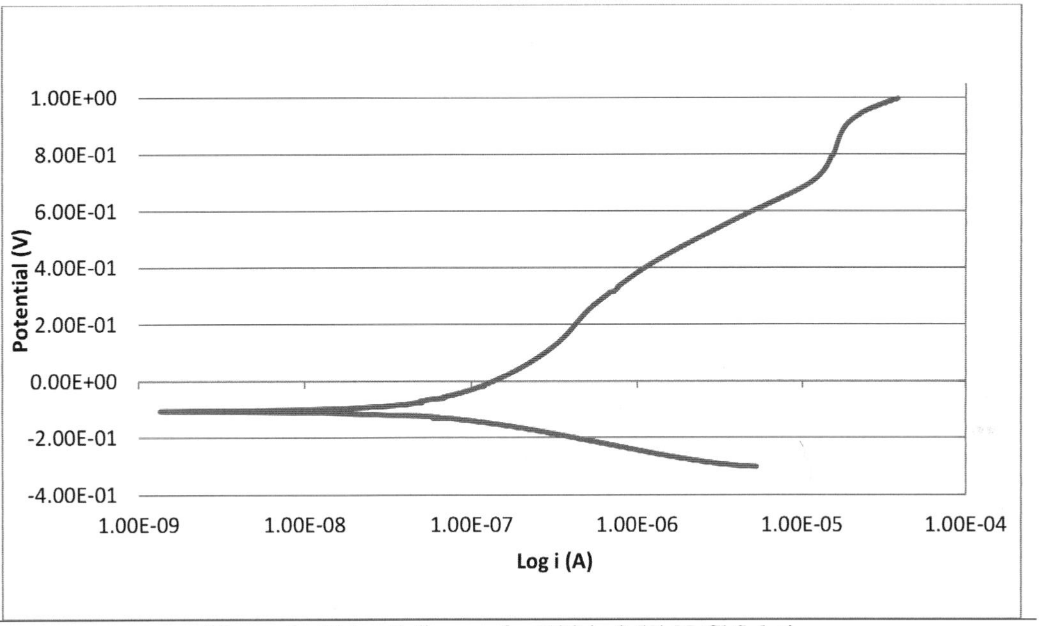

Figure 4. *'E-log i'* diagram for DSS in 3.5% NaCl Solution

Table 3. Corrosion data from Tafel analyses

E_{corr} (mV)	-106.0 mV
I_{corr} (nA)	55.80 nA
β_c	85.30×10^{-3}(mV/dec)
β_a	267.7×10^{-3} (mV/dec)
Corrosion Rate	8.439×10^{-3} (mpy)

3.3.3 Slow Strain Rate Test

Slow strain rate Stress-Strain curves for mild steel in seawater and air are shown in figure 5. The graph shows that air and seawater has almost similar UTS point and Strain to failure which indicating that no stress corrosion cracking took place in the allowed time. Figure 6 shows SSRT curve for mild steel in similar environments. Comparison of the two figures reveals that mild steel had a substantial effect of environmental conditions whereas it was not significant in case of DSS. Mild steel exhibited a loss of about 25% in elongation in seawater compared with air. The excellent corrosion and especially resistance to localized corrosion (pitting) is responsible for no loss of ductility in duplex stainless steel.

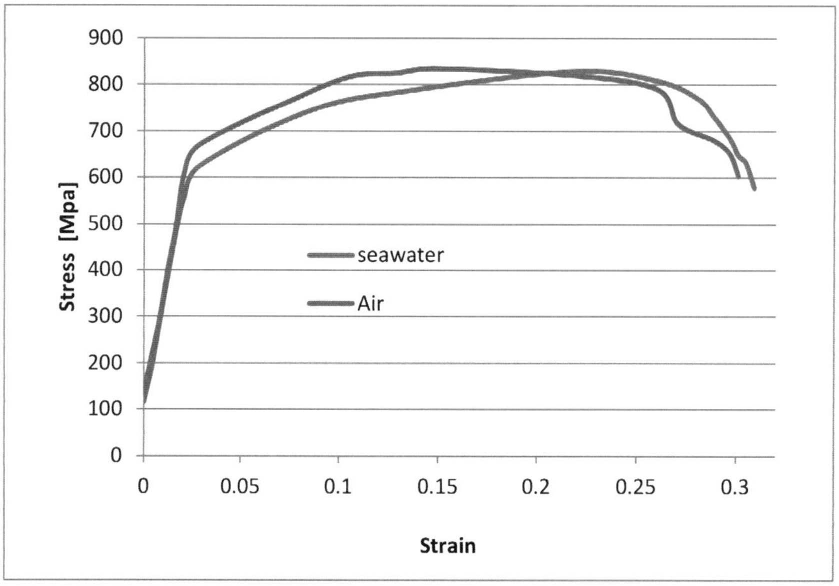

Figure 5. SSRT of DSS in air and seawater

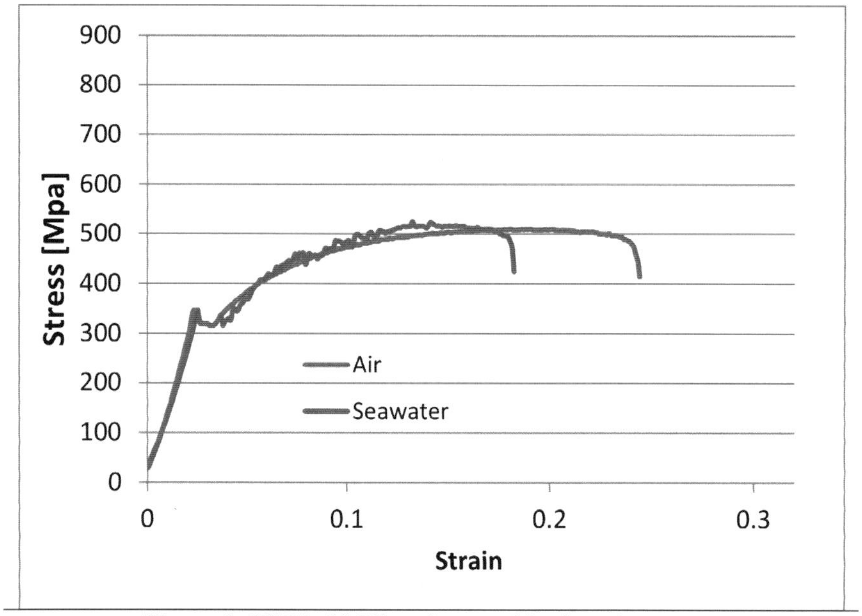

Figure 6. SSRT of 0.2% carbon steel in air and seawater

3.3.4 Fractography

Figure 7 reveals the fractured surface of DSS exposed to SSRT in seawater. The detailed examination of the fracture morphology showed a ductile fracture showing dimpled pattern.

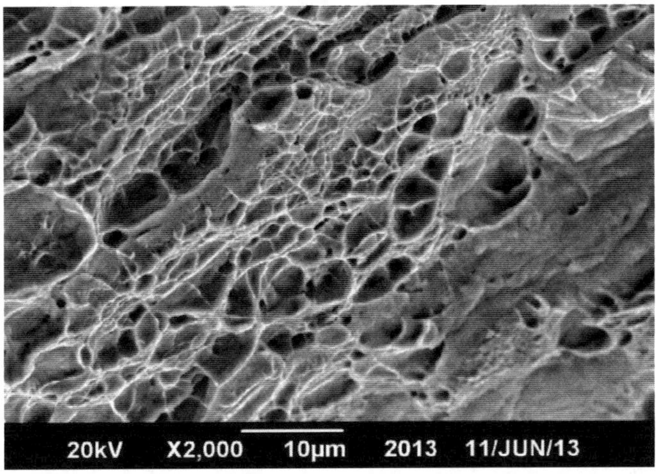

Figure 7. SEM fractograph of SSRT for DSS in 3.5% NaCl

4. Conclusions

- A higher volume fraction of ferrite increases hardness and gives a good uniform corrosion resistance as well as pitting.
- The DSS under investigation had about 60% higher UTS compared with mild steel.
- During slow strain rate testing in seawater, in comparison with DSS mild steel showed a loss of about 25% tensile elongation in seawater. The excellent corrosion resistance to localized corrosion (pitting) was presumably responsible for no loss of ductility in duplex stainless steel.

5. References

[1] Charles J 1991 Conf. proc. *Duplex Stainless steels '91*Beaune, Les éditions de physique 3–48

[2] Solomon H D, Devine TM and LuraR A 1983 *Duplex Stainless Steels* American Society for Metals, Metals Park OH 553–572

[3] Ahn YS and Kang JP 2000 *Mater Sci Technol* 16 382–387

[4] Nilsson J O, Wilson A 1993 *Mat Sci & Tech* 9(7) 545-554

[5] Garfias-Mesias L F, Sykes J M and Truck C D S 1996 *Corrosion Science38(12), 2267*

[6] Pohl M, Storz O and Glogowski T, January 2007 Materials Characterization 58(1) 65-71

[7] El-Sayed M Sherif 2012 *Int J Electrochem Sci* 7 2374-2388

[8] Tan H, Jiang Y, Sun T, Xu J and Li J 2009 *Materials Characterization* 60 1049-1054

[9] Fontana M G 1987 *Corrosion Engineering* McGraw-Hill

Solvothermal synthesis of luminescent bis-(8 hydroxy quinoline) cadmium complex nanostructures

M A Khan[1], Shafique Ahmed[1], Majid A[1], Ayaz Arif Khan[1], Mahboobullah[1], Asghar Hashmi[2], Tariq Iqbal[1], Azhar Ahmed[1] and Nasar Ahmed[1]
[1]Department of Physics, University of AJK, 13100, Muzaffarabad, Pakistan
[2]Department of Physics, Islamia University Bahawalpur, Pakistan

E-mail: rauf_ak@yahoo.com

Abstract. A facile solution– based route for the synthesis of Bis- (8- hydroxyquinoline) Cadmium (CdQ_2) complex nanorods, nanoflowers (bundles of nanorods) and nanosheets in an oleic acid- sodium oleate- ethanol- H_2O system at 50^0C -100^0C was reported. Scanning Electron Microscope (SEM) images indicated that longer time and higher temperature would result in nanoflowers, while lower temperature and shorter reaction time would be suitable for the formation of nanorods. However, a novel change in these structures was observed when the concentration of the surfactant (oleic acid) was reduced and we obtained 2- D nanosheets. Fourier- transform infrared (FTIR) spectroscopy was utilized to confirm that the samples were made up of CdQ_2. UV/ VIS spectroscopy was used to determine the different electronic transitions in CdQ_2 molecule. All the samples possessed excellent photoluminescence (PL) properties. Photoluminescence (PL) spectra showed a prominent peak around 500 nm which indicated a strong PL emission in the green region. This methodology could be extended for the controlled, large– scale preparation of other functional complexes, and the obtained nanostructures could be introduced as the basic building blocks for novel optoelectronic devices.

1. Introduction

It is well known that, at the nanoscale, materials exhibit fascinating optical, electronic and magnetic properties that differ drastically from their bulk counterparts [1-6]. Recently, one dimensional (1-D) conducting polymer and small molecule organic semiconductor nanomaterials have attracted much attention because of their unique physical properties and potential applications in fabrication of nano devices. However, the preparation techniques of organic nanomaterials are not diversiform compared with those of inorganic nanomaterials. The novel properties of nanostructured materials enable them to find potential applications in many new and promising fields such as nanofabrication, nanodevices, nanobiology, and nanotechnology [7- 12].

Currently, 8- hydroxyquinoline metal chelates are intensively investigated due to their excellent electro- and light- emitting properties, such as field emission, photoluminescence (PL) and electroluminescence (EL) [13- 16]. Here, we report the facile solvothermal synthesis of Bis- (8- hydroxyquinoline) Cadmium (CdQ_2) Complex nanorods, nanoflowers and nanosheets in an oleic acid (OA) - Sodium Oleate- ethanol-H_2O system at 50^0C-100^0C. SEM results show that the reaction time, reaction temperature, and the concentration of the surfactant play a crucial role in the morphology of the samples. Particularly, the reduction in the surfactant concentration results in the formation of (2-D) nanosheets. As for as we know, this significant change in the morphology of CdQ_2 samples has not yet been reported. The photoluminescence (PL) spectra reveal that the samples have excellent photoluminescence properties. These findings are very useful to fabricate optoelectronic devices based on CdQ_2 nanostructures [17, 18].

2. Experimental Section

2.1. Materials.

All the reagents in this work, including Cadmium chloride($CdCl_2.2½H_2O$), 8-hydroxyquinoline (C_9H_7NO), ethanol (C_2H_5OH), Sodium hydroxide (NaOH),and Oleic-acid ($CH_3(C_{11}H_{33}COOH)$, were of A.R grade and used without further purification. De-ionized water was used throughout the process.

2.2. Synthesis of the Samples.

In general synthesis, the reaction is mainly on the basis of the precipitation of $Cd^{2+}$and 8-hydroxyquinoline with a molar ratio 1:2 in a mixed ethanol–water (1:1) solvent at designed temperatures. In detail, NaOH, and oleic acid (OA) were orderly added into the mixed solvent with continuous stirring to produce a homogeneous normal emulsion. Then, the aqueous solution of Cd Cl $_{2.}2½H_2O$ was poured into the system.

Subsequently, the ethanol solution of 8-hydroxyquinoline was rapidly introduced into the solution to start the main reaction and the total volume of the mixture was kept at a certain value 30 mL. After stirring for about 15 min, the mixture was transferred into a 50 mL vessel, which was treated at a designated temperature for appropriate reaction time. Then, the system was cooled at room temperature naturally. The bottom precipitates were thoroughly washed with ethanol twice and then redispersed in ethanol for future use and characterization. The detailed experimental conditions for each sample are listed in table 1 and table 2, respectively.

2.3. Characterization

The sizes and morphologies of the samples were characterized by a Jeol JSM-6510LV Scanning Electron Microscope (SEM). Samples were prepared by placing a drop of a dilute ethanol dispersion of the product on the surface of Aluminium stub. FTIR spectra were conducted with a Perkin Elmer Spectrum 100 series FTIR Spectrometer. UV/ VIS-IR spectra were carried out by using Perkin-Elmer, LAMBDA 950 Spectrometer. PL spectra were taken by using Mini PL/ Raman system, photon system USA with Ne– Cu laser having wavelength of 248 nm.

3. Results and Discussions

In the synthesis process, shown in Figure 1, as the consequence of the addition of NaOH, sodium oleate (NaOA) is formed and then coordinate with Cd^{2+} to produce cadmium oleate, which releases Cd^{2+} slowly on the basis of the equilibrium equation to control the nucleation and growth process accurately. The other function of sodium oleate was to produce a buffer media jointly with the superfluous oleic acid which was lucrative for the precipitation between $Cd^{2+}$and 8-hydroxyquinoline [19]. On the other hand, the interactions such as van der Waals and the hydrophobic force between the alkyl chain and quinoline ring would have made it easier for 8-hydroxyquinoline molecules to collide with Cd^{2+} capped on the surface of sodium oleate to start the main reaction. At last, due to the low solubility of the complex, the final products deposited at the bottom of the container, giving convenience for collection and purification. In order to control the morphology and size of the product and understand the crystal growth behaviours of the purposeful molecule, we investigated the effects of the following factors, reaction time reaction temperatures, reactant concentration, and surfactant concentration. The table 1 and table 2 show the detailed experimental conditions for the synthesis of nanorods, nanoflowers and nanosheets. These morphologies are confirmed by the SEM images, shown in Figure 2 and Figure 3. It can be seen from Figure 2 and table 1 that longer time and higher temperatures (50^0C-100^0C) would result in the formation of nanoflowers, while lower temperatures (50^0C -55^0C) would be appropriate for the formation of nanorods.

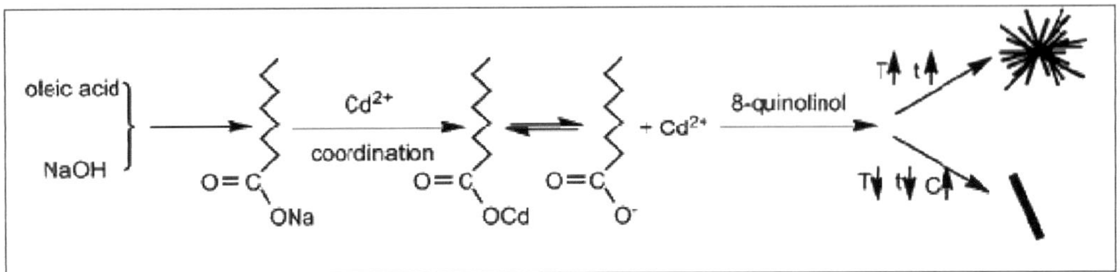

Figure 1. Reaction mechanism for the formation of Bis- (8- hydroxyquinoline) Cadmium (CdQ_2) complex nanorods and nanoflowers [Taken from Ref 17].

Table 1. Conditions for different morphologies and sizes of CdQ_2 nanorods and nanoflowers

Samples	Cd^{2+} (mmol.)	OA (ml.)	NaOH (g)	Temperature T(⁰C)	Time (hrs)	Morphology	Length (μm)	Average diameter (nm)
Sample-01	0.3	3.7	0.2	50	2.00	Nanorods	10.92	279
Sample-02	0.3	3.7	0.2	55	2.5	Nanorods	1.16	714
Sample-03	0.3	3.7	0.2	65	2.5	Honeycomb/ Nanoflowers	50	1000
Sample – 04	0.3	3.7	0.2	75	2.5	Nanoflowers	1.2	190
Sample – 05	0.3	3.7	0.2	100	2.00	Nanoflowers	55	882
Sample – 06	0.3	3.7	0.2	100	2.5	Nanoflowers	3.42	560

A novel change in the morphology of the CdQ2 samples has been observed when the concentration of the surfactant (OA) is changed. For instance, when the concentration of the surfactant (OA) was reduced from 3.7 ml to 3.5 ml, the morphology of the samples was remarkably changed from nanorods and nanoflowers to nanosheets. This significant change in the morphology of our samples has been confirmed by the SEM images, shown in Figure 3. Figure 3 and Table 2 indicate that this morphology remains consistent as we increase both the reaction time and temperature (65⁰C to 85⁰C). However, at lower temperature and at shorter reaction time (1.0- 2.0 hrs.), the nanosheets appear to be transparent whereas, at high temperatures (75⁰C to 85⁰C) and longer time (2.5- 3.0 hrs.), a little thicker nanosheets are obtained, as shown in Figure 3 (d- f).

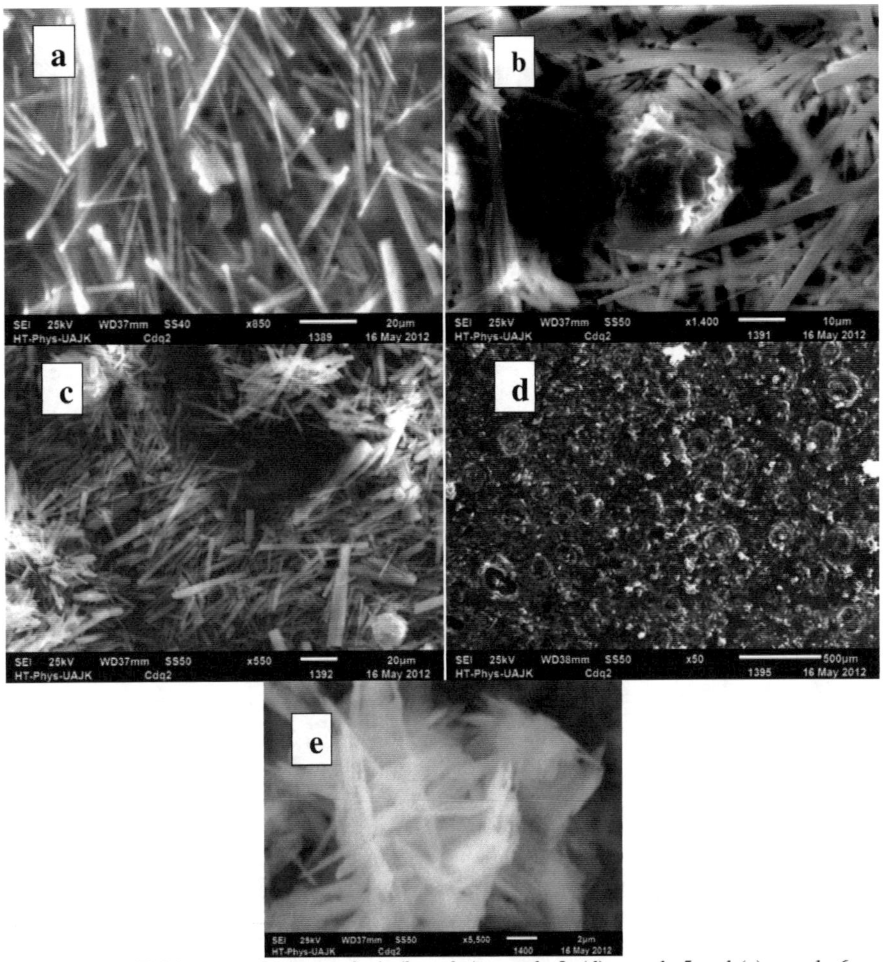

Figure 2. SEM images of (a) sample 1, (b and c) sample 2, (d) sample 5 and (e) sample 6.

Table 2. Conditions for morphology and different sizes of CdQ_2 nanosheets

Samples	Cd^{2+} (mmol.)	OA (ml.)	NaOH (g)	Temperature T(⁰C)	Time (hrs.)	Morphology	Length (µm)	Average Width (nm)
Sample - 07	0.3	3.5	0.2	65	1.00	Nanosheets	1	926
Sample-08	0.3	3.5	0.2	65	1.5	Nanosheets	7.24	788
Sample-09	0.3	3.5	0.2	65	2.00	Nanosheets	2.08	344
Sample-10	0.3	3.5	0.2	65	2.5	Nanosheets	1.04	1000
Sample-11	0.3	3.5	0.2	65	3.00	Nanosheets	4.72	313
Sample-12	0.3	3.5	0.2	75	2.5	Nanosheets	3.2	750
Sample-13	0.3	3.5	0.2	85	3.00	Nanosheets	2.48	800

Figure 3. SEM images of sample 07, sample 08, sample 09 (a, b, c), respectively, and sample 11, sample 12, sample 13 (d, e, f), respectively.

The component of the nanostructures has been identified with FTIR spectra. As indicated in Figure 4 and 5, (similar to congeneric compounds [20, 21]) the peak at 1571 cm^{-1} is an indicative peak of C=C aromatic stretching. Aromatic amine resonance peak of C-N-C at 1280 cm^{-1} and 1387 cm^{-1}, C=N stretching at 1425 cm^{-1} and aromatic C-O stretching at 1105 cm^{-1} are observed in the spectrum. The spectra show the characteristics peaks of aromatic ring stretching at 739 cm^{-1}, 740 cm^{-1}·793 cm^{-1} and 821 cm^{-1} which can be attributed to the existence of quinoline ring. The cadmium ion is a soft acid which tends to form a stable complex with quinoline by metal nitrogen bond. The Cd-O bond stretching was observed at 505 cm^{-1} while Cd-N bond stretching was observed at 490 cm^{-1}.

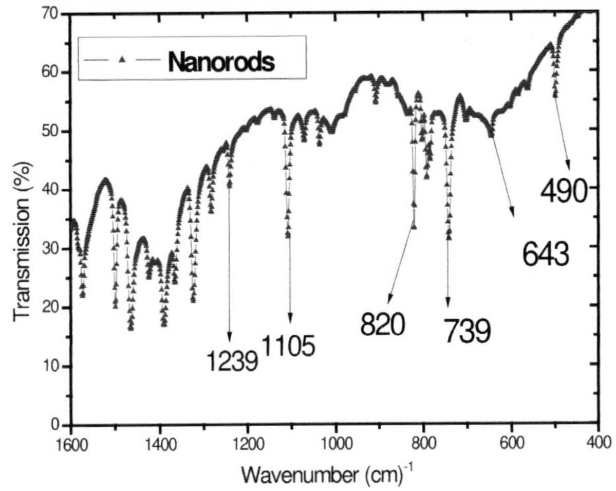

Figure 4. FTIR Spectra of CdQ_2 nanorods (sample 1).

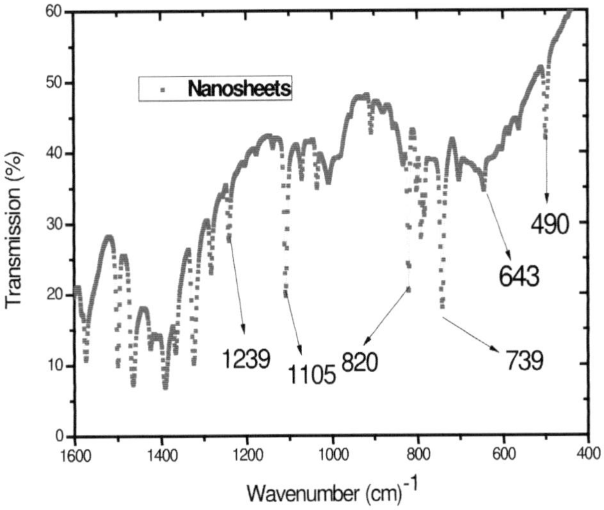

Figure 5. FTIR Spectra of CdQ_2 nanosheets (sample 09).

UV/ VIS spectra of our samples are shown in Figure 6.The solvent effect has been observed for the ethanol and water is at 278.43 nm and 262.88 nm, respectively. This is in good agreement with literature value [22]. The aromatic and hetero aromatic (quinoline) shows broad absorption band which consists of multiple peaks in the near UV region between 230-270 nm. E-band is the properties of aromatic structures. The acid produces the positive charge on the nitrogen shifts which shows absorption in the range of 270-290 nm. The nitrogen group is weak absorber in near UV- region ($n \rightarrow \pi^*$). Benzene shows a band at 256 nm. The absorption at 279 nm to 356 nm shows the presence of phenyl group in the molecule. The absorption range 226.49 nm 270 nm and 315 nm are observed in quinoline in E_1, E_2 and B-bands, respectively [19]. Thus we conclude that the broad UV- band is in the range of 250 nm -450 nm which clearly shows the absorption in the UV- region by CdQ_2. This may be the evidence of the fact that the emission from CdQ_2 may lie in the visible region.

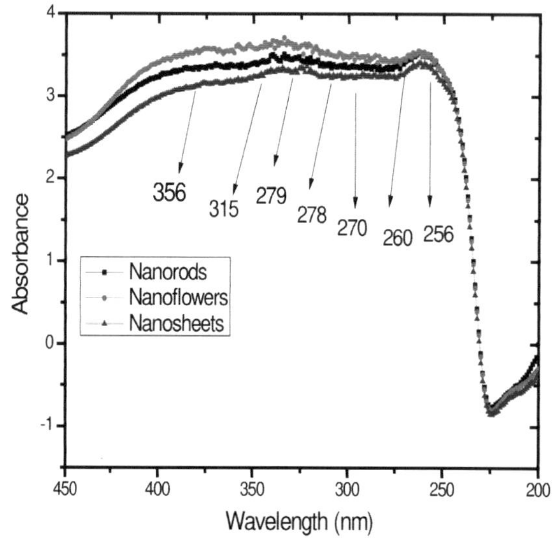

Figure 6. UV/VIS Spectra of CdQ_2 nanostructures.

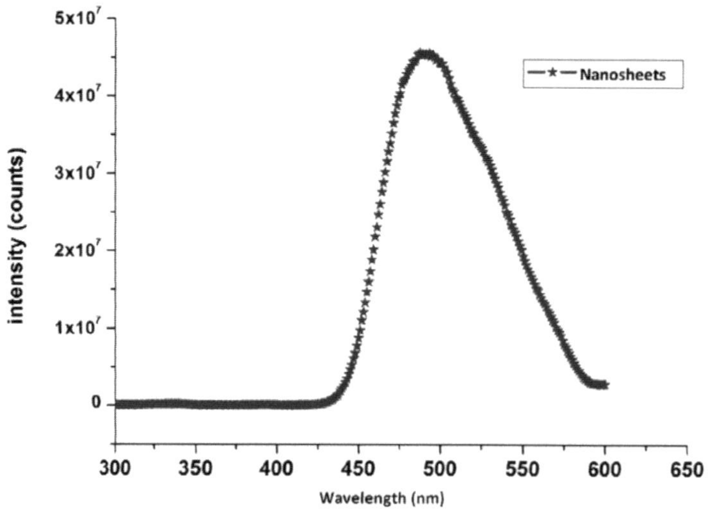

Figure 7. Photoluminescence (PL) spectra of CdQ_2 nanostructures

Photoluminescence (PL) is a very important characteristic for the 8-hydroxyquinoline metal chelates. Figure 7 is the room-temperature PL spectra of CdQ_2 nanostructures (nanosheets) dispersed in ethanol and irradiated by 350 nm of UV light, from which we can see that all the emissions are ranging from 430 to 600 nm and the maximum peak is located around 500 nm. Almost similar spectra are obtained for all samples (not shown). The PL spectra reveal that the CdQ_2 samples give a strong emission in the green region with a peak wavelength lies around 500 nm. This property can introduce applications in novel optoelectronic devices, particularly in organic light- emitting diodes.

4. Conclusions

In summary, Bis- (8-hydroxyquinoline) Cadmium (CdQ_2) complex nanostructures have been synthesized via solvothermal chemical route. Reaction time, reaction temperatures, reactant and surfactant concentration are the main influential factors to the size and morphology the final product. Various morphologies such as, nanorods, nanoflowers, and nanosheets have been achieved by changing different parameters. FTIR spectra confirm the formation of CdQ_2 nanomaterial. UV/ VIS spectra indicate a broad absorption band in the range of 250 nm -450 nm. Samples show a strong photoluminescence (PL) in the visible region (430 nm – 600 nm) with a maximum peak located at around 500 nm, indicating a strong green emission.

Acknowledgements. This work was supported by the Department of Physics and the High Tech Centralized Instrumentation lab of the University of AJK, Pakistan.

5. References

[1] Alivisatos A. P.1996 *Science 271* 933

[2] Murray C. B., Kagan C. R., Bawendi M. G. 2000 *Annu. Rev. Mater. Sci. 30* 545

[3] Xia Y., Yang P., Sun Y., Wu Y., Mayers B., Gates B., Yin Y., Kim F., Yan H. 2003 *Adv. Mater. 15* 53

[4] El-Sayed M. A. 2004 *Acc. Chem. Res. 37* 326

[5] Huang Y., Duan X. F., Lieber, C. M. 2005. *Acc. Chem. Res. 1* 142

[6] Lai E., Kim W., Yang P. D. 2008 *Nano Res.* 1 123

[7] Hu J. T., Odom T. W., Lieber C. M., 1999 *Acc. Chem. Res. 32* 435

[8] Cui Y., Wei Q. Q., Park H. K., Lieber C. M. 2001 *Science 293* 1289

[9] Liu Z., Robinson J. T., Sun X. M., Dai H. J. 2008 *J. Am. Chem. Soc., 130* 108

[10] Wang Z. L. 2008. *Nano Res.* 1 1

[11] Huang W., Kuhn J. N., Tsung C. K., Zhang Y., Habas S. E., Yang P., Somorjai G. A. 2008 *Nano. Lett.* 8 2027

[12] Sun, X. M.; Liu, Z.; Welsher, K.; Robinson, J. T.;Goodwin, A.; Zaric, S.; Dai, H. J. 2008 *Nano Res.* 203 212

[13] Chiu J. J., Wang W. S., Kei C. C., Cho C. P., Perng T. P., Wei P. K., Chiu S. Y. 2003 *App. Phys. Lett.,83* 4607

[14] Chiu J. J., Kei C. C., Perng T. P. Wang W. S. 2003 *Adv. Mater.* 15 1361

[15] Zhao Y. S., Di C. A., Yang W. S., Yu G., Liu Y. Q., Yao J. N. 2006 *Adv. Funct. Mater.*16 1985

[16] Cho C. P., Wu C. A., Perng T. P. 2006 *Adv. Funct. Mater.* 16 819

[17] Chen W., Peng Q., Li Y. D. *2008 Cryst Growth* Des 8 564

[18] Ouyang J. M., Zhang Z. M., Ling W. H., Liu H. Y., Lu Z.H. 1999. *Appl Surf Sci.* 151 67

[19] Irsching F. H., Brewer J. G. 1963 *Anal. Chem 35* 1630

[20] Halls M. D., Tripp C. P., Schlegel H. B. 2001.*Phys. Chem. 3* 2131

[21] Gavrilko T., Fedorovich. R., Dovbeshko G., Marchenko A., Naumovets A., Nechytaylo V., Puchkovska G.Viduta L., Baran J., Ratajczak H. 2004 *J. Mol. Struct. 704* 163

[22] Ayling G.M. 1974 *Appl. Spec. Rev.*, 8 1

Laser ablation characteristics of metallic materials: Role of Debye-Waller thermal parameter

M Z Butt
Rafi M. Chaudhri Chair, Centre for Advanced Studies in Physics, GC University, Lahore-54000, Pakistan

E-mail: mzbutt49@yahoo.com

Abstract. The interaction of a high intensity laser pulse with a solid target results in the formation of a crater and a plasma plume. The characteristics of both depend on physical properties of target material, environmental conditions, and laser parameters (e.g. wavelength, pulse duration, energy, beam diameter) etc. It has been shown for numerous metals and their alloys that plasma threshold fluence, plasma threshold energy, ablation efficiency, ablation yield, angular distribution of laser produced plasma (LPP) ions, etc. are a unique function of the Debye-Waller thermal parameter B or the mean-square amplitude of atomic vibration $\langle u^2 \rangle$ of the target material for given experimental conditions. The FWHM of the angular distribution of LPP ions, ablation yield, and ablation efficiency increase whereas plasma threshold fluence and plasma threshold energy decrease as B-factor of the target material increases.

1. Introduction

In 1993Butt et al [1] published a report on the correlations of macroscopic or bulk properties of cubic elements with their microscopic parameter Debye – Waller factor B measured at room temperature. They found that several general, thermal, mechanical and crystal defect properties of 20 cubic elements were a unique function of their room temperature B-factor. This motivated Butt and Chaudhary [2] to extend this study to the deformation behavior of cubic metals. They showed that for metals of a given cubic structure, the yield stress at a given temperature between 0 and 298K was a function of their room temperature B-factor; smaller the B-factor, higher is the yield stress. Similarly, temperature dependence of the yield stress of metals with a given cubic structure was also dependent on the room temperature B-factor; lower the B-factor, weaker is the temperature dependence of yield stress.

As far as solubility limit (SL) of one metal into another to form binary substitutional solid-solution is concerned, Butt and Kalsoom [3] observed that SL (wt. %) of different cubic metals in copper lattice is 100% if their B-factor is either equal to or less than that of copper. However, if $B_{solute} > B_{Cu}$, SL decreases progressively with the increase in B-value of solute metals. Greater the difference between the B-factors of solute and solvent metal, smaller is the solubility limit. The extremely small SL of Ag into Cu (0.1 wt. %), which is not explicable in terms of the well-known Hume-Rothery rules, can be attributed to the large difference (38.6%) in the B-factors of Ag and Cu.

In addition to several other correlations of bulk properties of cubic elements and cubic compounds with their B-factors in solid phase [4-13], it was interesting to find that solid→liquid and liquid→vapor phase transformations, characterized by melting and boiling temperatures (T_m and T_b) of crystalline materials, were also a unique function of room temperature B-factor. Butt and Jabeen [7], therefore, inferred that cubic elements and compounds possess lattice-vibration memory in the liquid →vapor phase transformation as well.

In laser – solid interaction experiments, the laser fluence and laser energy should have a critical minimum or threshold valueto generate plasma from the target solid by irradiation with laser of a given wave length. Both plasma threshold fluence (F_{th}) and plasma threshold energy (E_{th}) of the target solids for a laser of a given wavelength depends on the nature of the target. The laser produced plasma (LPP) plume, which may consist of electrons, ions, atoms, molecules and micro-sized particles, is formed as a result of the ablation of material from the target surface [14]. The angular distribution of LPP particles, whether ions or neutrals, has been studied for several metals, and their alloys [15–29]. The commonality in these investigations is that the angular distribution of LPP particles can be described either by the cosine power-law $F = F_o \mathrm{Cos}^n\theta$, where F is the flux measured in a direction making an angle θ with the normal to the target surface or by the full width at half maximum (FWHM) of the emission profile fitted by Gauss function. Under given experimental conditions, the value of exponent n or FWHM depends on the nature of the target. A number of researchers have attempted to correlate the exponent n and the FWHM of the angular distribution of particles in the LPP plume in terms of sublimation energy, atomic mass and melting temperature of the target solids. However, no consistent picture emerged from these attempts.

In this presentation, we will analyze the laser – solid interaction parameters, namely plasma threshed fluence and plasma threshed energy necessary for solid → plasma phase transformation, together with ablation efficiency, as a function of the Debye-Waller thermal parameter B or the mean-square amplitude of atomic vibration $<u^2>$ of the target materials under identical experimental conditions. Whether metallic materials possess lattice-vibration memory in the plasma state as well or not will also be reviewed.

2. Role of Debye-Waller thermal parameter

When radiations (X-rays, neutrons, electrons, and γ -rays) of wavelength λ are diffracted from a crystal, the intensity I of the diffracted beam at a Bragg angle θ_B is found to be less than that of the incident beam I_o due to lattice vibrations in the crystal by an exponential factor, $\exp(\frac{-2B\sin^2\theta_B}{\lambda^2})$, known as the Debye-Waller factor, i.e.$I = I_o \exp(\frac{-2B\sin^2\theta_B}{\lambda^2})$. In a mono atomic cubic crystal, the atomic vibrations are isotropic and the mean-square amplitude of atomic vibrations $<u^2>$ is related to B through the relation:$= \frac{8\pi^2<u^2>}{3}$. However, for non-cubic crystal structures, e.g. hexagonal close-packed, in which atomic vibrations are anisotropic, an average value of B_a and $B_c$is taken. Moreover, the B-factor value of a binary alloy is taken as mass weighted average of B_1 and B_2 of the constituent elements [17].

2.1. Laser thresholds for plasma generation

The effect of laser irradiation wavelength on plasma threshold of nine metals (Zn, Al, Ag, Cu, Ni, Fe, Cr, Mo, W) was studied by Cabalin and Laserna [30]. A nanosecond Q-switched pulsed Nd:YAG laser was operated at infrared (1064 nm), visible (532 nm), and ultraviolet (266 nm) wavelengths, and focused on the target in air at atmospheric pressure to generate plasma. They measured plasma threshold fluence (F_{th})and plasma threshold energy (E_{th}) of these metals for each laser wavelength (λ) cited above. For each laser wavelength, they tried to correlate the plasma threshold fluence (F_{th}) of target metals with their melting point (T_m) and boiling point (T_b). Although plasma thresholds show an increasing trend with T_m and T_b yet considerable scatter in each case (Figures 4 and 5 in [30]) clearly points to poor correlation.

Therefore, we will now re-analyze their data in terms of the Debye-Waller thermal parameter $B$of target metals. The data points in figure 1 denote the experimental values of plasma threshold fluence (F_{th}), taken from [30], and displayed here as a function of B-factor of target metals for three different laser irradiation wavelengths. The B-values were taken from [31–33]. One can readily note from figure

1 (a) and (b) for 266 and 532 nm wavelengths that the data points can be divided into two groups: One for body-centered cubic (bcc) metals (W, Mo, Cr, Fe) and the other for close-packed structure metals (Ni, Cu, Ag, Al, Zn). In each case, the data points are encompassed by a straight line obtained by least-squares fit method. However, figure 1(c) for1064 nm wavelength indicates a further sub-division in close-packed structure metals; hexagonal closed-packed (hcp) Zn separates itself from face-centered cubic (fcc) metals (Ni, Cu, Ag, Al). Nevertheless, one can again note quite good linear fit to the data points for bcc and fcc metals, setting aside single hcp Zn data point.

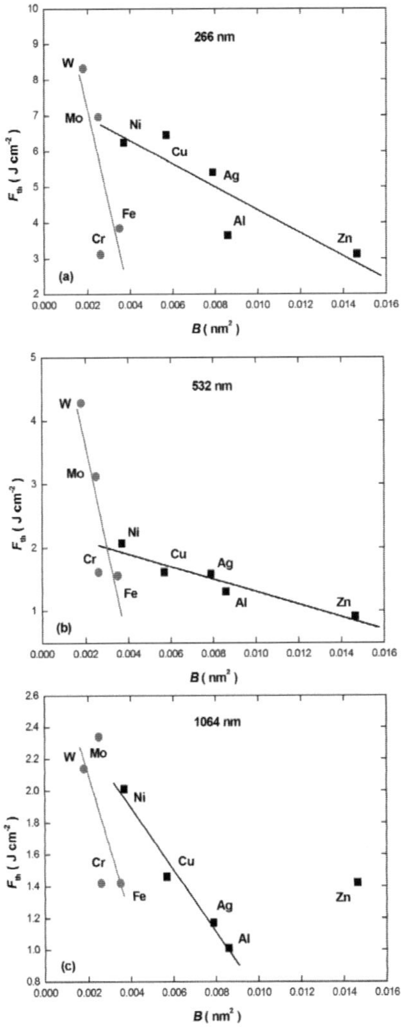

Figure 1. Correlation between the measured plasma threshold fluence and the room temperature *B*-factor of nine target metals for three different laser wavelengths: (a) 266 nm, (b) 532 nm, (c) 1064 nm.

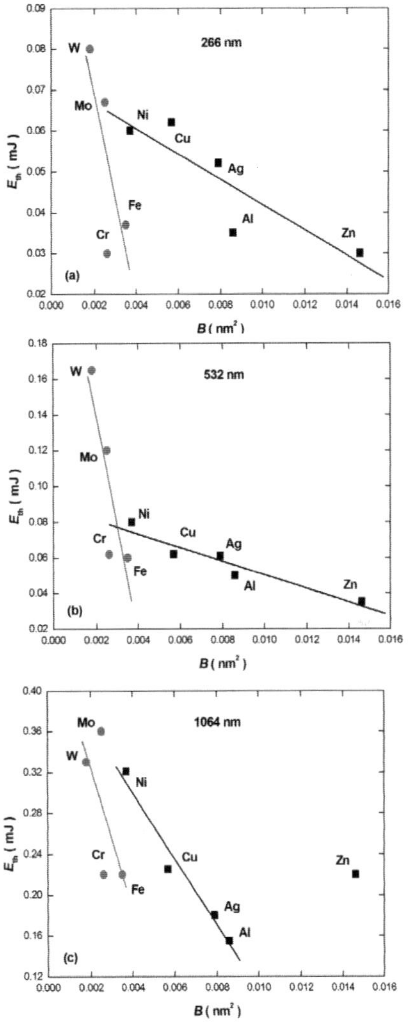

Figure 2. Correlation between the measured plasma threshold energy and the room temperature *B*-factor of nine target metals for three different laser wavelengths: (a) 266 nm, (b) 532 nm, (c) 1064 nm.

Similarly, the data points in figure 2 depicts the experimental values of plasma threshold energy (E_{th}) of nine metals reported by Cabalin and Laserna [30], for infrared (1064 nm), visible (532 nm) and ultraviolet (266 nm) wavelengths and displayed here as a function of B-factor of these metals. As far as the dependence of E_{th} on B is concerned (figure 2), it is a replica of that for F_{th}(figure 1).

2.2. Ablation Efficiency

In figure 3, ablation efficiency of pure Mo, Fe, Ni, Cu, Al and Pb metals in terms of ablated volume per unit of energy have been depicted as a function of the Debye-Waller thermal parameters B. The data points denote the measured values obtained by Semerok et al [34] in laser ablation experiments conducted in air at atmospheric pressure with sharply focused ns, ps, and fs laser pulses. The values of B were taken from [31–33]. Using least-squares fit method, a straight line is found to fit the data points

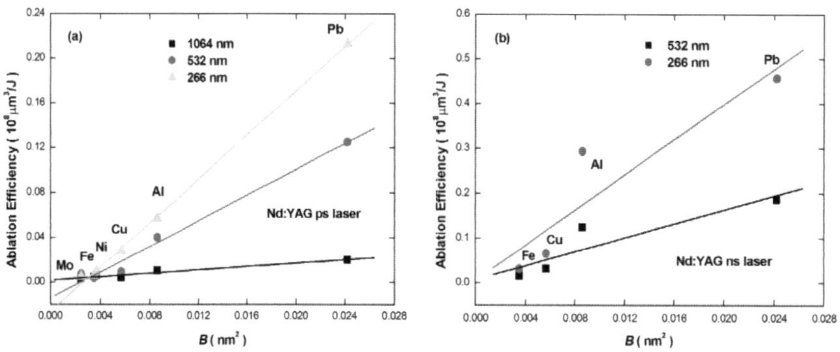

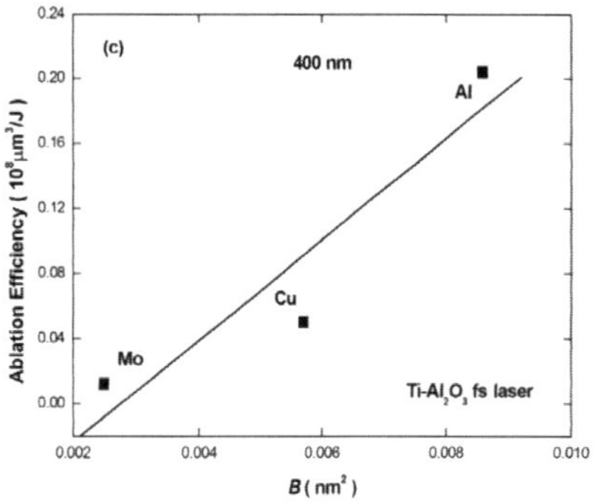

Figure 3. Ablation efficiency of some target metals as a function of their room temperature B-factor for (a) picosecond (b) nanosecond and (c) femtosecond Nd:YAG

in each case. The linear correlation coefficient values (0.909 – 0.999) being close to 1 points to an excellent linear dependence of ablation efficiency on the B-factor of target metals. Thus, under identical irradiation conditions, a metal with larger B-factor or higher mean-square amplitude of atomic vibrations $<u^2>$ will ablate more as compared to a metal with smaller B-factor and vice versa.

Similarly figure 4 illustrates the correlation between the ablation yield of several metallic targets in terms of atoms/shot and the room temperature Debye-Waller factor B. The data points denote the experimental values obtained by Thestrub et al [35] in laser ablation investigations carried out in a vacuum of 10^{-7} mbar operating the laser at 355 nm with pulse width of 6ns and fluence of 2.0 J/ cm^2. Using least-squares fit method, Ali et al [16] passed a straight line through the data points (figure 4) with a linear correlation coefficient $r = 0.94$. This shows that a metal with higher B-factor or larger mean-square amplitude of atomic vibrations $<u^2>$ will ablate more under given irradiation conditions, and hence the number of ablated atoms per shot will be larger.

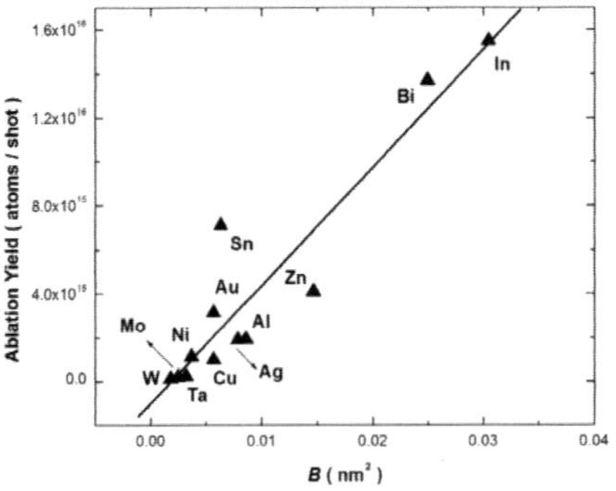

Figure 4. Ablation yield of twelve target metals as a function of their room temperature B-factor [16].

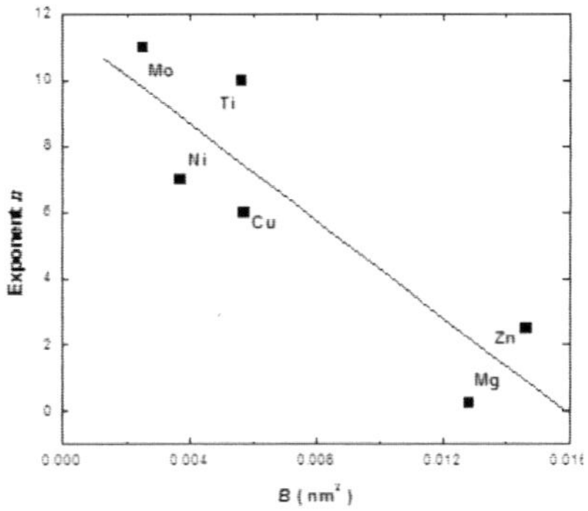

Figure 5. Dependence of exponent n of $\cos^n\theta$ distribution of LPP ions ejected from six target metals on their room temperature B-factor [16].

2.3. Angular distribution of LPP ions

Now we will review the angular distribution of LPP ions emitted from the metallic targets as a function of their room temperature B-factor.

Referring to figure 5, taken from [16], the data points denote the values of exponent n of $\cos^n\theta$ distribution of LPP ions for six target metals (Mo, Ni, Ti, Cu, Mg and Zn) as a function of their room temperature Debye-Waller thermal parameter B. The straight line fitted to the data points by least-squares fit method shows that as B increases n decreases. This shows that the plasma plume is broader for a metal with larger B-factoror higher mean-square amplitude of atomic vibrations $<u^2>$ and is narrower for a metal with smaller $<u^2>$.

Similarly, a linear n–B correlation is also observed [17]for metallic targets Ti, Al, and their binary alloys $Ti_{75}Al_{25}$ and $Ti_{60}Al_{40}$, as shown in figure 6(a). Moreover, figure 6(b) shows that the maximum average charge $(q_{av})^{max}$ of LPP ions emitted from these targets in a direction normal to the target surface also correlates well with their room temperature Debye-Waller thermal parameter B[17].Greater the B-factor, higher is the value of $(q_{av})^{max}$. It means that atoms are ejected to a greater extent from the target with higher B-factor or $<u^2>$-value, leading to higher value of $(q_{av})^{max}$ in the plasma plume.

As far as the angular distribution of various ionization states of ions is concerned, it can be seen from figure 7(a) and (b) , that both the parameters n and FWHM for ions of a given state (q^{+1} to q^{+4}) also show a linear dependence on the B- factor of Ti_xAl_{1-x} metallic targets with x = 0 to 1. For larger value of B-factor, the exponent n of $\cos^n\theta$ distribution of ions of a given state is smaller, and the FWHM is larger. This means that not only the angular distribution of overall LPP ion flux (accumulative charge) but also that of the individual ionization states are controlled by the B-factor of target material.

The observed increase in the value of exponent with the increase in ionization state of LPP ions emitted from a given target(figure 7) means that the angular distribution of LPP ions becomes narrower as the ionization state increase from +1to +4. It is because ions with higher ionization state possess more energy than lower ionization state since the acceleration of ions is proportional to their charge.

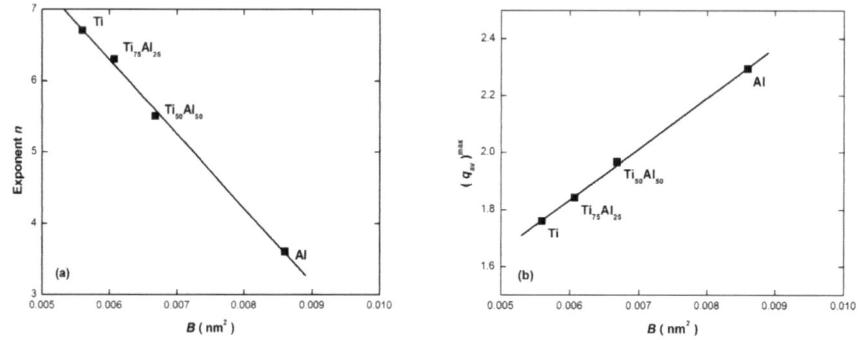

Figure6. Dependence of (a) exponent n of $\cos^n\theta$ distribution and (b) maximum average charge of LPP ions of metals Ti, Al, and their alloys $Ti_{50}Al_{50}$,$Ti_{75}Al_{25}$ on their room temperature Debye-Waller thermal parameter B [17].

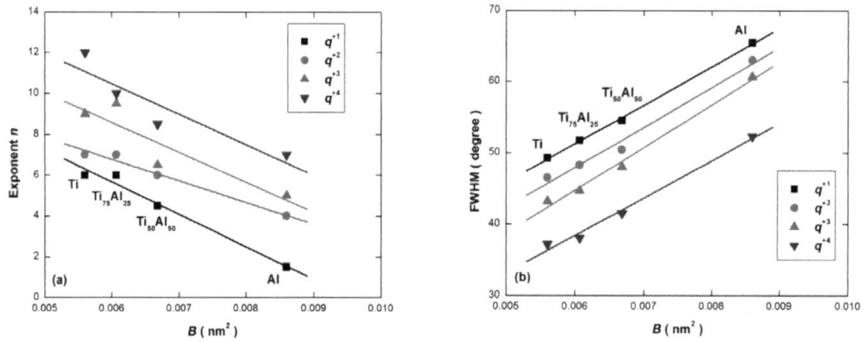

Figure 7. Dependence of (a) the exponent n of $\cos^n\theta$ distribution and (b) the FWHM of Gaussian distribution ofvarious ionization states on the room temperature Debye-Waller thermal parameter B of metals Ti, Al, and their alloys $Ti_{50}Al_{50}$,$Ti_{75}Al_{25}$[17]..

Finally, figure 8depicts the FWHM of Gaussion distribution of laser - ablated particles from Ti_xAl_{1-x} (x = 0, 0.25, 0.34, 0.5, 0.75, 1), W_xCu_{1-x} (x = 0, 0.6, 0.8, 0.9 and 1), and Mo_xCu_{1-x} (x = 0, 0.7, and 1) metallic targets as a function of their room-temperature B-factors [17]. A single straight line fitted to the entire data points shows that whatever the nature or composition of the target, the angular distribution of particles in the plasma plume is uniquely related to the B-factor. Thus, under given irradiation conditions, two metallic alloy targets of any composition but with identical B-factor value will give rise to identical angular distribution of ions in laser ablation plumes. In other words, angular distribution of LPP ions emitted from alloy targets, irrespective of the nature and composition of constituent metals, is B-equivalent.

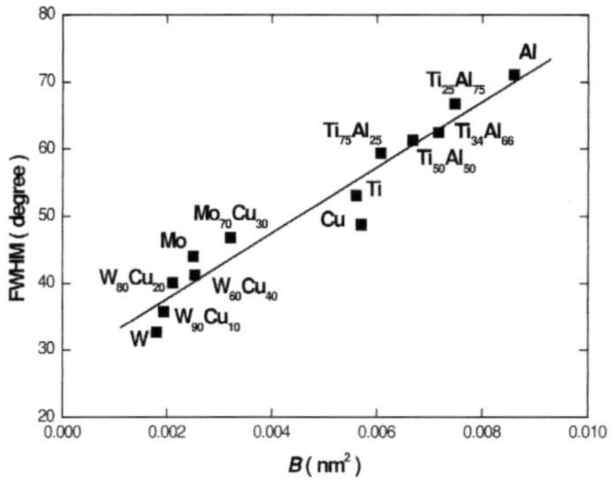

Figure 8. Dependence of the FWHM of Gaussian distribution of laser- ablated particle masses from Al, Ti, Cu, Mo, W and their alloys $Ti_{25}Al_{75}$, $Ti_{34}Al_{66}$, $Ti_{50}Al_{50}$, $Ti_{75}Al_{25}$, $W_{60}Cu_{40}$, $W_{80}Cu_{20}$, $W_{90}Cu_{10}$ and $Mo_{70}Cu_{30}$on the room temperature Debye-Waller thermal parameter *B* [17].

3. Conclusions

- The laser ablation parameters, namely plasma threshold fluence, plasma threshold energy,ablation efficiency, ablation yield, angular distribution of LPP ions, etc. are a unique function of the Debye-Waller thermal parameter *B* or the mean-square amplitude of atomic vibrations $\langle u^2 \rangle$ of the target material under given experimental conditions.
- Crystalline materials, whether metals or alloys, possess lattice-vibration memory not only in their solid → liquid and liquid → vapor phase transformations but also in the solid → plasma phase transformation as well as in the plasma state.
- Keeping in view the observed excellent correlations of the room temperature *B*-factor with laser ablation parameters as well as with numerous other properties of crystalline materials reported in the literature, one can speculate that *B*-factor of crystals is “equivalent” to the DNA of living organisms controlling their behavior in all respects.

4. References

[1] ButtN M, Bashir J and Khan M N1993*J. Mater. Sci.*281595

[2] Butt M Zand Chaudhary S A 1993 *Physical Review* (B)47 8418

[3] Butt M Z and Kalsoom U 1999 *Fizika A*8 75

[4] Niaz J, AnwarM, Nadeem A H and ButtM Z 1993*J. Nat. Sci. Math.*3689

[5] Butt M Zand Hameed T1993 *Science International* (Lahore) 5 313

[6] ButtN M, BashirJ, Khan M N and Willis B T M1994*J. Mater. Sci.Lett.*13 1440

[7] Butt M Zand Jabeen S 1996*J. Mater. Sci. Lett.*15 412

[8] KalsoomU, Butt M Z, and Khaleeq ur Rahman M 1998*J. Nat. Sci. Math.*38 261

[9] Butt M Z and Ahmad I 1999:*Czech. J. Phys.*49509

[10] Butt M Z 1999*Czech. J. Phys.*49 1177

[11] Butt M Z an Chaudhary S A 2000*Czech. J. Phys.*50 645

[12] Sirdeshmukh D B, Krishna P G and Subhadra K G 2003*J. Mater. Sci.*38 2001

[13] Ali D, Butt M Z and Naseem S 2013*Rad. Eff. Def. Solids*168 1

[14] Ali D 2010 *M.Phil. Thesis*, University of Engineering and Technology, Lahore, Pakistan

[15] Khaleeq-ur-Rahman M, Ali D and Butt M Z 2010*Vacuum*85 170

[16] Ali D, Butt M Z, and Khaleeq-ur-Rahman M 2011*App. Surf. Sci.* 2854

[17] AliD, Butt M Z and Butt S 2012*Mater. Chem. Phys.*137 147

[18] Ali D, Butt M Z, and Khaleeq-ur-Rahman M 2010 *J. Nat. Sci. Math.* 50 1

[19] Buttini E, Thum-Jager A and Rohr K 1998*J. Phys. D: Appl. Phys.*31 2165

[20] Thum-Jager A and Rohr K 1999*J. Phys. D: Appl. Phys.*32 2827

[21] Srivastava S N and Rohr K 2005*Nuclear Instrum. Methods B*237497

[22] MullerTh, Sinha B K and Rohr K P 2003*Phys. Rev. E*67 026415

[23] Laska L, Krasa J, Pfeifer M, RohlenaK, Gammino S, Torrisi L, Ando L and Ciavola, G2002 *Rev. Sci. Instrum.*73 654

[24] KonomiI, Motohiro T, and Asaoka T 2009*J. Appl. Phys.*106 013107

[25] KonomiI, Motohiro T, KobayashiT, and Asaoka T 2010*Appl. Surf. Sci.*256 4959

[26] Srivastava S N, Sinha B K and Rohr K P 2006*J. Phys. B: At. Mol. Opt. Phys.*39 3073

[27] Torrisi L, AndoL, Ciavola G, GamminoS andBarna A 2001*Rev. Sci. Instrum*72 68

[28] Antoni F, Fuchs C and Fogarassy E 1996 *Appl. Surf. Sci.*96-9850

[29] Torrisi L 2002*Appl. Surf. Sci.*195 8

[30] Cabalin L M and Laserna J J 1998*SpectrochimicaActa Part B*53 723

[31] Butt N M , Bashir J, Willis B T M and Heger G 1988*ActaCrystallographicaA*44 396

[32] Fox AG and Fisher R M 1988*Aust. J. Phys.* 41 461

[33] International Tables for X-rays Crystallography, vol. 3, Kynoch Press, Birmingham,1968, pp. 238–239.

[34] Semerok A, Chaleard C, Detalle V, Lacour J L, Mauchien P, Meynadier P, NouvellonC, Salle B, Palianov P, Perdrix M and Petite G 1999*Appl. Surf. Sci.*138–139 311

[35] Thestrup B, Toftmann B, Schou J, Doggett B and Lunney J G 2002 *Appl. Surf. Sci.*197-198 175

Growth and characterization of epilayer of SiC on Si (111) substrate using single solid source MBE

M Asghar[1], K Mahmood[1], M A Nawaz[1] and R Tsu[2]
[1]Department of Physics, The Islamia University of Bahawalpur, Pakistan
[2]Department of Electrical and Computer Engineering, UNC-Charlotte, NC USA

E-mail: mhashmi@iub.edu.pk

Abstract. 3C-SiC thin films have been grown by Molecular Beam Epitaxial (MBE) technique on a Si (111) substrate at 1050°C using a single solid source fullerene (C60). Fullerene molecules adsorb on the tissue like surface of substrate and then thermally decompose into amorphous carbon and produce 3C-SiC. The grown films were characterized by SEM/EDX, XRD, FTIR and Raman spectroscopy. XRD data confirmed the growth of thick, highly crystalline and epitaxial SiC. The grown film has amorphous surface but the inner core consists of nearly perfect crystalline lattice structure indicated by SEM. FTIR and Micro-Raman also confirmed the high crystalline growth of 3C-SiC . The method proves to be a milestone for low temperature and high crystalline growth of 3C-SiC.

1. Introduction

Silicon Carbide is very interesting semiconductor because it has some amazing properties such as high electron mobility, high thermal conductivity, resistance against radiations which can it is suitable candidate for high power, high frequency and high temperature devices [1, 2].

Mostly the epitaxial growth of SiC on Si substrate has been done by chemical vapour deposition (CVD) at very high substrate temperatures (1300-1600 °C) [3-7]. This high growth temperature results in high production price of Si-SiC heterostructure based electrical devices and several limitations regarding the quality of produced films due to thermally induced defects [8]. These problems can be minimized by using molecular beam due to low deposition temperature (T< 1100 °C) and clean growth ambient. Low growth temperature helps to decrease thermally induced defects especially the inter-diffusion across the Si-SiC interface which results in deterioration of Si-SiC heterostructure. On the other using Fullerene precursors can be advantageous because it do not introduce any impurity in the film. The key advantage of this method is low temperature growth (T~ 1000 °C) which is significantly less than other methods (T > 1050 °C). Therefore the growth of defect free SiC is still a matter of debate.

In this study we report the growth mechanism of 3C-SiC on Si (111) substrate using single solid source (fullerene) by MBE at 1050 °C. During growth period developments and surface reconstruction was monitored by Scanning Electron Microscopy (SEM)/EDEX. The crystallinity and defects were characterized by X-ray Diffraction (XRD), Fourier Transform Infrared spectroscopy (FTIR) and Raman Spectroscopy.

2. Experimental

We have grown SiC on Si (111) substrate. Firstly the substrate was cleaned according to the internationally accepted standard procedures. After cleaning we have loaded the substrate in MBE chamber for 3 hour growth process. During 1st hour C_{60} was deposited on the substrate keeping its temperature at 400 °C. In last 2 hour, we maintained the chamber pressure at 1.4×10^{-9} Torr, substrate temperature at 1050 °C and k-cell temperature, containing solid C_{60}, at 600 °C. The development during growth process and surface reconstruction was monitored by Scanning Electron Microscope (Hitachi 3000). The grown sample then characterized by many diagnostic techniques like X-ray Diffraction (Panaytical), Fourier Transform Infrared Spectroscopy (Tensor 27) and Raman Spectroscopy (Horiba having wavelength 325nm). All the measurements were performed on room temperature.

3. Results and Discussion

Fig. 1 shows the XRD pattern of grown film. The spectrum shows four peaks at angles 28.4° 35.6°, 58.8° and 75.5°, which are attributed to Si (111), 3C-SiC (111), Si (222) and 3C-SiC (222) [9, 10] respectively. From Gaussian fitting, FWHM of 3C-SiC (111) and (222) was found to be 0.22348° and 0.4812° respectively which is comparable with the values reported in literature for highly crystalline 3C-SiC [11]. Small values of FWHM is in agreement with thick and highly crystalline SiC film as reported earlier [11] which is further confirmed by the grain size calculated using Scherer's formula. The average grain size was found to be 23 nm. No peak from impurity or other phase demonstrated the highly epitaxial nature of grown film.

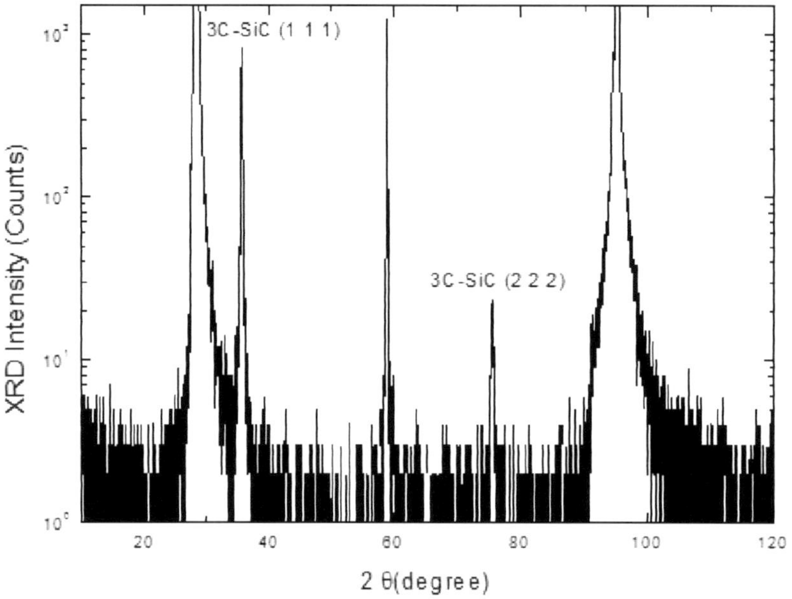

Figure 1. Typical XRD pattern of MBE grown SiC

FTIR in transmission mode has been performed to confirm crystallinity of the grown film and spectrum is shown in figure 2. FTIR (%transmission mode) shows SiC (TO) mode at 794cm^{-1} and in addition, we can see CO_2 and CO modes at 467 cm^{-1} and 1097 cm^{-1} respectively. The observation of C-O complexes yields the coexistence of C-O and Si-C on the Si substrate Peak at 796 cm^{-1} clearly indicates the presence of SiC crystals [12]. The observation of C-O can be attributed to the unpaired C_{60} on Si surface.

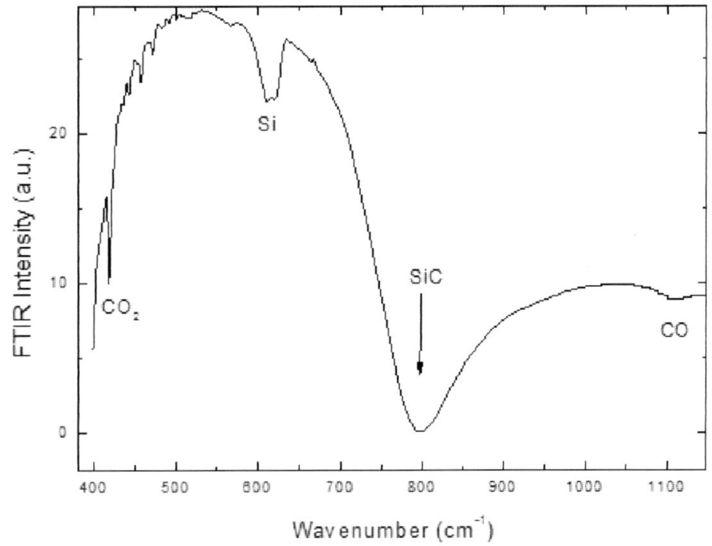

Figure 2. FTIR spectrum of 3C-SiC grown by single source MBE using C_{60} as source

Micro-Raman mapping (figure 3) shows a well known SiC (TO) mode at 794 cm^{-1} and SiC (LO) mode at 971 cm^{-1}. These peaks confirm the carbon rich growth of SiC [13]. Some weak modes associated with carbon/diamond and graphene are also observed at 1345 and 1600 cm^{-1} respectively.

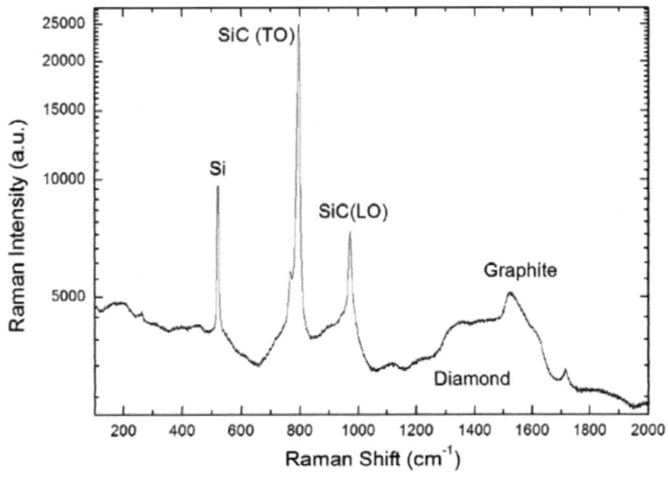

Figure 3. Raman scattering data of 3C-SiC grown by single source MBE

Large area SEM/EDEX depicts almost a semi-stoichiometric At% ratio of Si:C on SiC thin film as shown in Figure 4. It can be seen that the tissue like structure has extended itself over most of surface and C_{60} adsorption has increased to give out C-rich SiC with C:Si = 54:46).

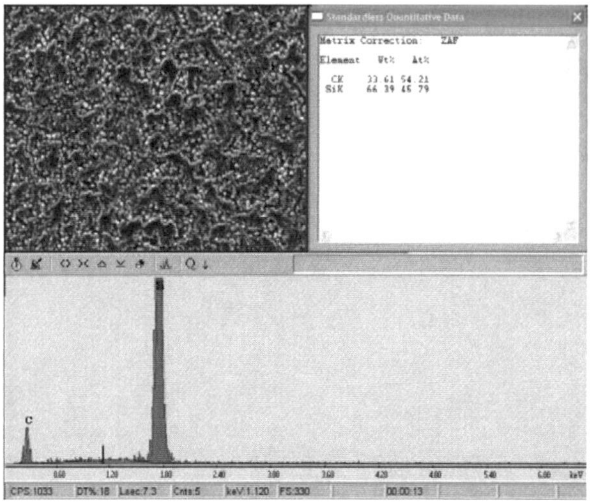

Figure 4. The SEM/EDX image of grown film clearly demonstrated the carbon rich growth of sample

SEM/EDEX images also revealed that Si substrate has tissue like structure (not shown here). This tissue like structure facilitates the adsorption of C_{60} molecule at high temperature. SiC nucleation first starts at the tip of this tissue like structure and with the passage of time extend themselves over the whole surface of the sample. Fullerene molecules adsorb at the surface and then thermally decompose into amorphous carbon and produces SiC.

4. Conclusion

In this paper, we have demonstrated the growth of Silicon Carbide (SiC) by Molecular Beam Epitaxy (MBE) on Si (111) substrate using single solid source fullerene (C60). The structural properties of grown film were confirmed by XRD, FTIR and Raman spectroscopy. XRD reveals that the grown film having cubic structure with (111) and (222) planes. FTIR and Raman data consists of a broad peak at 796 cm^{-1} which is due Si-C bonding. These peaks confirm the carbon rich growth of SiC also verified by EDX measurements. Some weak modes associated with carbon/diamond and graphene are also observed at 1345 cm^{-1} and 1600 cm^{-1} respectively.

5. References

[1] Asghar M, Iqbal F, Jokubavicius V, Wahab Q, Syvajarvi M 2012 *Physica B* 407 3038

[2] Asghar M, Iqbal F, Faraz S, Jokubavicius V, Wahab Q, Syvajaryi 2012 Physica B 407 3041.

[3] Li J. P, Steki J, Golecki I, Reidinger F, Wang L, X. Ning X. J, Pirouz P 1993 *Appl. Phys. Lett.* 62 3135.

[4] Ferro G, Monteil Y, Vincent H, Thevenot V 1996 *J. Appl. Phys.* 80 4691.

[5] Powell J.A, Larkin D. J, Matus L. G, Choyke W. J, Bradshaw J. L, Henderson L, Yoganathan M, Yang J, Pirouz P, *Appl. Phys. Lett.* 56 1442.

[6] Matsunami H 1993 *Physica B* 185 65.

[7] Karmann S, Suttrop W, Schöner A, Schadt M, Haderstroh C, Engelbrecht F, Helbig R, Pensl G1992 *J. Appl. Phys.* 72 5437.

[8] Morko H, Strite S, Gao G.B, Lin M. E, Sverdlov B, Burns M 1994 *J. Appl. Phys.* 73 1363.

[9] Wahab Q, Glass R. C, Ivanov I. P, Birch J, Sundgren J-E, Willander M 1993 *J. Appl. Phys.* 74 1663-1669.

[10] Madapura S, Steckl A. J, Loboda M 1999 *J. Electrochemical Soci.* 146 1197-1202.

[11] Chen J, Steckl A. J, Loboda M. J 1998 *J. Vac. Sci. Technol. B.* 16 3.

[12] Nakazawa H, Suemitsu M, Asami S 2000 *Thin Solid Films* 369 269-272.

[13] Wang Y. S, Li J. M, Lin L. Y, Zhang F. F 1999 *Appl. surf. sci.* 148 189-195.

Study of rheological, viscoelastic and vulcanization behavior of sponge EPDM/NR blended nano- composites

M Arshad Bashir[1,3], M Shahid[1], Riaz Ahmed[2], A G Yahya[2]
[1]School of Chemical & Materials Engineering, National University of Sciences and Technology, Islamabad, Pakistan
[2]Institute of Industrial control systems, P.O. Box 1398, Rawalpindi, Pakistan

E-mail: arshadbashir@scme.nust.edu.pk

Abstract. In this research paper the effect of blending ratio of natural rubber (NR) with Ethylene Propylene Diene Monomer (EPDM) were investigated. Different samples of EPDM/NR ratio were prepared to study the variation of NR in EPDM on rheology, curing characteristics, tangent δ, and viscosity variation during vulcanization of sponge nano composites.The main aim of present research is to develop elastomeric based sponge composites with the blending ratio of base elastomers along with the carbon nano particles for high energy absorbing and damping applications. The curing characteristics, rheology and viscoelastic nature of the composite is remarkably influenced with the progressive blending ratio of the base elastomeric matrix.

1. Introduction

Elastomeric based composites have attracted the attention of engineers and scientists for their versatile nature. They are light weight, easy processing, low cost and have high strength to weight ratio, most suitable for aerodynamics applications. Relatively, more impact energy absorption and damping efficiency make them unique for personal, building and armour protection[1].

Reinforcing fillers such as carbon black and silica are used during processing for the improvements in mechanical, thermal and aging characteristics[2]. Vulcanization and curing behavior including scorch safety, viscoelastic nature, damping efficiency are also influenced by the fillers. Concentration, size and morphology are the salient features of the fillers for the variation in the physical properties of finished elastomeric composites[3].

Elastomeric blends have gained much attention in the industry due to the improvements in the features as compared individual partner. Natural Rubber (NR) has high tensile strength as it crystallizes on stretching and resists thermal degradation due to existence of double bonds in molecular backbone. Oil resistance of NR is low as compared to other synthetic rubbers like EPDM, chloroprene rubber (CR) and nitrile butadiene rubber (NBR) etc[4]. To overcome this shortcomings NR is blended with other synthetic polymers. Mechanical properties and curing bahaviour of blended polymers are affected due to the difference in the polarity and interfacial interaction of the blended partners. The major important feature of blending is to attain the intermediate properties which are necessary in some applications[5].

The deformation and flow behavior of the elastomeric composite during vulcanization is known as rheology. The elastomers possess both viscous and elastic nature simultaneously so it is necessary to determine its rate of flow under various conditions of stress, temperature, time for the application of the stress and the elastic recovery on the removal of applied stress[6].

Viscoelasticity is an important property of elastomers. These materials exhibit both elastic and viscous behavior on deformation. Elstic part is recoverable while energy is dissipated in vicous part during stress application[7].

The novelty of present research is the significant execution of various NR blending ratio in EPDM in the presence of blowing agent and cabon nano particles on the sponge structure formation, curing rate index, viscosity variation during vulcanization, viscous and elastic torque with the variation of NR ratio in EPDM/NR sponge composites.

2. Materials

Blowing Agent (ACPW) was purchased from Qingdao Xiangsheng, Shandong, China. EPDM (KELTAN 4331) and NR received from Technical Rubber Products, China. Carbon Black (N330) from Continental Carbon USA. Mercaptobenzthiazole (MBTS) and Diphenylguanidine (DPG) were purchased from Dalian Richon Chemical Co.Ltd. China. Stearic acid, sulphur, and zinc oxide were supplied by Merck, Germany. Aromatic oil was purchased from International Petrochemicals (Pvt.) Ltd. Pakistan.

2.2 EPDM/NR Blends Preparation

Two-roll mixing mill was used for the preparation of EPDM/NR blends along with reinforcing fillers, accelerators, activators and processing aids according to ASTM 3182 [8]. Blending of the elastomers completed in two stages, the compounding of fillers and processing aids are carried in the first stage. In the second stage, accelerators, cross linkers and activators are incorporated into the previous mixture at the controlled temperature. Temperature control is very necessary at this stage so chilled water is passed through the rolls at a regulated flow rate to remove the excessive heat to avoid the premature vulcanization at the processing stage. Lateral shearing is caused the uniform distribution of the fillers within the blending mixture. Rubber sheet was cut diagonally and folded back at the nip frequently for uniform blending [1].

Table 1. Blending scheme of EPDM/NR

Ingredients (wt%)	Blended samples			
	S1	S2	S3	S4
EPDM	100	90	80	70
NR	0	10	20	30
Carbon black	5	5	5	5
Aromatic Oil	10	10	10	10
Stearic Acid	1	1	1	1
Activator	5	5	5	5
Blowing agent	5	5	5	5
MBTS	1	1	1	1
DPG	1	1	1	1
Cross linker	1.5	1.5	1.5	1.5

2.3 Rheological Characterization

Rheocheck profile-MD, Gibitre Instruments S.R.I, Italy was used to investigate tanδ, scorch time (ts_1 & ts_2), viscous torques at minimum and maximum viscosity according to ASTM 6204.

2.3.1 Curing behavior

Permanent cross linking between the polymer chains and ingredients is known as the curing or vulcanization. The elalstomeric blended composites converts into viscous liquid as the heating starts and at this stage lowest torque is measure by the moving die (Rheocheck-MD). Cross linking proceeds gradually with time and at the final stage of vulcanization maximum torque value is recorded. This effect is known as the cure rate index (CRI) and calculated by the relation

Cure Rate Index (CRI) = $100/t_{90}-t_{10}$ (Eq.1)[9]

Where t90 and t10 are the completion of 90% and 10% vulcanization respectively.

3. Results and Discussion

In figure-1, ts_1 is the time duration between the closure of die and increase torque up to 1dN.m, ts2 is the time for premature vulcanization up to 2 dN.m. Both ts_1 and ts2 are longer for pure EPDM (S1) and it shortens gradually with the increase in blending ratio of natural rubber in EPDM as displayed in Figure 1. This decrease in time durations is due to difference in reaction strategy of the both major components of polymer blends with the accelerators, activators, cross linkers and other ingredients which fastens the curing rate and shortens the scorch safety (ts_2) and ts_1 as well.

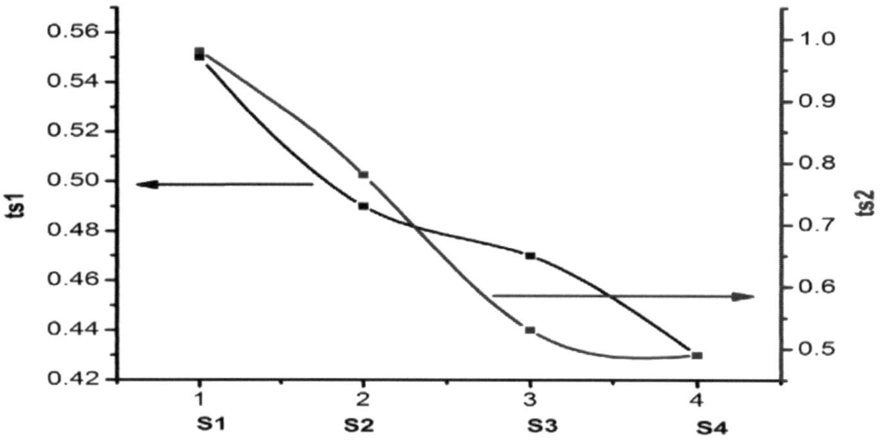

Figure1. Effect of blending ratio of Natural Rubber in EPDM on ts_1 and scorch time ts_2

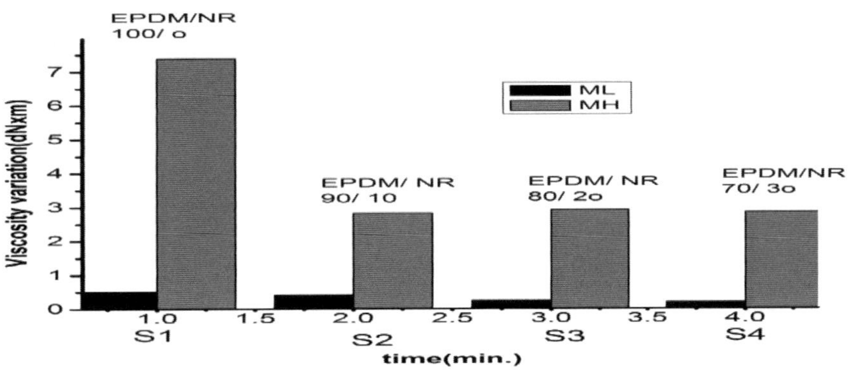

Figure 2. Viscosity variation of the blended composite with the incorporation of natural rubber component in EPDM

The variation in the viscosity of the blended composite, measured in torques at the initial and final stages of vulcanization as shown in Figure 2. Both lowest and maximum torques reduced with the augmentation of the blending of NR in EPDM which is due to the enhancement of viscous portion of the blended portion of NR [10].

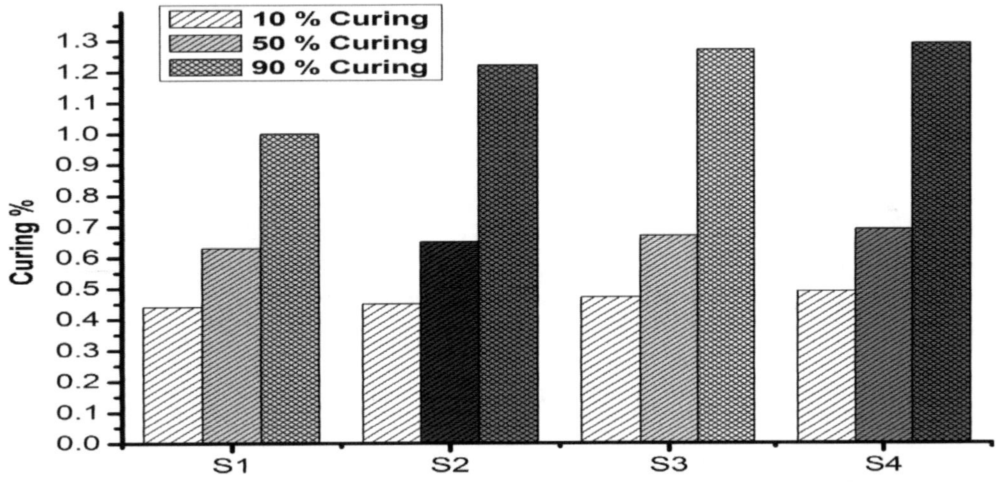

Figure 3. Variation in % curing of EPDM/NR blended composite with blending ratio of NR

The curing behaviour of the blended composite is found to be high as compared to pure EPDM(S1) as shown in Figure 3. This enhancement in the percent curing of the blended composite is due to the increase in blending part of NR due to more filler- polymer chains and chains network interactions of the composite.

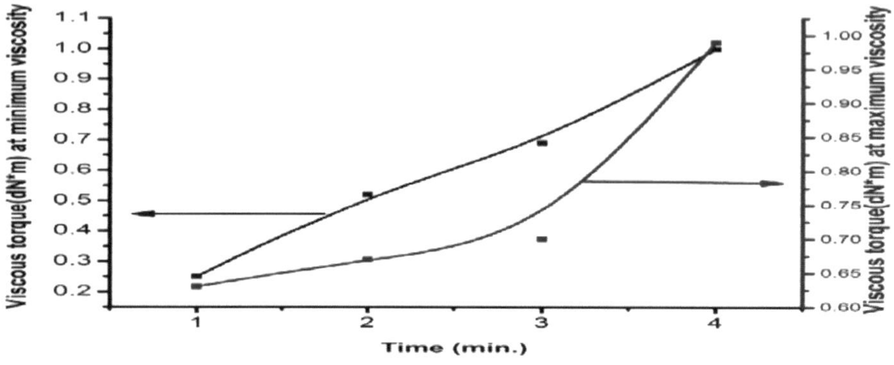

Figure 4. Variation in viscous torque at minimum and maximum viscosity with different loading of natural rubber in EPDM.

Polymers inherently possess both viscous and elastic part, the variation in viscous torques at minimum and maximum viscosity of the samples during vulcanization is displayed in Figure 4. The torque values increased gradually with the incorporation of NR component which is due the augmentation of viscous part of the blended NR component.

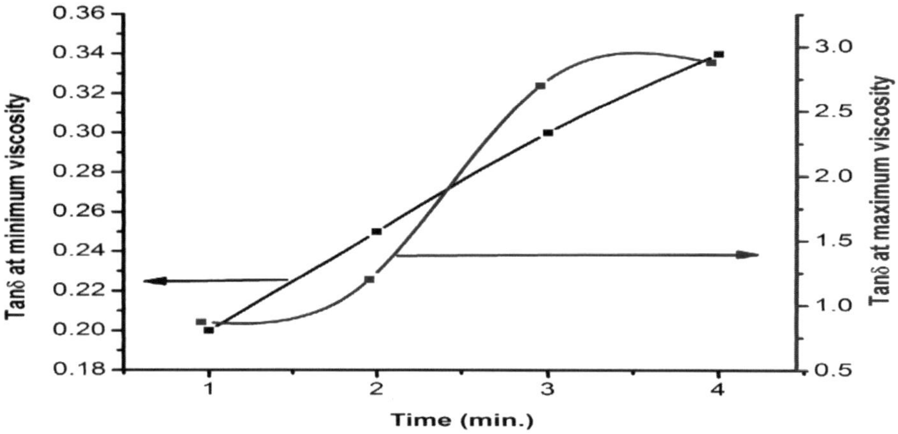

Figure 5. Effect of blending ratio of NR in EPDM on tan delta at maximum and minimum viscosity of the composite.

Tangent delta (tan δ) is the ratio of loss modulus to storage modulus, both tan δ values at minimum and maximum viscosities enhanced. The blending part causes the enhancement in loss modulus, reduction in storage modulus and elastic recovery of the blended composite as expressed in Figure 5. The blending of NR in EPDM increases in viscous part and reduces elastic component and as a result the tan δ values increases[11].

4. Conclusions

EPDM sponge blended with variant part of natural rubber studied the effect on the rheological, curing and visco-elastic properties of the composite. Curing rate, ts1, ts2, lowest torque, maximum torque diminish on the addition of natural rubber component in the EPDM. Viscous torque and tan delta of the blended composite augmented 3% and 550% respectively with the 30 % incorporation of natural rubber in the EPDM based blended composite. The enhancement in the % curing of the composite was also observed with the increase of blending ratio of NR in EPDM.

5. Acknowledgement:

The financial support of Higher Education Commission of Pakistan in this research is highly acknowledged by the authors.

6. References

[1] Arshad Bashir M, Iqbal N, Shahid M, Ahmed R. Structural, viscoelastic, and vulcanization study of sponge ethylene–propylene–diene monomer composites with various carbon black loadings. Journal of Applied Polymer Science 2013.

[2] Rattanasom N, Saowapark T, Deeprasertkul C. Reinforcement of natural rubber with silica/carbon black hybrid filler. Polymer Testing 2007;26:369-77.

[3] Guth E. Theory of filler reinforcement. Rubber Chemistry and Technology 1945;18:596-604.

[4] Lee HK, Ko SW. Structure and thermal properties of polyether polyurethaneurea elastomers. Journal of Applied Polymer Science 1993;50:1269-80.

[5] Morikawa A, Min K, White J. Flow Visualization of the Rubber Compounding Cycle in an Internal Mixer Based on Elastomer Blends. Int Polym Process 1989;4:23-31.

[6] Payne A. Rheology of elastomers. by 1958;1:11.

[7] McCrum NG, Read BE, Williams G. Anelastic and dielectric effects in polymeric solids: John Wiley & Sons Ltd; 1967.

[8] Norma AD. 3182,"Standard practice for rubber—materials, equipment, and procedures for mixing standard compounds and preparing standard vulcanized sheets". ASTM American Society for Testing and Materials 2007.

[9] Seema A, Kutty S. Thermal Degradation of Short Nylon-6 Fiber–Reinforced Styrene Butadiene Rubber Composite. International Journal of Polymeric Materials 2006;55:25-35.

[10] Gibala D, Hamed GR. Cure and Mechanical Behavior of Rubber Compounds Containing Ground Vulcanizates. Part I-Cure Behavior. Rubber Chemistry and Technology 1994;67:636-48.

[11] Katbab A, Nazockdast H, Bazgir S. Carbon black-reinforced dynamically cured EPDM/PP thermoplastic elastomers. I. Morphology, rheology, and dynamic mechanical properties. Journal of Applied Polymer Science 2000;75:1127-37.

International Symposium on Advanced Materials (ISAM 2013) IOP Publishing
IOP Conf. Series: Materials Science and Engineering **60** (2014) 012071 doi:10.1088/1757-899X/60/1/012071

Localized ageing in the heat affected zone of welded X5CrNiCuNb16-4 and X4CrNiSiTi14-7 sheets

S Sakhawat[1], A Falahati[2], H P Degischer[2], K Spiradek[3] and M Dománková[4]
[1]Pakistan Welding Institute, Islamabad, Pakistan
[2]Vienna University of Technology, Vienna Austria
[3]Austrian Institute of Technology, Seibersdorf, Austria
[4]Slovak University of Technology, Bratislava, Slovakia

Abstract. Localized ageing and corresponding microstructural developments as a result of welding heat in the heat-affected zone (HAZ) of two different precipitation-hardened Cr-Steels (X5CrNiCuNb16-4 and X4CrNiSiTi14-7) have been studied. The X5CrNiCuNb16-4 sheet was in solution annealed condition and X4CrNiSiTi14-7 sheet was in peak aged condition. The results showed that despite of initial heat treated condition, the fusion zone formed in both welded sheets has typical cast structure. The HAZ has different microstructure compared to fusion zone and base metal. The HAZ is found to be sensitive to the welding heat and was aged locally due to thermal effects of welding. This localized ageing forms regions in HAZ varying from over-aged to under aged, depending upon the initial ageing condition of the base sheet.

1. Introduction

Precipitation hardening martensitic stainless steels have been developed to overcome the limitation of fully austenitic and fully martensitic stainless steels on account of strength, ductility, toughness and high temperature performance capability. They are widely used in various industries such as nuclear, chemical, aircraft, and naval due to their favorable combination of mechanical properties, easy machinability, good weldability and adequate corrosion resistance. The unique ability of strength enhancement by age hardening makes these steels very attractive for high strength applications.

X5CrNiCuNb16-4 (a type of 17-4 PH) is very attractive to designers and engineers. This steel is strengthened by precipitation of copper rich phases in the martensitic matrix by aging treatment [1,2,3,4]. The second alloy type that is used in this study is X4CrNiSiTi14-7, which is a high tensile strength precipitation hardening stainless steel with martensitic microstructure. This alloy is hardened by G-phase precipitation[5,6].

These precipitates are sensitive to heat and significant changes in the microstructure of the weld and heat affected zone (HAZ) may occur as a result of welding. The high heat of welding produces stresses and local ageing in HAZ. This paper focuses on the microstructural investigation of the HAZ developed as a result of welding in different material condition, through optical metallography, EBSD, TEM and micro hardness measurements.

2. Material and experimental methods

Two types of precipitation hardening martensitic stainless steels are investigated in this work. They are X5CrNiCuNb16-4 (17-4PH) and X4CrNiSiTi14-7(15Cr-7Ni) sheets of thickness > 2mm. The chemical composition of these alloys is given in Table 1. The X5CrNiCuNb16-4 sheets were welded together in solution annealed condition and the X4CrNiSiTi14-7 sheets were welded together in peak age hardened condition. The welding was done by TIG welding technique, without using filler material and with two welding passes, one pass from each side. The first weld-pass was on the top and the second pass was from the bottom.

Table 1. Chemical composition of the alloys (wt %) balance Fe

Elements	X5CrNiCuNb16-4	X4CrNiSiTi14-7
Fe	Balance	Balance
C	0.024	0.038
Si	0.706	1.475
Mn	0.78	0.35
P	0.166	0.0258
S	0.0005	0.0006
Ni	5.04	7.03
Cr	15.04	13.61
Cu	3.3	0.6
Ti	0.016	0.33
Mo	0.115	0.79
Nb	0.26	0.006
Al	0.011	0.037
V	0.072	0.037

Metallography was done to characterize the microstructure. The metallographic samples were prepared in transversal direction. Conventionally used etchant Nital did not revealed microstructural details. The etchant WII (16g $FeCl_3.6H_2O$ in 50 ml H_2O dist. + 9g $(NH_4)2S_2O_8$ in 50 ml H_2O dist mix in 60 ml of 32% HCl) with etching time of 5-6 s was able to etch the samples. Metallography was carried out on the base metal and the welded area to characterize their microstructure. Vickers micro hardness profiles across the weld, in the HAZ and base material were measured according to EN 6507 in Vickers HV 1 and were performed on universal hardness tester model Emco Test M1C010.

The SEM-EBSD technique was used for resolving microstructural details of HAZ. The samples were mechanically ground and polished down to 1μm diamond paste. For good quality of EBSD patterns, it is important to remove the residual deformation in the surface layer of the specimens by electro-polishing. Different polishing solutions were tried for the purpose and finally a solution of 20 vol.% perchloric acid ($HClO_4$), 70 vol.% ethanol (C_2H_5OH) and 10 vol.% ethylene glycol monobutyl ether ($CH_3(CH_2)_3OCH_2CH_2OH$) at 20°C, with the voltage of 28V for 10 s was found successful for giving a good EBSD signals from the sample [7]. SEM-EBSD investigations were performed on FEI Quanta 200 FEGSEM operated at 20kV. EBSD scan were carried out in the middle of the specimen, with a step size of 50nm. During experiment the samples were maintained at standard 70° tilt by using special specimen holder.

The investigation of the HAZ of welded X5CrNiCuNb16-4 sheets has been performed with the transmission electron microscope CM20STEM and the high resolution electron microscope (HRTEM) Tecnai F20 operating at 200kV at Austrian Institute of Technology, Seibersdorf, Austria. Thin foils were prepared by mechanical grinding and subsequent electro polishing. The structure was analyzed in bright field and dark field imaging; and the precipitation was identified by electron diffraction, and with reference to the distance between the specific crystal planes in HRTEM.

Carbon extraction replicas were prepared for examination by TEM and analysis by EDX to identify the precipitates in X4CrNiSiTi14-7 alloy. The samples were etched in etchant consisting of 1-2 g of picric acid, 10 ml of ethanol, and 5 ml of water. The surface was coated by a C-layer and the replicas were removed from the sample surfaces electrochemically at 20 V using 3% nital as electrolyte. TEM observations were performed in a microscope JEOL 200CX operating at 200 kV.

3. Results and Discussions

The X5CrNiCuNb16-4 sheets were in precipitation free (solution annealed and rolled) condition before welding together. The microstructure in this condition was martensitic with an average grain size of about 25 μm. Figure 1 shows the macrostructure of cross section of the weldment showing the different regions developed. The first weld-pass was on the top and the second pass from the bottom. During welding the temperatures surpassed locally the solidus temperature, but the time of exposure to such high temperatures was very short, thus producing a temperature gradient in the weld and in the areas which were adjacent to the weld. As a result the microstructure varied from the fusion zone to the base material.

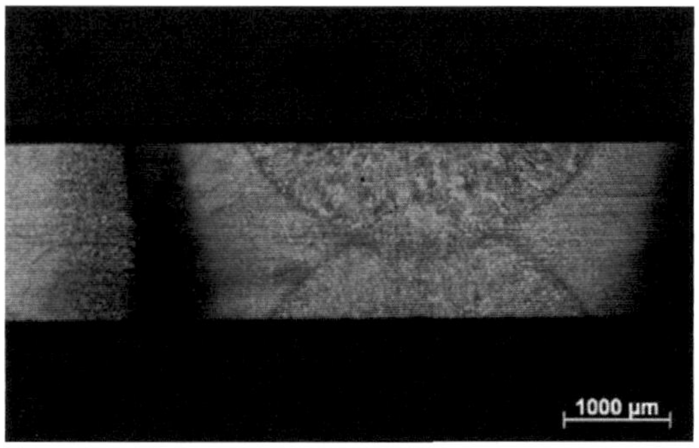

Figure 1. Macrograph of the cross-section of welded region (X5CrNiCuNb16-4)

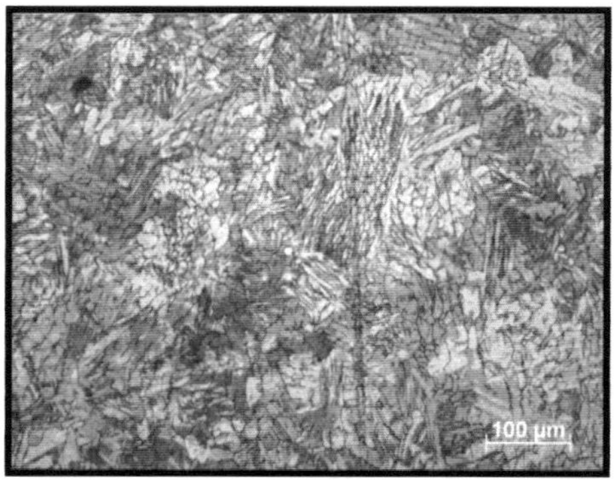

Figure 2. Microstructure of the weld zone (X5CrNiCuNb16-4)

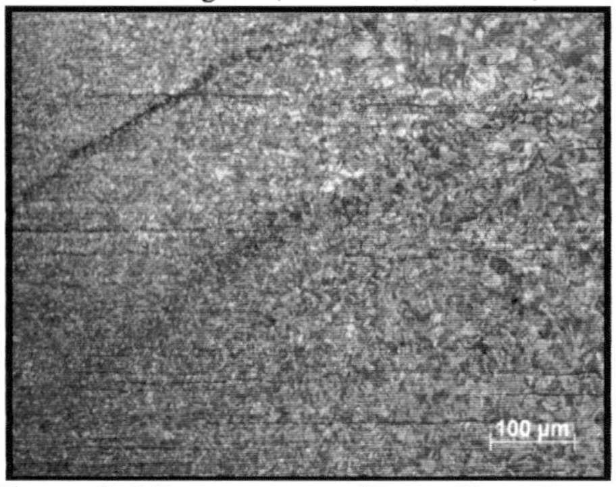

Figure 3. Microstructure of heat affected zone (X5CrNiCuNb16-4)

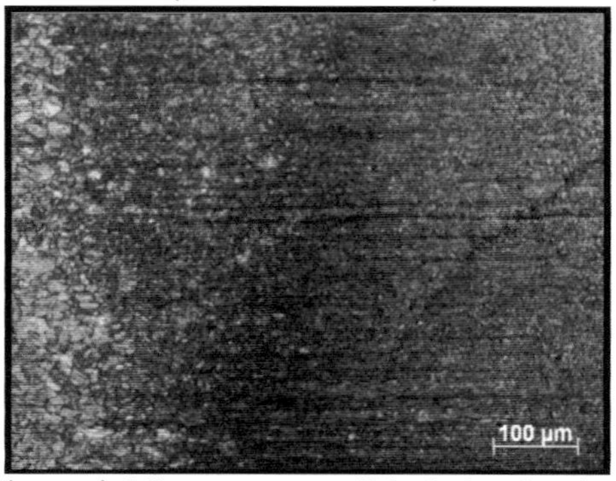

Figure 4. Microstructure of black tinted region (X5CrNiCuNb16-4)

The microstructures of the fusion zone is shown in Figure 2. The fusion zone is a typical cast structure, characteristic of solidification at rapid cooling. The HAZ is shown in Figure 3. As can be seen in Figure 1 that there is some black region in HAZ on both sides of the weld in the macrograph, which is shown in detail in Figure 4 depicting a strongly etched sample and remains unresolved by optical microscopy. Since this area appears black in metallographic images, hence termed as "black tinted region". There is a coarse grained structure adjacent to the fusion boundary in Figure 3. The black tinted region that remained unresolved by optical micrography was successfully resolved by EBSD. The microstructure is mostly martensitic with ca. 2% of austenite (Figure 5). The grain size has been found to be extremely small with 40% grains between 1-2 μm. The largest grain found in the area is less than 4 μm (Figure 6).

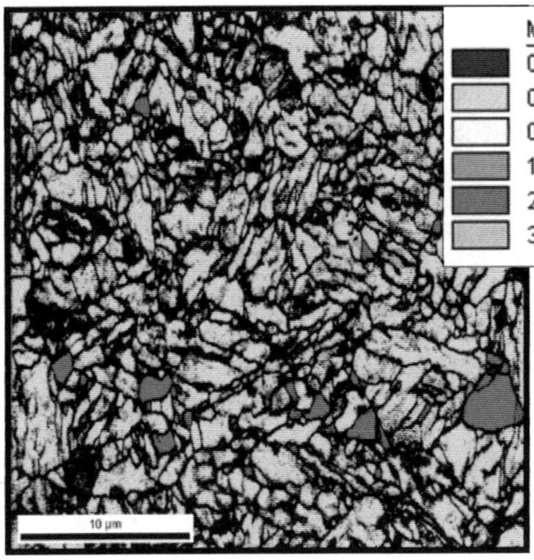

Figure 5. EBSD phase map (X5CrNiCuNb16-4), green= martensite, red= austenit

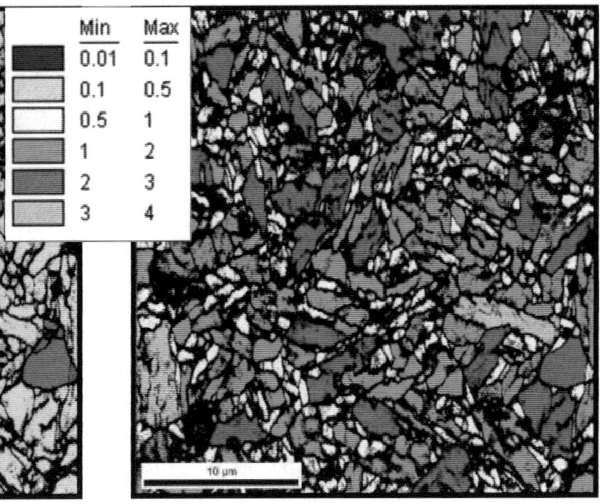

Figure 6. EBSD grain size map, grain size measurements as per attached scale (X5CrNiCuNb16-4)

Figure 7 presents the EBSD phase mapping of weld and HAZ. Big ferritic grains (>20μm) are mixed with small laths (<10 x 5 μm) present in the weld (Figure 7a). The grain size decreases and the small grains look rounder in Figure 7b. Very small grains (< 0.5 μm Φ) or highly distorted crystal structures, presumably both are found in the black region (Figure 7c). The grain size starts increasing outside dark region (Figure 7d). The possible explanation for this heterogeneity in HAZ is recrystallization, which occurs by the nucleation and growth of the new grains. The stored energy is the driving force for recrystallization in work hardened materials. During rapid heating of welding, the recrystallization temperatures may increase because it requires diffusion which takes time. During welding, the martensite in the coarse-grain region will transform into austenite and this reformed austenite subsequently transforms back into laths martensite during cooling with some retained austenite islands as suggested in literature [8].

Contains 3.3 wt% Cu which precipitates out upon ageing and produces precipitation hardening [1,2]. Before welding the X5CrNiCuNb16-4 sheets together, they were super saturated with Cu. During continuous cooling of the welded sheets in air, some regions of the HAZ have passed through temperatures, which were suitable to age these regions locally and as a result to precipitate Cu. The TEM investigations near the black-tinted region show a very fine and high density of spherical Cu precipitates of the size < 20 nm. One of the precipitate is shown in Figure 8. The HRTEM image in this figure shows some noncoherency of the precipitate with the matrix. Figure 9 shows the EDX spectra of matrix and precipitate and it clearly indicates that this precipitate is enriched with Cu. Hence the region close to the fusion line (points 6, 7, 12 and 13 in Figure 10) will be in solutionized condition, after which overaging (point 8 in Figure 10) might occur further away from the fusion line towards the region of peak hardness (point 3 and 16 in Figure 10) and thereafter follows the underaged region (point 1, 18 and 19 in Figure 10) until the base-material hardness is reached (about 1mm after max. hardness peak).

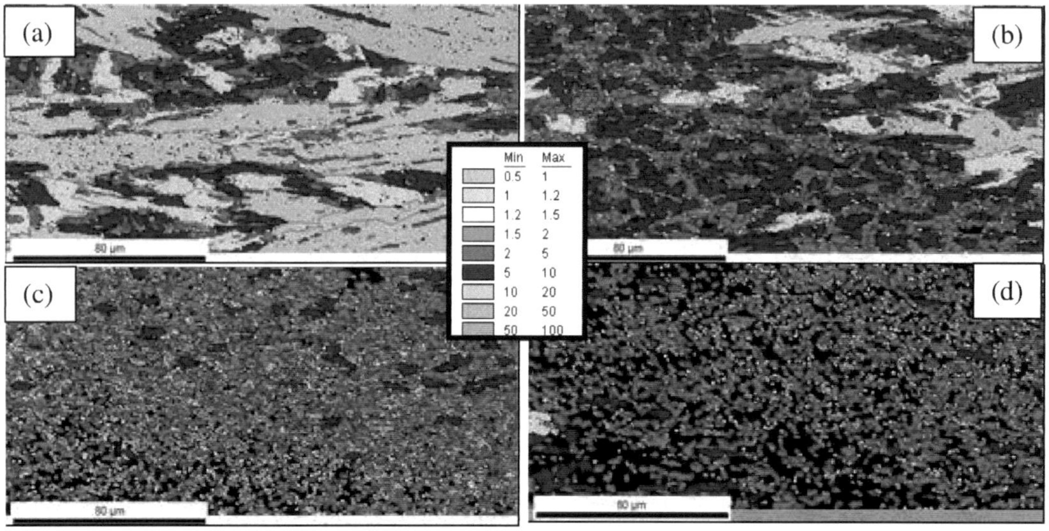

Figure 7. EBSD grain size mapping (X5CrNiCuNb16-4), a) weld, b) Weld and adjacent coarse grained HAZ, c) HAZ and black tinted region, d) outside black region. No EBSD signal obtained from black areas.

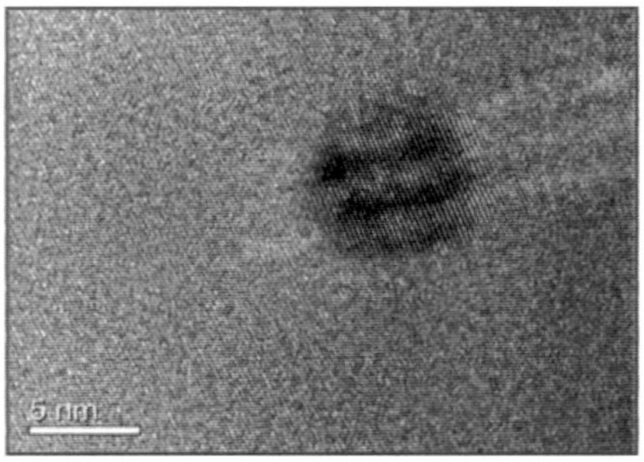

Figure 8. HRTEM, lattice fringe contrast, indicating some non-coherency of precipitate in X5CrNiCuNb16-4

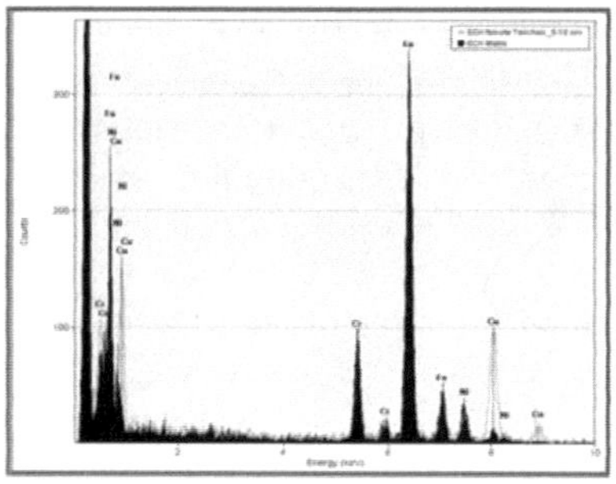

Figure 9. Overlapping EDX spectra of matrix and precipitate, clearly indicating the Cu enrichment of the precipitate shown in Fig.8.

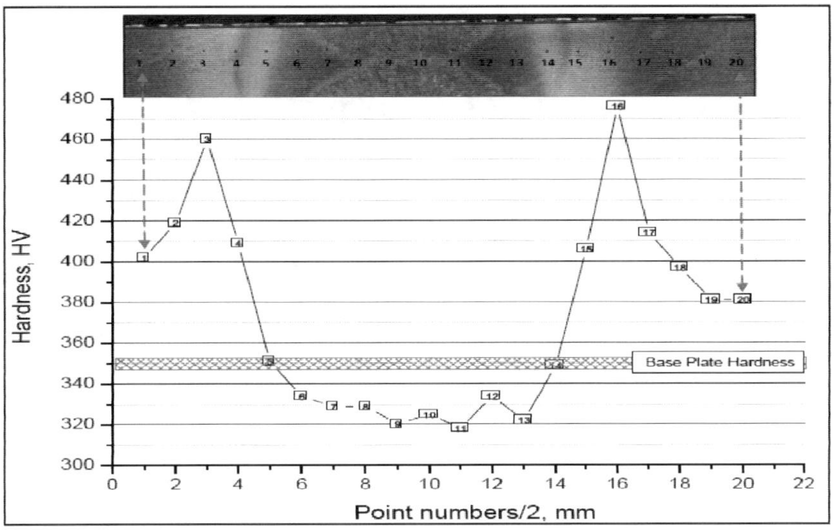

Figure 10. Micro hardness measurements across the weld and the HAZ(X4CrNiSiTi14-7)

The second material studied in this work was X4CrNiSiTi14-7. The sheets were in peak age hardened condition before welding together with a hardness value of 477 HV. Figure 11 shows the macrostructure of cross section of the weldment showing the different regions developed. The first weld-pass was on the top and the second pass from the bottom. The microstructures of the fusion zone and HAZ formed aftere welding X4CrNiSiTi14-7 aged sheets together are shown in Figure 12 and 13 respectively. The "black tinted region" was also found on both sides of the weld in the macrograph and is shown in Figure 14. The fusion zone, like as formed in X5CrNiCuNb16-4 solution annealed sheets, is a typical cast structure, characteristic of solidification at rapid cooling.

Figure 15 presents the phase mapping of weld, HAZ and . Big ferritic grains (>20μm) are mixed with small laths (<4μm) in the weld (Figure 15a). The grain size decreases in Figure 15b. Very small grains (<1μm Φ) or highly distorted crystal structures, presumably both are found in black the region (Figure 15c). The black region consists of very small grains (Figure 15d) mostly unresolved by EBSD.

Figure 11. Macrograph of the cross-section of welded region (X4CrNiSiTi14-7)

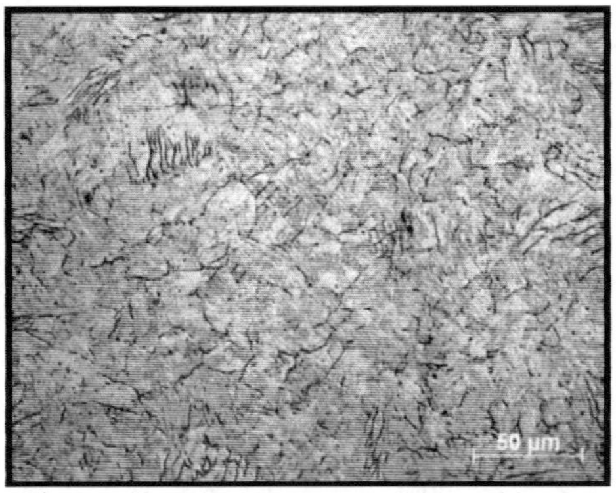

Figure 12. Microstructure of the weld zone (X4CrNiSiTi14-7)

Figure 13. Microstructure of heat affected zone (X4CrNiSiTi14-7)

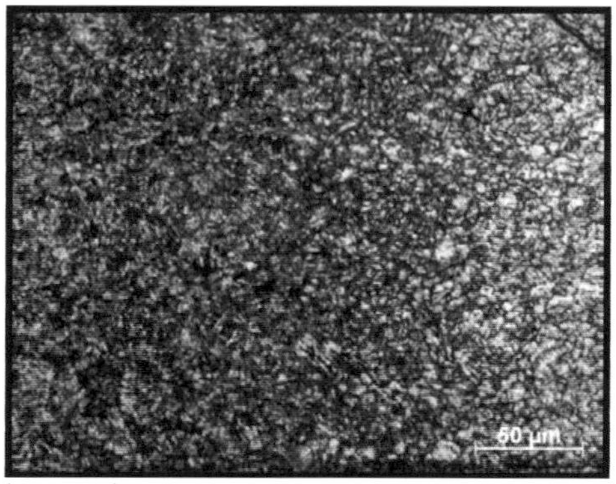

Figure 14. Microstructure of black tinted region (X4CrNiSiTi14-7)

Figure 10 shows the micro hardness profile done along cross sections of the weld, HAZ and base metal. The average hardness in the weld region is ~340 HV, whereas the hardness of the base material is ~350 HV. This small decrease in hardness indicates that melting and re-solidification in the weld zone has removed cold rolling effects. Since no prominent difference in hardness exists in between the weld zone and base material, this indicates that the martensite formed in the weld zone is precipitation free and is in untempered condition. The peak hardness (> 455HV) is about 0.5- 1mm outside of black tinted region in the HAZ on both sides of the weld seam. The weld seam has the lowest hardness values (320-335HV). This increase in hardness can be explained on the basis of precipitation hardening. The X5CrNiCuNb16-4 alloy

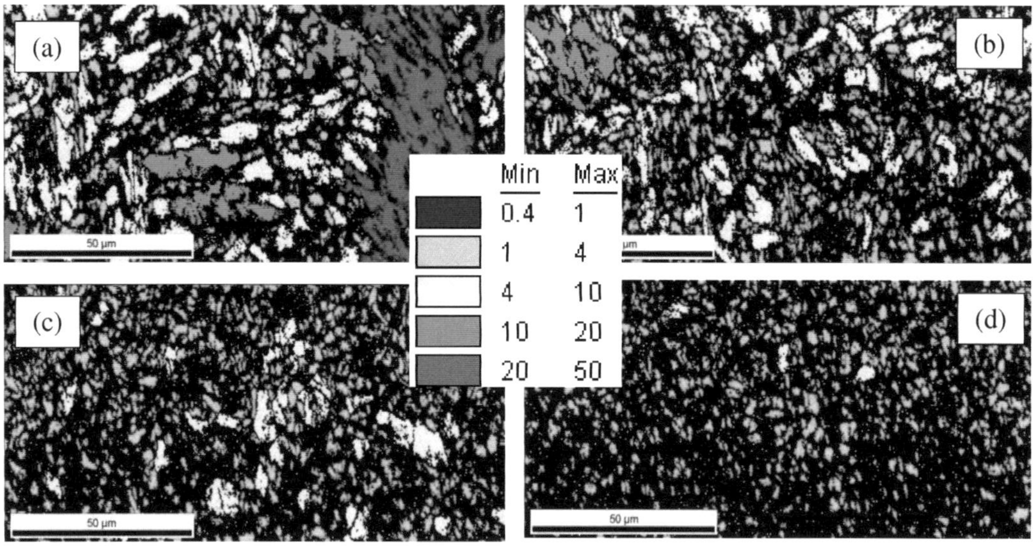

Figure 15. EBSD grain size mapping of X4CrNiSiTi14-7 weldment, a) weld, b) just outside weld, c) HAZ, d) black tinted region. No EBSD signal obtained from black areas (very distorted/ fine grained structure). (Color scale is shown on the right)

Figure 16 shows the micro hardness profile along cross sections of the weld, HAZ and base metal. The hardness of the weld seam (350-360HV) is lower than that of the base metal (~ 480 HV). Weld seam has the hardness of 350-360 HV. There is a hardness drop in HAZ near to the black tinted region, the lowest hardness value across the profile is at 1mm from the black tinted region in HAZ on both sides of the weld seam and the minimum hardness value is 317HV and then it increases again towards the base metal. This behavior can be explained on the basis of precipitation hardening. This high hardness of basematerial is due to the presence of fine G-phase precipitates [5,6] and shown in Figure 17. During welding the heat was sufficient to cause melting in the centre of weld zone. This melting also caused the dissolution of the G-phase precipitates present in the material. Since enough time was not available during re-solidification hence re-precipitation of G-phase did not takes place in the weld zone. As a result softening takes place due to the loss of precipitation hardening which was initially acting as a hindrance to dislocation motion. The microstructure is martensitic (precipitation free and in untempered condition) with some small islands of retained austenite. The weld became softer (340 Hv), which is very close to the hardness of the as-received condition with out ageing (335HV). This shows that weld zone will always be in solution annealed condition

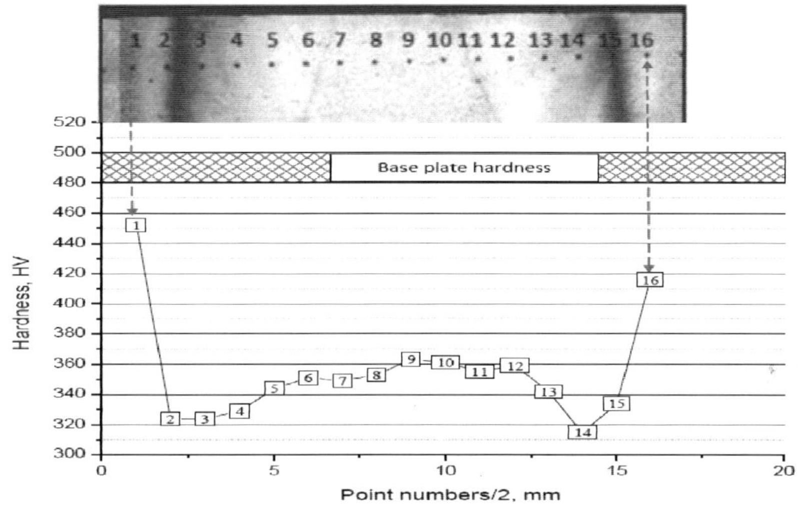

Figure 16. Micro hardness measurements across the weld and the HAZ(X4CrNiSiTi14-7)

despite of the initial heat treated condition of the base material. A similar condition exists in the region next to weld zone, where the temperatures were above the solution temperatures and were sufficient to dissolve the G-phase precipitates during heating, thus resulting in softening. Hence hardness values are also lower like in the weld zone (see for example points 5 and 12 in Figure 16). Beyond "near weld region" the temperatures were still high but lower than the solution temperature. The exposure to these temperatures causes overaging of the already aged base metal (points 2, 3 and 14 in Figure 16). The regions where the temperature in the heat-affected zone is not above the original ageing temperature, or for only a very short time, the hardness is not significantly affected [9].

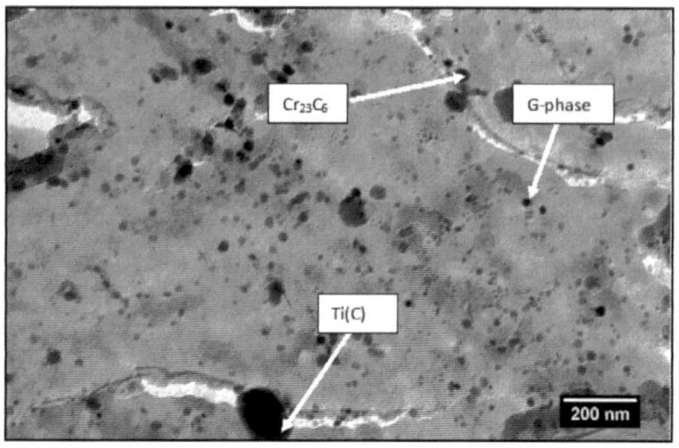

Figure 17. Transmission electron micrographs of X4CrNiSiTi14-7 showing precipitates

4. Conclusions

Microstructural changes in the HAZ of two types of precipitation hardening stainless steels have been characterized, and compared with the aging behavior.

During welding the HAZ is submitted to an "ageing" treatment producing strengthening and loss in ductility. Peak hardening is achieved at the borders of HAZ for X5CrNiCuNb16-4 sheets without heat treatment. In X4CrNiSiTi14-7 peak aged sheets, the hardness is reduced to the solution treated level close to the weld seam and outside where austenization did not occur, over-ageing takes place reducing the hardness further.

The "black tinted region" indicates very fine recrystallized regions representing either a precipitation strengthened region for an originally solution treated condition (X5CrNiCuNb16-4), or an overaged region (soft) for an initially peak aged sheet (X4CrNiSiTi14-7).

5. Acknowledgement

The work was partially supported by the Austrian FFG. The authors thank Higher Education Commission of Pakistan (HEC) for sponsoring PhD studies.

6. References

[1] Hsiao C N, Chiou C S and Yang J R 2002 *Materials Chemistry and Physics* 74 134-142

[2] Mirzadeh H, Najafizadeh A 2009 *Materials Chemistry and Physics* 116 119-124

[3] Murayama M, Katayama Y and Hono K 1999 *Metall. Mater. Trans* A 30A 345-353

[4] Wang J, Zou H, Cong L, Peng Y, Qiu S andShen B 2006 *Nuclear Engineering and Design* 236 2531-2536

[5] Sakhawat S, Falahati A and Domankova M 2010 *Prakt. Metallogr.* 47 500-516

[6] Murata Y, Ohashi S and Uematsu Y 1993 *ISIJ International* 33 711-720

[7] Jia N, Peng R L, Chai G C, Johansson S andWangY D 2008 *Materials Science and Engineering A*

[8] Bhaduri A K and Venkadesan S 1989 *Steel Research* 60 509-513

[9] Messler W R 2004 *Principles of welding* (Wiley VCH Weinheim) pp 540-542

Advanced materials and techniques for fibre-optic sensing

Philip J Henderson
Forman Christian College (a Chartered University),
Ferozepur Road, Lahore 54600, Pakistan

E-mail: philiphenderson@fccollege.edu.pk

Abstract. Fibre-optic monitoring systems came of age in about 1999 upon the emergence of the world's first significant commercialising company – a spin-out from the UK's collaborative MAST project. By using embedded fibre-optic technology, the MAST project successfully measured transient strain within high-performance composite yacht masts. Since then, applications have extended from smart composites into civil engineering, energy, military, aerospace, medicine and other sectors. Fibre-optic sensors come in various forms, and may be subject to embedment, retrofitting, and remote interrogation. The unique challenges presented by each implementation require careful scrutiny before widespread adoption can take place. Accordingly, various aspects of design and reliability are discussed spanning a range of representative technologies that include resonant microsilicon structures, MEMS, Bragg gratings, advanced forms of spectroscopy, and modern trends in nanotechnology. Keywords: Fibre-optic sensors, fibre Bragg gratings, MEMS, MOEMS, nanotechnology, plasmon.

1. Introduction

Fibre-optic sensors now provide a critical addition to our arsenal of monitoring technology with global sales reaching $1.9bn pa [1]. From an almost zero baseline a quarter of a century ago, fibre-optic monitoring systems have seen revolutionary levels of research, development and commercial exploitation due, in part, to: the many well-known potential advantages that optical sensors have over traditional technologies, the availability of high-quality glass optical fibres, and the introduction of solid-state sources and detectors. The market in fibre-optic sensors now exceeds 2.5% of the total sensors market [2], and may grow by as much as 20% pa over the next 5 years [1].

2. Early exploitations: Sensing within harsh environments

The success story began in the late 1980s when various global players sought to establish optical sensor technology within niche application-areas supported by its inherent advantages.

2.1. Body scanning, chemical synthesis and electrical switchgear

Development and commercialisation of rugged polychromatic monitoring systems [3-4], for example, was initially promoted through successful showcase monitoring of:

- Respiration rate, blood oxygenation and temperature of patients undergoing body scans.
- Pressure within the production vessel for an explosive monomer.
- Contact travel within a high-voltage circuit-breaker.
- Pressure, temperature and dielectric strength of SF_6 insulation.

2.2. Distribution transformers
The advent of Australian photonics in the early 1990s saw a massive initiative to promote fibre-optic sensors into the electric power industry, where a significant motivation was the need to measure "hot-spot" temperatures in distribution transformers.

2.2.1. Resonant microsilicon beams. A novel silicon undercutting process [5] creates a free-standing microsilicon beam that acts as one mirror of a reflective Fabry-Perot cavity, has high internal stress, and has a temperature-dependent resonant frequency.

2.2.2. Temperature sensing. Conversion of the microsilicon beam into a fibre-optic temperature sensor involves [6]:

- Placement under high vacuum: $<10^{-2}$mbar leads to resonance with a "Q-factor" exceeding 10^5.
- Optical excitation, and interferometric interrogation.
- Fourier-analysis of the interrogation signal, giving temperature resolutions of better than ±0.4°C.
- Attachment to the end of an optical fibre, and encapsulation within rugged packaging.

3. Fibre-optic microelectromechanical systems (MEMS)
While the focus is clearly on "optical" rather than "electrical", the section 2.2.1 technology is sometimes called a microelectromechanical system (MEMS) – more specifically it is an optical MEMS [7] or MOEMS. Section 2.2.2 could then be deemed to be describing a resonant fibre-optic MEMS sensor. The R&D of such resonant sensors peaked shortly after the mid-1990s with no significant commercialisation.

However, the underlying microsilicon technology is of major significance because it helped fuel revolutionary growth in non-resonant fibre-optically addressed MEMS [8] and silicon-on-insulator (hybrid silicon) photonics [9-12]. Often exploiting interferometric methods of signal transduction, such technologies are used in both sensing and interrogation. Most importantly, being based on the generic and abundant MEMS technology, the market for fibre-optic MEMS sensors is opened-up far beyond the original harsh-environment applications.

3.1. New trends in sensing
New forms of sensing structure are similar to those discussed in sections 2.2.1- 2.2.2, but they are not resonant, and thus do not need evacuation. A simple pressure (or strain) sensor is possible by configuring one end of the Fabry-Perot cavity as a diaphragm (or translatable) surface [13]. Reconfiguring the Fabry-Perot cavity enables other measurands – for example; temperature by the inclusion of a birefringent crystal [14], and refractive index by opening-up the cavity to the surroundings [15].

3.2. New trends in interrogation
While there are numerous initiatives towards fabricating individual MEMS components – such as mirrors [16], filters [17] and gratings [18] – the UK's collaborative AIMS project (2002-2005) was, perhaps, the first significant attempt to integrate all interrogation functions on to a single silicon-on-insulator (SOI) chip. This project developed an electrically-activated Fabry-Perot-based MEMS interrogating filter and associated SOI circuitry [19-20].

The processing of Fabry-Perot signals is a significant challenge impeding widespread acceptance due to the problems of the 2π-phase ambiguities and signal drift. Numerous approaches have been taken: the older ones being quadrature based methods [21] and the use of two interrogating wavelengths [22], and the newer ones being Fourier based methods [23] and low-coherence (white light) interferometry [24].

An exciting new alternative to Fabry-Perot based MEMS is the possibility of interrogation using microring resonators [25].

3.3. Comments on performance

Fibre-optic MEMS devices can be small – Opsens Inc., for example, is marketing pressure and temperature sensors with outside diameters respectively of 0.25 mm and 0.17 mm. Due to the small size of optical MEMS, the frequency response of pressure sensors can reach 80 kHz, and the response time of temperature sensors can be as short as 10 ms. As with the packaging of any Fabry-Perot device, it is important to use epoxy-free bonding to avoid signal distortions caused by adhesive expansion. Some form of low-coherence (white light) based interrogation seems to be the most acceptable solution to 2π phase ambiguities and signal drift. A number of other issues may need to be addressed such as fatigue-induced crack-growth, creep under constant stress, material deformation (due to annealing and other factors), anodic oxidation, and robustness to environmental factors (such as vibration).

4. Fibre-Bragg-grating

As a disruptive technology, the significance of the fibre-Bragg-grating (FBG) should not be underestimated – it immediately superseded more-complex sensing technologies and catapulted fibre-optic sensors into areas hitherto the domain of traditional technologies, notably thermometers and strain gauges. An FBG [26] is a narrow-band reflection filter created by inscribing a periodic variation in refractive index within the core of an optical fibre. This index variation is perturbed through a change of temperature or strain, leading directly to a change in the reflected so-called Bragg wavelength. Due to the encoding of the Bragg wavelength, FBGs may be used as temperature and strain sensors. Other parameters may be measured if they can be transduced to temperature or strain.

Backed by continuing collaborations and research output, the industry has matured with a number of global players positioned in the military, aerospace, civil engineering, energy, environmental, medical and other sectors [27-29].

4.1. FBG temperature sensing

Through early application of the FBG-temperature sensor to transformer “hot-spot” monitoring [30], the Australians quickly superseded their own work on resonant microsilicon sensors (sections 2.2.1-2.2.2)

4.2. FBG strain sensing

In terms of strain sensing, big breaks came in the form of two show-case collaborations, specifically the UK’s MAST (1995-98) and SHODOS (1996-1998) projects. While the former project monitored the flexing of composite yacht masts; the latter monitored long-term strain in civil engineering structures such as bridges, representing the technology’s first foray into Structural Health Monitoring. As sensors, FBGs may be retrofitted (as in SHODOS) or embedded, as in the exciting and highly active field of Smart Structures.

5. High-power systems

High-power systems are of importance in at least the following two situations. Powers up to about 1W are used in optically-powered hybrid systems, which feature fibre-optic activation of conventional electronic sensors [31]. At still-higher power, a 1.9W 1651nm Raman amplifier system with wavelength-tuneable output has been showcased by researchers at Gooch and Housego PLC for remote spectroscopic measurements of methane leaks.

6. Key new technologies and abilities

In addition to MEMS, there are a number of major developments that could help spark a second revolution in fibre-optic sensors:

- FBGs can now be inscribed at predetermined locations as the fibre is being drawn; thus avoiding any post-processing damage, and facilitating realisation of their great multiplexing capability.
- FBGs can now be written point by point, so that FBGS of arbitrary profile can be inscribed.
- Tilted fibre Bragg gratings may offer self-referencing, multiparameter physical sensing, as well as plasmonic applications.
- Plasmons, themselves, are waves that tend to propagate along the surface of metallic and certain dielectric materials. Such waves are highly sensitive to any changes in the refractive index of the material as well as the device's geometry, and may offer great potential in fibre-optic sensing applications.
- Optical fibre is available which has two mobile cores mounted in close proximity. Such fibre has unique transmission properties that may be useful in sensing.
- Nanotechnologies look set to offer enhanced sensing opportunities when used in conjunction with optical fibres.
- Nanogratings, in particular, may offer enhanced sensing and spectrometry.
- Fibre-modal interferometers offer the promise of highly sensitive sensors.
- Photonic crystal fibres may enhance many of the above opportunities.

7. Discussion and conclusions

At least the first generation of fibre-optic monitoring systems has come age of with the establishment of Bragg-grating based sensor technology. Such gratings can be inscribed at any locations at the time of fibre-manufacture (preserving its pristine strength), can have different Bragg wavelengths, can directly sense temperature and strain, and be subject to wavelength-demodulation based interrogation schemes. Some challenges remain such as simplifying the currently labour-intensive methods of installation, ensuring faithful strain transfer from the substrate to the grating, and efficiently separating out the dual temperature/strain dependencies. Despite these challenges, the FBG sensors are likely to remain the dominant contributor to the expected near-term 20% pa growth in the fibre-optic sensors market.

Paarticularly active within the market are interferometric sensors, for a variety of measurands, based on the Fabry-Perot cavity. Such devices are largely configured as reflective point-sensors; an important category of implementation. However, as such, they will not achieve the multiplexing convenience of the FBG.

Configurable in both sensing and interrogating forms, fibre-optic MEMS is one of a number of technologies that is likely to shake up the market in manner akin to the FBG. It holds the promise of cheap, mass-produced, miniaturized devices, principally by borrowing processing technology from the established silicon-on-insulator MEMS market. Sensors are typically based on the Fabry-Perot cavity, and a number of configurations are available for pressure and other measurands. Development of fibre-optic MEMS interrogators continues. Individual components have been made, but the completely integrated silicon-photonic interrogation system remains elusive due, in part, to silicon's indirect band-gap hindering its use as an emitter.

In the absence of a completely integrated solution, much interrogator development continues. Commercial activity is intense, as evidenced by the number of players, and the rate at which new/improved/cheaper interrogation systems are brought to the market place. Such systems range from bulk-optic affairs, incorporating spectrometer-type devices, to more miniaturized forms that include MEMS components, wavelength scanners and filters.

Also set to revolutionise the market in the next couple of decades are fibre-optic devices based on nanotechnology, photonic crystal fibres and surface plasmons.

8. References

[1] 2013 *Fiber Optic Sensors Global Market Forecast* (Aptos: ElectroniCast Consultants)

[2] 2013 *Global Markets and Technologies for Sensors* (Kirkland: Electronics.ca Publications)

Rallis I, Glomon L, Deakin A G, Spencer J W and Jones G R 2006 Polychromatic monitoring of complex biological systems *Sensor Review* 26 45–50

[3] 2008 *Chromatic Monitoring of Complex Conditions* ed G R Jones et al (London: Taylor & Francis)

[4] Wiszniewski W 1997 *United States Patent 5,682,053: Silicon Transducer with Composite Beam* (Alexandria: USPTO)

[5] Henderson P A, Henderson P H and Wiszniewski W R 1996 *Frequency-Encoded Temperature Sensor using Interferometric Microresonators: ACOFT'96 Proc. 21st Australian Conf. on Optical Fibre Technology* (Conrad Jupiters, Gold Coast, Queensland, 1-4 December, 1996) (Milsons Point: IREE Society) pp 282–285

[6] Piyawattanametha W and Qiu Z 2012 *Optical MEMS, Microelectromechanical Systems and*

a. *Devices* ed N Islam (New York: Intech) pp 291–330

[7] Pang C, Yu M, Zhang X M, Gupta A K and Bryden K M 2012 Multifunctional optical MEMS sensor platform with heterogeneous fiber optic Fabry-Perot sensors for wireless sensor networks *Sensors and Actuators* A 188 471–480

[8] Rickman A G, Reed G T and Namavar F 1993 *Silicon-on-Insulator Optical Rib Waveguide Circuits for Fiber Optic Sensors: Proc. SPIE 2071 Distributed and Multiplexed Fiber Optic Sensors III* ed A D Kersey and J P Dakin (Boston, 30 December, 1993) (Bellingham: SPIE) pp 190–6

[9] Noell W 2002 Applications of SOI-based optical MEMS *IEEE J. Sel. Topics Quant. Electron.* 8 148–54

[10] Jalali B and Fathpour S 2006 Silicon photonics *J. Lightwave Technology* 24 4600–4615

[11] Heck M J R, Bauters J F, Davenport M L , Doylend J K, Jain S, Kurczveil G, Srinivasan S, Tang Y and Bowers J E 2013 Hybrid silicon photonic integrated circuit technology *IEEE J. Sel. Topics Quant. Electron.* 19 6100117

[12] Ge Y and Wang M 2008 Optical MEMS pressure sensor based on a mesa-diaphragm structure *Optics Express* 16 21746–52

[13] Duplain G and Neste R V 2007 *United States Patent 7,265,847: Birefringent Optical Temperature Sensor and Method* (Alexandria: USPTO)

[14] Tian Y, Wang W, Wu N, Zou, Guthy C and Wang X 2011 A miniature fiber optic refractive index sensor built in a MEMS-based microchannel *Sensors* 11 1078–87

[15] Saitoh T, Nakamura K, Takahashi Y and Miyagi K 2005 *High-Speed MEMS Swept-Wavelength Light Source for FBG Sensor System: Proc. SPIE 5855 17th International Conference on Optical Fibre Sensors* Ed M Voet et al (Bruges, 23 May, 2005) (Bellingham: SPIE) pp 146– 9

[16] Beccherelli R, De Sio L, Umeton C, Donisi D, d'Alessandro A and Caponero M A 2008 *Fiber Bragg Grating Interrogation System Based on a Novel Integrated Optical Filter: IEEE Sensors* (Lecce, 26-29 October, 2008) (Washington: IEEE) pp 140-3

[17] Xiao G, Mrad N, Guo H, Zhang Z and Yao J 2008 A planar lightwave circuit based micro interrogator and its applications to the interrogation of multiplexed optical fiber Bragg grating sensors *Optics Communications* 281 5659–63

[18] Ang M T W, Reed G T, Ensell G J, Evans A G R, Foote P D, Pritchard A P and Barber D 2003 *Bragg Gratings Interrogating System using MEMS and Optical Circuits: Proc. SPIE 4997 Photonics Packaging and Integration III* (San Jose, 25 January, 2003) (Bellingham: SPIE) pp 232-41

[19] Keddie D P et al 2004 *Bragg Gratings Interrogation System using MEMS and Optical Circuits: Part 2, Component Development: SPIE Photonics West Invited Talk* (McEnery Convention Center, San Jose, 24-29 January, 2004)

[20] Kersey A D, Dandridge A and Tveten 1987 *Recent Advances in Demodulation/Multiplexing Techniques for Interferometric Fiber Sensors: Proc. SPIE 0734 Fibre Optics '87* ed L R Baker (London, 28 April, 1987) (Bellingham: SPIE) pp 261–9

[21] Xiao-qi N, Ming W, Xu-xing C, Yi-xian G and Hua R 2006 An optical fibre MEMS pressure sensor using dual-wavelength interrogation *Meas. Sci. Technol.* 17 2401–4

[22] Ge Y, Wang M, Chen X and Rong H, 2008 An optical MEMS pressure sensor based on a phase demodulation method *Sensors and Actuators* A 143 224–9

[23] Cong-fei W, Guang-long W, Jian-hui C and Yi D 2009 *The Signal Interrogation Technology of MEMS Optical Fiber Pressure Sensor: ICIA'09 Int. Conf. Information and Automation* (Zhuhai, 22-24 June 2009) pp 1285-8

[24] Yang J, Qiu C, Wang Q L, Wang M H and Yang J Y 2013 Tunable microring filter based on-chip interrogator for wavelength-modulated optical sensor *Key Engineering Materials* 562–565 265–7

[25] Hill K O and Meltz G 1997 Fiber Bragg grating technology fundamentals and overview *J. Lightwave Technology* 15 1263–76

[26] Todd M D, Nichols J M, Trickey S T, Seaver M, Nichols C J and Virgin L 2007 Bragg grating- based fibre optic sensors in structural health monitoring *Phil. Trans. R. Soc.* A 365 317-43

[27] Metje N, Chapman D N, Rogers C D F, Henderson P and Beth M 2006 Optical Fibre Sensors for remote monitoring of tunnel displacements – prototype tests in the laboratory *Tunnelling and Underground Space Technology* 21 417

[28] Glavind L, Olesen I B, Skipper B F and Kristensen M 2013 Fiber-optical grating sensors for wind turbine blades: a review *Optical Engineering* 52 030901-10

[29] Hammon T E and Stokes A D 1996 *Optical Fibre Bragg Grating Temperature Sensor Measurements in an Electrical Power Transformer using a Temperature Compensated Optical Fibre Bragg Grating as a Reference: 11th Int. Conf. on Optical Fiber Sensors* (Sapporo, Japan, 21 May, 1996) (Washington DC: Optical Society of America) pp 566-9

[30] Higgins A J and Hook I K 1995 *Optically Powered Isolated Sensors: Proc. Electronic Technology Directions to the Year 2000* (Adelaide, 23-25 May 1995) (Digital library: IEEE Xplore) pp 301-5

AUTHOR INDEX

AUTHOR INDEX